Skeletal Growth of Aquatic Organisms

Biological Records of Environmental Change

TOPICS IN GEOBIOLOGY

Series Editor: **F. G. Stehli,** University of Florida

Volume 1   SKELETAL GROWTH OF AQUATIC ORGANISMS: Biological Records
of Environmental Change
Edited by Donald C. Rhoads and Richard A. Lutz

Skeletal Growth of Aquatic Organisms

Biological Records of Environmental Change

Edited by

Donald C. Rhoads

Yale University
New Haven, Connecticut

and

Richard A. Lutz

Rutgers University
New Brunswick, New Jersey

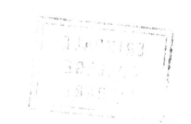

PLENUM PRESS · NEW YORK AND LONDON

Library of Congress Cataloging in Publication Data

Main entry under title:

Skeletal growth of aquatic organisms.

(Topics in geobiology; 1)
Includes index.
1. Aquatic animals–Growth. 2. Aquatic animals–Ecology. 3. Skeleton–Growth.
I. Rhoads, Donald C. II. Lutz, Richard A. III. Series.
QL120.S55   594'.04'7   79-25825
ISBN 0-306-40259-9

© 1980 Plenum Press, New York
A Division of Plenum Publishing Corporation
227 West 17th Street, New York, N.Y. 10011

Printed in the United States of America

Contributors

M. Ashton Geology Department, University of Newcastle Upon Tyne, Newcastle, England NE1 7RU

Edwin Bourget Groupe Interuniversitaire de Recherches Océanographiques du Québec (GIROQ), Département de Biologie, Faculté des Sciences et de Génie, Université Laval, Québec, Canada G1K 7P4

Joseph Gaylord Carter Department of Geology, University of North Carolina, Chapel Hill, North Carolina 27514

Robert M. Cerrato Department of Geology and Geophysics, Yale University, New Haven, Connecticut 06520

George R. Clark II Department of Geology, Kansas State University, Manhattan, Kansas 66506

Miles A. Crenshaw The Dental Research Center and Department of Pedodontics, School of Dentistry, University of North Carolina, Chapel Hill, North Carolina 27514

John F. Dillon Continental Oil Company, Albuquerque, New Mexico 87112

Richard E. Dodge Nova University Ocean Sciences Center, Dania, Florida 33004

R. Hewitt Geology Department, University of Birmingham, Birmingham, England B15 2TT

David Jablonski Department of Geology and Geophysics, Yale University, New Haven, Connecticut 06520. *Present address:* Department of Geological Sciences, University of California, Santa Barbara, California 93106

v

Michael J. Kennish Environmental Affairs Department, Jersey Central Power & Light Company, Morristown, New Jersey 07960

Richard A. Lutz Department of Oyster Culture, New Jersey Agricultural Experiment Station, Cook College, Rutgers University, New Brunswick, New Jersey 08903

P. J. W. Olive Department of Zoology, University of Newcastle Upon Tyne, Newcastle, England NE1 7RU

Giorgio Pannella Department of Geology, University of Puerto Rico, Mayaguez, Puerto Rico 00708

Donald C. Rhoads Department of Geology and Geophysics, Yale University, New Haven, Connecticut 06520

Gary D. Rosenberg Geology Department, Michigan State University, East Lansing, Michigan 48824. *Present address:* Geology Department, Indiana University/Purdue University, Indianapolis, Indiana 46202.

Danny M. Rye Department of Geology and Geophysics, Yale University, New Haven, Connecticut 06520

Raymond Seed Department of Zoology, University College of North Wales, Bangor, Wales, United Kingdom

D. J. Simmons Division of Orthopedic Surgery, Washington University Medical School, St. Louis, Missouri 63110

Michael A. Sommer II Department of Geology, Florida State University, Tallahassee, Florida 32306

Michael J. S. Tevesz Department of Geological Sciences, The Cleveland State University, Cleveland, Ohio 44115

Clifford L. Trump Groupe Interuniversitaire de Recherches Océanographiques du Québec (GIROQ), Département de Biologie, Faculté des Sciences et de Génie, Université Laval, Québec, Canada G1K 7P4 *Present address:* Applied Hydrodynamics Branch, Environmental Sciences Division, Naval Research Laboratory, Washington, D.C. 20375

J. Rimas Vaišnys Department of Geology and Geophysics, Yale University, New Haven, Connecticut 06520

Geerat J. Vermeij Department of Zoology, University of Maryland, College Park, Maryland 20742

Contents

Introduction · Skeletal Records of Environmental Change
Donald C. Rhoads and Richard A. Lutz

1. Perspective of This Volume 1
2. How Environmental Change Is Recorded by Skeletal Growth ... 3
3. Skeletal Growth Applied to Paleoecological and Ecological
   Problems 6
4. Plan of This Volume 14

I. The Mollusca

Chapter 1 · Shell Growth and Form in the Bivalvia
Raymond Seed

1. Introduction 23
2. General Features of the Bivalve Shell 24
3. Origin and Evolution of Bivalves 28
4. Shell Growth 32
5. Shell Shape and Body Form 45
6. Functional Morphology 52
   References 61

Chapter 2 · Environmental and Biological Controls of Bivalve
             Shell Mineralogy and Microstructure
Joseph Gaylord Carter

1. Introduction 69
2. Distribution of Bivalve Shell Mineralogy and Microstructure ... 71

vii

3. Evolution and Adaptive Significance of Shell Mineralogy and
   Microstructure 83
4. Shell Mineralogy and Structure as Records of Environmental
   Change 95
5. Conclusions 104
   References 107

Chapter 3 · Mechanisms of Shell Formation and Dissolution
Miles A. Crenshaw

1. Introduction 115
2. Shell Formation 116
3. Mechanisms of Shell Dissolution 125
   References 128

Chapter 4 · An Ontogenetic Approach to the Environmental
Significance of Bivalve Shell Chemistry

Gary D. Rosenberg

1. Introduction 133
2. Chemical Variations in the Shell 135
3. Recognition of the Original Elemental Composition of the
   Shell and Its Environmental and Physiological Significance 155
4. Summary 161
   References 162

Chapter 5 · Reconstructing Paleotemperature and Paleosalinity
Regimes with Oxygen Isotopes

Danny M. Rye and Michael A. Sommer II

1. Introduction 169
2. Oxygen Isotope Determinations 170
3. Isotopic Standards 172
4. Oxygen Isotope Geothermometry 176
5. Application of the "Internal" Thermometer 190
6. Summary 199
   References 200

Chapter 6 · Growth Patterns within the Molluscan Shell:
An Overview

Richard A. Lutz and Donald C. Rhoads

1. Introduction 203
2. Microgrowth Increments 206
3. Shell Structural Changes 230
    References 248

Chapter 7 · Shell Microgrowth Analysis: *Mercenaria mercenaria*
as a Type Example for Research in Population
Dynamics

Michael J. Kennish

1. Introduction 255
2. Shell Microgrowth Patterns in *Mercenaria mercenaria* 256
3. Application to Analysis of Population Dynamics 273
4. Population Dynamics of *Mercenaria mercenaria* in Barnegat
    Bay, New Jersey 274
5. Conclusions and Future Research Efforts 291
    References 292

Chapter 8 · Environmental Relationships of Shell Form and
Structure of Unionacean Bivalves

Michael J. S. Tevesz and Joseph Gaylord Carter

1. Introduction 295
2. Unionacean Form 296
3. Growth Banding 305
4. Ecological and Evolutionary Significance of Unionacean Shell
    Microstructure 309
    References 318

Chapter 9 · Molluscan Larval Shell Morphology: Ecological and
Paleontological Applications

David Jablonski and Richard A. Lutz

1. Introduction 323
2. Developmental Types in Marine Benthic Invertebrates 323
3. Recognition of Molluscan Developmental Types from Larval
    Shell Morphology 329

4. Ecological and Paleontological Applications 347
References 363

Chapter 10 · Gastropod Shell Growth Rate, Allometry, and Adult
Size: Environmental Implications
Geerat J. Vermeij

1. Introduction 379
2. Geometry of Shell Growth 379
3. Allometry and Adaptation 384
4. Growth Rate and Adult Size 387
5. Application to Environmental Reconstructions 390
References 391

Chapter 11 · Growth-Line Analysis as a Test for Contemporaneity
in Populations
John F. Dillon and George R. Clark II

1. Introduction 395
2. Analysis of a Recent Population: *Pecten maximus* 401
3. Guidelines for Future Research 413
References 414

Chapter 12 · Demographic Analysis of Bivalve Populations
Robert M. Cerrato

1. Introduction 417
2. Construction of Class-Frequency Histograms 418
3. Interpretation of Class-Frequency Histograms 420
4. Adaptiveness 453
5. Population Responses to Environmental Stress 455
References 463

II. Other Taxa

Chapter 13 · Barnacle Shell Growth and Its Relationship to
Environmental Factors
Edwin Bourget

1. Introduction 469
2. Shell Macrostructure 470

3. Shell Formation and Growth 471
4. Shell Growth Patterns 473
5. Ecological and Paleoecological Applications 485
    References 489

Chapter 14 · Skeletal Growth Chronologies of Recent and
              Fossil Corals
        Richard E. Dodge and J. Rimas Vaišnys

1. Introduction 493
2. General Features of Coral Growth 494
3. Applications of Coral Growth Features 498
4. Suggestions for Future Research 513
    References 514

Chapter 15 · Growth Patterns in Fish Sagittae
        Giorgio Pannella

1. Introduction 519
2. Morphology and Function of Sagittae 521
3. Calcification 524
4. Cyclical Growth Patterns 529
5. Growth Discontinuities 549
6. Growth Patterns and Ecology 552
7. Fossil Otoliths and Paleoecology 555
    References 556

Chapter 16 · Growth Lines in Polychaete Jaws (Teeth)
        P. J. W. Olive

1. Introduction 561
2. Evidence of Growth Lines in Recent Polychaete Jaws 563
3. Chemical Composition and Preservation Potential of
    Polychaete Jaws 572
4. Ecological and Paleoecological Applications 581
5. Conclusions 589
    References 590

Appendices

Introduction 595

Appendix 1 · Preparation and Examination of Skeletal Materials for Growth Studies

Part A · Molluscs

  1. Preparation of Acetate Peels and Fractured Sections for Observation of Growth Patterns within the Bivalve Shell .. 597

     Michael J. Kennish, Richard A. Lutz, and Donald C. Rhoads

  2. Study of Molluscan Shell Structure and Growth Lines Using Thin Sections 603

     George R. Clark II

  3. Techniques for Observing the Organic Matrix of Molluscan Shells 607

     George R. Clark II

  4. Study of Annual Growth Bands in Unionacean Bivalves ... 613

     Michael J. S. Tevesz and Joseph Gaylord Carter

Part B · Corals

Preparation of Coral Skeletons for Growth Studies 615

Richard E. Dodge

Part C · Fish

Methods of Preparing Fish Sagittae for the Study of Growth Patterns 619

Giorgio Pannella

Part D · Polychaetes

Preparation of Polychaete Jaws for Growth-Line Examination . 625

P. J. W. Olive

Appendix 2 · Bivalve Shell Mineralogy and Microstructure

Part A · Selected Mineralogical Data for the Bivalvia 627

  Joseph Gaylord Carter

Part B · Guide to Bivalve Shell Microstructures 645

Joseph Gaylord Carter

Appendix 3 · Quantitative Analysis of Skeletal Growth Records

Part A · Application of Normalized Power Spectra to the Analysis
of Chemical and Structural Growth Patterns 675

Gary D. Rosenberg, M. Ashton, R. Hewitt, and
D. J. Simmons

Part B · Study of Barnacle Shell Growth-Band Patterns Using
Time-Series Analysis 687

Clifford L. Trump and Edwin Bourget

Part C · Probabilistic Population Descriptions 699

J. Rimas Vaišnys and Richard E. Dodge

Index 725

Introduction

Skeletal Records of Environmental Change

DONALD C. RHOADS and RICHARD A. LUTZ

1. Perspective of This Volume 1
2. How Environmental Change Is Recorded by Skeletal Growth 3
   2.1. Ontogenetic Growth Records 3
   2.2. Demographic Parameters 5
3. Skeletal Growth Applied to Paleoecological and Ecological Problems 6
   3.1. Paleoecological Applications 7
   3.2. Ecological Applications 13
4. Plan of This Volume 14

1. Perspective of This Volume

A wealth of information about past and present-day environments is contained within the skeletons* of fossil and living organisms. Information
about an organism's dynamic environment is capable of being preserved
as a structural, morphological, or chemical change in the skeletal parts
of individual organisms. The demographic structure of a species population also contains much environmental information. Research into these
relationships has, in the past, been largely for the purpose of reconstructing paleoenvironments. The implications of this approach for ecological

---

* We use the term "skeleton" to include both endo- and exoskeletons and have broadened
its definition to include refractory accretionary structures such as polychaete jaws that
may be only partially mineralized.

DONALD C. RHOADS • Department of Geology and Geophysics, Yale University, New
Haven, Connecticut 06520.   RICHARD A. LUTZ • Department of Oyster Culture, New
Jersey Agricultural Experiment Station, Cook College, Rutgers University, New Brunswick,
New Jersey 08903.

1

work has recently interested many neontologists. The ecologist is frequently concerned with reconstructing environmental events after they have taken place. For instance, it may be desirable to assess the effects of major storms, salinity fluctuations, temperature changes, or pollution events on organisms after the event or change has taken place. In the absence of information about predisturbance rates of reproduction, recruitment, growth, and mortality, the ecologist, like the paleoecologist, is confronted with a problem of after-the-fact data acquisition. Skeletal growth records can address this kind of problem.

Our present knowledge of skeletal growth patterns is very limited, and therefore we are only beginning to understand how the ambient environment affects the physiology and, in turn, the structure, mineralogy, chemistry, and growth patterns of an organism's skeleton. Scleractinian corals and bivalved molluscs have historically been the subjects of extensive skeletal growth research, and yet current research on these two groups continues to yield important new insights into the coupling between the environment and the skeleton.

Over the past five years, bivalve and coral skeletal growth research has made impressive progress in identifying environmental causes for observed skeletal growth patterns. In addition, skeletal research of this type has been fruitfully extended to other taxa. These exciting new advances have prompted us to prepare this book as the first volume of the Plenum *Topics in Geobiology* series. We have brought together chapters that represent a cross section of current research that is dominated by work on aquatic invertebrates; one chapter (Chapter 15) is a vertebrate example. Most work has been done with molluscs, especially the class Bivalvia. Bivalves have received this attention because, as a group, they have a broad zoogeographical distribution, an extensive fossil record, and an excellent preservational record. Benthic molluscs therefore represent a "type" example for environmental growth analysis. Many of the techniques developed for their study can be extended to other taxa with little modification. Our contributors have prepared their reviews for nonspecialists and have attempted to identify a spectrum of environmental and paleoecological problems that can be approached with their respective organisms.

This volume will be successful if we have interested some of our readers to participate in the future development of skeletal research. To this end, we have included appendices that cover the following topics: (1) techniques of preparing skeletal materials for the study of ontogenetic growth patterns; (2) the complexities of bivalve shell mineralogy and microstructure; and (3) quantitative techniques for analyzing skeletal growth or mortality patterns related to environmental factors.

2. How Environmental Change Is Recorded by Skeletal Growth

A species is able to grow and reproduce as long as its functional range (biospace) is not exceeded by the ambient environment. Not all parts of the realized biospace promote equal growth or fecundity. Different combinations of niche parameters will be expressed as changed rates of growth, survivorship, or reproductive success. Suboptimal conditions can lead to ecological stress. When broadly defined in this way, ecological stress is probably responsible for many ontogenetic and demographic* patterns discussed in this volume.

Although we are addressing ontogenetic growth records and demographic parameters in general terms, we will illustrate these two sources of information in the following sections by drawing examples from the Bivalvia, in which ontogenetic and demographic skeletal features are well known.

2.1. Ontogenetic Growth Records

In the context of this volume, we define ontogenetic growth as the life history of an individual organism as it is preserved in mineralized, or otherwise refractory, tissues. The ontogenetic history of an organism can be preserved in the skeleton in several forms. For a change in the external environment to be recorded in the molluscan shell, the physiology of the organism must be altered. A metabolic response to an environmental change may, in turn, promote a biochemical response in the mantle and its associated pallial fluid (Chapter 3). Such an external environmental change can then be expressed in the shell. Many physiological changes are preserved in the shell as growth increments, growth discontinuities, or changes in shell allometry, structure, mineralogy, or chemistry, and these skeletal features can be observed on the whole shell or studied from specially prepared shells (Chapters 1, 4, 5, 6, 7, 8, and 11 and Appendix 1).

One of the features of the shell most frequently used in the study of ontogenetic growth is internal growth increments and patterns of increments. In molluscs, thin increments of calcium-carbonate-rich growth lay-

---

* Demography, taken literally, means writing about the people (Gr. dêmos, "the people," plus "to write"). The term was originally used to describe statistical studies of human populations: births, deaths, marriages, and other phenomena. We use demography in a broader sense: the statistical description of populations of any taxonomic group.

ers alternate with organic-rich layers. The time span represented by these increments varies among organisms and habitats, but lunar and solar days are commonly recorded in shallow-water species. These increments can be seen with a microscope in appropriately prepared cross sections of the shell (Appendix 1.A.1 and 1.A.2). No consistent nomenclature exists for the description of these growth features. They have been variously described as internal growth increments, microstructural or microscopic growth increments, or simply microgrowth increments. The prefix *micro*- is used to distinguish them from macroscopically visible growth "annulations" or bands seen on the external surfaces of many shells.*

Microscopic growth increments, observed in thin sections of shell material, are delimited by changes in the ratio of calcium carbonate to organic material. This is seen as a change in optical density of the shell. In acetate-peel replicas of sectioned and acid-etched shells, this compositional periodicity results in differential etching of the shell, and so the compositional change is observed in the peel as a topographic feature. Boundaries of increments can be gradational or abrupt. In the latter case, they may represent periods of nondeposition or an erosional unconformity produced through shell dissolution. These growth interruptions are sometimes called breaks, biochecks, or simply checks. They can be periodical, as in the case of shell resorption at each low tide, or aperiodical, related to occasional predator attack or disturbance caused by storm turbulence. Physiological changes may also be reflected in the shell as a change in shell microstructure—for example, nacreous, prismatic, or crossed lamellar structures (Chapters 2, 6, and 7)—or a change in shell mineralogy and chemistry (Chapters 2, 4, and 5), or both.

In summary, environmental change may be recorded in the skeleton as a change in shell shape and form, microgrowth increments, shell microstructure, mineralogy, or chemistry. A specific environmental change may involve one or several of these variables. At present, most skeletal growth research is attempting to relate observed changes in these shell variables to specific environmental factors.

In living organisms, ontogenetic growth studies involve correlating ecological stress factors such as temperature, salinity, dissolved oxygen, food concentration, substratum conditions, and water turbidity with the skeletal growth record. This may involve experimental manipulation of these variables in the laboratory or consist of correlating seasonally changing field conditions with growth of *in situ* populations. In some cases, transplantation of individual organisms from one habitat to another can

---

* Macroscopic growth bands observed on the surfaces of many shells may be a surface expression of closely spaced internal growth increments. Not all species have surface expressions of their internal increments.

provide interesting information about the effect of site-specific factors on ontogenetic growth.

Because it is not possible to experimentally determine ontogenetic responses of fossils to paleoenvironments, the paleoecologist must rely on his knowledge of skeletal growth responses of extant organisms. It is sometimes possible to extrapolate details of skeletal growth data from living organisms to fossils, especially when the comparison involves organisms of close taxonomic affinity. If extinct species have no close living representatives, the interpretation of growth patterns in fossils requires cautious *a priori* reasoning and analogy based on a general knowledge of environmental–skeletal relationships in Recent species.

The ontogenetic approach is not limited to the analysis of internal patterns of growth. One can compare allometric changes in overall external shell dimensions as measured over relatively large intervals of shell growth (usually annual)(Chapters 1, 8, and 10). In fossils in which the internal growth record is not preserved, analysis of external shell features is the only means of obtaining ontogenetic information. Also, shell sections representing a complete record of ontogenetic growth are difficult to obtain in three-dimensionally coiled shells (e.g., spired gastropods). Therefore, even in recent taxa or in well-preserved fossils with spired shells, ontogenetic information must be obtained from external shell features (see especially Chapter 10).

Another ontogenetic approach concerns the earliest growth stages: larval and early juvenile shells. The size, ornamentation, and shape of molluscan protoconchs and prodissoconchs can be used to infer larval developmental types, time of metamorphosis, and dispersal potential (Chapter 9). In some cases, once the larval developmental type has been determined from the features of the top shell, it is possible to make further inferences about adaptive strategies and, hence, the response of an organism to past, present, or future ecological stresses.

2.2. Demographic Parameters

Demographic records of environmental change are based on the response of populations to temporal and spatial gradients in environmental quality and ecological stress (Chapter 12). Such changes may be measured as changes in growth rate, recruitment, or survivorship as estimated from size or age-frequency distributions of skeletal parts.

When demographic analysis is applied to fossil populations, one must employ independent taphonomic* information to determine whether a

---

* Taphonomy is the study of sources of fossil remains and the manner in which they become buried within sediments.

specific fossil assemblage was formed by shells accumulating gradually
by age-dependent mortality or, on the other hand, if the death assemblage
was formed by a mass mortality. The contemporaneity or noncontemporaneity of death assemblages can be determined by comparing growth-
line patterns among specimens of a sampled death assemblage (Chapter
11). Once the paleoecologist has determined whether the fossil assemblage
was formed by a gradual accumulation of individual organisms of different
ages (a dynamic assemblage) or by a mass mortality (census or cohort
assemblage), he can construct population growth and survivorship curves
to make estimates of the seasonality of recruitment and rates of growth
and mortality.

If living populations are being studied, a one-time collection of a
population will reflect the instantaneous age structure of that population.
By making several such censuses, spaced over at least one year, population
recruitment, growth, and mortality can be assessed. Estimates of these
parameters made from size or age-frequency histograms can be checked
by using mark and recapture techniques. Population responses to specific
environmental change can be experimentally explored in the field by
making reciprocal transplants between habitats. Observed changes in demographic parameters are population measures of relative ecological
stress between habitats.

Because the paleontologist, through demographic analysis, is able to
make inferences about the dynamics of fossil populations, paleontology
and population biology enjoy a close kinship. The dimension of geological
time has been added to problems of island biogeography, population genetics, r and K selection, and community dynamics.

3. Skeletal Growth Applied to Paleoecological and Ecological Problems

A wide range of paleoecological and ecological problems can be approached at present with skeletal research. Because research is still at an
early stage of development, this range will undoubtedly expand in the
future. In the following sections, we have outlined applications that we
feel can be made at present with current knowledge and techniques. In
some of our examples, skeletal research has already proven to be fruitful;
in these cases, we have referred to chapters that explain these applications
in greater detail. In other examples, we have had to speculate as to the
future potential value of skeletal research because, to date, some problem
areas remain unexplored.

3.1. Paleoecological Applications

A paleoecological reconstruction can never be proven to be accurate but only more or less credible than an alternative reconstruction. The strength of a paleoenvironmental reconstruction is directly proportional to the number of pieces of independent data that support it. Herein lie both the strength and the weakness of skeletal growth data. On the one hand, several kinds of environmental variables may produce a similar growth record within the shells of a fossil population. Therefore, statements about specific causal events may, at best, be speculative. On the other hand, skeletal growth data, taken together with stratigraphic, sedimentological, and geochemical information, can provide a very important source of independent data for extablishing paleoecological relationships. In the following discussion of the application of skeletal growth to particular paleoecological problems, we have illustrated several examples by making specific reference to molluscan growth; our molluscan examples have broader implications for skeletal growth in other organisms.

3.1.1. Paleoclimatology and Paleolatitudes

Large annual changes in water temperature, runoff, primary production, and water turbulence are associated with temperate–boreal latitudes. In coastal marine habitats, climatic variability is at its extreme in areas that lie in the windward lee of continents. Temperature has a dominant effect on both soft-tissue and skeletal growth. Although important, the other seasonally related variables listed above appear to be of second-order importance relative to temperature in controlling seasonal growth rates of organisms. In many taxa, the skeletal growth record of extreme thermal variability is often recorded by marked seasonal clustering of internal growth increments; in boreal and temperate species, widely spaced increments frequently represent warm months and narrow increments, cold months. Seasonality may also be preserved in the skeleton as a change in mineralogy, chemistry, or micro- or ultrastructure (Chapters 2, 4, 5, 6, and 7). Because the skeletal growth record can reflect the thermal variability of a habitat, analysis of seasonal patterns over a broad biogeographical or stratigraphic range can potentially provide the paleoecologist with insights into climatic gradients in space and climatic change through time.

Seasonality exists even in the thermally equitable environments of the tropics and polar regions. In the tropics, seasonality exists in the form of annual changes in precipitation, wind mixing, or upwelling, and in polar regions, changes in ice cover and primary production may influence skeletal growth. Unfortunately, present information about how skeletal growth is related to these factors is sparse, and therefore we are unable to generalize about how nonthermal attributes of seasonality are expressed in the skeletal growth record. These aspects of skeletal growth should be the focus of future research.

At present, structural growth patterns can provide information only about relative changes in seasonality. However, oxygen isotope paleotemperature and geothermometry techniques can estimate absolute paloetemperatures. The isotopic approach is preferred for this reason, but there are constraints on the kind of skeletal material that can be used. Most important, the equilibrium oxygen isotope fractionation of the skeleton must not be diagenetically altered (Chapter 5).

Seasonality is also reflected in demographic features of populations. Reproduction, recruitment, population growth, and mortality are all seasonally dependent. In seasonably variable environments, age-frequency distributions are typically polymodal and very dynamic (Chapter 12).

Another potential source of paleoclimatic information exists in the geographic variability of the modal size of larval molluscan top shells. The size of molluscan larvae at the time of settlement and metamorphosis increases with increasing latitude and is thought to reflect the temperature dependence of larval development (Chapter 9). The top shell size–temperature relationship has been shown to exist in a few Recent species, but this relationship has not yet been explored for reconstructing paleolatitudes or paleoclimates.

3.1.2. Paleobathymetry

At a given latitude, seasonally variable properties of the water column decrease with increasing depth, especially below the seasonal thermocline. Shallow-water populations, especially those located in the intertidal zone, may therefore show more marked seasonal clustering of internal growth increments than deeper-water populations (Chapter 6). Also, shallow populations are known to experience high-frequency changes in the ambient environment such as tidal exposure and daily fluctuations in water temperature and light intensity. At greater depths, organisms may experience only low-frequency seasonal or annual fluctuations related to annual overturn or seasonal sedimentation of pelagic detritus to the bottom. Vertically migrating species, such as midwater fishes, which move across a thermocline or other depth-dependent change in water-column properties, may record such migratory behavior in the growth of their otoliths. These migratory growth patterns should appear very different from the otolith growth patterns of deep-water species that do not migrate vertically or from the growth patterns of saggitae of shallow-water species (Chapter 15).

Even abyssal species apparently experience cyclic growth. We have relatively little information about the growth of abyssal organisms, but a few molluscs have been studied, and their shells show thin and diffuse internal microgrowth increments. Abyssal barnacles show similar cyclic growth features (Chapter 13). Deep-water molluscan growth increments are not clustered into fortnightly, monthly, or seasonal periodicities as they are in shallow-water species (Chapter 6). The temporal value of growth increments in abyssal species is still a question, but this interesting phenomenon is at present being investigated by us (D.C.R. and R.A.L.) in the giant "clams" and "mussels" of the recently discovered Galapagos thermal sites (rift vents) on the Pacific abyssal seafloor.

Within the subtidal zone, one may be able to further discriminate between benthic habitats located above or below a seasonal thermocline on the basis of the degree of seasonal clustering of internal growth patterns or depth-related allometric changes in skeletal dimensions (Chapter 1).

3.1.3. Other Paleooceanographic Parameters

Earlier we alluded to the use of oxygen isotope paleothermometry to establish paleotemperatures. Recent advances in paleothermometry also permit estimates of paleosalinities. Most paleotemperatures have been based on whole (homogenized)-shell or single-mineral (either calcite or aragonite) analysis. To interpret whole-shell or single-mineral isotopic data, it has been necessary to estimate the oxygen isotopic composition of the ambient water of deposition. The $\delta^{18}O$ of a parcel of seawater can be highly variable, and therefore its isotopic composition is difficult to estimate. This problem may be circumvented by determining the isotopic fractionation between calcite and aragonite in bimineralogical species. This fractionation may yield the temperature of formation without making assumptions about the isotopic composition of the ambient water. In fact, the calcite–aragonite thermometer technique can be used to independently determine the isotopic equilibrium of the ambient water and its salinity (Chapter 5).

Ontogenetic variations of major and minor chemical constituents of the shell such as calcium or trace metals may in the future provide information about the properties of the ambient water. To extract this information, one must be able to recognize primary or biogenic shell compositions from diagenetically altered ones. Unaltered shell chemistry can be identified if the chemistry of the shell can be shown to change systematically with ontogeny. However, up to the present time, few workers have made complete ontogenetic descriptions of shell chemistry (Chapter 4). At present, research on shell chemistry and mineralogy is focused on how ontogenetic changes in shell composition are coupled with specific environmental factors (Chapters 2 and 4).

As suggested in Chapter 10, the mean maximum size of adult gastropod species, compared within a genus, may be closely tied to ambient nutrient levels. If this hypothesis is correct, one might be able to make inferences about relative nutrient levels or trophic resources by comparing adult sizes between local habitats or even between marine basins.

Ancient hydrographic features such as tidal fluctuations can be obtained from the study of internal growth increment patterns. This is best accomplished by studying fossil intertidal species, although some shallow subtidal organisms also record tidal phenomena. Intertidal species are forced to close their shells, or to clamp down on the substratum, to avoid desiccation. During the period of low tide when shell valves are closed, these organisms may respire anaerobically and secrete acidic metabolic end products. Metabolic acids will dissolve those parts of the shell that are in contact with the body fluids (Chapters 3 and 6). When an organism is again covered by water during the flood tide, the shell can open and aerobic metabolism and shell deposition are resumed. These small-scale shell-dissolution events produce striking growth-increment sequences, and detailed analyses of the clustering of the lunar periodicities can allow one to reconstruct the tidal frequency spectrum (Chapters 6 and 13).

3.1.4. Paleobiogeographic Barriers

The comparison of distributional patterns of benthic species with different modes of larval dispersal within, or between, basins may prove useful for establishing the time of appearance or disappearance of dispersal barriers (Chapter 9). For instance, during the early stages of continental separation, the shape of marine basins is commonly very long but narrow. The proportion of shelf is high relative to deep water. At this stage, larvae of all types, even those of short pelagic duration, may be dispersed across the narrow basin. As continental drift continues, and transbasinal distances increase, species with benthic and short-duration pelagic stages will no longer be shared between the separating shelves. This will lead to increasing faunal dissimilarity and promote endemism at each side of the basin. At great distances, and in the absence of intrabasinal islands, only species with long-distance dispersal capabilities will be shared. The convergence of continental masses might also be reflected in an increasing proportion of species with short-duration pelagic larval stages being shared on opposite sides of the basin as the oceanic barrier narrows.

It is now possible, in many cases, to identify the dispersal abilities of fossil molluscs from features of their larval top shells. Therefore, the temporal and spatial change in the frequency of larval dispersal types around the perimeter of an ancient basin may provide a relative chronology of changing ocean–continent relationships. Other types of biogeographical dispersal barriers such as island arcs, land bridges, or major deltas may similarly be evaluated with distributional patterns of larval top shells (Chapter 9).

3.1.5. Paleoenvironmental Reconstructions

Many paleoecological problems involve reconstructing past environmental gradients and habitat conditions. We have already seen how skeletal growth data can be used to reconstruct depth gradients, paleosalinities, and paleotemperatures. Other parameters such as relative changes in water turbidity, bottom stability, dissolved oxygen, and concentration of food can affect growth, frequency of reproduction, recruitment success, and survivorship. Of course, none of these variables leaves a unique imprint of its effect on skeletal growth. Also, in the context of ecological stress, these environmental factors will affect members of a community differently, depending on the phsyiological tolerance of each species for the time–space variation of each variable. Because the specific cause for a change in growth, reproduction, or mortality cannot be uniquely identified from the skeletal record alone, identification of specific causal factors will depend on the weight of independent evidence.

For instance, one might observe a difference in mean and maximum size, shell thickness, and shell shape of two fossil suspension-feeding mollusc populations, one collected from a sandstone and the other from a mudstone. Through stratigraphic and facies analysis, we determine that these two lithologies, and their contained fossils, were contemporaneous. The question then arises: are the observed morphotypes different species (or subspecies), or are they the same species with different growth rates, allometries, and survivorships? Ontogenetic and demographic analyses might show that the smaller shells of the mudstone grew less per unit time than did equal-aged specimens from the sandstone. Also, the mud population might have many more "disturbance" lines within their internal growth patterns than the sand population. Demographic data might show that recruitment and survivorship were lower in the mudstone than in the sandstone and that a clinal gradient of growth exists across the sandstone–mudstone facies transition. Further, we notice that the posterior region of the shell in the mud morph is extended relative to the sand-dwelling morph. We might also discover that this allometric difference is difficult to resolve in specimens sampled from the mixed sand–mud facies boundary.

With the foregoing observations, one could make a strong case for the two lithologically separated populations being ecotypes of the same species. In other words, this species is phenotypically plastic, and its shape, growth rate, and population structure reflect responses to different habitat conditions. In our hypothetical case, we might also be able to make more specific inferences about the nature of the stress factor(s). If skeletal analysis were extended to other species in the fossil assemblages, one might find that only heavily shelled epifaunal suspension feeders were "stunted" in the mudstone facies and that the growth rates of infaunal species were not affected. Independent data from sedimentary structures might further show that the mudstone was extensively bioturbated. On the basis of this additional evidence, one might hypothesize that the stress factors were related to substratum conditions. The quasi-fluid, bioturbated mud surface, and its associated turbidity, might be identified as operating selectively on heavily shelled epifauna. The bearing capacity or strength of such a surface would not provide firm support for heavily calcified species, and high levels of turbidity would be expected to clog ciliary feeding mechanisms. This reconstruction would be consistent with our earlier observation that the "stunted" mud morph had extended posterior relative to the sand morph. By analogy with Recent mud-dwelling suspension-feeding bivalves, one could show that this allometry serves to elevate the siphons above the mobile mud–water interface, a morphological solution to siphonal clogging.

This hypothetical case is, of course, an oversimplification, and actual reconstructions are usually based on more equivocal evidence. It should be obvious that this kind of *a posteriori* reasoning is fraught with the pitfalls of circular reasoning. However, the circularity can be avoided by drawing evidence from multiple independent sources.

Skeletal growth analysis can also be applied to anthropological reconstruction. Many ancient tribes spent at least part of the year fishing at aquatic habitation sites. To establish migrational patterns of these tribes, the anthropologist may wish to know whether the site was occupied permanently, or only seasonally, and the season of fishing may also be of interest. The season of death (Chapters 6, 7, and 15) can be determined from microgrowth patterns in shellfish and their associated calcified epifauna or from fish otoliths (sagittae) found in kitchen middens. From these patterns, it is possible to determine whether the prey populations were cropped within a single season (a census), or whether the assemblage represents a year-round fishing effort associated with a permanent habitation site.

3.2. Ecological Applications

The response of living organisms to changing environmental conditions is often explored by utilizing census sampling techniques; populations are sampled several times a year to define and follow recruitment patterns and population growth and mortality. Mark-and-recapture techniques are sometimes also employed to follow ontogenetic growth. These techniques are time-consuming and expensive because sampling must be carried out for at least one year. When completed, the results are applicable only to the year of study. This kind of approach is, of course, the only one available for organisms without skeletal parts. However, skeletonized organisms contain a great deal of ontogenetic and demographic information stored within their skeletal parts. Long-lived species may contain ontogenetic growth records that extend backward in time for several years.

3.2.1. After-the-Fact Pollution Studies

The neontologist is frequently confronted with the problem of reconstructing the effects of a pollution event after the pollution has taken place and in the absence of information about prepollution recruitment patterns and rates of growth and mortality. If the neontologist is able to work with calcified taxa, or the jaw structures of certain polychaetes (Chapter 16), many of the techniques described in the preceding section on paleoenvironmental reconstruction can be employed to study after-the-fact problems. The space–time dimensions of a pollution event can be defined in a high degree of detail using skeletal growth data (see, for example, Chapter 7).

3.2.2. Identification of Adaptive Strategies

Skeletal growth parameters can be used by both the neontologist and the paleontologist to recognize adaptive types. Opportunistic (pioneering) species have high intrinsic rates of population increase, high rates of ontogenetic growth, and high dispersal capability. Opportunists usually have short life spans and therefore reproduce at an early age. Mortality rates are also high. Equilibrium species have slower rates of population and ontogenetic growth. Birth rates are commonly balanced by death rates. All these life-history phenomena are capable of being deduced from ontogenetic shell growth, larval top shells of molluscs, and population shell growth parameters. Skeletal growth analysis can provide an efficient means of extracting this dynamic information. Once this is done, a sympatric association of species can be ordered into a relative spectrum of opportunism. With such a species ranking, one can make inferences about the role of each species in colonizing new, or newly disturbed, habitats (Chapters 9 and 12).

3.2.3. Fisheries Management

Fisheries management can also benefit from skeletal growth studies. Fish scales and otoliths have been used for some time to age fish, and these data have proven useful in the management of fisheries. Recent advances in otolith growth studies discussed in Chapter 15 should enhance the utility of this approach even further. The application of detailed skeletal growth analysis to shell fisheries has, as yet, not received the attention it deserves. The shell fisheries biologist is primarily interested in optimizing recruitment and maximizing the yield of soft-tissue biomass. Skeletal growth can reflect changes in soft-tissue biomass, but only indirectly. During cold months, some molluscs, for instance, experience negative growth of soft tissues, while the shell, although being deposited very slowly, experiences positive growth. In other molluscs, both soft and calcified tissues may experience negative winter growth (see, for example, Chapter 6). Poor correlations between changes in shell growth and soft-tissue biomass may also result when gametes are released. Soft-tissue biomass may decrease precipitously during spawning, while this reproductive event is recorded in the shell by only a few disturbance lines (Chapter 7). During the period of summer growth, increases in soft-tissue biomass are usually positively correlated with maximum rates of shell deposition.

Skeletal growth is probably best used in shellfish management to construct ontogenetic and population growth curves. The ontogenetic curve can be used to determine the age at which growth becomes asymptotic to the time (age) axis. The time or age of maximum yield, in terms of population growth, can also be determined from a demographic analysis of skeletal growth. Finally, the age of first reproduction, and the subsequent reproductive history, can often be deduced from spawning lines in the shell (Chapters 6 and 7). By relating these spawning lines to the seasonal clustering of internal growth patterns, one can plan the schedule of harvest so that the reproductive potential of the species is not compromised.

4. Plan of this Volume

Approximately three quarters of the chapters in this volume are in Section I, The Mollusca, and the class Bivalvia is the focus of most of these reviews. This taxonomic imbalance reflects the skewed nature of recent skeletal research. Because of the importance of the Bivalvia in skeletal research, Section I is introduced with three important background chapters.

Chapter 1 is a review of the gross morphology of the bivalve shell, its relationship to soft-part anatomy, and the effect of the environment on macroscopic features of shell growth and allometry. This macroscopic perspective is important since all subsequent chapters in Section I assume that the reader has a knowledge of the basic features of bivalve growth and morphology covered in Chapter 1.

Chapter 2, with its accompanying Appendix 2, is a review of bivalve shell mineralogy, microstructure, and architecture. The distribution and structural arrangement of calcite and aragonite in bivalve shells are a function of both genotype and environment. To recognize mechanical–structural adaptations in bivalves and to interpret environmental influences on shell construction, one must be able to distinguish among several common types of shell microstructure. The basic constructional units of shells may take several forms, for example, prisms, tablets, or blades. These basic units, generally aragonitic or calcitic, are associated with an organic matrix, and together they comprise shell layers. Several different kinds of shell layers may constitute a molluscan shell, and the arrangement of these layers defines its architecture. Time-parallel growth features, for example, growth increments, are superimposed on the shell's microstructure and architecture. The geometric relationships between the architecture of the shell layers and time-parallel growth features can sometimes be complex and confusing to the uninitiated. However, these two categories of shell features must be distinguished because they often reflect very different biological and environmental factors. The problem of their resolution is analogous to separating time-parallel depositional horizons from time-transgressive facies in stratigraphy. Thus, major shell layers are commonly time-transgressive, whereas growth features such as microgrowth increments are time-parallel (Chapter 6). Appendix 2.A outlines techniques for inferring original mineralogy for altered fossil shells, and provides a data base from which to evaluate the environmental and biological significance of bivalve shell mineralogy. Appendix 2.B is intended to help the reader recognize a variety of microstructures commonly encountered in bivalve shells. Chapter 6 shows the relationship between time-parallel growth structures and shell architecture.

The molluscan mantle, and the pallial fluid, are responsible for the formation of the microstructures described in Chapter 2 and Appendix 2, and for their arrangement into functional shell architectures. Unless the physiolgical basis of shell deposition and dissolution is well understood, one cannot go very far in interpreting how environmental changes may be recorded in shell growth. Chapter 3 provides this physiological perspective. Shells, like soft tissues, can experience both positive and negative growth. Under certain physiological conditions, part of the interior shell surface may be dissolved. In the process of shell dissolution and subsequent redeposition, primary shell architecture is sometimes reworked into distinctive secondary structures (see Chapter 6). These reworked shell layers provide an important datum for reconstructing an organism's physiological history (Chapters 6 and 7). Shell dissolution and reworking also have important implications for interpreting shell chemistry (Chapters 4 and 5).

Chapter 4 points out some of the sampling and operational problems that have, up to now, limited progress in recognizing the environmental significance of chemical periodicities observed in bivalve shells. The review points out, however, that analysis of ontogenetic variations in chemical composition may help clarify cause–effect relationships between environmental changes and patterns of shell chemistry. Allometric principles may also help to distinguish original unaltered shell compositional variations from diagenetic chemical alterations.

Although the theoretical basis for oxygen isotope geothermometry has been known since 1947, several problems have prevented this geochemical technique from being more widely applied. Chapter 5 reviews the history and development of $^{18}O/^{16}O$ paleothermometry, and the authors propose a new oxygen isotope geothermometer based on the isotopic fractionation between calcite and aragonite in bimineralogical shells. This calcite–aragonite thermometry can also provide an estimate of paleosalinity.

Internal ontogenetic growth patterns in marine bivalves have been the focus of considerable research. Detailed descriptions of these patterns, and how they may be used to reconstruct environmental change, as well as the physiology, metabolism, and reproductive history of a bivalve, are given in Chapters 6 and 7. The environmental significance of form and structure of freshwater bivalve shells is covered in Chapter 8.

The earliest growth stages of molluscs are often preserved in the umbonal or apical top shell of postmetamorphosed specimens. The size, shape, and ornamentation of the top shell can be related to larval developmental type. The larval developmental type, in turn, strongly influences the dispersal potential of a species and allows one to make inferences about a species's adaptive strategy. The ecological and paleoecological significance of bivalve and gastropod top shells is discussed in Chapter 9. Chapter 10 extends observations on the Gastropoda to postlarval growth stages and presents hypotheses on the relationship between environmental conditions and growth, allometry, and maximum adult size.

Analysis of skeletal growth in populations is covered in Chapters 11 and 12. Chapter 11 demonstrates how patterns of external growth banding on the shell can be used to establish whether a sample of a fossil species population, collected from a single horizon, represents a contemporaneous life assemblage. This is most easily done with mass mortalities. The establishment of contemporaneity or noncontemporaneity in a fossil population is very important for subsequent demographic analysis (Chapter 12). Chapter 12 is an introduction to the construction and interpretation of size (age)-frequency distributions and population growth and survivorship curves for both living and fossil molluscs. This chapter shows how comparisons of demographic parameters in space and time can provide useful insights into population responses to environmental change.

Section II presents ontogenetic growth studies in barnacles, corals, fish, and polychaetes. These four groups show ontogenetic growth patterns that are analogous to growth features described for the Mollusca. That we can compare ontogenetic growth structures among such phylogenetically distant organisms is important. Future work is certain to reveal that ontogenetic growth phenomena are taxonomically widespread and that environmental change is capable of being recorded in most skeletal structures. Striking growth patterns are observed within and on the surface of the tests of intertidal barnacles (Chapter 13). The study of these patterns, and their relationship to environmental factors, is limited at present to Recent species. However, these skeletal growth phenomena may be applied to the study of fossil Cirrepedia in the near future. Appendix 3.B discusses the manner in which barnacle skeletal growth patterns can be quantitatively related to external environmental factors using time-series analysis.

Chapter 14 is a radiographic study of internal growth banding in Recent corals from the Bermuda Platform. Growth banding patterns are used to reconstruct the effect of an airport land-fill project on coral growth and mortality in Castle Harbor, Bermuda. Appendix 3.C shows how hypotheses about environmental causes for observed coral growth and mortality patterns can be quantitatively stated and tested.

Sagittae (otoliths) have been used for some time to age fish. Chapter 15 attempts to define the underlying physiological and ecological factors that produce growth patterns in saggitae. Within this broader view of otolith growth, future research should provide important new insights into the coupling between otolith growth patterns and environmental change.

Polychaete jaw structures may, or may not, be highly mineralized, and these structures provide little structural support for soft parts. Therefore, one might argue that these structures are not truly skeletal. However, polychaete jaws, found in 13 important errant polychaete families, are refractory accretionary structures that allow these worms to "work on" or otherwise manipulate their environment. In addition, polychaetes are perhaps the most ubiquitous and abundant macrofaunal group in the marine environment. For these reasons, we have included Chapter 16. Despite the importance of polychaetes in benthic habitats, we have comparatively little detailed information about the ontogenetic growth history of many species. In those 13 families that possess accretionary jaw structures, scolecodonts contain an important, and as yet largely unutilized, source of information about reproduction and longevity.

The appendices are included for those who wish more detailed information about the preparation and analysis of skeletal growth records. Appendix 1 gives practical information on the preparation of skeletal materials for study. Appendix 2 is supplementary to Chapter 2, but also provides useful background information for many other chapters in Section II. Appendix 3 presents three quantitative approaches for relating environmental factors to observed skeletal growth patterns. The three examples are related to Chapters 4, 13, and 14, respectively. Although many of the examples are specifically applied to the three taxa covered in these chapters, many of the techniques can be generally applied.

ACKNOWLEDGMENTS. In 1975, we (D.C.R. and R.A.L.) had discussions with Eric Schneider, Director of the EPA Water Quality Laboratory, Narragansett, Rhode Island, about the potential of using skeletal growth analysis to monitor environmental change. These preliminary discussions subsequently resulted in our working with the EPA to explore the utility of using molluscan shell growth records in water pollution studies. The original scope of this project included preparation of a manual of molluscan skeletal growth analysis and techniques for EPA biologists. While preparing this manual, we decided that we would launch a separate project and invite specialists to contribute to a more comprehensive book on up-to-date advances in skeletal growth, extending the subject coverage to include other taxa. Without the initial funding and logistical support provided by the EPA, this volume could not have been realized. In particular, we wish to thank Eric Schneider, Gerald Pesch, and Robert Payne of the Narragansett Laboratory for their early interest in, and support of, our ideas. The preparation of parts of Chapters 6 and 9 and all of Chapter 12 was supported by EPA funds. We are also indebted to Drs. Robert Aller and Josephine Yingst, who prepared chapters for the original EPA manual. We have benefited from their work in the preparation and editing of this volume.

A major task in assembling a multiauthored volume is the review of manuscripts. This has been less of a task than we originally thought because the following reviewers were very responsive to our requests for prompt, yet thoughtful and critical, reviews: William Berry, A. Clarke, D. Crisp, D. Dean, R. M. Dillman, J. R. Dodd, R. Fairbanks, K. Fauchald, R. Green, W. J. Kennedy, M. Kennish, D. Lawson, J. Levinton, R. Linsley, C. MacClintock, A. L.,McAlester, E. McCoy, G. Pannella, S. Rachootin, R. Robertson, S. Saila, R. Scheltema, S. Stanley, C. Stearn, F. Stehli, R. Strathmann, J. D. Taylor, C. Thayer, I. Thompson, K. M. Towe, D. Townsend, R. D. Turner, R. Vaišnys, T. R. Waller, K. M. Wilbur, and S. Woodin.

Mr. Dwight Muschenheim assisted us with the many details of styling and proofreading of manuscripts. We also wish to thank our typists, who saw us through several drafts of edited copy: Betsy Dabakis and Doreen Orciari of the Department of Geology and Geophysics, Yale University, and Maxine Lane, Constance Williams, Stephanie Clifford, and Lois Lane of the Ira C. Darling Center, University of Maine.

This is Paper No. 5306 of the Journal Series, New Jersey Agricultural Experiment Station, Cook College, Rutgers University, New Brunswick, New Jersey.

I

The Mollusca

Chapter 1
Shell Growth and Form in the Bivalvia

RAYMOND SEED

1. Introduction 23
2. General Features of the Bivalve Shell 24
3. Origin and Evolution of Bivalves 28
4. Shell Growth 32
    4.1. Absolute Growth 33
    4.2. Allometric Growth 37
5. Shell Shape and Body Form 45
6. Functional Morphology 52
    References 61

1. Introduction

The Bivalvia constitute the second largest class within the phylum Mollusca. This exclusively aquatic group is exceedingly diverse, and includes species of considerable economic and ecological importance. Bivalves have had a long and supremely successful evolutionary history in which comparatively few lineages have become extinct. Although they are widely distributed, their apparent preference for shallow-water habitats makes them particularly amenable to detailed investigation.

To interpret the functional morphology and paleoecology of fossil bivalves, paleontologists depend to a great extent on a thorough knowledge of Recent taxa. Because many fossil bivalves have closely related

RAYMOND SEED • Department of Zoology, University College of North Wales, Bangor, Wales, United Kingdom.

living representatives, comparisons between Recent and fossil taxa can be made with perhaps a greater degree of confidence for the Bivalvia than is possible for virtually any other group of organisms. By paying particular attention to the ways in which living bivalves respond to environmental change, the paleontologist can obtain a considerable amount of valuable information concerning not only the functional nature of fossil taxa but also the environmental conditions of the ancient biospheres in which these taxa evolved. At the same time, however, considerable care must also be exercised with such an ecological approach because specific morphological structures may prove to have several distinct functions. This should serve as a caveat to those who may be too willing to generalize about bivalve form and function.

The presence of an accretionary exoskeleton in bivalves in which the total sequence of growth stages throughout ontogeny are recorded provides an opportunity for analyzing the effects of environmental change on shell growth and form. In this chapter, several aspects of bivalve growth as observed macroscopically (microscopic growth patterns are considered in Chapters 6, 7, and 11) are reviewed, and both their ecological and their paleoecological implications are briefly considered. First, however, the characteristic features of a generalized bivalve shell are described and the major evolutionary themes within the class Bivalvia summarized. Ontogenetic and environmentally induced variation in growth rate and body form are considered within the constraints that are imposed by an immutable* accretionary skeleton. The final section of this chapter is devoted to an analysis of the functional (adaptive) nature of shell morphology in bivalves that differ markedly in their respective modes of life.

2. General Features of the Bivalve Shell

The bivalve shell consists of three distinct parts—two laterally compressed calcareous valves united dorsally by means of an elastic ligament. This comparatively simple construction, which grows by the accretion of shell material around the valve margins, is, as we shall see later, subject to considerable ontogenetic, as well as phylogenetic, variation. Wainwright (1969) considers that the shape and surface texture of the bivalve shell are advantageous design features in a structure that is subject to continual mechanical stress. However, before reviewing shell growth and

---

* I use the term "immutable" only in the sense that the gross morphology cannot be changed once the shell is deposited; the bivalve shell is mutable, but only at the growing margins.

form, I shall first of all consider some of the basic terminology used for describing the morphology of bivalve shells.

Figure 1 illustrates the major morphological features and axes of orientation of a generalized bivalve mollusc. The areas of shell immediately surrounding the growth centers or *beaks* of each valve constitute the *umbones* (singular *umbo*). These generally raised, somewhat convex regions often bear traces of the larval shell or prodissoconch (see Chapter 9). They may be curved anteriorly (*prosogyrate*) or posteriorly (*opisthogyrate*), or they may be positioned so as to directly face each other (*orthogyrate*). Where the umbones are widely separated, the space between them is termed the *cardinal area*, but this may be reduced to an anterior heart-shaped *lunule* and a posterior *escutcheon* in shells with closely opposing umbones. Carter (1967a) discusses the structural and functional significance of these two differentiated regions, which are generally set apart from the rest of the shell by a change in surface sculpture.

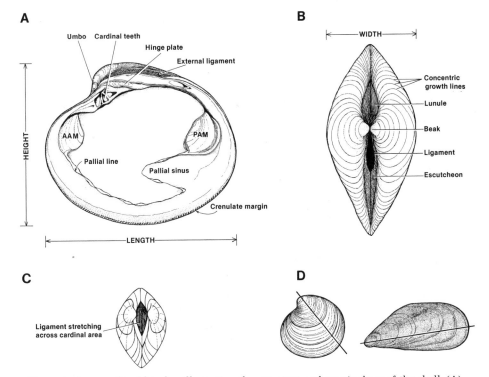

Figure 1. A generalized bivalve illustrating the structure and terminology of the shell. (A) Inside of right valve. (AAM, PAM) Anterior and posterior adductor muscle scars, respectively. (B) Dorsal view of shell. (C) Dorsal view of shell with extensive cardinal area. (D) Axis of maximum growth in a clam (*left*) and a mussel (*right*).

Dorsally, the valves are united by a three-layered elastic *ligament* that consists largely of the horny proteinaceous material conchiolin. The ligament is an integral part of the bivalve shell and is secreted by the underlying mantle tissue in much the same way as the shell valves. It functions like a spring and serves to open the valves when the adductor muscles relax. This it does via the energy stored in it during its elastic deformation whenever the valves are adducted. Where the ligament is external, it is generally positioned behind the umbones, but when an extensive cardinal area is present, it may extend on both sides of the umbones. Internal ligaments are usually more condensed and often set within shallow pits or *chondrophores* immediately below or to one side of the umbones. Because external ligaments are stretched and internal ligaments are compressed during shell closure, the terms *tensilium* and *resilium*, respectively, are sometimes applied to these two structures. The diversity, functional design, and mechanical efficiency of various types of bivalve ligaments are reviewed by Trueman (1969).

Situated on the inner surface of the dorsal region of the valves immediately beneath the umbones is the *hinge*, which functions during the opening and closing of the shell. Along the hinge is a series of *teeth*, the number and arrangement of which are important taxonomic features. Although the two valve surfaces interlock closely at this juncture, they are always separated by a neck of tissue, the mantle isthmus, which is continuously enlarging the teeth and deepening the opposing troughs as the shell grows. Hinge teeth ensure that the valves are correctly aligned when the shell closes; they also prevent rotational and shearing movements of one valve over the other. The *taxodont* condition, found in the Nuculacea and Arcacea, is probably the simplest and most primitive type of hinge. In this type, numerous small, uniform, straight or chevron-shaped teeth occur over the entire length of the hinge. The more common and more highly developed *heterodont* hinge consists of distinctly differentiated cardinal and lateral teeth, which diverge from a point immediately below the umbones. The taxodont and heterodont conditions, along with various other types of bivalve dentition, are detailed by Cox (1969). In some taxa, hinge teeth may be reduced or even absent.

Muscle scars present on the inside of the shell can sometimes provide valuable information concerning the structure of soft parts and the life habits of bivalves. Furthermore, since many of these scars are frequently well preserved in fossil material, they provide a potentially important paleoecological tool. Extending around the inside of the valves near the shell margin is the *pallial line* representing the point of attachment of the shell-secreting mantle lobes. In taxa with posterior siphons, the pallial line may be indented, forming a *pallial sinus* into which the siphons can be withdrawn by means of the siphonal retractor muscles (enlarged pallial muscles). The size of this sinus broadly reflects the strength and extrusibility of the siphons and the degree to which they are fused together, rather than their overall length (Kauffman, 1969). Valuable inferences about modes of feeding and burrowing depths can be made from the nature of the pallial sinus. The largest and perhaps most important muscle scars on the inner shell surface are those produced by the adductor muscles because the spatial orientation of many other features of the bivalve body can be related directly to these reference points. Since the adductors are responsible for valve closure, their action directly opposes that of the elastic ligament. Each muscle consists of two functionally dissimilar fibers, and the precise extent of these "quick" and "catch" components can occasionally be identified by careful inspection of the scar. Owen (1958) and Yonge (1953a) suggest that the adductor muscles probably originated through cross connections of pallial muscles at the ends of the mantle embayments in the ancestral bivalve. Typically, there are two adductor muscles present, one anterior and one posterior. This constitutes the *dimyarian* condition. These muscles may be similar in size (*isomyarian*), or the anterior muscle may show varying degrees of reduction (*anisomyarian* or *heteromyarian*). Alternatively, there may be only a single posterior adductor muscle present (*monomyarian*). Although the latter may be more or less centrally positioned, careful inspection shows that it is always set slightly posterior to the midline. Various other muscle scars such as those associated with the foot or byssal complex, and the small *cruciform* muscle characteristic of the Tellinacea, may also be identified by careful inspection of the shell, and these too furnish further valuable information concerning taxonomy and life habits. The arrangement and functional significance of many of these shell-attached muscles are described in some detail by Cox (1969) and Kauffman (1969).

Any study of comparative morphology requires a terminology related to growth axes. However, because bivalves lack a distinct head, confusion can easily arise over what constitutes the anterior–posterior axis of the body. For descriptive purposes, it is convenient to consider the hinge to be dorsal in position and to orient anterior and posterior regions with respect to the hinge axis (i.e., the axis along which the valves open and close). Previous workers have also defined the anterior–posterior axis by reference to lines drawn through the lower margins of the adductor muscles or through the mouth and anus. The former method is of little value in the case of monomyarian taxa, while the latter method clearly has limited potential for fossil material (though it is worth noting that the anus almost invariably opens over the posterior adductor muscle). Even the hinge axis may not always be easily determined, especially in bivalves with condensed ligaments, and Cox (1969) proposes that measurements of length and height (Fig. 1) should simply reflect the maximum shell dimensions between the anterior and posterior and the dorsal and ventral margins of the shell irrespective of the *exact* direction of measurement. Shell width is the maximum distance between the two valves measured normal to the plane of articulation (see the definition of *commissure* below). For sectioning purposes (see Appendix 1.A), an additional growth axis can be recognized. This is the "axis of maximum shell growth" and represents the maximum dimension between the umbo and the growing edge of the shell. Figure 1D shows this growth axis for two dissimilar bivalves.

Where one valve is a virtual mirror image of the other, the shell is described as *equivalve*; when the valves are dissimilar, as *inequivalve*. When regions of the shell anterior and posterior to the umbones are of approximately equal size, the shell is *equilateral*; when these regions differ appreciably, it is *inequilateral*. Externally, the bivalve shell can vary extensively with regard to color and degree of ornamentation. The line of junction between the two valves, but excluding the region of the ligament, is the *commissure*. Localized openings that remain between the valve margins when the adductor muscles are fully contracted are termed *gapes*, and these too can have considerable interpretive value. While the shells of most bivalves consist of two valves, various accessory plates or calcareous pallets may also be developed among those taxa that bore into rock or timber.

3. Origin and Evolution of Bivalves

Although bivalves are known to extend at least as far back as the Ordovician (e.g., Morris, 1967; Pojeta *et al.*, 1972; Stanley, 1977), their precise origins are far from clear, since they were already quite distinctive by the time they first appeared in the fossil record. A considerable and significant chapter in the evolution of bivalves thus appears to have occurred prior to shell calcification, and their early ancestry must therefore remain rather speculative. Morton and Yonge (1964) consider that the forerunners of bivalves probably had single, uncalcified limpetlike shells that were attached to the visceral mass by a series of muscles running down to the foot. Originally, several pairs of these muscles may have been present, as indeed they appear to be in *Babinka*, an Ordovician bivalve (McAlester, 1965; Stasek, 1972). However, these shell muscles have gradually become restricted to two pairs of pedal muscles in Recent taxa (four pairs in the more primitive Nuculacea). The development of the typical laterally compressed bivalve form led to the appearance of anterior and posterior embayments in the shell across which the characteristic adductor muscles developed by local fusions of the pallial muscles. The enclosure of the body between two compressed valves secreted by the underlying mantle folds led to the reduction of the head and its associated radula complex, while sense organs became arranged around the mantle periphery, where they could best sample the water currents entering the mantle cavity. Carter and Aller (1975) suggest that the immediate precursors of the bivalves may have been shell-less spiculose molluscs, while Runnegar and Pojeta (1974) propose a monoplacophoran ancestry.

We can only speculate as to the probable life style of these early bivalves. It is widely believed (e.g., Pojeta, 1971; Yonge, 1953a) that their early evolution occurred in littoral or shallow sublittoral deposits through which they moved by means of their muscular hydraulic foot. These early bivalves probably fed on organic deposits using extensions of their oral palps in much the same way as present-day protobranchs. In support of this infaunal origin is the observation that most modern taxa, irrespective of their life styles as adults, do possess features during their early ontogeny that would ideally suit them for a burrowing existence. Moreover, the earliest known bivalve, *Fordilla troyensis*, is thought to have been a prosogyrate burrower (Pojeta *et al.*, 1973). Tevesz and McCall (1976), however, believe that the earliest bivalves may in fact have been epifaunal suspension feeders rather than burrowers (also see Valentine and Gertman, 1972). Tevesz and McCall (1976) suggest that bivalves arose at small body size ($\approx$1 mm), a situation in which those features generally considered to be adaptations for burrowing can also be used for crawling over the surface of the substratum. This is particularly well illustrated among the tiny erycinaceans and the pediveliger stages of bivalves such as *Mytilus*. The development of the bivalve shell, which is also found in many small primitive crustaceans, probably afforded substantial protection from meiofaunal predators, better control of the internal environment, and more efficient circulation of water currents through the mantle cavity. Apart from those theories outlined above, a multiple-origin ancestry has also been proposed to explain the early evolution of bivalves (e.g., McAlester, 1966; Vokes, 1967).

Present-day bivalves fall into three groups. The primitive deposit-feeding protobranchs (e.g., *Nucula*, *Yoldia*), which probably represent a continuation of the ancestral stock, and the carnivorous septibranchs (e.g., *Cuspidaria*, *Poromya*), which are specialized for life in deeper nutrient-poor waters, are small groups compared with the more diverse lamellibranchs. By utilizing highly developed gills (ctenidia) for both respiration and feeding, the diverse and exceedingly successful lamellibranchs have freed themselves from the constraints of deposit feeding (although some taxa have in fact reverted to this life style) and have undergone considerable adaptive radiation in numerous aquatic habitats. The dependence of bivalves on gills and pelagic development, however, has effectively prevented them from becoming established in terrestrial habitats.

Two major evolutionary events have been largely responsible for the successful radiation of bivalves—the neotenous retention of the postlarval byssus complex and the development of siphons accompanied by varying degrees of mantle fusion. The byssus gland (a modified pedal mucous gland) originated as a postlarval organ used for temporary attachment during metamorphosis (Yonge, 1962). Once metamorphosis has been completed, the byssus gland is frequently lost, but its retention into the adult stages in many bivalves (by neoteny) was instrumental in the colonization of epifaunal habitats and permitted many diverse taxa to become independent of soft sediments. The colonization of hard surfaces using a system of proteinaceous byssal threads for attachment has had a profound influence on bivalve shell morphology. It has resulted in considerable shell asymmetry and the gradual reduction (anisomyarian) and eventual loss (monomyarian) of the anterior adductor muscle. Byssal attachment, which evolved polyphyletically in many different groups (e.g., Mytilacea, Arcacea, Anomiacea, Pteriacea), led, in turn, to the appearance of cemented taxa such as oysters (Ostreacea) and, perhaps most surprisingly, to free-swimming groups such as many of the scallops and file shells (Pectinacea), many of which have secondarily reverted to a more symmetrical shell form. Boring into the substratum confers considerable protection from water movement and predation and is known to have evolved by two independent routes, one of which was byssal attachment. Some byssate taxa, such as the fan shells (e.g., *Pinna*), have reinvaded sediments after having evolved many of the shell characteristics more typically associated with life on hard substrata (Yonge, 1953b). Reversion to an infaunal way of life has probably occurred many times during bivalve evolution, and this too may have been via neoteny, since the postlarval stages of byssate taxa still retain an active burrowing foot. Considerable evolutionary potential is thus stored in bivalve larvae (Stanley, 1972, 1977).

By contrast with byssally attached forms, bivalves inhabiting soft sediments are much more conservative in their shell shape. Nevertheless, considerable radiation in life habits has occurred, principally through the development of siphons and varying degrees of mantle fusion. Infaunal taxa are especially successful and account for well over three quarters of all living bivalves (Table I). Apart from the very specialized and rather aberrant Lucinacea (see Allen, J. A., 1958; Kauffman, 1969), deeply burrowing infaunal suspension-feeding bivalves were virtually unknown in the Paleozoic, and fossils from these strata rarely have a pallial sinus (Stanley, 1968). The development of siphons (Yonge, 1948, 1957) opened

Table I. Distribution of Marine Bivalve Genera among Major Life Habit Categories[a]

Substratum type	Method of attachment or locomotion	General life position		
		Epifaunal	Semiinfaunal	Infaunal
Soft substrata	Reclining	1	1	—
	Burrowing	—	(1)	166 (3)
	Byssally attached	21 (4)	3 (1)	(3)
Hard substrata	Boring	—	—	17 (1)
	Nestling	—	—	7
	Cemented	5	—	—
	Swimming	3 (2)	—	—

[a] From Stanley (1970). Parentheses indicate partial representation.

up many new habitats by permitting infaunal bivalves to penetrate much deeper into the sediment. Life deep in the substratum led to improved physical stability, reduced the risk of disinterment, and provided considerable protection from predators. An important comparison of infaunal and epifaunal environments is given by Levinton (1972). Siphon formation also provided for a much more orderly flow of water through the mantle cavity, and when accompanied by mantle fusion resulted in a more effective hydraulic system for both burrowing and siphon extension. The vast majority of suspension-feeding bivalves draw water (and food) into the mantle cavity through a posterior inhalent siphon (anterior in the Erycinacea). They may live near the surface and possess short siphons (e.g., Carditacea, Veneracea), or they may be deep-burrowing forms with either massive fused siphons (e.g., Myacea, Mactracea) or elongate shells (e.g., Solenacea). They can be sedentary, or they can be active burrowers. Each life style imposes its own suite of morphological features on the bivalve shell. The siphonate tellinaceans have reverted to the ancestral condition of surface deposit feeding and acquire their food either from the surface or from within the sediment itself. Unlike the more primitive protobranchs, however, these forms collect food by means of extremely long, mobile inhalent siphons that sweep the sediment like miniature vacuum cleaners. These active deposit-feeding siphonates are exceedingly successful, especially in rich organic muds to which they are adapted. Just as byssal attachment provided a route for the evolution of the boring habit, so too did certain adaptations for deep infaunal burrowing, because many of the features that evolved for deep burrowing are also equally adaptable to a life spent boring into hard substrata.

For many millions of years after their first appearance, bivalves remained at a relatively low level of species richness (Stanley, 1975a, 1977).

Their rapid expansion during the Ordovician thus appears to have been marked by the neotenous retention of the versatile byssus complex. Epibyssal attachment opened up the route for numerous diverse life styles including free swimming. Tevesz and McCall (1976) suggest that this Ordovician expansion followed the adoption of an infaunal habit by small primitively epifaunal bivalves. Moreover, apart from resulting in rapid taxonomic diversification, the invasion of this new adaptive zone could also explain the substantial increase in body size that is known to have accompanied this expansion. Recolonization of hard substrata by byssally attached fauna may thus have occurred later in the Ordovician at larger body size. Thayer (1979) considers that the diversification of immobile taxa living on hard surfaces, and mobile taxa on soft substrata, is possibly attributable to the increased amount of bioturbation brought about by the increase in the abundance and diversity of mobile deposit-feeding taxa during the Paleozoic. Perhaps one of the most striking features concerning the evolution of such a diverse group as the bivalves has been the repeated appearance of a comparatively restricted number of very successful shell morphologies. The fact that many primitive taxa have persisted while many modern taxa have reverted to more primitive life styles has led Stanley (1973, 1977) to conclude that competition, unlike predation, has been a comparatively weak selective force in bivalve evolution.

4. Shell Growth

Although the principal objective of this chapter is to examine the relationships between shell morphology and environmental change, I shall set the study of shell shape within the broader context of overall shell growth. This is considered appropriate, since the shape of any bivalve shell depends ultimately on growth processes. Furthermore, growth and size have, in themselves, several important ecological and evolutionary implications.

Growth has been extensively studied in bivalved molluscs not only because of the commercial importance of groups such as oysters, clams, and mussels, but also because their growth history can often be observed on the external surface of the valves in the form of disturbance rings. This latter feature, in particular, makes bivalves especially suitable for growth studies. Two principal methods have previously been used to represent bivalve growth: (1) the size of the whole organism can be related to age and (2) the growth of one shell parameter can be related either to that of another parameter or to the whole body. These methods will each be examined in turn.

4.1. Absolute Growth

By taking successive measurements of animals of known age, it is possible to construct a growth curve. The cumulative increase in body size with respect to time that is represented by such a curve is termed *absolute growth*. The percentage increase per unit of time provides a measure of *relative growth*. While the growth of living organisms is most appropriately measured in terms of increase in biomass or volume, the shell is such a prominent feature of the bivalve mollusc that growth is more frequently expressed as increase in shell length. This is also convenient for the paleoecologist, since it allows direct comparisons to be made between living and fossil taxa. In comparative investigations, however, caution should be exercised in using single linear parameters to measure growth rates. For example, although two individual organisms may be increasing in their total biomass at an approximately equal rate, their growth in terms of other parameters such as length, height, or width may vary quite considerably according to age or environmental conditions or both. Furthermore, there is also the additional problem that during winter, soft tissues may experience negative growth while shell continues to be deposited.

Several methods are available for assessing growth rates in bivalves. These methods include the use of modal size-frequency distributions, the use of annual disturbance lines where these are present on the shell, and the measurement of marked or segregated individuals. These methods are described in detail in Chapter 12. Other methods including X-ray techniques (particularly useful for boring species) and rates of incorporation of radioactive isotopes or fluorescent dyes into the shell are discussed by Wilbur and Owen (1964). Several of these techniques have been used to study growth in the mussel *Mytilus edulis*, but Seed (1976) concludes that probably the most reliable estimates of growth have been those obtained using a combination of methods. Recent analyses of internal shell growth increments in bivalves (Chapters 6, 7, and 11) appear to offer considerable potential for high-resolution growth studies in both Recent and fossil taxa (e.g., Clark, 1974; Lutz, 1976; Pannella and MacClintock, 1968). Microgrowth patterns within the shell record a variety of environmental changes and therefore have many ecological and paleoecological applications.

A habitat, by virtue of its resource-limiting environmental conditions, imposes a maximum size beyond which further growth can proceed only very slowly if at all. For example, when mussels that have more or less attained the maximum size imposed by a particular environmental regime are transplanted to more favorable conditions, they respond by a further period of enhanced growth (Seed, 1968) until the asymptote characteristic

Table II. Some Values for L_∞ (Maximum Attainable Length) and k (Growth Rate) (Constants in the von Bertalanffy Growth Equation) for *Mytilus edulis*[a]

Locality	L_∞	k	Authority	Comments
Danish Wadden Sea	77.60	0.5611	Thiesen (1968)	Mussels at low-water spring tides
Conway, North Wales	72.71	0.3426	Thiesen (1968), based on data in Savage (1956)	"Bank mussels"
	74.44	0.3927		"Channel mussels"
Greenland	77.5–283.9	0.022–0.162	Thiesen (1973)	Mussels from 12 widely varied sites
Menai Straits, North Wales	67.20	1.1378	G. Davies and P. J. Dare (personal communication)	Mussels suspended on ropes from rafts
Morecambe Bay, England	62.45	0.8103	Dare (1976)	Mussels near low-water mark of spring tides

[a] From Seed (1976).

of the new environmental regime is approached. The maximum attainable size (L_∞) under any set of environmental conditions can be approximated by means of the Ford–Walford plot (Beverton and Holt, 1957; Hancock, 1965) (see Chapter 12). This parameter is basic to many growth equations, one of which, the von Bertalanffy growth equation, has been widely used for various bivalve taxa. In this equation

$$L_t = L_\infty \left(1 - e^{-kt}\right)$$

where t = time, L_∞ = maximum size, and k is a growth constant. Equations of this type, however, are based on determinate* growth, while growth in many bivalves may not, in fact, be determinate, at least over their realized life span, and therefore may not cease at any particular fixed adult size. Despite criticisms (e.g., Knight, 1968), curve fitting by means of growth equations is perfectly acceptable providing that (1) data are relatively complete, i.e., that there is at least some evidence for asymptotic growth; and (2) it is appreciated that some uncertainty is always associated with such estimates. Table II compares the values of L_∞ and k among various populations of *Mytilus edulis* and illustrates low rates of growth associated with overcrowding, the low growth rates in Greenland com-

---

* *Determinate* growth indicates that some maximum attainable size exists for any given population. Species exhibiting *indeterminate* growth, by contrast, continue to grow throughout their entire life span.

pared with other North Atlantic locations, and the relatively large sizes attained by Arctic mussels.

One of the most striking features of bivalve growth is the degree to which it varies with respect to age and environmental conditions (Fig. 2). While patterns vary from species to species, bivalve growth curves are characteristically sigmoidal in shape, the initial exponential phase of growth being followed by a gradual decline as the animal ages. This decline in growth occurs to such an extent that in very old specimens, growth may virtually cease. Senility, however, is not necessarily the primary cause of reduced growth, because transplantation of old slow-growing animals to more favorable habitats generally results in renewed growth. Numerous environmental factors, e.g., food supply, temperature, substratum, salinity, light, depth, turbidity, population density (crowding), and exposure to high-energy environments, are known to influence growth rate in bivalves. The effects of these factors and others are discussed by Hallam (1965), Rhoads and Pannella (1970), Seed (1976), and Wilbur and Owen (1964). However, it is frequently difficult to isolate the important causal factors, especially since all variables are not independent of each other. Even individual specimens of similar size or age, or both, grown under apparently identical conditions can vary considerably in their

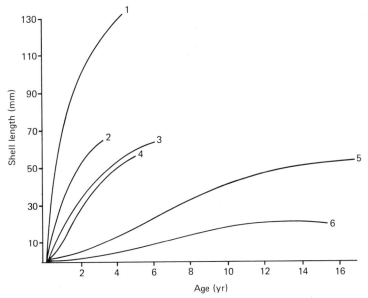

Figure 2. Growth curves for six populations of *Mytilus*. (1) *Mytilus californianus* (subtidal); (2) *Mytilus edulis*, River Boyne (subtidal); (3) *Mytilus edulis*, Chichester Harbour; (4) *Mytilus edulis*, Conway (subtidal); (5) *Mytilus edulis*, Filey Bay (high-shore); (6) *Mytilus edulis*, Robin Hood's Bay (high-shore). After Seed (1976).

growth rates, and there now seems to be some evidence to indicate that growth [and indeed survival (Newkirk et al., 1977)] is at least partly determined by genotype (Singh and Zouros, 1978; also see Murdock et al., 1975). Hallam (1965) considers that an examination of those factors causing stunting could be a potentially important interpretive technique, assuming that juveniles and stunted adults could be distinguished, by providing important data regarding paleoenvironmental conditions and the physical tolerances of many taxa. This is perhaps an overly optimistic view, especially for bivalves, the growth rates of which may vary within extremely broad limits even on a comparatively localized scale. Quite apart from the formidable task of disentangling all potentially important variables, the question of adaptation would also need to be considered. Some species, for example, appear to thrive in comparatively anaerobic sediments or in waters of exceedingly low salinity, i.e., in environments that are generally considered to be unsuitable for bivalves. Furthermore, body size is subject to natural selection and is known to change over evolutionary time (e.g., Bonner, 1968; Hallam, 1960).

Growth rate in several molluscs is known to vary with latitude (Newell, 1964). While populations from higher latitudes exhibit slower growth rates due to lower temperatures, longevity and ultimate size in these populations are frequently greater than in those from lower latitudes. Hallam (1967) suggests that where such data could be supplemented by growth-ring analyses, the prospects of establishing paleotemperature gradients may be quite promising, especially if fossil material were available from comparable horizons over a relatively large geographical area. Lutz and Jablonski (1978) (also see Chapter 9) also suggest that the size of veliger larvae at settlement is controlled by gradients in latitudinal temperature. Hence, the size of larval (or early juvenile) shells might also provide evidence for reconstructing paleotemperature gradients. However, considerable differences in size and age may occur even among bivalve populations inhabiting the same general geographical locality. For example, high-shore mussels on many exposed rocky coasts occupy a spatial refuge in which they are virtually free from predation (Paine, 1974; Seed, 1969, 1976). While very slow-growing, these high-shore populations frequently include many old, yet comparatively large, specimens when compared with faster-growing but more heavily predated populations in the low shore. Moreover, Gilbert (1973) has shown that at higher latitudes, longevity increases, whereas growth and maximum size decrease in Macoma balthica.

Body size has a number of important ecological and evolutionary implications. Fitness in any individual organism within a population can be measured broadly in terms of the extent to which it proves superior in propagating its genotype. Those individual organisms that propagate more of their genotypes by producing most offspring will have a proportionately greater influence on the course of evolutionary change. In most bivalves, fecundity is proportional to body size, and gamete production in any one year that reduces the capacity for growth has therefore to be weighed against increase in body size and the expectation of a higher fecundity during the following year. Because the amount of energy available to any population is generally limited, there sometimes exists a trade-off between reproductive effort and continued growth. Seed and Brown (1978) have recently shown that the different growth strategies adopted by subtidal *Modiolus modiolus* and intertidal *Cerastoderma edule* appear to operate to maximize reproductive efficiency within different frameworks of heavy mortality. We have already noted how the high shore can constitute a refuge from predation for certain intertidal species such as mussels. Some bivalves, e.g., *Modiolus modiolus* (Seed and Brown, 1978) and *Mytilus californianus* (Paine, 1976), also appear to be able to escape predation by growing rapidly to a size at which they are too large to be eaten by any of their major predators. These large specimens then make a reproductive contribution disproportionately greater than their numerical abundance would suggest. They may also influence species richness and community stability by diversifying the spatial structure of the environment through the addition of a vertical component that might not otherwise be present (Paine, 1976). Selection would therefore be expected to favor early rapid growth and delayed reproduction in those species that can escape predation by virtue of large size. Growth rate can also influence the outcome of competition between closely related species. Harger (1972) demonstrated that *Mytilus californianus* can virtually eliminate its major competitor, *M. edulis*, from the outer waveswept shores of the Pacific northwest coast of America by virtue of its mechanically stronger and faster-growing shell.

An area of particular paleoecological interest is the analysis of size-frequency distributions of living and fossil bivalve assemblages (e.g., Craig, 1967; Craig and Hallam, 1964; Craig and Oertel, 1966; Hallam, 1967, 1972; Johnson, 1965; Seed and Brown, 1975). This field of research is discussed in detail in Chapter 12, and it need only be noted at this point that size-frequency distributions require an adequate understanding of the interactions among the processes of growth, recruitment, and mortality before they can be properly evaluated.

4.2. Allometric Growth

Although it is important in any ecological study to establish the rate at which growth proceeds under different environmental conditions, it is clearly evident that body shape does not always change uniformly with an absolute increase in size of the whole animal. Changes in the relative proportions of the bivalve shell with increasing body size are the result of differential growth vectors operating at different points around the mantle edge. The study of the different ratios of growth between two parts of the body or between one part and the whole organism is termed *allometry*. The concept of allometry was first proposed by Huxley and Teissier (1936), and its contributions to developmental and evolutionary biology are critically reviewed by Gould (1966). One of the most significant effects of growth follows from the simple geometrical principle that while the area of a body increases as the square of its linear dimensions, volume increases as its cube. Thus, if geometrical similarity is maintained throughout growth, the area/volume ratio will progressively decline. However, because most exchanges between the external environment and the organism occur across body surfaces, relatively constant area/volume ratios are an adaptive necessity, and these constant ratios can be maintained only by changes in shape. Form is thus inextricably related to body size.

The relationship between any two parts of the body can be expressed by the power function

$$y = Ax^b$$

where y is a measure of one part and x that of another part or the whole body. A and b are constants. The exponent b in this equation is the growth coefficient and represents the relative growth rate of the two size variables under consideration. The constant A is calculated as the value of y when x is unity. This is the allometric equation, which, in view of its simplicity and the relative ease with which it can be interpreted, has found wide applicability among biologists. Although more sophisticated functions (e.g., polynomials) can be used to describe growth data, Gould (1966) questions whether such techniques add substantially to our understanding of biological processes. Rewritten in logarithmic form, the allometric equation becomes

$$\log y = \log A + b \log x$$

such that if y and x are allometrically related, they plot rectilinearly on logarithmic coordinates. The slope and intercept of such logarithmically transformed data are given by the constants b and A, respectively. When the two variables being considered have the same units of measurement, a value of unity for the exponent b is said to describe an isometric relationship in which the relative growth of the two variables is identical, thereby maintaining geometrical similarity with increasing size. Values of b greater than unity indicate that y is increasing relatively faster than x (positive allometry), while values of b less than unity indicate the reverse (negative allometry). If the dimensions of x and y differ, then different criteria for isometry and allometry will apply; e.g., where y is a weight or volume (L^3) and x is a length (L), then $b = 3$ corresponds to isometry. Growth coefficients for several combinations of size parameters for three species of bivalves are shown in Table III. The relative constancy of the allometric exponent over a wide size range indicates that while the two variables may themselves have different rates of increase, the ratio between these rates remains relatively constant.

The use of bivariate data in allometric studies is perhaps open to the criticism that only two variables are compared at any given time, and besides changing their proportions, animals also change their shape. Accepting that such data alone do not satisfactorily describe changes in body form, they do nonetheless enable simple and useful comparisons to be made of the variability in shell proportions among different bivalve populations and thereby provide valuable insight into the relationships between shell shape and environmental change in space or time. However, before any environmental control over shell shape can be demonstrated unequivocally, it is clearly necessary to establish how much variation is due simply to changes in body size, i.e., the extent to which shape varies during ontogeny.

Although shell growth in certain bivalves such as *Glycymeris* is virtually isometric (Thomas, 1976), most taxa exhibit some ontogenetic changes in shape. Marked alterations in the pattern of growth, as evidenced by relatively abrupt changes in the ratios of various shell dimen-

Table III. Coefficients of Allometry for Various Combinations of Size Parameters[a]

Dependent variable[b]	Independent variable[c]	Cerastoderma edule	Mytilus edulis	Modiolus modiolus
Shell weight	Tissue wet weight	1.074	0.957	0.968
Shell weight	Tissue dry weight	0.990	1.044	0.924[d]
Shell weight	Shell width	3.068	3.009	2.908[d]
Shell weight	Shell height	3.426[d]	3.868[d]	3.216[d]
Shell weight	Shell length	3.437[d]	3.300[d]	2.901
Shell width	Shell height	1.117[d]	1.286[d]	1.106[d]
Shell width	Shell length	1.120[d]	1.097[d]	0.998
Shell length	Shell height	0.997	1.172[d]	1.108[d]

[a] From Brown et al. (1976). The coefficient of allometry is coefficient b in the equation $y = Ax^b$, calculated as the average of the regression slope of log y on log x and of the regression slope of log x on log y. Values in italics indicate isometry.
[b] Variable y in the equation $y = Ax^b$.
[c] Variable x in the equation $y = Ax^b$.
[d] Both regression slopes were significant at $P < 0.001$.

sions (e.g., Johannessen, 1973; Ohba, 1959), may indicate that major changes in life habits have occurred during ontogeny. Usually, however, these ratios can be adequately described by a single allometric relationship (e.g., Mason, 1957; Nayar, 1955). Gradual allometric changes with increasing body size are more generally associated with maintaining physiologically favorable surface area/volume ratios, rather than with changing environmental conditions. Figure 3A compares the growth of various shell parameters relative to length in the mussel Mytilus edulis and shows that while a more or less linear relationship exists between length and width in this particular population, the rate of increase in shell height declines slightly in the larger size categories. Weight and volume, on the other hand, are related in typical exponential fashion. When these growth data are plotted on logarithmic axes (Fig. 3B), values for the constants A (intercept) and b (slope) in the allometric equation can be obtained. Apart from volume ($b = 3$), the values of b in these regressions depart significantly from isometry. Width and shell weight exhibit positive allometry, while height is negatively allometric with respect to shell length. Differential growth among these various shell parameters in Mytilus is thus reflected in a gradual change in shell shape, older ($=$ larger) mussels having relatively heavier, more elongate shells in which width can frequently exceed shell height (Fig. 4). Pohlo (1964) has shown that the ontogenetic changes in Tressus nuttalli can be related to differences in life style between the actively burrowing juveniles and the more sedentary adults, which live a more protected life deeper in the sediment. Positive allometric growth of the posterior region of the shell is here associated with the increase in space required for the enlarged siphons. Similar environmentally induced ontogenetic shape changes have also been recorded by Kristensen (1957) for Cardium ($=$Cerastoderma), by Johannessen (1973) for Venerupis, and by Stanley (1970) for Mercenaria. Brown et al. (1976) recently demonstrated that the species-specific changes in shape that occurred in Modiolus modiolus, Cerastoderma ($=$Cardium) edule, and Mytilus edulis appeared to have adaptive significance. Much greater emphasis was placed on shell growth than on tissue growth in the intertidal species (Mytilus and Cerastoderma) as compared to the subtidal population of Modiolus. Differences in the allometry of the linear dimensions, on the other hand, could be explained largely in terms of the relative degree of crowding experienced by these three bivalves. Because growth rate in bivalves is known to vary widely from one habitat to another, it should be remembered that specimens of similar size need not be of similar age. The differences in shell weight of mussels of similar length from different tidal levels (e.g., Baird and Drinnan, 1956; Seed, 1973) can be explained largely in terms of such age differences, the older, slower-growing specimens from high-shore populations having comparatively heavier

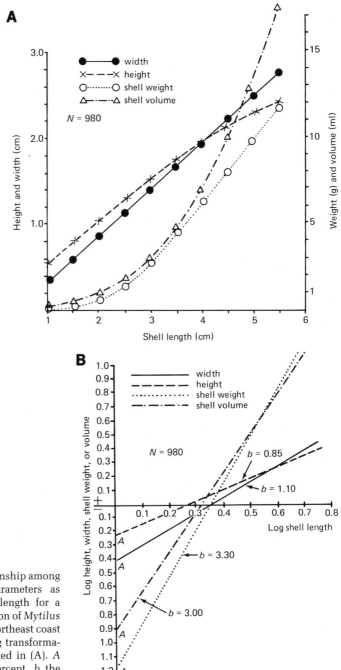

Figure 3. (A) Relationship among various growth parameters as functions of shell length for a high-shore population of *Mytilus edulis* (Filey Bay, northeast coast of England). (B) Log transformation of data presented in (A). *A* indicates the *y*-intercept, *b* the slope. After Seed (1973).

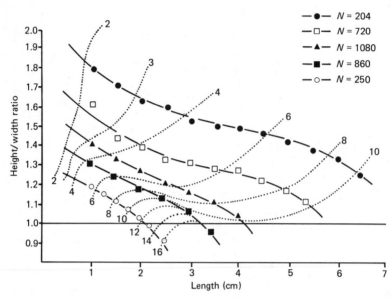

Figure 4. Relationship between height/width ratios and shell length in several populations of *Mytilus edulis*. (●) Whitby Harbour (M.L.W.S.); (□) Filey Bay (low-shore); (▲) Filey Brigg (midshore); (■) Filey Brigg (high-shore); (○) Robin Hood's Bay (high-shore). (· · · ·) Age contours (yr). Whitby Harbour and Filey Bay are low-energy environments, while Filey Brigg and Robin Hood's Bay are high-energy environments. After Seed (1968).

shells than their younger but faster-growing conspecifics lower in the intertidal region.

We have already seen that when weight and length are being compared, $b = 3$ corresponds to isometry. Some bivalves, however, show a progressive increase in their relative bulk density. Enhancing physical stability by increasing bulk density can be advantageous to those species that recline freely on the sediment surface or to shallow burrowers that are subject to frequent wash-out by currents. Increase in shell thickness or strength can also be an effective antipredator device. Heavier shells, on the other hand, can be detrimental to species living on very soft sediments into which they would readily sink, and the growth exponent (b) for two mud-dwelling bivalves, *Mulinia* (2.75 ± 0.82) and *Nucula* (2.57 ± 0.52), is therefore significantly less than 3 (Thayer, 1975). Bulk density in these bivalves thus decreases throughout ontogeny. In the wood-boring bivalve *Teredo*, for which the shell serves merely as a cutting tool, the growth exponent ($b = 1.37$) is reduced even further (Isham *et al.*, 1951).

There is an extensive literature concerning the potential influence of environmental change on shell proportions in bivalves. Crowding is known to affect shape in *Mytilus edulis diegensis* (Coe, 1946), *Mytilus edulis* (Seed, 1968, 1978), *Pinctada martensii* (Tanita and Kikuchi, 1957), *Venerupis semidecussata* (Ohba, 1956), and *Donax striatus* (Wade, 1967). Crowding also appears to be the principal cause of the morphological differences that exist among populations of *Modiolus demissus* that share a common gene pool (Lent, 1967). Mussels from similar but widely separated populations (relatively crowded salt marsh muds), on the other hand, were found to be remarkably alike in their shell proportions. The nature of the substratum is reported to have a marked effect on shape in *Venerupis pullastra* (Quayle, 1952), *Mya arenaria* (Swan, 1952), *Crassostrea virginica* (Gunter, 1938; Orton, 1936), *Cardium edule* (Purchon, 1939), and *Penitella penita* (Evans, 1968). Wave impact, trophic conditions, and water depth are also thought to influence bivalve shell shape (e.g., Clark, 1976; Eager, 1978; Fox and Coe, 1943; Orton, 1928). Nicol (1967) found that bivalves living below 5°C tended to be rather small, with thin chalky skeletons and subdued ornamentation. Fossil assemblages with similar shell characteristics might possibly be interpreted, therefore, as cold-water assemblages. Holme (1961a) showed that in *Venerupis rhomboides*, a linear relationship existed between height/length (H/L) ratios and water depth (possibly caused by hydrostatic pressure) when these were plotted on a logarithmic scale (Fig. 5). By means of this correlation, relatively accurate estimates of water depth could be made solely from a knowledge of shell proportions. Data such as these could clearly provide valuable information concerning the extent of shell transportation along the sea bed and the physical conditions of the original habitats of specimens from both beach and sub-Recent shell deposits. Eisma (1965) found

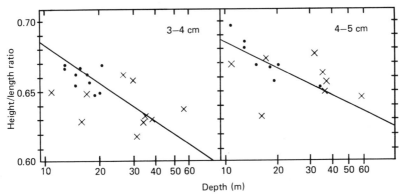

Figure 5. Regressions of median height/length (H/L) ratios of *Venerupis rhomboides* on water depth. (●) Samples from muddy sediments; (×) samples from "clean" sediments. Ratios for 3- to 4-cm and 4- to 5-cm length groups are plotted separately, since some change in proportions occurs with increasing length. After Holme (1961a).

a similar strong correlation between salinity and the number of radial ribs in *Cardium edule*. However, correlations between shell morphology and environmental parameters without any knowledge of their functional relationship has the obvious weakness that such correlations, however strong, may not be causal. Morphological features may be correlated with some factor other than the one chosen for correlation, and where this other factor does not correlate highly with the chosen parameter, the morphological correlation breaks down and interpretation based on morphology will be incorrect (Valentine, 1973). This is especially true for factors such as depth or latitude that are not in themselves physiological factors.

Considerable evidence is thus available to suggest that environmental change can have profound effects on the shell proportions of a number of bivalve molluscs. That such effects are indeed essentially phenotypic can be demonstrated experimentally by transplanting animals from one habitat to another and recording the resulting changes in shell form (e.g., Clark, 1976; Eager, 1978; Evans, 1968; Rhoads and Pannella, 1970; Seed, 1968). Figure 6 illustrates the changes that occur in shell proportions and rate of growth when two exposed shore mussel populations are transplanted to less crowded experimental cages at low tidal levels in a sheltered estuary. In a series of experiments using oysters, however, Hasuo (1958) concluded that ecological factors alone may not always explain the

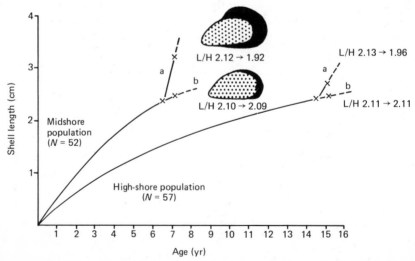

Figure 6. Changes in length/height (L/H) ratios and growth rate over *one* growing season in two exposed shore populations of *Mytilus edulis* after transplantation to cages at low intertidal levels in a sheltered estuary (Whitby Harbour, northeast coast of England). (a) Experimental animals; (b) controls. After Seed (1978).

observed differences in shell shape. Ontogenetic and phenotypic variation in shell morphology has obvious implications in taxonomic studies. Mistaken identification can easily arise from the assumption that shape is independent of body size. New shapes may simply result from the continuation of allometric trends to larger size, and differences in magnitude may be the only characters separating two otherwise identical forms (Gould, 1966). Fossil species, by nature, are morphological taxa, and because the assessment of morphological characters is somewhat arbitrarily determined, Wright (1972) suggests that paleontologists are occasionally guilty of overemphasizing morphological trivia. Characters commonly regarded as important taxonomically may eventually prove to be phenotypic. Only after the relationship between water depth and shell shape had been established (Holme, 1961a) was it fully appreciated that Venerupis rhomboides is a single, but highly variable, polymorphic species. Mytilus is another extremely plastic species in which shell shape varies with both age and environmental conditions. Such variability renders gross shell morphology a doubtful taxonomic character in this particular taxon (Seed, 1968, 1978). Phenotypic variation is probably far more widespread than is generally believed, especially among those species inhabiting less predictable habitats, such as the intertidal and shallow subtidal. In such species, the maintenance of wide variation may actually constitute a "functional insurance" for survival in a constantly changing environment (Eager, 1978).

5. Shell Shape and Body Form

Most allometric studies of bivalve shell shape have used bivariate data in which only two parameters (such as length, height, or width) are considered at any one time. Although useful for describing the changes in body proportions that occur either during ontogeny or as a result of environmental change, such an approach does not permit an adequate analysis of total body form. Our present understanding of shell geometry in bivalves is largely attributable to the work of Carter (1967b), Huxley (1932), Owen (1953a,b), Raup (1966), Stasek (1963a,b), and Thompson (1942).

In view of its accretionary mode of growth (i.e., by marginal increments), the bivalve shell has generally been considered in terms of a logarithmic spiral, the shape and growth of which have been analyzed in several distinct ways. One of the fundamental properties of logarithmic spiral growth (Fig. 7) is that it permits increase in size to occur without requiring any corresponding change in shape, a feature that, although it is approached by most bivalves once they are beyond their immediate

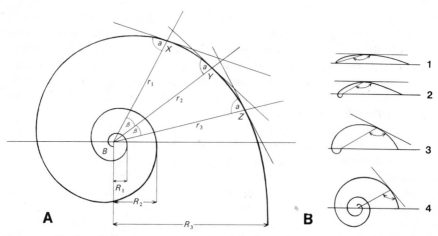

Figure 7. Bivalve shell geometry: properties of the logarithmic or equiangular spiral. (A) Triangles BXY and BYZ, in which the angle at $B(\beta)$ is the same, are always similar. Angle (a) formed between tangents and radii (r_1, r_2, r_3) is always constant. (B) Sections through beak and front edge of shell showing correlation between decrease in angle of equiangular spiral and increase of shell convexity. After Cox (1969) and Lison (1949).

postlarval stages, is rarely completely realized; most bivalves exhibit at least some degree of ontogenetic change during their shell growth.

Owen (1953b) considers that the direction of growth at any given point along the growing mantle edge can be resolved into radial and transverse components. The radial growth component, represented by lines radiating from the umbones, acts in the plane of the generating curve (represented by the mantle–shell margin), while the transverse component acts at right angles to this curve (Fig. 8). The ratio between these two components defines the spiral angle and thus the degree of concavity of the shell valves. Compare, for example, the exceedingly flat shells of tellins with the more obese shells found among cockles. The interaction of radial and transverse components is well illustrated in *Glycymeris*, in which a section passing through the umbones and the middle of the ventral shell margin describes a planospiral about the umbones and in the plane perpendicular to the generating curve. Where the radial and transverse components remain relatively constant, the resulting shell takes on the form of a logarithmic or equiangular spiral (see Thompson, 1942), in which growth proceeds without any concomitant change in shell shape. Flat shells, such as the upper valve of the scallop *Pecten maximus*, are generated when only the radial growth component is present.

The shape of the generating curve, which depends largely on the magnitude of the radial growth component around the shell margin, need

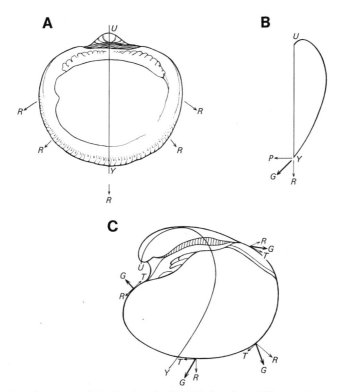

Figure 8. Growth vectors of the bivalve shell. (A) Left valve of *Glycymeris* sp. (B) Section of (A) through *UY*. (C) Left valve of *Glossus* sp. G is the direction of growth at shell margin, P the transverse component, R the radial component, T the tangential component, and *UY* the demarcation line or normal axis. After Owen (1953b).

not be symmetrical as it is in *Glycymeris* (Fig. 9A). Different functions associated with the anterior and posterior regions of the shell generally result in varying degrees of asymmetry about the anterior–posterior axis. In bivalves such as *Tivela*, marginal increments may be equally large in both anterior and posterior regions, thus producing a more or less elliptical shell, but in most taxa, these increments are greatest posteriorly. Such asymmetrical growth patterns are especially pronounced in the elongate shells of the razor clams, e.g., *Ensis* (Fig. 9C).

In addition to the radial and transverse growth components, many bivalves also have a third component that acts tangentially to, but in the same plane as, the generating curve. The effect of this tangential component is to produce a turbinate or twisted spiral, the growth of which is the result of two differential growth ratios, one acting perpendicular to the generating curve, the other (the lateral ratio) acting within the plane

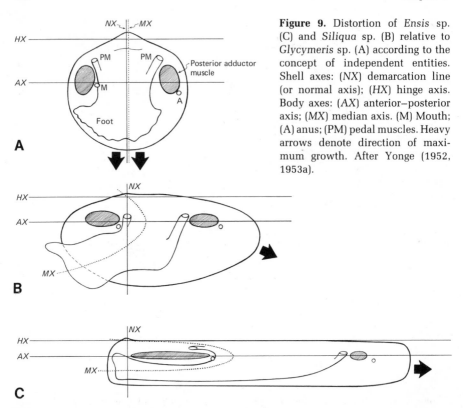

Figure 9. Distortion of *Ensis* sp. (C) and *Siliqua* sp. (B) relative to *Glycymeris* sp. (A) according to the concept of independent entities. Shell axes: (NX) demarcation line (or normal axis); (HX) hinge axis. Body axes: (AX) anterior–posterior axis; (MX) median axis. (M) Mouth; (A) anus; (PM) pedal muscles. Heavy arrows denote direction of maximum growth. After Yonge (1952, 1953a).

of the generating curve (Huxley, 1932; Owen, 1953b). This is best illustrated in shells with radial ornamentation because the ribs are displaced in the direction in which the tangential component acts. Thus, whereas the effects of interaction between radial and transverse components is to produce a planospiral about the umbones in a plane perpendicular to the generating curve, interaction between radial and tangential components produces a similar spiral, but within the plane of the generating curve. The form of the bivalve shell results from the interplay of these two systems of spirals (Owen, 1953b). In shells with a high spiral angle and no tangential component, there is a tendency for the opposing umbones, e.g., in cockles, to grind into one another when the shells open. Where the tangential component is large, the umbones become displaced either anteriorly (e.g., *Glossus*, *Arctica*) or more occasionally posteriorly (e.g., *Nucula*). Such spiral twisting may sometimes actually cause the ligament to split (Owen, 1953a). Raup (1966) and Stasek (1963b) examine some of the ways in which the problems caused by juxtaposition of the umbones can be circumvented.

For comparing shell form in different bivalves, extensive use has been made of the demarcation line (= normal axis) that starts at the umbo and passes through the region of maximum inflation of each valve (i.e., where the transverse-to-radial component is greatest). This line effectively divides the mantle–shell into anterior and posterior territories (Figs. 8 and 9). Where growth is not influenced by a tangential component, this line has the form of a planospiral, but when a tangential component is present, it becomes a turbinate spiral. The bivalve shell can thus be fully described by reference to the spiral angle of the normal axis, the form of the normal axis (i.e., whether a planospiral or turbinate spiral), and the outline of the generating curve. The demarcation line has been used extensively by Owen (1953b, 1958, 1959) and Yonge (1953a, 1955) for comparing shell forms as well as the interrelationship between the mantle–shell and the visceral mass.

While considerable importance has thus been attached to the demarcation line in describing bivalve growth and form, Lison (1949) puts greater emphasis on the directive spiral. This can be seen to greatest effect by reference to ribbed shells such as *Pecten*, in which only one rib follows a completely straight course (the directive plane) throughout its entire length (Fig. 10). The angle between the directive plane and the commissural plane is termed the *angle of incidence*. While the directive plane describes a planospiral (i.e., the directive spiral), all other radial elements are effectively turbinate spirals. The demarcation line and the directive spiral coincide only in symmetrical equivalve shells such as *Pecten* or *Glycymeris*. Knowing the angle of incidence, the angle of the directive spiral, and the shape of the generating curve (which he obtained graphically), Lison could effectively describe the form of any bivalve shell in mathematical terms. The angle of incidence also influences other shell features such as the development of the lunule and escutcheon and the position of the umbones (Cox, 1969). Carter (1967b) considers that Lison's often criticized and neglected work represents a major contribution to the

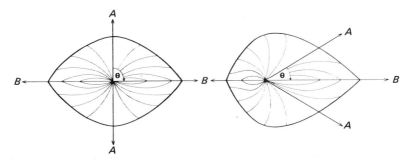

Figure 10. Bivalve shell geometry: Shells viewed dorsally. The directive plane (A) and the plane of commissure (B) are indicated by arrows. The angle between these planes (Θ) defines the angle of incidence. After Owen (1953b).

understanding of bivalve shell form and has not been accorded the attention that it obviously deserves.

Raup (1966) examined the influence of four different parameters on shell growth: (1) the shape of the generating curve (S); (2) the distance between the generating curve and the axis of coiling (D); (3) the rate of whorl expansion (W); and (4) the degree of whorl translation (T), i.e., the rate at which the whorl moves down the y axis. By altering each of these parameters except (S), Raup produced a broad array of computer-generated cones (Fig. 11). He noted, however, that only a limited number of

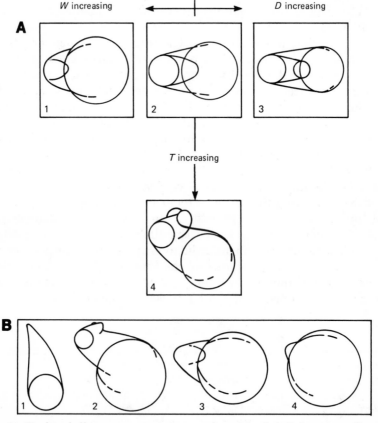

Figure 11. Bivalve shell geometry: parameters and models of shell form in molluscs. (A) Explanation of parameters used in construction of theoretical shell forms. The generating curve in all cases is circular. (1) Typical planispiral gastropod; (2) bellerophontid gastropod; (3) evolute ammonoid; (4) spired gastropod. (B) Models of shell form found in Bivalvia. (1) typical rudist with high whorl expansion (W) and high whorl translation (T); (2) *Diceras* sp. with relatively low W and intermediate T; (3) *Gryphaea* sp. with relatively low W and low T; (4) slightly equilateral shell of typical form with high W and low T. After Cox (1969) and Raup (1966).

architectural designs were ever actually found among bivalves. Most of the geometrical forms that were possible for logarithmic coiled shells were never utilized, suggesting that the evolutionary potential for this particular skeletal structure is perhaps rather restricted. Spiral growth thus seems to impose certain morphogenetic constraints on shell form in bivalves, in which only a finite number of solutions to particular functional problems appear to be possible (also see Kauffman, 1969; Trueman, 1964; Vermeij, 1970). Most taxa favor compact, rather than peaked, conical shells; flatter, laterally compressed valves probably improve the area/volume ratios of the adductor muscles and thus their strength and efficiency.

The mantle–shell and the visceral mass have been considered to be separate but interacting entities. This concept of "independent entities" proposed by Yonge (1952, 1953a, 1955) led to the use of a dual system of orientational axes (see Fig. 9). In symmetrical shells such as *Glycymeris*, shell and body axes are parallel, but in most bivalves, these axes become offset in various ways as the mantle–shell rotates about the fixed visceral mass. Stasek (1963a) criticizes this scheme and maintains that the body and shell are inseparable and evolve as an integrated unit. Comparable regions of the mantle–shell are always adjacent to comparable regions of the body. Differences among bivalves that vary markedly in body form are the result of changes in the relative proportions of the body and shell as a single entity. Stasek thus developed a uniform system of orientational terms and used transformational diagrams like those originally used by Thompson (1942) as a means of emphasizing the evolutionary unity between the mantle–shell and the rest of the body (Fig. 12). While both the concept of "independent entities" and the theory of transformation have led to a better understanding of bivalve growth and form, their use has caused some confusion concerning directional terminology. For example, in razor clams, is the region of the shell from which the foot emerges anterior in position, or is it ventral because the foot is usually considered ventral?

Various models have thus been proposed for analyzing total body form in bivalve molluscs. Nevertheless, the considerable difficulties inherent in quantifying the complex changes that accompany alteration in body form have so far proved a major obstacle to their more generalized use. Growth gradient studies and, perhaps more especially, multivariate techniques (see Blackith and Rayment, 1971), in which the complex interdependence of all variables are examined together, go a long way toward providing a more comprehensive approach to the study of changes in total body form. Although multivariate techniques have so far received comparatively little attention in bivalve growth studies (e.g., Brown *et al.*, 1976; Pohlo, 1964; Thomas, 1975), their use should ultimately supersede the simple allometric equation as a major quantitative technique for the explanatory analysis of allometry (Gould, 1966). However, the degree of

A

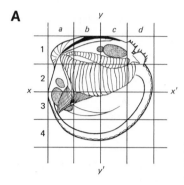

B

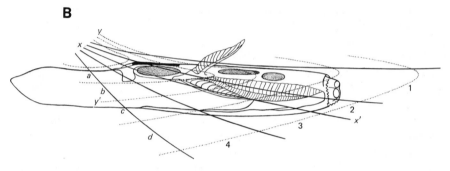

Figure 12. Deformation of *Solen* sp. (B) relative to *Clinocardium* sp. (A) using coordinate transformation. After Stasek (1963a).

analytical sophistication should always be related to the kinds of questions being asked, and for many practical purposes, bivariate analysis may in fact prove to be all that is required.

6. Functional Morphology

Most biologists would probably agree that all morphology is adaptive in one way or another. Seilacher (1970) proposes a framework for morphological interpretation in which any structure is examined in terms of the interaction of three important, yet quite distinct, influences: (1) *Historical*. We have already seen, for example, how the presence of an immutable accretionary skeleton can impose certain mechanical constraints on shell form. Such historical factors constitute the phylogenetic heritage of any evolving lineage. Furthermore, the functional requirements of bivalves may change during ontogeny, and features incorporated into the shell at one stage in development may limit subsequent morphological options; adult shells may therefore represent a compromise among different selective forces operating at different times throughout ontogeny. (2) *Functional*. These factors influence those features of morphology that arise directly from adaptation through natural selection. However, since the response of any one character can potentially influence the functional efficiency of other characters, it is important that all responses to selection pressure should be harmoniously integrated and suitably adapted to the particular life style of the organism. (3) *Structural.* These factors affect those morphological features that are determined by the inherent nature of the available materials and that need not in themselves necessarily be adaptive.

Raup (1972) suggests that two additional factors need to be considered—*chance* and *ecophenotypic effects,* the latter being of particular interest in the context of this volume. The importance of chance factors should not be underestimated, because any specific functional problem may have several possible solutions, and whichever solution is adopted need not necessarily be the outcome of selection; the observed morphology may, in fact, be the best one possible within the constraints imposed by the nature of the materials. Ecophenotypic factors determine the extent of phenotypic variation that is possible within any given genotype. Although such phenotypes could be considered functionally advantageous, this is not invariably the case. Densely crowded mussels, for example, develop elongate shells, yet such growth forms are probably functionally inferior to the taller, more expanded shells that develop in noncrowded conditions.

The interpretation of any morphological feature therefore depends on the relative contribution made by each of the aforenamed factors (Fig. 13). While bivalve morphology is thus ultimately controlled by several distinct factors, most work to date has concentrated on the functional (=adaptive) significance of particular morphologies and the evolutionary processes by which these were produced.

Bivalve molluscs provide excellent material for biologists and paleontologists whose interests lie in aspects of functional morphology. They

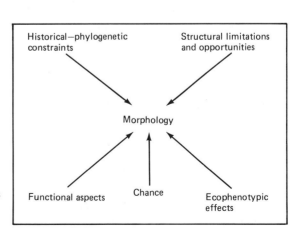

Figure 13. Principal factors controlling shell morphology. After Raup (1972).

exhibit extensive adaptive radiation, their shells frequently reflect environmental demands, and they have a long and successful evolutionary
history with comparatively few extinctions. Furthermore, convergence
and divergence both between and within members of phylogenetically
distant taxa has led to considerable homeomorphy (Stanley, 1970, 1975a).
This is especially well illustrated among the Mactracea (Fig. 14). Although
homeomorphy presents difficulties in recognizing major phylogenetic relationships within the Bivalvia, it does provide strong evidence for the
adaptive significance of shell morphology. Detailed studies of functional
morphology have been made on a number of diverse bivalve taxa, e.g.,
Veneracea (Ansell, 1961; Carriker, 1961; Pohlo, 1964; Purchon, 1955; Sellmer, 1967), Arcacea (Thomas, 1976, 1978), Carditacea (Yonge, 1969), Astartacea (Allen, J. A., 1968; Saleuddin, 1965), Lucinacea (Allen, J. A., 1958;
Kauffman, 1969), Mytilacea (Yonge, 1955), Tellinacea (Wade, 1969;
Yonge, 1949), Chamacea (Yonge, 1967), Pandoracea (Allen, M. F., and
Allen, 1955), Nuculacea (Yonge, 1939), Cyprinacea (Saleuddin, 1964),
Pholadacea (Purchon, 1955), and Erycinacea (Oldfield, 1955; Morton,
1973). In addition to work on selected taxa, several excellent comparative
studies of form, function, and evolution within the Bivalvia are also available (e.g., Allen, J. A., 1963; Eager, 1978; Kauffman, 1969; Stanley, 1968,
1970, 1972, 1975a). Attention has already been drawn to the two major
evolutionary events in bivalve evolution—adult retention of the byssus,
and siphon development. I shall conclude this chapter by briefly considering the impact that these two important events have had on bivalve
shell morphology.

Bivalve shell shape is remarkably plastic, and similar life styles in
many unrelated lineages reveal numerous instances of parallel or con-
vergent evolution. Consequently, it has generally been considered appropriate to group bivalves according to life styles and habitat preferences
(e.g., Kauffman, 1969; Kranz, 1974; Purchon, 1977; Schafer, 1972; Stanley,
1968, 1970). Although several environmental factors are undoubtedly important in controlling shell morphology (Fig. 15), water movement, via

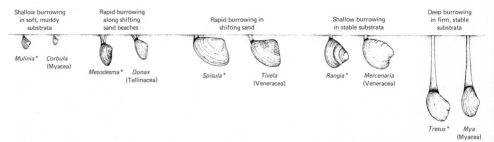

Figure 14. Adaptive divergence of the Mactracea (*) and convergence in form and habit with
unrelated burrowing taxa. After Stanley (1970).

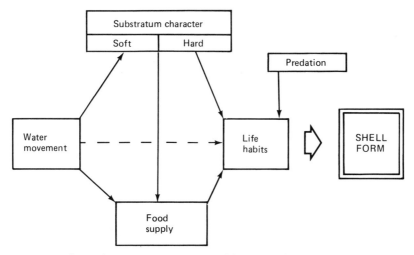

Figure 15. Interrelationships among environmental factors and bivalve life habits. Solid lines indicate major effects; the dashed line, minor effects. After Stanley (1970).

its effects on food supply and on the nature of the substratum, is arguably the most important single factor (Stanley, 1970).

Byssal attachment is widespread among many exceedingly diverse groups of bivalves (see Table I). Many of the shell features associated with epibyssal attachment to hard substrata can be considered as adaptations for improving physical stability. Epibyssate taxa such as mussels, in which the commissure plane is vertical, are usually anisomyarian with elongate, equivalve shells. Maximum shell width is situated close to the ventral shell margin, thereby lowering the center of gravity and providing a broad base for fixation (Stanley, 1972). Seed (1968) found that in *Mytilus*, many of these shell features that enhance physical stability become more pronounced in older specimens, strongly suggesting that they do indeed have adaptive value. The wedge-shaped profile of these anisomyarian shells effectively elevates the posterior current flow and may be beneficial to animals living in densely crowded conditions (Yonge and Campbell, 1968). Stasek (1966) considers that posterior expansion associated with the anisomyarian condition may, in fact, have evolved in the tropics as a method for improving feeding efficiency in comparatively nutrient-poor waters. The development of a powerful posterior adductor muscle places considerable strain on the ligament, and the small anterior adductor muscle is retained to prevent the ligament from splitting. Arc shells (Arcacea), which are among the least modified of byssally attached bivalves, have probably been compelled to retain the primitive isomyarian condition by virtue of their exceedingly weak ligament (Thomas, 1976).

Stanley (1972) suggests that a transitional period of infaunal or se-

miinfaunal byssal attachment probably preceded the ultimate colonization of hard substrata. Although originally abundant, these endobyssate
bivalves gradually declined with the appearance of epifaunal and free-
burrowing forms, and they now persist only in a few Recent taxa such as
Modiolus and *Brachidontes* (also see Thayer, 1979). The lineage from
Modiolus to *Mytilus* represents a gradual transition from an endobyssate
to an epibyssate way of life (Stanley, 1972). The adaptations accompanying this transition are illustrated in Fig. 16 and include a gradual reduction of the anterior region of the shell, the development of a steeper
hinge angle, a strengthening and change in position of the byssal retractor
muscle to pull the shell down more firmly against the substratum, and
a change in cross-sectional shape. Many of these features are even more
pronounced in *Mytilus galloprovincialis* than they are in *M. edulis* (Seed,
1978), supporting the belief that *M. galloprovincialis* is probably of comparatively recent origin, having evolved in the warmer, enclosed waters
of the Mediterranean. Taxa that are byssally attached in a nonvertical
position tend to be inequivalve because the effects of physical conditions
on the two valves may vary quite considerably. A recurrent stabilizing
strategy among these taxa is the development of shells with two marginal
points of contact with the substratum, one on either side of the byssal

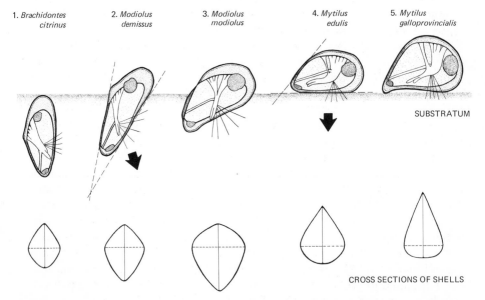

Figure 16. Grades of evolution in Recent Mytilidae. Heavy arrows in (2) and (4) indicate the
direction and relative force of the shell against the substratum. Horizontal dotted lines in
shell cross sections indicate positions of maximum shell width. After Stanley (1972) and
Seed (1978).

sinus (this is also encountered in some mytilids when the ventral margin appears incurved in lateral view). In the Pteriacea and Pectinacea, this is generally achieved by means of an elongate anterior auricle (See Fig. 17). Jingle shells (Anomiacea) are exceptional among byssate taxa in that the byssus forms a calcified plug that passes through an embayment in the flat lower valve. They are exceedingly variable in shape, thickness, and sculpturing, often closely following the contours of the substratum to which they are attached (Seed and Roberts, 1976). Cementation, which occurs, for example, in oysters (Ostreacea), is more commonly encountered in warmer waters where carbonate is more easily deposited (Nicol, 1964). Cemented bivalves typically have thick, well-ornamented shells for protection against predators. Taxa that recline freely on the surface of the sediment improve their physical stability by increasing bulk density. They may also have a convex lower valve to raise the commissure above the sediment surface and thus prevent clogging by suspended particles. Swimming, by clapping the valves together, has evolved among scallops and file shells (Pectinacea), probably from a mechanism originally designed for cleansing the mantle cavity of pseudofeces (Yonge, 1936). Scallops from high-energy environments are typically more heavily ribbed than the proficient smooth-shelled swimmers from deeper, less turbulent, waters. Whereas byssally attached pectinids are rather asymmetrical, swimming taxa have broad, lightweight shells with a large umbonal angle (Fig. 17). Bivalves that nestle in crevices are generally irregularly shaped due to their confined environment.

Most bivalve taxa are infaunal burrowers in which life habit, and therefore shell morphology, is determined principally by the nature of the substratum. Burrowing mechanisms are reviewed by Trueman (1966,

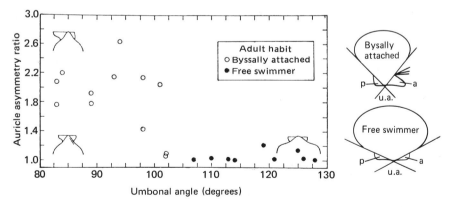

Figure 17. Graph of auricle asymmetry vs. umbonal angle (u.a.) for byssally attached and free-swimming members of the Pectinidae. The auricle assymetry ratio is the ratio between anterior (a) and posterior (p) auricle lengths. After Stanley (1970).

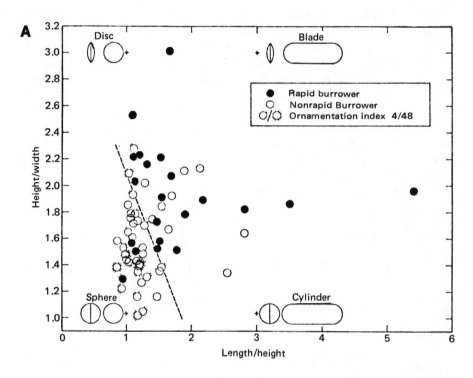

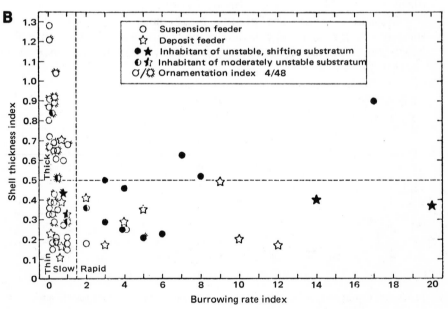

1975). Most advanced taxa (except the Lucinacea) orient themselves with posterior siphons uppermost and the hydraulic foot, which emerges more or less parallel to the long axis, directed forward and downwards into the substratum. Unlike byssally attached bivalves, most burrowing taxa retain the primitive isomyarian condition. Maximum shell width, which in epibyssate forms was located near the ventral shell margin, is here moved dorsally, thus leaving a knifelike leading edge to the shell for more efficient burrowing. Clam-shaped shells, which are more or less circular in lateral view (i.e., with low length/height ratios), burrow by means of a series of anterior–posterior rocking movements brought about by alternate contraction of the anterior and posterior adductor muscles. Stanley (1975b) simulated these burrowing movements experimentally using plaster casts of shells and showed that the prosogyrous shape and flat lunule were advantageous design features. Such shell rotations would obviously be impossible in elongate taxa such as razor shells in which the foot emerges from an anterior pedal gape. In these taxa, penetration of the substratum occurs almost parallel to the long axis via the pistonlike action of the foot within the elongate cylindrical shell. Stanley (1970) compared burrowing efficiency in several infaunal bivalves by means of the burrowing rate index (b.r.i.), where

$$\text{b.r.i.} = 100(\text{mass [g]})^{1/3}/\text{burrowing time}$$

For any given species, this index does not change appreciably during ontogeny, and it therefore remains independent of size. Most rapid burrowers were found to have streamlined shells that were either disklike, bladelike, or cylindrical in shape, rather than globular (Fig. 18A). Many carditids, however, are exceptionally good burrowers, but this is due to a powerful foot, rather than to their obese shape. Shells that are thick or highly ornamented, or both, hinder burrowing (Fig. 18B), although radial ribbing and divaricate (zigzag) ornamentation can sometimes be an asset, especially in taxa with more globular shells in which it offsets the disadvantages of shape. Shallow sand burrowers, which are subject to frequent disinterment, generally have thick shells, often with pronounced ribs or ridges for additional protection and stabilization. Radial ribbing also per-

←——

Figure 18. (A) Relationship of burrowing rate to gross shell shape and shell ornamentation. Empirical dashed line separates rapid-burrowing region from slow-burrowing region. (B) Relationship of burrowing rate to shell thickness, shell ornamentation, feeding type, and substratum character. From these data, one may draw the following conclusions: (1) Nearly all (19 of 22) thick-shelled species are slow burrowers. (2) Nearly all (18 of 21) rapid burrowers have very thin to moderately thick shells. (3) Nearly all (19 of 21) rapid burrowers are either (a) deposit feeders or (b) inhabitants of moderately unstable shifting substrata. (4) Only 1 of 16 deposit feeders is a slow burrower. (5) Nearly all (12 of 13) species with moderately coarse or very coarse ornamentation are slow burrowers. After Stanley (1970).

mits the valve edges to interdigitate and thus prevents lateral shearing; fine denticulations on the inner edges of the valves of certain bivalves (e.g., *Donax*) serve the same function. In shallow-burrowing arcids and carditids, the broad, flat posterior shell region lies parallel to and immediately beneath the sediment surface. Deep burrowers, by contrast, obtain substantial protection from the sediment itself and therefore tend to have thinner nonornamented shells. Many species (e.g., *Mya*) may lose the ability to burrow as adults once their required depth has been attained. Because they are not faced with problems of shear, many deep-burrowing taxa have more concentrated ligaments, while hinge teeth may be reduced or even absent. This allows the valves to rock about the dorsal–ventral axis, a movement that, together with a posterior gape, is necessary for the efficient accommodation of the elongate fleshy siphons. Vermeij and Veil (1978) found that infaunal bivalves with permanent posterior gapes become increasingly rare toward the tropics and attributed this to enhanced predation. Such latitudinal trends in predation and bivalve architecture (tropical taxa also have thicker, more highly ornamented shells) are therefore broadly in keeping with the parallel reduction in the proportion of gaping taxa from deep to shallow sediments. Carter (1968) reviews some of the strategies used by bivalves to avoid predation. Tellinids are unusual among burrowing infaunal taxa in that their thin, sometimes inequivalve shells lie horizontally in the sediment. Holme (1961b) suggests that this probably facilitates horizontal migration, so important to deposit feeders, which rapidly deplete their local food supplies. Coarse, shifting deposits of high-energy habitats, in which disinterment is a recurring hazard, are colonized only by active burrowers. Many taxa such as *Tivela* and *Donax* may develop comparatively thick shells for improved stability and protection, even though these features may slightly impair burrowing efficiency. Soft, muddy deposits more typical of bays and estuaries are rich in organic material and are largely dominated by deposit feeders (Tellinacea, Nuculacea). The turbidity associated with physically unstable fine-grained sediments, however, is detrimental to suspension feeders, and these taxa are therefore largely excluded from such habitats (Rhoads and Young, 1970). Mud-dwelling taxa have small, thin shells that reduce their bulk density and thereby prevent them from sinking too deeply into the soft sediment. These deposit feeders are mainly active, streamlined burrowers that rapidly exhaust their local food supplies. Bioturbation, caused by the burrowing activity of deposit feeders, may also be an important means of enhancing microbial turnover rates; this is mediated by the control of sediment chemistry through the bioturbation process (Rhoads *et al.*, 1978). By contrast, more cohesive sediments consisting of stable sandy mud are rich in suspension-feeding bivalves. Kranz (1974) demonstrated that certain features of shell morphology correlated well with the ability of bivalves to escape burial by sediment (e.g., anterior–ventral expansion indicates a relatively large foot and therefore good escape potential). More-over, because these escape responses could be shown to be influenced by several environmental factors, bivalves are potentially excellent indica-tors of sedimentary regimes. Boring into hard substrata, such as rock or wood, is an extremely specialized life style (e.g., Nair and Ansell, 1968; Purchon, 1955, 1956a,b; Turner, 1966; Yonge, 1955). Penetration may be accomplished mechanically (e.g., *Botula, Hiatella*) or chemically (e.g., *Lithophaga*), and the substratum may even be utilized as food (e.g., *Teredo*). The shells of boring taxa are often reduced and function merely as efficient cutting tools; accessory shell plates may also be developed. A condensed ligament that provides a flexible joint around which the two valves can rock about the dorsal–ventral axis has been instrumental in the evolution of an effective boring mechanism.

In conclusion, we have seen that the bivalve shell is subject to con-siderable phenotypic expression. Critical evaluation of such variation both during ontogeny and with respect to changing environmental con-ditions provides valuable information regarding the functional signifi-cance of various morphological features. Moreover, many of these fea-tures, such as overall shell proportions, cross-sectional shape, muscle scars, the presence of gapes and sinuses, or the ligament type, are fre-quently well preserved in fossil material and thus provide the paleon-tologist with a potentially valuable tool in paleoecological interpretation. Considerable progress in the study of functional morphology has been made over recent years. Further detailed research in this field should ultimately lead to an even better understanding of bivalve evolution and to the establishment of a more acceptable framework for interpreting the fossil record.

References

Allen, J. A., 1958, On the basic form and adaptations to habitat in the Lucinacea (Eulamel-libranchia), *Philos. Trans. R. Soc. London Ser. B* **241:**421–484.

Allen, J. A., 1963, Ecology and functional morphology of molluscs, *Oceanogr. Mar. Biol. Annu. Rev.* **1:**253–288.

Allen, J. A., 1968, The functional morphology of *Crassinella mactracea* (Linsley) (Bivalvia: Astartacea), *Proc. Malacol. Soc. London* **38:**27–40.

Allen, M. F., and Allen, J. A., 1955, On the habits of *Pandora inaequivalvis* (L.), *Proc. Malacol. Soc. London* **31:**175–185.

Ansell, A. D., 1961, The functional morphology of the British species of Veneracea (Eula-mellibranchia), *J. Mar. Biol. Assoc. U. K.* **41:**489–515.

Baird, R. H., and Drinnan, R. E., 1956, The ratio of shell to meat in *Mytilus* as a function of tidal exposure to air, *J. Cons. Perm. Int. Explor. Mer.* **22:**329–336.

Beverton, R. J. H., and Holt, S. J., 1957, On the dynamics of exploited fish populations, *Fish. Invest. Minist. Agric. Fish. Food (G. B.) Ser. II* **19:**1–533.

Blackith, R. E., and Rayment, R. A., 1971, *Multivariate Morphometrics*, Academic Press, London.

Bonner, J. T., 1968, Size change in development and evolution, *J. Paleontol.* **42:**1–15.

Brown, R. A., Seed, R., and O'Connor, R. J., 1976, A comparison of relative growth in *Cerastoderma* (=*Cardium*) *edule*, *Modiolus modiolus* and *Mytilus edulis* (Mollusca: Bivalvia), *J. Zool. (London)* **179**:297–315.

Carriker, M. R., 1961, Interrelation of functional morphology, behavior and autecology in early stages of the bivalve *Mercenaria mercenaria, J. Elisha Mitchell Sci. Soc.* **77**:168–241.

Carter, R. M., 1967a, On the nature and definition of the lunule, escutcheon, and corcelet in the Bivalvia, *Proc. Malacol. Soc. London* **37**:243–263.

Carter, R. M., 1967b, On Lison's model of bivalve shell form and its biological interpretation, *Proc. Malacol. Soc. London* **37**:265–278.

Carter, R. M., 1968, On the biology and paleontology of some predators of bivalved Mollusca, *Palaeogeogr. Palaeoclimatol. Palaeoecol.* **4**:29–65.

Carter, J. G., and Aller, R. C., 1975, Calcification in the bivalved periostracum, *Lethaia* **8**:315–320.

Clark, G. R., 1974, Growth lines in invertebrate skeletons, *Annu. Rev. Earth Planet. Sci.* **2**:77–99.

Clark, G. R., 1976, Shell convexity in *Argopecten gibbus*: Variation with depth in Herrington Sound, Bermuda, *Bull. Mar. Sci.* **26**:605–610.

Coe, W. R., 1946, A resurgent population of the Californian Bay mussel *Mytilus edulis diegensis, J. Exp. Zool.* **99**:1–14.

Cox, L. R., 1969, General features of Bivalvia, in: *Treatise on Invertebrate Paleontology, Part N, Mollusca 6, Bivalvia* (R. C. Moore, ed.), pp. N2–N129, University of Kansas Press, Lawrence.

Craig, G. Y., 1967, Size-frequency distributions of living and dead populations of pelecypods from Bimini, Bahamas, *J. Geol.* **75**:35–45.

Craig, G. Y., and Hallam, A., 1964, Size-frequency and growth-ring analyses of *Mytilus edulis* and *Cardium edule*, and their palaeoecological significance, *Palaeontology* **6**:731–750.

Craig, G. Y., and Oertel, G., 1966, Deterministic models of living and fossil populations of animals, *Q. J. Geol. Soc. London* **122**:315–355.

Dare, P. J., 1976, Settlement, growth, and production of the mussel, *Mytilus edulis* L., in Morecambe Bay, England, *Fish. Invest. Minist. Agric. Fish. Food (G. B.) Ser. II* **28**:1–25.

Eager, M. C., 1978, Shape and function of the shell: A comparison of some living and fossil bivalve molluscs, *Biol. Rev.* **53**:169–210.

Eisma, D., 1965, Shell characteristics of *Cardium edule* L. as indicators of salinity, *Neth. J. Sea Res.* **2**:493–540.

Evans, J. W., 1968, Growth rate of the rock-boring clam *Penitella penita* (Conrad) in relation to hardness of rock and other factors, *Ecology* **49**:619–628.

Fox, D. L., and Coe, W. R., 1943, Biology of the Californian sea-mussel *Mytilus californianus*. II. Nutrition, metabolism, growth and calcium deposition, *J. Exp. Zool.* **93**:205–249.

Gilbert, M. A., 1973, Growth rate, longevity and maximum size of *Macoma balthica* (L.), *Biol. Bull. (Woods Hole, Mass.)* **145**:119–126.

Gould, S. J., 1966, Allometry and size in ontogeny and phylogeny, *Biol. Rev.* **41**:587–640.

Gunter, G., 1938, Comments on the shape, growth and quality of the American oyster, *Science* **88**:546–547.

Hallam, A., 1960, A sedimentary and faunal study of the Blue Lias of Dorset and Glamorgan, *Philos. Trans. R. Soc. London Ser. B* **243**:1–44.

Hallam, A., 1965, Environmental causes of stunting of living and fossil marine benthonic invertebrates, *Palaeontology* **8**:132–155.

Hallam, A., 1967, The interpretation of size-frequency distributions in molluscan death assemblages, *Palaeontology* **10**:25–42.

Hallam, A., 1972, Models involving population dynamics, in: *Models in Paleobiology* (T. J. M. Schopf, ed.), pp. 62–80, Freeman Cooper, San Francisco.

Hancock, D. A., 1965, Graphical estimation of growth parameters, *J. Cons. Perm. Int. Explor. Mer.* **29**:340–351.

Harger, J. R., 1972, Competitive coexistence: Maintenance of interacting associations of the sea mussels *Mytilus edulis* and *M. californianus*, *Veliger* **14**:387–410.

Hasuo, M., 1958, The shell variation during growth in the Mie pearl oysters and those transplanted from Nagasaki Prefecture, *Bull. Natl. Pearl Res. Lab. (Kokuritsu Shinju Kenkyusho Hokoku)* **4**:318–324.

Holme, N. A., 1961a, Shell form in *Venerupis rhomboides*, *J. Mar. Biol. Assoc. U. K.* **41**:705–722.

Holme, N. A., 1961b, Notes on mode of life of the Tellinidae (Lamellibranchia), *J. Mar. Biol. Assoc. U. K.* **41**:699–703.

Huxley, J. S., 1932, *Problems of Relative Growth*, Dial Press, New York.

Huxley, J. S., and Teissier, G., 1936, Terminology of relative growth, *Nature (London)* **137**:780–781.

Isham, L. D., Moore, H. B., and Smith, F. G. W., 1951, Growth rate measurement of shipworms, *Bull. Mar. Sci. Gulf Caribb.* **1**:136–147.

Johannessen, O. H., 1973, Population structure and individual growth of *Venerupis pullastra* (Montagu) (Lamellibranchiata), *Sarsia* **52**:97–116.

Johnson, R. G., 1965, Pelecypod death assemblages in Tomales Bay California, *J. Paleontol.* **39**:80–85.

Kauffman, E. G., 1969, Form, function and evolution, in: *Treatise on Invertebrate Paleontology, Part N, Mollusca 6, Bivalvia* (R. C. Moore, ed.), pp. N129–N205, University of Kansas Press, Lawrence.

Knight, W., 1968, Asymptotic growth: An example of nonsense disguised as mathematics, *J. Fish. Res. Board Can.* **25**:1303–1307.

Kranz, P. M., 1974, The anastrophic burial of bivalves and its paleoecological significance, *J. Geol.* **82**:237–267.

Kristensen, I., 1957, Differences in density and growth in a cockle population in the Dutch Wadden Sea, *Arch. Neerl. Zool.* **12**:351–453.

Lent, C. M., 1967, Effect of habitat on growth indices in the ribbed mussel, *Modiolus (Arcuatula) demissus*, *Chesapeake Sci.* **8**:221–227.

Levinton, J., 1972, Stability and trophic structure in deposit-feeding and suspension-feeding communities, *Am. Nat.* **106**:472–486.

Lison, L., 1949, Recherches sur la forme et la mécanique du développement des coquilles des lamellibranches, *Mem. Inst. Sci. Nat. Belg.* **34**:1–87.

Lutz, R. A., 1976, Annual growth patterns in the inner shell layer of *Mytilus edulis* L., *J. Mar. Biol. Assoc. U. K.* **56**:723–731.

Lutz, R. A., and Jablonski, D., 1978, Larval bivalve shell morphology: A new paleoclimatic tool? *Science* **202**:51–53.

Mason, J., 1957, The age and growth of the scallop *Pecten maximus* (L.) in Manx waters, *J. Mar. Biol. Assoc. U. K.* **36**:473–492.

McAlester, A. L., 1965, Systematics, affinities, and life habits of *Babinka*, a transitional Ordovician lucinoid bivalve, *Palaeontology* **8**:231–246.

McAlester, A. L., 1966, Evolutionary and systematic implications of a transitional Ordovician lucinoid bivalve, *Malacologia* **3**:433–439.

Morris, N. J., 1967, Mollusca, Scaphopoda and Bivalvia, in: *The Fossil Record* (W. B. Harland et al., eds.), pp. 469–477, London Geological Society.

Morton, J. E., 1973, The biology and functional morphology of *Galeomma (Paralepida) takii* (Bivalvia: Leptonacea), *J. Zool. (London)* **169**:133–150.

Morton, J. E., and Yonge, C. M., 1964, Classification and structure of the Mollusca, in: *Physiology of Mollusca*, Vol. 1 (K. M. Wilbur and C. M. Yonge, eds.), pp. 1–58, Academic Press, London.

Murdock, E. A., Ferguson, A., and Seed, R., 1975, Geographical variation in leucine aminopeptidase in *Mytilus edulis* L. from the Irish coasts, *J. Exp. Mar. Biol. Ecol.* **19:**33–41.

Nair, N. B., and Ansell, A. D., 1968, Characteristics of penetration of the substratum by some marine bivalve molluscs, *Proc. Malacol. Soc. London* **38:**179–197.

Nayar, K. N., 1955, Studies on the growth of the wedge clam *Donax (Latona) cuneatus* Linnaeus, *Indian J. Fish.* **2:**325–348.

Newell, G. E., 1964, Physiological aspects of the ecology of intertidal molluscs, in: *Physiology of Mollusca,* Vol. 1 (K. M. Wilbur and C. M. Yonge, eds.), pp. 59–81, Academic Press, London.

Newkirk, G. F., Waugh, D. L., and Haley, L. E., 1977, Genetics of larval tolerance to reduced salinities in two populations of oysters, *Crassostrea virginica, J. Fish. Res. Board Can.* **34:**384–387.

Nicol, D., 1964, Lack of shell-attached pelecypods in Arctic and Antarctic waters, *Nautilus* **77:**92–93.

Nicöl, D., 1967, Some characteristics of cold water marine pelecypods, *J. Paleontol.* **41:**1330–1340.

Ohba, S., 1956, Effect of population density on mortality and growth in an experimental culture of a bivalve *Venerupis semidecussata, Biol. J. Okayama Univ.* **2:**169–173.

Ohba, S., 1959, Ecological studies in the natural population of a clam, *Tapes japonica,* with special reference to seasonal variations in the size and structure of the population and to individual growth, *Biol. J. Okayama Univ.* **5:**13–42.

Oldfield, E., 1955, Observations of the anatomy and mode of life of *Lasaea rubra* (Montagu) and *Turtonia minuta* (Fabricius), *Proc. Malacol. Soc. London* **31:**226–249.

Orton, J. H., 1928, On rhythmic periods of shell growth in *Ostrea edulis* with a note on fattening, *J. Mar. Biol. Assoc. U. K.* **15:**365–427.

Orton, J. H., 1936, Habit and shell-shape in the Portuguese oyster, *Ostrea angulata, Nature (London)* **138:**466–467.

Owen, G., 1953a, On the biology of *Glossus humanus* (L.) (Isocardia cor. Lam.), *J. Mar. Biol. Assoc. U. K.* **32:**85–106.

Owen, G., 1953b, The shell in the Lamellibranchia, *Q. J. Microsc. Sci.* **94:**57–70.

Owen, G., 1958, Shell form, pallial attachment and the ligament in the Bivalvia, *Proc. Zool. Soc. London* **131:**637–648.

Owen, G., 1959, Observations on the Solenacea with reasons for excluding the family Glaucomyidae, *Philos. Trans. R. Soc. London Ser. B* **242:**59–96.

Paine, R. T., 1974, Intertidal community structure: Experimental studies on the relationship between a dominant competitor and its principal predator, *Oecologia (Berlin)* **15:**93–120.

Paine, R. T., 1976, Size-limited predation: An observational and experimental approach with the *Mytilus–Pisaster* interaction, *Ecology* **57:**858–873.

Pannella, G., and MacClintock, C., 1968, Biological and environmental rhythms reflected in molluscan shell growth, *J. Paleontol.* **42:**64–80.

Pohlo, R. H., 1964, Ontogenetic changes of form and mode of life in *Tressus nuttalli* (Bivalvia: Mactridae), *Malacologia* **1:**321–330.

Pojeta, J., 1971, Review of Ordovician pelecypods, *U.S. Geol. Surv. Prof. Pap.* **695:**1–46.

Pojeta, J., Runnegar, B., Morris, N. J., and Newell, N. D., 1972, Rostroconchia: A new class of bivalved mollusks, *Science* **177:**264–267.

Pojeta, J., Runnegar, B, and Kříž, J., 1973, *Fordilla troyensis* Barrande: The oldest known pelecypod, *Science* **180:**866–868.

Purchon, R. D., 1939, The effect of the environment upon the shell of *Cardium edule, Proc. Malacol. Soc. London* **23:**31–37.

Purchon, R. D., 1955, Functional morphology of the rock boring lamellibranch *Petricola pholadiformis* Lmk, *J. Mar. Biol. Assoc. U. K.* **34:**257–278.

Purchon, R. D., 1956a, The structure and function of the British Pholadidae (rock-boring Lamellibranchia), *Proc. Zool. Soc. London* **124:**859–911.

Purchon, R. D., 1956b, A note on the biology of *Martesia striata* L. (Lamellibranchia), *Proc. Zool. Soc. London* **126**:245–258.

Purchon, R. D., 1977, *The Biology of the Mollusca*, Pergamon Press, Oxford.

Quayle, D. B., 1952, The rate of growth of *Venerupis pullastra* (Montagu) at Millport, Scotland, *Proc. R. Soc. Edinburgh Sect. B* **64**:384–406.

Raup, D. M., 1966, Geometric analysis of shell coiling: General problems, *J. Paleontol.* **40**:1178–1190.

Raup, D. M., 1972, Approaches to morphologic analysis, in: *Models in Paleobiology* (T. J. M. Schopf, ed.), pp. 28–44, Freeman Cooper, San Francisco.

Rhoads, D. C., and Pannella, G., 1970, The use of molluscan shell growth patterns in ecology and paleoecology, *Lethaia* **3**:143–161.

Rhoads, D. C., and Young, D. K., 1970, The influence of deposit-feeding organisms on sediment stability and community trophic structure, *J. Mar. Res.* **28**:150–178.

Rhoads, D. C., McCall, P. L., and Yingst, J. Y., 1978, Disturbance and production on the estuarine seafloor, *Am. Sci.* **66**:577–586.

Runnegar, B., and Pojeta, J., 1974, Molluscan phylogeny: The paleontological viewpoint, *Science* **186**:311–317.

Saleuddin, A. S. M., 1964, Observations on the habit and functional anatomy of *Cyprina islandica* (L.), *Proc. Malacol. Soc. London* **36**:149–162.

Saleuddin, A. S. M., 1965, The mode of life and functional anatomy of *Astarte* spp. (Eulamellibranchia), *Proc. Malacol. Soc. London* **36**:229–257.

Savage, R. E., 1956, The great spatfall of mussels in the River Conway Estuary in spring 1940, *Fish. Invest. Minist. Agric. Fish. Food (G. B.) Ser. II* **20**:1–21.

Schafer, W., 1972, *Ecology and Palaeoecology of Marine Environments* (G. Y. Craig, ed.; I. Oertel, trans.), Oliver & Boyd, Edinburgh.

Seed, R., 1968, Factors influencing shell shape in *Mytilus edulis* L., *J. Mar. Biol. Assoc. U. K.* **48**:561–584.

Seed, R., 1969, The ecology of *Mytilus edulis* L. (Lamellibranchiata) on exposed rocky shores. II. Growth and mortality, *Oecologia (Berlin)* **3**:317–350.

Seed, R., 1973, Absolute and allometric growth in the mussel *Mytilus edulis* L. (Mollusca: Bivalvia), *Proc. Malacol. Soc. London* **40**:343–357.

Seed, R., 1976, Ecology, in: *Marine Mussels: Their Ecology and Physiology* (B. L. Bayne, ed.), *International Biological Programme*, Vol. 10, pp. 13–65, Cambridge University Press.

Seed, R., 1978, The systematics and evolution of *Mytilus galloprovincialis* Lamarck, in: *The Genetics, Ecology, and Evolution of Marine Organisms* (J. A. Beardmore and B. Battaglia, eds.), pp. 447–466, Plenum Press, New York.

Seed, R., and Brown, R. A., 1975, The influence of reproductive cycle, growth and mortality on population structure in *Modiolus modiolus* (L.), *Cerastoderma edule* (L.) and *Mytilus edulis* (L.) (Mollusca: Bivalvia), *Proc. Eur. Mar. Biol. Symp.* 9 (H. Barnes, ed.), pp. 257–274, Aberdeen University Press.

Seed, R., and Brown, R. A., 1978, Growth as a strategy for survival in two marine bivalves, *Cerastoderma edule* (L.) and *Modiolus modiolus* (L), *J. Anim. Ecol.* **47**:283–292.

Seed, R., and Roberts, D., 1976, A study of three populations of saddle oÿsters (Family Anomiidae) from Strangford Lough, Northern Ireland, *Ir. Nat. J.* **18**:318–321.

Seilacher, A., 1970, Arbeitskonzept zur Konstruktionsmorphologie, *Lethaia* **3**:393–396.

Sellmer, G. P., 1967, Functional morphology and ecological life history of the gem clam *Gemma gemma* (Eulamellibranchia, Veneridae), *Malacologia* **5**:137–223.

Singh, S. M., and Zouros, E., 1978, Genetic variation associated with growth rate in the American oyster *Crassostrea virginica*, *Evolution* **32**:342–353.

Stanley, S. M., 1968, Post-Paleozoic adaptive radiation of infaunal bivalve molluscs—a consequence of mantle fusion and siphon formation, *J. Paleontol.* **42**:214–299.

Stanley, S. M., 1970, Relation of shell form to life habits in the Bivalvia, *Geol. Soc. Am. Mem.* **125**:1–296.

Stanley, S. M., 1972, Functional morphology and evolution of byssally attached bivalve mollusks, *J. Paleontol.* **46**:165–212.

Stanley, S. M., 1973, Effects of competition on rates of evolution with special reference to bivalve mollusks and mammals, *Syst. Zool.* **22**:486–506.

Stanley, S. M., 1975a, Adaptive themes in the evolution of the Bivalvia (Mollusca), *Annu. Rev. Earth Planet. Sci.* **3**:361–385.

Stanley, S. M., 1975b, Why clams have the shape they have: An experimental analysis of burrowing, *Paleobiology* **1**:48–58.

Stanley, S. M., 1977, Trends, rates and patterns of evolution in the Bivalvia, in: *Patterns of Evolution as Illustrated by the Fossil Record* (A. Hallam, ed.), pp. 210–250, Elsevier/North-Holland, Amsterdam.

Stasek, C. R., 1963a, Orientation and form in the bivalved Mollusca, *J. Morphol.* **112**:195–214.

Stasek, C. R., 1963b, Geometrical form and gnomonic growth in the bivalved Mollusca, *J. Morphol.* **112**:215–231.

Stasek, C. R., 1966, Views on the comparative anatomy of the bivalved Mollusca, *Malacologia* **5**:67–68.

Stasek, C. R., 1972, The molluscan framework, in: *Chemical Zoology 3* (M. Florkin and B. J. Scheer, eds.), pp. 1–44, Academic Press, New York.

Swan, E. F., 1952, Growth indices of the clam *Mya arenaria*, *Ecology* **33**:365–374.

Tanita, S., and Kikuchi, S., 1957, On the density-effect of the raft cultured oysters. I. The density effect within one plate, *Bull. Tokai Reg. Fish. Res. Lab.* **9**:133–142.

Tevesz, M. J. S., and McCall, P. L., 1976, Primitive life habits and adaptive significance of the pelecypod form, *Paleobiology* **2**:183–190.

Thayer, C. W., 1975, Morphological adaptations of benthic invertebrates to soft substrata, *J. Mar. Res.* **33**:177–189.

Thayer, C. W., 1979, Biological bulldozers and the evolution of marine benthic communities, *Science* **203**:458–460.

Thiesen, B. F., 1968, Growth and mortality of culture mussels in the Danish Wadden Sea, *Medd. Dan. Fisk. Havunders.* **6**:47–78.

Thiesen, B. F., 1973, The growth of *Mytilus edulis* L. (Bivalvia) from Disko and Thule district, Greenland, *Ophelia* **12**:59–77.

Thomas, R. D. K., 1975, Functional morphology, ecology and evolutionary conservation in the Glycymeridae (Bivalvia), *Paleontology* **18**:217–254.

Thomas, R. D. K., 1976, Constraints of ligament growth, form and function on evolution in the Arcoida (Mollusca: Bivalvia), *Paleobiology* **2**:64–83.

Thomas, R. D. K., 1978, Shell form and the ecological range of living and extinct Arcoida, *Paleobiology* **4**:181–194.

Thompson, D'A. W., 1942, *On Growth and Form*, Cambridge University Press.

Trueman, E. R., 1964, Adaptive morphology in paleoecological interpretation, in: *Approaches to Paleoecology* (J. Imbrie and N. Newell, eds.), pp. 45–74, Wiley, New York.

Trueman, E. R., 1966, Bivalve mollusks: Fluid dynamics of burrowing, *Science* **152**:523–525.

Trueman, E. R., 1969, Ligament, in: *Treatise on Invertebrate Paleontology, Part N, Mollusca 6, Bivalvia* (R. C. Moore, ed.), pp. N58–N64, University of Kansas Press, Lawrence.

Trueman, E. R., 1975, *The Locomotion of Soft-Bodied Animals: Contemporary Biology*, Arnold, London.

Turner, R. D., 1966, *A Survey and Illustrated Catalogue of the Teredinidae*, Museum of Comparative Zoology, Harvard University, Cambridge, Massachusetts.

Valentine, J. W., 1973, *Evolutionary Paleoecology of the Marine Biosphere*, Prentice-Hall, Englewood Cliffs, New Jersey.

Valentine, J. W., and Gertman, R. L., 1972, The primitive ecospace of the Pelecypoda, *Geol. Soc. Am. Abstr. Prog.* **4**:696.

Vermeij, G. J., 1970, Adaptive versatility and skeleton construction, *Am. Nat.* **104**:253–260.

Vermeij, G. J., and Veil, J. A., 1978, A latitudinal pattern in bivalve shell gaping, *Malacologia* **17**:57–61.

Vokes, H. E., 1967, Genera of the Bivalvia: A systematic and bibliographical catalogue, *Bull. Am. Paleontol.* **51**:105–394.

Wade, B. A., 1967, On taxonomy, morphology and ecology of the beach clam, *Donax striatus* *Bull. Mar. Sci.* **17**:723–740.

Wade, B. A., 1969, Studies on the biology of the West Indian beach clam *Donax denticulatus* L. 3. Functional morphology, *Bull. Mar. Sci.* **19**:307–322.

Wainwright, S. A., 1969, Stress and design in the bivalved mollusc shell, *Nature (London)* **224**:777–779.

Wilbur, K. M., and Owen, G., 1964, Growth, in: *Physiology of Mollusca*, Vol. 1 (K. M. Wilbur and C. M. Yonge, eds.), pp. 211–242, Academic Press, London.

Wright, A. D., 1972, The relevance of zoological variation studies to the generic identification of fossil brachiopods, *Lethaia* **5**:1–13.

Yonge, C. M., 1936, The evolution of the swimming habit in the Lamellibranchia, *Mem. Mus. Hist. Nat. Belg.* **3**:78–100.

Yonge, C. M., 1939, The protobranchiate Mollusca: A functional interpretation of their structure and evolution, *Philos. Trans. R. Soc. London Ser. B* **230**:79–147.

Yonge, C. M., 1948, Formation of siphons in Lamellibranchia, *Nature (London)* **161**:198.

Yonge, C. M., 1949, On the structure and adaptations of the Tellinacea, deposit-feeding Eulamellibranchia, *Philos. Trans. R. Soc. London Ser. B* **234**:29–76.

Yonge, C. M., 1952, Studies on Pacific coast molluscs. IV. Observations on *Siliqua patula* Dixon and on evolution within the Solenidae, *Univ. Calif. Berkeley Publ. Zool.* **55**:421–438.

Yonge, C. M., 1953a, The monomyarian condition in the Lamellibranchia, *Trans. R. Soc. Edinburgh* **62**:443–478.

Yonge, C. M., 1953b, Form and habit in *Pinna carnea* Gmelin, *Philos. Trans. R. Soc. London Ser. B* **237**:335–374.

Yonge, C. M., 1955, Adaptations to rock boring in *Botula* and *Lithophaga* (Lamellibranchia, Mytilidae) with a discussion on the evolution of this habit, *Q. J. Microsc. Sci.* **96**:383–410.

Yonge, C. M., 1957, Mantle fusion in the Lamellibranchia, *Pubbl. Stn. Zool. Napoli* **29**:151–171.

Yonge, C. M., 1962, On the primitive significance of the byssus in the Bivalvia and its effects in evolution, *J. Mar. Biol. Assoc. U. K.* **42**:113–125.

Yonge, C. M., 1967, Form, habit and evolution in the Chamidae (Bivalvia) with reference to conditions in the rudists (Hippuritacea), *Philos. Trans. R. Soc. London Ser. B* **252**:49–105.

Yonge, C. M., 1969, Functional morphology within the Carditacea (Bivalvia), *Proc. Malacol. Soc. London* **38**:493–527.

Yonge, C. M., and Campbell, J. I., 1968, On the heteromyarian condition in the Bivalvia with special reference to *Dreissena polymorpha* and certain Mytilacea, *Trans. R. Soc. Edinburgh* **68**:21–43.

Chapter 2

Environmental and Biological Controls of Bivalve Shell Mineralogy and Microstructure

JOSEPH GAYLORD CARTER

1. Introduction 69
2. Distribution of Bivalve Shell Mineralogy and Microstructure 71
   2.1. Mineralogy 71
   2.2. Microstructure Groups and Categories 78
3. Evolution and Adaptive Significance of Shell Mineralogy and Microstructure ... 83
   3.1. General Evolutionary Trends 84
   3.2. Evolution of Aragonitic Shell Layers 85
   3.3. Evolution of Calcitic Shell Layers 90
4. Shell Mineralogy and Structure as Records of Environmental Change 95
   4.1. Mineralogical Variations 96
   4.2. Structural Variations 99
5. Conclusions 104
   References 107

1. Introduction

Since the early descriptions of bivalve shell microstructure by Carpenter (1844) and of shell mineralogy by Prenant (1927) and Bøggild (1930), it has been generally recognized that these features are determined primarily through biological (i.e., evolutionary) processes. In fact, the available paleontological data suggest that mineralogical and microstructural evolution have followed definable, linear trends in this class, e.g., from ara-

JOSEPH GAYLORD CARTER ● Department of Geology, University of North Carolina, Chapel Hill, North Carolina 27514.

69

gonitic nacreous ancestors to aragonitic porcelaneous and calcitic foliated descendants (Douvillé, 1913; Newell, 1938; Taylor, J. D., 1973; Waller, 1972, 1978; Carter and Tevesz, 1978a,b). Studies of bivalve shell secretion have emphasized physiological controls of both mineralogy and microstructure, mediated by the composition of the extrapallial fluid and epitaxial or other mechanisms relating to the shell organic matrix (e.g., Watabe and Wilbur, 1960; Wilbur, 1960, 1964, 1972; Wilbur and Watabe, 1963; Nakahara and Bevelander, 1971; Erben and Watabe, 1974; Kitano et al., 1976; Duckworth, 1976). On the other hand, Lowenstam (1954a–c, 1963, 1964) and Dodd (1963, 1964) have noted that for certain bivalves, mineralogical evolution and shell aragonite/calcite ratios may also be related to environmental influences. More recently, even some aspects of shell microstructure have been attributed to environment-related modifications of secretory mechanisms (Davies and Sayre, 1970; Lutz, 1976a,b; Lutz and Rhoads, 1977). This chapter reviews current evidence for environmental and biological controls of shell mineralogy and microstructure and evaluates the utility of these features as records of environmental change.

To assess environmental influences on shell mineralogy and microstructure, it is first necessary to understand the range of these parameters in bivalve shells. Mineralogy and microstructure have both been extensively reviewed by Bøggild (1930), Kobayashi (1969, 1971), and Taylor et al. (1969, 1973). Appendix 2.A summarizes mineralogical data from these, and several other, literature sources for modern and fossil bivalves. Appendix 2.A also reviews procedures for determining the mineralogy of modern bivalves and outlines techniques for inferring original mineralogy from diagenetically altered shells. New data are provided for the Mytilacea, because representatives of this superfamily are emphasized in studies of environmental effects on shell mineralogy and microstructure. Also included here are mineralogical data for bivalve accessory calcification (e.g., tubes and burrow linings), which have been generally neglected in earlier studies, but which may demonstrate environmental influences. Appendix 2.B describes and illustrates the microstructure groups, categories, and varieties referred to in this chapter and provides a guide for their identification.

General Descriptive Terms for Shell Structure and Layering. Before outlining the distribution of mineralogy and microstructure in bivalve shells, it is necessary to define a general descriptive nomenclature for shell structure and shell layering. *Shell structure* is a general term referring to all aspects of shell construction except shell morphology. Shell structure encompasses shell architecture, microstructure, shell layering, and growth layering, including internal growth increments. *Shell architecture* refers to the major aspects of shell structure, i.e., the orientation of the largest units of shell microstructure with respect to shell form (e.g., horizontal nacreous laminae, concentric crossed lamellae and spatial relationships of the shell layers. *Shell microstructure* refers to the arrangement of the various basic structural units (e.g., tablets, rods, blades) in the microstructure varieties (Table I), as defined and illustrated in Appendix 2.B. Also included here are anaerobiosis-related changes in previously formed microstructures, local changes in the ratio of shell calcium carbonate to shell organic matrix, and fine-scale ultrastructural changes, some of which may be expressed in the shell as internal growth increments. Following the terminology of MacClintock (1967), a *shell layer* is a bed of shell material exhibiting a single microstructure or alternations of microstructures in successive *shell sublayers*. Some, but not all, shell layers and shell sublayers are also *growth layers* or *growth sublayers*, which are bounded by time-parallel growth surfaces.

2. Distribution of Bivalve Shell Mineralogy and Microstructure

2.1. Mineralogy

2.1.1. General Trends

Aragonite and calcite are the only two common forms of calcium carbonate in bivalve shells. A third variety, dahllite (a phosphatic calcium carbonate), was reported along with calcite in the larval shell of the pearl "oyster" *Pinctada martensi* (Watabe, 1956), but other bivalve larval shells appear to be generally aragonitic, even among species with largely calcitic adult shells (Stenzel, 1964) (also see Appendix 2.A, Table II). Calcified portions of ligaments, myostracal layers (layers deposited at sites of shell-muscle attachment), and periostracal calcification (secreted by the inner surface of the outer mantle fold) are also generally aragonitic in this class (Bøggild, 1930; Wada, 1961; Stenzel, 1962; Taylor, J. D., et al., 1969, 1973; Carter and Aller, 1975). Where they occur in the same shell, aragonite and calcite invariably maintain a sharp mutual boundary without microstructural intergradation (Bøggild, 1930; Taylor, J. D., et al., 1969). These two forms of calcium carbonate are generally separated as distinct shell layers, with the calcite more commonly comprising an outer shell

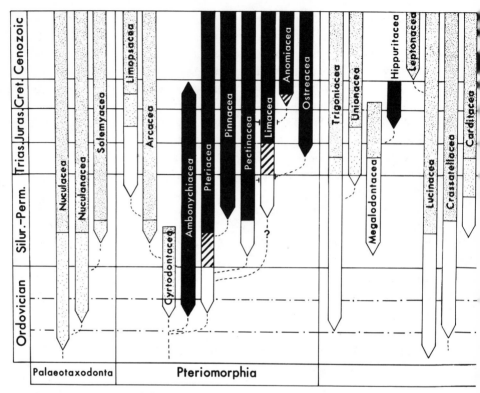

Figure 1. Diagrammatic summary of bivalve shell mineralogy (excluding accessory calcification). Mineralogical data are documented in Appendix 2.A, Tables I and II. This compilation assumes uniformity of shell mineralogy at the generic level, and on this basis, some
generic ranges allow extrapolation into the earlier fossil record prior to the earliest time of
actual documentation (indicated by separate symbols). Data from the literature have been

layer. Aragonite and calcite rarely occur in the same shell layer, and where
this occurs, the aragonitic and calcitic portions still maintain sharp mutual
boundaries (Appendix 2.B, Fig. 17) (Carter and Eichenberger, 1977).

Figure 1 illustrates the general distribution of shell mineralogy among
modern and fossil bivalve superfamilies, abstracted from a variety of literature sources and based, in part, on new mineralogical determinations
included in Table I of Appendix 2.A. Figure 1 represents considerable
interpolation of data between times of documented shell mineralogy and,
for some groups, also extrapolation into the earlier fossil record. Extrapolations of mineralogical data are based on the known geological time
ranges of genera (Cox *et al.*, 1969) and the assumption of mineralogical

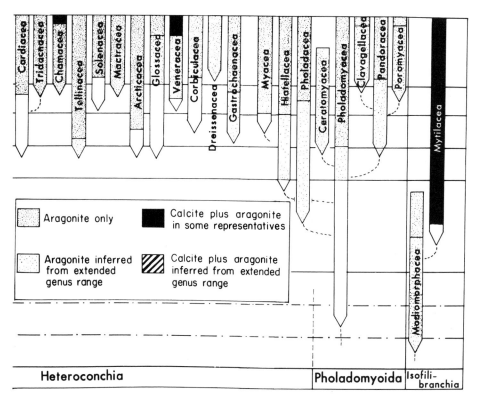

evaluated to exclude possible artifacts of diagenesis and contamination by sediment. Data judged uncertain from either of these points of view are excluded from this figure and from Appendix 2.A, Table I. The taxonomic arrangement of the superfamilies follows the outline by Newell (1969) as modified by Pojeta (1971).

uniformity at the generic level. Genera within the Pterioida and Hippuritacea follow this assumption with no known exceptions. But certain genera within the Mytilacea, Chamacea, Lucinacea, and Veneracea have added calcite to their shell during the course of species-level evolutionary differentiation. Consequently, extrapolations of calcitic mineralogy into their earlier fossil record are tenuous. Mineralogical data are rare for Ordovician fossils, and conclusions about the mineralogy of Lower and Middle Paleozoic bivalves are based largely on inferences from differential preservation and comparative shell microstructure.

Keeping in mind the limitations of the present data, Fig. 1 suggests unidirectional mineralogical evolution at the superfamily level. Thus, once calcite secretion evolves as a major shell layer in a superfamily, it is retained until the present day or, as in the Ambonychiacea and Hip-

puritacea, until extinction. This calls to mind the conclusion by Lowenstam (1954c) regarding linear trends in mineralogical evolution among
cephalopods and corals.

Paleontological data suggest that all major bivalve lineages defined
at the ordinal level are primitively entirely aragonitic, with the possible
exception of the Pterioida (Fig. 1). But even the Pterioida may have
evolved from wholly aragonitic arcoid ancestors, i.e., from the superfamily
Cyrtodontacea (see Pojeta, 1971; Carter and Tevesz, 1978b). This observation is compatible with the conclusion of J. D. Taylor (1973) that the
hypothetical monoplacophoran ancestors of the Bivalvia were entirely
aragonitic. If this is true, then mineralogical evolution in the Bivalvia has
resulted in the addition of calcite to certain representatives of all modern
subclasses except the Palaeotaxodonta and Anomalodesmata. These mineralogical changes have evolved in connection with a variety of life habits, including epifaunal, endolithic nestling, shallow burrowing, and even
deep burrowing. On the other hand, persistent calcitic shell layers (as
opposed to traces of calcite) have evolved only among epifaunal bivalves
and bivalves with an epifaunal ancestry (Kennedy and Taylor, 1968; Kennedy et al., 1969). The distribution of shell mineralogy among bivalves
secreting both aragonite and calcite is considered in more detail in the
following sections.

2.1.2. Mineralogy of the Pterioida and Hippuritacea

Calcite secretion is taxonomically more widespread among the Pterioida and Hippuritacea than in all other bivalve groups. Mineralogical
data are scarce for the presumed ancestors of these taxa (i.e., the Cyrtodontacea and Megalodontacea, respectively), but the available data suggest that they were characterized by entirely aragonitic shells (Kennedy
and Taylor, 1968; Carter and Tevesz, 1978b). Largely calcitic shells have
evolved in many pterioids such as scallops and oysters, probably through
gradual evolutionary expansion of calcitic shell layers at the expense of
middle and inner aragonitic layers. Thus, the Lower and Middle Paleozoic
pterioids generally possessed an outer calcitic layer and underlying middle and inner aragonitic layers (Pojeta, 1962; Taylor, J. D., et al., 1973;
Carter and Tevesz, 1978a,b). But by Upper Paleozoic time, some Pectinacea had evolved shells with expanded calcitic layers and reduced aragonitic layers, and largely calcitic shells became abundant by Middle
Mesozoic time (Newell, 1938; Newell and Boyd, 1970; Waller, 1972, 1978).
Largely calcitic shells have also evolved in the Pterioida in the Ostreacea
and in certain representatives of the Anomiacea (Bøggild, 1930; Chel'tsova,
1969; Taylor, J. D., et al., 1969). Among modern oysters, aragonite secre-

tion may be entirely restricted to the larval shell, ligament resilium, and myostracal deposits (Stenzel, 1962, 1963, 1964).

Calcite secretion in certain Hippuritacea may differ from that in the Pterioida in its late ontogenetic appearance. As described by Skelton (1974, p. 72), fibrous prismatic calcite first appears in the shell of the hippuritacean *Biradiolites angulosissimus* after half a whorl of growth. This ontogenetic appearance is especially interesting, considering that sporadic calcite appears ontogenetically in some Veneracea and Lucinacea (Carter, personal observations; Carter and Eichenberger, 1977). However, there is no evidence suggesting that this similarity is phylogenetically significant. To the contrary, the taxonomic distribution of late ontogenetic calcite in the Lucinacea and Veneracea suggests that this feature evolved independently in these two superfamilies. The reader is referred to Kennedy *et al.* (1970) for a discussion of phylogenetic relationships among the Hippuritacea, Veneracea, and another calcite-secreting superfamily of the subclass Heteroconchia, the Chamacea.

2.1.3. Mineralogy of the Veneroida

Until recently, the only veneroids known to secrete calcite were two or three cooler-water species of the genus *Chama*. In both *Chama pellucida* and *C. exogyra*, calcite is restricted to the outer layer, which comprises the shell's concentric ornament (Lowenstam, 1954b, 1963; Taylor, J. D., and Kennedy, 1969; Kennedy *et al.*, 1970). Lowenstam (1954b) detected a small amount of calcite in *C. semipurpurata* from Tokyo Bay, but his finding has yet to be confirmed by microstructural analysis.

In addition to the Chamacea, veneroid calcite occurs in certain representatives of the Lucinacea and Veneracea. Lucinacean calcite takes the form of extremely small patches within the largely aragonitic outer and middle shell layers of the Western Pacific *Anodontia bialata* (Appendix 2.B, Figs. 19–22). Preliminary observations of *A. bialata* suggest that calcite secretion is not ubiquitous among representatives of this species (Carter, personal observations). Calcite secretion in the Veneracea is restricted to the outer part of the outer prismatic shell layer of a few cooler-water species of *Protothaca*, *Saxidomus*, *Venerupis*, and *Petricola* (Carter and Eichenberger, 1977). Veneracean and lucinacean calcite is unusual in its sporadic occurrence and late ontogenetic appearance in the shell (Fig. 2). As discussed later in this chapter, ambient temperature may influence the initiation and abundance of calcite secretion in *Protothaca staminea*. However, the absence of calcite in certain cold-water populations of *P. staminea* suggests that biological controls are also important.

Figure 2. Shell of the veneracean *Venerupis crenata* Lamarck, Yale Peabody Museum 10115, Sydney Harbor, Australia, showing abundant mineralogical aberrations (sporadic calcitic patches) visible on the shell exterior. The calcitic patches appear as dark triangles and concentric bands concentrated in the shell posterior (i.e., toward the right in the upper photograph). Note the absence of calcite patches near the juvenile shell in the dorsal view (middle photograph). Scale bar: 5 mm.

2.1.4. Mineralogy of the Mytilacea and Modiomorphacea

The taxonomic distribution of mineralogy in the modern Mytilacea is extremely complex. Only one mytilid subfamily, the Lithophaginae, consistently shows outer-layer calcite, and only one modern species in the Mytilinae shows both outer and inner calcitic layers [i.e., *Mytilus californianus* (see Dodd, 1964)]. New mineralogical data for the Mytilacea, from Table II of Appendix 2.A, are integrated with data from Lowenstam

(1954b,c) and J. D. Taylor *et al.* (1969) and summarized diagrammatically in Fig. 5, which is discussed in detail in Section 4.1. For the moment, it may be pointed out that *Trichomya hirsuta* and *Botula fusca* show ontogenetic and other spatial variations in the distribution of calcite over the exterior of their subperiostracal shell layers (see notes 11 and 16 to Table II in Appendix 2.A). In addition, periostracal calcification in *Trichomya hirsuta* and *Gregariella coralliophaga* is aragonitic, even though this overlies partially or entirely calcitic shell layers.

Pojeta (1971) has suggested that the Mytilacea evolved from the Paleozoic superfamily Modiomorphacea. Mineralogical data for two Devonian Modiomorphacea indicate that their shells were probably entirely aragonitic (Carter and Tevesz, 1978a) (also see Appendix 2.A, Table I). These data and the occurrence of wholly aragonitic shells among the modern Mytilacea are compatible with the hypothesis that calcite secretion evolved after the evolutionary differentiation of this superfamily from the Modiomorphacea. If the linear trend of mineralogical evolution described above for bivalve superfamilies also applies to lower taxonomic levels, then the distribution of calcite among modern mytilacean genera and species may be helpful for assessing their phylogenetic relationships.

2.1.5. Mineralogy of Accessory Calcification

Accessory calcification includes plates attached to the shell margins, encrustations secreted over the exterior of the periostracum, tubes secreted by burrowing species, pallets secreted by the posterior mantle for closure of burrow apertures, and burrow linings secreted by boring species. These structures are especially common among representatives of the Gastrochaenacea, Clavagellacea, Mytilacea (subfamily Lithophaginae), and Pholadacea. Little has been published concerning their mineralogy, except for a few observations by Bøggild (1930), Pobeguin (1954), and J. D. Taylor *et al.* (1969). New observations of mineralogy of accessory calcification are summarized in Appendix 2.A, Table III.

Accessory calcification is generally aragonitic in the primarily tropical and subtropical superfamilies Gastrochaenacea and Clavagellacea (Appendix 2.A, Table III) (Carter, 1978). Some gastrochaenids show minor calcite in their middle or middle and posterior burrow linings, but this calcite is high in magnesium, and therefore probably represents contamination from associated encrusting calcareous red algae. As indicated by Chave (1954) and Dodd (1967), bivalve calcite is generally low in magnesium, whereas that secreted by calcareous red algae is commonly high in magnesium.

In comparison with the Gastrochaenacea and Clavagellacea, accessory calcification is mineralogically more variable in the Mytilacea and

Pholadacea. Lithophaginid burrow linings are commonly aragonitic, but the tube of one lithophaginid from the Pliocene (?) of Florida is entirely calcitic (Appendix 2.A, Table III). The posterior encrustations on the shells of many modern species of Lithophaga vary from aragonitic to aragonitic with minor calcite, or even largely calcitic. As described by Kühnelt (1930), the posterior encrustations in some species of Lithophaga consist of endogenous secretions plus debris eroded from the burrow walls.

Within the Pholadacea, representatives of the Pholadidae generally secrete aragonitic accessory structures, but the Teredinidae are mineralogically more variable. The data of Table III in Appendix 2.A and Bøggild (1930) suggest that pallets and posterior tubes in the Teredinidae are generally calcitic. But at least one teredinid, Uperotus clavus (Gmelin), is calcitic in the posterior of its tube and aragonitic anteriorly (Appendix 2.A, Table III). Pobeguin (1954) and J. D. Taylor et al. (1969) described only aragonite in teredinid tubes. However, J. D. Taylor et al. (1969) examined only Teredo navalis Linnaeus, and neither they nor Pobeguin (1954) indicated whether their samples came from the anterior or posterior portion of the tube. This discrepancy of mineralogical data for teredinid tubes may reflect spatial variation in tube mineralogy, as in Uperotus. Alternatively, Teredo tube linings may show a temperature influence on mineralogy, or possibly even mineralogical conversion of aragonite to calcite. The latter possibility is suggested by the account by Lowenstam (1954b) of apparent conversion of aragonite to calcite in serpulid worm tubes. Conversion of dry bivalve secretions from aragonite to calcite at room temperature has yet to be documented, but this possibility has probably not been investigated in connection with teredinid or other bivalve accessory calcification.

2.2. Microstructure Groups and Categories

Bivalve shell microstructures can be arranged into seven groups on the basis of their major structural arrangement, and into five categories on the basis of natural associations in major shell layers (Table I). The microstructure groups and categories are defined briefly in this section, with microstructure varieties described and illustrated in Appendix 2.B. The microstructure groups are defined independently of mineralogy, but as described below, they may generally associate with aragonite or calcite.

2.2.1. Microstructure Groups

Group I. Prismatic. Prismatic structures consist of generally parallel, columnar, adjacent structural units that do not strongly interdig-

Table I. Organization of Microstructure Terminology Used in Chapter 2 and Appendix 2.B[a]

Microstructure groups and their principal varieties	Microstructure categories and their constituent microstructures
I. Prismatic	I. Aragonitic prismatic
A. Simple prismatic	II. Calcitic prismatic
B. Fibrous prismatic	III. Nacreous (aragonitic)
C. Spherulitic prismatic	IV. Porcelaneous (aragonitic)
D. Composite prismatic	A. Aragonitic crossed lamellar
II. Spherulitic	B. Aragonitic crossed acicular
	C. Aragonitic complex crossed lamellar
III. Laminar	D. Aragonitic crossed-matted/lineated
A. Nacreous	E. Aragonitic homogeneous
B. Regularly foliated	V. Foliated (calcitic)
IV. Crossed	A. Regularly foliated
A. Crossed lamellar	B. Calcitic crossed lamellar
B. Crossed acicular	(= crossed foliated)
C. Complex crossed lamellar	C. Calcitic complex crossed lamellar
D. Crossed-matted/lineated	(= complex crossed foliated)
V. Homogeneous	
A. Homogeneous *s.s.*	**Microstructures rarely encountered in the Bivalvia**
B. Granular	I. Spherulitic
VI. Isolated spicules or spikes	II. Isolated spicules or spikes
VII. Isolated crystal morphotypes	III. Isolated crystal morphotypes

[a] See Appendix 2.B for microstructure definitions and illustrations.

itate along their mutual boundaries. The prism columns are generally longer than they are wide, but rather short calcitic prisms occur in certain Pectinacea, Ostreacea, and Anomiacea. Varieties of prismatic structure include simple, fibrous, spherulitic, and composite prismatic. The reader may note the distinction between spherulitic prisimatic and spherulitic structure (Appendix 2.B). All varieties of prismatic structure presently described are known to occur as aragonite and calcite, but composite prismatic structure is almost always aragonitic (Appendix 2.B, Figs. 1–15 and 19).

Group II. Spherulitic. Spherulitic structure consists of densely packed spherical to subspherical aggregations of elongate structural subunits, with the latter radiating in all directions from a central nucleation site (Appendix 2.B, Figs. 16 and 18). Spherulitic structure commonly grades into spherulitic prismatic structure through elongation of the larger spherulites toward the depositional surface. Major shell layers consisting of spherulitic structure are rare in the Bivalvia, and are at present known only in aragonite. Because isolated spherulites do not comprise a compact shell layer, they are not included here but are regarded as isolated crystal morphotypes (see Group VII below).

Group III. Laminar. Laminar microstructures consist of sheets of strongly flattened structural units deposited uniformly parallel or oblique to the general depositional surface. This group includes aragonitic nacreous structure (mother of pearl) and calcitic regularly foliated structure (Appendix 2.B, Figs. 23–27). Calcitic regularly foliated structure is one of several microstructures presently included in the foliated structure category (defined in Section 2.2.2, Category V, below).

Group IV. Crossed. Crossed microstructures consist of elongate structural subunits organized into adjacent, interdigitating aggregations referred to as "first-order lamellae." The first-order lamellae show the following properties: (1) they may assume a variety of shapes, but their elongate structural subunits are not parallel with the surface of deposition; (2) within each first-order lamella, the elongate structural subunits are arranged generally in parallel or, figuratively speaking, on the surfaces of stacked cones; and (3) adjacent first-order lamellae show two or more dip directions of their elongate structural subunits relative to the surface of deposition. Of the microstructure varieties included in this group, crossed acicular and crossed-matted/lineated structures are at present known only in aragonite (Appendix 2.B, Figs. 32 and 42–44), but crossed lamellar and complex crossed lamellar structures occur in aragonite or calcite (Appendix 2.B, Figs. 28–31 and 33–41). Crossed foliated and complex crossed foliated structures are merely calcitic varieties of crossed lamellar and complex crossed lamellar structures, respectively (Appendix 2.B, Figs. 30–31).

Group V. Homogeneous. Homogeneous structure consists of irregularly shaped but generally more or less equidimensional crystallites, or aggregations of these crystallites, which show no regular arrangement except for possible separation into growth layers. The present definition of homogeneous structure differs from that of Bøggild (1930) in its independence from optical crystallographic criteria, and in its diagnosis, which is based on scanning electron microscopy criteria rather than on features observed with optical microscopy. The homogeneous structure group includes homogeneous structure *sensu stricto* (s.s.) and granular structure, which are differentiated by the size of their major structural units. In homogeneous structure s.s., the major structural units are generally smaller than 5.0 μm in diameter. In granular structure, the major structural units are generally larger than 5.0 μm in diameter. Both varieties of homogeneous structure are known in aragonite (Appendix 2.B, Figs. 45–58) but only homogeneous structure s.s. is at present also known in calcite.

Group VI. Isolated Spicules or Spikes. Isolated spicules or spikes are variably shaped calcified structures sparsely distributed within the periostracum or within a tissue layer. Calcified supporting rods occur in the gills of certain Trigoniacea and Unionacea (Atkins, 1938), but their mineralogy is at present unknown. Calcified periostracal spikes and granules are at present known only in aragonite (Appendix 2.A, Table II; Appendix 2.B, Fig. 49).

Group VII. Isolated Crystal Morphotypes. This group includes all sparsely distributed, irregularly oriented structural units that do not comprise a persistent, major shell layer, and that show morphologies typical of inorganically precipitated crystal forms, including single crystals, twins, and more or less isolated spherulites. Aragonitic and calcitic isolated crystal morphotypes are commonly associated with voids originally filled with extrapallial fluid but isolated from close proximity with the mantle epithelium (Appendix 2.B, Fig. 50) (also see Wind and Wise, 1976).

2.2.2. Microstructure Categories

The microstructure categories described herein represent an alternative organization of the more common microstructure groups and their structural varieties (Table I). Microstructure Groups VI and VII defined above are not included here because they are rarely encountered and do not comprise major shell layers. The spherulitic structure group (Group II) is also excluded here because of its rarity in bivalve shells. Each of the categories described below represents a natural association of microstructures that commonly intergrade within aragonitic or calcitic shell layers.

Category I. Aragonitic Prismatic. All varieties of aragonitic prismatic structure.

Category II. Calcitic Prismatic. All varieties of calcitic prismatic structure.

Category III. Nacreous. All varieties of aragonitic nacreous structure. Because nacre is invariably aragonitic in the Bivalvia, the term "nacreous" is used here with the implication of aragonitic mineralogy.

Category IV. Porcelaneous. All varieties of aragonitic crossed and homogeneous structures (e.g., crossed lamellar, crossed acicular, complex crossed lamellar, crossed-matted/lineated, homogeneous *sensu stricto*, and granular). As with the term nacreous in the preceding category, use of the term "porcelaneous" carries an implication of aragonitic mineralogy.

Category V. Foliated. Any calcitic shell layer consisting of adjacent, flattened calcitic blades or laths, regardless of their structural organization. This category includes calcitic regularly foliated structure, crossed foliated structure, and calcitic varieties of complex crossed lamellar structure (e.g., irregular complex crossed foliated and cone complex crossed foliated).

The three aragonitic categories (I, III, and IV) are naturally separated from the two calcitic categories (II and V) by the fact that microstructures with differing mineralogy do not show microstructural intergradations.

In addition, microstructures of differing mineralogy are almost always separated into distinct shell layers in the Bivalvia (Bøggild, 1930; Taylor, J. D., et al., 1969). Some microstructural intergradation occurs among the three aragonitic and between the two calcitic categories, but this is generally limited to contacts between adjacent shell layers. By comparison, microstructural intergradations are common among the microstructure varieties of individual categories, both vertically and laterally within individual shell layers. Intergradations among crossed foliated, complex crossed foliated, and regularly foliated structures are so common in the Pectinacea that many investigators do not distinguish among these varieties of foliated structure.

The present nacreous, porcelaneous, and foliated categories as defined in this chapter correspond generally to the "nacreous," "porcelaneous," and "foliated" microstructures, respectively, of most previous workers (e.g., Hatchett, 1799; Carpenter, 1844; Douvillé, 1913; Schmidt, 1924; Oberling, 1964; Taylor, J. D., et al., 1969, 1973). The reader may note, however, that Lucas (1952) excluded homogeneous structure from his category of porcelaneous structure. In addition, certain previous workers have included calcitic fibrous prisms in their application of the term "foliated" structure where these occur in the same shell valve as true foliated structure.

2.2.3. Taxonomic Distribution of Microstructure Categories

2.2.3a. Aragonitic Prismatic Category. Aragonitic prismatic structures probably occur in all bivalves, if only in their myostracal layers. Aragonitic prismatic structures are especially common as outer layers (Appendix 2.B, Figs. 1, 5, 6, and 10–12), but they also commonly comprise thin sublayers within middle and inner nacreous and aragonitic complex crossed lamellar layers (Appendix 2.B, Fig. 37) (also see Lutz and Rhoads, 1977). Major inner layers of aragonitic prismatic structure are uncommon, but have been described for certain Mytilacea, Lucinacea, Carditacea, Crassatellacea, Arcticacea, Glossacea, and Pandoracea (Taylor, J. D., et al., 1969, 1973) (Appendix 2.B, Fig. 3).

2.2.3b. Calcitic Prismatic Category. Calcitic prismatic structures are associated almost exclusively with outer shell layers. Middle- or inner-layer calcitic prisms are at present known in only a few Pectinacea, e.g., the fibrous prisms in *Propeamussium dalli*, and in the inner layer of the mytilacean *Mytilus californianus* (Dodd, 1964; Carter, personal observations of *Propeamussium*). Among veneroid bivalves, in which calcitic mineralogy is rare, calcitic prisms may occur as a major outer shell layer (in certain cooler-water *Chama*) or as more or less isolated patches within a largely aragonitic outer or middle shell layer [as in certain Veneracea and Lucinacea (Fig. 2 and Appendix 2.B, Figs. 13–15 and 19–22)]. Outer-layer calcitic prisms are especially common among representatives of the Pterioida and Hippuritacea (Taylor, J. D., et al., 1969, 1973; Skelton, 1974, 1976; Waller, 1972, 1976a,b, 1978; Carter and Tevesz, 1978a,b).

2.2.3c. Nacreous Category. Nacre is generally regarded as one of the most primitive microstructures in the Bivalvia (Taylor, J. D., 1973), and it is accordingly widely distributed among many ancient superfamilies (e.g., the Nuculacea, some Nuculanacea, Cyrtodontacea, Ambonychiacea, Pteriacea, Pinnacea, some Pectinacea, Modiomorphacea, most Mytilacea, Trigoniacea, and Pholadomyacea) (Appendix 2.B, Figs. 23–25). Only a few of the anatomically more specialized and later-evolving bivalve super-families are represented by nacreous species (e.g., the Unionacea, some Myacea, Clavagellacea, Pandoracea, and Poromyacea) (Strachimirov, 1972; Taylor, J. D., et al., 1969, 1973; Carter and Tevesz, 1978a,b).

2.2.3d. Porcelaneous Category. Bivalves losing their ancestral na-creous structure have generally replaced this with porcelaneous structure (Appendix 2.B, Figs. 28, 29, and 32–48). Porcelaneous structures are known in certain representatives of all superfamilies tabulated in Fig. 1 except the Nuculacea, Ambonychiacea, Pinnacea, Ostreacea, and Unionacea (Taylor, J. D., et al., 1969, 1973; Carter and Tevesz, 1978a,b). However, porcelaneous structures are relatively rare in the Pteriacea and Phola-domyacea (Bøggild, 1930; Newell, 1938; Taylor, J. D., et al., 1969, 1973; Newell and Boyd, 1975).

2.2.3e. Foliated Category. Foliated structures in the Bivalvia are ap-parently restricted to the Pterioida, in which they commonly associate with porcelaneous and calcitic simple prismatic structures. In some Pec-tinacea, Anomiacea, and Ostreacea, foliated structure comprises the entire shell except for minor calcitic prismatic layers, the larval shell, and ar-agonitic myostracal deposits (Taylor, J. D., et al., 1969; Waller, 1972, 1978) (Appendix 2.B, Figs. 26, 27, 30, and 31). Although foliated structure prob-ably does not co-occur with nacreous structure in any modern pterioid, this association may have been common in representatives of the Paleo-zoic pectinacean family Pterinopectinidae (Carter, personal observations of two Middle Devonian representatives of *Pseudaviculopecten* from the Hamilton Group of central New York). As discussed below, the fossil record of the Pectinacea suggests that largely foliated shells evolved in this superfamily through gradual expansion of outer and inner foliated layers at the expense of middle porcelaneous layers.

3. Evolution and Adaptive Significance of Shell Mineralogy and Microstructure

This section reviews and interprets from a unified historical and func-tional perspective the major aspects of mineralogical and microstructural evolution in the Bivalvia. Because of space restrictions, and also because of limitations of the available data base, emphasis is placed on the microstructure categories, rather than on individual microstructure varieties. The five microstructure categories defined above are ideally suited for this kind of summary because their constituent microstructure varieties have tended to evolve as recurrent associations in many bivalve lineages. It is important for the reader to note that many of the present inferences of mineralogical and microstructural evolution and their functional interpretations are largely speculative, because of the paucity of data for Lower and Middle Paleozoic bivalves and because much has yet to be learned about the mechanical properties of individual microstructure varieties. However, these speculations are at least compatible with the data on mineralogy, microstructure, and mechanical properties provided by all available literature sources, including, but not limited to, Bøggild (1930), Newell (1938, 1954), Newell and Boyd (1970), Kobayashi (1969, 1971), Kennedy et al. (1969), J. D. Taylor et al. (1969, 1973), J. D. Taylor and Layman (1972), Currey (1976, 1977), Currey and Taylor (1974), Waller (1972, 1976b, 1978), Carter (1976), Carter and Aller (1975), and Carter and Tevesz (1978a,b).

3.1. General Evolutionary Trends

J. D. Taylor (1973) suggested that the most primitive molluscan microstructures are aragonitic simple prisms and nacre, because of their occurrence in the Monoplacophora (the presumed ultimate shelled molluscan ancestors of the Bivalvia) and because these two structural arrangements can be explained largely on the basis of the principles of group growth of spherulites or other inorganic crystal morphotypes. Taylor also hypothesized that crossed lamellar and foliated microstructures evolved through microstructural or microstructural and mineralogical modifications of nacre, respectively.

Alternatively, molluscan microstructural evolution can be viewed from the perspective of increasing sophistication of the calcification process, beginning with largely unicellular controls of microstructure and culminating in tissue-grade organization of shell microstructure, in which microstructure orientation is integrated with shell mechanical design (Carter, 1979). Beedham and Trueman (1968) and Carter and Aller (1975) hypothesized that the most primitive molluscs secreted aplacophoranlike aragonitic cuticular spicules. The studies by Hoffman (1949) and Salvini-Plawen (1972) suggest that spicule calcification in the Aplacophora is essentially unicellular, i.e., with individual mantle cells secreting one spicule at a time. If the earliest shelled molluscs evolved from aragonitic spiculose ancestors, then their shell plates probably evolved through a simple modification of the mechanism of spicule-type cuticular calcification in which part of the mantle epithelium became specialized for subcuticular shell layer secretion. In their most primitive form, these initial shell plates may have consisted of no more than a relatively thin, flexible aragonitic spherulitic prismatic layer, with each prism secreted by a single or at most a few mantle cells. This primordial microstructure grade would be genetically comparable to that in modern articulate brachiopods, in which the mantle retains largely unicellular controls over the formation of its prismatic structure (Williams, 1966; Jope, 1971). Despite the similarity between many brachiopod and bivalve shell microstructures, brachiopods have apparently never evolved complex microstructures that are uniformly directionally oriented with respect to both the plane of the shell layer and the shell margins. Brachiopod crossed bladed structure is comparable to bivalve crossed lamellar structure in its internal complexity, but unlike the latter, its constituent structural units show variable dip directions with respect to the shell margins (Armstrong, 1969; Carter, 1979).

By the present hypothesis, molluscs evolving increased integration of their shell microstructure with shell mechanical design would have been preadapted for the evolution of uniformly oriented microstructures such as concentric crossed lamellae. This is compatible with the occurrence of relatively nondirectional structures such as sheet nacre and columnar nacre in primitive molluscs such as the Monoplacophora, and with the general replacement of these nacreous structures with concentric crossed lamellar structure in many lineages in the Bivalvia (Douvillé, 1913; Schmidt, 1959; Erben et al., 1968; Taylor, J. D., 1973; Carter and Tevesz, 1978a,b) (see Appendix 2.B for microstructure definitions and illustrations).

As described in more detail in the following sections, the available paleontological data support the contention of J. D. Taylor (1973) that bivalves are primitively aragonitic and that crossed lamellar and other porcelaneous structures have evolved from ancestral nacreous structure. However, these data also suggest that foliated structures evolved through microstructural modification of calcitic prismatic structure, rather than through combined mineralogical and microstructural modifications of nacre (see below and Waller, 1976a,b).

3.2. Evolution of Aragonitic Shell Layers

If, as suggested above, the earliest molluscan shell plates evolved through modification of ancestral spicule-type cuticular calcification, then these plates may have shown aragonitic spherulitic prismatic structure. Spiculelike cuticular calcification appears to be generally aragonitic in the Aplacophora, Polyplacophora, and Bivalvia (Lowenstam, 1963; Beedham and Trueman, 1968; Haas, 1972; Carter and Aller, 1975). The hypothesis of ancestral aragonitic mineralogy is also compatible with the view that molluscs evolved from turbellarianlike ancestors, because aragonitic spicule-type calcification is regarded as primitive in the Turbellaria (Rieger and Sterrer, 1975) (see Vagvolgyi, 1967, for a discussion of molluscan origins). Ancestral outer-layer spherulitic prismatic structure is suggested by its presence in the girdle cuticle spicules of the polyplacophoran *Acanthochitona* and in the periostracal spikes of the pandoracean bivalve *Laternula* (Plate 16 and Fig. 3 in Haas, 1972; Fig. 4 in Aller, 1974). In addition, this is the most widely distributed microstructure in molluscan outer shell layers, having been described or illustrated for certain Monoplacophora, nautiloid Cephalopoda, Gastropoda, Polyplacophora, and Bivalvia [e.g., in *Neopilina* (Plate 1 in Erben *et al.*, 1968), in *Nautilus* (Fig. 1 in Meenakshi *et al.*, 1974), in *Pleurotomaria* (Plate 5, Fig. 5 in Erben and Krampitz, 1970), in *Chiton* (Plate 2, Fig. 1 in Haas, 1972), and in *Spisula* (Appendix 2.B, Fig. 1)]. In the Bivalvia, aragonitic spherulitic prismatic structure comprises part of the outer shell layer in certain representatives of at least 15 superfamilies, 5 of which have Paleozoic origins [i.e., the Nuculacea, Cyrtodontacea, Lucinacea, Modiomorphacea, and Mytilacea (Carter, personal observations of Devonian *Nuculoidea opima*; Carter, 1976; Carter and Tevesz, 1978a,b)]. Unfortunately, the hypothesis of ancestral outer-layer aragonitic spherulitic prismatic structure cannot at present be tested against direct observations of Cambrian molluscs because of preservational problems. However, there is evidence for both aragonitic and calcitic original shell mineralogy in certain Ordovician cephalopods, gastropods, and bivalves, indicating that several molluscan lineages became mineralogically diverse early in their evolutionary history (see Fig. 1) (Bøggild, 1930, pp. 299 and 323).

3.2.1. Outer Aragonitic Shell Layers

The most ancient bivalves for which microstructure data are at present available for outer aragonitic layers are certain Middle Devonian representatives of the Nuculacea, Cyrtodontacea, Modiomorphacea, Crassatellacea, and Pholadomyacea. These bivalves show various combinations of spherulitic prismatic, fibrous prismatic, and simple prismatic structure in outer layers inferred to have been aragonitic on the basis of differential preservation (see Appendix 2.A, Table I) (Carter, personal observations of Middle Devonian *Cypricardella* and *Grammysia*; Carter and Tevesz, 1978a,b). None of these bivalves shows any of the varieties of composite prismatic structure that characterize outer aragonitic layers in certain modern Nuculacea, Trigonioida, Veneroida, Myoida, and Pholadomyoida (Bøggild, 1930; Taylor, J. D., et al., 1969, 1973; Carter, 1976).

Aragonitic composite prismatic structure may have evolved at least by Carboniferous time, if the outer layer of the trigoniacean *Schizodus* was microstructurally comparable to modern *Neotrigonia* (Newell and Boyd, 1975; Carter, personal observations of *Neotrigonia gemma*). However, fundamental differences in the ultrastructure and mode of formation of the varieties of composite prismatic structure suggest that they evolved independently in modern nuculaceans, in the Trigonioida, and in various superfamilies and families in the Veneroida, Myoida, and Pholadomyoida. In this regard, the reader may note the genetic difference between the *denticular* composite prisms of modern *Nucula* and certain Cardiacea and Tellinacea and the *nondenticular* composite prisms of the Trigonioida (both the Trigoniacea and Unionacea), certain Veneroida (certain Tellinacea and some representatives of many other superfamilies), Myoida (several superfamilies), and Pholadomyoida (certain Pandoracea) (Bøggild, 1930; Taylor, J. D., et al., 1969, 1973; Carter, personal observations) (also see Appendix 2.B for microstructure definitions and illustrations). According to J. D. Taylor and Layman (1972), certain composite prismatic structures may have evolved for increased resistance to abrasion in the context of increasingly active burrowing life habits. Their data indicate higher microhardness values for composite prisms than for several other aragonitic microstructures analyzed (Fig. 3). However, mechanical data are not yet available for comparisons among the varieties of composite prismatic structure or for comparisons between microhardness measurements and actual resistance to abrasion.

The available paleontological data for the Isofilibranchia suggest a general evolutionary trend from aragonitic outer layers in the Paleozoic Modiomorphacea to aragonitic, combined aragonitic and calcitic, and calcitic outer layers in their likely descendants, the Mytilacea (see Appendix 2.A, Table II) (see Pojeta, 1971, for a discussion of phylogenetic relationships in the Isofilibranchia). Similar aragonitic spherulitic prismatic and irregular simple prismatic structures occur in the outer layers of certain Modiomorphacea and Mytilacea. However, mytilaceans may also show calcitic equivalents of these structures and/or calcitic fibrous prismatic structure in their outer layers (Appendix 2.A, Table II). Possible environmental influences on mytilacean mineralogy and shell architecture are discussed in Section 4.

3.2.2. Middle and Inner Aragonitic Shell Layers

Like aragonitic spherulitic prismatic structure, nacreous structure is taxonomically widespread in the Mollusca, in this case occurring in certain representatives of all shelled molluscan classes with the possible

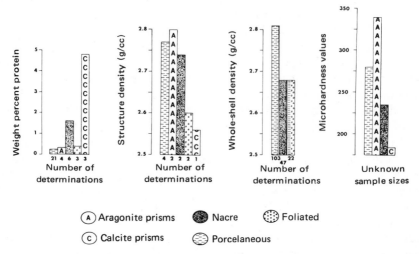

Figure 3. Shell microstructure mechanical data abstracted from J. D. Taylor and Layman (1972) and Carter (1976; unpublished data for whole-shell density). The patterns within each histogram represent one of the five microstructure categories defined in this chapter, with the exception of the whole-shell density graph. The latter shows, from left to right, aragonitic porcelaneous + aragonitic prismatic shell layers, aragonitic nacreous + calcitic prismatic shell layers, and predominantly calcitic foliated + aragonitic porcelaneous shell layers. The data of J. D. Taylor and Layman (1972) have been pooled for certain microstructures that belong in the same microstructure category. The numbers below each histogram bar indicate the number of determinations represented. Other notes are as follows: *Weight percent protein.* Data abstracted from Table 4 of J. D. Taylor and Layman (1972). These data are based on organic nitrogen content. *Structure density.* Data abstracted from Table 5 of J. D. Taylor and Layman (1972); density measurements (g/cc). *Whole-shell density.* Whole-shell density measurements are based on air-dried specimens and are calculated using shell valve weight and displacement volume (g/cc). *Microhardness.* These data are based mostly on dried specimens and were determined using an Akashi microhardometer with a diamond pyramid indenter (Taylor, J. D., and Layman, 1972, p. 76). Aragonitic crossed lamellar and calcitic foliated structures are at present excluded from this tabulation because they show vectorial (i.e., directionally variable) microhardness values (Taylor, J. D., and Layman, 1972, p. 81). Figure 9 of J. D. Taylor and Layman (1972) shows microhardness values of about 293 and 183 for aragonitic crossed lamellar and calcitic foliated structures, respectively. The reader should note that *only* composite prisms are plotted for the category of aragonite prisms for the sake of microstructural uniformity. The microhardness values of J. D. Taylor and Layman (1972) for aragonitic prisms other than composite prisms were significantly lower, averaging only about 220 microhardness units.

exception of the Scaphopoda and Polyplacophora (Bøggild, 1930; Taylor, J. D., 1973). Nacre has long been regarded as the most primitive middle- and inner-layer microstructure in the Bivalvia because of its association with several anatomically primitive and/or ancient superfamilies, most notably the Nuculacea, Pteriacea, Trigoniacea, and Pholadomyacea (Douvillé, 1913; Bøggild, 1930; Taylor, J. D., et al., 1969, 1973; Taylor, J. D., 1973). In addition, recent studies indicate the presence of nacreous structure in certain Lower or Middle Paleozoic representatives of the Nuculanacea, Cyrtodontacea, Ambonychiacea, Pectinacea, and Crassatellacea (Taylor, J. D., et al., 1973; Carter and Tevesz, 1978a,b; Carter, personal observations of the Middle Devonian crassatellacean *Cypricardella tenuistriata*).

Paleontological data suggest that porcelaneous shell layers have replaced nacreous ones at least once within each bivalve subclass (Bøggild, 1930; Taylor, J. D., et al., 1969, 1973; Taylor, J. D., 1973; Strachimirov, 1972; Carter and Tevesz, 1978a,b). Previous investigators have noted that this evolutionary change appears paradoxical from the perspective of mechanical adaptation, because nacreous structures are generally stronger than porcelaneous ones in tests of compression, bending, and resistance to breakage on impact (Taylor, J. D., and Layman, 1972; Currey and Taylor, 1974; Currey, 1976). According to Currey (1977), the aragonitic tablets and organic matrix in sheet nacre show an optimal thickness, spacing, and mutual arrangement to provide for maximum dissipation of fracture energy prior to complete structural failure. On the other hand, the high strength of nacreous structure is probably not required in connection with all combinations of bivalve life habit, environment, shell form, and shell thickness. This idea is compatible with the observation that middle and inner nacreous layers occur exclusively in the Bivalvia only in association with relatively thin-shelled, largely unornamented, intermediate-sized and larger bivalves inhabiting exposed, rocky substrates in turbulent, high-energy environments (e.g., *Mytilus edulis*). Most, if not all, other combinations of life habit, environment, shell form, and shell thickness in this class are represented by many species with porcelaneous and/or foliated shell layers, and in many instances, also by species with nacreous layers (Taylor, J. D., and Layman, 1972; Carter, 1976). Comparative microhardness data suggest that porcelaneous structures may be generally more resistant to abrasion than nacreous structures (Fig. 3) (also see Taylor, J. D., and Layman, 1972; Currey, 1976). However, direct abrasion experiments have yet to be conducted to verify whether microhardness measurements provide an accurate indication of relative resistance to abrasion for shell microstructures.

In addition to the possible advantage of resistance to abrasion, the concentric crossed lamellar variety of porcelaneous structure may have evolved for concentric deflection of marginal, radial fractures. Concentric crossed lamellar structure commonly comprises a middle shell layer between outer prismatic or foliated, and inner foliated or complex crossed lamellar layers (Bøggild, 1930; Taylor, J. D., et al., 1969, 1973). In this position, the crossed lamellar structure can intercept radial fractures propagating from the shell margins toward the central part of the shell, thereby deflecting some of these fractures along the concentric boundaries between the adjacent first-order crossed lamellae. This concentric fracture deflection insures that much of the fracture energy will be dissipated near the shell margins, thereby minimizing the hazard of fracture propagation toward the central part of the shell overlying the viscera. The capacity for concentric crossed lamellar structure to deflect radial fractures can be demonstrated by attempting to propagate radial fractures through the outer radial prismatic and middle crossed lamellar layers of the limacean *Ctenoides scabra* (Carter, personal observations).

In a few rather elongate, thin-shelled representatives of the Arcacea, such as *Trisidos tortuosa*, the middle concentric crossed lamellar shell layer is supplemented by an inner radial crossed lamellar layer (Carter, personal observations). Although extremely rare in the Bivalvia, this combination of concentric and radial crossed lamellar layers is relatively common in the Gastropoda (Bøggild, 1930). According to Currey and Kohn (1976), the superposition of one concentric and two radial crossed lamellar layers in the gastropod *Conus* enhances the shell's capacity for absorption and dissipation of fracture energy, thereby minimizing the hazard of complete structural failure.

Other than possible resistance to abrasion, little is known at present about the mechanical or other advantages provided by the varieties of porcelaneous structure other than crossed lamellar structure (Table I). However, it may be noted that crossed-matted/lineated structure is restricted to the Arcacea, where it apparently represents a transition between irregular complex crossed lamellar and radial crossed lamellar structures in the inner shell layer (see microstructure definitions and Appendix 2.B, Figs. 42–44).

3.3. Evolution of Calcitic Shell Layers

As noted earlier in this chapter, probably all subclasses in the Bivalvia are primitively aragonitic, and calcitic outer shell layers have evolved in many representatives of the Pteriomoprhia, Heteroconchia, and Isofilibrachia (see Fig. 1). Within the Pteriomorphia, many pterioids have evolved foliated layers that have largely or entirely replaced ancestral middle and inner nacreous and porcelaneous layers. In some Pectinacea and Anomiacea, and apparently in all modern Ostreacea, this replacement has resulted in the evolution of shells that are entirely calcitic with the exception of the ligament resilium, myostracal deposits, and larval shell (Stenzel, 1962, 1963, 1964; Waller, 1972, 1978; Taylor, J. D., 1973; Carter and Tevesz, 1978a). The following sections summarize ideas regarding the possible adaptive significance of calcitic shell layers in these and other bivalves. First, however, it is important to note that not all additions of calcite to a shell, and not all evolutionary changes in calcitic shell layers, need be related to mechanical or other advantages of calcitic mineralogy *per se*. For example, evolutionary increases in the relative area of secretion of the outer calcitic layer in the Pinnacea may be adaptive for enhancing shell-margin flexibility, a property that almost certainly derives from the high organic matrix content and microstructure of the simple prismatic layer, rather than from its calcitic mineralogy (Taylor, J. D., and Layman, 1972; Currey and Taylor, 1974; Currey, 1976; Carter and Tevesz, 1978b). In addition, the possibility exists that some mineralogical "evolution" reflects largely environmental influences on the physiological controls of shell mineralogy (see Section 4.1).

3.3.1. Mechanical Advantages of Calcite Secretion

The available mechanical data suggest that calcite secretion generally does not enhance shell strength or resistance to abrasion (Fig. 3) (Taylor, J. D., and Layman, 1972; Currey and Taylor, 1974; Currey, 1976). This is not surprising, considering that inorganic calcite is softer than aragonite and, unlike aragonite, tends to break along well-defined cleavage planes (Table II). On the other hand, this cleavage property of calcite may be adaptive for enhancing fracture localization in connection with certain varieties of foliated structure. As noted by J. D. Taylor and Layman (1972), the calcitic blades in oyster foliated layers tend to break near the area of application of stress, rather than propagate fracture energy, and fracture damage, to distant parts of the shell. Consequently, whereas oyster foliated

Table II. Selected Physical Properties of Inorganic Aragonite and Calcite[a]

	Density	Hardness (Mohs' scale)	Cleavage, fracture, and tenacity
Aragonite	2.93 g/cc	3.5–4.0	Imperfect cleavage on 010 plane; subconchoidal, brittle
Calcite	2.71 g/cc	3.0	Perfect rhombohedral cleavage on $10\bar{1}1$ plane

[a] From Berry and Mason (1959).

layers may be more easily damaged than nacreous and porcelaneous layers, they are more likely to localize the area of damage resulting from strong point impacts.

It is also possible that calcite secretion is adaptive for enhancing crack-stopping in conjunction with aragonitic shell layers. In this case, the mechanical advantage would derive from the contribution of differing mineralogy toward increasing both the microstructural discontinuity and difference in mechanical properties between adjacent aragonitic and calcitic shell layers. As noted by J. D. Taylor and Layman (1972), the capacity for a composite material to arrest fractures can be enhanced by the superposition of layered materials with differing mechanical properties. In addition, the mineralogical contrast between adjacent shell layers may insure that they are not microstructurally intergradational along their mutual boundary, thereby increasing the probability that vertical fractures will be deflected along their horizontal interface. These indirect mineralogical effects may well enhance the capacity of the nacreous structure in nacroprismatic pinnaceans such as *Atrina serrata* for deflecting vertical fractures along the horizontal interlayer boundary with the calcitic prismatic layer (Fig. 4).

3.3.2. Other Possible Advantages of Calcite Secretion

In addition to indirect mechanical advantages, calcite secretion may have evolved for minimizing shell density. Mineralogy is probably the most important contributing factor to shell density in most bivalves. Porosity may significantly affect shell density in the case of ostreaceans and hippuritaceans with abundant chalky deposits, vacuoles, or large chambers within their shell layers, but such features are largely restricted to these two superfamilies (Chel'tsova, 1969; Stenzel, 1971; Masse and Philip, 1972; Skelton, 1976). The percentage of organic matrix may also significantly affect shell density in certain isolated instances. However, the structure density data compiled by J. D. Taylor and Layman (1972) indicate that despite the wide range of values of percentage of organic matrix among the major shell microstructures, calcitic shell layers are still generally less dense than aragonitic ones by at least 0.14 g/cc (see Fig. 3). Ideally, calcitic mineralogy can account for a density decrease of about 7.5% in comparison with aragonitic shells, based on the densities of inorganic calcite and aragonite (Table II). The measurements of whole-shell density summarized in Fig. 3 suggest that slightly over half this possible 7.5% density decrease accompanied the evolution of prominent calcitic outer shell layers in the Pteriacea, with the remaining density decrease coming about through the subsequent evolution of largely calcitic foliated shells in certain Pectinacea, Anomiacea, and Ostreacea.

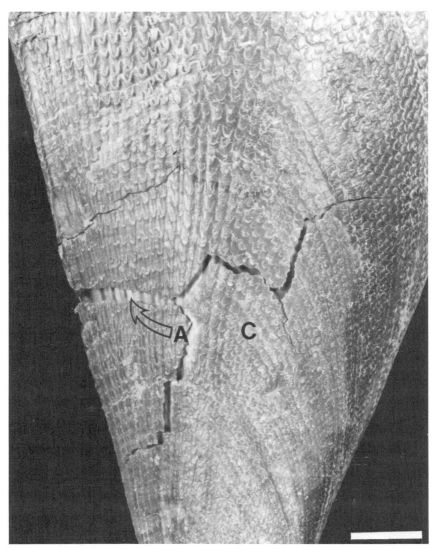

Figure 4. Fractures in the outer calcitic simple prismatic layer in the pinnacean *Atrina serrata* (Sowerby), University of North Carolina 5191, Sanibel, Florida. The underlying aragonitic nacreous layer (visible at the point of the arrow from A) is largely intact, despite extensive fracture propagation within the calcitic prismatic layer (C) and separation of the two shell layers along their mutual boundary. The fractures were probably incurred during dislodgement of the animal from its natural semiinfaunal living position or during subsequent transport in the shallow subtidal and intertidal zones. Scale bar: 10 mm.

The ecological distribution of largely calcitic foliated shells in the Pectinacea is compatible with the hypothesis that they evolved for low density and fracture localization in connection with increasing speciali-zation for active swimming. Some pectinaceans evolved minor foliated layers in conjunction with calcitic prismatic structure by Middle Devon-ian time (Carter, personal observations of *Pseudaviculopecten scabridus*). However, largely foliated pectinaceans did not become abundant until late Paleozoic or early Mesozoic time, when many representatives of this group were evolving increased shell muscular and shell morphological specializations for prolonged, active swimming (Newell, 1938; Newell and Boyd, 1970; Waller, 1978; Stanley, 1972). Among modern pectina-ceans, the more actively swimming species in the Pectinacea are largely or entirely foliated, whereas the more sedentary species (e.g., the Spon-dylidae and Plicatulidae) retain prominent inner aragonitic layers (Taylor, J. D., *et al.*, 1969; Waller, 1972, 1978). Among other swimming pterioids, the Limacea are not morphologically specialized for prolonged, active swimming, and they likewise retain prominent inner aragonitic layers (Taylor, J. D., *et al.*, 1969). According to Yonge (1936) and Gilmour (1967), the Limacea possess only a moderate swimming ability that they use pri-marily as a defensive measure when routed from their byssus nests.

Although the Mesozoic oysters were nonswimmers, the circumstan-ces of their sedentary life habits suggest that they too may have benefited from low shell density. According to Stenzel (1971, pp. N1071–N1072), the Jurassic "*Gryphaea*, its gryph-shaped descendants, and its homeo-morphs lived mostly on sea bottoms composed of soft, waterlogged oozes with small fragments of shells, and their distinguishing features are ad-aptations to this environment. . . . It is now believed that the animals lived partly sunk into the soft ooze lying with their valve commissures nearly horizontal. . . . In this position, the animal's weight, shell and all, was about equal to the weight of the ooze it displaced so that the *Gryphaea* essentially floated in or on the soft ooze." Although the early ostreaceans may not have actually floated in their substratum, their low shell density would certainly have reduced the hazard of their valve commissures sink-ing below the sediment–water interface. Natural selection for low shell density is also compatible with their occasional evolution of porous chalky deposits and vacuolization (Chel'tsova, 1969; Stenzel, 1971).

On the other hand, the retention of highly porous or vacuolated cal-citic layers in connection with sedentary life habits in the Hippuritacea and in many modern oysters may also be taken to suggest natural selection for crack-stopping mechanisms, economy of secretion, or rapidity of shell layer thickness increase. As explained by Weber *et al.* (1969) for echi-noderms, fenestration of a skeleton can maximize its strength-to-weight ratio, thereby fulfilling its strength and volumetric requirements with a minimum amount of material. These authors also noted that fenestrate structures provide large shock-resistance capabilities because local impact-loading can be transmitted rapidly to a large area, and much energy can be dissipated in local fracture of the framework elements. However, incorporation of porosity into a skeleton beyond a certain point can lower its strength-to-weight ratio, suggesting, in such instances, that low weight rather than strength is at a premium (Currey, 1975; Wainwright et al., 1976). In ostreaceans such as *Gryphaea* faced with the hazard of sinking below the sediment–water interface, the higher molar volume of calcite relative to aragonite may have contributed to the effect of vacuolization or chalky layers in permitting a rapid rate of shell layer thickness increase for the lower valve. It is also possible that the greater molar volume of calcite contributes to its metabolic economy of secretion in terms of energy expended per unit volume of shell material. However, the problem of energy efficiency of calcite vs. aragonite secretion requires study of the energy priorities of various mantle tissues, and data on comparative energy of secretion over a broad range of environmental conditions. According to a hypothesis proposed by Duckworth (1976), the uniform aragonitic mineralogy of myostracal deposits reflects the *higher* metabolic energy required for calcite secretion because of the necessity for excluding magnesium from the precipitating solutions.

As a final note, there is little evidence to support the hypothesis that calcite secretion evolved to reduce the hazard of exterior shell dissolution. Such an advantage is expected because of the greater thermodynamic stability of calcite in comparison with aragonite, considering similarly sized crystals at normal marine and freshwater temperatures and pressures (Berner, 1971). However, calcite is uncommon in freshwater molluscs, and calcite secretion is entirely unknown in bivalves inhabiting only freshwater environments, except in certain instances of shell regeneration [e.g., in the unionacean *Elliptio complanatus* (Wilbur and Watabe, 1963)] (for shell mineralogy data for the Corbiculacea and Unionacea, see Taylor, J. D., et al., 1969, 1973). Instead of evolving calcite secretion, these bivalves have tended to evolve persistent periostracal layers and thick, internal organic laminae to reduce the hazard of deep exterior dissolution by acidic freshwater environments (Taylor, J. D., et al., 1969) (also see Chapter 8).

4. Shell Mineralogy and Structure as Records of Environmental Change

In view of the abundant evidence for phylogenetic controls, and considering the mechanisms proposed by many workers for physiological influences on mineralogy and shell structure, environmental influences on these parameters are almost certainly mediated through genetically determined biological processes. Therefore, mineralogy and shell structure can be regarded as records of environmental change only insofar as environmental influences can be separated from biological controls. Several kinds of mineralogical, microstructural, and architectural variations have been attributed to environmental influences. These variations are summarized here and evaluated for their application to paleoenvironmental studies.

4.1. Mineralogical Variations

Three kinds of mineralogical variations in bivalve shells have been attributed to environmental controls: (1) addition of calcite to shells of cooler-water species; (2) increase in the whole-shell calcite/aragonite ratio among cooler-water and/or higher-salinity populations of a species; and (3) increase in the abundance of mineralogical aberrations (i.e., isolated patches of calcite) among cooler-water populations of a species. Lowenstam (1954b,c) described several bivalve genera that illustrate apparent temperature-related additions of calcitic shell layers to cooler-water species. But he also noted that biological influences on shell mineralogy may transcend environmental influences in certain other bivalves. Most of his examples of apparent calcite layer addition come from the superfamilies Chamacea and Mytilacea. Lowenstam (1954b,c) indicated that the relatively cool-water *Chama pellucida* and *C. semipurpurata* show substantial calcite or a trace of calcite, respectively, in their whole-shell preparations. In contrast, the warmer-water species of *Chama* are wholly aragonitic. J. D. Taylor and Kennedy (1969) and Kennedy *et al.* (1970) verified Lowenstam's determination of calcite in the outer layer of *C. pellucida*, but they did not reexamine *C. semipurpurata*. Kennedy *et al.* (1970) noted the occurrence of calcite in the cool-water *C. exogyra*, but they speculated that this species may be synonymous with *C. pellucida*.

Lowenstam's other examples of apparent temperature-related calcite secretion deal with several genera in the Mytilacea, i.e., *Volsella*, *Choromytilus*, *Brachidontes*, and *Mytilus*.* His major conclusions have been summarized and evaluated by Kennedy *et al.* (1969). Inasmuch as the

---

* Taxonomic revisions of the mytilid bivalves considered by Lowenstam (1954b,c) include the following: *Brachidontes recurvum* is now *Ischadium recurvum* (Rafinesque); *Brachidontes bilocularis* is now *Septifer bilocularis* (Linnaeus); *Cuneolus tippana* is now *Lycettia tippana* (Conrad); *Choromytilus* is now a subgenus of *Perna*; *Volsella* is now *Modiolus*; and *Volsella tulipa* is now *Modiolus americanus* (Leach). These revisions are based on Olsson (1961), Soot-Ryen (1969), and Abbott (1974).

taxonomy of the Mytilacea has been revised since the publication of the works of Lowenstam (1954b,c), and because more information is now available for this superfamily, e.g., J. D. Taylor *et al.* (1969) and Appendix 2.A, these data are integrated here and evaluated within a current taxonomic framework.

The present analysis indicates that the Myilinae show the strongest apparent temperature control of calcite secretion among the four subfamilies of the family Mytilidae (Fig. 5). The Mytilinae show a strong positive correlation between calcite secretion and species preference for cooler ambient temperature. In comparison, positive correlations between calcite secretion and species preference for cooler temperature are weaker for the Modiolinae and Crenellinae and nonexistent for the Lithophaginae.

Within the Mytilinae, evidence for temperature control of calcite secretion within individual genera is provided by representatives of *Mytilus* and *Perna* (= Lowenstam's "*Choromytilus*"). The tropical *Mytilus arciformis* is entirely aragonitic, whereas the cooler-water *M. edulis* and *M. californianus* show outer or outer and inner calcitic layers, respectively (also see Dodd, 1964). As noted previously by Lowenstam (1954b,c), the tropical *Perna palliopunctatus* is entirely aragonitic, whereas the cooler-water *P. chorus* secretes both aragonite and calcite (also see Taylor, J. D., *et al.*, 1969).

Certain species of *Modiolus* (subfamily Modiolinae) show a correlation between cooler ambient temperature and calcite secretion, but other representatives of this subfamily show exceptions to the expected temperature correlation. In keeping with the expected trend, *Modiolus pseudotulipus* is a warm-water species secreting only aragonite, and the cooler-

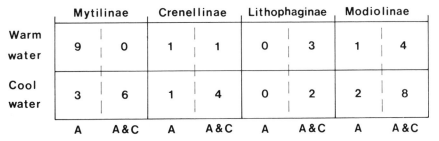

	Mytilinae		Crenellinae		Lithophaginae		Modiolinae	
Warm water	9	0	1	1	0	3	1	4
Cool water	3	6	1	4	0	2	2	8
	A	A&C	A	A&C	A	A&C	A	A&C

Figure 5. Temperature distribution of wholly aragonitic (A) and aragonitic plus calcitic (A&C) species within each of the four mytilid subfamilies. *Warm water* refers to species with a tropical distribution, in some cases rarely occurring in warm-temperature water as well. *Cool water* refers to species (1) with a warm-temperate and cooler-water distribution; or (2) with a primarily warm-temperate distribution, but rarely occurring in the tropics as well; or (3) with a cold-temperate and Arctic or Antarctic distribution. This figure incorporates data from Lowenstam (1954a–c, 1963), J. D. Taylor *et al.* (1969), and Table II of Appendix 2.A, including a total of 45 species. Each number indicates the number of species in each temperature + mineralogy category within one of the four subfamilies.

water *M. modiolus* secretes aragonite and calcite. But the warm-water *M. americanus* secretes aragonite and calcite, and the cooler-water *Dacrydium vitreum* is entirely aragonitic (Appendix 2.A, Table II). The reader may consult Appendix 2.A for examples of apparent temperature control of shell mineralogy, and for exceptions to the expected mineralogy–temperature correlation, among the Crenellinae and Lithophaginae.

In summary, the present data suggest that some mytilid subfamilies exercise more biological control over their mineralogy than others, with certain Mytilinae showing the greatest apparent environmental influence and hence the greatest potential for paleotemperature analysis. Inference of a tropical vs. nontropical environment is reasonably well founded when considering the presence or absence of calcite in a modern representative of the Mytilinae. On the other hand, a correct inference of general temperature based on mineralogy is to be expected for only 50–80% of the modern species considered for the Crenellinae and for about 70% of the modern species considered for the Modiolinae (Fig. 5).

Another kind of environment-related mineralogical variation, whole-shell calcite/aragonite ratio within a species, has been described for *Mytilus* by Lowenstam (1954a–c), Dodd (1963, 1964, 1966), Eisma (1966), and Davies (1966). These studies have shown that calcite/aragonite ratios in *M. edulis* and *M. californianus* may reflect a variety of parameters, including temperature, salinity, spawning season, shell size, shell thickness, water turbulence (which may influence shell thickness), degree of intertidal exposure, growth retardation, and stage of ontogenetic development. Additionally, geographically isolated populations of *M. edulis* may show genetically determined physiological differences in their response to certain environmental factors (Dodd, 1966; Eisma, 1966). Smaller shells of *M. californianus* do not show a correlation between ambient temperature and calcite/aragonite ratio, but larger individuals of this species show a stronger temperature correlation than do larger individuals of *M. edulis* (Dodd, 1963, 1964). Despite these complexities, Malone and Dodd (1967) have demonstrated that calcite/aragonite ratios in *M. edulis* can be useful for assessing paleotemperature and paleosalinity, and they and Dodd (1964) have pointed out the advantages of utilizing this approach to paleotemperature analysis in conjunction with geochemical studies.

According to Waller (1972), certain Pectinidae show a rough positive correlation between the degree of development of their aragonitic crossed lamellar shell layer and environmental temperature. However, pectinid shell calcite/aragonite ratios have yet to be evaluated for application to quantitative paleotemperature analyses.

The third kind of mineralogical variation, mineralogical aberrations appearing as isolated patches of calcite within a largely aragonitic shell layer, has been described by Carter and Eichenberger (1977). As in the case of pectinid mineralogical variations, these require further study to evaluate their utility for assessing paleotemperature and paleosalinity. Preliminary data show a low, but statistically significant, negative correlation between abundance of calcitic mineralogical aberrations visible on the exterior of *Protothaca staminea* and February open-ocean surface-water temperature (Fig. 6). However, the variability of these data and especially the absence of calcite in some cold-water *P. staminea* suggest that other environmental factors (salinity?) and possibly biological controls are also influential. Considering that veneracean calcite is largely restricted to the outer shell layer, where it can be quantified by inspection of the shell exterior, these mineralogical aberrations may provide readily accessible data for future studies of environmental and biological controls of shell mineralogy.

4.2. Structural Variations

4.2.1. Combined Mineralogical–Structural Variations

All the examples of mineralogical variation described in the preceding section could alternatively be described as combined mineralogical-

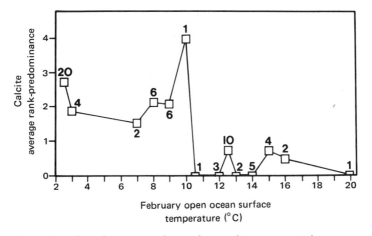

Figure 6. Regression of predominance of outer-layer calcite against February open-ocean surface-water temperature for *Protothaca staminea* (Conrad). The mineralogical data are from Carter (personal observations); the temperature data are from Sverdrup *et al.* (1942). The number adjacent to each square indicates the number of samples averaged to obtain the indicated calcite rank-predominance value. A rank-predominance value of 0 represents no outer-layer calcite visible on the adult shell exterior. A rank-predominance value of 6.0 represents abundant outer-layer calcite. Kendall's tau rank-order correlation statistic between calcite rank-order predominance and water temperature is -0.37 ($P < 0.001$) for the 67 samples.

structural variations. Virtually all mineralogical variations in bivalve shells are associated with at least minor changes in shell architecture or microstructure, or both. For example, in bimineralogical veneraceans such as *Protothaca staminea*, the calcitic prisms of the outer shell layer show larger diameters and more widely diverging structural subunits than the aragonitic prisms of the same general shell layer (Carter and Eichenberger, 1977). Similar structural contrasts are presented by the calcitic and aragonitic portions of the outer layer in *Venerupis crenata* (Appendix 2.B, Figs. 12–15). In *Mytilus edulis* and *M. californianus*, variations in whole-shell calcite/aragonite ratios correspond with changes in the relative thickness and/or area of secretion of the nacreous and calcitic prismatic shell layers (Dodd, 1963, 1964). In other mytilaceans, additions of calcite secretion to a shell may represent additions of an entirely new shell layer.

Considering the minor contribution to shell construction provided by the calcitic patches in certain Veneracea, and by the thin, outer calcitic layers in certain Mytilacea, it is possible that their environment-related mineralogical–structural variations are not subject to strong biological controls. On the other hand, the likelihood of strong biological control related to adaptive shell mechanical design would appear to be greater in the case of *M. edulis* and *M. californianus*, because their mineralogical variations can result in major changes in shell architecture. Unfortunately, except for evidence of possible subspecies or physiological races in *M. edulis* (Dodd, 1962, 1966), little is known at present about the interpopulation variability of biological controls related to mineralogical–structural variation in any bivalve. Further study of this aspect of mineralogical–structural variation should be undertaken to insure that mineralogical data for species such as *M. edulis* and *M. californianus* can be confidently utilized as records of environmental change.

4.2.2. Structural Variations Not Accompanying Mineralogical Change

Previous workers have suggested relationships between environmental factors and three kinds of structural variations not related to mineralogical change: (1) microstructural sublayers; (2) changes in the area of secretion of adjacent aragonitic shell layers; and (3) variations in the concentration of the shell's organic matrix.

The most common microstructural sublayers in bivalve shells consist of sheets or lenses of aragonitic homogeneous and/or irregular simple prismatic structure within nacreous or porcelaneous layers. These sublayers are especially common in the inner layers of many mytiloids, veneroids, and myoids, where they occasionally comprise a prominent component of the shell (see Appendix 2.B, Fig. 37) (Bøggild, 1930; Taylor, J. D., *et al.*, 1969, 1973). According to Lutz and Rhoads (1977), the homogeneous sublayers in the inner nacreous layer of the mytilid *Geukensia demissa* reflect anaerobic respiration induced by certain conditions of ambient temperature and/or oxygen availability, and they presumably represent dissolution–destruction of previously deposited microstructures. Many bivalves change from aerobic to anaerobic respiration during valve closure, during which time they utilize their shell calcium carbonate to buffer their acidic waste products of anaerobiosis (Collip, 1920; Dugal, 1939; Crenshaw and Neff, 1969; Taylor, A. C., and Brand, 1975; Wilkes and Crenshaw, 1979). Davies and Sayre (1970) suggested that this dissolution–destruction may provide a permanent record of physical stress and related environmental factors, and Lutz (1976a) and Lutz and Rhoads (1977) emphasized the potential paleoenvironmental significance of homogeneous sublayers in *G. demissa*. Unlike homogeneous sublayers, irregular simple prismatic sublayers, which also occur in the inner layer of *G. demissa*, have been described as representing modifications of shell secretion brought about by periods of transitional aerobic–anaerobic respiration (Lutz and Rhoads, 1977). These prismatic sublayers are structurally similar to the irregular simple prismatic myostracal layers deposited by most or all bivalves at sites of shell-muscle attachment.

Despite the apparent relationship between homogeneous and irregular simple prismatic sublayer formation and environmental parameters, the utility of these sublayers as records of environmental change is limited at present by the complexity of factors that may bring about bivalve anaerobic respiration. Studies of mytilid ecology and physiology by Coleman (1973), Bayne *et al.* (1976), and Widdows (1976) suggest that acclimatization, behavioral strategies such as subaerial mantle ventilation, and physiological differences between semiisolated populations may influence a bivalve's response to environmental factors, and therefore also influence the environmental significance of its anaerobic respiration. In addition, the possibility has yet to be explored that certain homogeneous and irregular simple prismatic sublayers are not related to anaerobic respiration, but reflect natural selection for certain mechanical advantages related to composite shell layers, such as absorption of fracture energy.

A less common type of microstructural sublayer consists of columnar nacreous structure within a sheet nacreous layer (see Figs. 23 and 25 and the microstructure guide in Appendix 2.B for illustrations and definitions of columnar and sheet nacre). Columnar nacreous sublayers occur along with homogeneous and irregular simple prismatic sublayers in the inner layer of *G. demissa*, and like the latter sublayer type, they are presumed to correlate with modifications of shell secretion accompanying transitional aerobic–anaerobic respiration (Lutz and Rhoads, 1977). This idea is not incompatible with the hypothesis of Wise (1970) that nacre tablet stacking modes associate with different rates of vertical shell layer accretion. However, it appears unlikely that anaerobiosis is the only major contributing factor to variation in stacking mode between and within nacreous shell layers. According to Wise (1970), columnar nacre can accrete faster than sheet nacre because its vertical stacking exposes more growing tablet edges per unit area of depositional surface. He based this hypothesis on the general correlation between columnar nacre and low aperture expansion rates in the Gastropoda and Cephalopoda, and between sheet nacre and high aperture expansion rates in the Bivalvia. Additional support for Wise's hypothesis comes from the distribution of columnar nacreous layers in the Bivalvia, which generally associate with more rapidly accreting parts of the shell, e.g., on the curved shell margins in certain Cyrtodontacea and Trigoniacea (see microstructure data in Taylor, J. D., et al., 1969; Carter and Tevesz, 1978b). According to the hypothesis of Wise (1970), the columnar nacreous sublayers in G. demissa may represent parts of the inner shell layer that have increased in thickness rapidly in order to restore a dissolution-pitted shell interior to its original smooth topography. However, it is unknown at present whether nacre tablet stacking mode is the cause or result of differential vertical shell layer accretion rate. In any event, the capacity of some bivalves to control both the orientation and stacking mode of their nacre tablets is suggested by consistent differences between the nacreous structures of certain morphologically similar representatives of the Pinnacea. Unlike most or all species of Atrina, species of Pinna show a unique row stack nacreous structure that may have evolved to complement the other adaptations of this genus for flexible closure in the shell posterior (Yonge, 1953; Wise, 1970; Carter and Tevesz, 1978b). The reader may consult Wada (1961, 1972) and Lutz (1976b) for further discussion of biological and environmental influences on nacre tablet size, shape, thickness, and internal texture.

A fourth type of microstructural sublayer consists of porous chalky deposits within the foliated layers of many oysters. These were briefly mentioned above in the context of possible natural selection for economy or rapidity of secretion and low shell density. Margolis and Carver (1974) described these deposits as consisting of calcitic "prisms" (actually blades) oriented more or less perpendicular to the depositional surface. Margolis and Carver proposed that chalky deposits are secreted during periods of accelerated aerobic respiration in order to quickly rid the body of excess calcium. According to Korringa (1951), oyster shell chalky deposits represent an economical shell material that can be used to smooth out interior shell contours or modify these to improve the shell's hydrodynamic properties in relation to ventilating mantle currents. If, as Korringa suggests, chalky deposits offer a faster maximum rate of shell layer thickness increase than foliated structure, then chalky deposits and foliated structures might be regarded as functionally analogous to columnar and sheet nacreous structures, respectively. The reader is referred to Korringa (1951) for a more complete history of ideas concerning the possible environmental significance of oyster chalky deposits.

Changes in the area of secretion of adjacent aragonitic shell layers have yet to be thoroughly evaluated for their utility as records of environmental change independently of growth band and internal growth increment variations. Such changes are commonly described or illustrated in connection with studies of shell growth in the veneroids *Mercenaria mercenaria*, "*Cardium*" (= *Cerastoderma*) *edule*, and *Spisula solidissima*. Radial, vertical sections through the shells of these and many other veneroids show intertonguing of the outer and middle shell layers, or periodic expansion and contraction of the area of secretion of only the outer (commonly prismatic) shell layer (see Plate 8 in Farrow, 1972; also see, among the more recent studies, Kennish and Olsson, 1975; Kennish, 1978; Jones *et al.*, 1978). It is unknown at present whether these architectural variations reflect merely extension and retraction of the marginal mantle epithelium or fundamental changes in the pattern of shell secretion of certain mantle zones. In his study of shell repair by the unionacean *Anodonta*, Beedham (1965) indicated that the normal relationships between shell layers and mantle secretory zones do not appear to be specific and unchangeable. Further study of architectural variations among wholly aragonitic shell layers in veneroids such as *Mercenaria mercenaria* may contribute significantly to our understanding of biological vs. environmental controls of architectural variations in bimineralogical bivalves.

Finally, variations in the concentration of the shell's organic matrix have been related to the formation of internal growth increments and to environmental salinity, among other factors. Lutz and Rhoads (1977) hypothesized that daily and subdaily internal growth increments originate through concentration of the shell's organic matrix during anaerobiosis-related dissolution of previously deposited calcium carbonate. Although anaerobiosis is almost certainly responsible for certain internal growth increments, especially within inner shell layers, the possibility still exists that most internal growth increments are not directly related to anaerobiosis. Alternative modes of formation include variations in the proportion of secreted organic matrix vs. calcium carbonate, or changes in microporosity or ultrastructure of the deposited mineral that occur independently of anaerobiosis (see the discussions by Gordon and Carriker, 1978; Wilkes and Crenshaw, 1979; Watabe and Dunkelberger, 1979). A relationship between environmental salinity and concentration of the shell's organic matrix relative to calcium carbonate has been suggested for certain Anthracosiacea by Kolesnikov (1970). However, it must be noted that other factors besides salinity may influence the proportion of shell organic matrix, including incidence of mantle irritation and shell repair, ontogenetic variation in the thickness of the periostracum, and possibly also the deposition of thick, interior organic layers, as in certain unionaceans, to reduce the hazard of deep exterior shell dissolution (Wilbur, 1964; Beedham, 1965; Saleuddin, 1967; Tolstikova, 1972, 1974) (also see Chapter 8). Thus, as in the case of mineralogical variations, much has yet to be learned about the phylogenetic history of particular bivalve lineages and contributory environmental factors before data on concentration of the shell's organic matrix can be confidently applied to paleoenvironmental studies.

5. Conclusions

The data summarized in this chapter are compatible with the hypothesis that the major variations in shell mineralogy, architecture, and microstructure among bivalve species are largely biologically controlled, and have evolved for a variety of advantages, including resistance to breakage on impact, resistance to abrasion, resistance to deep exterior shell dissolution, fracture localization, fracture deflection, dissipation of fracture energy, flexibility, low density, volumetric or energy economy of secretion, and variable relative or absolute rates of vertical shell layer accretion. The more obvious evolutionary trends in bivalve shell mineralogy and structure consist of: (1) the replacement of ancestral nacreous layers with porcelaneous layers in certain representatives of all bivalve subclasses; (2) the evolution of outer-layer calcite secretion in the Pterioida, Hippuritacea, and certain Chamacea; (3) the evolution of foliated structure from calcitic prismatic structure in certain early Pterioida, followed by the evolution of largely calcitic foliated shells in certain Pectinacea, Anomiacea, and in most or all Ostreacea; (4) the evolution of prominent calcitic simple prismatic layers and relatively reduced nacreous layers in many Pteriacea and Pinnacea; (5) the evolution of row stack nacre from ancestral sheet nacre in certain Pinnacea; (6) the evolution of chalky deposits and vacuolated layers in many Ostreacea; (7) the evolution of vacuolated layers and larger chambers in many Hippuritacea; (8) the evolution of sporadic calcite secretion in a few Veneracea and at least one lucinacean; and (9) the evolution of thick organic laminae within the nacreous structure of certain Unionacea. Current speculations regarding the adaptive significance of these and other mineralogical and structural variations among bivalve species are summarized in Table III. With the possible exception of certain outer calcitic shell layers in the Mytilacea, and sporadic calcite secretion in certain Veneracea, the taxonomic distribution of calcite secretion among modern bivalve species is controlled by biological rather than environmental factors.

Table III. Summary of Possible Adaptive Significance of Selected
Microstructures and Structural Arrangements

Shell structural or mineralogical feature	Possible adaptive significance
1. Calcitic simple prisms as in *Pinna*	High flexibility
2. Aragonitic composite prismatic, complex crossed lamellar, and homogeneous structures	High resistance to abrasion
3. Aragonitic concentric crossed lamellar structure	High resistance to abrasion; fracture deflection
4. Superposed concentric and radial crossed lamellar layers	High resistance to abrasion; dissipation of fracture energy
5. Sheet nacreous structure	High resistance to breakage on impact; relatively low maximum rate of vertical shell layer accretion.
6. Columnar nacreous structure	High resistance to breakage on impact; relatively high maximum rate of vertical shell layer accretion
7. Row stack nacreous structure	High resistance to breakage on impact; directional flexibility
8. Foliated structure	Fracture localization, low density, volumetric economy of secretion; relatively low maximum rate of vertical shell layer accretion
9. Chalky deposits (within foliated layers)	Fracture localization, low density, volumetric economy of secretion; relatively high maximum rate of vertical shell layer accretion
10. Thick organic laminae in unionacean shells	Resistance to deep exterior shell dissolution in connection with acidic freshwater environments; substratum for subsequent calcification
11. Superposition of aragonitic and calcitic shell layers	Mineralogical difference may enhance mechanical differences between the layers and insure their microstructural discontinuity, thereby promoting fracture deflection along their mutual boundary.

Environmental factors are known to associate with a variety of mineralogical and structural variations within bivalve species (Table IV). Some of these variations, e.g., the combined mineralogical–structural variations in *Mytilus edulis*, show strong potential as records of environmental and paleoenvironmental change. However, most of the variations listed in Table IV have yet to be developed as useful records of environmental change through the quantification of their contributory environmental and biological influences. Of particular importance here are possible effects of salinity and physiological races, both geographic and temporal, which remain poorly understood even for *M. edulis*. Realization of the full potential of mineralogical and structural variations in bivalve

Table IV. Summary of Environmental and Biological Controls of Selected Mineralogical and Microstructural Variations

Mineralogical or microstructural variation	Possible contributing factors
1. Addition of calcite to cooler-water species of a genus, as in the Mytilinae	Ambient temperature and possibly other environmental factors; high probability of strong biological (genetic) influence
2. Whole-shell calcite/aragonite ratios among populations of a species, as in *Mytilus edulis* and *M. californianus*	Ambient temperature; environmental salinity; water turbulence; degree of intertidal exposure; stage of ontogenetic development; spawning season; shell size; shell thickness; growth retardation; physiological races
3. Changes in the proportional area of secretion of adjacent aragonitic and calcitic shell layers, as in some Pectinidae	Ambient temperature; other possible contributing factors, both biological and environmental, have yet to be analyzed.
4. Mineralogy of tube linings, as in certain Teredinidae	Possibly ambient temperature; other possible contributing factors, both biological and environmental, have yet to be analyzed.
5. Isolated patches of calcite within a largely aragonitic outer shell layer, as in *Protothaca staminea*	Ambient temperature; stage of ontogenetic development; physiological races; other environmental factors, especially salinity, have yet to be analyzed.
6. Homogeneous sublayers, as in *Geukensia demissa*	Anaerobiosis-related dissolution of previously deposited calcium carbonate; ambient temperature; salinity; dissolved oxygen; possible biological (genetic) controls related to shell mechanical design and physiological races; acclimatization; behavioral strategies such as subaerial mantle ventilation
7. Irregular simple prismatic sublayers, as in *G. demissa*	As above, except possibly related to shell secretion during transitional aerobic–anaerobic respiration, rather than anaerobiosis-related dissolution. These sublayers may also reflect temporary attachment by the mantle epithelium to the shell interior (R. Lutz, personal communication).
8. Columnar nacreous sublayers, as in *G. demissa*	May be related to shell secretion during transitional aerobic–anaerobic respiration; these sublayers probably reflect, or bring about, higher rates of vertical shell layer accretion than adjacent parts of the same shell layer depositing sheet nacre; possibly strong biological controls.
9. Chalky deposits	Biological controls related to higher rates of vertical shell layer accretion than adjacent parts of the same shell layer depositing foliated structure; may also reflect accelerated aerobic respiration or natural selection for low density and economy of secretion.
10. Ratio of shell organic matrix to calcium carbonate in certain Anthracosiacea and presumably in other freshwater bivalves	Salinity; incidence of mantle irritation and shell repair; stage of ontogenetic development; biological controls related to deposition of thick interior organic laminae to reduce the hazard of deep exterior shell dissolution.

· shells as records of environmental change lies in: (1) determining variations among those listed in Table IV, or other possible variations, which
show maximum environmental and minimum biological influence; and
(2) studying mineralogical and structural variations in modern bivalves
from a combined environmental and evolutionary perspective in order
to justify, or limit, our interpretation of similar variations in the fossil
record.

ACKNOWLEDGMENTS. I thank Dr. Richard Lutz, Dr. Donald Rhoads, Dr.
Steven Stanley, Dr. Kenneth Towe, and Dr. Thomas Waller for their helpful comments regarding various versions of this chapter and the accompanying Appendix 2. Paul Belyea, Jack Hall, and Robert VanGundy assisted in proofreading the manuscript, and Jim Dischinger drafted Fig. 5.
Funding for this research was provided by the Department of Geology of
the University of North Carolina at Chapel Hill.

References

Abbott, R. T., 1974, *American Seashells*, 2nd ed., Van Nostrand, New York, 663 pp.
Aller, R. C., 1974, Prefabrication of shell ornamentation in the bivalve *Laternula*, *Lethaia*
   7:43–56.
Armstrong, J. D., 1969, The crossed bladed fabrics of the shells of *Terrakea solida* (Etheridge
   and Dun) and *Steptorhynchus pelicanensis* Fletcher, *Palaeontology* **12**:310–320.
Atkins, D., 1938, On the ciliary mechanisms and interrelationships of lamellibranchs. Part
   VII. Latero-frontal cilia of the gill filaments and their phylogenetic value, *Q. J. Microsc.
   Sci.* **80**:345–436.
Bayne, B. L., Bayne, C. H., Carefoot, T. C., and Thompson, R. J., 1976, The physiological
   ecology of *Mytilus californianus* Conrad, *Oecologia* **22**:229–250.
Beedham, G. E., 1965, Repair of the shell in species of *Anodonta*, *Proc. Zool. Soc. London*
   145:107–124.
Beedham, G. E., and Trueman, E. R., 1968, The cuticle of the Aplacophora and its evolu-
   tionary significance in the Mollusca, *J. Zool. (London)* **154**:443–451.
Berner, R. A., 1971, *Principles of Chemical Sedimentology*, McGraw-Hill, New York, 240
   pp.
Berry, L. G., and Mason, B., 1959, *Mineralogy*, W. H. Freeman, San Francisco, 612 pp.
Bøggild, O. B., 1930, The shell structure of the mollusks, *K. Dan. Vidensk. Selsk. Skr. Na-
   turvidensk. Math. Afd. 9 Raekke* **2**:231–326.
Carpenter, W. B., 1844, On the microscopic structure of shells, *Br. Assoc. Adv. Sci. Rep.*
   1844:1–24.
Carter, J. G., 1976, The structural evolution of the bivalve shell, with notes on the phylo-
   genetic significance of crossed lamellar structures, Ph.D. dissertation, Yale University,
   New Haven, Connecticut, 255 pp.
Carter, J. G., 1978, Ecology and evolution of the Gastrochaenacea, with notes on the evolution
   of the endolithic habitat, *Peabody Mus. Nat. Hist. Yale Univ. Bull.* **41**, 92 pp.

Carter, J. G., 1979, Comparative shell microstructure of the Mollusca, Brachiopoda and Bryozoa, in: Scanning Electron Microscopy/1979 (O. Johari, director), Vol. II, pp. 439–446 and p. 456, Chicago Press Corporation, Chicago.

Carter, J. G., and Aller, R. C., 1975, Calcification in the bivalve periostracum, Lethaia 8:315–320.

Carter, J. G., and Eichenberger, N. L., 1977, Ontogenetic calcite in the Veneracea (Mollusca: Bivalvia), Geol. Soc. Am. Abstr., Annual Meeting in Seattle, Washington, p. 922.

Carter, J. G., and Tevesz, M. J. S., 1978a, Shell microstructure of a Middle Devonian (Hamilton Group) bivalve fauna from central New York, J. Paleontol. 52:859–880.

Carter, J. G., and Tevesz, M. J. S., 1978b, The shell structure of Ptychodesma (Crytodontidae; Bivalvia) and its bearing on the evolution of the Pteriomorphia, Philos. Trans. R. Soc. London Ser. B 284:367–374.

Chave, K. E., 1954, Aspects of the biogeochemistry of magnesium. 1. Calcareous marine organisms, J. Geol. 62:266–283.

Chel'tsova, N. A., 1969, Znachenie Mikrostruktury Rakoviny Melovykh Ustrits dlia Ikh Sistematiki, Akademia Nauk, SSSR (Academy of Sciences, U.S.S.R.), Moscow, 82 pp.

Coleman, N., 1973, Water loss from aerially exposed mussels, J. Exp. Mar. Biol. Ecol. 12:145–155.

Collip, J. B., 1920, Studies on molluscan celomic fluid: Effect of change in environment on the carbon dioxide content of the celomic fluid: Anaerobic respiration in Mya arenaria, J. Biol. Chem. 45:23–49.

Cox, L. R., et al., 1969, Bivalvia, Part N, Mollusca 6, Vols. 1 and 2, in: Treatise on Invertebrate Paleontology (R. C. Moore, ed.), Geological Society of America and University of Kansas Press, Lawrence, 952 pp.

Crenshaw, M. A., and Neff, J. M., 1969, Decalcification at the mantle–shell interface in molluscs, Am. Zool. 9:881–885.

Currey, J. D., 1975, A comparison of the strength of enchinoderm spines and mollusc shells, J. Mar. Biol. Assoc. U.K. 55:419–424.

Currey, J. D., 1976, Further studies on the mechanical properties of mollusc shell material, Proc. Zool. Soc. London 180:445–453.

Currey, J. D., 1977, Mechanical properties of mother of pearl in tension, Proc. R. Soc. London Ser. B 196:443–463.

Currey, J. D., and Kohn, A. J., 1976, Fracture in the crossed-lamellar layer of Conus shells, J. Mater. Sci. 11:1615–1623.

Currey, J. D., and Taylor, J. D., 1974, The mechanical behavior of some molluscan hard tissues, J. Zool. (London) 173:395–406.

Davies, T. T., 1966, Effect of environmentally induced growth rate changes in Mytilus edulis shell, Geol. Soc. Am. Spec. Pap. 87:42.

Davies, T. T., and Sayre, J. G., 1970, The effect of environmental stress on pelecypod shell ultrastructure, Geol. Soc. Am. Abstr., Southeastern Section, pp. 204–205.

Dodd, J. R., 1962, Shell mineralogy and structure as evidence for two subspecies of Mytilus edulis on the Pacific coast of North America, Geological Society of America Abstracts, Annual Meeting in San Diego, California, 1961, p. 20.

Dodd, J. R., 1963, Paleoecological implications of shell mineralogy in two pelecypod species, J. Geol. 71:1–11.

Dodd, J. R., 1964, Environmentally controlled variation in the shell structure of a pelecypod species, J. Paleontol. 38:1065–1071.

Dodd, J. R., 1966, The influence of salinity on mollusk shell mineralogy: A discussion, J. Geol. 74:85–89.

Dodd, J. R., 1967, Magnesium and calcium in calcareous skeletons: A review, J. Paleontol. 41:1313–1329.

Douvillé, H., 1913, Classification des lamellibranches, *Bull. Soc. Geol. Fr. Ser.* 4 **12**:419–467.

Duckworth, D. L., 1976, A model for the physiological control of mineralogy and trace element composition of biogenic carbonate, *Geol. Soc. Am. Abstr.*, Annual Meeting, Denver, Colorado, p. 844.

Dugal, L. P., 1939, The use of calcareous shell to buffer the product of anaerobic glycolysis in *Venus mercenaria*, *J. Cell. Comp. Physiol.* **13**:235–251.

Eisma, D., 1966, The influence of salinity on mollusk shell mineralogy: A discussion, *J. Geol.* **74**:89–94.

Erben, H. K., and Krampitz, G., 1970, Ultrastruktur und Aminosäuren-Verhältnisse in den Schalen der rezenten Pleurotomariidae (Gastropoda), *Biomineralisation* **6**:12–31.

Erben, H. K., and Watabe, N., 1974, Crystal formation and growth in bivalve nacre, *Nature (London)* **248**:128–130.

Erben, H. K., Flajs, G., and Siehl, A., 1968, Über die Schalenstruktur von Monoplacophoren, *Akad. Wiss. Lit. Mainz 1968* **(1)**:1–24.

Farrow, G. E., 1972, Periodicity structures in the bivalve shell: Analysis of stunting in *Cerastoderma edule* from the Thames estuary, *Palaeontology* **15**:61–72.

Gilmour, T. H. J., 1967, The defensive adaptations of *Lima hians* (Mollusca, Bivalvia), *J. Mar. Biol. Assoc. U.K.* **47**:209–221.

Gordon, J., and Carriker, M. R., 1978, Growth lines in a bivalve mollusk: Subdaily patterns and dissolution of the shell, *Science* **202**:519–521.

Haas, W., 1972, Untersuchungen über die Mikro- and Ultrastruktur der Polyplacophorenschale, *Biomineralisation* **5**:3–52.

Hatchett, C., 1799, Experiments and observations on shell and bone, *Philos. Trans. R. Soc. London* **18**:554–563.

Hoffman, S., 1949, Studien über das Integument der Solenogastren nebst Bemerkungen über die Verwandtschaft zwischen den Solenogastren und Placophoren, *Zool. Bidr. Uppsala* **27**:293–427.

Jones, D. S., Thompson, I., and Ambrose, W., 1978, Age and growth rate determinations for the Atlantic surf clam *Spisula solidissima* (Bivalvia: Mactracea), based on internal growth lines in shell cross-sections, *Mar. Biol.* **47**:63–70.

Jope, M., 1971, Constituents of brachiopod shells, in: *Comprehensive Biochemistry* (M. Florkin and E. H. Stotz, eds.), Vol. 26C, pp. 749–782, Elsevier, Amsterdam.

Kennedy, W. J., and Taylor, J. D., 1968, Aragonite in rudists, *Proc. Geol. Soc. London* **1645**:325–331.

Kennedy, W. J., Taylor, J. D., and Hall, A., 1969, Environmental and biological controls on bivalve shell mineralogy, *Cambridge Philos. Soc. Biol. Rev.* **44**:499–530.

Kennedy, W. J., Morris, N. J., Taylor, J. D., 1970, The shell structure, mineralogy and relationships of the Chamacea (Bivalvia), *Palaeontology* **13**:379–413.

Kennish, M. J., 1978, Effects of thermal discharges on mortality of *Mercenaria mercenaria* in Barnegat Bay, New Jersey, *Environ. Geol.* **2**:223–254.

Kennish, M. J., and Olsson, R. K., 1975, Effects of thermal discharges on the microstructural growth of *Mercenaria mercenaria*, *Environ. Geol.* **1**:41–64.

Kitano, Y., Kanamori, N., and Yoshioka, S., 1976, Influence of chemical species on the crystal type of calcium carbonate, in: *The Mechanisms of Mineralization in the Invertebrates and Plants* (N. Watabe and K. M. Wilbur, eds.), pp. 191–202, University of South Carolina Press, Columbia.

Kobayashi, I., 1969, Internal microstructure of the shell of bivalve molluscs, *Am. Zool.* **9**:663–672.

Kobayashi, I., 1971, Internal shell microstructure of recent bivalvian molluscs, *Sci. Rep. Niigata Univ. Ser. E* **2**:27–50.

Kolesnikov, C. M., 1970, Paleobiochemistry of fossil organisms as exemplified by Mesozoic freshwater mollusks, *Paleontol. J.* **4**:39–47.

Korringa, P., 1951, On the nature and function of chalky deposits in the shell of *Ostrea edulis* Linnaeus, *Proc. Calif. Acad. Sci. Ser.* 4 **27**:133–159.

Kühnelt, W., 1930, Bohrmuschelstudien I, *Paleobiologica* **3**:51–91.

Lowenstam, H. A., 1954a, Environmental relations of modification compositions of certain carbonate-secreting marine invertebrates, *Proc. Natl. Acad. Sci. U.S.A.* **40**:39–48.

Lowenstam, H. A., 1954b, Factors affecting the aragonite:calcite ratios in carbonate-secreting marine organims, *J. Geol.* **62**:284–322.

Lowenstam, H. A., 1954c, Status of invertebrate paleontology. XI. Systematic, paleoecologic and evolutionary aspects of skeletal building materials, *Bull. Mus. Comp. Zool. Harv. Univ.* **112**:287–317.

Lowenstam, H. A., 1963, Biologic problems relating to the composition and diagenesis of sediments, in: *The Earth Sciences: Problems and Progress in Current Research* (T. W. Donnely, ed.), pp. 137–195, Rice University Semicentennial Publication, University of Chicago Press.

Lowenstam, H. A., 1964, Coexisting calcites and aragonites from skeletal carbonates of marine organisms and their strontium and magnesium contents, in: *Recent Researches in the Fields of Hydrosphere, Atmosphere, and Nuclear Geochemistry* (Y. Miyake and T. Koyama, eds.), pp. 373–404, Maruzen, Tokyo.

Lucas, G., 1952, Étude microscopique et pétrographique de la coquille des lamellibranches, in: *Traité de Paléontologie* Vol. II (J. Piveteau, ed.), pp. 246–261, Masson, Paris, 790 pp.

Lutz, R. A., 1976a, Geographical and seasonal variation in the shell structure of an estuarine bivalve, *Geol. Soc. Am. Abstr.*, Annual Meeting, Denver, Colorado, p. 988.

Lutz, R. A., 1976b, Annual growth patterns in the inner shell layer of *Mytilus edulis* L., *J. Mar. Biol. Assoc. U.K.* **56**:723–731.

Lutz, R. A., and Rhoads, D. C., 1977, Anaerobiosis and a theory of growth line formation, *Science* **198**:1222–1227.

MacClintock, C., 1967, Shell structure of patelloid and bellerophontoid gastropods (Mollusca), *Peabody Mus. Nat. Hist. Yale Univ. Bull.* **22**:1–140.

Malone, P. G., and Dodd, J. R., 1967, Temperature and salinity effects on calcification rate in *Mytilus edulis* and its paleoecological implications, *Limnol. Oceanogr.* **12**:432–436.

Margolis, S. V., and Carver, R. E., 1974, Microstructure of chalky deposits found in shells of the oyster *Crassostrea virginica*, *Nautilus* **88**:62–65.

Masse, J. P., and Philip, J., 1972, Observations sur la croissance et l'ontogenèse du test des Radiolitidae (Rudistes): Conséquences phylogénétiques et paléoécologiques, *C. R. Acad. Sci. Ser. D* **274**:3202–3205.

Meenakshi, V. R., Martin, A. W., and Wilbur, K. M., 1974, Shell repair in *Nautilus macromphalus*, *Mar. Biol.* **27**:27–35.

Nakahara, H., and Bevelander, G., 1971, The formation and growth of the prismatic layer of *Pinctada radiata*, *Calcif. Tissue Res.* **7**:31–45.

Newell, N. D., 1938, Late Paleozoic pelecypods: Pectinacea, *Kansas State Geol. Surv. Bull.* **10**:1–23 ("1937").

Newell, N. D., 1954, Status of invertebrate paleontology, 1953. V. Mollusca: Pelecypoda, *Mus. Comp. Zool. (Harv. Univ.) Bull.* **112**:161–172.

Newell, N. D., 1969, Classification of Bivalvia, in: *Treatise on Invertebrate Paleontology, Part N, Mollusca 6, Bivalvia* (R. C. Moore, ed.), pp. N205–N224, Geological Society of America and University of Kansas Press, Lawrence.

Newell, N. D., and Boyd, D. W., 1970, Oyster-like Permian Bivalvia, *Bull. Am. Mus. Nat. Hist.* **143**:217–282.

Newell, N. D., and Boyd, D. W., 1975, Parallel evolution in early trigoniacean bivalves, *Bull. Am. Mus. Nat. Hist.* **154:**53–162.

Oberling, J. J., 1964, Observations on some structural features of the pelecypod shell, *Mitt. Naturforsch. Ges. Bern.* **20:**1–63.

Olsson, A. A., 1961, *Panamic–Pacific Pelecypoda*, Paleontological Research Institution, Ithaca, New York, 572 pp.

Pobeguin, T., 1954, Contribution a l'étude des carbonates de calcium précipitation du calcaire par les végétaux comparison avec le monde animal, *Ann. Sci. Nat. Bot. Ser. 11* **14:**31–109.

Pojeta, J., Jr., 1962, The pelecypod genus *Byssonychia* as it occurs in the Cincinnatian at Cincinnati, Ohio, *Paleontogr. Am.* **4:**169–216.

Pojeta, J., Jr., 1971, Review of Ordovician pelecypods, *U.S. Geol. Surv. Prof. Pap.* **695:**1–46.

Prenant, M., 1927, Les formes minéralogiques du calcaire chez les êtres vivants, et le problème de leur déterminisme, *Biol. Rev.* **2:**365–393.

Rieger, R. M., and Sterrer, W., 1975, New spicular skeletons in Turbellaria, and the occurrence of spicules in marine meiofauna, *Z. Zool. Syst. Evolutionsforsch.* **13:**207–278.

Saleuddin, A. S. M., 1967, The histochemistry of the mantle during the early stage of shell repair, *Proc. Malacol. Soc. London* **37:**371–380.

Salvini-Plawen, L. v., 1972, Zur Morphologie und Phylogenie der Mollusken: Die Beziehungen der Caudofoveata und der Solenogastres als Aculifera, als Mollusca, und als Spiralia, *Z. Wiss. Zool. (Leipzig)* **184:**205–394.

Schmidt, W. J., 1924, *Die Bausteine des Tierkörpers in polarisiertem Lichte*, Friedrich Cohen, Bonn, 528 pp.

Schmidt, W. J., 1959, Bemerkungen zur Schalenstruktur von *Neopilina galathea*, *Galathea Rep.* **3:**73–77.

Skelton, P. W., 1974, Aragonitic shell structures in the rudist *Biradiolites*, and some palaeobiological inferences, *Geol. Mediterraneenne* **1:**63–74.

Skelton, P. W., 1976, Functional morphology of the Hippuritidae, *Lethaia* **9:**83–100.

Soot-Ryen, T., 1969, Superfamily "Mytilacea," in: *Treatise on Invertebrate Paleontology, Part N, Mollusca 6* (R. C. Moore, ed.), pp. N271–N281, Geological Society of America and University of Kansas Press, Lawrence.

Stanley, S. M., 1972, Functional morphology and evolution of byssally attached bivalve molluscs, *J. Paleontol.* **46:**165–212.

Stenzel, H. B., 1962, Aragonite in the resilium of oysters, *Science* **136:**1121–1122.

Stenzel, H. B., 1963, Aragonite and calcite as constituents of adult oyster shells, *Science* **142:**232–233.

Stenzel, H. B., 1964, Oysters: Composition of larval shell, *Science* **145:**155–156.

Stenzel, H. B., 1971, "Oysters," in: *Treatise on Invertebrate Paleontology, Part N, Mollusca 6* (R. C. Moore, ed.), pp. N953–N1224, Geological Society of America and University of Kansas Press, Lawrence.

Strachimirov, B., 1972, Recherches morphologiques et microstructurales sur la coquille de *Corbula (Varicorbula) gibba* du Tortonien et du Tchokrakien en Bulgarie, *Proc. Int. Geol. Congr.*, 24th session, Montreal, **7:**41–47.

Sverdrup, H. U., Johnson, M. W., and Fleming, R. H., 1942, *The Oceans*, Prentice-Hall, Englewood Cliffs, New Jersey, 1087 pp.

Taylor, A. C., and Brand, A. R., 1975, A comparative study of the respiratory responses of the bivalves *Arctica islandica* (L.) and *Mytilus edulis* L. to declining oxygen tension, *Proc. R. Soc. London Ser. B* **190:**443–456.

Taylor, J. D., 1973, The structural evolution of the bivalve shell, *Palaeontology* **16:**519–534.

Taylor, J. D., and Kennedy, W. J., 1969, The shell structure and mineralogy of *Chama pellucida* Broderip, *Veliger* **11:**391–398.

Taylor, J. D., and Layman, M., 1972, The mechanical properties of bivalve (Mollusca) shell structures, *Palaeontology* **15**:73–87.

Taylor, J. D., Kennedy, W. J., and Hall, A., 1969, The shell structure and mineralogy of the Bivalvia: Introduction: Nuculacea–Trigonacea, *Bull. Br. Mus. (Nat. Hist.) Zool. Suppl.* **3**:1–125.

Taylor, J. D., Kennedy, W. J., and Hall, A., 1973, The shell structure and mineralogy of the Bivalvia. II. Lucinacea–Clavagellacea: Conclusions, *Bull. Br. Mus. (Nat. Hist.) Zool.* **22**:253–294.

Tolstikova, N. V., 1972, Microstructural variability of shells in *Unio tumidus* (Bivalvia), *Zool. Zh.* **51**:1565–1569.

Tolstikova, N. V., 1974, Microstructural characteristics of freshwater bivalves (Unionidae), *Paleontol. J.* **8**:55–60.

Vagvolgyi, J., 1967, On the origin of the molluscs, the coelom, and coelomic segmentation, *Syst. Zool.* **16**:153–168.

Wada, K., 1961, Crystal growth of molluscan shells, *Bull. Natl. Pearl Res. Lab. (Kokuritsu Shinju Kenkyusho Hokoku)* **7**:703–828.

Wada, K., 1972, Nucleation and growth of aragonite crystals in the nacre of some bivalve molluscs, *Biomineralisation* **6**:141–159.

Wainwright, S. A., Biggs, W. D., Currey, J. D., and Gosline, J. M., 1976, *Mechanical Design in Organisms*, Wiley, New York, 423 pp.

Waller, T. R., 1972, The functional significance of some shell microstructures in the Pectinacea (Mollusca: Bivalvia), *Proc. Int. Geol. Congr.*, 24th session, Montreal, **7**:48–56.

Waller, T. R., 1976a, The development of the larval and early postlarval shell of the bay scallop, *Argopecten irradians*, *Am. Malacol. Union Inc. Bull.* **1976**:46.

Waller, T. R., 1976b, The origin of foliated-calcite shell microstructure in the subclass Pteriomorphia (Mollusca: Bivalvia), *Am. Malacol. Union Inc. Bull.* **1975**:57–58.

Waller, T. R., 1978, Morphology, morphoclines and a new classification of the Pteriomorphia (Mollusca: Bivalvia), *Philos. Trans. R. Soc. London Ser. B* **284**:345–365.

Watabe, N., 1956, Dahllite identified as a constituent of prodissoconch of *Pinctada martensii* (Dunker), *Science* **124**:630.

Watabe, N., and Dunkelberger, D. G., 1979, Ultrastructural studies on calcification in various organisms, in: *Scanning Electron Microscopy/1979* (O. Johari, director), Vol. II, pp. 403–416, Chicago Press Corporation, Chicago.

Watabe, N., and Wilbur, K. M., 1960, Influence of the organic matrix on crystal type in molluscs, *Nature (London)* **188**:334.

Weber, J., Greer, R., Voight, B., White, E., and Roy, R., 1969, Unusual strength properties of echinoderm calcite related to structure, *J. Ultrastruct. Res.* **26**:355–366.

Widdows, J., 1976, Physiological adaptation of *Mytilus edulis* to cyclic temperatures, *J. Comp. Physiol.* **105**:115–128.

Wilbur, K. M., 1960, Shell structure and mineralization in molluscs, in: *Calcification in Biological Systems* (R. F. Sognnaes, ed.), pp. 15–40, American Association for the Advancement of Science, Washington, D.C.

Wilbur, K. M., 1964, Shell formation and regeneration, in: *Physiology of Mollusca*, Vol. 1 (K. M. Wilbur and C. M. Yonge, eds.), pp. 243–282, Academic Press, New York.

Wilbur, K. M., 1972, Shell formation in mollusks, in: *Chemical Zoology, Vol. VII, Mollusca* (M. Florkin and B. Scheer, eds.), pp. 103–154, Academic Press, New York.

Wilbur, K. M., and Watabe, N., 1963, Experimental studies on calcification in molluscs and the alga *Coccolithus huxleyi*, *Ann. N.Y. Acad. Sci.* **109**:82–112.

Wilkes, D. A., and Crenshaw, M. A., 1979, Formation of a dissolution layer in molluscan shells, in: *Scanning Electron Microscopy/1979* (O. Johari, director), Vol. II, pp. 469–474, Chicago Press Corporation, Chicago.

Williams, A., 1966. Growth and structure of the shell of living articulate brachiopods, *Nature (London)* **211**:1146–1148.

Wind, F. H., and Wise, S. W., Jr., 1976, Organic vs. inorganic processes in archaeogastropod mineralization, in: *The Mechanisms of Mineralization in the Invertebrates and Plants* (N. Watabe and K. M. Wilbur, eds.), pp. 369–387, University of South Carolina Press, Columbia.

Wise, S. W., Jr., 1970, Microarchitecture and mode of formation of nacre (mother of pearl) in pelecypods, gastropods, and cephalopods, *Ecologae. Geol. Helv.* **63**:775–797.

Yonge, C. M., 1936, The evolution of the swimming habit in the Lamellibranchia, *Mem. Mus. Hist. Nat. Belg.* **3**:77–100.

Yonge, C. M., 1953, Form and habit in *Pinna carnea* Gmelin, *Philos. Trans. R. Soc. London Ser. B* **237**:335–374.

Chapter 3

Mechanisms of Shell Formation and Dissolution

MILES A. CRENSHAW

1. Introduction 115
2. Shell Formation 116
   2.1. Physiological Aspects of Shell Formation 116
   2.2. Shell Deposition 121
3. Mechanisms of Shell Dissolution 125
   3.1. Shell Dissolution 125
   3.2. Metabolic Aspects of Shell Dissolution 126
   References 128

1. Introduction

Molluscan shell formation has been the subject of frequent reviews (e.g., Wilbur, 1964, 1972; Wilbur and Simkiss, 1968; Grégoire, 1972). In this chapter, I shall not attempt another comprehensive and balanced review of the earlier literature. Instead, the focus will be on the limitations of some experimental approaches that have been used and the interpretation of some recent experimental results, especially regarding the soluble organic shell matrix.

Structural and biochemical evidence shows that shell formation is not a uniform, unidirectional process. Growth is not uniform over the entire shell surface, and the entire outer mantle epithelium should not be expected to be uniform in its functions. Additionally, it has been shown that shell can function as an alkali reserve and is dissolved during periods of anaerobic acid production.

MILES A. CRENSHAW • The Dental Research Center and Department of Pedodontics, School of Dentistry, University of North Carolina, Chapel Hill, North Carolina 27514.

Shell dissolution has not been as extensively studied as shell formation. The recent hypothesis advanced by Lutz and Rhoads (1977) that
periodic shell dissolution may account for growth-line formation in molluscan shells has sparked new interest in the patterns and extent of anaerobic shell dissolution. Research in progress may disprove the conclusions drawn here before this chapter is published. Nevertheless, a
summary of the evidence to date and its implication is justified.

Shell dissolution occurs as a result of anaerobic acid production by
molluscs. The metabolic pathways and end products of molluscan anaerobiosis have recently been reviewed (de Zwaan and Wijsman, 1976; de
Zwaan et al., 1976; de Zwaan, 1977).

2. Shell Formation

It is convenient to divide processes involved in molluscan shell formation into two categories. The first is the transfer of materials to the
growing surface of the shell. The second is the assembly of calcium carbonate crystals and organic matrix into a highly organized structure.

2.1. Physiological Aspects of Shell Formation

The molluscan shell-forming system is usually depicted as a series
of compartments (Fig. 1). The primary compartments are the outer mantle
epithelium, the extrapallial space, and the inner shell surface. The extrapallial fluid is enclosed between the outer mantle epithelium and the
inner surface of the shell. This extracellular fluid is considered to be the
medium from which shell deposition takes place, and its composition is
controlled by the metabolic activity of the outer mantle epithelium (Wilbur, 1964, 1972; Wilbur and Simkiss, 1968).

The primary compartments of the shell-forming system are intercon-
nected with secondary compartments that are indirectly involved in shell
formation. The principal secondary compartment is the blood sinus between the two epithelial layers of the mantle. Materials derived from food
or the medium bathing the inner surface of the mantle pass through this
blood space to the outer epithelium of the mantle.

2.1.1. The Mantle

The highest rate of shell deposition takes place at or near the growing
shell margin (Wilbur and Jodrey, 1952; Zischke et al., 1970; Wheeler et

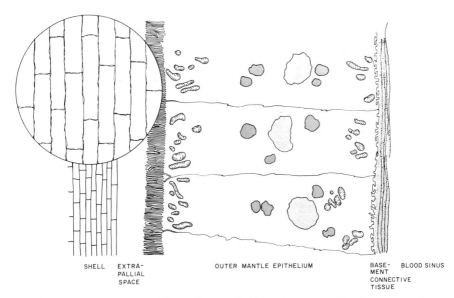

SHELL   EXTRA-   OUTER MANTLE EPITHELIUM   BASE-   BLOOD SINUS
        PALLIAL   MENT
        SPACE   CONNECTIVE
                                                           TISSUE

Figure 1. Compartments of the molluscan shell-forming system. Not drawn to scale.

al., 1975; Wheeler and Wilbur, 1977). One would therefore expect a higher portion of the metabolic activity associated with net shell formation to be in the margin of the outer mantle epithelium that is apposed to the shell margin. The central (proximal) zone of the mantle is more closely associated with the acid–base metabolism of the animal. These two functional zones are separated by the pallial attachment, where it is present, or, in the absence of a pallial attachment, by the transition from tall columnar mantle cells in the marginal zone to more cuboidal cells in the central zone.

Direct and indirect observations indicate that there is greater metabolic activity in the marginal zone than in the central zone of the mantle. Jodrey and Wilbur (1955) found that the marginal zone had a greater endogenous respiratory rate than the central zone of the mantle from *Crassostrea virginica.* A parallel difference is often evident in the ultrastructure of the two mantle zones in certain bivalve and gastropod species (Fig. 2). The cells of the marginal zone have numerous mitochondria, well-developed endoplasmic reticula, and Golgi apparatus. The mitochondria are much less numerous in the central zone. Here the endoplasmic reticulum is less extensive and less well developed. Golgi figures are difficult to locate in the cells of the central zone (Kawaguti and Ikemoto, 1962; Tsujii, 1968, 1976; Saleuddin, 1970; Kniprath, 1971; Crenshaw and Travis, unpublished observations).

Mantle Cells

Inner Shell Surfaces

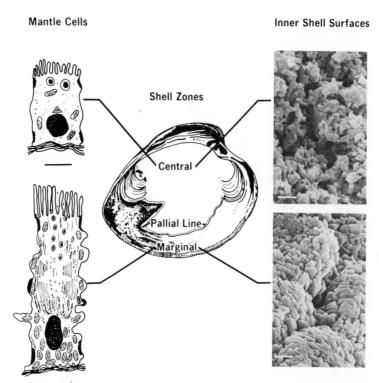

Figure 2. Zonation in the shell of *Mercenaria mercenaria*. The central and marginal zones are separated by the pallial line. *Left:* The composite drawings of the cells were made from electron micrographs and illustrate the relative sizes of the cells and abundance of cellular organelles in the corresponding zones in the outer mantle epithelium. The scale bar between the cells represents 3 μm. *Right:* The scanning electron micrographs were taken of the inner surface of the same valve after the animal was out of water for 3 hr. The shell was treated with 2.5% sodium hypochlorite to remove the organic matrix before it was coated with gold. Scale bars: upper, 0.5 μm; lower, 0.1 μm.

The correlation of cell ultrastructure and function in the mantle is strengthened by the shell-regeneration studies of Tsujii (1976). In these studies, a small piece of the shell overlying the central zone of the mantle of *Anodonta* sp. was removed, and cytological changes in the outer epithelium of this area were followed. During the period of shell repair, the cells in this area assumed an ultrastructure very much like that of the cells in the marginal zone. After shell repair was completed, the cells reverted to the cytology typical of the central zone of the epithelium on control mantles.

The movement of calcium and bicarbonate through the mantle does not appear to be a limiting factor in shell mineralization. Rates of isotopic exchange between the mantle and the medium indicate high rates of both calcium (Jodrey, 1953; Wilbur, 1964; Wheeler *et al.*, 1975) and bicarbonate (Wheeler *et al.*, 1975) movement through the mantle. Direct measurements of the transepithelial movement of calcium appear to have been made only on the central zones of mantles from freshwater bivalves. Both the inner and outer epithelia are highly permeable to the bidirectional flux of calcium (Istin and Maetz, 1964).

Structurally, the central zone of the outer mantle epithelium is well suited for the passive paracellular movement of calcium into and out of the extrapallial fluid (Machin, 1977). The intercellular junctions in this zone are "gap" junctions (Neff, 1972) that allow the free diffusion of calcium. In addition, studies in which lanthanum was used as a tracer for the paracellular routes of calcium (Whittembury and Rawlins, 1971; Machen *et al.*, 1972) showed that this tracer permeated the entire length of the paracellular channels in the central zone of the outer mantle epithelium. In contrast, lanthanum did not penetrate the paracellular space in the marginal zone (Crenshaw and Travis, unpublished data). This preliminary work corroborates earlier evidence of the paracellular movement of calcium through the central zone (Neff, 1972) and indicates that calcium must pass through the cells of the marginal zone to reach the shell through the mantle.

Most of the evidence for the functional zonation of the outer mantle epithelium is presumptive, and this hypothesis must be tested with comparative studies. The evidence available indicates that the cells in the marginal zone have an active aerobic metabolism and a high synthetic capacity and that they control calcium movement through this zone of the mantle. In contrast, the cells of the central zone probably have a lower aerobic respiratory rate and a lower synthetic capacity and allow calcium to pass freely between the cells. If these differences prove to be valid, the marginal zone may be more closely associated with net shell formation, and the central zone may be more closely associated with anaerobic shell dissolution.

2.1.2. The Extrapallial Fluid

Conceptually, the extrapallial fluid is that fluid space between the cells of the outer mantle epithelium and the inner shell surface through which components must pass to be incorporated into the shell. This extracellular fluid may also contain materials that are not incorporated into the shell. The composition of the extrapallial fluid may be determined by the direct access of the external medium to this compartment. However, the periostracum, as a hydrophobic barrier (Hunt, 1970; Grégoire, 1972), may restrict the free exchange of seawater into the pallial space, especially where the periostracum is complete from the periostracal groove to the outer shell surface. When a pallial attachment is also present, the extra-pallial fluid may be divided into a marginal and a central fluid space.

A primary purpose for analyzing the extrapallial fluid has been to describe the physiochemical environment in which shell formation takes place. The two analyses of the extrapallial fluid that have been reported (Crenshaw, 1972a; Wada and Fujinuki, 1976) were on samples of fluid taken from areas inside the pallial line. In both cases, the samples were judged to be saturated or slightly supersaturated with respect to the shell mineral. Significant changes in the composition of the extrapallial fluid were observed only when the valves closed; the animal became anaerobic and dissolved previously deposited shell (Crenshaw, 1972a).

The aforementioned studies have contributed little to the understanding of shell formation. The samples were taken from the central compartment rather than from the marginal compartment of the extrapallial fluid. During shell deposition, the epithelium may move in very close proximity to the inner shell surface, so that only a small fraction of the total extrapallial fluid separates the two surfaces. This distance may, in fact, be so small that the transfer of materials to the inner shell is essentially direct.

2.1.3. Carbonic Anhydrase

One aspect of the physiology of shell calcification that has received little attention is the fact that calcification is an acid-forming process:

$$Ca^{2+} + HCO_3^- \rightleftarrows CaCO_3 + H^+$$

Once mineralization has started, the hydrogen ions must be removed from the calcifying medium for the process to continue. Wheeler (1975) proposed a mechanism of hydrogen ion removal that may explain the role of the enzyme carbonic anhydrase in shell mineralization. This enzyme has frequently been implicated in shell mineral deposition because inhibitors of carbonic anhydrase inhibit calcification (cf. Wilbur, 1972). Wheeler (1975) found that at least 70% of the carbonic anhydrase activity was in the membrane fraction of homogenates of the mantle from *Crassostrea virginica*. He also found that the bicarbonate fluxes across the mantle were approximately equal in both directions. The carbonic anhydrase inhibitor acetazolamide inhibited the flux away from the shell, but did not affect the flux toward the shell. From this evidence, Wheeler (1975) proposed that carbonic anhydrase catalyzes the reaction of hydrogen ions, produced by mineral formation, with additional bicarbonate to produce carbon dioxide, which then could readily diffuse away from the mineralization sites.

2.2. Shell Deposition

Any general theory of molluscan shell formation must account for: (1) the nucleation of the calcium carbonate crystals of the shell; (2) the determination of the calcium carbonate polymorph deposited; and (3) the control of the crystal orientation and the direction of crystal growth. Evidence for the mechanisms by which these processes are accomplished is incomplete. It is possible that all the parameters are determined by the nucleation process and are expressed by postnucleation growth (Pashley, 1970).

The organic matrix of molluscan shells is considered to have a primary role in shell mineralization (Wilbur, 1964, 1972; Wilbur and Simkiss, 1968; Grégoire, 1972). There are two primary hypotheses of the mechanisms whereby the matrix controls the deposition of calcium carbonate. The "matrix" hypothesis assumes that the organic matrix is deposited as a substrate that nucleates crystals, a process similar to epitaxy (Wilbur, 1964, 1972; Wilbur and Simkiss, 1968; Grégoire, 1972). The "compartment" hypothesis supposes that crystal nucleation occurs within preexisting, empty compartments of organic matrix, and that these compartments control crystal growth (Bevelender and Nakahara, 1969). Neither of these hypotheses has been expressed in precise enough detail to allow direct experimental testing.

The "compartment" hypothesis has generally fallen into disfavor. Conceptually, this hypothesis offers no explanation of the structural order observed in shell (Towe, 1972). Additionally, the experimental basis for the hypothesis (Bevelander and Nakahara, 1969) appears to have been due in part to inadequate fixation (Towe and Hamilton, 1968; Crenshaw and Ristedt, 1975, 1976) or shell dissolution (Lutz and Rhoads, 1977) (also see Chapter 6) or both. Parallel sheets of matrix have been observed in immature nacre, but they were separated by several layers of nacreous tablets, rather than by one layer of tablets (Wise, 1970; Erben, 1972), as required by the "compartment" hypothesis.

2.2.1. The Insoluble Organic Matrix

Experimentally, the matrix of shells is seen as the organic material into which the mineral is embedded or as the residue left after the mineral has dissolved. There is no reason, *a priori*, to assume that the entire organic residue has a function in mineral deposition (Towe, 1972). In fact, Wheeler *et al.* (1975) found that the protein/mineral ratio varied in the shell of *Argopecten irradians*. The role of some components of the organic matrix may be more closely associated with functionally important mechanical properties of the shell than with crystal formation.

In view of problems related to the definition of the functional organic matrix, it is not surprising that analyses for the primary components of the residue left after demineralization of shell have contributed little to our understanding of how the matrix controls the mineral phase of the shell. The utility of these analyses is also limited by their being incomplete. Although it is known that shell matrix contains significant quantities of carbohydrate (Grégoire, 1967, 1972), the majority of analyses have been restricted to amino acids (Wilbur and Simkiss, 1968; Hunt, 1970; Grégoire, 1972). Additionally, the amino acid compositions have most frequently been determined on acid hydrolysates using automatic analyzers. This method precludes the detection of most amino acid derivatives that may have been present in the protein (Glazer et al., 1975).

Determination of the primary composition of a macromolecular assemblage, such as shell matrix, is an important first step in the understanding of its function. However, any description of the mechanism whereby this assemblage may function requires a knowledge of the primary structure of each component and interactions among components. Some significant contributions toward this end have been made by investigators at the University of Liège. They have separated and analyzed three components from nacreous matrix and correlated these fractions with structural components of the matrix. The striking feature from the amino acid analyses is the proponderance of amino acids having hydrophobic side chains (for reviews, see Grégoire et al., 1955; Grégoire, 1967, 1972). These results indicate that the dominant characteristic of the insoluble nacreous matrix may be its hydrophobicity. This property could affect an ordering in the solution layers immediately adjacent to the matrix sheets (Shultz and Asumaa, 1970). The increase in order (decrease in entropy) would tend to reduce the activation energy required for initial crystal formation (Neuman and Neuman, 1958).

2.2.2. The Soluble Organic Matrix

A significant fraction of the organic matrix from shell is soluble in agents used to remove the mineral (Voss-Foucart et al., 1969; Crenshaw, 1972b; Weiner and Hood, 1975; Krampitz et al., 1976; Weiner et al., 1977). Analyses of this soluble matrix have been primarily limited to amino acid determinations. The soluble matrix from Mercenaria mercenaria, however, contains about 20% carbohydrates (Crenshaw, 1972b).

Using gel filtration and cellulose acetate electrophoresis (conditions that do not denature the shell proteins), Crenshaw (1972b) found only one component in the soluble matrix fraction. With the more precise method of sodium dodecyl sulfate electrophoresis, Weiner et al. (1977) demonstrated that the soluble matrix of bivalves, gastropods, and cephalopods was heterogeneous. However, the electrophoretic bands of those molluscs accounted for only about 10% of the protein loaded onto the gel. Therefore, the question remains, is the soluble fraction heterogeneous or is it one primary component that is 90% pure?

Aspartic acid, glycine, and serine account for at least half the amino acids in the acid hydrolysates of the soluble matrix (Crenshaw, 1972b; Weiner and Hood, 1975). The β-carboxyl group of aspartic acid appears to be amidated rather than being a free acidic group in the soluble matrix of *Mercenaria mercenaria* (Crenshaw, 1972b). This amide nitrogen of asparagine may be involved in the carbohydrate–peptide linkage of the glycoprotein. When alkali-induced β-elimination of the carbonhydrate was tried in order to detect the presence of the O-glycosidic linkage to serine (Neuberger *et al.*, 1966), no evidence was found of this linkage in the soluble matrix of *M. mercenaria*. Treatment of the soluble matrix with 0.5 N NaOH did not cause any change in absorbance at 241 nm, and β-alanine was not detected in hydrolysates of the alkali-treated material (Crenshaw, unpublished data). However, an N-(aspartyl)-acylglycosylamine has not been isolated. The isolation of such a compound is necessary to prove that asparagine is involved in the carbohydrate–peptide linkage. The compositional data obtained to date represent a partial characterization of the major component of the soluble matrix, and it is this type of data that is needed to correlate structure and function.

Attempts to describe the function of the soluble matrix have also been undertaken. The soluble matrix from *M. mercenaria* specifically binds calcium. The stoichiometry of this binding is one mole of calcium bound for each two moles of ester sulfate in the soluble matrix. The specificity of the calcium binding is dependent on the conformation of the matrix protein (Crenshaw, 1972b). Because these results were so similar to those obtained for a porcine cartilage glycoprotein (Woodward and Davidson, 1968), Crenshaw (1972b) postulated that calcium binding by the soluble matrix was accomplished by the same mechanism as with the cartilage protein, that is, chelation by ester sulfate groups from two adjacent polysaccharide side chains. This mechanism of binding, however, has not been proven.

Alternative mechanisms by which calcium may be bound by the soluble matrix have been proposed. From their amino acid analyses and determinations of the frequency at which aspartyl residues occur in the peptide chain of soluble matrices, Weiner and Hood (1975) proposed that the β-carboxyl groups of aspartic acid bind calcium so that the interatomic distances closely approximate those in aragonite and calcite crystals. In this manner, the soluble matrix would act as a crystal nucleator. This hypothesis is very much like the one proposed earlier by Matheja and Degens (1968). However, if the β-carboxyl groups of aspartyl residues are amidated (Crenshaw, 1972b), this mechanism is not possible. Neither Matheja and Degens (1968) nor Weiner and Hood (1975) used methods that could distinguish between aspartic acid and asparagine.

There appears to be a definite structural relationship between the soluble and the insoluble matrices in some species. Grégoire (1967, 1972) has shown that when interlamellar sheets of insoluble matrix are obtained by the decalcification of nacre, the intercrystalline matrix collapses onto the interlamellar sheets to form polygonal outlines corresponding to the domains of individual nacreous tablets. He termed these outlines on the interlamellar matrix "crystal imprints." Crenshaw and Ristedt (1975, 1976) used histochemical methods to localize a soluble, calcium-binding, sulfated polysaccharidic component in the central region of the crystal imprints in the interlamellar matrix from the septal nacre of *Nautilus pompilius*. They used cetylpyridinium chloride and formaldehyde in low-ionic-strength EDTA to fix the soluble component *in situ* as the mineral was removed. When the cetylpyridinium chloride was omitted, or high-ionic-strength solutions were used, pores were then evident in the central region of the crystal imprints.

Iwata (1975) reported a different distribution of soluble matrix components in the nacreous interlamellar membranes from bivalves, gastropods, and a cephalopod. When he decalcified with EDTA or HCl, he observed the patterns of pores that Grégoire (1967) has shown to be characteristic of the molluscan classes. When chromium sulfate was used to simultaneously decalcify and fix the matrix, the pores were not seen. Iwata (1975) suggested that the pores were filled with a soluble matrix similar to that isolated from *M. mercenaria* (Crenshaw, 1972b). The different localization of the soluble matrix in nacreous interlamellar membranes may be accounted for by the different methods used. Iwata (1975) used uranyl acetate staining or metal shadowing, which is more sensitive for the detection of pores, while the calcium and ruthenium red staining used by Crenshaw and Ristedt (1975, 1976) was more selective.

Crenshaw and Ristedt (1975, 1976) advanced the hypothesis that the calcium-binding soluble matrix nucleated crystals by "ionotropy." This term was taken from the work of Thiele (1967), who showed that calcium-binding macromolecules can induce ordered crystal formation. Crystal induction in these model systems is accomplished by selective concentration of ions around calcium-binding sites. Preliminary experiments (D. Manyak, personal communication) have shown that the organic matrix from the tube lining of the shipworm, *Bankia gouldi*, that has been decalcified with low-ionic-strength EDTA, cetylpyridinium chloride, and glutaraldehyde will recalcify in an artificial extrapallial fluid (Crenshaw, 1972a). This new hypothesis of ionotropic nucleation should be considered to be a redefinition of the "active site" portion of the "matrix" hypothesis (Wilbur, 1972), in which the nucleation process requires a less precise order in the calcium-binding groups than was previously suggested (Wilbur and Simkiss, 1968).

In summary, our knowledge of the processes involved in shell formation is very incomplete. Experiments should be designed with some knowledge of where and when net shell deposition is taking place. Very little information is available about the synthesis of the organic matrix and how this synthesis is controlled. If the matrix has a primary role in shell formation, then information regarding its synthesis is of paramount importance. In the future, analyses of the organic matrix must be more complete in terms of the primary components and must show how different fractions are assembled in relation to one another and to the mineral phase of the shell.

3. Mechanisms of Shell Dissolution

3.1. Shell Dissolution

Shell formation is often considered to be a unidirectional deposition of mineral and matrix. However, it has been known for over 50 years that some bivalves dissolve previously deposited shell when they become anaerobic. This shell solution has been evidenced by increased concentrations of calcium and carbon dioxide in body fluids and soft tissues of animals that had been removed from the water (Collip, 1920, 1921; Dotterweich and Elssner, 1935; Dugal, 1939). Dugal (1939) found that shell dissolution in *Mercenaria mercenaria* was caused by an anaerobically produced acid that had a pK similar to that of lactic acid. Crenshaw and Neff (1969) showed that the acid was succinic acid. They found that lactic acid accounted for only 2% of the total acid and that its concentration did not change significantly with time.

Crenshaw and Neff (1969) used analytical techniques that were not available to earlier investigators and were thereby able to infer that shell dissolution began in minutes, rather than days, after the onset of anaerobiosis induced by removing the animals from water. Their analyses of body fluids and soft tissues of *M. mercenaria* showed that there was a stoichiometric relationship between the increase in succinic acid and calcium concentrations. However, analyses of the extrapallial fluid indicated that succinic acid accounted for only 80% of the calcium change in this compartment (see Chapter 6, Fig. 4). This apparent discrepancy may be accounted for by the carbonate and bicarbonate derived from the shell and by ionic exchange with the mantle epithelium.

Shell dissolution is not restricted to subaerially exposed molluscs. Bivalves kept in aerated sea water close their valves periodically and undergo shell dissolution. Crenshaw and Neff (1969) observed this phenomenon by simultaneously measuring the oxygen tension and pH with indwelling microelectrodes and by sampling the extrapallial fluid for calcium (see Chapter 6, Fig. 3). Wijsman (1975) also found that the pH of the extrapallial fluid in *Mytilus edulis* decreased when submerged animals closed their valves from time to time. Gordon and Carriker (1978) observed a much more regular periodicity in the valve closure and concomitant decreases in pH. These observations show that submerged bivalves can also undergo periodic anaerobic shell dissolution. This type of "voluntary" shift to anaerobiosis has been observed in other marine bivalves (Hammen, 1976).

The dissolution of the shell that occurs during anaerobiosis appears to occur primarily inside the pallial line. The chalky appearance of the inner shell surface that characteristically develops with this dissolution is not found outside the pallial line in *M. mercenaria* even after extended periods out of water (Dugal, 1939; Wilkes and Crenshaw, 1979). Scanning electron micrographs reveal that crystals inside the pallial line have irregular edges and are poorly organized with large voids on the inner surface (see Fig. 2). Outside the pallial line, the crystals have sharp edges, are well organized, and fill available space (Fig. 2).* When *Geukensia demissa* is removed from water for as long as 24 hr, shell dissolution also appears to be confined to the area of the shell inside the pallial line (Wilkes and Crenshaw, 1979). Shell dissolution outside the pallial line is not evident even when *G. demissa* is placed in a moist nitrogen atmosphere for 8 hr (Crenshaw, unpublished data). These observations reinforce the hypothesis that there is a functional zonation in the mantle. If dissolution does not generally take place outside the pallial line, ideas concerning the relationship between microgrowth increment formation and thickness and anaerobiosis (Lutz and Rhoads, 1977; Gordon and Carriker, 1978) (also see Chapter 6) may have to be revised or refined.

3.2. Metabolic Aspects of Shell Dissolution

It has been known for some time that certain molluscs can survive for extended periods without a supply of oxygen that is adequate to sup-

---

* One must also bear in mind, however, that the microstructure in the two areas is different. Inside the pallial line, it is predominantly homogeneous and/or complex crossed lamellar, while it is predominantly crossed lamellar and prismatic outside the pallial line.

port aerobic respiration (Newell, 1964; Galtsoff, 1964). This euryoxic* capacity allows certain molluscs to occupy habitats that have great fluctuations in oxygen concentrations, such as the intertidal zone and bottom layers that are periodically anoxic (Theede, 1973). Of course, not all molluscs are euryoxic. The Pectinidae, for example, appear to be strictly aerobic (stenoxic) (Giese, 1969). Here, we shall be primarily interested in the events associated with short periods such as those imposed on bivalves that are exposed by the ebbing tide within each tidal cycle.

During periods of oxygen deprivation, the energy requirements of an animal are supplied by anaerobic glycolysis. The acidic end products of this metabolism must be neutralized to maintain the constant pH required for normal function. For molluscs, the shell is an alkali reserve that is dissolved to neutralize the acid produced. Simpson and Awapara (1966) found that the primary acidic end product of the anaerobic catabolism of glucose by bivalves was succinic acid rather than lactic acid, which is the end product in vertebrates. Further investigations showed that alanine was the other primary end product of glucose degradation in *Rangia cuneata* (Stokes and Awapara, 1968; Chen and Awapara, 1969). The formation of succinate and alanine, for instance, from glucose is advantageous because less total acid is produced than if two moles of lactic acid were formed from one mole of glucose. Subsequent work by a number of investigators proved that the succinate metabolic pathway was common to a number of bivalves and that its utilization could be measured by the accumulation of succinate, alanine, glutamate, and propionate (de Zwaan and Wijsman, 1976; de Zwann et al., 1976). The relative importance of any one end product is not constant, for it can vary with time, species (de Zwann and Wijsman, 1976; de Zwaan et al., 1976), and tissue type (Chaplin and Loxton, 1976).

Most biochemical studies on the anaerobic metabolism of bivalves have been done, of necessity, on isolated tissue in anoxic media. This approach precludes the detection of behavioral adaptations to the intertidal environment. Several bivalve species use oxygen from the air when they are subaerially exposed (Kuenzler, 1961; Boyden, 1972; Bayne et al., 1976). The significance of this aerial respiration can be illustrated by considering the metabolism of *Cerastoderma* (= *Cardium*) *edule*. Gäde (1975) reported that this bivalve accumulates glycolytic end products typical of the succinate pathway when exposed to a nitrogen atmosphere. However, Widdows et al. (1979) found that these anaerobic end products did not accumulate in aerially exposed animals because they obtained sufficient

---

* This term, introduced by Rhoads and Morse (1971), provides a more appropriate description of Metazoans tolerating a wide range of oxygen concentrations than the term "facultative anaerobic," introduced by Hammen (1969). See Hammen (1976) for discussion.

oxygen directly from air to support aerobic metabolism. These contrasting results serve to illustrate that an animal may be able to metabolize glycogen via the succinate pathway, but it does not necessarily indicate that this pathway is used by aerially exposed animals.

In summary, several aspects of shell dissolution requiring further experimental attention are apparent. How common is this phenomenon among bivalves? Direct evidence for shell dissolution has been reported only in those bivalve species that close their valves tightly during exposure to air. What happens to the organic matrix associated with the shell when the mineral phase is dissolved? It may be left on the inner shell surface, as suggested by Lutz and Rhoads (1977) (also see Chapter 6) and Gordon and Carriker (1978), or it may be ingested into lysosomes and digested (Bevelander and Nakahara, 1966). Do all tissues of those bivalve species that obtain oxygen from air obtain enough oxygen to continue aerobic respiration when the animal, as a whole, is aerobic? For example, the outer mantle epithelium could be anaerobic, but the amount of succinate pathway end products from this tissue would not be detected in an analysis of the whole animal.

ACKNOWLEDGMENTS. Without intending to imply approval or disapproval of viewpoints expressed in this chapter, I should like to thank J. G. Carter, R. M. Dillaman, R. A. Lutz, D. Manyak, H. Ristedt, R. D. Roer, K. M. Towe, H. Waite, T. R. Waller, N. Watabe, A. P. Wheeler, and K. M. Wilbur for their fruitful discussions of this material. I should also like to thank R. A. Lutz for critically reviewing the manuscript, Diane Baker for typing it, and the editors for their patience. The preparation of this manuscript was supported, in part, by NIH Grant DE 02668.

References

Bayne, B. L., Bayne, C. J., Carefoot, T. C., and Thompson, R. S., 1976, The physiological ecology of *Mytilus californianus* Conrad. 2. Adaptations to low oxygen tension and air exposure, *Oecologia (Berlin)* **22**:229–250.

Bevelander, G., and Nakahara, H. 1966, Correlation of lysosomal activity and ingestion by mantle epithelium, *Biol. Bull.* **131**:76–82.

Bevelander, G., and Nakahara, H., 1969, An electron microscopic study of the formation of the nacreous layer in the shell of certain bivalve molluscs, *Calcif. Tissue Res.* **3**:84–92.

Boyden, C. R., 1972, Aerial respiration of the cockle *Cerastoderma edule* in relation to temperature, *Comp. Biochem. Physiol.* **43A**:697–712.

Chaplin, A. E., and Loxton, J., 1976, Tissue differences in the response of the mussel, *Mytilus edulis*, to experimentally induced anaerobiosis, *Biochem. Soc. Trans.* **4**:437–441.

Chen, C., and Awapara, J., 1969, Effect of oxygen on the end-product of glycolysis in *Rangia cuneata*, *Comp. Biochem. Physiol.* **31**:395–401.

Collip, J. B., 1920, Studies on molluscan celomic fluid: Effect of change in environment on the carbon dioxide content of the celomic fluid: Anaerobic respiration in *Mya arenaria*, *J. Biol. Chem.* **45**:23–49.

Collip, J. B., 1921, A further study of the respiratory processes in *Mya arenaria*, and other marine molluscs, *J. Biol. Chem.* **49**:297–310.

Crenshaw, M. A., 1972a, The inorganic composition of molluscan extrapallial fluid, *Biol. Bull.* **143**:506–512.

Crenshaw, M. A., 1972b, The soluble matrix from *Mercenaria mercenaria* shell, *Biomineralisation* **6**:6–11.

Crenshaw, M. A., and Neff, J. M., 1969, Decalcification at the mantle–shell interface in molluscs, *Am. Zool.* **9**:881–885.

Crenshaw, M. A., and Ristedt, H., 1975, Histochemical and structural study of nautiloid septal nacre, *Biomineralisation* **8**:1–15.

Crenshaw, M. A., and Ristedt, H., 1976, The histochemical localization of reactive groups in the septal nacre from *Nautilus pompilius* L., in: *The Mechanisms of Mineralization in the Invertebrates and Plants* (N. Watebe and K. M. Wilbur, eds.), pp. 335–367, University of South Carolina Press, Columbia.

de Zwaan, A., 1977, Anaerobic energy metabolism in bivalve molluscs, *Annu. Rev. Oceanogr. Mar. Biol.* **15**:103–187.

de Zwaan, A., and Wijsman, T. C. M., 1976, Anaerobic metabolism in Bivalvia (Mollusca). I. Characteristics of anaerobic metabolism, *Comp. Biochem. Physiol.* **54B**:313–324.

de Zwaan, A. K., Kluytmans, J. H. F., and Zandee, D. I., 1976, Facultative anaerobiosis in molluscs, *Biochem. Soc. Symp.* **41**:133–168.

Dotterweich, H., and Elssner, E., 1935, Die Mobilisierung des Schalenkalkes für die Reaktionsregulation der Muscheln (*Anodonta cygnea*), *Biol. Zentralbl.* **55**:138–163.

Dugal, L. P., 1939, The use of calcareous shell to buffer the product of anaerobic glycolysis in *Venus mercenaria*, *J. Cell. Comp. Physiol.* **13**:235–251.

Erben, H. K., 1972, Über die Bildung und das Wachstum von Perlmutt, *Biomineralisation* **4**:15–46.

Gäde, G., 1975, Anaerobic metabolism of the common cockle, *Cardium edule*. I. Utilization of glycogen and accumulation of multiple end-products, *Arch. Int. Physiol. Biochim.* **83**:879–896.

Galtsoff, P. S., 1964, The American oyster, *Crassostrea virginica*, Fisheries Bulletin Vol. 64, U.S. Government Printing Office, Washington, D.C.

Giese, A. C., 1969, A new approach to the biochemical composition of the molluscs' body, *Oceanogr. Mar. Biol. Annu. Rev.* **7**:175–229.

Glazer, A. N., DeLange, R. J., and Sigman, D. S., 1975, *Chemical Modification of Proteins*, pp. 13–21, North-Holland, Amsterdam.

Gordon, J., and Carriker, M. R., 1978, Growth lines in a bivalve mollusk: Subdaily patterns and dissolution of the shell, *Science* **202**:519–521.

Grégoire, C., 1967, Sur la structure des matrices organiques des coquilles de mollusques *Biol. Rev.* **42**:653–688.

Grégoire, C., 1972. Structure of the molluscan shell, in: *Chemical Zoology*, Vol. VII, *Molluscs* (M. Florkin and B. Scheer, eds.), pp. 45–102, Academic Press, New York.

Grégoire, C., Duchateau, G., and Florkin, M., 1955, La trame protidique des nacres et des perles, *Ann. Inst. Oceanogr.* **31**:1–36.

Hammen, C. S., 1969, Metabolism of the oyster, *Crassostrea virginica*, *Am. Zool.* **9**:309–318.

Hammen, C. S., 1976, Respiratory adaptations: Invertebrates, in: *Estuarine Processes I* (M. Wiley, ed.), pp. 347–355, Academic Press, New York.

Hunt, S., 1970, *Polysaccharide–Protein Complexes in Invertebrates*, pp. 250–262 and 270–274, Academic Press, New York.

Istin, M., and Maetz, J., 1964, Permeabilité au calcium du manteau de lamellibranches d'eau douce etudiée des isotopes ^{45}Ca et ^{47}Ca, *Biochim. Biophys. Acta* **88**:225–227.

Iwata, K., 1975, Ultrastructure of the conchiolin matrices in molluscan nacreous layer, *J. Fac. Sci. Hokkaido Univ. Ser. 4* **17**:173–229.

Jodrey, L. H., 1953, Studies on shell formation. III. Measurement of calcium deposition in shell and calcium turnover in mantle tissue using the mantle–shell preparation and Ca^{45}, *Biol. Bull.* **104**:398–407.

Jodrey, L. H., and Wilbur, K. M., 1955, Studies on shell formation. IV. The respiratory metabolism of the oyster mantle, *Biol. Bull.* **108**:346–358.

Kawaguti, S., and Ikemoto, N., 1962, Electron microscopy on the mantle of a bivalve *Fabulina nitidula*, *Okayama Univ. Biol. J.* **8**:21–30.

Kniprath, E., 1971, Die Feinstruktur des Drusenpolsters von *Lymnaea stagnalis*, *Biomineralisation* **3**:1–11.

Krampitz, G., Engels, J., and Cazaux, C., 1976, Biochemical studies on water-soluble proteins and related components of gastropod shells, in: *The Mechanism of Mineralization in the Invertebrates and Plants* (N. Watabe and K. M. Wilbur, eds.), pp. 155–193, University of South Carolina Press, Columbia.

Kuenzler, E. S., 1961, Structure and energy flow of a mussel population in a Georgia salt marsh, *Limnol. Oceanogr.* **6**:191–204.

Lutz, R. A., and Rhoads, D. C., 1977, Anaerobiosis and a theory of growth line formation, *Science* **198**:1222–1227.

Machen, T. E., Erlij, D., and Wooding, F. B. P., 1972, Permeable junctional complexes: The movement of lanthanum across rabbit gallbladder and intestine, *J. Cell Biol.* **54**:302–312.

Machin, J., 1977, Role of integument in molluscs, in: *Transport of Ions and Water in Animals* (B. L. Gupta, R. B. Moreton, J. L. Oschman, and B. L. Wall, eds.), pp. 735–762, Academic Press, New York.

Matheja, J., and Degens, E. T., 1968, Molekulare Entwicklung mineralisations-fähiger organischer Matrizen, *Neves Jahrb. Geol. Palaeontol. Abh.* **4**:215–229.

Neff, J. M., 1972, Ultrastructure of the outer epithelium of the mantle in the clam, *Mercenaria mercenaria*, in relation to calcification of the shell, *Tissue Cell* **4**:591–600.

Neuberger, A., Gottschalk, A., and Marshall, R. D., 1966, Carbohydrate–peptide linkages in glycoproteins and methods for their elucidation, in: *Glycoproteins* (A. Gottschalk, ed.), pp. 273–295, Elsevier, Amsterdam.

Neumann, W. F., and Neuman, M. F., 1958, *The Chemical Dynamics of Bone Mineral*, pp. 23–38, University of Chicago Press, Chicago.

Newell, G. E., 1964, Physiological aspects of the ecology of intertidal molluscs, in: *Physiology of Mollusca*, Vol. I (K. M. Wilbur and C. M. Yonge, eds.), pp. 59–81, Academic Press, New York.

Pashley, D. W., 1970, Recent developments in the study of epitaxy, in: *Recent Progress in Surface Science III* (J. F. Danielli, A. C. Riddiford, and M. D. Rosenberg, eds.), pp. 291–332, Academic Press, New York.

Rhoads, D. C., and Morse, J. W., 1971, Evolutionary and ecologic significance of oxygen-deficient marine basins, *Lethaia* **4**:413–428.

Saleuddin, A. S. M., 1970, Electron microscopic study of the mantle of normal and regenerating *Helix*, *Can. J. Zool.* **48**:409–416.

Shultz, R. D., and Asumaa, S. K., 1970, Ordered water and the ultrastructure of the cellular plasma membrane, in: *Recent Progress in Surface Science III* (J. F. Danielli, A. C. Riddiford, and M. D. Rosenberg, eds.), pp. 291–332, Academic Press, New York.

Simpson, J. W., and Awapara, J., 1966, The pathway of glucose degradation in some vertebrates, *Comp. Biochem. Physiol.* **18**:537–548.

Stokes, T. M., and Awapara, J., 1968, Alanine and succinate as end-products of glucose degradation in the clam, *Rangia cuneata, Comp. Biochem. Physiol.* **25**:883–892.

Theede, H., 1973, Comparative studies on the influence of oxygen deficiency and hydrogen sulfide on marine bottom invertebrates, *Neth. J. Sea Res.* **7**:244–252.

Thiele, H., 1967, Geordnete Kristallisation, *J. Biomed. Mater. Res.* **1**:213–231.

Towe, K. M., 1972, Invertebrate shell structure and the organic matrix concept, *Biomineralisation* **4**:1–14.

Towe, K. M., and Hamilton, G. H., 1968, Ultrastructural and inferred calcification of mature and developing nacre in bivalve mollusks, *Calcif. Tissue Res.* **1**:306–318.

Tsujii, T., 1968, Studies on the mechanism of shell and pearl formation. X. The submicroscopic structure of the epithelial cells on the mantle of pearl oyster, *Petria martensii, Rep. Fish. Mie Univ.* **6**:41–58.

Tsujii, T., 1976, An electron microscopic study of the mantle epithelial cells of *Anodonta* sp. during shell regeneration, in: *The Mechanisms of Mineralization in the Invertebrates and Plants* (N. Watabe and K. M. Wilbur, eds.), pp. 339–353, University of South Carolina Press, Columbia.

Voss-Foucart, M. F., Laurent, C., and Grégoire, C., 1969, Sur les constituants organiques des coquilles d'etherides, *Arch. Int. Physiol. Biochim.* **77**:901–923.

Wada, K., and Fujinuki, T., 1976, Biomineralization in bivalve molluscs with emphasis on the chemical composition of the extrapallial fluid, in: *The Mechanisms of Mineralization in the Invertebrates and Plants* (N. Watabe and K. M. Wilbur, eds.), pp. 175–190, University of South Carolina Press, Columbia.

Weiner, S., and Hood, L., 1975, Soluble protein of the organic protein matrix of mollusk shells: A potential template for shell formation, *Science* **190**:987–989.

Weiner, S., Lowenstein, H. A., and Hood, L., 1977, Discrete molecular weight components of the organic matrices of molluscan shells, *J. Exp. Mar. Biol. Ecol.* **30**:45–51.

Wheeler, A. P., 1975, Oyster mantle carbonic anhydrase: Evidence for plasma membrane-bound activity and for a role in bicarbonate transport, Ph.D. thesis, Duke University, Durham, North Carolina.

Wheeler, A. P., and Wilbur, K. M., 1977, Shell growth in the scallop *Argopecten irradians.* II. Processes of shell growth, *J. Mollusc Stud.* **43**:155–161.

Wheeler, A. P., Blackwelder, P. L., and Wilbur, K. M., 1975, Shell growth in the scallop *Argopecten irradians.* I. Isotope incorporation with reference to diurnal growth, *Biol. Bull.* **148**:472–482.

Whittembury, G., and Rawlins, F. A., 1971, Evidence of a paracellular pathway for ion flow in the kidney proximal tubule: Electronmicroscopic demonstration of lanthanum precipitate in the tight junction, *Pfluegers Arch.* **330**:302–309.

Widdows, J., Bayne, B. L., Livingstone, D. R., Newell, R. I. E., and Donkin, P., 1979, Physiological and biochemical responses of bivalve molluscs to exposure to air, *Comp. Biochem. Physiol.* **62A**:301–308.

Wijsman, T. C. M., 1975, pH fluctuations in *Mytilus edulis* L. in relation to shell movements under aerobic and anaerobic conditions, in: *The Biochemistry, Physiology and Behavior of Marine Organisms in Relation to Their Ecology* (H. Barnes, ed.), pp. 139–149, University of Aberdeen Press, Aberdeen, Scotland.

Wilbur, K. M., 1964, Shell formation and regeneration, in: *Physiology of Mollusca I* (K. M. Wilbur and C. M. Yonge, eds.), pp. 243–282, Academic Press, New York.

Wilbur, K. M., 1972, Shell formation in mollusks, in: *Chemical Zoology,* Vol. VII, *Molluscs* (M. Florkin and B. Scheer, eds.), pp. 103–145, Academic Press, New York.

Wilbur, K. M., and Jodrey, L. H., 1952, Studies on shell formation. I. Measurement of the rate of shell formation using ^{45}Ca, *Biol. Bull.* **103**:269–276.

Wilbur, K. M., and Simkiss, K., 1968, Calcified shells, in: *Comprehensive Biochemistry 26A* (M. Florkin and E. H. Stotz, eds.), pp. 229–295, Elsevier, Amsterdam.

Wilkes, D. A., and Crenshaw, M. A., 1979, Formation of a dissolution layer in molluscan shells, in: *Scanning Electron Microscopy/1979/II* (O. Johari and R. P. Becker, eds.), pp. 469–474, SEM, O'Hare, Illinois.

Wise, S. W., Jr., 1970, Microarchitecture and deposition of gastropod nacre, *Science* **167**:1486–1488.

Woodward, C., and Davidson, E. A., 1968, Structure–function relationships of protein polysaccharide complexes: Specific ion-binding properties, *Proc. Natl. Acad. Sci. U.S.A.* **60**:201–205.

Zischke, J. A., Watabe, N., and Wilbur, K. M., 1970, Studies on shell formation: Measurement of growth in the gastropod *Ampullarius glaucus, Malacologia* **10**:423–439.

Chapter 4

An Ontogenetic Approach to the Environmental Significance of Bivalve Shell Chemistry

GARY D. ROSENBERG

1. Introduction 133
2. Chemical Variations in the Shell 135
    2.1. Calcium and Organic Matter 135
    2.2. Minor and Trace Elements 139
3. Recognition of the Original Elemental Composition of the Shell and Its
    Environmental and Physiological Significance 155
4. Summary 161
    References 162

1. Introduction

Bivalved molluscs are common constituents of fossil assemblages from the early Paleozoic to the Recent, but their chemical evolution has yet to be described. Not one evolutionary trend has been established beyond reasonable doubt for the distribution of any element within any bivalve taxon. We do not know how the chemical compositions of Recent bivalves compare with those of their fossil ancestors, nor do we have adequate knowledge of how the chemical composition of a species changes from habitat to habitat or throughout ontogeny.

Variations in the concentration of calcium, magnesium, strontium, barium, boron, sulfur, amino acids, sodium, and other minor and trace constituents within the bivalve shell have been variously described as

GARY D. ROSENBERG ● Geology Department, Michigan State University, East Lansing, Michigan 48824.   *Present address:* Geology Department, Indiana University/Purdue University, Indianapolis, Indiana 46202.

133

being species-specific indicators of growth rate or records of a species's response to changes in temperature, salinity, oxygen/carbon dioxide ratio, tides, or illumination. However, generalizations relating shell chemical composition to environmental or physiological events have been compromised by a lack of clear-cut repeatable trends and conflicting data. Species-specific effects, complicated interactions of dependent environmental variables, and diagenesis have all been invoked to explain the confusion.

As Raup and Stanley (1971) wrote in the preface to their book, *Principles of Paleontology*: "Our omission of topics that might together be called 'biogeochemistry' may draw criticism, but we believe that the many isotopic and trace element approaches undertaken during the past two decades have thus far contributed little to general paleontologic knowledge largely because of the thorny problems imposed by diagenetic alteration." When we consider what has, and has not, been done in the field of bivalve biogeochemistry, this statement is a bold one indeed. Diagenesis is frequently the mechanism used to explain apparently random chemical data. However, so little is known about the original chemical composition of extant or fossil bivalve shells that it is at present not an easy matter to evaluate chemical diagenesis.

In this chapter, I will attempt to summarize, and explain shortcomings of, past work on bivalve shell chemistry. Difficulties in interpreting this work arise from (1) methods of sampling the shell; (2) data sets that are often conflicting or are of limited applicability; (3) interpretations based on invalid assumptions; and (4) a lack of criteria for distinguishing diagenetic alterations from variations in shell chemistry related to an organism's response to the environment.

Most data on shell chemistry are derived from whole-shell, or whole-layer, analyses, or from (apparently) randomly located sampling positions within the shell. While there is no *a priori* reason to believe that such analyses should not be typical of a species, it so happens that they apparently are not. Data from the few workers who have studied ontogenetic changes in shell chemistry show that (1) analyses of parts of the shell are not necessarily typical of the whole shell and (2) a whole-shell analysis represents the elemental concentration incorrectly; elements are not necessarily distributed uniformly throughout the shell. Thus, any similarities or differences in average chemical composition among species may merely reflect inadequate sampling procedures.

The environmental and evolutionary significance of the chemical composition of bivalve shells is poorly understood; not one single bivalve species has been completely or adequately described chemically. There is evidence that the Ca, Sr, Mg, and organic matter concentrations within the shell change throughout ontogeny. The shell may be composed of trace amounts of sulfates and hydroxides, as well as metal carbonates other than $CaCO_3$, complicating predictions of elemental distribution based on stoichiometric considerations alone. The amino acid concentration varies from layer to layer. Our knowledge, however, is fragmentary, and detailed maps of these various shell components are lacking. Distributions along dorsal–ventral and anterior–posterior shell axes are poorly defined. Variations from specimen to specimen, from population to population, and from species to species have not been adequately quantified.

In the following review, I will discuss some of the scanty, but important, data on the ontogenetic variation of elements within the bivalve shell. I will attempt to demonstrate that neglect of ontogenetic principles is the reason fractional and whole-shell analyses have led to tenuous correlations between environmental variables and shell chemistry. Finally, I will demonstrate how allometric variations in shell composition may be used to test the hypothesis that diagenesis has obscured original elemental distributions within the shell.

2. Chemical Variations in the Shell

2.1. Calcium and Organic Matter

Calcium and organic matter are considered together because they are the major shell components. Ironically, two opposing and incorrect assumptions are found throughout past research regarding their relative distribution. First, the fluctuations in one are assumed to be inversely proportional to the other. This assumption is especially common in discussions of bivalve growth increment composition, as will be seen below. Second, Ca is sometimes assumed to be distributed uniformly throughout the shell (e.g., Pilkey and Goodell, 1964). This assumption is most apparent every time a researcher reports trace- and minor-element ratios relative to Ca. The Ca concentration is almost never measured, and the variability of the ratio relative to fluctuations in Ca is never assessed. However, the alternation of $CaCO_3$ and organic layers implies that $CaCO_3$ can vary locally from 0 to 100%. Even across a transect of the entire shell, the nonuniform distribution of Ca has been known for many years. Fox and Coe (1943) found that young *Mytilus californianus* specimens have less than 87% (ash dry weight) inorganic matter, while older specimens have more than 90%. Certain other species also show increasing Ca concentration with age (Vinogradov, 1953).

In the living Pacific coast bivalve *Chione undatella*, Rosenberg (1972, 1973) found that, with age, Ca concentration increased (a maximum of

3%) in 10 specimens, was uniform in 5 specimens, and decreased 2% in 1 specimen.*

The variations in Ca throughout ontogeny in *Chione* may be a result of two factors. First, variations may reflect real ontogenetic differences among individual specimens. Second, the line of section along which Ca was measured could have deviated from the medial dorsal–ventral axis. Calcium (and other element) incorporation rates vary with position along the ventral margin (see Figs. 1 and 7A), and any deviation in the position of the section could result in *apparent* ontogenetic variations. If the observed ontogenetic change in calcium concentration is determined to be real, this observation may support the observations of Hare and Abelson (1965), who determined (based on whole-shell analyses) that "phylogenetically advanced" bivalves have a higher Ca/organic matter ratio than "primitive" species. For example, the Veneridae were found to contain only 10% (by weight) of the shell protein found in the Mytilidae. It seems reasonable that a phylogenetic trend of increasing Ca concentration should parallel an ontogenetic one. However, this conclusion requires substantiation. Phenotypic differences between the venerids and mytilids will also affect the concentration of Ca within their shells. These phenotypic differences must be separated from true phylogenetic and evolutionary differences.

If the trends differing from the mode are real, they may also be explained by allometric variations in carbonate incorporation along the ventral margin. Wheeler *et al.* (1975) found that carbonate deposition in *Argopecten irradians* was highest at the midventral margin and declined dorsally, anteriorly, and posteriorly (Fig. 1). Furthermore, they state that "the rate of [carbonate] incorporation at the ventral shell edge did not change with increase in shell size." This means that while the rate of increase in size declines with age (and may vary seasonally and with other factors), the rate of carbonate secretion may remain constant. It follows that the Ca content in a unit area of the shell is a function of two variables that can be considered independent, within limits; a unit change in Ca secretion need not result in a *uniformly* proportional change in growth rate. Consequently, we might predict that ontogenetic patterns of Ca concentration could vary among specimens living in different habitats and growing at different rates.

Moreover, the Ca and organic concentration in the shell need not fluctuate inversely. Radiographs of shells (Wilbur, 1964; Omori *et al.* 1974) show the distribution of X-ray-opaque and X-ray-translucent shell material. Regions of slow growth are sometimes X-ray-opaque and some-

---

* It is interesting that the last specimen was taken from a habitat in which the clams grew the largest, as measured along the dorsal–ventral axis.

Figure 1. Variation of the rate of incorporation of [^{14}C]carbonate with position along the shell margin of *Agropecten irradians*. The values show the mean counts per minute per milligram of shell sample for a 4-hr incubation for 10 animals. The medium contained 20 μCi/liter [^{14}C]bicarbonate. In addition to the rate of incorporation, the rate of growth must also be known to accurately predict concentration of Ca or carbonate within the shell. Also see Fig. 7A. Modified from Wheeler *et al.* (1975) with the permission of the *Biological Bulletin (Woods Hole, Massachusetts)* and the authors; figure provided by K. M. Wilbur.

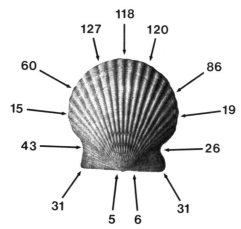

times translucent. It is not yet clear that the absolute concentration of Ca is the same in all translucent zones or in all opaque zones.

Thus, while CaCO$_3$ and organic matter are the two most prominent components of the bivalve shell, one has reason to ask whether there is a gradation in composition between the two extremes. This is a complication that is generally passed over in discussing shell composition. As indirect evidence, consider that basophilic and eosinophilic "matrices" may be distributed with no apparent direct relationship to each other (Omori *et al.*, 1974). The exact role of each in increasing or retarding mineralization has yet to be defined. However, if hematoxylin-staining areas were always Ca-rich and eosin-staining areas were always organic-rich, as Okada (1943) proposed for vertebrate and invertebrate accretionary tissues, one could claim with greater justification that the bivalve shell consists of only two major components. Shell metabolism could thus be completely explained in terms of the Ca-rich and organic-rich end-members. Moreover, Rosenberg (1972) described Ca fluctuations that were superimposed on an ontogenetic trend of increasing Ca concentration, as measured along a dorsal–ventral axis of the shell of *Chione undatella*. In the study of Rosenberg (1972), the Ca concentration was found to increase markedly after a period of gradual decline in *some* specimens. Furthermore, Ca variation was not necessarily in phase with structural growth patterns [also see Rosenberg and Jones (1975) and Fig. 2].

Even at a fine scale of resolution, it has not been proven that growth increments are always uniform alternations of carbonate- and organic-rich layers; the terms used by Pannella (1975), "organic lines" and "organic layers," are only interpretations. No one has yet measured the difference in Ca concentration among structural components of all increments. Although electron micrographs show organic lamellae alternating with in-

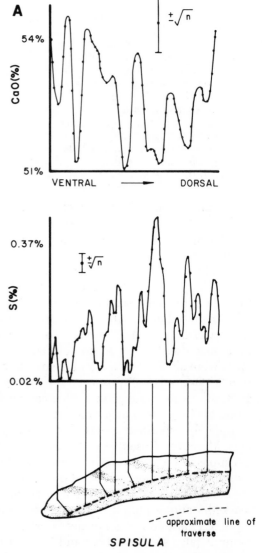

Figure 2. (A) Distribution of Ca and S at the base of the outer shell layer in *Spisula* sp. as measured with the electron microprobe. Data smoothed by recombination of frequencies with the highest amplitudes in a Fourier transform of the original data series (also see Appendix 3.A). (B) S and Ca tidal rhythms in the outer shell layer of *Cardium edule*. Three repeated microprobe traverses (A, B, C and A[1], B[1], C[1]) of the same tidal series are related to patterns of narrow (N) and broad (B) growth increments. Data were smoothed as described in (A) above. (a–k, m–z) Cycles of concentration that are repeated in the three traverses. Note that Ca, S, and structural growth cycles do not maintain a constant phase difference in either *Spisula* or *Cardium*. After Rosenberg and Jones (1975).

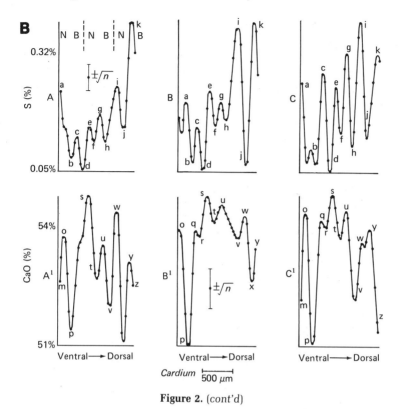

Figure 2. (cont'd)

organic layers (e.g., Pannella, 1975), it is not clear that these organic lamellae are entirely or *uniformly* free of Ca (also see Taylor *et al.*, 1969).
In addition, considering that both long- and short-term fluctuations in Ca
concentration are present in bivalve shells, it would be interesting to know
if there are also ontogenetic variations in the amino acid composition.
Previous research has established that amino acid content varies among
young and old specimens, among shell layers, and among shells of different species (Akiyama and Wyckoff, 1970; Bricteux-Grégoire, 1968;
Grégoire, 1972; Hare, 1962, 1963; Florkin, 1966; Wada, 1966; Weiner et
al., 1977; Wilbur and Simkiss, 1968). Although these researchers do not
describe ontogenetic amino acid variations in single shells, the systematic
variability that they note gives one reason to hope that growth trends in
the concentration of amino acids can also be quantified.

2.2. Minor and Trace Elements

Numerous minor and trace elements have been found in a wide range
of bivalve shells (Clarke, F. W., and Wheeler, 1922; Vinogradov, 1953;

Graham, 1972; Ohta, 1977; Brooks and Rumsby, 1965; Segar *et al.*, 1971; Milliman, 1974; Wolf *et al.*, 1967; Kobayashi, 1975). However, the taxonomic distribution of only a few elements has been studied thoroughly, and our knowledge of their distribution within shells, as well as their environmental and evolutionary significance, is fragmentary. One or another of these elements may yet prove to be of use as an indicator of paleotemperature, paleosalinity, or physiological state, although our present understanding suggests numerous complications that can be resolved only by ontogenetic studies.

For example, the distribution of elements such as Mg, Fe, Sr, P, S, Na, and Ca has been measured normal to the outer and inner shell surfaces (i.e., across shell layers), and claims have been made that observed differences are typical of whole layers (Kobayashi, 1975; Estes, 1972). However, given the possibility of ontogenetic variations within each layer, it is not clear what the differences are along all possible axes of section (Fig. 3). The determination of elemental distributions throughout ontogeny within single layers is thus necessary.

Second, trace and minor elements are associated with various shell structural components (also see Dodd, 1967). Such elements may be bound directly to conchiolin or may be present in organometallic pigments (Comfort, 1951; Fox, 1966). Alternatively, they may substitute for Ca or C within the inorganic fraction as carbonates, hydroxides, oxides, or sulfates, or they may simply be adsorbed onto the shell surface. Unfortunately, the differentiation of components can be difficult at low concentrations; this could explain why high-Mg calcite, rather than dolomite, has been reported in other invertebrates using X-ray analyses (Wyckoff, 1972). The problem of recognizing different chemical shell components means that stoichiometric relationships determined from calculations of total concentration of the element may not be maintained. Physiologically, this could mean that the shell is a much more complex and dynamic exoskeleton than previously thought. In addition, distribution of an element within a shell could appear to be random and show neither ontogenetic regularity nor significant correlation with environmental changes unless the components in which the element exists are considered separately. On the other hand, this is all the more reason to question whether the mere presence of an element in the shell can be assumed to have physiological or environmental implications. Four examples—Mn, S, B, and Pb—will be discussed.

Manganese, like other trace metals, is bound within pigments or is part of the mineralogical structure of carbonates and oxides (Comfort, 1951; Fox, 1966; Horiguchi, 1959a,b; Horiguchi and Tsujii, 1967). One of the most interesting recent technical breakthroughs in the chemical analysis of shells has resulted from electron paramagnetic resonance studies.

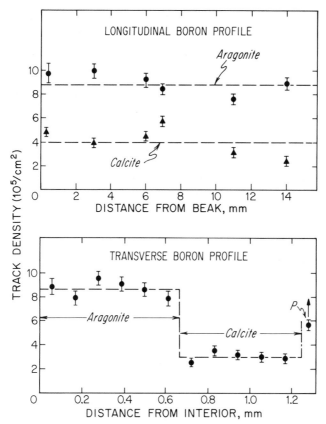

Figure 3. Distribution of boron in the aragonitic (inner) and calcitic (outer) shell layers of *Mytilus edulis* as measured by neutron irradiation–alpha emission. Track density is proportional to concentration. The average absolute concentration of B in aragonite is approximately 9 ppm. (A) Transect within each shell layer beginning at the beak and moving toward the posterior margin at right. (B) Transect extending from the interior growth surface of the inner aragonitic nacreous layer (left) to the periostracum (P) at right. The differences in B concentration among the shell layers will vary with the position at which the transect is taken due to ontogenetic variations within each layer. After Furst *et al.* (1976) with the permission of *Geochimica et Cosmochimica Acta* and the authors.

Two papers have shown that Mn^{2+} can substitute for Ca^{2+} in the aragonite lattice of *Mya arenaria* (White *et al.*, 1977) or in the calcite lattice of *Mytilus edulis* (Blanchard and Chasteen, 1976). The former paper also stated that "substitution of Mn^{2+} for Ca^{2+} in (aragonite) apparently does not appear to be possible in normal geological processes." The latter study showed that the amount of Mn^{2+} substituting for Ca^{2+}, rather than total Mn, was correlated with the tide level at which the animal lived. The work requires additional confirmation, since few specimens were studied.

If this relationship is real, it is still not clear why specimens growing above the mean low water level (MLW) should have a higher Mn^{2+}/Mn_{total} ratio than those growing below the MLW. The duration of an organism's exposure to either the atmosphere or well-oxygenated intertidal waters might be important factors to consider. In any event, one would think that the relationship between shell Mn^{2+} and oxygenation would be complicated by those variables (e.g., temperature, nutrition, salinity) that affect the total Mn concentration. Only an ambiguous correlation between total shell Mn concentration and salinity has been established (Horiguchi, 1959a,b; Girardi and Merlini, 1963; Schelski, 1964; Horiguchi and Tadashi, 1967; Harvey, 1969; Crisp, 1975, 1976; Eisma et al., 1976). Mn_{total} tends to increase inversely with salinity, but the relationship is by no means definitive in all species. Manganese concentration may be independent of salinity in same species; Chipman and Schommers (1968) suggested that the activity of shell-surface bacteria was important in controlling the concentration of Mn within the shell.

Sulfates are another group of compounds indicating the complexity of chemical distributions associated with shell structural components. Sulfated compounds were detected long ago in the shells of bivalves (Clarke, F. W., and Wheeler, 1922; Vinogradov, 1953). Despite their more recent detection in bivalves (Kaplan et al., 1963; Travis, 1968) and the recognition that sulfate groups may be involved in the initiation of mineralization (Crenshaw, 1972; Wada and Furuhashi, 1970), the importance of their presence in the shell is not universally accepted (Gunatilaka, 1975). Travis (1968) provided evidence for the presence of Ba and Sr sulfates and carbonates in Mytilus. Kaplan et al. (1963) measured up to 0.26% shell S (as sulfate) in the genus Chlamys. Rosenberg and Jones (1975) found tidal variations in S concentration in Cardium edule with the electron microprobe (see Fig. 2B). The electron microprobe measures total S, not only sulfate S. Sulfur is also present in the shell amino acids, cystine and methionine (Hare, 1969), and in the chondroitin sulfate fraction of protein-polysaccharides. To the extent that amino acids and chondroitin sulfate are intimately bound with carbonate, and not immediately combusted during analysis, they may also be measured by the microprobe. Nevertheless, it is interesting to note that while S_{total} rhythms could be detected, they were not in phase with the Ca rhythms (Rosenberg and Jones, 1975) (also see Fig. 2). This provides additional evidence that the fluctuation of one chemical component (carbonate) is not necessarily inversely proportional to the fluctuation of another (S_{total}).[*]

---

[*] An inverse fluctuation might be expected if S_{total} were an index of organic matter abundance, according to some models (see the previous section).

Boron is an example of an element that may substitute for C rather than Ca. Furst *et al.* (1976) studied the B concentration in Recent mollusc shells (see Fig. 3). Boron appears to be concentrated in the inorganic shell fraction, and is more abundant in aragonite than in calcite. While B has a small ionic radius relative to that of Ca, and therefore might be expected to be more abundant in calcite than in aragonite, Furst *et al.* (1976) suggested that B substitutes for C rather than for Ca. The B concentration within either the calcitic or the aragonitic shells examined seemed to increase with salinity. However, the authors admitted that unassessed effects of temperature and nutrition on B concentration could complicate the relationship between salinity and B concentration. Cook (1977) agreed with the possibility that B was substituting for C. He also observed that the B concentration decreased with (geological) age in Holocene oyster shells from Queensland, but doubted that this was evidence for a lower salinity in the Holocene environment.

Lead is an example of an element that can be adsorbed onto the shell. Ferrel *et al.* (1973) determined that Pb concentration in *Crassostrea virginica* was proportional to its concentration in seawater. However, A. W. Clarke and J. H. Clarke (1974) and J. H. Clarke *et al.* (1976) studied Pb accumulation in shells from dead *Corbicula manillensis* in rivers. They claimed that Pb was adsorbed onto shells within the death assemblage in proportion to the Pb concentration in the rivers; populations closer to the source of Pb pollution were reported to have higher Pb concentrations than those farther away from the Pb source. However, the amount of Pb adsorbed onto the shells before death and the rate of Pb accumulation on dead shells within each environment were poorly determined. Nevertheless, the fact that shells are able to adsorb cations should be an important consideration to those who would calculate shell composition from stoichiometric relationships, or who would suggest that the presence of an element implies that it is structurally bound within the shell.

In conclusion, the observations discussed in this section are an enumeration of a list of cause–effect relationships between the environment and shell chemistry. An important variable that is seldom considered in these kinds of studies is the organism's ontogeny. At different times during its life, the organism's metabolism will respond differently to the same physiological or environmental variable.

2.2.1. Magnesium

Magnesium and strontium have been the most thoroughly studied elements in the bivalve shell. The cause–effect relationships between environmental change and shell concentrations of these elements have been actively pursued. A good example is the study of Moberly (1968) on the

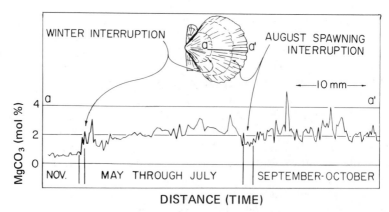

Figure 4. Distribution of magnesium along a transect (a–a') in the outer shell layer of *Aequipecten irradians*, as measured with the electron microprobe. Note the gradual increase in Mg concentration with age. After Moberly (1968).

distribution of Mg in the outer calcitic layer of *Aequipecten irradians* from New Jersey (Fig. 4). In his electron-microprobe analysis of ontogenetic variation of shell Mg, Moberly (1968) attempted to establish cause–effect relationships, rather than to consider ontogenetic variations for their own merit. He claimed that Mg concentration increased with the growth rate, not with temperature, as would be expected.* That is, the Mg concentration seemed to decrease during the growth slowdown before spawning in late summer–early fall, and the concentration seemed to increase just after the winter interruption in growth. However, Moberly (1968) inaccurately stated that a general relationship exists between growth rate and Mg abundance because his Mg data show a gradual but clear increase with age and reduced growth rate. Moreover, his criteria are not clear for distinguishing between prespawning and winter slowdowns on the basis of external growth increments. One would also like to know how these relationships vary from specimen to specimen.

Additionally, it is interesting that Moberly (1968) finds the Mg concentration to be locally as high as 4 mol% within both *Aequipecten* and *Crassostrea*. The determinations of other workers of Mg concentrations in closely related species in similar environments yield values much less than 1 mol% Mg (Turekian and Armstrong, 1960; Lowenstam, 1963). However, it must be emphasized that the studies cited are based on whole-shell determinations. A Mg concentration of 4 mol% is near the upper limit for "low-Mg" calcite. Bivalves are unusual in that they typically have very low-Mg calcite (and aragonite) shells relative to those of other

---

* This contradiction is in reference to a relationship between temperature and Mg concentration that will be discussed more fully later.

invertebrates (Chave, 1954; Lowenstam, 1963; Dodd, 1967; Wolf et al., 1967; Wyckoff, 1972). This observation prompted Lowenstam (1963) to state that bivalves had evolved an enhanced ability to discriminate against Mg.* This may provide an adaptive advantage because $MgCO_3$ is more soluble than $CaCO_3$ (Milliman, 1974). However, the determination by Moberly (1968) of locally high Mg concentrations in *Aequipecten*, and the fact that most other determinations have been based on whole-shell analyses (obscuring localized maximum abundances), prompts one to ask just how low in Mg bivalve shells really are. At least it underscores the importance of determining ontogenetic variations in elemental concentration.

The importance of this ontogenetic theme is also seen in studies that compare species or shell layers of similar mineralogy and that attempt to subtract crystallographic effects from environmental or evolutionary influences on Mg concentration. F. W. Clarke and Wheeler (1922) noted that the Mg concentration in biogenic calcites is greater than in aragonites. The ionic radius of Mg^{2+} is smaller than that of Ca^{2+}. Because there is less room for the metal ion in the calcite lattice, Mg^{2+} substitutes for Ca^{2+} in the calcite lattice more readily than in aragonite. This relationship among mineralogy, cation substitution, and concentration has been emphasized for Mg more than for any other minor or trace element (Bøggild, 1930; Vinogradov, 1953; Chave, 1954; Harriss, 1965; Dodd, 1965, 1966; Wolf et al., 1967; Gunatilaka, 1975). Nevertheless, from time to time, exceptions do appear. Kado (1960) reported that Mg concentration in the entire calcitic prismatic region of *Pinctada martensii* is less than that in the aragonitic nacreous region. Turekian and Armstrong (1960) found that "those genera (of bivalves and gastropods) which have species with some calcite in the shell are generally high in Mg, and this obtains even when a species of such a genus has no calcite in it at all." However, they analyzed only one "calcitic" bivalve genus (*Pecten*), all the species of which are composed of at least 80% calcite. In addition, the data were obtained from an undefined region of the shell, and the authors do not take possible ontogenetic variation into account. All their examined *Pecten* species contained more Mg than any of the all-aragonitic bivalve genera. The applicability of their data to bivalves in general is thus not clear, but this would be an interesting subject for future work.

Even research aimed at establishing a temperature effect on Mg concentration has led to uncertain conclusions except in those few studies that have taken ontogeny into account. F. W. Clarke and Wheeler (1922) originally made "one very curious discovery . . . namely that in certain

---

* Magnesium is uniformly at least 3 times as abundant by weight as Ca in seawater (Goldberg, 1957).

groups of organisms the proportion of $MgCO_3$ is dependent upon or determined by temperature." They recognized the positive correlation between temperature and Mg concentration as a "tendency" with "probable exceptions." Thus, the statement by Dodd (1967) that "the validity of Clarke and Wheeler's findings outlining all the major trends in Mg distribution has not been significantly altered by more recent work" should not be taken to mean that Mg concentration and temperature, or any other environmental or physiological parameter for that matter, can be simply correlated. For example, it is difficult to analyze the data of Turekian and Armstrong (1960) for temperature relationships because there is no independent information regarding temperature ranges within the life habitat of their specimens. However, we can say by examination of their data that some warm-water species have a higher Mg concentration than closely related or unrelated cold-water species, but others do not. On the basis of whole-shell analyses, Gunatilaka (1975) reported that the Mg concentration within the shells of cold- and warm-water calcitic species of bivalves was independent of temperature. Pilkey and Goodell (1963) were similarly unsuccessful in correlating Mg concentration in bivalves with shells composed of low-Mg calcite and temperature. Rucker and Valentine (1961) analyzed "wedge-shaped cross sections [of *Crassostrea virginica* shell]" starting just behind the hinge region "so that all periods of growth except the earliest" were represented in their analyses. It is not surprising to find that their attempts to correlate Mg concentration with environmental temperature and salinity measurements, averaged over the entire period of shell growth, failed. Dodd (1965, 1966), however, in his ontogenetic approach, found a weak but repeatable positive correlation between environmental temperature (more appropriately, season) and Mg concentration in the outer calcitic layer of individual living *Mytilus* specimens. Dodd (1965) studied the Mg concentration in the last-formed portion of shells collected at different times of the year (Fig. 5A). He assumed that the ventral margin had been secreted just prior to collection. However, one must be aware that bivalve growth can be discontinuous during certain times of the year, so that the seasonal timing of chemical variations determined in such a manner can be uncertain. Moreover, his approach was not truly ontogenetic, because he synthesized populations, not single shells. It would be interesting to know whether these limitations have any bearing on the inability of Dodd (1966) to find preserved seasonal variations in Mg concentration in Pleistocene *Mytilus*.

The summary by Milliman (1974) of data relating Mg concentration (and other minor- and trace-element concentrations) to temperature could be misinterpreted because he does not qualify the discussion with the original researchers' statements concerning limitations and uncertainty of their data. Thus, the correlation between temperature and Mg concen-

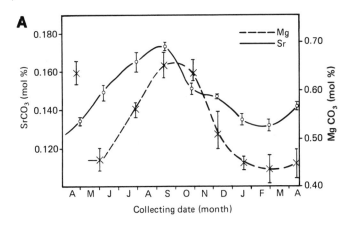

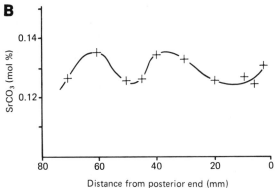

Figure 5. (A) Variation in strontium concentration (—) and magnesium concentration (– –) with collecting date in the last-formed portion of outer (calcitic) prismatic layer of *Mytilus edulis (diegensis)* from California. Data were obtained by emission spectrography. Compare with the presentation by Milliman (1974) of the same data. After Dodd (1965) with the permission of *Geochimica et Cosmochimica Acta* and the author. (B) Distribution of Sr in the outer (calcitic) prismatic shell layer of *Mytilus californianus* (X-ray fluorescence). Strontium maxima are interpreted as representing summer growth and minima as representing winter growth. Note that there is no well-defined trend in Sr concentration with age in the specimen. After Dodd (1965) with the permission of *Geochimica et Cosmochimica Acta* and the author. (C) Distribution of Sr in *Mytilus californianus* from the Pleistocene of Torrey Pines, California. Note prominent ontogenetic decrease in Sr in contrast to living *Mytilus* (B). This difference could be due to diagenesis, ontogeny, temperature, or other factors. After Dodd (1966) with the permission of the Geological Society of America and the author.

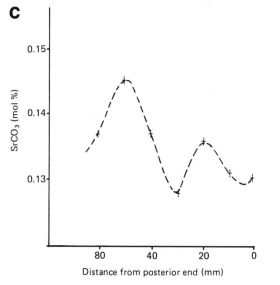

tration has passed from a hypothesis to a trend accepted without quali-
fication. In the words of Wolf *et al.* (1967), "more research is required on
the species and subspecies level before paleotemperature reconstructions
can be accepted as reliable." Chave (1954) also regarded the temperature
relationships of Mg concentration in bivalves as more uncertain than in
any other group he studied.

In view of the uncertain relationship between the Mg/Ca ratio and
temperature, it is premature for Berlin and Khabakov (1973, 1974) and
Yasamanov (1973) to determine environmental temperatures to within
0.1°C using Mg concentration in bivalves and other molluscs of Cretaceous
and Jurassic age. These workers ignore the variance of their own data and
have fitted lines to very few temperature–Mg concentration determinations.
Additionally, they ignore the possibility that different species concentrate
Mg differently in response to changes in temperature. Zolotarev (1974)
is more circumspect, since he admits that salinity, growth rate, mineralogy, and temperature could affect Mg concentration (Fig. 6). He measured the Mg concentration within calcitic portions of *Mizihopecten yessoensis*, *Swiftopecten swifti*, and *Crenomytilus grayanus* and found that
the Mg concentration increased in all three species with increasing temperature, but to different extents. The concentration also decreased with
age and declining growth rate. This stands in contrast to the results of
Moberly (1968) for *Aequipecten* (see Fig. 4) and of Estes (1972) for *Mercenaria*. Differences in Mg concentration among species must therefore
be considered, but the determinations of Zolotarev (1974) would be more
convincing if he had more clearly evaluated the repeatability of his determinations in different specimens (Fig. 6).

The relationship between salinity and the Mg/Ca ratio in the environment has also been studied (Leutwein and Waskowiak, 1962; Dodd,
1967). Milliman (1974) is misleading when he states that the Mg concentration in bivalve shells increases with decreasing salinity, or that Mg (or
Sr) tends to be more concentrated in freshwater than in marine shells.
Eisma *et al.* (1976) compared *Cardium*, *Macoma*, *Mya*, and *Mytilus* in
different salinity regimes and defined only weak trends even within a
single species. Lerman (1965b) found Mg concentration in *Crassostrea*
calcite to be independent of the Mg/Ca ratio in the environment, although
Lorens and Bender (1976) found that the Mg/Ca ratio in *Mytilus edulis*
grown in laboratory tanks increased exponentially with an increasing Mg/
Ca ratio in the water. They believed that this represented a breakdown,
with increasing Mg concentration in the medium, of the physiological
mechanism that discriminates against Mg concentration in the shell. In
the same species, Dodd (1965) found that Mg concentration in the calcitic
fraction increased with decreasing salinity (and probably Mg concentra-

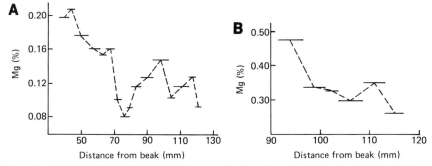

Figure 6. Distribution of magnesium from the beak to the posterior margin in the calcitic shell layer of *Mytilus yessoensis*, as measured by flame photometry. A slow-growing specimen (A) shows a lower average Mg content than a fast-growing specimen (B). The Mg content also declines with age (reduced growth rate) in both these *Mytilus* specimens. This ontogenetic relationship is different from that described for *Aequipecten*, in which Mg increases with age (see Fig. 4). After Zolotarev (1974) with the permission of the American Geological Institute.

tion). In the brackish water mussel, *Brachiodontes recurvus*, Davies (1972) determined that low concentrations of Mg within the shell were positively correlated with low salinity (Mg concentration). However, specimens collected from areas of highest salinity had intermediate Mg concentrations. Therefore, shell Mg was not simply correlated with environmental salinity. Perhaps the work of Lorens and Bender (1976) should be extended to several other species, and experiments carried out over each organism's entire physiological range and ontogeny to determine how each species discriminates against Mg at different environmental concentrations, and at different times during its life span.

2.2.2. Strontium

Many of the problems associated with skeletal magnesium are similar for strontium and barium (barium will not be discussed here). Few workers have defined ontogenetic variations in the concentrations of these elements. Those trends that have been described show increasing concentration with age in some species and decreasing concentration in others, although changes in concentration are sometimes close to the limits of measurement error. Some of the data have been used in an attempt to establish a cause–effect relationship between the environment and shell Sr at the expense of ignoring ontogenetic variations in distributions.

The analyses by Dodd (1965) of Sr in *Mytilus* (see Fig. 5) are among the best available. Dodd's Sr analyses were made simultaneously with his Mg analyses of *Mytilus* collected at various times of the year. Thus, the same limitations of their ontogenetic interpretation apply for Sr as discussed previously for Mg. Our concern here, however, is more with the temperature significance of the data. Dodd found that the concentration of Sr increased in the outer calcitic shell layer with increasing temperature.* Strontium concentration decreased in the aragonitic nacreous layer with increasing temperature (season). Dodd regressed the Sr/Ca ratio against environmental temperature measured approximately at the time of collection. From least-squares fits to the data, Dodd determined equations for temperature from the Sr/Ca ratio. However, other seasonably dependent variables such as illumination, nutrition, water circulation, and mussel growth rate may also affect the Sr/Ca ratio within the shell. Therefore, it is of questionable wisdom to extend the temperature regression to other populations of these species (*Mytilus edulis* and *Mytilus californianus*), and premature to imply that the equations can be used with other species, or with fossil specimens of *M. edulis* or *M. californianus*.

On the basis of whole-shell analyses, Thompson and Chow (1955) found that the Sr/Ca ratio in more than 30 bivalve species was independent of temperature. Similarly, Gunatilaka (1975) reported that the average Sr/Ca ratio in temperate bivalve species is similar to the ratio for warm-water species. The claim of Zolotarev (1974) that the Sr concentration in *Mizihopecten yessoensis* increases from 0 to 10°C, and then decreases above 10°C, is not substantiated by his data, which have a large uncertainty at all temperatures. Similarly, Hallam and Price (1968) claimed that the Sr concentration in the aragonitic shell of *Cardium edule* is inversely proportional to temperature, as it is in *Mytilus* aragonite (Dodd, 1965). However, Hallam and Price (1968) state that there is no correlation between Sr concentration and growth rate, although their data actually do show an increasing Sr level with increasing age (i.e., reduced growth rate) in two specimens.

Swan (1956) originally proposed an inverse relationship between growth rate and Sr concentration based on examination of *Modiolus*, *Mytilus*, and *Mya*. Swan's data are often considered to be applicable to all species. However, Dodd (1965, 1966) and Stanton and Dodd (1970) report that the Sr/Ca ratio decreases with (ontogenetic) age in Recent, Pleistocene, and Pliocene *Mytilus* (see Fig. 5B and C). Crisp (1975) found

---

* As with Mg, the correlation is more with seasonal change than with temperature change alone.

that Sr concentration generally increases with age and also varies with position along the ventral shell margin in freshwater bivalves (Fig. 7).*

In contrast to the ontogenetic observations of Crisp (1975), Nelson (1961) claimed that the Sr/Ca ratio increases with increasing growth rate in the freshwater bivalve *Anodonta*. Nelson (1967) interpreted Quaternary paleoclimates on the basis of Sr concentration in bivalve shells found in Indian middens. He found increasing Sr concentration with increasing amount of shell added by *Anodonta* (Nelson, 1961). This index is misleading, however, because in another study, Nelson (1967) notes that in six species of freshwater bivalves (including *Anodonta*), the Sr concentration increases with age. In the latter case, some bivalves actually showed ontogenetic patterns in Sr concentration opposite the general trend. For example, one specimen of *Anodonta* showed increasing concentrations of Sr with age, while another specimen showed highest Sr concentrations in the middle years of growth (Fig. 8). At the very least, the variation and uncertainty in these data stand in contrast to Nelson's (1961) earlier positive correlation between growth rate and Sr concentration in *Anodonta*. Thus, the need to evaluate ontogenetic variability among individual specimens of a species population is clear.

Odum (1957) recognized that the Sr concentration in the bivalve shell is proportional to the Sr concentration in the environment. The average Sr concentration in bivalves is low relative to that in other invertebrates (Lowenstam, 1964; Ueda *et al.*, 1973). Lorens and Bender (1976) and Lerman (1965a,b) found that the Sr concentration in the shells of *Mytilus* and *Crassostrea* increased with an increasing Sr/Ca ratio in the environment. Similarly, Muller (1968) reported that the Sr concentration in shells of *Anodonta*, *Sphaerium*, and *Pisidium* of Lake Constance was positively correlated with high Sr/Ca levels in the lake. Also, Nelson (1962) found that relatively high concentrations of ^{90}Sr were incorporated into *Unio* shells that were located close to the release site of radioactive wastes in the Clinch and Tennessee River system. All these observations are consistent with the original note of Odum (1957), and would seem to indicate that bivalve shells can reflect Sr levels in the environment. However, in *Brachiodontes recurvus* from the Rappahannock River Estuary, specimens from areas of highest salinity had intermediate levels of Sr and Mg when compared with specimens living in less saline water (Davies, 1972). In

---

* Compare Fig. 7A with Fig. 1, which shows variations in carbonate incorporation along the shell margins in *Argopecten*. Crisp (1975) also stated that the faster-growing areas (4, 5, 6 in Fig. 7A) incorporate less Sr and Mn than the slower-growing areas and that the ontogenetic distribution of Sr and Mn differed among shell layers (Fig. 7B and C). In turn, the distributions differed among species (Fig. 7D). Clearly, the distinction among ontogenetic, growth-rate, and shell-layer effects must be made to completely describe the distribution of elements in a bivalve species.

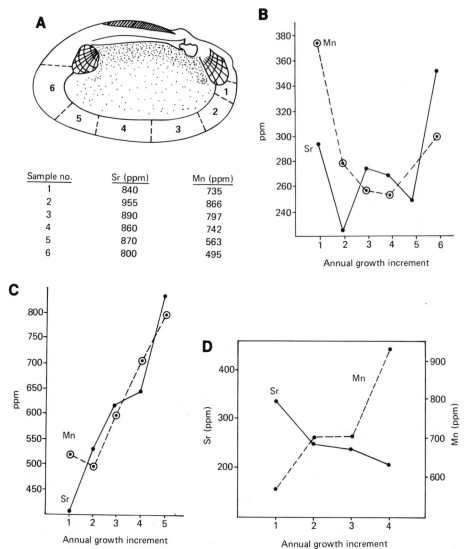

Figure 7. Distribution of strontium (——) and manganese (– –) in the shells of two freshwater bivalves as measured by X-ray fluorescence. (A) Variation in Sr and Mn concentration within the ventralmost annual increment of the outer shell layer of *Lampsilis* sp. from the Jefferson River, Cardwell, Montana. Note that the ontogenetic chemical patterns could appear to be very different depending on the sampling transect through the shell. (B) Sr and Mn in successive annual growth increments within the outer nacreous layer of *Lampsilis* sp. from Lake Huron. (C) Sr and Mn in successive annual growth increments within the inner nacreous layer of *Lampsilis* sp. from Lake Huron. (D) Sr and Mn in successive annual growth increments of the outer nacreous layer of *Anodonta corpulenta* from a small pond in Indiana. Note that ontogenetic trends vary among different species and among different shell layers within the same species. After Crisp (1975) with the permission of the author.

environments in which the Sr/Ca ratio in the water decreased, the ratio increased in the shell. These data are from a different species than those studied by Nelson (1962, 1967), but it would be facile to allude to "species-specific" effects to explain the differences. What is important is the clear warning from the work of Davies (1972) that the response of a species over its entire physiological range must be considered before correlations can be established between shell and environmental concentrations of Sr.

The need for circumspection can be seen from the salinity studies of Eisma *et al.* (1976) and Rucker and Valentine (1961). These authors found that the correlation was weak between salinity and shell Sr or Mg within a single species. Rucker and Valentine (1961) noted a multiple postive correlation between salinity and Sr + Mg concentration in the shell. Such multiple-correlation work has not been extensively pursued by other workers.

It should be clear from the foregoing discussion that correlations between any environmental or physiological variable and shell Sr have not been clearly defined over the entire life span or physiological range of any species. Thus, we must regard as tenuous the conclusions drawn by Nelson (1967) and Lee and Wilson (1969) for Quaternary paleoclimates

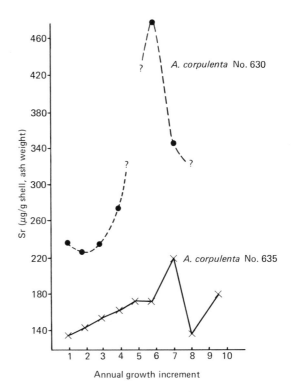

Figure 8. Distribution of strontium in the inner nacreous layer of two specimens of *Anodonta corpulenta*. Note that the Sr concentration does not increase uniformly with age in both specimens. Also compare this figure with Fig. 7. Analyses were made with flame spectrophotometry and emission spectrography. After Nelson (1967).

based on Sr composition of midden shells. Nelson (1967) found that 1000- to 2000-year-old shells had 50–500% more Sr than contemporary specimens. He claimed that the destruction of forests by the European agricultural economy led to decreased soil CO_2, reduced dissolution of alkaline earths, and reduced concentrations of dissolved minerals in rivers. Thus, the more contemporary bivalves supposedly reflect reduced levels of Sr in river water.

Lee and Wilson (1969) measured the Sr concentration in several species of freshwater bivalves sampled in middens at different stratigraphic levels. In three middens, they found similar variations in Sr concentrations from level to level, although the differences were within the limits of analytical error. They claimed that populations of shells with increased amounts of Sr marked periods of increased aridity (higher Sr concentrations in the water). However, neither Nelson nor Lee and Wilson acknowledged the alternative environmental and physiological influences on Sr concentration.

Finally, the ionic radius of Sr^{2+} is larger than that of Ca^{2+}, and on this basis, it can be predicted that Sr should be more abundant in biogenic aragonite than in biogenic calcite. Consequently, Harriss (1965) compared the concentrations of Sr, Mg, Fe, and Mn in molluscan calcites and aragonites, based on the results of various workers (Turekian and Armstrong, 1960; Odum, 1957; Chave, 1954; Rucker and Valentine, 1961; Pilkey and Goodell, 1963). Strontium concentration in molluscan aragonite and calcite is similar, in contrast to the case of Mg, in which the shell concentration tends to be controlled by skeletal mineralogy. Harriss (1965) concluded that biological control over the concentration of Sr is more significant than biological control over Mg (although the mineralogical effect on Sr^{2+} may also be less than on Mg^{2+} because the Ca^{2+} and Sr^{2+} ions are more similar in size). However, the analyses of Harriss (1965) are not species-by-species comparisons, and if the original data (e.g., Turekian and Armstrong, 1960) are examined, some calcitic species (of *Pecten*) are seen to have slightly less Sr than some completely aragonitic species. Harriss (1965) asks whether we can predict the concentration of Sr in the shell of any bivalve species on the basis of a consideration of the relative sizes of ions and their consequent ability to substitute for one another in different crystal lattices. The problem arises because the comparison is between all aragonites and all calcites within the phylum Mollusca. However, the evolutionary response of organisms to physical characteristics of the environment occurs at the level of populations of individual specimens, and different populations of a species respond differently to similar environmental variables. Thus, it would be more important first to ask how the distribution of Sr within individual specimens varies during ontogeny in order to compare those variations among species, and then to ask why in some species the prediction holds and in others it does not.

3. Recognition of the Original Elemental Composition of the Shell and Its Environmental and Physiological Significance

From the preceding discussion, it should be clear that we cannot infer life-habitat parameters of a fossil bivalve on the sole basis of a fractional or a whole-shell compositional analysis. On the basis of such analyses, one might find statistically significant chemical differences among populations of a living species, but at present there is no way of deciding whether the differences in fossil or living populations are coincidental or reflect real habitat differences.

Thus far, I have said little about diagenesis. Although many researchers have described some of the changes in shell chemistry related to diagenesis, and because many of the chemical changes are independent of structural diagenesis (e.g., Curtis and Krinslay, 1966; Land, 1967; Estes, 1972; Crisp, 1975; Kamiya, 1975; Walls et al., 1977), I believe that the original distribution of an element within the shell must be known before one can describe subsequent alterations in distribution or average concentration. Because such distribution has rarely been described in a systematic, ontogenetic way, descriptions of the specific events in diagenesis are as uncertain as claims of original shell composition in fossils. Few fossil species are represented by living specimens; without criteria for recognizing original composition, neither diagenesis nor fossil metabolism can be described adequately.

Diagenesis of a shell begins while the organism is alive, and factors weathering the shell may be the same factors that effect changes in an organism's shell-secreting metabolism (Rosenberg, 1972). Thus, the few ontogenetic patterns that have been discussed earlier in this chapter suggest that the original shell composition of living specimens can be preserved despite weathering.

However, the immediate onset of diagenesis would seem to make the role of diagenesis all the more compelling with regard to fossil bivalve shell chemistry. This has led to the belief that despite the difficulty of describing the process of diagenesis, one can judge the extent of diagenesis and use it for crude age-dating. Ohta (1977) found that the Na concentration declines exponentially, and the Mn concentration increases exponentially, with geological age between some stratigraphically separated fossil bivalve assemblages. Both elements level off at about 10 million years b.p. Schoute-Vanneck (1960) roughly dated middens of Late Stone and Iron Age, based on the declining ratio of conchiolin/CaCO$_3$ in bivalves of increasing geological age. Hare (1969) proposed that amino acid degradation could also be used for age-dating providing that time- and temperature-dependent diagenetic effects on degradation could be quantified. But none of this proves that all ontogenetic chemical variations are altered, or lost, during weathering. There is hope, if only because such preserved ontogenetic trends have rarely been searched for in fossils.

Thus, it is essential to establish criteria for recognizing original shell chemical composition. Such criteria may help separate effects of diagenesis from original patterns of chemical composition regardless of the subtlety of original compositional variations. One approach to defining original shell composition is to describe ontogenetic trends in elemental composition. Only a few workers have taken on this tedious task, and the ontogenetic comparisons between living and fossil species are based on very few specimens (Nelson, 1967; Stanton and Dodd, 1970; Estes, 1972; Crisp, 1975; Immega, 1976). The assumption is that similar ontogenetic patterns in living and fossil specimens are evidence of original shell composition in the fossils. Obviously, this assumption does not provide for the possibility of evolutionary or phenotypic differences in ontogenetic patterns. In addition, ontogenetic variations are sometimes small and may be close to the limits of resolution. Moreover, the variation in the patterns among individual members of a single species population has never been properly assessed, and conclusions separating original from altered chemical distributions based on a few ontogenetic comparisons are thus tenuous. However, one can decide whether supposed ontogenetic variations are real based on allometric relationships, as will be shown below.

The first step is to recognize that chemical and structural growth increments are basic units of bivalve growth. As a working hypothesis, which requires testing, one can state that increments of similar composition or structure have similar environmental or physiological significance. Incremental growth patterns are often cyclical and are superimposed on long-term ontogenetic changes. Shell structural cycles have been matched with metabolic changes within soft tissues and the extrapallial fluid (see Chapter 6), and these changes in turn are often related to environmental cycles associated with tides, seasons, illumination, and reproduction among other factors. Chemical growth rhythms have been so little studied, however, that correlations with environmental rhythms are tentative (Stanton and Dodd, 1970; Rosenberg and Jones, 1975). Some of the most interesting problems in molluscan physiology are related to how long- and short-term ontogenetic chemical variations in shell composition are coupled to compositional changes in the chemistry of the extrapallial fluid and soft tissues. Such studies are bound to be difficult, given the effort necessary to make complete shell chemical analyses, the difficulty of sampling the extrapallial fluid without altering its composition (Wilbur, 1964), and the difficulty of studying soft-tissue chemistry of a single bivalve throughout its entire ontogeny. However, if variations in shell chemistry can be matched with changes in chemistry of the extrapallial fluid and soft tissues, we will be much closer to understanding cause–effect relationships among environmental change, physiological responses, and shell composition. For example, Wada (1961) has determined that the Ca concentration in the extrapallial fluid may be inversely proportional to the deposition rate. At present, however, one cannot simply infer that the Ca concentration in the extrapallial fluid is inversely proportional to the calcium concentration in the shell because the rate of incorporation of carbonate into the growing margin of the shell may remain constant despite changes in the rate of increase in shell size with age (Wheeler et al., 1975).

In vitro experiments on the precipitation of aragonites and calcites (Lerman, 1965a,b; Kitano et al., 1971, 1974) suggest that changes in trace-element concentrations in the shell could be out of phase with changes in the extrapallial fluid. Kitano et al. (1974) have shown that Mg in solution favors precipitation of aragonite. As mentioned earlier, Mg tends to be excluded from the aragonite lattice. Therefore, reduction in Mg concentration in the shell could be related to an increase in Mg in the extrapallial fluid. However, matters are likely to be even further complicated by the fact that the Mg concentration in precipitated crystals is also influenced by the concentration of organic compounds and other trace elements in solution (Kitano et al., 1974).

Masumoto et al. (1934) have shown that increases in shell weight of Crassostrea (= Ostrea) gigas are greatest in February, May, and September and least in January, April, and August, paralleling trends in ash content of the soft tissues. The most encouraging studies of time-varying compositional changes within the soft tissues come from Bryan (1973). Bryan measured concentrations of trace metals in the kidney, digestive gland, gonad/foot, gills, and mantle of Pecten maximus and Chlamys opercularis. He found seasonal variations in Co, Cu, Fe, Mn, Ni, Pb, and Zn most prominent in the kidney and digestive glands. The concentrations in most cases were highest in the autumn and winter months. There was considerable variation among individual specimens, so the seasonal curves of concentration are close to or within the statistically predicted range of uncertainty. However, seasonal variations were repeatable over the more than 2-year period of observation.

Bryan (1973) also related changes in soft-tissue metal content to changes in food supply. Concentration of the metals was generally highest when phytoplankton productivity was lowest, and tended to drop in the spring when primary productivity increased. Bryan believed that as a result of incorporation of trace metals into phytoplankton cells, periods of population growth resulted in reduced concentrations of metals in the water and dilution of the metals throughout the phytoplankton population. Thus, the metals were effectively less available to the bivalves.

Surely this is not the only possible cause–effect relationship between environmental change and compositional cycles in soft tissues (especially the mantle), so that the discussion of Bryan (1973) will undoubtedly be amended by future work. The ultimate goal is to correlate soft-tissue (especially mantle) chemical cycles and skeletal rhythms in a given species. Then one could also look for environmental cycles to match with observed periodicities in the composition of the shell, soft tissue, and extrapallial fluid. Structural growth rhythms have been correlated with daily and tidal, as well as illumination, cycles (Evans, 1975; Pannella, 1976). However, temperature, oxygen/carbon dioxide levels, carbonate saturation, food supply, concentration of trace elements, and pH are also cyclic environmental parameters.

Lutz and Rhoads (1977) (also see Chapter 6) have suggested that cycles of aerobic–anaerobic metabolism (and consequent changes in shell deposition) are timed to fluctuations in tide level. These workers believe that at high tide, bivalves secrete calcium-carbonate-rich layers while they are feeding. The bivalves record low tide with an organic-rich layer that reflects shell closure and partial shell dissolution to buffer the metabolic end products of anaerobic respiration. From the previous discussion on Ca distribution in the shell, it is not yet clear that bivalve growth increments are *uniform* alternations of Ca- and organic-rich layers (i.e., the variation is as important as the average difference). That is, nontidal causes of incrementation are superimposed on tidal causes. For example, the pumping rate and valve movement in some bivalves are also correlated with oxygen consumption (Hamwi and Haskin, 1969). Oxygen consumption, in turn, may be related to cyclic oxygen/carbon dioxide levels in the water that reflect photosynthetic/respiratory activity in some coastal habitats; the cycles are both daily and seasonal, as well as independent of the tide (Schmalz and Swanson, 1969; Jackson, 1972). Schmalz and Swanson (1969) found that in shallow lagoonal habitats, CO_2 levels are minimal in the late afternoon following depletion by algal and vascular plant photosynthesis. The CO_2 level in the water is maximal just before dawn. Carbonate saturation and pH are 180° out of phase with the CO_2 levels. All these factors may influence shell calcification, as well as the concentration of Ca in the shell, perhaps by the same physiological mechanism proposed by Lutz and Rhoads (1977).

All environmental and physiological variables affecting shell depositional rhythms need not be understood before an attempt is made to establish original shell compositional rhythms in fossil bivalves. In fact, it would be misleading to state that all chemical rhythms must maintain a constant phase difference with structural cycles in the shell, and that all structural or chemical periodicities in the shell have simple cause–effect relationships with environmental factors, changes in soft tissue composition, or metabolism.

As a second working hypothesis, we can predict that chemical growth rhythms within the shell will vary allometrically if they have not been altered by diagenesis. Simply stated, just as structural growth increments vary in thickness with growth rate, the wavelength of chemical rhythms will also vary with ontogeny. For example, as a result of a decline in growth rate with age, growth increments decrease in thickness in mature specimens. The wavelength of chemical rhythms will similarly decrease with age. If such an ontogenetic change in chemical rhythms can be found in a fossil, one has further evidence of preservation of initial compositional changes (Fig. 9, B–B' vs. A–A').

Second, shell growth rate is highest along the axis of maximum growth (see Chapter 1 for illustration of this axis). Growth increments are widest along this axis, narrowing anteriorly and posteriorly. Chemical rhythms will consequently have a longer wavelength along this axis than along any other (Fig. 9, A–A' vs. C–C' and D–D').

The outer surface of the shell, with its greater curvature, grows more rapidly than the inner surface, and consequently, growth increments in the outer shell layer are wider than those in the inner shell layer. Chemical growth rhythms in the outer layer will therefore have a longer wavelength than those in the inner layer (Fig. 9 E–E' vs. F–F'). Evidence that allometric chemical variations do exist (at least in teeth) is presented in Appendix 3.A.

It must be emphasized that chemical and structural rhythms need not maintain a constant phase difference with each other (see previous discussion on Ca and S rhythms); chemical and structural rhythms may be independent, but chemical rhythms must demonstrate allometric principles nonetheless. The absence of such allometric changes in fossils would be a clear indication of diagenetic alteration. One could argue that the inverse is not necessarily true. One could suppose that anything can happen during diagenesis; narrow increments could chemically alter differently than wide ones, and diagenesis could thus induce *apparent* allometric variations. However, the idea that weathering could systematically *order* rather than disorder *all* chemical rhythms to produce pseudoallometric variations in all elements must itself be demonstrated.

A final means of recognizing original shell chemical composition in fossils is derived from evolutionary considerations. In a wide variety of invertebrate taxa, there is an evolutionary tendency toward an increase in size (Newell, 1949). In addition, larger species within such lineages

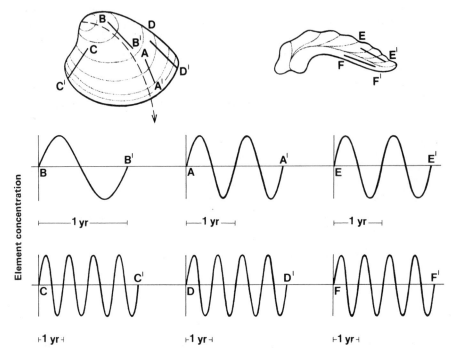

Figure 9. Diagrams illustrating how allometric principles provide criteria for filtering original chemical variations in bivalve shells from diagenetic noise. Different areas of the bivalve shell grow at different rates; the wavelength of chemical periodicities will consequently vary with growth rate. Growth rate declines with age (from B–B' to A–A') and is maximal along the axis of maximum growth (arrow) and declines anteriorly and posteriorly (C–C' and D–D'). Growth rate is also more rapid in the outer shell layer (E–E') than in the inner shell layer (F–F'). For simplicity, the sine waves below the shells represent one chemical cycle of known periodicity (say, annual). Note, however, that the temporal significance of the cycles need not be known. Fourier spectra of all cycles will be shifted to shorter wavelengths in areas of reduced growth.

tend to live longer than smaller species (Bonner, 1965). The life span determines the maximum (or fundamental) periodicity of an organism's biological rhythms (Rosenberg, 1976). Individual organisms that live 12 months cannot record in their shell biological rhythms that are years in duration. Thus, evolutionary lineages that show increasing size through time should record within their shells a history of increasing fundamental periods of biological rhythms.

A quantum increase in mean life span occurred toward the end of the Precambrian with the evolution of multicellular organisms. The life span of multicellular organisms is measured in days, weeks, and years, whereas the life span of unicellular organisms is measured in a maximum of hours and days. We note that if it is true that the length of the day and month has increased since the Precambrian (see Rosenberg and Runcorn, 1975; Rosenberg, 1977), then the length of the day and month did not approach Recent values until about the same time as the origin of multicellular, relatively long-lived organisms. It is interesting that daily rhythms are basic to the physiology of organisms and that circadian rhythms cannot be shortened in the laboratory to less than about 17 hr by altering light–dark cycles. Since the length of the day at the time of origin of the multicellular organisms was about 17 hr, one wonders whether the evolving length of day and month provided a threshold to multicellular evolution. Tappan and Loeblich (1971) have observed that multicellular organisms evolved only after the solar energy declined in intensity (e.g., by means of atmospheric filtering of ultraviolet light). The discussion herein would also imply that there is a maximum circadian frequency of energy input above which the metazoa could not originate.

There are two implications of species longevity for this discussion. First, within any bivalve lineage that shows an evolutionary increase in body size and life span, one should observe a shift in chemical periodicities preserved within the shells of these bivalves to lower and lower frequencies through time, paralleling the ontogenetic trend. Of course, this trend will be most easily detected in temperate, shallow-water bivalves, the growth of which reflects seasonal, as well as daily, physiological changes. Second, because the life span is so important in determining the spectrum of frequencies recorded within an organism's shell, some rhythms may not be in phase with environmental cycles. This does not necessarily mean that the rhythms are independent of environmental phasing, and certainly does not imply the existence of internal, controlling clocks independent of the environment. It does mean that we should avoid the pitfall of trying to link *all* chemical rhythms within living and fossil shells with synchronous environmental cycles.

4. Summary

The extent to which shell chemistry of bivalves can be used to make environmental and evolutionary interpretations is, at present, unclear. In the past, such determinations have involved measuring the concentration of an element within the whole shell or part of the shell and plotting these data against temperature, salinity, or mineralogy in an attempt to establish cause–effect relationships between environmental variables (or physiology) and shell composition. However, shell composition is complex, varying with position in the shell and with ontogenetic age. The distribution of elements within a shell must be fully described before interpretations can be made.

Allometric variations in ontogenetic patterns of rhythmic growth can be used to discriminate original chemical growth sequences from diagenetic alterations. At this point, we still do not know whether all ontogenetic patterns of variation in shell composition are correlated with single environmental variables, many environmental variables, or a combination of physiological and environmental variables.

The description of chemical ontogeny and its allometric variations promises to be a long and difficult task with present analytical methods. Ontogenetic variations can be of small amplitude, perhaps variable within a species, and may be readily altered by diagenesis. Despite these problems, defining chemical variations throughout ontogeny is a worthy goal because, at present, there seem to be few other criteria for recognizing original shell chemistry. Furthermore, ontogenetic variations in chemistry can be predicted from allometric variations in structural growth increment patterns, although the two may not be in phase. Even if chemical ontogenetic patterns prove to be highly variable within some species, and hence of little use in establishing environmental relationships, this knowledge is, in itself, important.

ACKNOWLEDGMENTS. I thank Dr. Richard A. Lutz and Dr. Donald C. Rhoads and their reviewers for considerable editorial assistance. Dr. Norman Newell and Dr. Helen Loeblich have encouraged some of my evolutionary thinking relating to growth rhythms. However, errors remaining in the manuscript are a result of my own inability to grasp the complexities of the subject of bivalve shell chemistry. In addition, some of my evolutionary ideas have proven to be very controversial, and if they are (sooner or later) proven to be false, it isn't because I wasn't warned many times by chronobiologists and paleontologists alike.

Professor K. M. Wilbur, Duke University, kindly sent the photo appearing as Figure 1 in this contribution. I also thank Dr. Ralph Moberly, University of Hawaii, and Dr. Marian Furst, California Institute of Technology, for sending me original figures from their manuscripts, which are cited herein.

References

Akiyama, M., and Wyckoff, R. W. G., 1970, Total amino acid content of fossil *Pecten* shells, *Proc. Natl. Acad. Sci. U.S.A.* **67**:1097–1100.

Berlin, T. S., and Khabakov, A. V., 1973, Calcium–magnesium ratios, chlorine content and mineral composition in shells of recent pelecypod molluscs, *Geochem. Int.* **10**:939–946.

Berlin, T. S., and Khabokov, A. V., 1974, Calcium and magnesium paleotemperature determinations for carbonate fossils and country rocks, Geochem. Int. **11**:427–433.

Blanchard, S. C., and Chasteen, N. D., 1976, Electron paramagnetic resonance spectrum of a seashell, J. Phys. Chem. **80**:1362–1367.

Bøggild, O. B., 1930, The shell structure of the molluscs, K. Dan. Vidensk. Selsk. Skr. Naturvidensk. Math. Afd. **2**:419–431.

Bonner, J. T., 1965, Size and Cycle, Princeton University Press, Princeton, New Jersey, 201 pp.

Bricteux-Grégoire, S., 1968, Prism conchiolin of modern or fossil molluscan shells, Comp. Biochem. Physiol. **24**:567–572.

Brooks, R. R., and Rumsby, M. G., 1965, The biogeochemistry of trace element uptake by some New Zealand bivalves, Limnol. Oceanogr. **10**:521–528.

Bryan, G. W., 1973, The occurrence and seasonal variation of trace metals in the scallops Pecten maximus and Chlamys opercularis, J. Mar. Biol. Assoc. U.K. **53**:145–166.

Chave, K. E., 1954, Aspects of the biogeochemistry of magnesium. 1. Calcareous marine organisms, J. Geol. **62**:266–283.

Chipman, W., and Schommers, E., 1968, Role of surface associated organisms in the uptake of radioactive manganese by the clam Tapes decussatus: Radioactivity in the sea, International Atomic Energy Agency Publ. 24, 11 pp.

Clarke, A. N., and Clarke, J. H., 1974, A static monitor for lead in natural and waste waters, Environ. Lett. **7**:251–260.

Clarke, F. W., and Wheeler, W. C., 1922, The inorganic constituents of marine invertebrates, U.S. Geol. Surv. Prof. Pap. **124**:1–62.

Clarke, J. H., Clarke, A. N., Wilson, D. J., and Freauf, J. J., 1976, Lead levels in freshwater mollusk shells, J. Environ. Sci. Health Part A **All**:65–78.

Comfort, A., 1951, The pigmentation of molluscan shells, Biol. Rev. **26**:285–301.

Cook, P. J., 1977, Loss of boron from shells during weathering and possible implication for the determination of salinity, Nature (London) **268**:426–427.

Crenshaw, M. A., 1972, The soluble matrix from Mercenaria mercenaria shell, Biomineralisation **6**:6–11.

Crisp, E. L., 1975, The skeletal trace element chemistry of freshwater bivalves, Ph.D. dissertation, University of Indiana, Bloomington, 187 pp.

Crisp, E. L., 1976, Skeletal trace element chemistry of freshwater bivalves, Fed. Soc. Am. Proc. **8**:15–16 (abstract).

Curtis, D. D., and Krinslay, D., 1966, The detection of minor diagenetic alteration in shell material, Geochim. Cosmochim. Acta **29**:71–84.

Davies, T. T., 1972, The effect of environmental gradients in the Rappahannock River Estuary on the molluscan fauna, in: Environmental Framework of Coastal Plain Estuaries (B. W. Nelson, ed.), Geol. Soc. Am. Mem. **133**:263–290.

Dodd, J. R., 1965, Environmental control of strontium and magnesium in Mytilus, Geochim. Cosmochim. Acta **29**:385–398.

Dodd, J. R., 1966, Diagenetic stability of temperature-sensitive skeletal properties in Mytilus from the Pleistocene of California, Geol. Soc. Am. Bull. **77**:1213–1224.

Dodd, J. R., 1967, Magnesium and strontium in calcareous skeletons: A review, J. Paleontol. **41**:1313–1329.

Eisma, D., Mook, W. G., and Das, H. A., 1976, Shell characteristics, isotopic composition and trace element contents of some euryhaline molluscs as indicators of salinity, Palaeogeogr. Palaeoclimatol. Palaeoecol. **19**:39–62.

Estes, E. L., III, 1972, Diagenetic alteration of Mercenaria mercenaria as determined by laser microprobe analysis, Ph.D. dissertation, University of North Carolina, Chapel Hill, 103 pp.

Evans, J., 1975, Growth and micromorphology of two bivalves exhibiting non-daily growth lines, in: *Growth Rhythms and the History of the Earth's Rotation* (G. D. Rosenberg and S. K. Runcorn, eds.), pp. 119–134, John Wiley, London.

Ferrel, R. E., Carville, T. E., and Martinez, J. D., 1973, Trace metals in oyster shells, *Environ. Lett.* **4**:311–316.

Florkin, M., 1966, *A Molecular Approach to Phylogeny*, Elsevier, Amsterdam, 176 pp.

Fox, D. L., 1966, The pigmentation of molluscs, in: *Physiology of Mollusca*, Vol. 2 (K. M. Wilbur and C. M. Yonge, eds.), pp. 249–274, Academic Press, New York.

Fox, D. L., and Coe, W. R., 1943, Biology of the California sea mussel (*Mytilus californianus*). II. Nutrition, metabolism, growth, and calcium deposition, *J. Exp. Zool.* **93**:205–249.

Furst, M., Lowenstam, H. A., and Burnett, D. S., 1976, Radiographic study of the distribution of boron in recent mollusc shells, *Geochim. Cosmochim. Acta* **40**:1381–1386.

Girardi, F., and Merlini, M., 1963, Studies on the distribution of trace elements in a mollusk from a freshwater environment by activation analysis, Ispra, Italy, European Atomic Energy Community, EUR 474e.

Goldberg, E. D., 1957, The biogeochemistry of trace metals, in: *Treatise on Marine Ecology and Paleoecology* (J. W. Hedgpeth, ed.), *Geol. Soc. Am. Mem.* **67**:345–358.

Graham, D. L., 1972, Trace metal levels in intertidal mollusks of California, *Veliger* **14**:365–372.

Grégoire, C., 1972, Structure of the molluscan shell, in: *Chemical Zoology*, Vol. VII, *Mollusca* (M. Florkin and B. T. Scheer, eds.), pp. 45–102, Academic Press, New York.

Gunatilaka, A., 1975, The chemical composition of some carbonate secreting marine organisms from Connemara, *Proc. R. Ir. Acad. Sect. B* **75**:543–556.

Hallam, A., and Price, N. B., 1968, Environmental and biochemical control of strontium in shells of *Cardium edule*, *Geochim. Cosmochim. Acta* **32**:319–328.

Hamwi, A., and Haskin, H. H., 1969, Oxygen consumption and pumping rates in the hard clam *Mercenaria mercenaria*: A direct method, *Science* **163**:823–824.

Hare, P. E., 1962, Variations in the composition of the organic matrix of some modern calcareous shells, *Geol. Soc. Am. Spec. Pap.* **68**:191–192.

Hare, P. E., 1963, Amino acids in the proteins from aragonite and calcite in the shells of *Mytilus californianus*, *Science* **139**:216–217.

Hare, P. E.. 1969, Geochemistry of proteins, peptides, and amino acids, in: *Organic Geochemistry* (G. Eglinton and M. T. J. Murphy, eds.), pp. 438–463, Springer-Verlag, New York.

Hare, P. E.. and Abelson, P. H., 1965, Amino acid composition of some calcified proteins, *Yearb. Carnegie Inst. Wash., D.C.* **64**:223–231.

Harriss, R. C., 1965, Trace element distribution in molluscan skeletal material. I. Magnesium, iron, manganese, and strontium, *Bull. Mar. Sci.* **15**:265–273.

Harvey, R. S., 1969, Uptake and loss of radionuclides by the freshwater clam *Lampsilis radiata*, *Health Phys.* **17**:149–154.

Horiguchi, Y., 1959a, Biochemical studies in *Pteria (Pinctada) martensii* (Dunker) and *Hydriopsis schlegelii* (v. Martens). VII. On the separation and purification of sulfumucopolysaccharide and detection of its component sugars, *Rep. Fac. Fish. Perfect Univ. Mie* **3**:399–406.

Horiguchi, Y., 1959b, Biochemical studies on *Pteria (Pinctada) martensii* (Dunker) and *Hydriopsis schlegelii* (v. Martens). VIII. Trace components in the shells of shellfish (Part I), *Bull. Jpn. Soc. Sci. Fish.* **25**:392–396.

Horiguchi, Y., and Tsujii, T., 1967, Studies on the production of black pearls by irradiation with radioactive rays. III. Relationship between the coloration obtained by x-ray irradiation and the manganese contents in the shells of several shellfish, *Bull. Jpn. Soc. Sci. Fish.* **33**:5–11.

Immega, N., 1976, Environmental influence on trace element concentrations in some modern and fossil oysters, Ph.D. dissertation, University of Indiana, Bloomington, 194 pp.

Jackson, J. B. C., 1972, The ecology of molluscs of *Thalassia* communities, Jamaica, W. Indies. II. Molluscan population variability along an environmental stress gradient, *Mar. Biol.* **14**:304–337.

Kado, Y., 1960, Studies on shell formation in mollusca, *J. Sci. Hiroshima Univ. Ser. Bl* **19**(4):163–210.

Kamiya, H., 1975, Study of the internal structure of some fossil molluscan shells and their diagenetic alteration, *J. Geol. Soc. Jpn.* **81**:431–436.

Kaplan, I. R., Emery, K. O., and Rittenberg, S. C., 1963, The distribution and isotopic abundance of sulfur in recent marine sediments of southern California, *Geochim. Cosmochim. Acta* **27**:297–332.

Kitano, Y., Kanomori, N., and Oomori, T., 1971, Measurement of distribution coefficients of Sr and Ba between carbonate precipitation and solution—Abnormally high values of distribution coefficients measured at early stages of carbonate formation, *Geochem. J.* **4**:183–206.

Kitano, Y., Kanomori, N., and Yoshioka, S., 1974, Influence of chemical species on the crystal type of CaCO₃, in: *The Mechanisms of Mineralization in the Invertebrates and Plants* (N. Watabe and K. M. Wilbur, eds.), pp. 191–202, University of Carolina Press, Columbia.

Kobayashi, I., 1975, Preliminary study on the distribution of some elements in the shell of some bivalvian molluscs by the electron microprobe analysis, *Sci. Rep. Niigata Univ. Ser. E* **3**:41–50.

Land, L. S., 1967, Diagenesis of skeletal carbonates, *J. Sediment. Petrol.* **37**:914–930.

Lee, G. F., and Wilson, W., 1969, Use of chemical composition of freshwater clam shells as indicators of paleohydrologic conditions, *Ecology* **50**:990–997.

Lerman, A., 1965a, Paleoecological problems of magnesium and strontium in biogenic calcites in light of recent thermodynamic data, *Geochim. Cosmochim. Acta* **29**:977–1002.

Lerman, A., 1965b, Strontium and magnesium in water and in *Crassostrea* calcite, *Science* **150**:745–751.

Leutwein, F., and Waskowiak, R., 1962, Geochemische Untersuchungen an rezenten marinen Molluskenschalen, *Neues Jahrb. Mineral.* **99**:45–78.

Lorens, R. B., and Bender, M. L., 1976, The physiological exclusion of Mg^{++} from *Mytilus edulis* calcite, *Geol. Soc. Am. Abstr. Proc.* **8**:986–987.

Lowenstam, H. A., 1963, Biologic problems relating to the composition and diagenesis of sediments, in: *The Earth Sciences* (T. W. Donnelly, ed.), pp. 137–195, Rice University, Houston.

Lowenstam, H. A., 1964, Strontium–calcium ratio of skeletal aragonites from the recent marine biota at Palau and from fossil gastropods, in: *Isotopic and Cosmic Chemistry* (H. Craig, S. L. Miller, and G. J. Wasserburg, eds.), pp. 114–132, North Holland, Amsterdam.

Lutz, R. A., and Rhoads, D. C., 1977, Anaerobiosis and a theory of growth line formation, *Science* **198**:1222–1227.

Masumoto, B., Masumoto, M., and Hibino, M., 1934, Biochemical studies of Magaki (*Ostrea gigas* Thunberg). I. The seasonal variation in the chemical composition of *Ostrea gigas*, *J. Sci. Hiroshima Univ. Ser. A* **4**:47–56.

Milliman, J. D., 1974, *Marine Carbonates*, Springer-Verlag, Berlin, 313 pp.

Moberly, R., Jr., 1968, Composition of magnesium calcites of algae and pelecypods by electron microprobe analysis, *Sedimentology* **11**:61–82.

Muller, G., 1968, Exceptionally high strontium concentration in freshwater onkolites and mollusk shells in Lake Constance, in: *Recent Developments in Carbonate Sedimentology, Central Europe* (G. Muller and G. M. Friedman, eds.), pp. 116–127, Springer-Verlag, Berlin.

Nelson, D. J., 1961, The strontium and calcium relationships in Clinch and Tennessee River mollusks, in: Radioecology (V. Schultz and A. W. Klement, Jr., eds.), pp. 203–211, Reinhold, New York.

Nelson, D. J., 1962, Clams as indicators of strontium-90, Science 137:38–39.

Nelson, D. J., 1967, Microchemical constituents in contemporary and pre-Columbian clam shell, in: Quarternary Paleoecology, Proceedings of the VIIth Congress of the International Association for Quarternary Research (E. J. Cushing and H. E. Wright, Jr., eds.), pp. 185–204, Yale University Press, New Haven, Connecticut.

Newell, N. D., 1949, Phyletic size increase—an important trend illustrated by fossil invertebrates, Evolution 3:103–124.

Odum, H. T., 1957, Biogeochemical deposition of strontium, Publ. Inst. Mar. Sci. Univ. Texas 4:39–106.

Ohta, N., 1977, Relation between shell materials and environment, Sekko To Sekkai (Gypsum and Lime) 146:41–46.

Okada, M., 1943, Studies on the periodic pattern of hard tissues in animal body, Shanghai Evening Post, Medical Edition of September, 1943, pp. 26–31.

Omori, M., Kobayashi, I., Shibata, M., Mano, K., and Kamiya, H., 1974, On some problems concerning calcification and fossilization of taxodontid bivalves, in: The Mechanisms of Mineralization in the Invertebrates and Plants (N. Watabe and K. M. Wilbur, eds.), pp. 403–426, University of South Carolina Press, Columbia.

Pannella, G., 1975, Paleontological clocks and the history of the Earth's rotation, in: Growth Rhythms and the History of the Earth's Rotation (G. D. Rosenberg and S. K. Runcorn, eds.), pp. 253–284, John Wiley, London.

Pannella, G., 1976, Tidal growth patterns in recent and fossil mollusk bivalve shells: A tool for the reconstruction of paleotides, Naturwissenschaften 63:539–543.

Pilkey, O. H., and Goodell, H. G., 1963, Trace elements in recent mollusk shells, Limnol. Oceanogr. 8:137–148.

Pilkey, O. H., and Goodell, H. G., 1964, Comparison of the composition of fossil and recent mollusk shells, Geol. Soc. Am. Bull. 75:217–228.

Raup, D., and Stanley, S., 1971, Principles of Paleontology, W. H. Freeman, San Francisco, 370 pp.

Rosenberg, G. D., 1972, Patterned growth of the bivalve Chione undatella Sowerby relative to the environment, Ph.D. dissertation, University of California, Los Angeles, 220 pp.

Rosenberg, G. D., 1973, Calcium concentration in the bivalve Chione undatella Sowerby, Nature (London) 244:155–156.

Rosenberg, G. D., 1976, The biological clock paradox, Address to the Third International Conference on the Study of Time, Alpbach, Austria, July 1–10, 1976.

Rosenberg, G. D., 1977, The development of a chronobiologic time scale, J. Interdiscip. Cycle Res. 8:211–214.

Rosenberg, G. D., and Jones, C. B., 1975, Approaches to chemical periodicities in molluscs and stromatolites, in: Growth Rhythms and the History of the Earth's Rotation (G. D. Rosenberg and S. K. Runcorn, eds.), pp. 223–242, John Wiley, London.

Rosenberg, G. D., and Runcorn, S. K. (eds.), 1975, Growth Rhythms and the History of the Earth's Rotation, John Wiley, London, 538 pp.

Rucker, J. B., and Valentine, J. W., 1961, Salinity response of trace element concentration in Crassostrea virginica, Nature (London) 190:1099–1100.

Schelske, C. L., 1964, Ecological implications of radioactivity accumulated by molluscs, Ecology 45:149–150.

Schmalz, R. E., and Swanson, F. J., 1969, Diurnal variations in the carbonate saturation of seawater, J. Sediment. Petrol. 39:255–267.

Schoute-Vanneck, C. A., 1960, A chemical aid for the relative dating of coastal shell middens, S. Afn. J. Sci. **56**:67–70.

Segar, D. A., Collins, J. D., and Riley, J. P., 1971, The distribution of the major and some minor elements in marine animals. Part II. Molluscs, J. Mar. Biol. Assoc. U. K. **51**:131–136.

Stanton, R. J., and Dodd, J. T., 1970, Paleoecologic techniques—comparison of faunal and geochemical analyses of Pliocene paleoenvironments, Kettlemen Hills, California, J. Paleontol. **44**:1092–1121.

Swan, E. F., 1956, The meaning of strontium–calcium ratios, Deep Sea Res. **4**:71.

Tappan, H., and Loeblich, A. R., Jr., 1971, Geobiologic implications of fossil phytoplankton evolution and time–space distribution, in: Palynology of the Late Cretaceous and Early Tertiary (R. Kosanke and A. T. Cross, eds.), Geol. Soc. Am. Spec. Pap. **127**:247–340.

Taylor, J. D., Kennedy, N. J., and Hall, A., 1969, The shell structure and mineralogy of the Bivalvia: Introduction: Nuculacea–Trigonacea, Bull. Br. Mus. (Nat. Hist.) Zool. Suppl. **3**:1–125.

Thompson, T. G., and Chow, T. J., 1955, The strontium–calcium atom ratio in carbonate secreting marine organisms, in Pap. Mar. Biol. Oceanogr., Suppl. to Deep Sea Res. **3**:20–39.

Travis, D. F., 1968, The structure and organization of, and the relationship between, the inorganic crystals and the organic matrix of the prismatic region of Mytilus edulis, J. Ultrastruct. Res. **23**:183–215.

Turekian, K., and Armstrong, R. L., 1960, Magnesium, strontium, and barium concentrations and calcite–aragonite ratios of some Recent molluscan shells, J. Mar. Res. **18**:133–151.

Ueda, T., Suzuki, Y., and Nakamura, R., 1973, Accumulation of strontium in marine organisms. I. Strontium and calcium contents, CF and OR values in marine organisms, Bull. Jpn. Soc. Sci. Fish. **39**:1253–1262.

Vinogradov, A. P., 1953, The elementary chemical composition of marine organisms, Mem. Sears Found. Mar. Res. **2**:1–647.

Wada, K., 1961, Crystal growth of molluscan shells, Bull. Natl. Pearl Res. Lab. (Kokuritsu Shinju Kenkyusho Hokoku) **7**:703–828.

Wada, K., 1966, Taxonomic modification of amino acid composition in the proteins of molluscan shells, Bull. Jpn. Soc. Sci. Fish. **32**:253–259.

Wada, K., and Furuhashi, T., 1970, Studies on the mineralization of the calcified tissue in molluscs. XVII. Acid polysaccharide in the shell of Hydriopsis schlegeli, Bull. Jpn. Soc. Sci. Fish. **36**:1122–1126.

Walls, R. A., Ragland, P. C., and Crisp, E. L., 1977, Experimental and natural early diagenetic mobility of strontium and magnesium in biogenic carbonates, Geochim. Cosmochim. Acta **41**:1731–1737.

Weiner, S., Lowenstam, H. A., and Hood, L., 1977, Discrete molecular weight components of the organic matrices of mollusk shells, J. Exp. Mar. Biol. Ecol. **30**:45–51.

Wheeler, A. P., Blackwelder, P. L., and Wilbur, K. M., 1975, Shell growth in the scallop Argopecten irradians. I. Isotopic incorporation with reference to diurnal growth, Biol. Bull. (Woods Hole, Mass.) **148**:472–482.

White, L. K., Szabo, A., Carkner, P., and Chasteen, N. D., 1977, An electron paramagnetic study of Mn II in the aragonite lattice of a clam shell, Mya arenaria, J. Phys. Chem. **81**:1420–1424.

Wilbur, K. M., 1964, Shell formation and regeneration, in: Physiology of Mollusca (K. M. Wilbur and C. M. Yonge, eds.), pp. 243–282, Academic Press, New York.

Wilbur, K. M., and Simkiss, K., Calcified shells, in: Comprehensive Biochemistry, Vol. 26A (M. Florkin and E. H. Stotz, eds.), pp. 229–295, Elsevier, New York.

Wolf, K. H., Chilingar, G. V., and Beals, F. W., 1967, Elemental composition of carbonate skeletons, minerals and sediments, in: Carbonate Rocks, Developments in Sedimentology, Vol. 9B (G. V. Chilingar, H. J. Bissel, and R. W. Fairbridge, eds.), pp. 23–150, Elsevier, Amsterdam.

Wyckoff, R. W. G., 1972, *The Biochemistry of Animal Fossils*, Scientichnica, Bristol, 127 pp.

Yasamanov, N. A., 1973, Temperature of the habitat of Jurassic and Cretaceous brachiopods, cephalopods, and pelecypods in the western Transcaucasian Basin, *Geochem. Int.* **10**:583–590.

Zolotarev, V. N., 1974, Magnesium and strontium in the shell calcite of some modern pelecypods, *Geochem. Int.* **11**:347–353.

Chapter 5

Reconstructing Paleotemperature and Paleosalinity Regimes with Oxygen Isotopes

DANNY M. RYE and MICHAEL A. SOMMER II

1. Introduction 169
2. Oxygen Isotope Determinations 170
3. Isotopic Standards 172
4. Oxygen Isotope Geothermometry 176
    4.1. Theoretical Basis 176
    4.2. Historical Background 177
    4.3. Natural Variations in the $\delta^{18}O$ (SMOW) of Water 180
    4.4. Meteoric Waters: Rain, Snow, and Ice 181
    4.5. Ocean Water $\delta^{18}O$ (SMOW) Variations with Salinity 183
    4.6. Ocean Water $\delta^{18}O$ (SMOW) Variations with Time 185
    4.7. Other Environmental and Biological Parameters 186
    4.8. Aragonite–Calcite Thermometry (Independent δ_w and δ_c Determinations) 189
5. Application of the "Internal" Thermometer 190
    5.1. Results 190
    5.2. Estimation of Paleosalinity 196
    5.3. Discussion of Results 197
6. Summary 199
    References 200

1. Introduction

Naturally occurring oxygen is composed of the stable isotopes ^{16}O, ^{17}O, and ^{18}O. In 1950, Nier (1950) determined that these isotopes comprise, respectively, 99.759, 0.0374, and 0.2039% of the total atmospheric oxygen.

DANNY M. RYE • Department of Geology and Geophysics, Yale University, New Haven, Connecticut 06520.   MICHAEL A. SOMMER II • Department of Geology, Florida State University, Tallahassee, Florida 32306.

Thus, the ratio of ^{18}O to ^{16}O in air is about 1:489; however, in nature, this ratio can vary by about 10%. Many chemical and physical processes occurring in nature are accompanied by oxygen isotope fractionations. These fractionations are often temperature-dependent. The oxygen isotope paleotemperature technique makes use of the temperature-dependent oxygen isotope fractionation between $CaCO_3$ and H_2O. Because the differences in $^{18}O/^{16}O$ between two samples can be determined far more accurately than the actual isotopic abundances, we will consider in this chapter only the isotopic differences between samples, not the absolute ratio of ^{18}O to ^{16}O.

Since the development of the oxygen isotope paleotemperature method by Harold Urey and his associates in the early 1950s, the use of oxygen isotope data obtained from carbonate shells of marine organisms has contributed greatly to our knowledge of paleoenvironments. The objectives of this chapter are threefold:

1. To describe the early development of the oxygen isotope paleothermometer method. In addition to describing the development of the technique, we will introduce the nomenclature used by isotope geochemists and discuss possible errors and potential problems with the method.
2. To describe how and when the method can be applied to fossil molluscan species. To evaluate the molluscan data, it will be necessary to describe the variations in the oxygen isotopic composition of natural water and to describe some of the advances in our understanding of paleoenvironments obtained from the study of foraminifera. These studies will make it clear that the basic problem with the oxygen isotope paleotemperature technique is that the effects of temperature usually cannot be distinquished from those of salinity.
3. To describe a technique utilizing calcite–argonite oxygen isotope thermometry that will allow the determination of both paleotemperature and paleosalinity. We will include a practical example of how this technique can be applied to reconstructing paleotemperature and paleosalinity from the molluscan fossil record.

2. Oxygen Isotope Determinations

Because the nomenclature used in oxygen isotope paleothermometry evolved with the technique, it is necessary that we establish a consistent nomenclature before describing the theoretical basis and the development of the oxygen isotope paleotemperature technique. The parameters in the paleotemperature equation are a function of the way in which the samples were analyzed in the original paleotemperature experiments. Therefore, it is also important that we introduce the techniques used to analyze the oxygen isotopic composition both of $CaCO_3$ and of the water in which the $CaCO_3$ was precipitated. As we shall see, either the measurements must be carried out in the same way as the original experiments or corrections must be applied to the data to put them in a form compatible with the paleotemperature equation.

When analyzing the oxygen isotopic composition of shell material, carbon dioxide is the most suitable gas for use in the mass spectrometer. During a measurement, the unknown $^{18}O/^{16}O$ ratio of CO_2 from a sample is compared with the known $^{18}O/^{16}O$ ratio of a standard CO_2 sample by switching back and forth between sample and standard. This procedure ensures that conditions in the mass spectrometer are the same for both gases. The results are reported in the δ notation (see, for example, Epstein, 1953):

$$\delta = [(R_{sample}/R_{standard}) - 1] \times 1000$$

where R = mass 46/mass 44 = $^{12}C^{18}O^{16}O/^{12}C^{16}O^{16}O$ = $^{18}O/^{16}O$. Therefore, δ is defined as the per ml deviation* of the sample $^{18}O/^{16}O$ ratio from that of the standard.

The usual technique for obtaining CO_2 from carbonates for mass spectrometric analysis is to first remove any organic skeletal material by roasting the sample at about 400°C either in a helium atmosphere or under vacuum for about 30 min. After the organic material is removed, the carbonate is reacted with 100% phosphoric acid at 25.3°C. (Many analysts now use 50°C to accelerate the reaction process.) The evolved CO_2 is then cleaned of any contaminant gases using fractional freezing, and collected (see McCrea, 1950; Urey et al., 1951; Epstein et al., 1953).

After collection, the CO_2 is introduced into a ratio mass spectrometer for comparison with a standard CO_2. The standard CO_2 gas is usually the CO_2 gas liberated from a standard carbonate† by reacting it with 100% phosphoric acid at the same temperature at which the standard gas was evolved.

The standard way of reporting the ^{18}O measurement when analyzing carbonates for paleotemperature work is in δ notation, where R_{sample} is the mass 44/mass 46 ratio of the CO_2 from the sample and R_{std} is the mass 44/

---

* We will use the symbol ‰ to mean per mil (parts per thousand deviation).
† This standard carbonate is known as a working standard; the isotopic δ values relative to the working standard are converted to δ values relative to an international standard.

mass 46 ratio of the CO_2 from the PDB-1 standard carbonate.[*] We will refer to this value as δ_c. That is,

$$\delta_c = \left(\frac{{}^{18}O/{}^{16}O \text{ of } CO_2 \text{ gas from sample}}{{}^{18}O/{}^{16}O \text{ of } CO_2 \text{ gas from PDB-1 standard)}} - 1 \right) \times 1000$$

As we shall discuss later, when oxygen isotope values are used as a paleothermometer, the isotopic composition of the water in which the animal grew must also be determined. The isotopic composition of the water is analyzed by measuring the isotopic composition of CO_2 gas that has been equilibrated with the water at 25.3°C (Epstein and Mayeda, 1953). When used in the paleotemperature technique, the isotopic composition of the water is expressed as the per mil deviation of the equilibrated CO_2 relative to the PDB-1 standard CO_2 gas. We will refer to this value as δ_w. That is,

$$\delta_w = \left(\frac{{}^{18}O/{}^{16}O \text{ of } CO_2 \text{ equilibrated with } H_2O}{{}^{18}O/{}^{16}O \text{ of } CO_2 \text{ from PDB-1 standard}} - 1 \right) \times 1000$$

It is important to note that neither the isotopic composition of carbonates nor the isotopic composition of water is reported on a true Peedee Belemnite (PDB) scale for paleotemperature work. The PDB scale for carbonate would be the oxygen isotopic composition of carbonate in the sample $CaCO_3$ relative to the oxygen isotopic composition of the PDB-1 $CaCO_3$. The oxygen isotopic composition of H_2O on the PDB scale would be the oxygen isotopic composition of H_2O relative to the oxygen isotopic composition of the PDB-1 $CaCO_3$. We will discuss this further in the following section.

3. Isotopic Standards

There are two internationally accepted reference standards used to report variations in oxygen isotope ratios: Peedee Belemnite (PDB) and Standard Mean Ocean Water (SMOW). The PDB standard is a sample of belemnite guard from the Peedee Formation in North Carolina. The original supply is now exhausted, so that secondary standards are now used, but as mentioned before, the values are reported relative to PDB-1.

The SMOW standard was originally a hypothetical water sample with an isotope ratio of oxygen similar to that of an average sample of ocean

---

[*] Further discussion of standards may be found in the next section.

water. It was defined by Craig (1961a) in terms of a National Bureau of Standards reference water, NBS-1, as follows:

$$(^{18}O/^{16}O) \text{ SMOW} = 1.008 \, (^{18}O/^{16}O) \text{ NBS-1}$$

More recently, a large quantity of water called SMOW was prepared by R. Weiss and H. Craig for the International Atomic Energy Agency (Vienna). According to Friedman and O'Neil (1977), the SMOW is apparently within analytical limits identical to the originally defined SMOW.

To relate the SMOW and the PDB scales, the oxygen isotope fractionation factors (α)* for the CO_2–H_2O equilibration and for the $CaCO_3$–CO_2 acid reaction must be known. Any uncertainty in these fractionation factors will result in an uncertainty in the relationship between PDB and SMOW. To relate $\delta^{18}O$ values of calcite on the PDB and SMOW isotopic scales, the following expressions are used†:

$$\delta^{18}O \, (\text{SMOW}) = 1.03086 \, \delta^{18}O \, (\text{PDB}) + 30.86$$

and

$$\delta^{18}O \, (\text{PDB}) = 0.97006 \, \delta^{18}O \, (\text{SMOW}) - 29.94$$

These relationships are based on the following fractionation factors $(\alpha\text{'s})$ at 25°C $[\alpha = (^{18}O/^{16}O)$ in species $A/(^{18}O/^{16}O)$ in species B]:

$$\alpha = 1.0412 \quad \text{for } CO_2/H_2O$$

$$\alpha = 1.01025 \text{ for } H_3PO_4\text{-liberated } CO_2/CaCO_3$$

$$\alpha = 1.00022 \text{ for } H_3PO_4\text{-liberated } CO_2 \text{ from PDB-1}/CO_2$$
$$\text{in equilibrium with SMOW}$$

Now let us examine the relationship between the $\delta^{18}O$ (PDB) and δ_c values and the $\delta^{18}O$ (SMOW) and δ_w values. The relationship between the oxygen isotope standards and the CO_2 gases normally prepared from them is shown in Fig. 1. The relationship between the CO_2 gases normally prepared and the standards is shown because in the paleotemperature equation, neither the SMOW nor the PDB scale is used, which has led to

---

* $\alpha = R_A/R_B$, where R is the ratio of any two isotopes and A and B refer to chemical compounds A and B.

† To convert δ values from one standard to another, Craig (1957) has given the equation $\delta(x - A) = \delta(x - B) + \delta(B - A) + 10^{-3} \delta(x - B)\delta(B - A)$, where $\delta(x - A)$ is the δ value of the unknown on the A scale, $\delta(x - B)$ is the δ value of the unknown on the B scale, and $\delta(B - A)$ is the δ value of the standard for the A scale on the B scale.

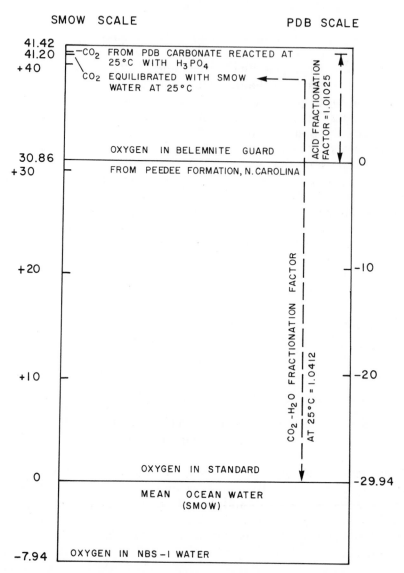

Figure 1. Relationship among PDB, SMOW, and other oxygen standards. Note that a water with a $\delta^{18}O$ (SMOW) value of 0‰ has a δ_w value of -0.22‰; that is, CO_2 gas in equilibrium with SMOW water is 0.22‰ lighter than CO_2 gas produced by reaction of PDB-1 carbonate with H_3PO_4 at 25°C. After Friedman and O'Neil (1977).

a great deal of confusion. As we have shown earlier, both δ_c and δ_w are defined in terms of the CO_2 gases produced in the extraction technique relative to the CO_2 produced from the PDB-1 $CaCO_3$ standard.

When reporting oxygen isotope data for carbonates, it is important to specify whether δ_c or $\delta^{18}O$ (PDB) is used, because when CO_2 is liberated from carbonate reacted with 100% phosphoric acid, only two thirds of the oxygen is incorporated into the CO_2. There is a resultant kinetic isotope fractionation with the CO_2 being about 10‰ enriched in ^{18}O relative to the original oxygen in the carbonate. As long as one is interested only in isotopic differences between different samples of the same carbonate, the kinetic isotope fractionation will be the same for both the sample and the standard, and the measured isotopic difference between CO_2 gases will be the same as the true isotopic difference between $CaCO_3$ samples. However, in comparing $^{18}O/^{16}O$ ratios of carbonates with those for other minerals, or other carbonates, it is important to correct the carbonate analyses for residual oxygen not extracted during the H_3PO_4 treatment. We must therefore know the exact values of the kinetic fractionation factors for each type of carbonate mineral. Fortunately, the two polymorphs (aragonite and calcite) that must be considered when analyzing molluscan material apparently have the same kinetic isotopic fractionation factors (Tarutani et al., 1969). Therefore, for our purposes, the measured value of δ_c is the same as the $\delta^{18}O$ value of the carbonate on the PDB scale. The δ_w value, however, is not the $\delta^{18}O$ value of water on the PDB or the SMOW scale.

In his paper defining the SMOW scale, Craig (1961a) stated that "the $\delta^{18}O$ value of Chicago PDB-1 carbonate standard (CO_2 from reaction with 100% H_3PO_4 at 25°C) is +0.22 per mil on the SMOW scale." The correct statement should have been that the $\delta^{18}O$ value of the CO_2 produced by reaction of 100% H_3PO_4 with PDB-1 calcite at 25°C is +0.22‰ relative to CO_2 equilibrated at 25°C with SMOW water. This statement means that water with a $\delta^{18}O$ value of 0‰ on the SMOW scale has a δ_w value of −0.22‰ (see Fig. 1).

From the preceding discussion, it follows that the relationship among $\delta^{18}O$ (SMOW), δ_w, $\delta^{18}O$ (PDB), and δ_c is:

$$\delta^{18}O \text{ (SMOW)} = 1.03086\, \delta_c + 30.86$$

$$\delta_c = \delta^{18}O \text{ (PDB)} = 0.97006\, \delta^{18}O \text{ (SMOW)} - 29.94$$

and

$$\delta^{18}O \text{ (SMOW)} = 1.00022\, \delta_w + 0.22$$

$$\delta_w = 0.99978\, \delta^{18}O \text{ (SMOW)} - 0.22$$

4. Oxygen Isotope Geothermometry

4.1. Theoretical Basis

The theoretical basis for oxygen isotope geothermometry is given by Urey (1947) and Bigeleisen and Mayer (1947), who also discuss the general principles of isotopic exchange reactions. The equilibrium constant (K) for isotopic exchange is temperature-dependent, and it commonly varies with temperature in a simple manner. For low temperatures ($\approx <300°$ Kelvin), ln K is usually approximated by $A + B/T$ (where T is degrees Kelvin and A and B are constants), which, for small temperature ranges, can be approximated by ln $K = C + DT$ (where T is degrees Kelvin and C and D are constants). At higher temperatures, ln K is usually approximated by $A' + B'/T^2$ (where T is in degrees Kelvin and A' and B' are constants). However, we are usually interested in the fractionation factor (α), rather than in the equilibrium constant. The fractionation factor α is defined as the ratio of the numbers of any two isotopes in chemical compound A divided by the corresponding ratio for another chemical compound B, that is, $\alpha_{A-B} = R_A/R_B$, where $R = {}^{18}O/{}^{16}O$, for instance. In terms of quantities actually measured in the laboratory (δ values), this expression becomes

$$\alpha_{A-B} = (1 + \delta_A/1000)/(1 + \delta_B/1000)$$

If the isotopes are distributed over all possible positions in compounds A and B, the α value is related to the equilibrium constant K by the relationship $\alpha = K^{1/n}$, where n is the number of atoms exchanged. For simplicity, isotope-exchange reactions are written so that only one atom is exchanged. For example:

$$\tfrac{1}{2}C^{16}O_2 \text{ (gas)} + H_2{}^{18}O \text{ (liquid)} \rightleftarrows \tfrac{1}{2}C^{18}O_2 \text{ (gas)} + H_2{}^{16}O \text{ (liquid)}$$

so that α will be equal to K:

$$K = \alpha = (C^{18}O^{18}O)^{1/2}(H_2{}^{16}O)/(C^{16}O^{16}O)^{1/2}(H_2{}^{18}O)$$

$$= ({}^{18}O_{CO_2})({}^{16}O_{H_2O})/({}^{16}O_{CO_2})({}^{18}O_{H_2O})$$

$$= ({}^{18}O/{}^{16}O_{CO_2})/({}^{18}O/{}^{16}O_{H_2O})$$

Because ln K or ln α is proportional to $1/T^2$ or $1/T$, a useful expression is ln α. Note that ln α is just ln $(1 + \delta_A/1000) - $ ln $(1 + \delta_B/1000)$, and for small δ (≤ 10), this is approximately $(\delta_A - \delta_B)/1000$. Therefore, 1000·

$\ln \alpha = \delta_A - \delta_B$, which is proportional to $1/T^2$ or $1/T$. We often write the expression $\delta_A - \delta_B$ as Δ_{A-B} or just Δ.

Craig (1965) has shown that over a small temperature range, the equation $1000 \ln \alpha_{A-B} = A' + B'/T^2$ can be approximated by $t°C = a + b(\delta_A - \delta_B) + c(\delta_A - \delta_B)^2$, where $t°C$ is degrees centigrade and a, b, and c are constants. We can see, therefore, that the $1000 \ln \alpha_{A-B}$ or Δ_{A-B} $(\delta_A - \delta_B)$ value between any two materials that contain oxygen and that commonly coexist at equilibrium in nature can in theory be used as an oxygen isotope geothermometer. Generally, these will be two minerals; however, the carbonate paleothermometer scale is based on $CaCO_3$ and ocean water. As we shall see in later sections, this represents a major problem, because in dealing with fossil material, we cannot determine the $\delta^{18}O$ value for water that was present at the time the $CaCO_3$ was precipitated.

For any two materials to be useful as an oxygen isotope thermometer, they must satisfy the following requirements:

1. They must form together in isotopic equilibrium.
2. The equilibrium oxygen isotopic fractionation factor must exhibit regular temperature variations that are large enough to be readily measurable.
3. Equilibrium must be "frozen in"; that is, both materials must remain unaltered through geological time.
4. The temperature dependence of the fractionation factor between the two materials must be determined.

In this chapter, we shall present two types of geothermometers. The first of these is the standard oxygen isotope paleothermometer; the second is a geothermometer based on the oxygen isotope fractionation between calcite and aragonite.

4.2. Historical Background

In 1947, Harold Urey (1947) presented a paper concerning the thermodynamics of isotopic systems and suggested that variations in the temperature of water from which $CaCO_3$ precipitates would lead to measurable variations in the $^{18}O/^{16}O$ ratio of calcium carbonate. The development of precise mass spectrometry (Nier, 1947; McKinney et al., 1950) and careful sample-preparation procedures (McCrea, 1950) made possible the accurate measurement of these variations in the isotopic abundances of oxygen in $CaCO_3$ to a precision of 0.1–0.2‰. McCrea (1950) provided the first link between theory and the actual establishment of a paleotemperature curve by carrying out experiments on inorganically precipitated $CaCO_3$.

He clearly demonstrated that there is a temperature-dependent isotope fractionation of the $^{18}O/^{16}O$ ratio between $CaCO_3$ and water. The real break-through occurred when Epstein et al. (1951) published a carbonate–water isotopic temperature scale using $CaCO_3$ precipitated by living molluscs grown in controlled experiments or grown in well-characterized natural environments. In 1953, they repeated the experiment because of experi-mental errors in the earlier work that resulted from the roasting technique to remove organic impurities. As a result of the new experiments, Epstein et al. (1953) proposed the paleotemperature equation, where $t = 16.5 - 4.3\delta + 0.14\delta^2$ (Fig. 2), in which t is the temperature in °C and δ is the per mil difference between the ratio of masses $46(^{12}C^{18}O^{16}O)$ to $44(^{12}C^{16}O^{16}O)$ of the sample CO_2 and the PDB-1 standard CO_2 corrected for the isotopic composition of the water in which the animal grew. They called this δ value the δ-corrected value, and in the nomenclature we have used herein, it is equal to $\delta_c - \delta_w$. This form of the equation was very useful for mod-ern samples in which the δ_w value could be measured or in which the δ_w value could be assumed to be close to 0‰, but it obscured the fact that two measurements must be made to get a unique temperature.

Epstein and Mayeda (1953) showed that there could be large variations in the ^{18}O content of natural waters. For this reason, they for-mulated the paleotemperature scale in the form

$$t = 16.5 - 4.3(\delta_1 - A) + 0.14(\delta_1 - A)^2$$

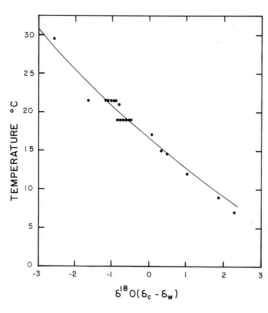

Figure 2. Isotopic temperature scale and original data points of Epstein et al. (1953). Temperature is in de-grees Celsius. The δ values are the δ-corrected values, which are equal to $\delta_c - \delta_w$. After Epstein et al. (1953).

where t is the temperature in °C, A equals δ_w, and δ_1 equals δ_c in our notation. Clearly, t depends on both A and δ_1 and A is an unknown quantity when paleooceans are considered. They showed that there was enough variation in the isotopic composition of modern surface ocean water that one could make errors at least as large as 30°C in the temperature if the wrong A value were used.

Craig (1965) made a further small adjustment in the paleotemperature equation. He noted that no correction was made for the ^{18}O content of the original CO_2 that was equilibrated with the H_2O, and he noted that there were many small instrumental errors for which corrections could be made. Making those corrections, he obtained a slightly different equation in which

$$t°C = 16.9 - 4.2(\delta_c^+ - \delta_w^+) + 0.13(\delta_c^+ - \delta_w^+)^2$$

The δ_c^+ and δ_w^+ terms refer to the true δ values of the CO_2 with no instrumental bias.

Clearly, if the δ_w^+ value is not known, severe limitations are placed on the usefulness of the paleotemperature equation. There are, in addition, some possible problems with the 1953 experiments.

Epstein et al. (1953) analyzed Haliotis cracherodii, H. rufescens, Kelletia Kellettii, Strombus gigas, Macoma sp., Mytelles californianus, and an unidentified calcareous worm. They analyzed both aragonitic and calcitic polymorphs of $CaCO_3$. Epstein and his co-workers were careful to note which sections of the shells contained which polymorph before they started their experiments. However, there are two parts of their procedure that might have led to small errors, and these errors in turn probably led to their fractionation curve being somewhat biased toward the calcitic polymorph of $CaCO_3$.

The first of the potential problems occurred in their tank experiments. The rims of the shells of large abalones were notched or drilled. The damaged areas were then allowed to regenerate in a controlled-temperature environment. After an animal regenerated its shell, the isotopic composition of CO_2 equilibrated with water in which the animal grew and of CO_2 from $CaCO_3$ were measured and a δ-corrected $(\delta_c - \delta_w)$ value established at the temperature of the tank. They did not check to make certain the animal regenerated its shell with the original polymorph. Recent studies (Wilbur, 1972) have shown that the polymorph secreted in the regenerated shell may not be the polymorph normally deposited in that region of the shell. It is therefore possible that Epstein et al. (1953) did not analyze the polymorph of $CaCO_3$ that the animal would normally secrete at the shell margin.

The second potential problem was in the roasting technique. To analyze the $CaCO_3$, they had to remove organic material. They did this by roasting the samples at 470°C in a helium atmosphere. A temperature of 470°C is likely to accelerate the inversion of metastable aragonite to calcite. During the inversion, it is likely that exchange could take place. One of their criteria for a good roasting procedure was that calcite and aragonite δ_c values were equal for animals grown at a given temperature. It is therefore likely that any errors introduced in the roasting procedure would tend to bias the values toward the calcite polymorph values.

We think that the experimental problems to which we have just alluded caused only minor discrepancies with the fractionation curve and that the curve is, for most practical purposes, correct. This conclusion is substantiated by the fact that the inorganic calcite–water oxygen isotope fractionation curve determined by O'Neil et al. (1969) is essentially identical to the paleotemperature curve.

There are, however, some important problems associated with applying the oxygen isotope calcium carbonate–water fractionation curve:

1. When examining fossil material, it must be assumed or be demonstrated that the original isotopic composition is preserved.
2. It must be assumed that the calcium carbonate was precipitated in isotopic equilibrium with water.
3. The most serious problem may be that the δ_w value is not known. Therefore, an assumption has to be made concerning the oxygen isotopic composition of the water.

For many years, these assumptions, although acknowledged, were essentially ignored. Some attempts were usually made to establish that the shell material was original. It was usually assumed that the δ_w values of ancient oceans were similar to the average open ocean values of today. Temperatures were calculated usually assuming a $\delta^{18}O$ (SMOW) value or δ_w value of 0‰. The results were considered valid if the temperatures were reasonable (between 0 and 30°C) and the results were more or less comparable among individual specimens analyzed from a fossil assemblage. The problem with this approach is that oftentimes only values that agreed with preconceived ideas were accepted. In the following sections, we will evaluate each of these problems.

4.3. Natural Variations in the $\delta^{18}O$ (SMOW) of Water

To understand the natural variations that occur in the oxygen isotopic composition of natural surface waters, it is necessary to look at the natural water cycle. This cycle has been compared to a multiple-stage distillation

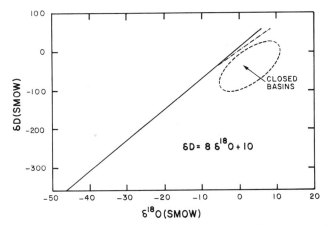

Figure 3. Deuterium and oxygen isotopic variations in rivers, lakes, rain, and snow. Data are expressed as per-millage enrichments relative to SMOW. Note that meteoric water follows the relationship $\delta D = 8\,\delta^{18}O + 10$. After Craig (1961b).

column with reflux of the condensate to the reservoir (Epstein and Mayeda, 1953). The oceans correspond to the reservoir, and the ice fields at the poles correspond to the highest stages of the column. In all processes concerning the water cycle, the hydrogen isotopes (hydrogen and deuterium) are fractionated in proportion to the oxygen isotopes, because a similar difference in vapor pressures exists between H_2O and HDO in one case and $H_2{}^{16}O$ and $H_2{}^{18}O$ in the other. Therefore, the hydrogen and oxygen isotope distributions are correlated in meteoric water, that is, water that has gone through the meteorological cycle of evaporation, condensation, and precipitation. Craig (1961b) has given the following relationship (Fig. 3):

$$\delta D = 8\delta^{18}O + 10$$

where both δ values are relative to SMOW.

4.4. Meteoric Waters: Rain, Snow, and Ice

The comparison of D/H and $^{18}O/^{16}O$ ratios shows that atmospheric precipitation normally follows a Rayleigh process at liquid–vapor equilibrium. The atmospheric Rayleigh process explains why, at higher altitudes and·latitudes, fresh water becomes progressively lighter isotopically, whereas tropical samples show very small depletion relative to open ocean water (Fig. 4). However, because the processes determining the

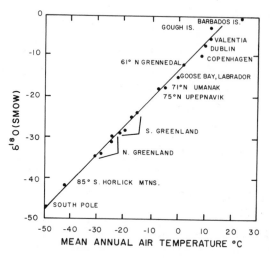

Figure 4. Mean annual $\delta^{18}O$ of precipitation as a function of the mean annual air temperature at the earth's surface. Note that $\delta^{18}O$ values are progressively lighter as the mean annual temperature becomes lower. After Dansgaard (1964).

isotopic content of precipitation are much more complex than a simple Rayleigh distillation, there is often only qualitative agreement, rather than quantitative agreement. When water evaporates from the surface of the ocean, the water vapor is enriched in H and ^{16}O relative to D and ^{18}O because $H_2^{16}O$ has a higher vapor pressure than $HD^{16}O$ and $H_2^{18}O$. Under equilibrium conditions at 25°C, the fractionation factors (αH_2O–vapor) for evaporating water are 1.0092 for $^{18}O/^{16}O$ and 1.074 for D/H (Craig and Gordon, 1965). However, under natural conditions, the actual isotopic composition of water vapor is significantly more negative than the predicted equilibrium values due to kinetic effects (Craig and Gordon, 1965). Vapor leaving the surface of the ocean cools as it rises, and rain forms when the dew point is reached. During precipitation from the moist air mass, the vapor is continuously depleted in the heavy isotopes, since the rain leaving the system is enriched in ^{18}O and D. If the air mass moves poleward or to higher altitudes and becomes cooler, additional rain forms having less ^{18}O than the first rain (Fig. 5). When water is cooled from 25°C (equator region) to about −15°C (polar region), there is a hundredfold decrease in water vapor content. Thus, the air masses that arrive at higher latitudes have lost about 99% of their moisture, and the residual water vapor has a $\delta^{18}O$ (SMOW) of −45‰. A snowfall from this vapor would have a $\delta^{18}O$ (SMOW) of ≃ −35‰. The relationship among the $\delta^{18}O$ of precipitation, the percentage of vapor remaining, and the temperature of precipitation is shown in Fig. 6. As a result of this process, the $\delta^{18}O$ of fresh water is lighter as the poles are approached and at higher elevations.

In contrast to condensation, evaporation takes place mostly under kinetic conditions, especially when the relative humidity is significantly less than 100%. In the Sahara, for instance, Fontes and Gonfiantini (1967)

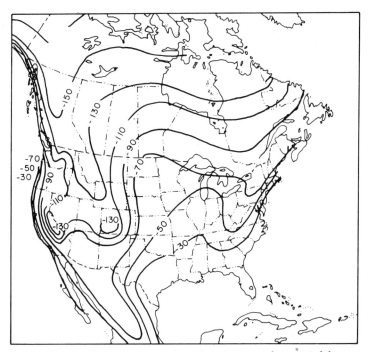

Figure 5. Map of North America showing how the D/H ratios of meteoric lake, stream, river, and spring waters vary over the land surface. Contours are δD values relative to SMOW. Note how δD values get lighter both at higher altitudes and at higher latitudes. After Sheppard *et al.* (1969); generalized from data by Friedman *et al.* (1964), and Dansgaard (1964).

have measured $\delta^{18}O$ values as heavy as $+30‰$ and δD values as heavy as $+129‰$. Kinetic fractionation during evaporation from the ocean is responsible for the fact that the meteoric water line, shown in Fig. 2, does not pass through the point $\delta D = 0$, $\delta^{18}O = 0$.

4.5. Ocean Water $\delta^{18}O$ (SMOW) Variations with Salinity

The isotopic composition of ocean water has been discussed in detail by Redfield and Friedman (1965), Craig and Gordon (1965), and Broecker (1974). Ocean water with a salinity of 35 parts per thousand exhibits a $\delta^{18}O$ range of less than 1‰. However, evaporation processes strongly affect both the isotopic composition and the salinity. Evaporation will increase the salinity and will cause a preferential depletion in the lighter isotopes of oxygen that became enriched in the vapor phase. For this reason, highly

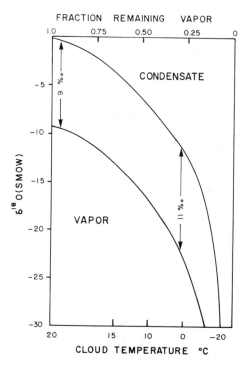

Figure 6. $\delta^{18}O$ in cloud vapor and condensate plotted as a function of the remaining vapor in the cloud for a Rayleigh process. The temperature of the cloud in degrees Celsius is shown on the lower axis. $\delta^{18}O$ values are relative to SMOW. The increase in fractionation with decreasing temperature is taken into account. After Dansgaard (1964).

saline surface waters generally have high ^{18}O content. This effect is well illustrated by surface water from the Gulf of Mexico. Low salinities, caused by fresh-water and melt-water dilution, are correlated with a low ^{18}O concentration (Epstein and Mayeda, 1953; Redfield and Friedman, 1965). An example of this dilution effect is demonstrated in a $\delta^{18}O$ vs. salinity diagram from North Atlantic surface water (Fig. 7). Those samples with salinities of about 36 parts per thousand have $\delta^{18}O$ values about 1‰ higher than samples of SMOW, whereas water taken close to Greenland with a salinity of 16 parts per thousand has an oxygen isotopic composition of $-11‰$ (SMOW). This translates into an error of approximately $30°C$ in the temperature calculated from the paleotemperature curve if 0‰ were used instead of $-11‰$. All the North Atlantic surface water (NASW) falls along a regression slope, on a $\delta^{18}O$ (SMOW) vs. salinity plot (Fig. 7), suggesting that these waters are mixtures of normal ocean water with fresh water that has a fixed salinity and $\delta^{18}O$ value. For a full discussion of $\delta^{18}O$ vs. salinity for the world's oceans, see Craig and Gordon (1965) and Broecker (1974) (other important effects include the formation of sea ice). The important point to be made here is that in near-shore environments where salinity can be very low, the $\delta^{18}O$ of the water can be very light, whereas in environments of high evaporation and high salinity, the $\delta^{18}O$ of the water can be very heavy. Therefore, it is extremly important in

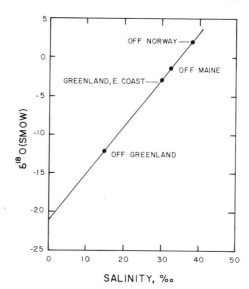

Figure 7. Relationship between $\delta^{18}O$ and salinity of surface water from the North Atlantic Ocean. The standard is SMOW; salinity is given in parts per thousand. Data from Epstein and Mayeda (1953) and Craig and Gordon (1965).

looking at fossil material to be able to recognize paleoenvironments (from the fossil record) if we are to apply the paleotemperature curve in a meaningful way. Clearly, if one is dealing with estuarine environments or desert environments, δ_w values of the ambient water could be significantly different from the δ_w value of the average ocean. If open-ocean conditions can be deduced, we can assume that the $\delta^{18}O$ and the salinity of the water were fairly constant. However, the δ_w value still cannot be known with certainty, and at best only approximate relative temperatures can be determined.

4.6. Ocean Water $\delta^{18}O$ (SMOW) Variations with Time

In addition to local variation, an important question concerning isotopic composition of ocean water is how constant its isotopic composition has been throughout geological history. If all the continental ice sheets in the world were melted, the $\delta^{18}O$ values of the oceans would be lowered by about 0.5‰ (Craig, 1965). On the other hand, estimates for the maximum enrichment of the oceans in ^{18}O during the Pleistocene glaciation include 0.4‰ (Emiliani, 1955), 1.0% (Craig, 1965), and 1.4‰ (Shackleton, 1967). At least throughout Phanerozoic time, the isotopic composition of ocean water has probably fluctuated between approximately -0.6 and $+1.4$‰ (SMOW) due to the amount of light oxygen stored as the result of the presence of the varying amounts of polar ice. If the δ_c values from Pleistocene animals are assumed to result from temperature alone and the ice storage effect is ignored, the temperatures calculated from the paleotemperature equation (assuming $\delta_w = 0$) will be as much as approximately 6°C too low.

The problem of neglecting the isotopic composition of the water in which the shell material grew or assuming it has remained equal throughout time was pointed out by Emiliani (1955, 1958, 1964, 1966) in a number of classic papers. Emiliani pioneered the work of paleoclimatology based on the $\delta^{18}O$ of biogenic carbonate—primarily foraminifera collected from deep-sea sediments. As Emiliani pointed out in discussing his work on Pleistocene sediments, the isotopic signal observed in foraminifera from the Pleistocene is actually due to two main features: growth temperatures and the isotopic composition of sea water. The increase in $^{18}O/^{16}O$ ratio of foraminifera during glacial times reflects an increase in both the ^{18}O content of seawater because of the ^{16}O enrichment in glacial ice and the isotopic fractionation during a decrease in ocean temperature. Emiliani argued that two thirds of the total isotopic signal was due to a temperature change and one third to ice storage. Subsequently, Shackleton (1967) and Shackleton and Opdyke (1973) have argued that the proportions of the effects should be reversed—that two thirds of the isotopic signal is due to ice storage and one third to temperature. Most investigators now recognize that Shackleton's conclusions are most likely correct. These effects account for approximately 1.4‰ of the variations observed in foraminifera from Pleistocene cores, and clearly the ^{18}O content of oceans has varied through time. The results have relegated paleotemperature determinations to "isotopic temperature," which should *not* be confused with "real" temperatures.

4.7. Other Environmental and Biological Parameters

Biological factors play an important role in the recording of temperature in calcareous skeletal material. A study of the Bermuda biota (Epstein and Lowenstam, 1953) shows that marine animals may deposit shell over only a portion of the total temperature range, rather than over the entire ambient range. Therefore, the analyses of such shell material will yield different temperatures for different species. Thus, only through studies of seasonal variations in a number of species can one obtain an accurate estimate of local temperature range from the paleotemperature equation even if the isotopic composition of the water is correctly assumed.

Figure 8A is a reproduction of the Epstein and Lowenstam (1953) data from a Recent *Strombus gigas*. Subsamples were obtained by grinding successive layers of shell, parallel to growth lines from, quoting Epstein and Lowenstam (1953, p. 430): "a rectangular piece of shell, whose length was the thickness of the lip of the shell and which was approximately

5mm., in width and breadth, [which] was cut from the lip of the shell.'' Temperatures were obtained using the Epstein and Mayeda (1953) form of the paleotemperature equation [$t°C = 16.5 - 4.3 (\delta_1 - A) + 0.14 (\delta_1 - A)^2$]. Their data clearly indicate the presence of variations in the growth of the gastropod shell, which in turn closely reflect seasonal variations in the Bermuda surface water temperature ($\sim$17–28°C).

Figure 8B shows the data of Epstein and Lowenstam (1953) for the bivalve *Chama macerophylla*. Again quoting from Epstein and Lowenstam (1953, p. 432): ''In the case of *Chama macerophylla*, a cross-sectional

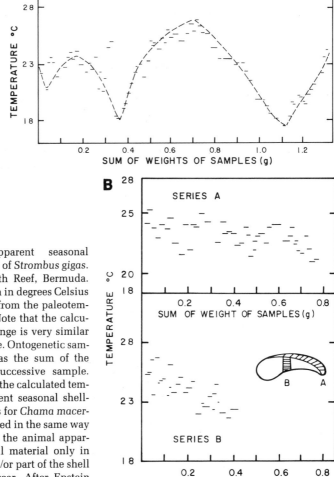

Figure 8. (A) Apparent seasonal growth temperatures of *Strombus gigas*. Collected from North Reef, Bermuda. Temperature is given in degrees Celsius and was calculated from the paleotemperature equation. Note that the calculated temperature range is very similar to the observed range. Ontogenetic sampling is presented as the sum of the weights for each successive sample. Note the cyclicity of the calculated temperature. (B) Apparent seasonal shell-growth temperatures for *Chama macerophylla* data presented in the same way as in (A). Note that the animal apparently produces shell material only in summer months and/or part of the shell is dissolved each year. After Epstein and Lowenstam (1953).

piece of shell approximately 5mm. in thickness was cut across the complete half-shell extending from the hinge to the lip. . . ." Series A represents the data from "samples cut in succession parallel to the growth lines from the outside 5 mm. layer." Series B results are for "samples cut from a rectangular piece of shell covering the thickness of the shell and approximately 5 mm. in the other dimension." Ontogenetic sampling was done by cutting successive layers of shell parallel to growth lines. The results indicate that this species deposits shell material during the warm part of the year at temperatures above 21°C. Epstein and Lowenstam point out that the shell may be partially dissolved and reprecipitated during the summer months, but conclude that the data probably resulted from the animal's growing its shell almost exclusively during the summer months.

Epstein and Lowenstam (1953) further pointed out that the oxygen isotope data suggest that in general, the majority of the Bermuda bivalves appear to grow primarily during warm seasons, whereas the gastropods generally appear to grow in both winter and summer.

This study clearly demonstrates the necessity of sampling successive (time-parallel) growth bands when analyzing shell material and the need to analyze a wide taxonomic range of organisms. For instance, if an entire shell were to be ground up, the "average" temperature obtained would be heavily biased toward maximum growth temperatures.

In nature, we can often find animals that do not precipitate $CaCO_3$ in isotopic equilibrium with their environment. The equilibrium deposition of $CaCO_3$ theoretically should be closely approximated in many cases. It is generally assumed that carbonate deposited is continually bathed with water that has the same isotopic composition as the water in the surrounding environment and therefore may effectively exchange its oxygen with the surrounding water during the process of deposition. However, in many organisms, or parts of organisms, the water in contact with the $CaCO_3$ may not have the same isotopic composition as the ambient sea water. Such an organism will produce $CaCO_3$ that is not in equilibrium with sea water and will display what Urey *et al.* (1951) termed a "vital" effect.

Many organisms display "vital" effects. For instance, coelenterates and some algae fractionate oxygen isotopes in their metabolism (Lowenstam and Epstein, 1954), and echinoderms show marked fractionation of both carbon and oxygen isotopes (Weber and Raup, 1966). In addition, Keith *et al.* (1964) have demonstrated that the $\delta^{18}O$ values of argonite from ligament fibers from pelecopods are very different from the $\delta^{18}O$ values of the shell material.

4.8. Aragonite–Calcite Thermometry (Independent δ_w and δ_c Determinations)

In principle, there is an approach to circumvent the problem of our inability to directly measure the δ_w value for fossil material. If two (or more) minerals are precipitated in isotopic equilibrium with one another, then the isotopic fractionation between each mineral pair can be used to calculate the temperature of formation. Barring coincident slopes in the mineral–water fractionation curves, an assemblage of n phases can yield $n - 1$ independent temperature determinations (Epstein and Taylor, 1967). Therefore, if the oxygen isotopes are measured for two different minerals from the same animal, it may be possible to determine a growth temperature. If the temperature can be determined, then the δ_c value and the calcite–water oxygen isotope fractionation curve can be used to determine the δ_w value. In this way, absolute temperatures and estimated salinities from the δ_w value can be obtained. However, it remains to be shown that this method of "internal" thermometry can be applied to biological systems.

Longinelli (1966) attempted to use $\Delta^{18}O$ (phosphate–calcite) thermometry. Longinelli and Nuti (1973), however, have shown that the $\Delta^{18}O$ (phosphate–H_2O) temperature curve and the $\Delta^{18}O$ (calcite–H_2O) temperature curve have essentially the same slope. Therefore, the $\Delta^{18}O$ (phosphate–calcite) value is not a function of temperature. More recently, Sommer and Rye (1978) have developed an empirical aragonite–calcite oxygen isotope fractionation curve as a function of temperature.

To establish an aragonite–calcite oxygen isotope fractionation curve, we (Sommer and Rye, 1978) collected aragonitic and calcitic deep-water benthic foraminifera from piston core tops. We then measured the δ_c values of both aragonite and calcite and established an empirical relationship between the apparent isotopic fractionation of aragonite–calcite and the measured *in situ* temperatures. The data are reproduced in Fig. 9. On the basis of these data, we proposed that the empirical equation $\Delta^{18}O$ (aragonite–calcite) $= 1.39 - 0.038t°C$ was valid over the temperature range of 0–25°C.

One of the early studies in which unreasonable paleotemperatures were obtained from $\delta^{18}O$ values of $CaCO_3$ in well-preserved, diagenetically unaltered fossils is found in Tourtelot and Rye (1969). They obtained δ_c values on well-preserved cephalopods and bivalves found in late Campanian and Maestrichtian sedimentary deposits of North America. In general, they found that the ammonite genus *Baculites* gave reasonable temperatures for the Cretaceous seaway if normal-salinity (normal δ_w) water was assumed. However, many of the δ_c values for the bivalve genera *Inoceramus* and *Ostrea* resulted in unreasonably high temperatures if nor-

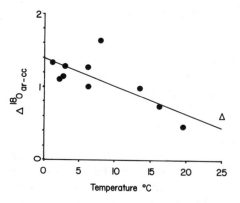

Figure 9. Graph of oxygen isotope fractionation data of aragonite–calcite from foraminiferal pairs vs. *in situ* temperature. (△) An inorganic oxygen isotope aragonite–calcite fractionation data point from Tarutani *et al.* (1969). After Sommer and Rye (1978).

mal-salinity water was assumed. Tourtelot and Rye (1969) concluded that the low δ_c values obtained in many *Inoceramus* samples were the result of some "vital" effect and not the result of the animal living in low-salinity water (low-δ_w water).

In the following section, we suggest an alternative explanation for the Tourtelot and Rye (1969) data that is based on the aragonite–calcite oxygen isotope geothermometer. This treatment of the data suggests that during the Late Campanian, the salinities in the North American seaway may have been lower than in normal sea water. We present these results as an example of how the "internal" oxygen isotope geothermometer may be useful in establishing paleotemperatures and paleosalinities. We picked these particular data because they are the only data set available at this time appropriate for the application of aragonite–calcite thermometry.

5. Application of the "Internal" Thermometer

5.1. Results

In an attempt to establish paleotemperatures for the sea in which the Pierre Shale of Late Cretaceous age (Late Campanian and Maestrichtian) was deposited, Tourtelot and Rye (1969) investigated the δ_c values of the shells from the various molluscan species preserved in the Pierre Shale. Tourtelot and Rye took great care to make certain that the shell material had retained its original isotopic values. They were able to demonstrate that: (1) All specimens in which the preservation of original composition was investigated had cyclical arrangements of δ_c values that were consistent with the interpretation of seasonal temperature variations (Fig. 10).

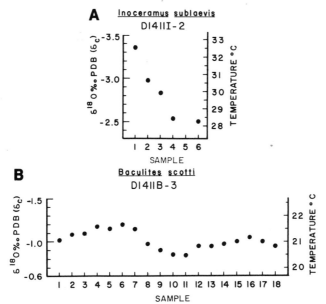

Figure 10. Oxygen isotope compositions in aragonite from *Inoceramus sublaevis* (A) and *Baculites scotti* (B), from Oral, South Dakota. Temperatures were calculated from the Epstein *et al.* (1953) paleotemperature equation assuming $\delta^{18}O$ (SMOW) was equal to 0‰, and are given in degrees Celsius. The sample numbers for the *Inoceramus* are those of samples obtained by flaking off sequential layers of aragonite, where sample 1 is the oldest. The sample numbers for the *Baculites* are those of sequential samples taken from the living chamber of the baculite by drilling out material parallel to growth lines. Sample 1 is the oldest. Note that the *Baculites* specimen shows cyclicity suggesting seasonal variation of only 1.5°C between the "isotopic" temperatures of 20.5 and 22°C, whereas the *Inoceramus* specimens suggest a much higher apparent growth temperature and wider range in temperature. After Tourtelot and Rye (1969).

(2) None of the specimens had arrangements of δ_c or $\delta^{13}C$ (PDB)* values that would indicate the effects of postdepositional exchange. (3) The isotopic composition of the fossils and the enclosing concretions could not be interpreted as having resulted from exchange processes. (4) Aragonite was completely preserved in both the ammonites and the bivalves.

The data of Tourtelot and Rye (1969) are summarized in Table I and Fig. 11. The temperatures in Fig. 11 were calculated assuming $\delta^{18}O$ (SMOW) was 0‰ (normal-salinity water) using the Epstein *et al.* (1953) paleotemperature equation.

There are several noteworthy systematic trends in the Tourtelot and Rye (1969) data that we reproduce in Table I and Fig. 11:

$$* \ \delta^{13}C \ (\text{PDB}) = \left(\frac{{}^{13}C/{}^{12}C \text{ of sample}}{{}^{13}C/{}^{12}C \text{ of PDB-1 CaCO}_3} - 1 \right) \times 1000$$

Table I. Summary of Data on Fossils and Enclosing Concretions[a]

Sample No.	Locality and fossil	Sample	Shell aragonite δ18O	Shell aragonite δ13C	Shell calcite δ18O	Shell calcite δ13C	Concretion δ18O	Concretion δ13C
D14111-2	Oral *Inoceramus sublaevis*	Average of 5	-2.84	+2.22	—	—	—	—
		Single sample	—	—	-3.66	+0.11	—	—
		Adjacent to aragonite	—	—	—	—	-6.83	-15.95
		5 cm from aragonite	—	—	—	—	-2.01	-18.26
D1411B-2	*Baculites scotti* Colorado	Average of 6	-0.72	-3.36	—	—	-0.69	-21.74
D7151	*Inoceramus convexus*	Average of 9	-3.90	+4.34	—	—	-6.99	-4.19
		Average of 10	—	—	-4.57	+2.35	—	—
D4930	Chadron *Ostrea* sp. Plum Creek	—	—	—	-6.04	+0.40	-6.68	-41.77
D2899	*Baculites* cf. *compressus* Tennessee	—	-2.13	-1.31	—	—	-4.95	-18.92
CCl-1	*Inoceramus vanuxemi*	—	-1.75	+1.69	—	—	-5.03	-12.71
CCE-1	*Eutrephoceras* sp. Alberta	—	-1.95	-3.78	—	—	-6.10	-13.53
D4131	*Baculites compressus* Saskatchewan	—	-2.69	-1.97	—	—	-4.63	-12.79
D4114	*Inoceramus* sp. Manitoba	—	-4.60	+4.10	—	—	-1.66	-4.80
D4112	*Inoceramus* sp. Northwest Territory	Average of 4	-3.16	+1.80	—	—	-1.41	-18.32
69201	*Inoceramus* sp.		—	—	-5.56	-0.67	—	—
	Ostrea sp.	Growing on *Inoceramus* shell	—	—	-5.05	+2.77	-1.02	-17.90

[a] After Tourtelot and Rye (1969).

1. If normal-salinity [$\delta^{18}O$ (SMOW) = 0‰] oceans are assumed, the δ_c values of both aragonite and calcite from the *Inoceramus* and *Ostrea* specimens yield unreasonably high temperatures for many of the samples if the paleotemperature equation is used.
2. Most of the associated *Baculites* yield reasonable temperatures if normal-salinity sea water is assumed.
3. In the sample in which both *Ostrea* and *Inoceramus* were analyzed, the two have similar δ_c values.
4. In general, the δ_c values obtained in *Inoceramus* and *Ostrea* are lighter than the values obtained in *Baculites* and other cephalopods.

Because δ_c values of both the *Inoceramus* and *Baculites* specimens are probably original, the differences between the δ_c values can be interpreted only in terms of a salinity difference or a "vital" effect. Tourtelot and Rye (1969) argued that because *Inoceramus* was benthic and both juvenile and adult nektic *Baculites* are found preserved together with the *Inoceramus*, the two genera must have lived together. They conclude that the difference in δ_c values must be due to a metabolic "vital" effect on

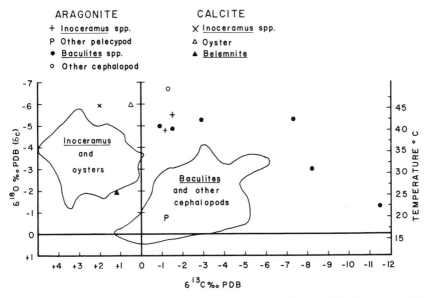

Figure 11. Summary of oxygen and carbon isotope compositions of fossils collected by Tourtelot and Rye (1969). Averages are plotted where there is more than one analysis of a specimen. Note that the oxygen isotope compositions of *Inoceramus* and the other pelecypods are characteristically relatively light and the carbon isotope compositions relatively heavy when compared to those of the *Baculites* and other cephalopods. After Tourtelot and Rye (1969).

the δ_c values of the *Inoceramus* specimen. This argument does not account for some *Inoceramus* specimens that do not show a metabolic "vital" effect. We recognize that the *Inoceramus* specimens analyzed represent several different species and that there may be species-specific effects. We do not, however, believe that this is the reason that some *Inoceramus* specimens have δ_c values the same as or similar to those of associated cephalopods whereas others do not.

Figure 12 presents isotopic data for calcite and aragonite from an *Inoceramus* specimen (D715I) collected by Tourtelot and Rye (1969). Included in this figure are the calculated paleotemperatures based on normal ocean water salinity [$\delta^{18}O$ (SMOW) $= 0$] and the paleotemperature curve of Epstein *et al.* (1953). Clearly, both polymorphs give paleotemperatures that are too high, and calcite and aragonite yield different temperatures. In all cases, aragonite is isotopically heavier than calcite. However, both polymorphs show isotopic variations that suggest that they may reflect seasonality. The fact that aragonite has a δ_c value that is consistently heavier than the calcite value suggests that the aragonite–calcite isotope thermometer may be applicable to resolving this problem.

To apply the aragonite–calcite fractionation curve properly, it is important to compare δ_c values only from aragonite and calcite that have been deposited at the same time. Unfortunately, this kind of detailed comparison cannot be made for the Tourtelot and Rye data. However, we noted earlier that the aragonite and calcite data for this specimen show a cyclic pattern and that this cyclicity could be caused by seasonality of deposition. Averaging these "seasonal" data should give a close approximation to the average δ_c values reflective of the average yearly temperature. By averaging the data in this way, we should be able to eliminate the problem

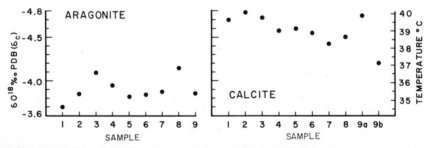

Figure 12. Oxygen isotope composition of aragonite and calcite of *Inoceramus convexus* D715I, from Pueblo, Colorado. Isotopic temperatures are given on the right axis and were calculated using the paleotemperature equation assuming the δ^{18} (SMOW) of the water was 0‰. Sequential sampling of the aragonite was obtained by flaking of successive layers. Ontogenetic sampling was obtained on the calcite by grinding samples parallel to growth lines. In each case, sample 1 is the oldest. Both aragonite and calcite suggest seasonal cyclicity. However, aragonite has consistently higher δ_c values. After Tourtelot and Rye (1969).

of giving too much weight to the temperature at which maximum growth took place. For instance, if the animal produced more shell material in the warm months than in the cool months, a bulk sample would yield a δ_c value that would overestimate the average yearly temperature.

There are two *Inoceramus* specimens from the Tourtelot and Rye collection that meet all the qualifications for being useful in aragonite–calcite geothermometry: samples D715I and D14111-2 (see Table I).

Specimen D715I was collected near Pueblo, Colorado, from sediments that contained the ammonite *Baculites scotti*, a stratigraphic zone fossil of the western interior ammonite sequence. Specimen D14111-2, collected near Oral, South Dakota, was also associated with *B. scotti*. Stratigraphically, the two *Inoceramus* specimens were collected from the equivalent of the Red Bird Silty member of the Pierre Shale. This member was deposited during the Judith River regression. A reconstruction of the Judith River strand line (Fig. 13) (Gill and Cobban, 1972) shows that the two *In-*

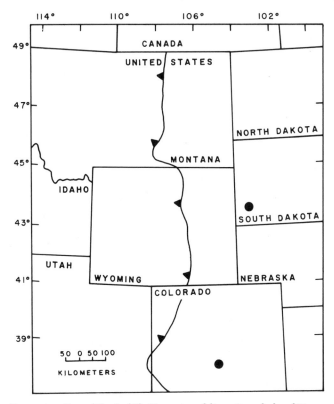

Figure 13. Reconstruction of the Judith River strand line. Sample localities are shown relative to the shoreline. After Gill and Cobban (1972).

oceramus specimens were collected from localities that are at least 180 km from the Judith River strand line.

If we employ the aragonite–calcite oxygen isotope curve, the data from the *Inoceramus* specimen D1411I-2 yield a paleotemperature of 14°C. If we then use this temperature and the isotopic composition of the calcite, we obtain a $\delta^{18}O$ (SMOW) value of $-4.2‰$ from the O'Neil *et al.* (1969) calcite–water curve.* The data from the *Inoceramus* specimen D715I yield a temperature of 19°C and a $\delta^{18}O$ (SMOW) value of the water of $-4.0‰$.

These results suggest that the two *Inoceramus* specimens lived in colder and lower-salinity water (low δ_w) than the Cretaceous *Baculites*.† One might argue, however, that the *Inoceramus* δ_c values represent a "vital" effect. This argument is unlikely, because if the δ_c values of $CaCO_3$ were caused by a metabolic effect, the inner aragonitic and outer calcitic layer in these two samples would have to have been affected in the same way by nearly the same amount; otherwise, the fractionation between calcite and aragonite would not indicate a reasonable temperature. That both polymorphs should be similarly affected by a physiological effect appears unlikely, because any kinetic isotope effect produced by the animal during the precipitation of $CaCO_3$ is unlikely to have been the same in the regions of the animal where the stable-phase calcite was being precipitated and in regions where the metastable-phase aragonite was being precipitated. Also, if a "vital" effect were responsible for light δ_c values, one might expect that this effect would be observed in all *Inoceramus* specimens. However, an *Inoceramus* collected in Tennessee (Sample CCI-1 in Table I) had a δ_c value essentially identical to that of a cephalopod (Sample CCE-1 in Table I) collected from the same deposit. Furthermore, an *Ostrea* specimen attached to an *Inoceramus* shell collected in the Northwest Territory, Canada, had a δ_c value very similar to that of its *Inoceramus* host (Sample 69201 in Table I). In addition, the data of Lowenstam and Epstein (1954) suggest that many *Inoceramus* shells have δ_c values that reflect equilibrium with near-"normal" salinity water [$\delta^{18}O$ (SMOW) $\cong 0‰$].

5.2. Estimation of Paleosalinity

To estimate the paleosalinity from the $\delta^{18}O$ (SMOW) values of the water in which the two *Inoceramus* specimens lived, it is necessary to

---

* This curve is based on $1000 \ln \alpha = (2.78 \times 10^6/T^2) - 3.39$, where T is degrees Kelvin. This curve is used rather than the Epstein *et al.* (1953) $CaCO_3$ curve because the latter curve might be slightly biased toward aragonite–water fractionation values.

† The aragonite–calcite geothermometer technique cannot be applied to the *Baculites* specimens because the shells are entirely aragonite. However, if the *Baculites* and the *Inoceramus* lived in the same water (at the same temperature), the δ_c value for the aragonite from both animals should be the same.

establish the $\delta^{18}O$ (SMOW) value of the fresh water added to this Creta-ceous seaway. Most of the fresh-water input was probably from the moun-tains to the west and entered the seaway by means of large deltas at the western margin of the basin (Weimer, 1970).

In principle, the $\delta^{18}O$ (SMOW) of rainwater could be estimated if the latitude and the mean elevation of the watershed were known. During the Late Cretaceous, the temperature gradient between the poles and the equator was probably about half the present gradient (Donn and Shaw, 1977). The smaller temperature gradient probably resulted in the $\delta^{18}O$ (SMOW) of rainwater at 50°N and at sea level being about -2 to -5‰, whereas present-day values at this latitude and elevation are approxi-mately -5 to -10‰. The orographic effect is much more difficult to estimate because we do not know the elevation of the mountains to the west, the prevailing weather patterns, or the proximity of the mountains to the western edge of the Cretaceous shoreline. However, Keith et al. (1964) analyzed well-preserved aragonite from several Cretaceous fresh-water molluscs from this area. They obtained δ_c values between -9.4 and -15.7‰; the average value was -11.6 ± 2.1‰.

We might infer from the freshwater-mollusc data that a reasonable $\delta^{18}O$ (SMOW) value for fresh water derived from the mountains was some-where between -10 and -15‰. If we accept these values for the $\delta^{18}O$ (SMOW) of fresh water added to the Cretaceous seaway, we can calculate the salinity of the water in which the two Inoceramus specimens (D715I and D1411I-2) lived. Assuming that a salinity of 35 parts per thousand is equivalent to a 0‰ $\delta^{18}O$ (SMOW) ocean, the calculated paleosalinities for the two specimens fall between 20 and 24.5 parts per thousand. The implication is that if the Inoceramus specimens are preserved in the en-vironment in which they lived, the Upper Cretaceous seaway had lower than normal salinity values at the time of the Judith River regression.

5.3. Discussion of Results

It would be easy to conclude from the Inoceramus data that the inland Cretaceous sea was brackish. However, Baculites specimens, found in the same concretion layers as the Inoceramus specimens, have been inter-preted as living in normal-salinity seawater (Tourtelot and Rye, 1969). If both organisms deposited their shells in isotopic equilibrium with water over the ambient temperature range,* one or both of these fossils is (are) not preserved in the environment in which it (they) lived. There are sev-

---

* The cyclicity in the data from both Inoceramus and Baculites suggests that seasonality has been preserved in both organisms.

eral explanations: (1) *Inoceramus* was not entirely a benthic organism. Some of them may have lived in the water column attached to floating objects in estuarian environments. After death, they could have been transported to their preservational sites. (2) The *Baculites* experienced postmortem transport. (3) The *Baculites* came into low-salinity water only to breed or to feed, or both, but spent most of their lives in the open ocean.

The possibility that *Inoceramus* may have attached itself by abyssus to floating objects is difficult to evaluate. However, *Inoceramus* specimens have been collected from the Sharon Springs member of the Pierre Shale, and the Sharon Springs member was deposited in a basin with anoxic bottom conditions (Byers, 1976). The implication is that either the *Inoceramus* lived just above the anoxic mud or they were transported to their site of preservation.

Waage (1964) has suggested that fossil-rich strata that are repeated throughout the Fox Hills Formation represent recurrent mass mortalities. The same origin may also apply to Pierre Shale fossil assemblages, which underlie the Fox Hills Formation. Waage (1964) further suggested that the mass mortalities of the Fox Hills fauna were the result of occasional exposure to high turbidity and reduced salinity. He suggested that these conditions might have been associated with periodic riverine influx of sediment and fresh water during flood runoff. It follows that if *Inoceramus* was an attached form, it is possible that some of them could have been transported during large storms from the brackish water of an estuarine environment to the site of their deaths and preservation. However, in view of the large number of individual *Inoceramus* specimens found throughout the Pierre Shale, the unbroken nature of many of the shells, and the large size of the specimens analyzed (15 × 12 cm, for D715I), we think it unlikely that the *Inoceramus* specimens analyzed by Tourtelot and Rye (1969) were transported to their preservational sites.

That the *Baculites* floated to their preservational site after death seems unlikely if *Inoceramus* is preserved at its life site. If *Inoceramus* is preserved at its life site, then the data imply that the entire Cretaceous seaway was brackish during the Judith River regression, and the *Baculites* would have had to float or swim in from the open ocean, presumably from the region of the present Gulf Coast. We think it unlikely that the large number of *Baculites* preserved in the Pierre Shale could have floated into the Cretaceous seaway over that distance without concomitant mixing of ocean water into the seaway. Such transport and mixing would not permit brackish water conditions to exist in the seaway for very long.

If it is the case that the *Baculites* were not preserved in their usual life site, a more likely possibility for their presence in the Pierre Shale fossil assemblage is that the *Baculites* migrated into the brackish Cretaceous seaway for breeding purposes. If this were the case, juveniles and adults would be preserved together, but the oxygen isotopic composition of the juvenile and adult shells would be quite different. If the adult *Baculites* spent most of their lives in the open ocean, and came into brackish water to breed, than juvenile *Baculites* and the early ontogenetic chambers of the adult *Baculites* should have low δ_c values reflecting this lower-salinity water.

The number of specimens available for which both aragonite and calcite have been analyzed from the Tourtelot and Rye (1969) paper is too small to establish any definite paleosalinity patterns for the Upper Cretaceous of the western interior of the United States. However, we submit that their data permit the interpretation that *Inoceramus* lived in salinities ranging from normal (the Tennessee specimen) to brackish (the Cretaceous Seaway specimens) and that the cephalopods associated with the bivalves spent most of their lives in normal-salinity ocean water. *Inoceramus* probably lived near shore and in estuarine environments, whereas the ammonites spent most of their adult lives in the open seas, but may have moved into a brackish seaway to spawn.

6. Summary

It has been the aim of this chapter to familiarize the reader with the oxygen isotope paleotemperature technique and to present a new technique that may solve some of the problems associated with the application of the paleotemperature technique.

First, we presented the oxygen isotope paleotemperature technique developed by Epstein *et al.* (1953). In our discussion of the development of this technique, it was made apparent that the paleotemperature equation is subject to large errors when fossil material is being considered because temperature effects oftentimes cannot be separated from salinity effects. A review of the hydrologic cycle revealed that salinity variations observed in sea water can lead to errors as large as 30°C. In addition, we showed that the $\delta^{18}O$ (SMOW) value of average seawater has varied through time, and that the variation from the Pleistocene to the present may yield errors as large as 6°C.

In addition to the problem of not being able to separate temperature effects from salinity effects, we pointed out that in living organisms, there can be large "vital" effects. These "vital" effects are caused by kinetic isotope fractionations. We also pointed out that some animals do not produce shell material over the entire ambient-temperature range. These problems suggest that any isotopic studies carried out on fossils should be done ontogenetically and on several different kinds of shells. These kinds of detailed studies can give valuable information on the biological and/or physical parameters that may have acted on individual organisms within an assemblage of organisms. However, these studies would be good mostly for comparison of relative parameters, because the δ_w value cannot be known with certainty.

Next, we showed that in principle the fact that we do not know the δ_w values in the past can be circumvented if we can find two phases that were precipitated together in isotopic equilibrium and are both preserved. We showed that calcite and aragonite oxygen isotope fractionations may be useful in this regard.

Finally, we presented an example of how the aragonite–calcite oxygen isotope geothermometer may be used to calculate the growth temperature and the $\delta^{18}O$ (SMOW) value of the water in which the animal secreted its shell. This $\delta^{18}O$ (SMOW) value can then in some cases be used to estimate the salinity of the ambient water.

ACKNOWLEDGMENTS. We thank Karl K. Turekian for discussions concerning the manuscript. We also thank W. Graustein and R. Lutz for their critical comments on the manuscript.

References

Bigeleisen, J., and Mayer, M. G., 1947, Calculation of equilibrium constants for isotopic exchange reactions, *J. Chem. Phys.* **15**:261–267.

Broecker, W., 1974, *Chemical Oceanography*, Harcourt Brace Jovanovich, New York, 214 pp.

Byers, C. W., 1976, Depositional environments of the Pierre Shale, Cretaceous of eastern Wyoming, *Geol. Soc. Am. Abstr.* **8**:800–801 (abstract).

Craig, H., 1957, Isotopic standards for carbon and oxygen and correction factors for mass-spectrometric analysis of carbon dioxide, *Geochim. Cosmochim. Acta* **12**:133–149.

Craig, H., 1961a, Standard for reporting concentrations of deuterium and oxygen-18 in natural waters, *Science* **133**:1833–1834.

Craig, H., 1961b, Isotopic variations in meteoric waters, *Science* **133**:1702–1703.

Craig, H., 1965, The measurement of oxygen isotope paleotemperatures, in: *Stable Isotopes in Oceanographic Studies and Paleotemperatures*, Spoleto, July 26–30, 1965, No. 3 (E. Tongiorgi, ed.), pp. 1–24, Consiglio Nazionale delle Ricerche, Laboratorio di Geologica Nucleare, Pisa.

Craig, H., and Gordon, L., 1965, Deuterium and oxygen-18 variations in the oceans and marine atmosphere, in: *Stable Isotopes in Oceanographic Studies and Paleotemperatures*, Spoleto, July 26–30, 1965, No. 9 (E. Tongiorgi, ed.), pp. 1–122, Consiglio Nazionale delle Ricerche, Laboratorio di Geologica Nucleare, Pisa.

Dansgaard, W., 1964, Stable isotopes in precipitation, *Tellus* **16**:436–468.

Donn, W. L., and Shaw, D. M., 1977, Model of climate evolution based on continental drift and polar wandering, *Bull. Geol. Soc. Am.* **88**(3):390–396.

Emiliani, C., 1955, Pleistocene temperatures, *J. Geol.* **63**:538–578.

Emiliani, C., 1958, Paleotemperature analysis of core 280 and Pleistocene correlations, *J. Geol.* **66**:264–275.

Emiliani, C., 1964, Paleotemperature analyses of the Caribbean cores A254–BR-C and CP-28, *Bull. Geol. Soc. Am.* **75**:129–144.

Emiliani, C., 1966, Paleotemperature analysis of Caribbean core P6304-8 and P6304-9 and a generalized temperature curve for the past 425,000 years, *J. Geol.* **74**(2):109–126.

Epstein, S., and Lowenstam, H. A., 1953, Temperature–shell growth relations of Recent and interglacial Pleistocene shoal-water biota from Bermuda, *J. Geol.* **61**:424–438.

Epstein, S., and Mayeda, T., 1953, Variations in O^{18} content of waters from natural sources, *Geochim. Cosmochim. Acta* **27**(4):213–224.

Epstein, S., and Taylor, H. P., Jr., 1967, Variation of O^{18}/O^{16} in minerals and rocks, in: *Researches in Geochemistry*, Vol. 2 (P. H. Abelson, ed.), pp. 29–62, John Wiley and Sons, New York and London.

Epstein, S., Buchsbaum, R., Lowenstam, H. A., and Urey, H. C., 1951, Carbonate–water isotopic temperature scale, *Bull. Geol. Soc. Am.* **62**:417–426.

Epstein, S., Buchsbaum, R., Lowenstam, H. A., and Urey, H. C., 1953, Revised carbonate–water isotopic temperature scale, *Bull. Geol. Soc. Am.* **64**:1315–1326.

Fontes, J. C., and Gonfiantini, R., 1967, Fractionnement isotopique de l'hydrogène dans l'eau de cristallisation du gypse, *C. R. Acad. Sci.* **2650**:4–6.

Friedman, I., and O'Neil, J. R., 1977, Compilation of stable isotope fractionation factors of geochemical interest, in: *Data of Geochemistry*, 6th ed. (M. Fleischer, ed.), *U.S. Geol. Surv. Prof. Pap.* 440 KK.

Gill, J. R., and Cobban, W. A., 1972, Stratigraphy and geologic history of the Montana group and equivalent rocks, Montana, Wyoming and North and South Dakota, *U.S. Geol. Surv. Prof. Pap.* **776**:1–37.

Keith, M. L., Anderson, G. M., and Eichler, R., 1964, Carbon and oxygen isotopic composition of mollusk shells from marine and fresh-water environments, *Geochim. Cosmochim. Acta* **28**:1757–1786.

Longinelli, A., 1966, Ratios of oxygen-18; oxygen-16 in phosphate and carbonate from living and fossil marine organisms, *Nature (London)* **211**:923–929.

Longinelli, A., and Nuti, S., 1973, Revised phosphate–water isotopic temperature scale, *Earth Planet. Sci. Lett.* **19**:373–376.

Lowenstam, H. A., and Epstein, S., 1954, Paleotemperatures of the post-Aptian Cretaceous as determined by the oxygen isotope method, *Geology* **62**(3):207–248.

McCrea, J. M., 1950, On the isotopic chemistry of carbonates and a paleotemperature scale, *J. Chem. Phys.* **18**(6):849–857.

McKinney, C. R., McCrea, J. M., Epstein, S., Allen, H. A., and Urey, H. C., 1950, Improvements in mass spectrometers for the measurement of small differences in isotope abundance ratios, *Rev. Sci. Instrum.* **21**:724–730.

Nier, A. O., 1947, A mass spectrometer for isotope and gas analysis, *Rev. Sci. Instrum.* **18**:398–411.

Nier, A. O., 1950, A redetermination of the relative abundances of the isotopes of carbon, nitrogen, oxygen, argon and potassium, *Phys. Rev.* **77**:789–793.

O'Neil, J. R., Clayton, R. N., and Mayeda, T. K., 1969, Oxygen isotope fractionation in divalent metal carbonates, *J. Chem. Phys.* **51**:5547–5548.

Redfield, A. C., and Friedman, I., 1965, Factors affecting the distribution of deuterium in the ocean, in: *Symposium on Marine Geochemistry, Univ. R. I. Occ. Publ.* **3**:149–168.

Shackleton, N. J., 1967, Oxygen isotope analysis and Pleistocene temperatures re-assessed, *Nature (London)* **215**(5096):15–17.

Shackleton, N. J., and Opdyke, N. O., 1973, Oxygen isotope and paleomagnetic stratigraphy of equatorial Pacific core V28-V38: Oxygen isotope temperatures and ice volumes on a 10^5 and 10^6 year scale, in: *CLIMAP Program, Quat. Res.* (Wash. Univ. Quat. Res. Cent.) **3**(1):39–55.

Sheppard, S. M. F., Nielsen, R. L., and Taylor, H. P., 1969, Oxygen and hydrogen isotope ratios of clay minerals from Porphyry Copper deposits, *Econ. Geol.* **64**:755–777.

Sommer, M. A., II, and Rye, D. M., 1978, Oxygen and carbon isotope internal thermometry using benthic calcite and aragonite foraminifera pairs, in: *Short Papers of the Fourth International Conference, Geochionology, Cosmochionology, Isotope Geology,* U.S. Geol. Surv. Open File Rep. No. 78-701, pp. 408–410.

Tarutani, T., Clayton, R. N., and Mayeda, T. K., 1969, The effect of polymorphism and magnesium substitution on oxygen isotope fractionation between calcium carbonate and water, *Geochim. Cosmochim. Acta.* **33**:487–496.

Tourtelot, H. A., and Rye, R. O., 1969, Distribution of oxygen and carbon isotopes in fossils of late Cretaceous age, western interior region of North America, *Bull. Geol. Soc. Am.* **80**:1903–1922.

Urey, H. C., 1947, The thermodynamic properties of isotopic substances, *J. Chem. Soc.* **1947**:562–581.

Urey, H. C., Lowenstam, H. A., Epstein, S., and McKinney, C. R., 1951, Measurement of paleotemperatures and temperatures of the Upper Cretaceous of England, Denmark and the Southeastern United States, *Bull. Geol. Soc. Am.* **62**:399–416.

Waage, K. M., 1964, Origin of repeated fossiliferous concretion layers in the Fox Hills Formation, *Bull. Kansas Geol. Surv.* **169**:541–563.

Weber, J. N., and Raup, D. M., 1966, Fractionation of the stable isotopes of carbon and oxygen in marine calcareous organisms—the Echinoidea. Part I. Variation of C^{13} and O^{18} content within individuals, *Geochim. Cosmochim. Acta* **30**(7):681–703.

Weimer, R. J., 1970, Rates of deltaic sedimentation and intrabasin deformation, upper Cretaceous of the Rocky Mountain Region, *Society of Economic Paleontologists and Mineralogists Spec. Publ.* **15**:270–292.

Wilbur, K. M., 1972, Shell formation in mollusks, in: *Chemical Zoology,* Vol. VII (M. Florkin and B. T. Scheer, eds), pp. 103–145, Academic Press, New York.

Chapter 6

Growth Patterns within the Molluscan Shell
An Overview

RICHARD A. LUTZ and DONALD C. RHOADS

1. Introduction	203
2. Microgrowth Increments	206
2.1. Formation of Microgrowth Increments	208
2.2. Summary of Documented Microgrowth Increment Patterns	218
2.3. Ecological and Paleoecological Applications	223
3. Shell Structural Changes	230
3.1. Formation of Structural Changes	230
3.2. Documented Growth Patterns and Their Ecological, Paleoecological, and Phylogenetic Applications	238
References	248

1. Introduction

Growth patterns on the external surface of the molluscan shell (Fig. 1) have long been the subject of both biological and paleontological research (Isely, 1913, 1931; Mossop, 1921, 1922; Orton, 1923, 1926; Gutsell, 1931; Weymouth *et al.*, 1931; Weymouth and Thompson, 1931; Moore, 1934, 1958; Newcombe, 1935, 1936; Davenport, 1938; Tang, 1941; Matteson, 1948; Shuster, 1951; Fairbridge, 1953; Savilov, 1953; Haskin, 1954; Mason, 1957; Lubinsky, 1958; Sastry, 1961; Craig and Hallam, 1963; Merrill *et al.*, 1965; Clark, 1968, 1969a,b, 1974, 1975; Olsen, 1968; Andrews, 1972; Seed, 1973, 1976; Thiesen, 1973; for an earlier compendium, see Weymouth, 1923) (also see Chapters 8, 11, and 12 of this volume for dis-

RICHARD A. LUTZ ● Department of Oyster Culture, New Jersey Agricultural Experiment Station, Cook College, Rutgers University, New Brunswick, New Jersey 08903. DONALD C. RHOADS ● Department of Geology and Geophysics, Yale University, New Haven, Connecticut 06520.

Figure 1. The geoduck, *Panope generosa* (Gould). Distinct growth patterns are visible on the external surface of the shell. Note wheelbarrow for scale. Origin of photograph unknown.

cussions of external growth patterns). Excellent discussions on the interpretation of such growth patterns, as well as the effects of various environmental variables on shell growth and surface morphology, are presented by Hedgepeth (1957), Moore (1958), Hallam (1965), and Clark (1974). Despite these extensive studies, the usefulness of external shell growth patterns in ecological and paleoecological studies has been limited.* The principal difficulty encountered in interpreting shell surface growth features arises from an inability to distinguish spawning and disturbance lines from annual patterns (Lutz, 1976a). Problems associated with distinguishing these growth features have been reduced over the past two decades by the discovery of periodic patterns of growth within the shells, i.e., internal growth patterns, of numerous Recent and fossil molluscs (Barker, 1964, 1970; Berry and Barker, 1968, 1975; House and Farrow, 1968; Pannella and MacClintock, 1968; Rhoads and Pannella, 1970; Farrow, 1971, 1972; Evans, 1972; Cunliffe, 1974; Pannella, 1975; Lutz, 1976a; Lutz and Rhoads, 1977; Gordon and Carriker, 1978; Jones et al., 1978). In addition to facilitating interpretation of the temporal significance and causes for observed surface morphological features, such internal growth patterns have proven useful in: (1) geophysical studies for defining Phanerozoic changes in the rate of the earth's rotation (Berry and Barker,

---

* Limitations discussed do not apply to external growth patterns of all species. Many bivalves (e.g., certain pectinids) have periodic external shell patterns that are potentially extremely useful in both ecological and paleoecological studies (see Chapter 11).

1968, 1975; Pannella and MacClintock, 1968; Pannella *et al.*, 1968; Pannella, 1972, 1975; Dolman, 1975; Rosenberg and Runcorn, 1975; Weinstein and Keeney, 1975; Whyte, 1977); (2) ecological and paleoecological studies for assessing the effects of various biological and environmental stresses (Pannella and MacClintock, 1968; Rhoads and Pannella, 1970; Farrow, 1972; Kennish and Olsson, 1975; Kennish, 1977a–c, 1978) (also see Chapter 7); and (3) archeological studies for reconstructing settlement patterns of prehistoric hunter–gatherers (Coutts, 1970, 1975; Coutts and Higham, 1971; Koike, 1973, 1975; Ham and Irvine, 1975; Clark, 1977, 1979).

Although internal shell growth patterns may potentially be found within all classes of molluscs, the Bivalvia have been by far the most universally studied. This apparent research bias is largely a reflection of the relative ease with which a complete record of bivalve growth can be obtained. Such an ontogenetic record is revealed by sectioning or fracturing the shell along a plane passing from the oldest part of the shell, the umbo, to the growing margin along the axis of maximum growth (Pannella and MacClintock, 1968; Rhoads and Pannella, 1970) (see Chapter 1 for illustration of maximum axis of growth and Appendix 1.A.1 for details of preparation procedures). Complete ontogenetic records are difficult, if not impossible, to obtain from tightly coiled or otherwise torqued shells (e.g., gastropods) using current methods (see Chapter 10 for discussion).

The primary purpose of this chapter is to provide a comprehensive review of the types of internal growth patterns thus far described within molluscan (primarily bivalve) shells. We will also speculate on the origins of these growth patterns. Realized, as well as potential, ecological and paleoecological applications are discussed at some length.* In the following discussion, internal growth patterns have been assigned to one of two categories: (1) microgrowth increments and their sequential changes or (2) shell structural changes. Growth patterns within the former category are associated with alternating regions containing variable concentrations of organic material, while those patterns within the latter category are associated with changes in shell microstructure (e.g., nacreous, prismatic, crossed lamellar; see Appendix 2.B) within a given shell layer or shifts in the relative position of shell layer boundaries on the depositional surface.† When viewed in acetate-peel or thin-section preparations (see Ap-

---

* For an in-depth discussion of the manner in which many of these internal patterns can be utilized in studies of population dynamics, see Chapter 7.

† While microgrowth increments may be considered "structural" features, we have avoided the use of the term "structure" when referring to such increments in order to avoid confusion. In this chapter, we use the term structure in reference to shell structural types (i.e., shell microstructures) outlined in Appendix 2.B or in reference to the arrangement of shell layers. The term structure is employed in a slightly different manner in Chapter 2.

pendix 1.A.1,2), microgrowth increment patterns often appear superimposed on shell structural changes.

It should gradually become evident to the reader of this chapter that most, if not all, growth patterns within the molluscan shell are a reflection of complex interactions between environmental variables and the organism's physiology. A change in the external environment will often induce physiological (particularly respiratory) changes that result in altered concentrations of metabolic end products within the extrapallial fluid. Such fluid chemistry changes can, in turn, have profound effects on the composition and structure of the shell. In the following discussion, we will consider in detail the types of internal growth patterns, resulting from such shell compositional and structural changes, that various workers have described.

2. Microgrowth Increments

Microgrowth increments within the molluscan shell were first described in detail by Barker (1964) in his thin-section observations of four genera of bivalves (*Mactra, Mercenaria, Anadara,* and *Chione*). Since

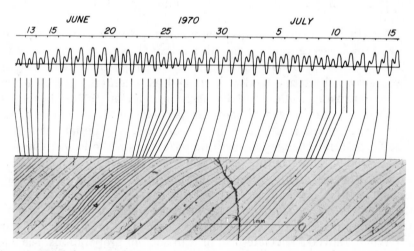

Figure 2. Microgrowth increments in the outer shell layer of the intertidal cockle *Clinocardium nuttalli* collected at Charleston, Oregon, compared with tidal predictions for the same period for Empire, Oregon. The horizontal line drawn through the tidal curves marks the intertidal position at which the specimen was collected. The optical micrograph is of an acetate peel of a polished and etched, radially sectioned shell valve. Diurnal, semidiurnal, and fortnightly patterns are apparent. Slightly modified from figures that appeared in Evans (1972) and Clark (1974). Reprinted with permission. Copyright © 1972 by the American Association for the Advancement of Science.

Barker's original study, similar "microscopic periodicity structures" have been found in numerous other molluscs, and the micro- and ultrastructures of such increments have been described in detail (for a review, see Clark, 1974). It is now generally accepted that many, if not all, microgrowth increments within the molluscan shell are a reflection of variations in the relative proportions of organic material (conchiolin) and calcium carbonate (aragonite or calcite). Alternation of calcium-carbonate-rich regions and organic-rich regions or lines has been well documented for numerous Recent and fossil species through optical- and electron-microscope studies of thin sections, acetate peels, and polished and etched surfaces of the shell (see Appendix 1.A.1–3 for preparation techniques). These growth lineations were originally interpreted as reflections of solar time (House and Farrow, 1968; Pannella and MacClintock, 1968; Pannella et al., 1968; Farrow, 1971). More recent studies, however, have revealed a complex relationship between incremental shell growth and lunar and solar cycles (Clark, 1974, 1975; Pannella, 1975, 1976; Thompson, 1975; Whyte, 1975). Although a one-to-one correspondence has not been established, the formation of microgrowth increments in bivalves is highly correlated with shell valve movements (Thompson, 1975; Gordon and Carriker, 1978).* Because the valves of many species are usually closed during low tide and open during high tide, a high positive correlation also exists between the number of increments and the number of tides to which an organism has been subjected (Evans, 1972, 1975) (Fig. 2). While valve-movement rhythmicity is usually most pronounced in intertidal specimens, subtidal specimens of at least one species [*Mercenaria mercenaria* (L.)] exhibit biological rhythms in relative synchrony with the tidal cycle (Thompson, 1975).† There has been general agreement among growth-line workers that when the valves are open and the organism is actively pumping, shell material is deposited that is rich in calcium carbonate relative to the adjacent shell material (which contains higher concentrations of organic material). Until recently, however, hypotheses attempting to explain the origin of layers relatively rich in organic content have been based on the assumption that calcium carbonate and organic material were *deposited* at variable rates during daily or tidal cycles (Pannella and MacClintock, 1968; Evans, 1972, 1975; Pannella, 1975, 1976), although a few references were made to the possibility that shell disso-

---

* Difficulty in establishing a one-to-one correspondence between the formation of increments and valve movements arises as a result of the lack of clearly defined microgrowth increments, when viewed in shell section, near the shell margin. The absence of growth lines in this region has been reported by several workers (Thompson, 1975; Gordon and Carriker, 1978) and may be related to the lack of polymerization of the organic matrix of the shell near the growing margin (Gordon and Carriker, 1978).

† Gordon and Carriker (1978), in their laboratory study of *Mercenaria mercenaria*, did not find circadian locomotor (valve-movement) rhythms reported by Thompson (1975).

lution might play a role in the formation of organic-rich layers (Rhoads and Pannella, 1970; Wilbur, 1972; Rhoads, 1974; Thompson, 1975). In 1977, Lutz and Rhoads (1977) published a hypothesis of growth-line formation that proposed that organic concentrations in various regions of the shell were simply residues left behind as a result of dissolution of shell material during periods of anaerobiosis. Independent quantitative evidence in partial support of this hypothesis has recently been presented by Gordon and Carriker (1978). In the following section, we summarize the mechanism of growth-line formation as outlined by Lutz and Rhoads (1977).

2.1. Formation of Microgrowth Increments

Prior to describing the physiologically controlled mechanism of growth line formation, it is necessary that we provide a brief background based on recent studies of molluscan anaerobiosis and shell organic–inorganic relationships. In the following sections, we consider the implications of these studies in detail.

2.1.1. Anaerobic Respiration in Molluscs

The ability of molluscs to respire anaerobically has been well documented. Excellent reviews of this subject have been presented by von Brand (1944), Hochachka and Mustafa (1972), and de Zwann and Wijsman (1976). The biochemical pathways operating during anaerobiosis, at least in bivalves, appear different from those described in vertebrates. In contrast to vertebrate pathways, in which quantities of lactic acid are produced during anaerobic metabolism, the major end products of molluscan anaerobic respiration are succinic acid and alanine (Stokes and Awapara, 1968; de Zwann and Zandee, 1972; de Zwann and van Marrewijk, 1973). Early studies (Collip, 1920, 1921) suggested that calcium carbonate from the shell served to buffer the acidic products of anaerobic metabolism, and visible corrosion of the growth surfaces of *Mercenaria mercenaria* was observed after prolonged periods of shell closure (Dugal, 1939). This hypothesis has been confirmed through studies of chemical changes in the composition of the extrapallial fluid (Crenshaw and Neff, 1969; Gordon and Carriker, 1978) (also see Figs. 3 and 4) and measurement of ^{45}Ca deposition and solution (Crenshaw and Neff, 1969); at least a portion of the succinic acid produced during anaerobic metabolism is neutralized by dissolution of $CaCO_3$ from the shell. Furthermore, measurements of the oxygen tension and succinate levels in the extrapallial fluid have demonstrated that *M. mercenaria* becomes anaerobic when the valves are closed (Figs. 3 and 4). As alluded to by Wilbur (1972), periodic valve

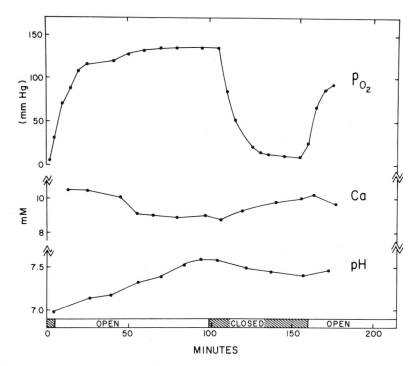

Figure 3. Oxygen tension (P_{O_2}), calcium (Ca) concentration, and pH of the extrapallial fluid of one *Mercenaria mercenaria* specimen with respect to the opening and closing of the shell valves. Valve position is indicated by the bar at the bottom of the graph. Reprinted with permission from Crenshaw and Neff (1969).

closure should therefore result in an alternation of shell deposition and decalcification. Any theory on growth-line formation must account for decalcification at the interface between the mantle and shell and its effect on preservation of the complete record of growth found within the shells of numerous species (Berry and Barker, 1968; Pannella and MacClintock, 1968; Rhoads and Pannella, 1970; Farrow, 1971, 1972; Kennish and Olsson, 1975; Kennish, 1978).

2.1.2. Organic–Inorganic Relationships

Recent shell structure research has contributed greatly to our understanding of micro- and ultrastructural relationships between organic material and inorganic crystals of calcium carbonate within the shells of bivalves, gastropods, and cephalopods (Watabe and Wilbur, 1960, 1961; Oberling, 1964; Hudson, 1967, 1968; Wise and Hay, 1968a,b; Kennedy et al., 1969; Mutvei, 1969, 1970, 1972; Taylor, J. D., et al., 1969; Grégoire, 1972; Crenshaw and Ristedt, 1975) (also see Appendix 2.B). While struc-

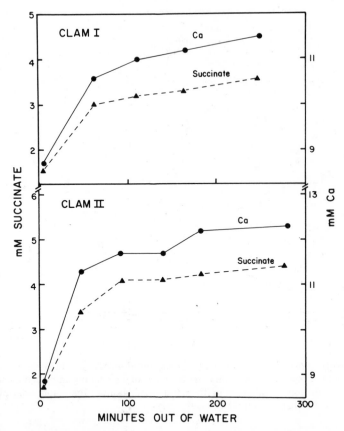

Figure 4. Concentration of calcium and succinate in the extrapallial fluids of two clams (*Mercenaria mercenaria*) with respect to time out of water. Reprinted with permission from Crenshaw and Neff (1969).

tures such as interlamellar, intercrystalline, and interprismatic organic matrices are universally recognized, the structural organization of intra-crystalline and intraprismatic organic material has been the subject of some controversy (Watabe, 1963, 1965; Towe and Hamilton, 1968a,b; Travis, 1968; Travis and Gonsalves, 1969; Mutvei, 1970; Towe, 1972; Towe and Thompson, 1972). Sometimes this organic material assumes the form of coherent matrices or "envelopes" located within prisms or aragonitic nacreous tablets. However, some workers (Towe and Hamilton, 1968a,b; Towe, 1972; Towe and Thompson, 1972) have suggested that such organic matter, when present, may represent impurities in the car-bonate in the form of trapped proteins or polypeptides. It was originally suggested that such intracrystalline organic matter was water-soluble

(Crenshaw, 1972), but subsequent studies have refuted this hypothesis and indicate that the water-soluble fraction appears to be adsorbed on crystal surfaces (Erben, 1974; M. A. Crenshaw, personal communication).

2.1.3. Lutz–Rhoads Hypothesis

With the foregoing discussion serving as a background, we offer the following as an explanation of growth-line formation in the molluscan shell. During aerobic metabolism, molluscs deposit calcium carbonate in the form of aragonite or calcite, together with organic material, resulting in shell construction. Aerobic metabolism is usually associated with periods of active pumping during high tide in well-oxygenated waters. As the concentration of dissolved oxygen falls, such as in the internal microenvironment created by the organism during periods of shell closure, anaerobic respiratory pathways are employed, and the level of succinic acid (or other acidic end products) within the extrapallial fluid rises. The acid produced is gradually neutralized by the dissolution of shell calcium carbonate, leading to increased levels of Ca^{2+} and succinate (or other end products) within the extrapallial and mantle fluids (Fig. 4) (also see Crenshaw and Neff, 1969). As a result of this decalcification, the ratio of relatively acid-insoluble organic material to calcium carbonate increases at the interface between the mantle and shell* One need not invoke the complication of an increased concentration of organic material in a given volume, although a collapse of unsupported matrix structures or a movement of the mantle as a compensatory response to the increased distance between mantle and shell could result in increased concentrations of freed organic material in specific regions of the extrapallial fluid. With the opening of the valves and the resumption of aerobic metabolism (see Fig. 3), the deposition of calcium carbonate and organic material within an area containing an insoluble residue of organic material should result in a localized increase in the ratio of organic material to calcium carbonate within the specific shell region. The end product of this process, from a strictly structural viewpoint, is one growth increment.†

---

* It is also conceivable that the ratio of organic material to calcium carbonate at the mantle–shell interface could increase as a result of cessation of calcium carbonate deposition, with continued deposition of organic material. This hypothesis does not require shell dissolution and might be invoked to explain growth-line formation, as well as to explain some of the ultrastructural observations described elsewhere in this chapter. Evidence in support of this hypothesis, however, is unavailable at present.

† The presence of intraprismatic and intracrystalline coherent organic matrices would generally be more compatible with this hypothesis than would reincorporation into the shell of organic inclusions freed during decalcification, although the latter interpretation is certainly tenable.

2.1.4. Supporting Evidence

As mentioned earlier, evidence in partial support of the hypothesis discussed above has recently been provided by Gordon and Carriker (1978) based on measurements of extrapallial fluid pH,* rates of shell dissolution during valve closure [from the data of Crenshaw and Neff (1969)], thickness of subdaily growth lines, and duration of valve closure of *Mercenaria mercenaria*. In their study, Gordon and Carriker measured thicknesses of organic-rich "subdaily growth striations" ranging from 0.45 to 0.9 μm. Crenshaw and Neff (1969) calculated that at measured rates of calcium carbonate dissolution, an organism with a shell weight of 100 g would lose about 2 mg of shell per hour. Shell weights of *M. mercenaria* specimens used by Gordon and Carriker (1978) ranged from 40 to 80 g, with internal shell surface areas of 44–64 cm². Assuming the dissolution rate to be uniform over the entire internal surface, Gordon and Carriker calculated the thickness of shell that would be removed for every hour the shell is closed:

(shell loss per hour/unit weight) × (total weight/surface area)

× (1/density) = thickness lost per hour

For example, one specimen with valves weighing 64 g and an internal surface area of 56 cm² would lose 7.8×10^{-2} μm/hr (using a density of 2.93 g/cm³ for aragonite). Experimental organisms in the study of Gordon and Carriker (1978) kept their valves closed for 2.5–12 hr. At the calculated rate of shell dissolution, the thickness of shell removed would be 0.2–0.94 μm. These values are sufficiently close to the measured widths (0.45–0.9 μm) of subdaily lines to suggest a causative relationship between the dissolution and subdaily line thickness.

As a result of the measurements and calculations discussed above, Gordon and Carriker (1978) conclude that:

> To account for subdaily striations, then, it is only necessary to envision continuous and simultaneous secretion of organic matrix and calcium carbonate during the aerobic shell-building part of the animal's growth cycle. In the anaerobic period, increasing acid in the extrapallial fluid dissolves a portion of newly deposited shell. Some of the associated matrix may also dissolve, but at least part of it is sufficiently insoluble to resist attack by metabolic acids and remains behind as a residue to be covered by a new layer of calcified material during the next cycle of aerobic deposition. Because the matrix at this point is hardened by polymerization of the protein, it maintains its structural integrity during and after decalcification. As a result, the width of residual matrix provides a record of the length of time that the shell was exposed to metabolic acids.

---

* Since calibration buffers of suitable ionic strength were not used in their study, Gordon and Carriker (1978) actually measured electrode potential rather than hydrogen ion concentration.

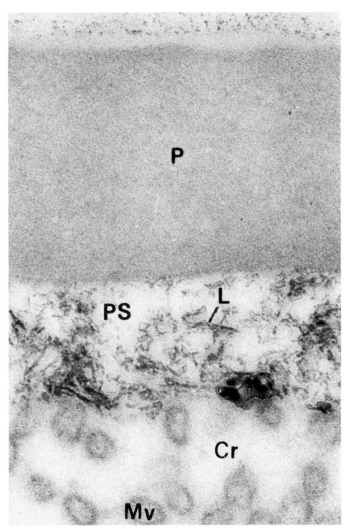

Figure 5. Electron micrograph of a vertical section of the pallial region of *Pinctada radiata*. The live specimen was relaxed in a solution of magnesium sulfate prior to fixation in glutaraldehyde and osmium tetroxide. Calcium carbonate crystals (Cr) and fragments of "electron-dense lamellae" (L) are seen scattered throughout the pallial space (PS). (P) Periostracum; (Mv) microvilli of mantle cells. Unstained section ×110,000. Reprinted with permission from Nakahara and Bevelander (1971).

Thus, results and conclusions from the study of Gordon and Carriker (1978) lend considerable support to the hypothesis of growth-line formation outlined above.

Further evidence suggesting the importance of respiratory changes in producing microgrowth increments is provided by a reinterpretation of the works of Bevelander and Nakahara (Bevelander and Nakahara, 1969; Nakahara and Bevelander, 1971). In their electron-microscope study of the formation and growth of the prismatic layer of *Pinctada radiata* (Leach), Nakahara and Bevelander (1971) envisioned fragments of "electron-dense lamellae" at the internal boundary of future prisms (Figs. 5 and 6) as migrating through the extrapallial fluid and ultimately forming envelopes within which crystal nucleation and growth occur. If the micrographs they presented (Figs. 5 and 6) are reinterpreted and the process of prism formation they described is envisioned as occurring in reverse, we can observe quite vividly the process of gradual shell destruction. As calcium carbonate slowly dissolves, relatively insoluble organic envelopes, matrices, or inclusions remain at the mantle–shell interface, presenting the appearance of an electron-dense lamella (Figs. 5 and 6). That this organic residue is subsequently incorporated into the shell, as proposed by our hypothesis, is strongly suggested by the presence of "intraprismatic organic strands" approximately paralleling the inner growth surface of the prisms (Fig. 6c). While details of the treatment of the studied pearl oysters prior to anesthetization were not presented by Nakahara and Bevelander (1971), one can reasonably assume that laboratory conditions and/or exposure to air prior to or during the placement of the organisms in magnesium sulfate could have induced valve closure and subsequent anaerobiosis.

Results of an earlier study conducted by Bevelander and Nakahara (1969) on the formation and growth of nacre have been interpreted as

---→

Figure 6. Electron micrographs of the pallial region of *Pinctada radiata*. The live specimens were relaxed in a solution of magnesium sulfate prior to fixation in glutaraldehyde and osmium tetroxide. All sections have been stained with uranyl acetate and lead citrate. (A) Vertical section of a mantle cell (bottom of micrograph), pallial space (Ps), lamella (L), and portion of a prism (Pr). This region was located proximal to the outer fold of the mantle. (Mv) Microvilli. × 27,000. (B) Vertical section in a region of pallial space showing a portion of a prism (Pr) containing scattered calcium carbonate crystals (Cr), interprismatic wall (IW), lamella (L), and tips of microvilli (Mv). Decalcified section × 76,000. (C) Vertical section of mantle cells (bottom of micrograph), pallial space (Ps), lamella (L), and portion of three prisms (Pr). Microvilli of mantle cells extend into the pallial space. The prisms are separated from one another by substantial interprismatic walls (IW); "intraprismatic organic strands" (IS) and "remnants of envelopes and ground substance" (Nakahara and Bevelander, 1971) are seen between the walls. Decalcified section × 6700. All micrographs reprinted with permission from Nakahara and Bevelander (1971).

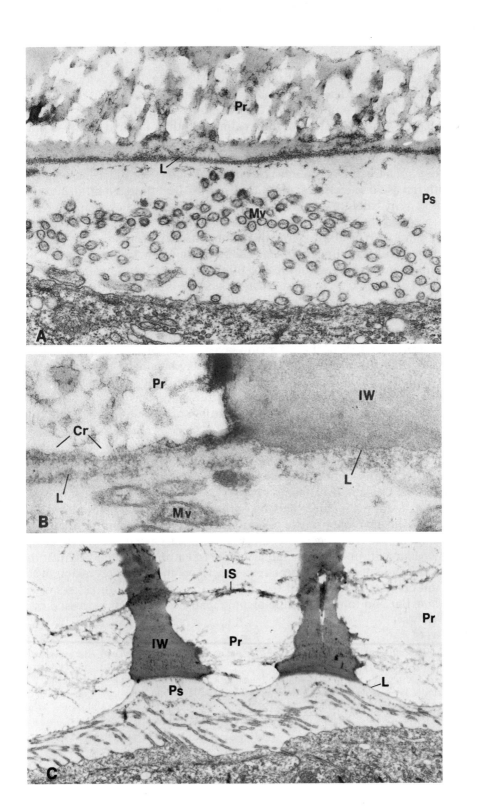

illustrative of shell dissolution (Erben, 1972, 1974; J. D. Taylor, personal communication). That their organic "compartments" (Fig. 7) at the mantle–shell interface may indeed be reflections of shell dissolution is suggested by an examination of the treatment of their studied bivalves prior to fixation. For a period of 1 hr before fixation, the molluscs were placed in a refrigerator at 5°C. As implied earlier (also see Crenshaw and Neff, 1969), such treatment is almost certainly sufficient to induce anaerobiosis, with release of succinic acid into the extrapallial fluid. Neutralization of this acid by calcium carbonate from the shell would be expected to result in selective shell destruction in a manner virtually identical to that seen in Figure 7. While microgrowth increments have seldom, if ever, been reported in nacre, they do occur. Nonreflected growth lines* within the middle nacreous layer† of the Atlantic ribbed mussel, Geukensia (= Modiolus) demissa (Fig. 8) show clustering patterns similar to those reported in a number of species (Pannella and MacClintock, 1968; Evans, 1972, 1975; Clark, 1974; Pannella, 1975, 1976). Mechanisms of growth-line formation within nacreous structures may or may not be precisely analogous to the process within the prismatic layer of Pinctada. If intracrystalline organic matrices are present, the process may be almost identical to that occurring in Pinctada prisms. In light of the current controversy over the presence or absence of such organic structures (for a discussion, see Towe, 1972), we suggest an alternative mechanism of growth-line formation in nacre.

It has been proposed that nacreous crystal nucleation occurs on the surface of an organic substratum and that during growth these crystals are enclosed within tight-fitting organic envelopes (Erben, 1972, 1974; Mutvei, 1972; Erben and Watabe, 1974). As a result of the lateral fusion of the crystals to form a mature nacreous lamina, adjacent lateral and distal portions of the envelopes merge to become intercrystalline and interlamellar organic matrices, respectively. If crystal nucleation and growth occasionally occur within organic compartments remaining after decalcification, and subsequent growth proceeds as described above, lateral fusion of adjacent crystals should result in a marked increase in the thickness of interlamellar and intercrystalline organic matrices. Such an increased con-

---

* See Pannella and MacClintock (1968) for a detailed description of nonreflected growth lines within the bivalve shell.

† L. F. Gainey, Jr., and M. J. Greenberg (personal communication) have indicated that this middle layer is equivalent to the "inner nacreous layer" described by Blackwell et al. (1977). Also see Lutz and Rhoads (1978) for a critique of the study by Blackwell et al. (1977), as well as a detailed description of the shell structure of Geukensia demissa (Dillwyn).

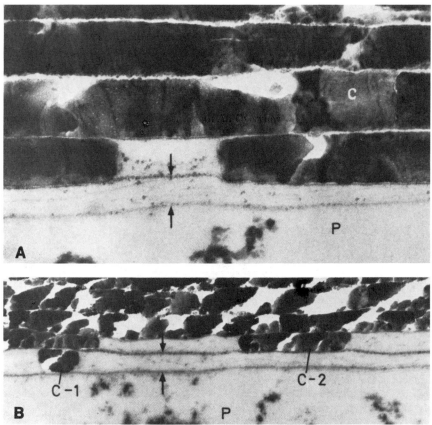

Figure 7. (A,B) Electron micrographs showing portions of the pallial space (P) and formed shell (C) of *Mytilus exustus*. The live specimens were placed in a refrigerator at 5°C for 1 hr prior to fixation in glutaraldehyde and osmium tetroxide. Such prefixation treatment may have induced anaerobiosis, with release of succinic acid into the extrapallial fluid. "Compartments" are delimited by organic lamellae (arrows). (C-1, C-2) Incomplete (partially dissolved?) aragonitic tablets in adjacent nacreous laminae. (A) ×46,000; (B) ×20,000. Both micrographs reprinted with permission from Bevelander and Nakahara (1969).

centration of conchiolin relative to calcium carbonate could explain growth patterns observed on acetate peels of polished and etched sections of *G. demissa* nacre. Evaluation of the above alternation of shell dissolution and deposition in the perspective of the current controversy associated with the *template* and *compartment* hypotheses (Bevelander and Nakahara, 1969; Nakahara and Bevelander, 1971; Mutvei, 1972; Towe, 1972; Erben and Watabe, 1974) suggests that while compartments may

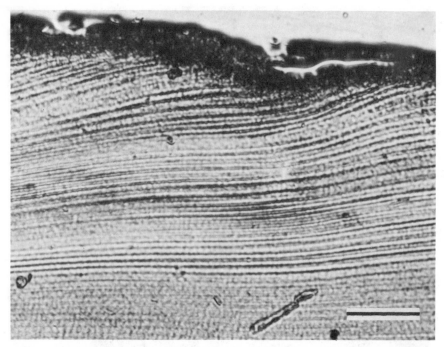

Figure 8. Optical micrograph of an acetate peel of a polished and etched longitudinal shell section showing nonreflected growth lines within the middle nacreous layer of *Geukensia demissa*. Note the pronounced clustering of microgrowth increments. Scale bar: 100 μm. Reprinted with permission from Lutz and Rhoads (1977). Copyright © 1977 by the American Association for the Advancement of Science.

exist from a strictly structural viewpoint, they are not primary features, but rather are "residual" in nature, and should be considered of only incidental importance in the formation of nacre. A similar point of view concerning the presence (or lack) of preformed matrix layers or compartments has been expressed by Gordon and Carriker (1978).

2.2. Summary of Documented Microgrowth Increment Patterns

The following discussion is intended to provide a current literature review* of microgrowth increment patterns described to date within the molluscan shell. Detailed descriptions and micrographs of the various types of patterns may be found in Chapter 7. For the purposes of this review, we have assigned documented microgrowth increment patterns to one of the following five temporal categories: (1) semidiurnal and diur-

---

* For earlier compendia, see Rhoads and Pannella (1970) and Clark (1974).

nal; (2) fortnightly; (3) monthly; (4) annual; (5) semiperiodic or random events.*

2.2.1. Semidiurnal and Diurnal

As mentioned above, Barker (1964) has been credited with the discovery of microgrowth increments (and their sequential patterns) within the molluscan shell. As a result of his examination of several species of bivalves, Barker (1964) suggested that these small-scale internal growth features might reflect semidiurnal or diurnal periodicities, or both. Since 1964, considerable research has been conducted to test the validity of this hypothesis.

The first convincing evidence that diurnal periodicities were indeed recorded within the molluscan shell was provided by Pannella and MacClintock (1968). In their study of *Mercenaria mercenaria*, these workers counted from 360 to 370 lines and from 720 to 725 lines for growth periods of 368 and 723 days, respectively. These results strongly suggested that such microscopic growth increments were formed with a solar periodicity. "Solar" daily increments were also described by Pannella and MacClintock (1968) in *Tridacna squamosa*. More recently, other workers have examined and analyzed growth patterns in *M. mercenaria* and *T. squamosa* in greater detail, increasing our understanding of factors responsible for the formation of diurnal (and semidiurnal) increments in these and other species of molluscs (MacClintock and Pannella, 1969; Rhoads and Pannella, 1970; Thompson and Barnwell, 1970; Evans, 1972, 1975; Pannella, 1972, 1975,† 1976; Cunliffe and Kennish, 1974; Kennish and Olsson, 1975; Thompson, 1975; Kennish, 1976, 1977b, 1978; Gordon and Carriker, 1978). It is now generally accepted that such microgrowth increments are largely reflections of complex interactions between lunar and solar cycles (Clark, 1974, 1975; Pannella, 1975, 1976; Thompson, 1975; Whyte, 1975). At about the same time that Pannella and MacClintock (1968) reported the results of their study of growth patterns in *M. mercenaria*, House and Farrow (1968) independently suggested that similar microscopic features within the shell of *Cerastoderma* (= *Cardium*) *edule* were formed daily. Since these early studies, periodic microgrowth increments have been found in many other Recent and fossil species of

---

* A few workers (Pannella and MacClintock, 1968; Kennish and Olsson, 1975) (also see Chapter 7) offer a description of bidaily patterns in *Mercenaria mercenaria*. Such patterns have not been described in detail in other species and will not be discussed in this section. Similarly, growth patterns reflecting a semiannual cycle (for a discussion of this type of periodicity, see Clark, 1974) have not been well documented and are not considered in this discussion.

† Pannella (1975) also described daily increments within the shell of the closely related venerid, *Mercenaria campechiensis*.

molluscs. Pannella *et al.* (1968) reported the presence of well-preserved "daily" growth increments in 12 species of molluscs from several geological periods (Middle Devonian to Upper Miocene). Barker (1970) prepared thin sections of the shells of over 98 species of bivalves and described numerous subannual structures within these shells that he felt might be related to daily or tidal cycles within the environment. Other detailed studies providing evidence for the presence of diurnal and/or semidiurnal periodicity structures within molluscan shells include those of Rhoads and Pannella (1970), Farrow (1971, 1972), Evans (1972, 1975) (also see Fig. 2), Pannella (1975), Whyte (1975), and Lutz and Rhoads (1977).

2.2.2. Fortnightly

Microgrowth patterns recurring at approximately fortnightly intervals have been described in a large number of molluscs, particularly those inhabiting intertidal and shallow subtidal environments (Barker, 1964; Berry and Barker, 1968; Pannella and MacClintock, 1968; Pannella *et al.*, 1968; House and Farrow, 1968; Rhoads and Pannella, 1970; Evans, 1972, 1975; Clark, 1974; Kennish and Olsson, 1975; Kennish, 1976, 1977b; Pannella, 1975, 1976) (also see Chapter 7). Barker (1964), in his detailed examination of "microtextural variation" in the bivalve shell, described "3rd-order layers" composed of microgrowth increments cyclically varying in thickness and hypothesized that such "layers" might recur at intervals of approximately 15 days. Later, Pannella and MacClintock (1968), and more recently Pannella (1975), published detailed micrographs of acetate peels of sectioned *Mercenaria mercenaria* and *Tridacna squamosa* showing fortnightly "clusters" of microgrowth increments within the outer shell layers of these two Recent bivalves.* These workers described similar "fortnightly cycles of deposition" within the shells of numerous fossil specimens† and presented micrographs of "fortnightly" patterns within *Mercenaria campechiensis ochlockoneenis* (Upper Miocene), *Crassatella mississippiensis* (Upper Eocene), *Limopsis striatopunctatus* (Upper Cretaceous), and *Conocardium* sp. (Upper Pennsylvanian). Similar clustering (fortnightly?) patterns of microgrowth increments have been observed within the middle nacreous layer of the Recent mytilid *Geukensia demissa* (Fig. 8) (also see Lutz and Rhoads, 1977). Evans (1972,

---

* Kennish and Olsson (1975) and Kennish (1976, 1977b) provide additional discussions of fortnightly patterns of microgrowth increments in *Mercenaria mercenaria* (also see Chapter 7).
† The "fortnightly" periodicity of patterns in these fossil specimens was inferred from their structural similarity to those patterns that Pannella and MacClintock (1968) observed within the shells of *Mercenaria mercenaria*.

1975) presented a micrograph (see Fig. 2) of an acetate peel of *Clinocardium nuttalli* that illustrates a fortnightly pattern of microgrowth increments within the outer shell layer of this cockle. Pannella (1976) published a similar micrograph of *Limopsis striatopunctatus* (Upper Cretaceous) showing an inferred "fortnightly" pattern of growth. Dolman (1975), in his Fourier analysis of microgrowth periodicities in *Cerastoderma* (= *Cardium*) *edule*, recognized weakly developed fortnightly growth cycles in this species. Pannella (1975) photographically illustrated fine structures* within the shell of an unidentified Middle Devonian cephalopod and suggested that observed clusters of these structures might reflect a fortnightly periodicity. Earlier brief reviews of fortnightly patterns within the molluscan shell are presented by Rhoads and Pannella (1970) and Clark (1974).

2.2.3. Monthly

Numerous workers have provided evidence for the presence of monthly microgrowth patterns within the shells of various species of molluscs (House and Farrow, 1968; Pannella and MacClintock, 1968; Pannella *et al.*, 1968; Rhoads and Pannella, 1970; Farrow, 1972; Kennish and Olsson, 1975; Kennish, 1976, 1977b) (also see Chapter 7). House and Farrow (1968), in their analysis of microgrowth increments within the shell of *Cerastoderma edule*, observed "a regular bunching" of growth increments that occurred with a frequency of 29 days. Farrow (1972), in a later study of this same bivalve, showed that this monthly periodicity could be related to prolonged exposure due to very low high tides at every other neap tide for part of the year. Pannella and MacClintock (1968) reported a similar (synodic) monthly periodicity in *Mercenaria mercenaria* characterized by groupings of fortnightly "clusters" in pairs, with one fortnightly cluster more pronounced than the other. This internal pattern is also expressed, at least on some parts of the outer surface of this species, by concentric undulations with a wavelength periodicity of 29 days (also see Kennish and Olsson, 1975; Kennish, 1976, 1977b; Chapter 7). Pannella *et al.* (1968) reported the presence of "synodic-month patterns" in numerous fossil molluscs ranging in geological age from the Middle Devonian to Upper Miocene. The "monthly" periodicity of patterns within the shells of these fossil specimens was inferred from the structural similarity of patterns to those patterns that Pannella and MacClintock (1968) observed within the shells of *Mercenaria mercenaria*. Rhoads and Pannella (1970) photographically illustrated "monthly clustering" patterns of microgrowth increments formed during the autumn months in *Nucula prox-*

---

* Such fine structures were shell surface topographic features and need not correspond with microgrowth increments within the shell.

ima from 6 m of water in Long Island Sound. More recently, Kahn and Pompea (1978) have suggested that septal formation within the cephalopod *Nautilus pompilius* may take place at monthly intervals* and reported the presence of similar "monthly" structures within the shells of numerous fossil cephalopods (Upper Ordovician to Upper Miocene). Possible factors responsible for the formation of monthly growth patterns in molluscs are discussed by Pannella and MacClintock (1968), Clark (1974), and Pannella (1976).

2.2.4. Annual

Annual patterns of microgrowth increments have been observed within the shells of the majority† of specimens that have been examined and analyzed in detail. Such patterns are generally a result of seasonal variations in growth that, in turn, usually result in the formation of relatively thick microgrowth increments during certain seasons (e.g., northern hemisphere, high latitude, summer) and relatively thin increments during others (e.g., northern hemisphere, high latitude, winter). These annual patterns constitute the "1st-order layers" described by Barker (1964) in his detailed examination of *Mercenaria mercenaria*, *Spisula* (= *Mactra*) *solidissima*, *Chione cancellata*, and *Anadara ovalis*. Pannella and MacClintock (1968), and later Rhoads and Pannella (1970), Kennish and Olsson (1975), and Kennish (1976, 1977b) (also see Chapter 7), presented detailed micrographs illustrating these patterns within *M. mercenaria*. Similar annual patterns were described in detail by House and Farrow (1968) and Farrow (1971, 1972) in their examination of microgrowth increments in *Cerastoderma edule*. Spawning patterns, while sometimes occurring with an annual periodicity in certain species of bivalves, are considered in this chapter to be more accurately described as semiperiodic events and will therefore be considered in the next section. Further discussions of annual growth patterns within the molluscan shell may be found in Rhoads and Pannella (1970) and Clark (1974), and in Chapter 7.

2.2.5. Semiperiodic or Random Events

Many patterns are encountered within the shells of molluscs that do not occur with definite cyclic periodicities. "Subdaily" growth incre-

---

* Similarly, Martin *et al.* (1977), in their study of *Nautilus macromphalus*, interpreted observed periodic weight fluctuations, occurring at intervals of approximately 30 days, to indicate the formation of new septa.

† Annual patterns have generally not been recognized in deep-water molluscan species; in such species, major growth patterns have been interpreted to mark spawning events (Rhoads and Pannella, 1970).

ments described by numerous workers may be examples of such aperiodic features (Pannella and MacClintock, 1968; Kennish and Olsson, 1975; Kennish, 1976, 1977b; Gordon and Carriker, 1978) (also see Chapter 7). Distinct microgrowth patterns also result from damage to the shell margin (see Chapter 7 for detailed illustrations and discussions of these patterns). Notching of the shell with a small file, for instance, causes formation of a distinct disturbance line and microgrowth pattern (see Pannella and MacClintock, 1968; Kennish and Olsson, 1975; Kennish, 1976, 1977b) (also see Chapter 7). Shell damage caused by predator attacks may result in similar growth patterns. Storm events are often recorded in the shell and can sometimes be recognized by the presence of associated silt particles that are trapped between the mantle and the shell during the storm and are then subsequently incorporated into the shell (Barker, 1974; House and Farrow, 1968; Rhoads and Pannella, 1970; Kennish and Olsson, 1975) (also see Chapter 7). Certain pollution events, such as the discharge of heated effluent waters from nuclear or fossil fuel power plants, may also result in the formation of microgrowth patterns within the shell (Kennish and Olsson, 1975) (also see Chapter 7). Rhoads and Pannella (1970) demonstrated that the transplantation of organisms to different environments will often result in marked differences in the pattern of microgrowth increments before and after transplantation. Spawning events, which may be considered semiperiodic, have been found to be reflected in microgrowth patterns within almost all molluscan shells that have been examined in detail. Even bivalves (e.g., *Malletia*) from depths as great as 4970 m have patterns that have been interpreted to mark spawning events (Fig. 9) (also see Rhoads and Pannella, 1970). Pannella and MacClintock (1968), and later Rhoads and Pannella (1970), Kennish and Olsson (1975), and Kennish (1976, 1977b, 1978) (also see Chapter 7), presented detailed descriptions and illustrations of the appearance of spawning patterns in a few species of shallow-water bivalves. For a further detailed discussion of the types of semiperiodic and random events that can be recorded within the molluscan shell, see Chapter 7.

2.3. Ecological and Paleoecological Applications

Ecology has been defined as the study of relationships between organisms and their environment. In functioning ecosystems, it is possible to make direct observations of these relationships and their temporal changes. Organism–environment relationships cannot always be directly measured, however, but may require an indirect or deductive approach; for instance, one may wish to assess the effect of a storm, pollution event, change in salinity, temperature, or other factor on a population after the

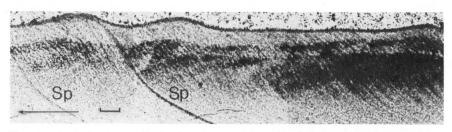

Figure 9. Optical micrograph of a thin section of the outer shell layer of *Malletia* sp. from 4970 m on the lower continental rise. Two major breaks in shell deposition have been interpreted by Rhoads and Pannella (1970) to mark spawning events (Sp). Arrow indicates direction of growth. Scale bar: 100 μm. Thin section YPM-IP 28157. Reprinted with permission from Rhoads and Pannella (1970).

event has taken place. In the absence of data about predisturbance rates of growth, death, and reproduction, one is totally dependent on indirect techniques. This kind of after-the-fact problem is common in paleoecology and promises to be an increasingly important approach in ecological studies. In this section, we will consider the various ways in which detailed analyses of molluscan microgrowth patterns have been applied to ecological and paleoecological studies, and we will speculate on their future applications.

Microgrowth increments have proven especially useful in recognizing how seasonally variable environmental factors affect molluscan growth. For instance, through analysis of patterns of "daily" microgrowth increments, Farrow (1971) found that a substantial portion of a population of the shallow subtidal cockle *Cerastoderma edule* from the Thames estuary in England stopped growing during winter because of subzero temperatures. Similarly, Tevesz (1972) observed that *Gemma gemma* Totten grew very little in the winter; growth increments were very closely spaced, and the inner shell layer had a brownish hue. During the summer, *G. gemma* grew rapidly; microgrowth increments were widely spaced, and the inner shell layer was clear and translucent in appearance. Through an examination of microgrowth increments in numerous acetate peels, Evans and LeMessurier (1972) were able to demonstrate striking winter growth rate differences between two sympatric species of bivalves. They found winter growth of the rock-boring clam *Penitella penita* to be approximately 75% of the summer growth rate, while the growth rate of the cockle *Clinocardium nuttalli* which inhabited a neighboring mud flat, decreased during the winter, relative to the summer, by a factor of as much as 19. For further examples and discussions of the usefulness of microgrowth increments for reconstructing seasonal variations in molluscan growth, see House and Farrow (1968), Pannella and MacClintock (1968), and Rhoads and Pannella (1970).

Microgrowth increment patterns can also supply detailed information on the age of individual specimens at the time of death, as well as the season of their death. In those species for which the periodicities of microgrowth patterns have been established, the age at death is readily obtained by simply counting the total number of patterns. The periodicity with which the counted patterns occur (e.g., daily, fortnightly, annual) will determine the precision of the age estimate. The season of death is determined by relating the position of the last increment at the margin of the shell to the seasonal growth pattern (Fig. 10). For example, a margin preceded by a complete summer depositional record represents death early in the fall. A margin that follows a long period of winter growth represents late winter or early spring death. Often, however, when the shell margin is preceded by a few days of growth slowdown, comparison of several specimens may be necessary to determine whether growth-rate changes reflect a moribund condition prior to death or are related to seasonal changes (Rhoads and Pannella, 1970). By counting the number of microgrowth increments (or patterns of increments), it is also often possible to relate the season of death to absolute age at death. Furthermore, the identification of the season of death and age at time of death in a mollusc assemblage may be used to distinguish a mass mortality from a death assemblage produced by natural mortality. As emphasized by

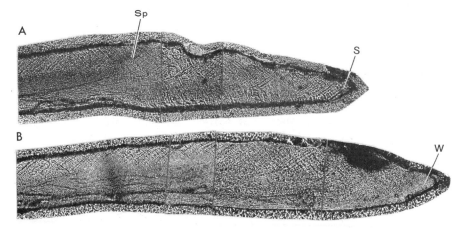

Figure 10. Summer and winter death recorded in microgrowth increment patterns of *Gemma gemma* Totten. (A,B) Optical micrographs of acetate peels of polished and etched shell sections. (A) Summer death (S). The shell edge is preceded by relatively large increments characteristic of summer growth. In the earlier part of the shell, the thin increments are probably related to spawning (Sp). Acetate peel YPM-IP 28158. (B) Winter death (W). The shell edge is preceded by thin increments marking a period of slow growth. Thicker bandings in the early part the shell mark the preceding summer's growth. Acetate peel YPM-IP 28163. Reprinted with permission from Rhoads and Pannella (1970).

Rhoads and Pannella (1970), this information may be used to determine seasons of peak environmental stress. This technique can be especially useful when applied to drilled shells. By determining the season of death of drilled members of a population, one may identify the season of peak predator activity (Rhoads and Pannella, 1970).

Microgrowth increment patterns can also frequently be utilized to determine the age at sexual maturity and season of reproduction by relating the position of spawning breaks to absolute age and the seasonal pattern of growth. An illustration of how growth patterns have been used for such purposes is provided by Rhoads and Pannella (1970) in their analysis of microgrowth patterns in the shell of *Gemma gemma*.

Variations in environmental parameters including food supply, substratum type, salinity, dissolved oxygen, turbidity, temperature, and population density can influence the growth of bivalves. Hallam (1965) reviews these various environmental parameters as causes of stunting and dwarfing of living and fossil marine benthic invertebrates. Several studies conducted over the past few years have used microgrowth increments to define the effects on bivalve growth of various environmental perturbations, such as those outlined by Hallam (1965). Rhoads and Pannella (1970), for example, through careful examination of microgrowth patterns on both acetate peels and thin sections, have shown that *Mercenaria mercenaria* grows faster in sandy sediments than in mud when other variables are eliminated. Also, these workers analyzed microgrowth increment patterns within *Astarte castanea* Say and *M. mercenaria* to demonstrate that growth rate (and shell structural) changes occur when specimens of these species are transplanted from a natural subtidal environment to a laboratory holding tank without substratum (Fig. 11). Farrow (1971) used microgrowth increments within the shell of *Cerastoderma edule* to illustrate that high-density populations of this species had a much shorter growing season than low-density populations. An inverse relationship between individual cockle size and cockle population density was also noted. In a subsequent study, Farrow (1972) used microgrowth increments within the outer shell layer of *C. edule* to demonstrate that individual specimens living in the high intertidal zone were stunted relative to low intertidal specimens. The higher-shore cockles were situated near the high-water mark and, consequently, were aerially exposed for several days during neap tides. Following neap tide deceleration, there was a resumption of vigorous growth. Many of the high-intertidal cockles were some two thirds the size of specimens in the lower intertidal zone, where growth was more continuous.

Microgrowth patterns can also provide evidence about the biology of deep-water molluscs. Preliminary data of Rhoads and Pannella (1970) indicate that such patterns can shed light on problems related to deep-

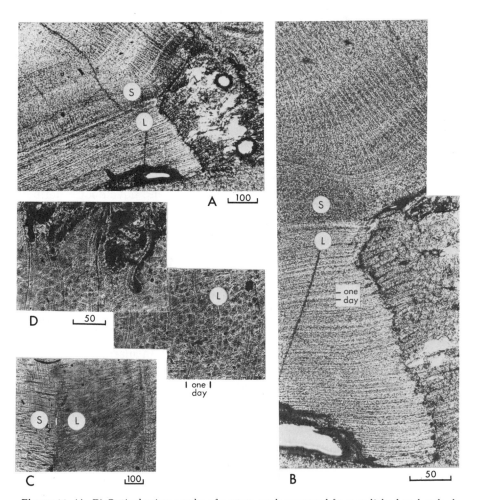

Figure 11. (A–D) Optical micrographs of acetate peels prepared from polished and etched shell sections. (A) Acetate peel of an *Astarte castanea* Say specimen transplanted from a subtidal environment (S) to a laboratory holding tank without substratum (L). Periostracal extensions and changes in shell structure and growth rate mark the transplant event. (B) Enlargement of (A). A daily growth increment is shown. Acetate peel YPM-IP 28159. (C,D) Acetate peels of a *Mercenaria mercenaria* (Linnaeus) specimen transplanted from a subtidal environment (S) to a laboratory holding tank without substratum (L). (C) Changes in shell structure and growth rate mark the transplant event; (D) enlargement of periostracal extensions showing a large daily growth increment. Acetate peel YPM-IP 28161. Scale bars in micrometers. Reprinted with permission from Rhoads and Pannella (1970).

water periodicity in growth and reproduction, as well as yield information
on temporal variations in the deep-water environment. Furthermore,
study of the relationship of microgrowth patterns to bathymetry has
yielded information on the effect of water depth on calcification (Rhoads
and Pannella, 1970). Preliminary results suggest that microgrowth patterns within the shells of fossil molluscs may be of use in paleobathymetric
studies for defining relative depth gradients in Phanerozoic marine environments. As discussed by Rhoads and Pannella (1970), a bathymetric
change from shallow to deep water is accompanied by decreased surface
wave motion at the bottom, decreased light penetration, and decreased
variation in temperature, salinity, and food supply. Depth variation in the
fluctuation of the physical–chemical environment should be reflected in
features of microgrowth patterns.* Microgrowth increments of deep-
water species are relatively regular in thickness and have poorly defined
boundaries that are delimited by color variation. Shallow-water bivalves,
especially those living in the boreal intertidal zone, show sharply delimited microgrowth increment boundaries, and these increments show differences in thickness related to tidal and seasonal periodicities. Data available at present (i.e., Rhoads and Pannella, 1970) are limited to species
from extremes in bathymetric range. Further work on the relationship of
microgrowth patterns to water depth is required for species from intermediate water depths, especially on either side of the seasonal thermocline.

Clark (1974), in his review of growth lines in marine invertebrate
skeletons, summarizes a few potential ecological uses of microgrowth increment patterns that have not been discussed above. For example, he
points out that the presence of disturbance lines, or variation in the spacing of periodic lines, can be used as an index of environmental variability,
and high variability argues for relatively shallow water. Similarly, the
presence of annual patterns of microgrowth increments suggests a climate
with well-defined seasons, and the presence of increments occurring with
a tidal periodicity implies a habitat in or near the intertidal zone. Clark
(1974) summarizes his review by stating that "virtually any approach
which can be used to demonstrate that growth lines form today as reflections of the environment can be turned about to make interpretations
of past environments from fossil lines . . . ." Along these same lines, Pannella and MacClintock (1968) have suggested that seasonal and climatic
variations of microgrowth patterns can be useful in the study of latitudinal
distributions of fossil assemblages. Similarly, Hall (1975) found that the
number of microgrowth increments in "biochecks and bands of fast

---

* Pannella (1976) has more recently extended and refined these ideas, providing evidence
  that microgrowth patterns within the shells of fossil molluscs can be used to reconstruct
  tidal curves of paleoseas.

growth" within the shells of *Tivela stultorum* and *Callista chione* varied with age, latitude, and water depth, and has suggested that counts of microgrowth increments in similar shell regions of fossil specimens might be of use in reconstructing both paleolatitudes and paleoclimate.

In all the examples discussed in the preceding paragraphs, observed changes in microgrowth increment patterns result from natural environmental perturbations. With the increased potential environmental impact of many of man's activities (e.g., dredge-spoil dumping, oil spills, thermal discharges, disposal of chemical and radioactive wastes), the number of environmental perturbations of anthropogenic origin has been steadily increasing, and these anthropogenic changes in the environment can also be recorded in skeletal growth. One of the best examples of how such pollution events can be manifested within the molluscan shell comes from a recent study conducted by Kennish and Olsson (1975). These workers analyzed in detail microgrowth increments within the shell of *Mercenaria mercenaria* to define the effects of thermal discharges from a nuclear power plant on the growth of this species. They found that clams from within a mile radius of the mouth of the discharge canal of the nuclear plant had a much higher number of shell-growth interruptions, thinner shells, and slower summer growth rates than did clams farther from the plant. By counting growth increments back from the shell margin, they were able to determine that many of the growth breaks* were associated with rapidly changing water temperatures that resulted from abrupt shutdowns and startups of the power plant.† Kennish and Olsson (1975) also suggested that the thermal effluent may be adversely affecting physiological functions other than growth. At the station nearest the effluent, no spawning breaks were observed within the shells, while such breaks were seen in specimens from all control sites.

In summary, the fact that microgrowth increment patterns reflect environmental changes has many applications to ecological and paleoecological problems. As we have attempted to emphasize, comparison of growth patterns among specimens collected from different biotopes may be used to identify clinal gradients in growth. Recognition of growth gradients (both spatial and temporal) in Recent and fossil assemblages may greatly assist both the ecologist and the paleoecologist in recognizing environmentally related stress gradients. When these and the other appli-

---

* Such breaks were associated with relatively unique patterns of microgrowth increments within the shell. See Chapter 7, Plate IIe–g, for photographic illustrations of this type of break (and associated microgrowth increment patterns) in *Mercenaria mercenaria*.

† The growth rate of *M. mercenaria* generally increases with increasing temperatures and peaks between 20 and 24°C; Haskin (communicated to Kennish and Olsson, 1975) found decreased growth of *M. mercenaria* above 26°C. The thermal effluent raised water temperature in areas around the nuclear plant 3–5°C above ambient.

cations discussed above are combined, the molluscan shell might well be interpreted as a long-term continuous environmental recorder.

3. Shell Structural Changes

At this point, it is imperative that the reader recognize that we are no longer discussing microgrowth increment patterns. A distinction was made, in the introduction to this chapter, between patterns of microgrowth increments and patterns associated with shell structural changes. We thus emphasize once again that shell structural changes, which are considered in the following discussion, are associated either with changes in shell microstructure (see Appendix 2.B for a guide to shell microstructures) within a given layer or shifts in the relative position of shell layer boundaries on the depositional surface.* As stated earlier, we use the general term "structure" in reference to shell structural types (i.e., shell microstructures) or in reference to the arrangement of shell layers.

In Chapter 2, Carter emphasizes that the type of structure deposited within the molluscan shell is strongly biologically (genetically) controlled and more or less independent of environmental influences. It is also pointed out in Chapter 2 that notable exceptions to this generalization exist—in particular, structural changes associated with changes in (1) employed respiratory pathways and (2) shell mineralogy. In the following discussion we will consider these exceptions in detail and summarize a few other apparent exceptions that may or may not be related to metabolic changes. Before discussing the growth patterns themselves and their applications, we will outline hypothesized mechanisms responsible for the formation of certain structural changes within the shell.

3.1. Formation of Structural Changes

For over 50 years workers have recognized that structural changes observed within the shells of certain bivalves may reflect changes in environmental conditions (Weymouth, 1923; Dodd, 1964; Davies and Sayre, 1970; Kennish and Olsson, 1975; Lutz and Rhoads, 1977). Only recently, however, have mechanisms responsible for some of these changes been considered in detail. In 1970, Davies and Sayre (1970) suggested that various conditions of environmental stress could lead to shell dissolution and, furthermore, that such dissolution events were recorded in the ultrastructure of the shell. More recently, Lutz and Rhoads (1977) have hy-

---

* We include here the addition of a shell layer (e.g., addition of a calcitic layer to cooler-water representatives of a largely aragonitic species).

pothesized that extensive reworking of original shell microstructure occurs as a result of environmentally controlled alternating periods of shell deposition and dissolution associated with utilization of aerobic and anaerobic pathways, respectively. Evidence that such shell-destructive mechanisms may be operating in bivalves during periods of environmental stress has been provided by Lutz (1976b, 1977) and Lutz and Rhoads (1977, 1978) through a detailed analysis of seasonal and geographical variation in the shell structure of a Recent mytilid (*Geukensia demissa*). In the following paragraphs we will summarize the results obtained by these workers to provide an example of the manner in which environmentally induced metabolic changes might drastically alter the original structure of a bivalve shell. Preparation procedures for analysis of described structural changes may be found in Appendix 1.A.1–3.

The structure of the inner shell layer of *G. demissa* varies not only with season, but also with latitude. If specimens from relatively cold-water (high-latitude) environments, such as the Gulf of Maine, are examined, the seasonal sequence of events occurring on the growth surface of the inner shell layer is summarized as follows: During the warm summer months (June through September), very regular hexagonal nacreous tablets are arranged in steplike patterns characteristic of bivalve nacre (Fig. 12A). As water temperatures decline, the nacreous tablets become smaller and less regular,* showing visible signs of corrosion in the form of marked pitting and hollow crystals (Erben, 1972; Wada, 1972), as well as increased proportions of fine-grained structures (Fig. 12B–D). An increased stacking of nacreous tablets (i.e., increased proportions of columnar nacre) is also apparent in various areas of the inner layer growth surface (Fig. 13). During the colder months of the year (January to March, with water temperatures below 3°C), shell corrosion becomes visible to the naked eye, the entire inner shell surface often presenting a chalky white appearance. Ultrastructurally, this surface appears uniformly fine-grained (Fig. 12E). Similar macroscopically visible corrosion has been reported in *Mercenaria mercenaria* after long periods of valve closure resulting from extended periods of anaerobiosis (Dugal, 1939). The ability of *G. demissa* to respire anaerobically for extended periods of time has been described (Kuenzler, 1961; Lent, 1967, 1968, 1969), as has the relative increased efficiency, in this species, of some of the citric acid cycle enzymes in an anaerobic direction (Hammen and Lum, 1966). The observed shell corrosion may well be a reflection of buffering of acidic end products from anaerobic metabolism during the colder months, when oxygen transport into the cells should theoretically be reduced relative to that occurring at higher

---

* A similar irregularity of nacreous crystals as a result of dissolution during reduced winter temperatures (as low as 8°C) was found by Wada (1972) in his examination of growth surfaces of *Pinctada martensii* and *Pinna attenuata*.

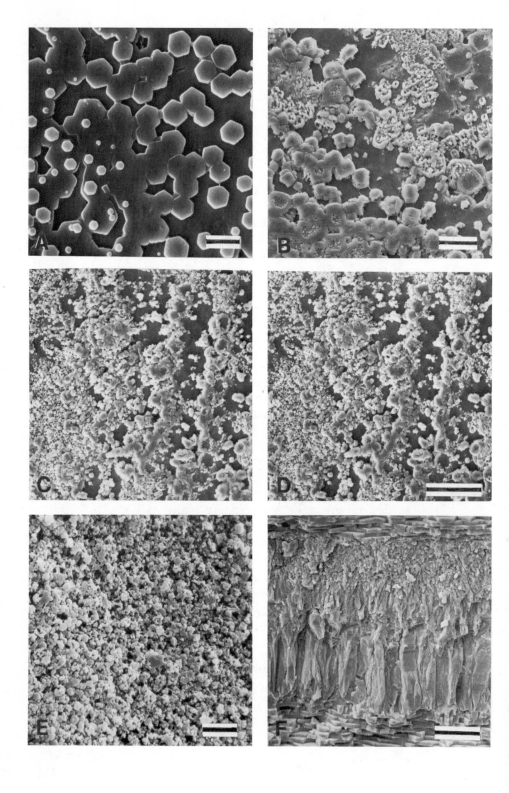

temperatures (Lange *et al.*, 1972). Wilbur (1972) has suggested that during periods of "adverse environmental conditions," shell decalcification may predominate over growth. The gradation in fractured, as well as polished and etched, vertical shell sections of *G. demissa* nacreous laminae into fine-grained structures (suggestive of massive corrosion) instead of regular prisms (Fig. 12F) tends to support this view. As water temperatures increase during the spring, the sequence of events described above is reversed. Examination of growth surfaces during transition periods (spring and fall) between normal nacreous deposition (summer) and drastic corrosion (winter) reveals differential dissolution of calcium carbonate and organic material, with corroded aragonitic tablets above and below exposed sheets of inter-lamellar organic matrices (Fig. 13). Despite such shell dissolution, successive monthly samples often indicate a net gain of shell material during these "transition" periods. We therefore suggest that the observed growth surfaces are a reflection of alternating periods (not necessarily rhythmic) of shell deposition and destruction. Here, alternating periods of aerobic and anaerobic metabolism, respectively, which have already been demonstrated to occur in at least one species [*M. mercenaria* (Crenshaw and Neff, 1969; Gordon and Carriker, 1978)], could easily provide the driving forces.

In the preceding paragraph we have described changes occurring primarily on the growth surface itself. When the shell is viewed in section, using fracture, acetate-peel, or thin-section techniques (see Appendix 1.A. 1,2), one is able to observe how these changes are recorded within the shell. Figures 12F and 15B depict, respectively, a fractured section and an acetate peel of a polished and etched longitudinal section of the inner layer of a *G. demissa* shell valve from the Gulf of Maine (Damariscotta River, Lincoln County, Maine). In these sections one observes a gradation (progressing toward the inner shell layer growth surface) of nacreous laminae into fine-grained "homogeneous" structure that gradually becomes

←

Figure 12. Scanning electron micrographs of *Geukensia demissa* (Dillwyn). (A–E) Inner shell layer growth surface as seen during various months of the year. (A) Regular hexagonal nacreous tablets arranged in steplike patterns characteristic of bivalve nacre (August sample). Scale bar: 10 μm. (B) Small, irregular nacreous tablets showing visible signs of corrosion in the form of marked pitting and increased proportions of fine-grained structures (November sample). Scale bar: 5 μm. (C,D) Extremely irregular nacreous tablets with considerable amounts of fine-grained structures (December sample). The stereo pair was taken with a 6° angular displacement between exposures. Scale bar: 10 μm. (E) Fine-grained structures reflective of extensive shell dissolution during the colder months of the year (February sample). Scale bar: 10 μm. (F) Vertical fracture through the inner shell layer. Nacreous tablets grade into fine-grained structures at the top of the micrograph, and the prisms grade into nacre at the bottom. The most recently deposited crystals are at the bottom of the micrograph. Scale bar: 10 μm. Micrographs A, B, E, and F reprinted with permission from Lutz and Rhoads (1977). Copyright © 1977 by the American Association for the Advancement of Science.

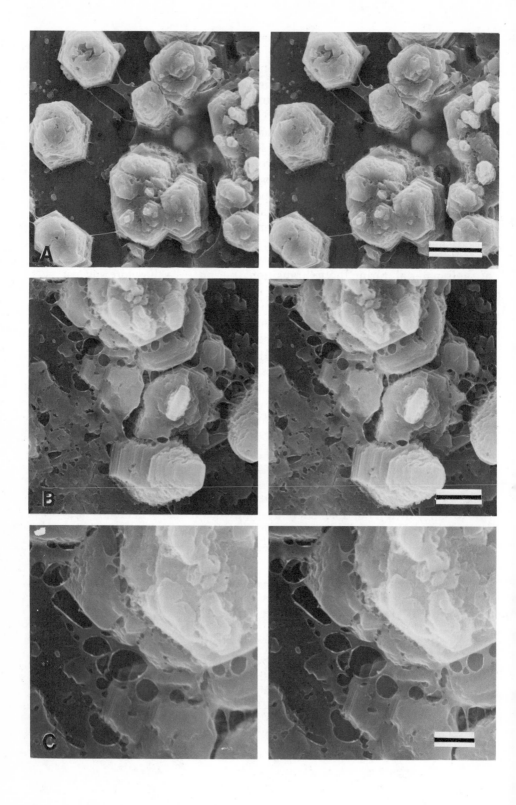

increasingly coarse-grained and subsequently grades into simple aragonitic prisms. In turn, these simple prisms, which are strikingly similar in appearance to myostracal prisms, grade into nacreous tablets. It is also interesting to note that at the bottom of Figure 12F, the first few nacreous tablets appear to be stacked upon one another, while subsequently deposited tablets assume a more steplike [*Treppen* (see Schmidt, 1923; Wise, 1970a,b)] pattern. These repetitive changes result in a series of distinct sublayers within the inner shell layer of mature specimens.

We have described above the sequence of seasonally related changes occurring on the growth surface and within the shell of G. *demissa* specimens from relatively cold-water (high-latitude) environments. Seasonal changes occurring on the inner shell layer surface of specimens from warmer-water environments (e.g., south of Cape Cod on the east coast of North America) are somewhat more complex. In these populations, one still sees alternating sublayers associated with cold-water shell deposition and dissolution, but one also encounters irregular prismatic sublayers that are associated with extremely warm summer temperatures (Lutz and Rhoads, 1978). These irregular prismatic sublayers may be associated with the increased utilization of anaerobic pathways during the hot summer months when oxygen solubility is markedly reduced. However, in all the specimens examined to date from south of Cape Cod (Lutz, 1976b, 1977; Lutz and Rhoads, 1978), we have never observed homogeneous structures in "warm-water" sublayers. This may well indicate decreased shell-destructive processes at relatively high (relative to extremely low) temperatures. When viewed in longitudinal section, such complex seasonal variation often results in an alternation of three distinct types of sublayers within the inner layer: (1) irregular prismatic ("warm-water" sublayer); (2) homogeneous + irregular prismatic ("cold-water" sublayer); and (3) nacre. These three types of sublayers are generally present in the majority of specimens sampled from relatively temperate environments (e.g., Cape Cod to Cape Hatteras) (Figs. 14 and 15C). In subtropical environments (e.g., south of Cape Hatteras), homogeneous structures are absent, resulting in an alternation of only irregular prisms and nacre (Fig. 15D).

Structural patterns described in the preceding discussion are summarized in Fig. 15. The micrographs in this figure depict acetate peels of the inner shell layer of G. *demissa* specimens from four geographically

----

Figure 13. Scanning electron micrographs of the inner shell layer growth surface of Geukensia demissa (Dillwyn) showing natural shell dissolution. Stereo pairs were taken with a 6° angular displacement between exposures. Note the marked pitting and stacked appearance of nacreous tablets. The differential solubility of calcium carbonate and organic matrices is apparent. Scale bars: (A) 5 μm; (B) 2 μm; (C) 1 μm. Reprinted with permission from Lutz and Rhoads (1977). Copyright © 1977 by the American Association for the Advancement of Science.

Figure 14. Optical micrograph of an acetate peel prepared from a polished and etched anteroposterior longitudinal section of a *Geukensia demissa* shell from Wachapreague, Virginia. The micrograph is taken in the region of the umbo. Three types of sublayers are apparent: (1) nacre (n); (2) irregular prismatic (p); (3) irregular prismatic (p) + homogeneous (h). Note the decreased percentage of nacre in the most recently formed regions (bottom of micrograph) of the shell. Scale bar: 100 μm.

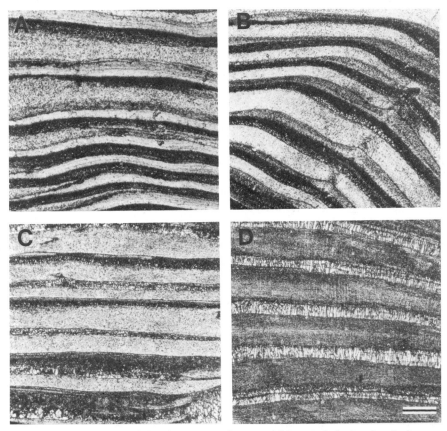

Figure 15. Optical micrographs of acetate peels prepared from polished and etched anteroposterior sections of the inner shell layer of *Geukensia demissa* specimens from four latitudinally separated populations. All micrographs were taken in the region of the umbo. The direction of growth is toward the left. (A) Specimen sampled from St. Peter's Bay, Prince Edward Island. (B) Specimen sampled from the Damariscotta River, Lincoln County, Maine. (C) Specimen sampled from Wachapreague, Virginia. (D) Specimen sampled from Cape Kennedy, Florida. Scale bar (applicable to all four micrographs): 100 μm.

separated populations from Prince Edward Island to Florida. Marked differences in the percentages of various structural types are obvious.

The mechanism described above for the formation of structural changes has important implications. For example, in paleotemperature studies based on oxygen isotope analyses, it is important to realize that truncated seasonal paleotemperature curves may reflect a net loss of shell material during certain times of the year (see Chapter 5). Similarly, shell dissolution may account for certain apparent ontogenetic discontinuities in elemental (or chemical) distribution patterns within the shells of individual specimens (see Chapter 4). We hope that the implications of shell dissolution and redeposition discussed above will eventually result in the realization that the shell is far more than a simple accretionary deposit.

3.2. Documented Growth Patterns and Their Ecological, Paleoecological, and Phylogenetic Applications

In the previous section, we proposed a hypothesis for the mechanisms responsible for the formation of *certain* structural changes within the molluscan shell. At this point, it is important for the reader to realize that not all the patterns that will be discussed here necessarily reflect periodic or aperiodic respiratory changes. Alternate hypotheses for the formation of structural growth patterns are discussed by Carter in Chapter 2.* The reader should also realize that relatively few studies have been conducted in which patterns of structural change within molluscan shells have been adequately documented and analyzed. Thus, while it may appear that we have summarized the results of only a few relatively specific studies, we have, in fact, presented a comprehensive review of the status of current knowledge.

Environmentally controlled variation in the mineralogy of the molluscan shell is discussed in Chapter 2 and will therefore not be considered in detail here. Worth mentioning, however, is some early work that is relevant to subsequent studies of shell structural variation. In 1954, Lowenstam (1954) showed that the relative amounts of aragonite and calcite composing the shells of certain molluscs can reflect environmental conditions at the time of shell formation. Since aragonite and calcite are present in distinct structural units in molluscan shells (Bøggild, 1930), the variation in shell mineralogy found by Lowenstam (1954) suggested to Dodd (1963, 1964) that variation in the structure of the shell should also occur. As a result, Dodd conducted two detailed studies of the effect of temperature on the shell mineralogy and structure of the Recent mytilid

---

* As mentioned in Chapter 2, sublayers of simple aragonitic prisms possibly result from temporary mantle attachment (Taylor, J. D., *et al.*, 1969). While this might be viewed as an alternate hypothesis of structural growth pattern formation, it may not be independent of the mechanisms outlined in this chapter. It is conceivable, for example, that mantle attachment itself may be a direct consequence of anaerobiosis. Personal observations on *Arctica islandica* specimens that have been kept for extended periods of time under anoxic conditions have suggested that the entire surface of the mantle inside the pallial line may be attached to the shell during anaerobiosis. If this is indeed the case, the entire surface of the mantle might be acting, during such periods, as a "muscle" that attaches the shell to the soft tissues. If so, the genetic term "myostracal prisms," which has been used by J. D. Taylor *et al.* (1969, 1973), may well appropriately characterize simple prisms occurring within sublayers in the inner shell layer of this (see Fig. 19) and other species of molluscs. Before the term "myostracal prisms" can be employed, however, it must be demonstrated that the entire mantle surface can be appropriately termed a "muscle"; until evidence for this is provided, it is probably best to refer to such simple aragonitic prisms as "myostracal-like" prisms.

Mytilus californianus. Significant results obtained by Dodd (1963, 1964) for this species were as follows: (1) the percentage of aragonite in the shell varies with temperature; (2) the extent and development of the inner calcitic prismatic layer are strongly dependent on growth temperature; (3) wedges of nacreous structure that often project into the inner prismatic layer form annually, making age determination of the shell possible; (4) growth rates determined from these wedges are directly proportional to mean temperature of the growth locality; and (5) quantitative paleotemperature determination based on shell-structure variation in this species is possible. Moreover, Dodd suggested that the shell structure of other species of molluscs, particularly those having shells composed of a combination of calcite and aragonite, could potentially be used for determining paleotemperatures.

Since the pioneering work of Dodd (1963, 1964), several workers have examined the effects of temperature (or latitude) on the shell structure of various species of bivalves. Blackwell *et al.* (1977) considered the possibility that the structure of the "middle" aragonitic prismatic layer of *Geukensia demissa* varied with latitude, but concluded that the observed variation reflected subspecific differences and not a climatic gradient coinciding with the geographical range of this species. More recently, however, Lutz (1976b, 1977) and Lutz and Rhoads (1977, 1978) have shown that the structure of the inner shell layer of this species varies with both season and latitude. The extent of this variation can be seen in Fig. 15. It should be apparent from a close examination of the micrographs in this figure that there is a relatively sharp decrease in the percentage of homogeneous structure with decreasing latitude (or increasing mean temperature). In light of this, it may eventually be possible to develop a quantitative approach [similar to that developed by Dodd (1964) for *M. californianus*] for the analysis of growth patterns within the shells of *G. demissa* and other closely related Recent and fossil species. Such an approach may prove useful in ecological studies for defining the effect of thermal stress gradients (e.g., power-plant waste-heat gradients) on bivalve shell growth and in paleoecological studies for defining both spatial and temporal paleotemperature patterns. Furthermore, analysis of structures adjacent to the inner shell layer surface of *G. demissa* from death assemblages* should prove useful in ecological, paleoecological, and ar-

---

* In analyzing the shells of specimens from death assemblages, one needs to distinguish postmortem shell dissolution from shell dissolution that occurred while the organism was alive and respiring anaerobically. In the case of *G. demissa*, this is readily done by comparing growth surfaces of shell layers on either side of the pallial line. During anaerobiosis, shell structural changes in this species are apparent only within that portion of the shell inside the pallial line; evidence of dissolution of nacre on both sides of the pallial line (i.e., in both the middle nacreous layer and the inner layer of the shell) would indicate postmortem diagenesis.

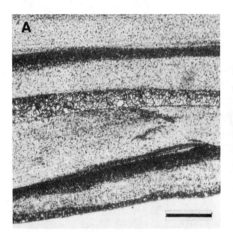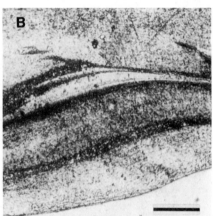

Figure 16. Optical micrographs of acetate peels of polished and etched anteroposterior longitudinal sections of *Geukensia demissa* shells sampled at different times of the year. Examination of structures adjacent to the inner shell layer surface should prove useful for defining the season of death of individual specimens. (A) Specimen sampled on August 4, 1976, at Wachapreague, Virginia. Note irregular prismatic structure at the growth surface (bottom of micrograph. Scale bar: 50 μm. (B) Specimen sampled during mid-October at Crisfield, Maryland. Note nacre at the growth surface (bottom of micrograph). Scale bar: 50 μm.

cheological studies for defining the season of death of individual specimens (Fig. 16).

Other studies in which structural changes have been observed to be associated with thermal stress include those of Kennish and Olsson (1975), Kennish (1976, 1977a,b, 1978) (also see Chapter 7), and Farrow (1972). In their studies of the effects of thermal discharges on the microstructural growth and mortality of *M. mercenaria*, Kennish and Olsson (1975) and Kennish (1976, 1977a,b, 1978) found that "crossed lamellar structure"* replaced prismatic structure in the outer shell layer of this species throughout the duration of high-temperature stress periods. Farrow (1972) reported a somewhat similar phenomenon in *Cerastoderma edule*, in which a greater prominence of "crossed lamellar structure" in the outer shell layer was observed to be associated with winter reduction in growth rate. Farrow further observed that the lamellae became "markedly deflected" if an actual "stoppage" of growth took place.

Bryan (1969) described drastic changes within the shell of the gastropod *Nucella lapillus* that occurred subsequent to addition of oil-spill removers ("detergents") to the organisms' natural environment (following

* For an interpretation of regions containing "crossed lamellar structure" in the outer shell layer of *Mercenaria mercenaria* and *Cerastoderma edule*, see Chapter 2.

a large spill of Kuwait crude oil). When N. *lapillus* shells were sectioned longitudinally or transversely, it was found that "the shell edges had often been temporarily sealed by continuing the inner nacreous layer of the shell to the outer surface. Later growth of the shell was laid down on this nacreous layer partition and this produced a growth mark and a line of weakness in the shell" (Bryan, 1969, pp. 1074–1075).

Many of the studies in which shell structural changes have been documented have been primarily concerned with the usefulness of observed growth patterns for determining the age of examined specimens. Weymouth (1923) commented on the presence of "translucent regions" within the inner shell layer of *Tivela stultorum* and provided evidence that such regions were formed with an annual periodicity (during winter) and, hence, could be of value in estimating the age of individual specimens. More recently, Rhoads and Pannella (1970) have demonstrated the annual nature of pigmented growth bands within the inner "homogeneous" layer of the hard-shelled clam, *M. mercenaria*. Similar annually formed pigmented bands have been found by Jones *et al.* (1978) within the inner shell layer of *Spisula solidissima* and were used by these workers to determine the age and growth rate of this species in both inshore and offshore populations (for a more detailed discussion of growth patterns in *S. solidissima*, see Chapter 2). Lutz (1976a) has presented evidence that structural patterns within the inner nacreous layer of *Mytilus edulis* reflect annual cycles of growth and has suggested that careful examination of these patterns can provide a relatively accurate estimate of the age of individual mussels (Fig. 17). The ultrastructure of the patterns described by Lutz (1976a) is shown in Figure 18. Relatively thick aragonite crystalline laminae abruptly decrease in size to form a zone of fine laminae that gradually increase in thickness toward the inner shell surface. It is interesting to note that the zone of fine laminae was observed to form only during the spring of each year when water temperatures were rising (Fig. 18C). If, as suggested by Lutz (1976a), the thickness of nacreous laminae is temperature-dependent, one would expect to observe within the shell a *gradual* decrease in the thickness of nacreous tablets, rather than the abrupt decrease seen in Fig. 18A and C. We suggest that the observed structural patterns within *M. edulis* may reflect anaerobiosis-related shell dissolution during the late winter and early spring (perhaps related to gametogenesis). If shell-destructive processes are indeed operating during this period, nacreous laminae may actually be destroyed, and one would expect to encounter only relatively thick, previously deposited nacreous laminae at the inner shell surface (e.g., Fig. 18B). With increased utilization of aerobic pathways during the late spring, net shell growth takes place, albeit perhaps irregularly at first, resulting in the formation of thin, irregular nacreous laminae that gradually increase in thickness (Fig. 18C

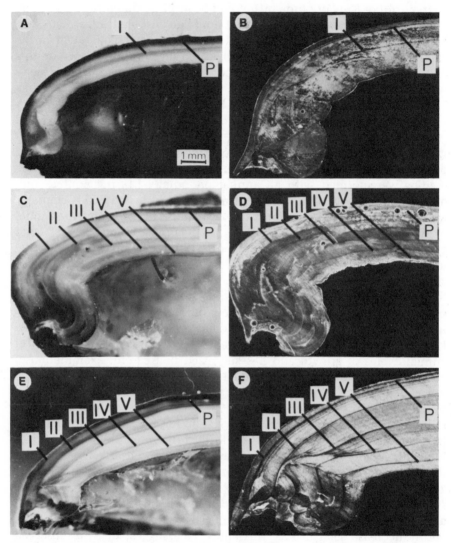

Figure 17. Finely polished and etched anteroposterior longitudinal shell sections (left) of *Mytilus edulis* L. with corresponding acetate peels (right). Roman numerals designate annual growth patterns. The scale for all shell sections appears in (A). All specimens were sampled from the Damariscotta River, Lincoln County, Maine. (A) 1½-year-old specimen from a constantly submerged (rafted) population (larval settlement, May, 1971; sampling date, January 4, 1973). (B) Acetate peel of umbonal region of (A). (C) Specimen sampled from a constantly submerged population on the underside of a floating platform that was launched in July, 1967 (sampling date, August 4, 1972). (D) Acetate peel of umbonal region of (C). (E) Specimen sampled from an intertidal population during the month of November. (F) Acetate peel of umbonal region of (E). Reprinted with permission from Lutz (1976a).

and D) as anaerobic pathways are employed less and less. Patterns morphologically very similar to those described above in M. edulis have been found by Hudson (1968; personal communication)* within the inner shell layer of the Jurassic mytilid Praemytilus strathairdensis. Hudson (1968) speculates on the annual origin of these "sublayer junctions." If such patterns are indeed analogous to those encountered within M. edulis, they should be of considerable assistance in reconstructing growth rates of P. strathairdensis in various paleoenvironments. Finally, Wada (1961, 1972) has commented on the presence of somewhat similar structural changes in various bivalves (particularly Pinctada). He comments on seasonal variations in the size of aragonite crystals, as well as the effects of extrapallial fluid pH, viscosity, and impurities on nacre tablet size, shape, and texture.

In the remainder of this chapter, we would like to discuss briefly several of the phylogenetic and paleoecological implications of our ideas on the relationships between bivalve metabolism and shell structure. For purposes of illustration, we will discuss many of the implications in light of the patterns within the shell of G. demissa described above.

It has been proposed that the stacked "mode of deposition" of nacre is a primitive trait among molluscan species that has been lost by degrees in bivalves during their evolution of a "new (bivalve) shell form" (Wise, 1970a,b). More recently, J. D. Taylor (1973; personal communication) has suggested that the "vertical component" of nacre is more closely related to geometry of the shell rather than the antiquity of the lineage, arguing that bivalves with low expansion rates (high convexity) have better-developed stacked nacre than forms with high rates of expansion. A similar point of view has been expressed by Carter (see Chapter 2), who suggests that the major variations in nacre tablet-stacking mode reflect adaptive secretory strategies for controlling shell-accretion rate (also see Wise, 1970b). One may alternatively view the various configurations of nacre within the bivalve shell as reflecting changing metabolic pathways (Lutz and Rhoads, 1977). Assuming that atmospheric oxygen levels increased from the late Precambrian to present levels in the Ordovician (Berkner and Marshall, 1965; Cloud and Nelson, 1966; Brinkman, 1969; Rhoads and Morse, 1971; Crimes, 1974), we suggest that aerobic respiration pro-

---

* In his paper, Hudson (1968, p. 173), in referring to "sublayer junctions" within the inner nacreous layer of Praemytilus strathairdensis, states that aragonite laminae "tend to grade into those above [i.e., older], and have a sharp junction with those below." Plate 33, Fig. 2, of his paper, however, shows just the reverse, with crystalline laminae grading into those below (i.e., younger) and having a sharp junction with those above. In personal communications (see Lutz, 1976a), Dr. Hudson has indicated that the text statement is in error and that the structural patterns within the inner shell layer of P. strathairdensis are very similar to those found within the shells of M. edulis.

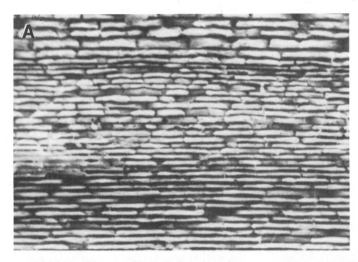

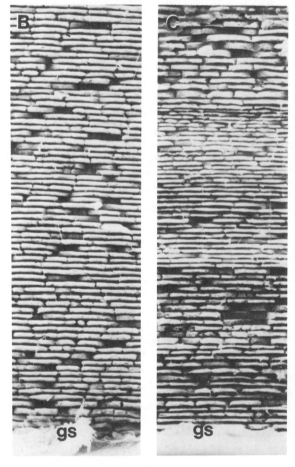

gressively increased over this same period of geological time and resulted in an increased shell deposition/shell dissolution ratio and, hence, increased proportions of sheet nacre (see Appendix 2.B)* in certain bivalves. Moreover, in *G. demissa*, during "transition" periods described in the previous section, increased stacking of crystals† actually results in structural changes, during which nacre subtly grades into simple aragonitic prisms, or the reverse (see Figs. 12F and 13). Physiological and/or other mechanisms responsible for the increased stacking of nacreous tablets (i.e., increased proportions of columnar nacre) and for this gradation of nacre into simple prisms warrant additional research.‡ Similar bands of simple aragonitic prisms are found alternating with complex crossed lamellar and homogeneous structures within the inner shell layer of *Arctica islandica* (Fig. 19), an organism known to respire anaerobically during

* Schmidt (1923) used the terms *Backsteinbau*, *Treppen*, and *Vertikalschichtung* for, respectively, brick-wall, stepped, and columnar (or lenticular) patterns of nacre (also see Lutz and Rhoads, 1977). Schmidt's terms were defined by him (and are used by most workers) in only two dimensions. In the microstructure guide presented in Appendix 2.B, the various stacking modes of nacre are classified according to a three-dimensional scheme. "Sheet" nacre includes *Backsteinbau* and *Treppen* in various vertical two-dimensional views. "Row stack" nacre appears as *Backsteinbau* or *Treppen* in some vertical two-dimensional views, and as *Vertikalschichtung* in others. Finally, "columnar" nacre appears as *Vertikalschichtung* in all vertical two-dimensional views.
† A similar stacking of nacreous tablets may be seen in micrographs presented by Wise (1970b) of the surfaces of *Pinna carnea* and *Nucula annulata* [= *Nucula proxima* (Hampson, 1971)].
‡ Similar structural changes (i.e., sublayers of simple aragonitic prisms) within the shells of numerous molluscs, such as the Pennsylvanian gastropod *Shansiella carbonara* (Batten, 1972), the Jurassic bivalves *Lithiotis problematica*, *Cochlearites loppianus*, and *Lithiopedalion* sp. from the "Lithiotis" facies (Loriga and Neri, 1976; Benini and Loriga, 1977; Benini, personal communication), the Recent bivalves *Modiolus modiolus*, *Cyclinella saccata*, and *Pholadomya candida* (Taylor, J. D., *et al.*, 1969, 1973), and numerous species within the superfamily Chamacea (Kennedy *et al.*, 1970), may result from periods of anaerobiosis.

←

Figure 18. Scanning electron micrographs of polished and etched anteroposterior longitudinal sections of the inner nacreous shell layer of *Mytilus edulis* L. All specimens were sampled from the Damariscotta River, Lincoln County, Maine. (A) Annually formed ultrastructural crystalline pattern. The most recently formed aragonite crystalline laminae are at the bottom of the micrograph. × 2800. (B) Aragonite crystalline laminae adjacent to the inner shell surface (gs) of a specimen sampled May 16, 1974. × 1400. (C) Aragonite crystalline laminae adjacent to the inner shell surface (gs) of a specimen sampled June 18, 1974. × 1400. (D) Aragonite crystalline laminae adjacent to the inner shell surface (gs) of a specimen sampled July 2, 1974. Although the actual junction between thick and thin laminae is above the upper border of the micrograph, a gradual increase in the thickness of crystal laminae with decreasing distance from the inner shell growth surface (gs) is apparent. × 1400. Reprinted with permission from Lutz (1976a).

Figure 19. Scanning electron micrograph of a vertical fracture through the inner shell layer of *Arctica islandica* near the umbo. Bands of simple aragonitic prisms are seen alternating with complex crossed lamellar structures. Scale bar: 40 μm.

extended aperiodic burrowing activities (Taylor, A. C., 1976).* This suggests that such aragonitic prisms may represent a shell microstructure that reflects low-oxygen environments. Such a view is compatible with the phylogenetic tree presented by J. D. Taylor *et al.* (1973), in which the relationships among bivalve superfamilies, their possible histories, and

---

* Thompson and Jones (1977) have recently suggested that these bands of simple aragonitic prisms may occur with an annual periodicity. These workers have arrived at age estimates, based on counts of such "annual" bands, as high as 150 years for certain specimens. If these simple prismatic sublayers are related to the aperiodic burrowing activity of this species, such age estimates may be unrealistically high.

their shell structures have been summarized. The earliest ancestral forms in which shell structures can be recognized are composed of nacre and simple prisms of aragonite. It has been suggested (Taylor, J. D., 1973) that the simple-prism-and-nacre combination originally arose spontaneously as a consequence of the precipitation of calcium carbonate contemporaneously with organic matrix under a certain set of physicochemical conditions. Subsequently, because of some selective advantage in this structural combination (Taylor, J. D., and Layman, 1972), this condition became stabilized.* In the foregoing discussion, we imply that simple aragonitic prisms represent an altered form of nacre resulting from alternating processes of shell deposition and dissolution. Assuming this to be the case, we suggest that the evolutionary scheme proposed by Wise (1970a,b) be extended backward one step. We propose a gradual change (not necessarily evolutionary in a genetic sense) within certain ancestral molluscan lines from simple aragonitic prisms to columnar [*Vertikalschichtung* of Schmidt (1923)], then row stack, and, finally, sheet [*Treppen* and *Backsteinbau* of Schmidt (1923)] nacre. It is also interesting to note that if such structural changes have occurred, aragonitic (myostracal) prisms associated with the sites of muscle attachment may represent vestiges of an early ancestral shell structure. Inasmuch as such structural changes may be correlated with oxygen tension, a detailed analysis of relationships between ambient oxygen concentrations and various shell structural types within certain Recent and fossil molluscan superfamilies may eventually prove useful in determining dissolved oxygen gradients in present, as well as ancient, marine environments (Lutz and Rhoads, 1977).

ACKNOWLEDGMENTS. We thank G. R. Clark II, G. D. Rosenberg, G. Pannella, J. G. Carter, and M. J. Kennish for discussions and critical reviews of the manuscript; M. Castagna, J. N. Kraeuter, M. D. Bertness, D. K. Muschenheim, J. G. Carter, and A. Sesona for sampling specimens of *Geukensia demissa*; W. C. Phelps, P. Duckett, K. Genthner, V. Clark, and R. Noddin for preparation of specimens; W. K. Sacco and R. D. Smith for assistance with photographic reproductions; A. S. Pooley and E. Tveter Gallagher for technical assistance with the scanning electron microscopy; G. Bevelander, H. Nakahara, and G. R. Clark II for kindly providing figures utilized in this chapter; and S. E. Hurlburt for many heartfelt smiles. Much of the research summarized in this chapter was supported by Environmental Protection Agency grant R804-909-010 and NOAA Sea Grants 04-6-158-44056, SGI-77-17, and 04-7-158-44034. This is Paper No. 5305 of the Journal Series, New Jersey Agricultural Experiment Station, Cook College, Rutgers University, New Brunswick, New Jersey.

---

* For an alternative view on the evolution of shell structures, see Chapter 2.

References

Andrews, J. T., 1972, Recent and fossil growth rates of marine bivalves, Canadian Arctic, and Late-Quaternary Arctic marine environments, *Palaeogeogr. Palaeoclimatol. Palaeocol.* **11**:157–176.

Barker, R. M., 1964, Microtextural variation in pelecypod shells, *Malacologia* **2**:69–86.

Barker, R. M., 1970, Constituency and origins of cyclic growth layers in pelecypod shells, Ph. D. thesis, University of California, Berkeley, 265 pp.

Batten, R. L., 1972, The ultrastructure of five common Pennsylvanian pleurotomarian gastropod species of eastern United States, *Am. Mus. Novit.* **2501**:1–34.

Benini, C. A., and Loriga, C. B., 1977, *Lithiotis* Gümbel, 1871 e *Cochlearites* Reis, 1903 I°: Revisione morfologica e tassonomica, *Boll. Soc. Paleontol. Ital.* **16**:15–60.

Berkner, L. V., and Marshall, L. G., 1965, On the origin and rise of oxygen concentration in the earth's atmosphere, *J. Atmos. Sci.* **22**:225–261.

Berry, W. B. N., and Barker, R. M., 1968, Fossil bivalve shells indicate longer month and year in Cretaceous than present, *Nature (London)* **217**:938–939.

Berry, W. B. N., and Barker, R. M., 1975, Growth increments in fossil and modern bivalves, in: *Growth Rhythms and the History of the Earth's Rotation* (G. D. Rosenberg and S. K. Runcorn, eds.), pp. 9–25, John Wiley & Sons, London.

Bevelander, G., and Nakahara, H., 1969, An electron microscopic study of the formation of the nacreous layer in the shell of certain bivalve molluscs, *Calcif. Tissue Res.* **3**:84–92.

Blackwell, J. F., Gainey, L. F., Jr., and Greenberg, M. J., 1977, Shell ultrastructure in two subspecies of the ribbed mussel, *Geukensia demissa* (Dillwyn), *Biol. Bull.* (Woods Hole, Mass.) **152**:1–11.

Bøggild, O. B., 1930, The shell structure of the mollusks, *K. Dan. Vidensk. Selsk. Skr. (Copenhagen)* **2**:232–325.

Brinkman, R. T., 1969, Dissociation of water vapor and the evolution of oxygen in the terrestrial atmosphere, *J. Geophys. Res.* **74**:5355–5367.

Bryan, G. W., 1969, The effects of oil-spill removers ("detergents") on the gastropod *Nucella lapillus* on a rocky shore and in the laboratory, *J. Mar. Biol. Assoc. U. K.* **49**:1067–1092.

Clark, G. R., II., 1968 Mollusk shell: Daily growth lines, *Science* **161**:800–802.

Clark, G. R., II., 1969a, Shell characteristics of the family Pectinidae as environmental indicators, Ph.D. thesis, California Institute of Technology, Pasadena, 101 pp.

Clark, G. R., II., 1969b, Daily growth lines in the bivalve family Pectinidae, *Geol. Soc. Am. Abstr. Programs* **1**(7):34–35.

Clark, G. R., II., 1974, Growth lines in invertebrate skeletons, *Annu. Rev. Earth Planet. Sci.* **2**:77–99.

Clark, G. R., II., 1975, Periodic growth and biological rhythms in experimentally grown bivalves, in *Growth Rhythms and the History of the Earth's Rotation* (G. D. Rosenberg and S. K. Runcorn, eds.), pp. 103–117, John Wiley & Sons, London.

Clark, G. R., II., 1977, Seasonal growth variations in bivalve shells and some applications in archeology, *J. Paleontol.* **51**(Suppl. to No. 2):7.

Clark, G. R., II., 1979, Seasonal growth variations in the shells of recent and prehistoric specimens of *Mercenaria mercenaria* from St. Catherine's Island, Georgia, *Am. Mus. Nat. Hist. Anthropol. Pap.* **56**:161–179.

Cloud, P. E., Jr., and Nelson, C. A., 1966, Phanerozoic–Cryptozoic and related transitions—New evidence, *Science* **154**:766–770.

Collip, J. B., 1920, Studies on molluscan celomic fluid: Effect of change in environment on the carbon dioxide content of the celomic fluid: Anaerobic respiration in *Mya arenaria*, *J. Biol. Chem.* **45**:23–49.

Collip, J. B., 1921, A further study of the respiratory processes in Mya arenaria and other marine mollusca, J. Biol. Chem. **49:**297–310.

Coutts, P. J. F., 1970, Bivalve growth patterning as a method of seasonal dating in archaeology, Nature (London) **226:**874.

Coutts, P. J. F., 1975, The seasonal perspective of marine-oriented prehistoric hunter-gatherers, in: Growth Rhythms and the History of the Earth's Rotation (G. D. Rosenberg and S. K. Runcorn, eds.), pp. 243–252, John Wiley & Sons, London.

Coutts, P. J. F., and Higham, C., 1971, The seasonal factor in prehistoric New Zealand, World Archaeol. **2**(3):266–277.

Craig, G. Y., and Hallam, A., 1963, Size-frequency and growth-ring analyses of Mytilus edulis and Cardium edule, and their palaeoecological significance, Palaeontology **6:**731–750.

Crenshaw, M. A., 1972, The soluble matrix from Mercenaria mercenaria shell, Biomineralisation **6:**6–11.

Crenshaw, M. A., and Neff, J. M., 1969, Decalcification at the mantle–shell interface in molluscs, Am. Zool. **9:**881–885.

Crenshaw, M. A., and Ristedt, H., 1975, Histochemical and structural study of nautiloid septal nacre, Biomineralisation **8:**1–8.

Crimes, T. P., 1974, Colonisation of the early ocean floor, Nature (London) **248:**328–330.

Cunliffe, J. E., 1974, Description, interpretation, and preservation of growth increment patterns in shells of Cenozoic bivalves, Ph.D. thesis, Rutgers University, New Brunswick, New Jersey, 171 pp.

Cunliffe, J. E., and Kennish, M. J., 1974, Shell growth patterns in the hard-shelled clam, Underwater Nat. **8:**20–24.

Davenport, C. B., 1938, Growth lines in fossil pectens as indicators of past climates, J. Paleontol. **12:**514–515.

Davies, T. T., and Sayre, J. G., 1970, The effect of environmental stress on pelecypod shell ultrastructure, Geol. Soc. Am. Abstr. Programs **2**(3):204–205.

de Zwann, A., and van Marrewijk, W. J. A., 1973, Anaerobic glucose degradation in the sea mussel Mytilus edulis L., Comp. Biochem. Physiol. B **44:**429–439.

de Zwann, A., and Wijsman, T. C. M., 1976, Anaerobic metabolism in Bivalvia (Mollusca): Characteristics of anaerobic metabolism, Comp. Biochem. Physiol. B **54:**313–324.

de Zwann, A., and Zandee, D. I., 1972, The utilization of glycogen and accumulation of some intermediates during anaerobiosis in Mytilus edulis L., Comp. Biochem. Physiol. B **43:**47–54.

Dodd, J. R., 1963, Paleoecological implications of shell mineralogy in two pelecypod species, J. Geol. **71:**1–11.

Dodd, J. R., 1964, Environmentally controlled variation in the shell structure of a pelecypod species, J. Paleontol. **38:**1065–1071.

Dolman, J., 1975, A technique for the extraction of environmental and geophysical information from growth records in invertebrates and stromatolites, in: Growth Rhythms and the History of the Earth's Rotation (G. D. Rosenberg and S. K. Runcorn, eds.), pp. 191–222, John Wiley & Sons., London.

Dugal, L. P., 1939, The use of calcareous shell to buffer the product of anaerobic glycolysis in Venus mercenaria, J. Cell. Comp. Physiol. **13:**235–251.

Erben, H. K., 1972, Über die Bildung und das Wachstum von Perlmutt, Biomineralisation **4:**15–46.

Erben, H. K., 1974, On the structure and growth of the nacreous tablets in gastropods, Biomineralisation **7:**14–27.

Erben, H. K., and Watabe, N., 1974, Crystal formation and growth in bivalve nacre, Nature (London) **248:**128–130.

Evans, J. W., 1972, Tidal growth increments in the cockle *Clinocardium nuttalli, Science* **176**:416–417.

Evans, J. W., 1975, Growth and micromorphology of two bivalves exhibiting nondaily growth lines, in: *Growth Rhythms and the History of the Earth's Rotation* (G. D. Rosenberg and S. K. Runcorn, eds.), pp. 119–134, John Wiley & Sons, London.

Evans, J. W., and LeMessurier, M. H., 1972, Functional micromorphology and circadian growth of the rock-boring clam, *Penitella penita, Can. J. Zool.,* **50**:1251–1258.

Fairbridge, W. S., 1953, A population study of the Tasmanian "commercial" scallop, *Notovola meridionalis* (Tate) (Lamellibranchiata, Pectinidae), *Aust. J. Mar. Freshwater Res.* **4**:1–41.

Farrow, G. E., 1971, Periodicity structures in the bivalve shell: Experiments to establish growth controls in *Cerastoderma edule* from the Thames estuary, *Palaeontology* **14**:571–588.

Farrow, G. E., 1972, Periodicity structures in the bivalve shell: Analysis of stunting in *Cerastoderma edule* from the Burry Inlet (South Wales), *Palaeontology* **15**:61–72.

Gordon, J., and Carriker, M. R., 1978, Growth lines in a bivalve mollusk: Subdaily patterns and dissolution of the shell, *Science* **202**:519–521.

Grégoire, C., 1972, Structure of the molluscan shell, in: *Chemical Zoology VII: Molluscs* (M. Florkin and B. T. Scheer, eds.), pp. 45–102, Academic Press, New York.

Gutsell, J. S., 1931, Natural history of the bay scallop, *Fish. Bull. (U.S.)* **46**:568–632.

Hall, C. A., Jr., 1975, Latitudinal variation in shell growth patterns of bivalve molluscs: Implications and problems, in: *Growth Rhythms and the History of the Earth's Rotation* (G. D. Rosenberg and S. K. Runcorn, eds.), pp. 163–175, John Wiley & Sons, London.

Hallam, A., 1965, Environmental causes of stunting in living and fossil marine benthonic invertebrates, *Palaeontology* **8**:132–155.

Ham, L. C., and Irvine, M., 1975, Techniques for determining seasonality of shell middens from marine mollusc remains, *Syesis* **8**:363–373.

Hammen, C. S., and Lum, S. C., 1966, Fumarate reductase and succinate dehydrogenase activities in bivalve mollusks and brachiopods, *Comp. Biochem. Physiol.* **19**:775–781.

Hampson, G. R., 1971, A species pair of the genus *Nucula* (Bivalvia) from the eastern coast of the United States, *Proc. Malacol. Soc. London* **39**:333–342.

Haskin, H. H., 1954, Age determination in mollusks, *Trans. N. Y. Acad. Sci.* **16**:300–304.

Hedgepeth, J. W., 1957, *Treatise on Marine Ecology and Paleoecology,* Vol. 1, Ecology, Geol. Soc. Am. Mem. **67,** Vol. 1, 1296 pp.

Hochachka, P. W., and Mustafa, T., 1972, Invertebrate facultative anaerobiosis, *Science* **178**:1056–1060.

House, M. R., and Farrow, G. E., 1968, Daily growth banding in the shell of the cockle, *Cardium edule, Nature (London)* **219**:1384–1386.

Hudson, J. D., 1967, The elemental composition of the organic fraction, and the water content, of some recent and fossil mollusc shells, *Geochim. Cosmochim. Acta* **31**:2361–2378.

Hudson, J. D., 1968, The microstructure and mineralogy of the shell of a Jurassic mytilid (Bivalvia), *Palaeontology* **6**:327–348.

Isely, F. B., 1913, Experimental study of the growth and migration of fresh-water mussels, *Rep. U.S. Comm. Fish. 1913,* Appendix 3, 24 pp.

Isely, F. B., 1931, A fifteen year growth record in fresh-water mussels, *Ecology* **12**:616–619.

Jones, D. S., Thompson, I., and Ambrose, W., 1978, Age and growth rate determinations for the Atlantic surf clam *Spisula solidissima* (Bivalvia: Mactracea), based on internal growth lines in shell cross-sections, *Mar. Biol.* **47**:63–70.

Kahn, P. G. K., and Pompea, S. M., 1978, Nautiloid growth rhythms and dynamical evolution of the Earth–Moon system, *Nature (London)* **275**:606–611.

Kennedy, W. J., Taylor, J. D., and Hall, A., 1969, Environmental and biological controls on bivalve shell mineralogy, *Cambridge Philos. Soc. Biol. Rev.* **44:**499–530.

Kennedy, W. J., Morris, N. J., and Taylor, J. D., 1970, The shell structure, mineralogy and relationships of the Chamacea (Bivalvia), *Palaeontology* **13:**379–413.

Kennish, M. J., 1976, Monitoring thermal discharges: A natural method, *Underwater Nat.* **9:**8–11.

Kennish, M. J., 1977a, Growth increment analysis of *Mercenaria mercenaria* from artificially heated coastal marine waters: A practical monitoring method, *Proceedings of the XII International Society of Chronobiology Conference* (Washington, D.C.), pp. 663–669.

Kennish, M. J., 1977b, Effects of thermal discharges on mortality of *Mercenaria mercenaria* in Barnegat Bay, New Jersey, Ph. D. thesis, Rutgers University, New Brunswick, New Jersey, 161 pp.

Kennish, M. J., 1977c, Effects of thermal discharges on mortality of *Mercenaria mercenaria* in Barnegat Bay, New Jersey, *J. Paleontol.* **51**(Suppl. to No. 2):17.

Kennish, M. J., 1978, Effects of thermal discharges on mortality of *Mercenaria mercenaria* in Barnegat Bay, New Jersey. *Environ. Geol.* **2:**223–254.

Kennish, M. J., and Olsson, R. K., 1975, Effects of thermal discharges on the microstructural growth of *Mercenaria mercenaria*, *Environ. Geol.* **1:**41–64.

Koike, H., 1973, Daily growth lines of the clam, *Meretrix lusoria*: A basic study for the estimation of prehistoric seasonal gathering, *J. Anthropol. Soc. Nippon (Zinruigaku Zasshi)* **81:**122–138.

Koike, H., 1975, The use of daily and annual growth lines of the clam *Meretrix lusoria* in estimating seasons of Jomon Period shell gathering, in: *Quaternary Studies* (R. P. Suggate and M. M. Cresswell, eds.), pp. 189–193, The Royal Society of New Zealand, Wellington.

Kuenzler, E. J., 1961, Structure and energy flow of a mussel population in a Georgia salt marsh, *Limnol. Oceanogr.* **6:**191–204.

Lange, R., Staaland, H., and Mostad, A., 1972, The effect of salinity and temperature on solubility of oxygen and respiratory rate in oxygen-dependent marine invertebrates, *J. Exp. Mar. Biol. Ecol.* **9:**217–229.

Lent, C. M., 1967, Effect of habitat on growth indices in the ribbed mussel *Modiolus (Arcuatula) demissus* (Dillwyn), *Chesapeake Sci.* **8:**221–227.

Lent, C. M., 1968, Air-gaping by the ribbed mussel, *Modiolus demissus* (Dillwyn): Effects and adaptive significance, *Biol. Bull. (Woods Hole, Mass.)* **134:**60–73.

Lent, C. M., 1969, Adaptations of the ribbed mussel, *Modiolus demissus* (Dillwyn), to the intertidal habitat, *Am. Zool.* **9:**283–292.

Loriga, C. B., and Neri, C., 1976, Aspetti paleobiologici e paleogeografici della facies a "Lithiotis" (Giurese Inf.), *Riv. Ital. Paleontol.* **82:**651–706.

Lowenstam, H. A., 1954, Factors affecting the aragonite–calcite ratios in carbonate secreting marine organisms, *J. Geol.* **62:**284–322.

Lubinsky, I., 1958, Studies on *Mytilus edulis* L. of the "Calanus" expeditions to Hudson Bay and Ungava Bay, *Can. J. Zool.* **36:**869–881.

Lutz, R. A., 1976a, Annual growth patterns in the inner shell layer of *Mytilus edulis* L., *J. Mar. Biol. Assoc. U. K.* **56:**723–731.

Lutz, R. A., 1976b, Geographical and seasonal variation in the shell structure of an estuarine bivalve, *Geol. Soc. Am. Abstr. Programs* **8:**988.

Lutz, R. A., 1977, Annual structural changes in the inner shell layer of *Geukensia* (. = *Modiolus*) *demissa*, *Proc. Natl. Shellfish. Assoc.* **67:**120.

Lutz, R. A., and Rhoads, D. C., 1977, Anaerobiosis and a theory of growth line formation, *Science* **198:**1222–1227.

Lutz, R. A., and Rhoads, D. C., 1978, Shell structure of the Atlantic ribbed mussel, *Geukensia demissa* (Dillwyn): A reevaluation, *Am. Malacol. Union Inc. Bull.* **1978:**13–17.

MacClintock, C., and Pannella, G., 1969, Time of calcification in the bivalve mollusk Mercenaria mercenaria (Linnaeus) during the 24-hour period, *Geol. Soc. Am. Abstr. Programs* **1**(7):140.

Martin, A. W., Catala-Stucki, I., and Ward, P. D., 1977, Growth rate and reproductive behavior of *Nautilus macromphalus*, *Geol. Soc. Am. Abstr. Programs* **9**:1086.

Mason, J., 1957, The age and growth of the scallop, *Pecten maximus* (L.), in Manx waters, *J. Mar. Biol. Assoc. U. K.* **36**:473–492.

Matteson, M. R., 1948, Life History of *Elliptio complanatus* (Dillwyn), *Am. Midl. Nat.* **40**:690–723.

Merrill, A. S., Posgay, J. A., and Nichy, F. E., 1965, Annual marks on shell and ligament of sea scallop *(Placopecten magellanicus)*, *Fish. Bull. (U.S.)* **65**:299–311.

Moore, H. B., 1934, On "ledging" in shells at Port Erin, *Proc. Malacol. Soc.* **21**:213–217.

Moore, H. B., 1958, *Marine Ecology*, Wiley, New York, 493 pp.

Mossop, B. K. E., 1921, A study of the sea mussel (*Mytilus edulis* Linn.), *Contrib. Can. Biol. Fish.* **2**:17–48.

Mossop, B. K. E., 1922, The rate of growth of the sea mussel (*Mytilus edulis* L.) at St. Andrews New Brunswick, Digby Nova Scotia, and Hudson Bay, *Trans. R. Can. Inst.* **4**:3–21.

Mutvei, H., 1969, On the micro- and ultrastructure of the conchiolin in the nacreous layer of some recent and fossil molluscs, *Stockholm Contrib. Geol.* **20**:1–17.

Mutvei, H., 1970, Ultrastructure of the mineral and organic components of molluscan nacreous layers, *Biomineralisation* **2**:48–72.

Mutvei, H., 1972, Ultrastructural relationships between the prismatic and nacreous layers in *Nautilus* (Cephalopoda), *Biomineralisation* **4**:80–88.

Nakahara, H., and Bevelander, G., 1971, The formation and growth of the prismatic layer of *Pinctada radiata*, *Calcif. Tissue Res.* **7**:31–45.

Newcombe, C. L., 1935, Growth of *Mya arenaria* L. in the Bay of Fundy region, *Can. J. Res. Sect. D* **13**:97–137.

Newcombe, C. L., 1936, Validity of concentric rings of *Mya arenaria* L. for determining age, *Nature (London)* **137**:191–192.

Oberling, J. J., 1964, Observations on some structural features of the pelecypod shell, *Mitt. Naturforsch. Ges. Bern.* **20**:1–63.

Olsen, D., 1968, Banding patterns of *Haliotis rufescens* as indicators of botanical and animal succession, *Biol. Bull. (Woods Hole, Mass.)* **134**:139–147.

Orton, J. H., 1923, On the significance of "rings" on the shells of *Cardium* and other molluscs, *Nature (London)* **112**:10.

Orton, J. H., 1926, On the rate of growth of *Cardium edule*. I. Experimental observations, *J. Mar. Biol. Assoc. U.K.* **14**:239–279.

Pannella, G., 1972, Palaeontological evidence on the Earth's rotational history since Early Precambrian, *Astrophys. Space Sci.* **16**:212–237.

Pannella, G., 1975, Palaeontological clocks and the history of the earth's rotation, in: *Growth Rhythms and the History of the Earth's Rotation* (G. D. Rosenberg and S. K. Runcorn, eds.). pp. 253–284, John Wiley & Sons, London.

Pannella, G., 1976, Tidal growth patterns in Recent and fossil mollusc bivalve shells: A tool for the reconstruction of paleotides, *Naturwissenschaften* **63**:539–543.

Pannella, G., and MacClintock, C., 1968, Biological and environmental rhythms reflected in molluscan shell growth, *J. Paleontol.* **42**:64–80.

Pannella, G., MacClintock, C., and Thompson, M. N., 1968, Paleontological evidence of variations in length of synodic month since late Cambrian, *Science* **162**:792–796.

Rhoads, D. C., 1974, Organism–sediment relations on the muddy seafloor, *Oceanogr. Mar. Biol. Annu. Rev.* **12**:263–300.

Rhoads, D. C., and Morse, J. W., 1971, Evolutionary and ecologic significance of oxygen-deficient marine basins, *Lethaia* **4**:413–428.

Rhoads, D. C., and Pannella, G., 1970, The use of molluscan shell growth patterns in ecology and paleoecology, *Lethaia* **3**:143–161.

Rosenberg, G. D., and Runcorn, S. K. (eds.), 1975, *Growth Rhythms and the History of the Earth's Rotation*, John Wiley & Sons, London, 559 pp.

Sastry, A. N., 1961, Studies on the bay scallop, *Aequipecten irradians concentricus* Say, in Alligator Harbor, Florida, Ph.D. thesis, Florida State University, Tallahassee, 118 pp.

Savilov, A. I., 1953, The growth and variation in growth of the White Sea invertebrates *Mytilus edulis, Mya arenaria,* and *Balanus balanoides, Tr. Inst. Okeanol.* **7**:198–213, 252–258.

Schmidt, W. J., 1923, Bau und Bildung der Perlmuttermasse, *Zool. Jahrb. Abt. Anat. Ontog. Tiere* **45**:1–148.

Seed, R., 1973, Absolute and allometric growth in the mussel, *Mytilus edulis* L. (Mollusca: Bivalvia), *Proc. Malacol. Soc. London* **40**:343–357.

Seed, R., 1976, Ecology, in: *Marine Mussels* (B. L. Bayne, ed.), pp. 13–65, Cambridge University Press, Cambridge.

Shuster, C. N., Jr., 1951, On the formation of mid-season checks in the shell of *Mya, Anat. Rec.* **111**:127.

Stokes, T. M., and Awapara, J., 1968, Alanine and succinate as end-products of glucose degradation in the clam *Rangia cuneata, Comp. Biochem. Physiol.* **25**:883–892.

Tang, S. F., 1941, The breeding of the escallop *Pecten maximus* (L.) with a note on the growth rate, *Proc. Liverpool Biol. Soc.* **54**:9–28.

Taylor, A. C., 1976, Burrowing behaviour and anaerobiosis in the bivalve *Arctica islandica* (L.), *J. Mar. Biol. Assoc. U.K.* **56**:95–109.

Taylor, J. D., 1973, The structural evolution of the bivalve shell, *Palaeontology* **16**:519–534.

Taylor, J. D., and Layman, M., 1972, The mechanical properties of bivalve (Mollusca) shell structures, *Palaeontology* **15**:73–87.

Taylor, J. D., Kennedy, W. J., and Hall, A., 1969, The shell structure and mineralogy of the Bivalvia: Introduction, Nuculacea–Trigonacea, *Bull. Br. Mus. (Nat. Hist.) Zool.*, Suppl. 3, 125 pp.

Taylor, J. D., Kennedy, W. J., and Hall, A., 1973, The shell structure and mineralogy of the Bivalvia. II. Lucinacea–Clavagellacea: Conclusions, *Bull. Br. Mus. (Nat. Hist.) Zool.* **22**:253–294.

Tevesz, M. J. S., 1972, Implications of absolute age and season of death data compiled for Recent *Gemma gemma, Lethaia* **5**:31–38.

Thiesen, B. F., 1973, The growth of *Mytilus edulis* L. (Bivalvia) from Disko and Thule District, Greenland, *Ophelia* **12**:59–77.

Thompson, I., 1975, Biological clocks and shell growth in bivalves, in: *Growth Rhythms and the History of the Earth's Rotation* (G. D. Rosenberg and S. K. Runcorn eds.), pp. 149–161, John Wiley & Sons, London.

Thompson, I. L., and Barnwell, F. H., 1970, Biological clock control and shell growth in the bivalve *Mercenaria mercenaria, Geol. Soc. Am. Abstr. Programs* **2**:704.

Thompson, I. L., and Jones, D. S., 1977, The ocean quahog, *Arctica islandica,* "tree" of the North Atlantic shelf, *Geol. Soc. Am. Abstr. Programs* **9**:1199.

Towe, K. M., 1972, Invertebrate shell structure and the organic matrix concept, *Biomineralisation* **4**:1–14.

Towe, K. M., and Hamilton, G. H., 1968a, Ultrastructure and inferred calcification of the mature and developing nacre in bivalve mollusks. *Calcif. Tissue Res.* **1**:306–318.

Towe, K. M., and Hamilton, G. H., 1968b, Ultramicrotome-induced deformation artifacts in densely calcified material, *J. Ultrastruct. Res.* **22**:274–281.

Towe, K. M., and Thompson, G. R., 1972, The structure of some bivalve shell carbonates prepared by ion-beam thinning, *Calcif. Tissue Res.* **10**:38–48.

Travis, D. F., 1968, The structure and organization of, and the relationship between, the inorganic crystals and the organic matrix of the prismatic region of *Mytilus edulis*, *J. Ultrastruct. Res.* **23**:183–215.

Travis, D. F., and Gonsalves, M., 1969, Comparative ultrastructure and organization of the prismatic region of two bivalves and its possible relation to the chemical mechanism of boring, *Am. Zool.* **9**:635–661.

von Brand, T., 1944, Anaerobiosis in invertebrates, *Biodynamica* **4**:1–328.

Wada, K., 1961, Crystal growth of molluscan shells, *Bull. Natl. Pearl Res. Lab. (Kokuritsu Shinju Kenkyusho Hokoku)* **7**:703–828.

Wada, K., 1972, Nucleation and growth of aragonite crystals in the nacre of some bivalve molluscs, *Biomineralisation* **6**:141–159. .

Watabe, N., 1963, Decalcification of thin sections for electron microscope studies of crystal–matrix relationships in mollusc shells, *J. Cell Biol.* **18**:701–703.

Watabe, N., 1965, Studies on shell formation. XI. Crystal–matrix relationships in the inner layers of mollusc shells, *J. Ultrastruct. Res.* **12**:351–370.

Watabe, N., and Wilbur, K. M., 1960, Influence of the organic matrix on crystal type in molluscs, *Nature (London)* **188**:334.

Watabe, N., and Wilbur, K. M., 1961, Studies on shell formation. IX. An electron microscope study of crystal layer formation in the oyster, *J. Cell Biol.* **9**:761–772.

Weinstein, D. H., and Keeney, J., 1975, Palaeontology and the dynamic history of the Sun–Earth–Moon system, in: *Growth Rhythms and the History of the Earth's Rotation* (G. D. Rosenberg and S. K. Runcorn, eds.), pp. 377–384, John Wiley & Sons, London.

Weymouth, F. W., 1923, The life history and growth of the pismo clam (*Tivela stultorum* Mawe), *Fish Bull. Calif.*, No. 7, pp. 1–120.

Weymouth, F. W., and Thompson, S. H., 1931, The age and growth of the Pacific cockle *Cardium corbis* Martyn, *Bull. Bur. Fish. Wash.* **46**:633–641.

Weymouth, F. W., McMillin, J. C., and Rich, W. H., 1931, Latitude and relative growth in the razor clam, *Siliqua patula*, *J. Exp. Biol.* **8**:228–249.

Whyte, M. A., 1975, Time, tide and the cockle, in: *Growth Rhythms and the History of the Earth's Rotation* (G. D. Rosenberg and S. K. Runcorn, eds.), pp. 177–189, John Wiley & Sons, London.

Whyte, M. A., 1977, Turning points in Phanerozoic history, *Nature (London)* **267**:679–682.

Wilbur, K. M., 1972, Shell formation in mollusks, in: *Chemical Zoology. VII. Molluscs* (M. Florkin and B. Scheer, eds.), pp. 103–145, Academic Press, New York.

Wise, S. W., 1970a, Microarchitecture and deposition of gastropod nacre, *Science* **167**:1486–1488.

Wise, S. W., 1970b, Microarchitecture and mode of formation of nacre (mother-of-pearl) in pelecypods, gastropods, and cephalopods, *Ecologae Geol. Helv.* **63**:775–797.

Wise, S. W., and Hay, W. W., 1968a, Scanning electron microscopy of molluscan shell ultrastructures. I. Techniques for polished and etched sections, *Trans. Am. Microsc. Soc.* **87**:411–418.

Wise, S. W., and Hay, W. W., 1968b, Scanning electron microscopy of molluscan shell ultrastructures. II. Observations of growth surfaces, *Trans. Am. Microsc. Soc.* **87**:419–430.

Chapter 7

Shell Microgrowth Analysis
Mercenaria mercenaria as a Type Example for Research in Population Dynamics

MICHAEL J. KENNISH

1. Introduction 255
2. Shell Microgrowth Patterns in *Mercenaria mercenaria* 256
   2.1. Cyclical Growth Patterns 258
   2.2. Growth Breaks 269
3. Application to Analysis of Population Dynamics 273
4. Population Dynamics of *Mercenaria mercenaria* in Barnegat Bay, New Jersey ... 274
   4.1. Size and Age Distributions, Growth Rates, and Recruitment Patterns 275
   4.2. Mortality Patterns 282
5. Conclusions and Future Research Efforts 291
   References 292

1. Introduction

The molluscan shell has been the subject of intensive research in ecology and paleoecology for many years. Much of this research has centered around the examination of macroscopic growth features on the surfaces of shells of marine bivalves and the utilization of these growth bands in the investigation of environmental and paleoenvironmental problems in marine ecosystems. Recent studies of microscopic growth features within the shells of various molluscan taxa, however, have revealed a more detailed and higher-resolution record of growth. These internal shell microgrowth patterns in molluscs have proven to be valuable for reconstructing habitat conditions in both ancient and extant marine environments. Such patterns also yield information about population dynamics that is difficult or impossible to obtain by other methods.

MICHAEL J. KENNISH • Environmental Affairs Department, Jersey Central Power & Light Company, Morristown, New Jersey 07960.

A molluscan species ideally suited for shell microgrowth research the coastal marine bivalve *Mercenaria mercenaria* (Linné), which ranges from the Gulf of St. Lawrence to the Gulf of Mexico (Saila and Pratt, 1973). This infaunal bivalve lives in the intertidal and shallow subtidal zones. During the past 15 years, the shell microstructure of *M. mercenaria* has been extensively studied (Barker, 1964; Pannella and MacClintock, 1968; Rhoads and Pannella, 1970; Cunliffe, 1974; Cunliffe and Kennish, 1974; Kennish and Olsson, 1975; Kennish, 1976, 1977a,b, 1978; Lutz and Rhoads, 1977; Gordon and Carriker, 1978). These workers have documented the remarkably well preserved record of growth retained within the shell of this clam.

Because shell growth of *M. mercenaria* is strongly influenced by environmental conditions, shell microgrowth patterns effectively reflect conditions in the organism's ambient environment. External environmental factors affect a bivalve's physiology and hence its shell microgrowth patterns. Physiological events such as reproduction are also recorded. Analysis of the shell microstructure, therefore, can yield important data on the population dynamics of the species.

This chapter consists of a detailed description of patterns of growth increments and growth breaks in *M. mercenaria* and the environmental and physiological significance of each pattern. I will give a detailed example of how these microgrowth patterns can be used to approach an ecological problem; the investigation of the population dynamics of *M. mercenaria* in a lagoonal-type estuarine system. The objective of this chapter is to demonstrate how the ontogenetic shell growth record of *M. mercenaria* can serve as a "type" example for ecological and paleoecological research using other molluscan species.

2. Shell Microgrowth Patterns in *Mercenaria mercenaria*

The shell of *M. mercenaria* is composed primarily of calcium carbonate (aragonite) and conchiolin. A vertical section through a valve reveals four shell layers: (1) the inner homogeneous layer; (2) the pallial myostracum (pallial muscle-scar layer); (3) the middle layer; and (4) the outer layer (Pannella and MacClintock, 1968) (Fig. 1). Shell microgrowth patterns are most conspicuous in the outer layer, where shell growth reaches a maximum and growth-increment widths exceed those in the inner and middle layers.

Three stages of growth (youth, maturity, and old age) are evident in valve cross sections of this species (Cunliffe, 1974). In youth, which persists for approximately 2 years, the outer shell layer terminates in con-

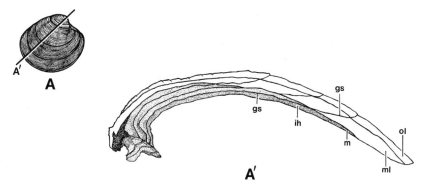

Figure 1. Valve of *Mercenaria mercenaria* (Linné). (A) Outer surface view. Approximately × 0.5. (A′) Valve cross section showing former growth surfaces (gs) and the four shell layers; inner homogeneous (ih), pallial myostracum (m), middle (ml), and outer (ol). Approximately × 2. After Pannella and MacClintock (1968). Reproduced at 60%.

centric ridges at the outer shell surface, with growth increments inter-secting the outer shell surface nearly at right angles. When the mature stage of growth is attained, the concentric ridges disappear and growth increments become recurved in shape. The beginning of the mature stage of growth corresponds closely with the onset of sexual maturity and lasts for 3–8 years. The final stage of growth is old age, and at this time crossed lamellar shell structure completely replaces prismatic shell structure in the outer shell layer, growth increments become thin and perpendicular to the outer shell surface, and numerous growth breaks develop.

To study shell microgrowth patterns of *M. mercenaria* throughout the three ontogenetic stages outlined above, valves of specimens are sectioned perpendicular to the shell surface from the umbo to the ventral margin along the maximum axis of growth (see Chapter 1 for a description of this axis; also see Appendix 1.A.1). Next, acetate-peel replicas of the acid-etched cross sections of the valves are made according to the method outlined in Appendix 1.A.1. Microgrowth patterns in the peels are studied microscopically. Shell microgrowth patterns can also be investigated in thin sections of shell material, but thin sections require considerably more preparation time and therefore place practical limits on sample size (see Appendix 1.A.2).

Using acetate peels and thin sections, Pannella and MacClintock (1968), Rhoads and Pannella (1970), Cunliffe and Kennish (1974), Kennish and Olsson (1975), and Kennish (1978) analyzed shell microgrowth pat-terns of *M. mercenaria*. They recognized two categories of microgrowth patterns based on cyclical and interrupted growth incrementation. The first category, cyclical growth patterns, includes subdaily, daily, bidaily, fortnightly, lunar-monthly, and annual types (Table I). These patterns result from variable rates of calcium carbonate deposition or dissolution or both (Lutz and Rhoads, 1977; Gordon and Carriker, 1978) (see Chapter 6 for a detailed discussion). Environmental conditions (Clark, 1974) and a biological-clock mechanism in the clam (MacClintock and Pannella, 1969; Thompson and Barnwell, 1970; Thompson, 1975) appear to control the formation of such cyclical growth patterns. The second category, growth breaks, reflects periods of environmental and physiological stress. Seven types of growth breaks are documented, and they develop from both random and periodic events (Table II). These types consist of growth breaks caused by freeze shock (winter), heat shock (summer), thermal shock, shell-margin abrasion, spawning, neap tides, and storms.

2.1. Cyclical Growth Patterns

Cyclical growth patterns, viewed in cross section, appear as an alternation of calcium-carbonate-rich layers and organic-rich lines (Plate Ia–c). Recent studies indicate that this repetitive sequence originates from alternating periods of shell deposition and dissolution during aerobic and anaerobic respiration, respectively (Lutz and Rhoads, 1977; Gordon and Carriker, 1978) (see Chapter 6). When the valves of M. mercenaria are open and water is being pumped over the gills, the organism respires aerobically and the mantle deposits calcium carbonate and organic matrix simultaneously. During periods of shell closure, anaerobic metabolism occurs, and the concentration of succinic acid within the extrapallial fluid increases. Dissolution of shell calcium carbonate neutralizes the succinic acid produced during anaerobiosis, resulting in an increase in the concentration of calcium ions and succinate in the extrapallial and mantle fluids (Crenshaw and Neff, 1969). This carbonate-dissolution process leaves behind a conchiolin residue that forms organic-rich lines in the shell microstructure. With the return to aerobic conditions, shell decalcification ceases and a new phase of shell-building is initiated. Further details of this process are discussed in Chapter 6.

From a structural viewpoint, alternating periods of shell deposition and decalcification result in growth increments in the shell of M. mercenaria. Barker (1964), Pannella and MacClintock (1968), Rhoads and Pannella (1970), Cunliffe (1974), Cunliffe and Kennish (1974), Thompson (1975), Kennish and Olsson (1975), and Kennish (1976, 1977a, 1978) have documented daily growth increments in the shell of this species indicative of a circadian rhythm (Plate Ia–c). The existence of a circadian rhythm in shell growth of M. mercenaria has been confirmed by mark-and-recapture experiments. For example, Rhoads and Pannella (1970) investigated shell

Table I. Cyclical Microgrowth Patterns in *Mercenaria mercenaria*[a]

Pattern	Periodicity	Description
Subdaily	< 1 day	This pattern is observed as diffuse organic-rich lines surrounded by calcium-carbonate-rich regions. The diffuse organic-rich lines form during intervals of anaerobiosis and shell decalcification within a 24-hr period. The calcium-carbonate-rich regions form during episodes of shell secretion.
Daily	1 day	A daily growth increment comprises the daily microgrowth pattern. This increment appears as a calcium-carbonate-rich region between successive organic-rich lines. In contrast to the diffuse organic-rich lines in the subdaily pattern, diurnal organic-rich lines are well developed. The daily growth increment forms in response to a circadian locomotor rhythm in the species. When the valves of the organism are open and water is freely pumped, aerobic metabolism occurs and calcium-carbonate-rich regions of the shell are deposited. When the valves of the clam are closed, anaerobic respiratory pathways are utilized and shell decalcification occurs, resulting in organic-rich growth lines. This alternating sequence of shell deposition and dissolution takes place on a daily basis.
Bidaily	2 days	A bidaily growth pattern consists of pairs of growth increments—one thick increment followed by a thin one. The cause of this couplet is unknown.
Fortnightly	14 days	Fortnightly patterns appear as a cluster of six to eight thin daily growth increments followed by a cluster of six to eight thick daily growth increments. Alternating neap and spring tides generate this pattern.
Lunar-month	29 days	Monthly patterns are clusters of tidal cycles with an overall periodicity of 29 days. This pattern is manifested on the external shell surface as concentric ridges formed by the thickened outer shell layer. The synodic month produces this pattern.
Annual	340–380 days	The yearly cycle is composed of a long sequence of thin daily increments followed by a long sequence of thick daily increments. Seasonal temperature changes are mainly responsible for this cyclical pattern, with low winter temperatures yielding thin daily increments and high summer temperatures yielding thick daily increments.

[a] After Kennish and Olsson (1975).

Table II. Growth Breaks in *Mercenaria mercenaria*[a]

Type of break	Description
Freeze-shock break	This break is characterized by a deep V-shaped notch in the outer shell layer and an increase in organic material that extends from the outer layer to the inner homogeneous layer. Daily growth increments gradually decrease in thickness as the break is approached and gradually increase in thickness going away from it. Crossed lamellar shell structure replaces prismatic shell structure on one or both sides of the break. The gradual decrease in increment thickness approaching the break reflects the gradual decrease in water temperature associated with the onset of winter. The break itself often marks the first freeze of winter, and the gradual increase of increment thickness following the break indicates the slow recovery from the freeze shock. More than one freeze-shock break may be present per winter season.
Heat-shock break	A heat-shock break is structurally similar to a freeze-shock break, but is characteristically developed during the summer months. It apparently occurs when the organism is shocked by excessively high water temperatures. More than one heat-shock break may be present per summer season.
Thermal-shock break	A thermal-shock break appears as a sudden break in the normal pattern of growth-increment addition. There is usually no slowdown of growth prior to the break, massive slowdown following the break, and crossed lamellar shell structure throughout the stressed period. This break was first described in specimens from Barnegat Bay, New Jersey, which were affected by rapid fluctuations in water temperatures associated with sharp changes in operations of the Oyster Creek Nuclear Generating Station.
Abrasion break	An abrasion break is structurally similar to a thermal-shock break but shows less slowdown of growth following the break. Abrasion breaks are frequently produced by filing or sanding the valve margins during transplantation experiments.
Spawning break	In this break, the normal pattern of daily growth-increment addition is abruptly interrupted by a break in growth and immediately followed by a sequence of thin, indistinct increments. The sudden occurrence of the break corresponds to the triggering of the spawning event when the proper temperature is attained. The thin increments associated with the break reflect the period during which the clam stops feeding. Crossed lamellar shell structure is generally absent.
Neap-tide break	This break in growth is preceded by a rapid decrease in the thickness of growth increments and is followed by a rapid increase in the thickness of the growth increments. Such a break, when present, coincides with neap tides. Crossed lamellar shell structure is usually absent.

(Continued)

Table II. (Continued)

Type of break	Description
Storm break	A storm break appears as a break in the normal depositional pattern of shell growth, which is abruptly followed by recovery. The thickness of growth increments prior to and following the break is not significantly altered. Silt grains are occasionally incorporated into the outer shell layer. Crossed lamellar shell structure is generally lacking.

[a] After Kennish and Olsson (1975).

microgrowth patterns of specimens that had been notched at the ventral margin of the shell, placed in sediment-filled plastic trays, and transplanted to intertidal and subtidal zones in Connecticut and Massachusetts. Individual specimens retrieved several months after transplantation were found to have faithfully recorded the transplant sequence on a daily basis. Similarly, Cunliffe (1974, personal communication) and Kennish (1977a,b, 1978) transplanted several thousand notched clams to intertidal and subtidal zones in New Jersey. Specimens subsequently recovered and studied contained approximately the same number of growth increments in the outer shell layer as the number of days that had elapsed since the day of transplantation. When a daily growth increment was missing in the outer shell layer, it corresponded to a growth break in the shell. Pannella and MacClintock (1968) demonstrated the ability of M. mercenaria to add an increment of shell each day, even under strong adverse conditions. Clams kept in tanks for several months without any food deposited a thin growth increment every day. Other specimens placed on a table for 7 days secreted seven thin growth increments.

A daily growth increment in M. mercenaria appears as a calcium-carbonate-rich region between successive, well-developed organic-rich lines. The character of daily growth increments is strongly influenced by environmental conditions through physiological responses of the bivalve (see Chapter 6). Water temperature, salinity, tides, water depth, substratum, storms, food supply, spawning events, and the individual specimen's age affect the thickness of daily growth increments. Of these factors, water temperature and age appear to exert the major controlling influence on growth.

The rate of shell growth of M. mercenaria peaks between 20 and 25°C (Ansell, 1968; Kennish and Olsson, 1975) and decreases rapidly below 10°C and above 30°C. Growth rates reach maximum levels in clams between 1 and 3 years of age and minimum levels in clams greater than 5 years of age (Ansell, 1968; Kennish, 1977b). Studies by a number of workers show that seasonal growth rates are highest in the summer when water temperatures are elevated (Pannella and MacClintock, 1968; Rhoads and

Pannella, 1970; Cunliffe, 1974; Kennish and Olsson, 1975; Kennish, 1978). Evidence of differences in the seasonal growth rates of M. mercenaria exists in the shell microstructure of specimens collected in Barnegat Bay, New Jersey, a temperate East Coast estuary with an annual water temperature range of -1.4 to $28.0°C$. During the summer, these clams form thick daily growth increments ranging from 15 to 150 μm (Plate Ia–c), whereas during the winter, they form thin daily growth increments ranging from 1 to 50 μm (Plate Id–f). Daily growth increments in the spring and fall are intermediate in thickness between summer and winter increments and range from 10 to 75 μm in thickness.

Occasionally, daily growth increments are grouped into a bidaily microgrowth pattern. Although formed in the shell of M. mercenaria during most seasons of the year, the bidaily pattern is not as common as subdaily, daily, fortnightly, and annual types. Bidaily patterns consist of an alternating sequence of one thick and one thin growth increment formed during a 48-hr period (Plate Ig). The cause of these couplets is unknown.

A subdaily microgrowth pattern consists of one or more diffuse organic-rich lines within a daily growth increment (Plate Ih). Subdaily lines represent organic (conchiolin) residues left behind as a result of shell dissolution during periods of shell closure and anaerobiosis (Gordon and Carriker, 1978) (also see Chapter 6 for a discussion). Data from Gordon and Carriker (1978) indicate a causal relationship between rhythms of valve movement in the clam and the appearance of the diffuse organic-rich lines. The rhythms of valve movement range from 5.4 to 12.8 hr in different specimens.

The subdaily pattern is commonly encountered in intertidal clams, but has also been found in subtidal specimens. Kennish and Olsson (1975) attributed the formation of subdaily lines in some specimens to semidiurnal low tides. This suggests that such a pattern is associated with valve closure during low tide.

In addition to semidiurnal events, the shell of M. mercenaria contains a record of spring and neap tides, as well as the lunar month. Spring- and neap-tide events appear as a clustering of daily growth increments into a fortnightly pattern composed of six to eight thick increments followed by six to eight thin increments (Plate Ii). The series of thick increments is formed during spring tides and the series of thin increments during neap tides. Although the fortnightly microgrowth pattern occurs in the shells of nearly all specimens of this species, it is, like the subdaily pattern, more conspicuous in specimens from the intertidal zone, which are most strongly influenced by the tides.

Groupings of pairs of tidal clusters, with an overall periodicity of 29 days, comprise the lunar-month microgrowth pattern (Plate Ij). This pat-

Text continues on p. 269.

Plate I. (a–c) Optical micrographs of acetate peels showing summer microgrowth patterns in the outer shell layer of *Mercenaria mercenaria* (Linné). Calcium-carbonate-rich regions (light bands) represent periods of shell deposition, whereas the organic-rich lines (dark striations) reflect periods of shell dissolution. Note the thick daily growth increments. (a, b) Scale bars: 100 μm; (c) scale equals that in (b). (d–f) Winter microgrowth patterns. Calcium-carbonate-rich regions are much thinner than those deposited in the summer (see a–c), indicating that shell growth occurs at a reduced rate during the winter season. (d, e) Scales equal that in (b); (f) scale bar: 100 μm. (g) Bidaily microgrowth pattern. This pattern consists of a couplet: one thick growth increment that follows one thin growth increment (arrows). The couplet forms during a 48-hr period. The cause of this periodicity is unknown. Scale equals that in (b). (h) Subdaily microgrowth pattern. Diffuse organic-rich lines (arrows) occur within daily growth increments. These lines form during intervals of valve closure and shell decalcification within a 24-hr period. Subdaily lines can be distinguished from heavier, daily organic-rich lines. Semidiurnal tides may produce the subdaily microgrowth pattern. Scale equals that in (b). (i) Fortnightly microgrowth pattern. The sequence of thin daily growth increments (n) corresponds to shell growth during neap tides, whereas the series of thick daily growth increments (s) corresponds to shell growth during spring tides. Scale bar: 100 μm. Modified from Pannella and MacClintock (1968). Photograph courtesy of C. MacClintock. (j) Lunar-monthly microgrowth pattern. Undulations (s) on the outer shell surface reflect a 29-day periodicity. The synodic month causes this microgrowth pattern. Scale bar: 500 μm. Modified from Pannella and MacClintock (1968). Photograph courtesy of C. MacClintock.

Plate II. (a–c) Freeze-shock breaks. Deep V-shaped notches in the outer shell layer (arrows) are accompanied by thin daily growth increments, crossed lamellar shell structure, and increased organic material. (a) Scale bar: 100 μm; (b, c) scales equal that in (a). (d) Heat-shock break. Thin daily growth increments surround a break in growth (arrow) caused by high water temperatures in the summer. This pattern is structurally similar to the seasonal freeze-shock break (see a–c). Scale bar: 100 μm. (e–g) Thermal-shock breaks. A sudden, but transient, temperature change may produce breaks (arrows) delineated by the absence of growth deceleration prior to the breaks. Compare these breaks to those in (a–d). (e–g) Scales equal that in (d). (h, i) Abrasion breaks. These breaks in growth (arrows) were generated by breaking the periostracal seal at valve margins with fine sandpaper. Shell repair is marked by crossed lamellar shell structure and thin increments of growth. (h) Scale bar: 200 μm;

(i) scale bar: 100 μm. (j) Spawning break. The slight decreases in the thickness of growth increments (horizontal arrows) on both sides of the break (downward-pointing arrow) are related to the physiological stress of spawning. Crossed lamellar shell structure is not associated with the break. Scale equals that in (a). (k) Storm break. Seafloor erosion and high bottom turbidity associated with storms may result in sediment particles becoming trapped between the organism's mantle and shell. These grains can be permanently incorporated into the shell. Note the presence of this break (arrow) with no change in the thickness of growth increments on either side of the break. Scale equals that in (a).

Plate III. (a, b) Season of death: early spring. Shell microgrowth patterns near the edge of the shell in specimens that die in the early spring include thin growth increments and growth breaks (arrows) associated with crossed lamellar shell structure (dark areas). (a) Scale bar: 200 μm; (b) scale bar: 100 μm. (c) Season of death: late spring. Daily growth increments at the shell edge are thicker than those of clams that die in the early spring (see a and b). Crossed lamellar shell structure is absent. Scale equals that in (b). (d, e) Season of death: summer. Heat-shock breaks (arrows) and crossed lamellar shell structure occur near the shell margin. Scales equal that in (b). (f, g) Season of death: summer. Spawning breaks located near the shell edge (arrows) were formed during the summer season. Scales equal that in (b). (h) Season of death: summer. A summer storm break (arrow) is near the shell edge. Scale equals that in (a). (i, j) Season of death: summer. Thick daily growth increments diagnostic of rapid summer growth precede the shell margins. Scales equal that in (b).

Plate IV. (a–d) Season of death: fall. Daily growth increments near the valve margins are of thickness intermediate between those of winter and summer. Fall microgrowth patterns are differentiated from spring patterns by being preceded by summer patterns. A gradual reduction in the thickness of growth increments at the shell margin suggests a moribund condition prior to death. A growth break occurs in micrograph (d) (arrow). (a) Scale bar: 100 μm; (b–d) scales equal that in (a). (e–h) Winter microgrowth patterns at the ventral valve margins include thin growth increments and freeze-shock breaks (downward-pointing arrows). Horizontal arrows point to crossed lamellar shell structure. Scale in (e) equals that in (a); (f) scale bar: 200 μm; (g, h) scales equal that in (f). (i) Exterior view of the right-hand valve of the northern quahog, M. mercenaria. True height (h) of the organism; (– – –) dimension of height (h') used by Kennish (1978). Approximately ×0.5, reproduced at 78%.

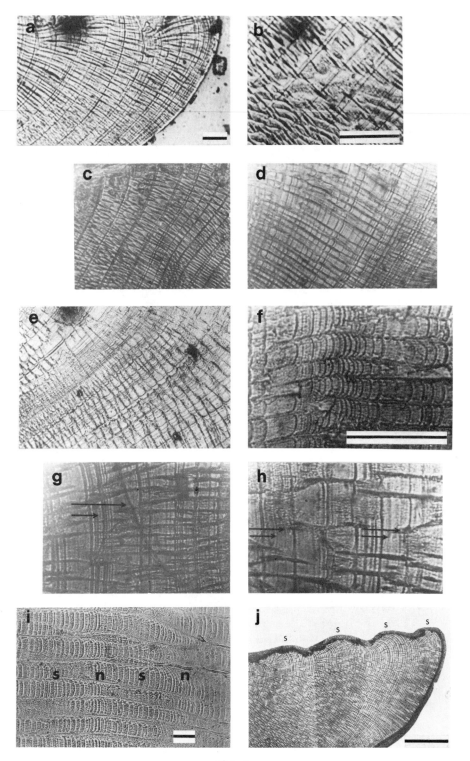

Plate I

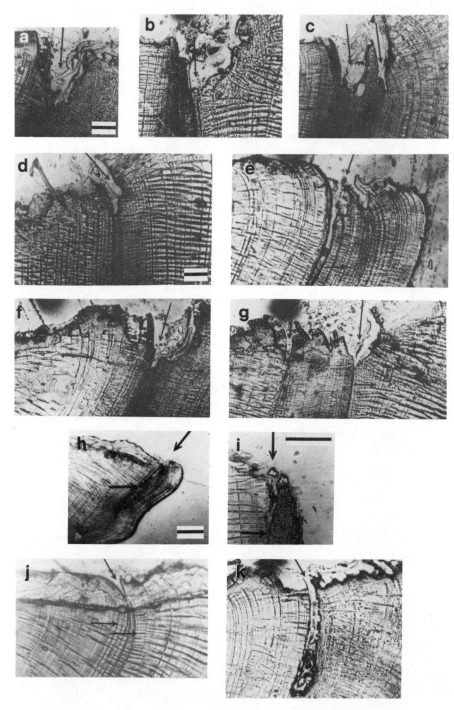

Plate II

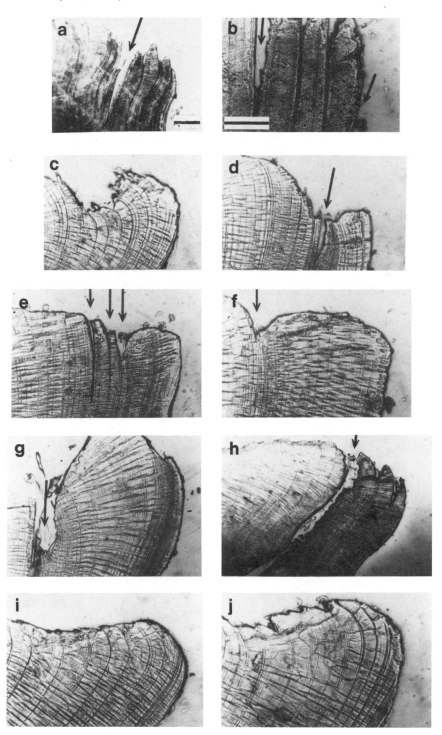

Plate III

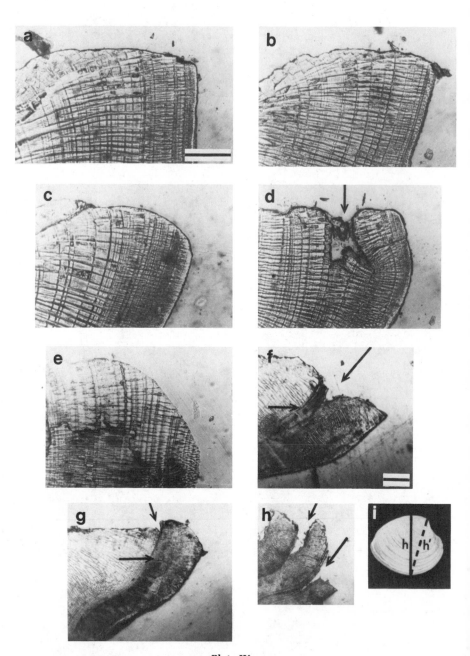

Plate IV

tern has a periodicity equivalent to that of the synodic month, and it is manifested on the external shell surface by concentric ridges formed by the thickened outer shell layer. The lunar-month microgrowth pattern occurs infrequently in M. mercenaria.

Annual periodicities in water temperature, food supply, light input, and perhaps other environmental variables result in annual cycles in the thickness of growth increments. This annual periodicity is strongly developed in M. mercenaria. Low water temperature and a reduced food supply in the winter result in slow shell growth (thin daily growth increments), whereas high water temperature and an abundant food supply in the summer accelerate growth (thick daily growth increments). Freeze- and heat-shock breaks are also commonly associated with seasonal periodicities.

2.2. Growth Breaks

Research on transplanted and natural populations of M. mercenaria indicates that extremes in environmental conditions, rapid changes in environmental conditions, and other physiological stresses can preclude shell secretion for a period of 24 hr or more. During such a period, the valves of the organism remain closed, anaerobic respiratory pathways are employed, and shell decalcification takes place (Crenshaw and Neff, 1969; Lutz and Rhoads, 1977; Gordon and Carriker, 1978) (see Chapter 6). Some clams withdraw their mantle when stressed, producing V-shaped notches in the outer shell layer (Plate IIa–c). These notches have been termed "growth breaks" (Cunliffe, 1974; Kennish and Olsson, 1975; Kennish, 1977a, 1978).

Some growth breaks are extensive, leaving U-shaped notches in the outer shell layer that can be observed on the external shell surface as concentric "growth" lines, linear depressions, or rings. When viewed in shell cross section, crossed lamellar shell structure and an increased concentration of shell organic material are usually associated with the region of the breaks. These microstructural changes may enable the bivalve to strengthen its shell during periods of severe environmental or physiological stress (MacClintock, 1967). Other kinds of depositional breaks are difficult to distinguish on the exterior of the shell because they represent only a few microns of change in the microstructure. These are generally not associated with crossed lamellar shell structure or a high organic content.

It is nearly impossible to identify the diverse origins of growth breaks by examining only the exterior of the shell. By studying the shell microstructure, however, the origin of many growth breaks can be established. Each type of growth break can be identified in the microstructure because each one is preceded and followed by a unique pattern of daily growth increments and shell structural changes.

Of the seven types of growth breaks known to occur in M. mercenaria, four are associated with periodic events (freeze shock, heat shock, spawning, and neap tides) and three with semiperiodic or random events (thermal shock, abrasion, and storms). Shock breaks are particularly abundant and noticeable in the shell of this species; they are caused by temperature stress. For example, freeze- and heat-shock breaks, although developed during different seasons of the year, have similar characteristics. Both occur after a gradual decrease in overall shell convexity and consist of very thin increments prior to and following the break. Also, a pronounced U- or V-shaped notch is developed in the outer shell layer bordered by a region of high organic content; crossed lamellar shell structure transgresses and replaces prismatic shell structure on one, or each, side of the break (Plate IIa–d).

The thin increments preceding a freeze-shock break represent a decline in shell deposition in the fall and winter as water temperature drops below the optimum for growth in the species. The depositional break generally arises at the time of the first freeze of the year, but may also take place later in the winter season. The gradual increase in the thickness of growth increments subsequent to the break is indicative of physiological recovery from the period of temperature stress.

In contrast to a freeze-shock break, thin growth increments preceding a heat-shock break result from the gradual rise in water temperature above the optimum for growth in the species. The depositional break occurs during the period of maximum summer temperatures. The increase in the thickness of growth increments after the break, similar to that of a freeze-shock break, represents physiological recovery from the temperature stress.

Freeze- and heat-shock breaks do not develop at exactly the same time each year because seasonal water temperatures vary from one year to the next. Within a given population, however, these breaks occur synchronously in a large number of specimens. This synchronous effect allows the determination of age, growth rate, and mortality patterns of the population.

Rapid short-term changes in water temperature can also create breaks in accretionary shell growth of M. mercenaria. Kennish and Olsson (1975) and Kennish (1976) have shown that shells of clams at the mouth of Oyster Creek in Barnegat Bay, New Jersey, contain thermal-shock breaks due to sudden fluctuations in water temperature (up to 5°C) associated with thermal discharges from the Oyster Creek Nuclear Generating Station. Thermal-shock breaks form specifically when the station undergoes abrupt shutdowns, massive load reductions, and rapid renewal of operations following shutdown or load-reduction periods.

A thermal-shock break appears as a V-shaped notch in the outer shell layer that is accompanied by crossed lamellar shell structure extending from the middle shell layer into the outer shell layer. Thin increments of growth follow the break. The thermal-shock break can be distinguished from freeze- or heat-shock breaks by its sudden occurrence, the lack of growth deceleration prior to the break, the massive slowdown in growth following the break, and its formation during all seasons of the year (Plate IIe–g).

By counting daily growth increments backward from the valve margins of freshly killed specimens, it is possible to accurately date the time of formation of thermal-shock breaks. These events are represented in most members of a population and therefore serve as useful reference marks in the correlation of other ecological stresses that take place during the life span of the population. Thus, thermal-shock breaks are of value in the study of the population dynamics of clams subjected to thermal discharges.

Random anthropogenic effects, other than thermal discharges, may also produce growth breaks in M. mercenaria. For example, the abrasion of shells of specimens during transplantation experiments often interrupts shell deposition sufficiently to cause a disturbance band (Pannella and MacClintock, 1968; Rhoads and Pannella, 1970; Kennish, 1978). Shell abrasion results in a growth break similar in morphology to the thermal-shock break, with a sequence of thin increments of growth following a V-shaped notch (Plate IIh,i).

The shell of M. mercenaria also preserves a record of biological events, for example, spawning periods. Depending on geographical location, spawning in the species ranges from late spring to early fall as water temperature increases above 20°C (Ansell, 1968). Carriker (1961) noted maximum frequency of spawning in clams in Little Egg Harbor, New Jersey, between 24 and 26°C. Spawning is synchronous, stimulated in part by water temperature and by a hormone carried in the sperm (Nelson and Haskin, 1949).

At the time of spawning, the bivalve suddenly stops feeding, and its growth slows down substantially. This sequence is expressed in the shell as an abrupt break in growth preceded by little or no reduction in the thickness of growth increments and is followed by a series of thin growth increments that reflects reproductive stress (Plate IIj). Recovery from the stress is rapid; the thickness of growth increments quickly returns to the thickness observed prior to spawning. Breaks caused by spawning are less severe than freeze-, heat-, and thermal-shock breaks, with notches in the shell being thinner and shallower in the outer shell layer. Crossed lamellar shell structure may or may not be associated with reproductive breaks.

Spawning breaks in *M. mercenaria* do not develop until the bivalve attains sexual maturity at an age of 2 years (Rhoads and Pannella, 1970). Some specimens have more than one reproductive break per year, and males display more spawning breaks than females (Pannella and Mac-Clintock, 1968). Thus, within the same population, males should be smaller than females of the same age.

Cunliffe (1974) described clams that had spawning breaks coincident with neap tides. His work supported an earlier observation by Carriker (1961) that showed mass spawning of *M. mercenaria* during neap tides. Spawning at this time in the fortnightly tidal cycle serves to retain larvae in the estuary because there is less exchange of water with the open ocean during neap tides than during spring tides.

Extreme neap tides may cause a growth break at times other than spawning, especially in intertidal and tide pool clams. The neap-tide break is preceded by a group of increments decreasing in thickness toward the break and followed by a series of increments increasing in thickness away from the break. Notches in the outer shell layer are generally less deeply incised than those associated with other types of breaks, and crossed lamellar shell structure is usually not associated with the region of the break.

Another type of growth break identified in the shell of this species is the storm break. The features of the storm break vary, depending on the severity of the storm and the depth of the clam below sea level. In most cases, the thickness of growth increments during pre- and poststorm periods is about the same. The break appears suddenly, and it may contain silt grains trapped deep within an identation in the outer shell layer (Plate IIk). The silt grains become trapped between the mantle and shell during the storm and are incorporated in the microstructure when growth resumes after the storm.

Other types of breaks, some of anthropogenic origin, could occur in the shell microstructure of *M. mercenaria*, but have not yet been documented. Floods, sedimentation events, attacks by predators, disease, and changes in water chemistry may also produce structural changes in the shell. Variations in turbidity and substrate stability associated with dredging operations are also capable of affecting shell secretion. Additional experimentation on this species should continue, therefore, to determine the causal relationship between specific anthropogenic, environmental, and physiological events and microgrowth patterns.

3. Application to Analysis of Population Dynamics

Microgrowth patterns in the shell of M. mercenaria accurately record environmental and physiological conditions throughout the ontogeny of the organism, enabling applications to various ecological and paleoecological problems. A number of applications of this type have been made in the past. For example, Barker (1964) investigated changes in the thickness of growth increments with respect to latitude in extant M. mercenaria, as did Cunliffe (1974). Cunliffe (1974) also described the paleoecology of the Choptank and St. Marys Formations (Middle Miocene age) in Maryland by examining growth increments in fossil M. mercenaria. Rhoads and Pannella (1970) studied the influence of substratum, water depth, turbidity, and temperature on juvenile M. mercenaria by analyzing the shell microstructure in transplanted specimens. Kennish and Olsson (1975) and Kennish (1976, 1977a) evaluated the impact of thermal discharges from the Oyster Creek Nuclear Generating Station on shell microgrowth patterns of clams in Barnegat Bay, New Jersey.

Although a substantial volume of data on shell microgrowth patterns has been gathered on the behavior of this bivalve, there has been no attempt to determine its population dynamics by utilizing shell microstructure. In general, research on the population dynamics of an organism requires the analysis of size-frequency histograms of a population. This usually demands a great deal of time and effort on the part of the investigator (see Chapter 11). By implementing shell microgrowth analysis, however, these problems can be mitigated or eliminated because of the life-history data stored within the shell of each clam.

Studies of population dynamics include defining the age and growth rates of organisms and their mortality and recruitment patterns (Hallam, 1972). In the case of M. mercenaria, these data are obtainable by microstructural analysis of the shell, if the daily growth increment is used as the fundamental unit of study. For instance, counts of daily growth increments from the umbo to the ventral valve margin yield data on the absolute age of the animal. If the thicknesses of daily growth increments are measured, daily and yearly growth rates and size distributions can be established. Superimposing growth rates on the size distributions of life assemblages of clams supplies information on the frequency and success of recruitment (Kennish, 1978).

By adapting these methods of analysis to dead specimens, it is possible to determine the absolute age at death, as well as the growth rates of specimens in death assemblages (Rhoads and Pannella, 1970). These data can be used to generate life tables that accurately depict mortality of a population. In addition to providing data on the absolute age at death, growth-increment counting in combination with the interpretation of microgrowth patterns at the valve margin can yield season-of-death data (Rhoads and Pannella, 1970; Tevesz, 1972). For example, a ventral margin following a complete sequence of winter growth increments and growth breaks implies spring death (Plate IIIa–c). A heat-shock break (Plate IIId, e), a spawning break (Plate IIIf, g), and a summer storm break (Plate IIIh) adjacent to the growing edge of the shell signify summer death, as do thick growth increments at the outer edge preceded by a freeze-shock break located 160–200 increments dorsally (Plate IIIi,j). A ventral margin preceded by a long period of summer growth increments and growth breaks suggests fall death (Plate IVa–d). Thin growth increments located at the shell edge approximately 160–200 increments ventrally from a heat-shock break denote winter death (Plate IVe). A freeze-shock break found at the ventral margin also connotes winter death (Plate IVf–h).

Analysis of microgrowth patterns at the valve margin can be of value in differentiating between types of mortality in *M. mercenaria* populations (i.e., natural vs. census or catastrophic mortality). A population experiencing natural mortality will consist of specimens with a high degree of variability in the terminal microgrowth patterns of their shells, indicating that the organisms died during different seasons of the year. Census (mass) mortality, in contrast, will be manifested by great similarity in the terminal microgrowth patterns of component shells of a death assemblage, reflecting widespread mortality during a single season of the year.

It is feasible to use shell microgrowth patterns to define different generations of clams within a death assemblage. Some growth breaks (i.e., freeze-shock and heat-shock breaks) occur synchronously in a large part of a population, whereas others (i.e, thermal-shock and storm breaks) can be dated precisely by alternate methods (Kennish, 1978). In effect, it is often possible to match and date these growth breaks across many specimens in a death assemblage, employing each break as a microgrowth datum for purposes of correlation. Those specimens with a significant correlation in their growth breaks have a high probability of belonging to the same generation of clams (also see Chapter 10).

4. Population Dynamics of *Mercenaria mercenaria* in Barnegat Bay, New Jersey

In a previous study (Kennish, 1978), I examined size and age distributions, growth rates, and recruitment and mortality patterns of *M. mercenaria* by implementing the methods of shell microgrowth analysis described above. This work was conducted in Barnegat Bay, New Jersey, a shallow tidal basin located midway on the eastern New Jersey coast (Fig. 2). Results of this study are summarized below.

4.1. Size and Age Distributions, Growth Rates, and Recruitment Patterns

I sampled death assemblages of M. mercenaria at four sites in Barnegat Bay (sites 1, 3, 4, and 7 in Fig. 2) in the summer of 1974 and collected life and death assemblages of the bivalve from nine sites in the summer of 1976 (sites 1–9 in Fig. 2) (Kennish, 1978). Size measurements on individual specimens were made from the umbo to the ventral valve margin at a slight angle to the dorsal–ventral plane for the purpose of investigating

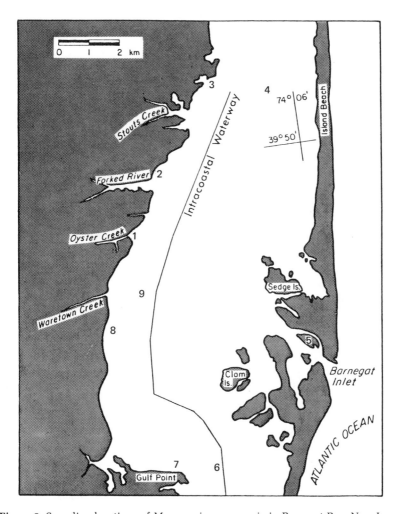

Figure 2. Sampling locations of Mercenaria mercenaria in Barnegat Bay, New Jersey.

the entire growth record of the clam (Plate IVi). The axis measured only approximates the height of the organism; therefore, it was labeled h' to differentiate it from true shell height (h).

Figures 3 and 4 illustrate the size distributions of all M. mercenaria collected in the summers of 1974 and 1976, and Table III summarizes these distributions. The size-frequency distributions of all death assemblages and most of the life assemblages consist of unimodal curves with negative skewness. These indicate a great similarity in clam sizes throughout the estuary except at Barnegat Inlet (site 5), where juveniles comprise both the life and death assemblages. Barnegat Inlet, however, is a transitional environment between the Atlantic Ocean and Barnegat Bay, and environmental conditions in the inlet are quite different from those within

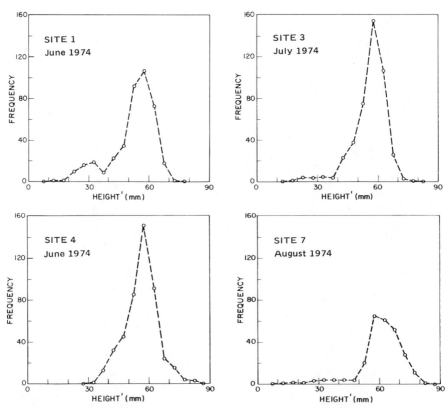

Figure 3. Size (h')-frequency distributions for death assemblages of Mercenaria mercenaria collected at sites 1, 3, 4, and 7 in Barnegat Bay in 1974. Height' represents a dimension of the shell from the umbo to the ventral margin at a slight angle to the dorsal–ventral plane (see Plate IVi).

the estuary. These differences may account for spatial variations in recruitment and growth of the species in the area of study.

Research on extant and fossil invertebrate assemblages has shown that growth rates, recruitment, and mortality primarily control the form of size-frequency distributions of invertebrate populations (Craig and Hallam, 1963; Craig and Oertel, 1966; Craig, 1967; McKerrow et al., 1969; Schmidt and Warme, 1969; Levinton and Bambach, 1970; Hallam, 1972) (also see Chapter 11). The two major factors influencing the shape of size-frequency distributions of death assemblages are growth rates and mortality (Hallam, 1972). Evidence of seasonal recruitment rarely exists in death assemblages because annual fluctuations in recruitment tend to average out through the accumulation of many generations of clams; this contributes to the formation of unimodal size-frequency distributions. Recruitment patterns are preserved best in mass mortality assemblages, in which the size–age distribution of the life assemblage is "frozen" at a time surface (Sheldon, 1965).

To test these principles on M. mercenaria in Barnegat Bay, I sectioned and analyzed the valves of 278 specimens from death assemblages gathered in 1974 at sites 1, 3, 4, and 7 (Kennish, 1978). The valves ranged from 14.0 to 76.5 mm in h'. Daily growth-increment and seasonal growth-break counts were made on 277 valves to obtain absolute growth rates and on 278 valves to obtain absolute mortality patterns.

Tables IV and V show the mean and absolute age–h' relationships of M. mercenaria at each site, and Table VI gives the pooled age–h' values of 277 specimens. Most of the clams are less than 7 years of age and 70.0 mm in h'. Growth rates of the specimens are uniform at all four sites, with growth being more rapid during youth than during maturity and old age. The bivalve undergoes a gradual, exponential decline of growth rate with increasing age; growth is nearly linear during the juvenile life stage.

This type of ontogenetic growth conforms best to the following exponential growth equation:

$$(\text{differential form}) \ dN/dt = N_{max}r_0 - r_0N_t$$

$$(\text{integrated form}) \ N_t = N_{max}(1 - ce^{-r_0t})$$

where dN/dt is the rate of shell growth, N_{max} is the asymptotic h' or maximum h' attainable, r_0 is the maximum intrinsic rate of natural increase, N_t is the h' at a given time t, c is the integration constant defining the complement of that proportion of N_{max} that specifies the initial N_0, and e is the base of the natural logarithm (Kennish, 1977b). This exponential growth equation predicts growth curves for M. mercenaria that are realistic in comparison to other exponential functions.

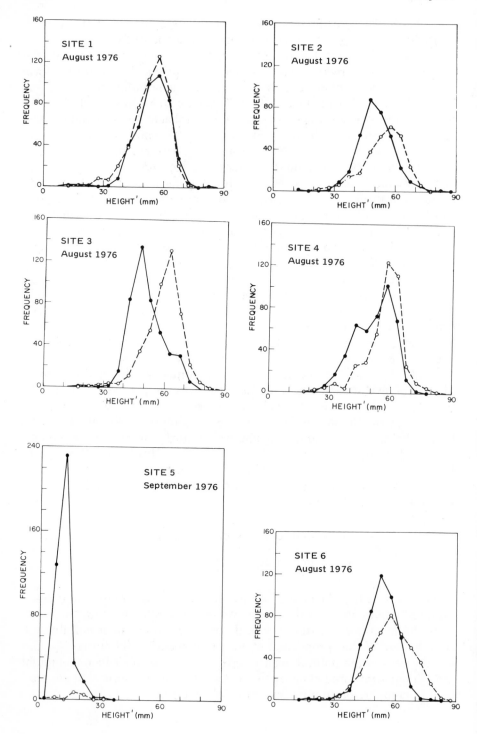

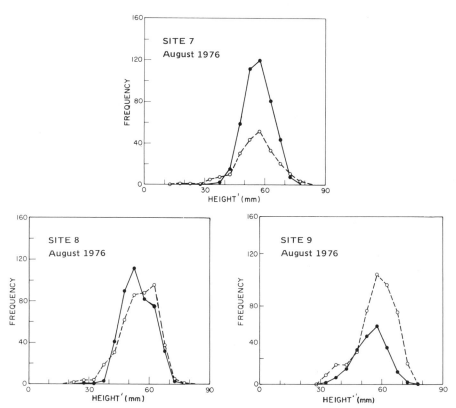

Figure 4. Size (h')-frequency distributions for life and death assemblages of *Mercenaria mercenaria* collected at sites 1–9 in Barnegat Bay in 1976. (○----○) Death assemblages; (●—●) life assemblages. Height' represents a dimension of the shell from the umbo to the ventral margin at a slight angle to the dorsal–ventral plane (see Plate IVi).

A comparison of the observed growth rates of specimens within death assemblages of clams at sites 1, 3, 4, and 7 (Table V) with the corresponding size-frequency distributions at these locations (Fig. 3) indicates that most specimens in these samples range from 40 to 70 mm in h' corresponding to ages of 4–9 years. Mortality is highest in larger and older clams, and this reflects the effect of high rates of death during the gerontic stage of life. Thus, the unimodal size-frequency distributions of negative skewness in the death assemblages result from an interplay of growth rates that decrease with age and mortality that increases with age. The similar size distributions of the death assemblages suggest comparable growth and mortality patterns throughout the estuary.

Most *M. mercenaria* in the life assemblages, as in the death assemblages, range from 40 to 70 mm in h'. Most live clams in Barnegat Bay,

Table III. Size Statistics for Life and Death Assemblages of *Mercenaria mercenaria* in Barnegat Bay

Site (Fig. 2)	Year	Life assemblage				Death assemblage			
		Number of clams	Mean h' (mm)^a ±95% C.I.	S.D. for mean	Median h' (mm)^a	Number of clams	Mean h' (mm)^a ±95% C.I.	S.D. for mean	Median h' (mm)^a
1	1974	No data	—	—	—	405	51.75 ± 1.06	10.81	54.50
3	1974	No data	—	—	—	446	55.38 ± 0.74	7.91	57.00
4	1974	No data	—	—	—	464	55.89 ± 0.73	7.97	56.50
7	1974	No data	—	—	—	261	60.47 ± 1.29	10.56	62.00
1	1976	435	54.56 ± 0.75	7.97	55.20	499	53.36 ± 0.76	8.69	54.45
2	1976	342	50.33 ± 0.91	8.52	49.60	280	54.16 ± 1.11	9.47	55.48
3	1976	446	51.07 ± 0.78	8.38	49.45	445	59.17 ± 0.78	8.37	60.60
4	1976	441	51.24 ± 0.88	9.41	52.70	408	56.45 ± 0.87	8.95	57.80
5	1976	416	11.84 ± 0.37	3.85	11.10	16	17.52 ± 3.44	6.45	16.85
6	1976	450	52.52 ± 0.70	7.54	52.63	413	57.78 ± 1.02	10.53	57.80
7	1976	436	56.26 ± 0.63	6.64	56.25	214	55.25 ± 1.30	9.63	56.30
8	1976	440	54.24 ± 0.70	7.51	53.77	436	54.62 ± 0.84	8.89	55.60
9	1976	208	54.20 ± 1.04	7.63	55.00	438	57.27 ± 0.85	9.07	58.60

^{a} Height h' represents a dimension of the shell from the umbo to the ventral margin at a slight angle to the dorsal–ventral plane (see Plate IVi).

Table V. Mean Age–Height' Relationships Observed in Death Assemblages of Clams at Sites 1, 3, 4, and 7

Age	Site 1 (N^a = 101)			Site 3 (N^a = 74)			Site 4 (N^a = 64)			Site 7 (N^a = 38)		
	N^b	Mean h' (mm)^c	S.D.	N^b	Mean h' (mm)^c	S.D.	N^b	Mean h' (mm)^c	S.D.	N^b	Mean h' (mm)^c	S.D.
1	101	12.39	3.76	74	11.36	4.16	64	7.04	2.71	38	12.72	5.15
2	89	23.00	4.49	68	23.51	4.96	64	17.70	4.24	35	25.61	6.19
3	65	32.91	4.49	63	35.01	4.89	63	29.54	5.22	25	37.30	6.42
4	50	42.36	5.26	44	43.78	4.15	54	39.55	5.23	19	46.97	5.25
5	28	49.37	6.27	19	49.89	3.01	28	47.75	5.81	12	55.58	7.05
6	16	57.61	7.76	9	56.18	3.94	10	55.11	4.05	6	62.13	8.88
7	6	59.52	5.41	6	63.60	1.70	7	63.30	4.56	3	65.80	3.04
8	0	—	—	4	70.60	1.75	3	69.13	5.95	0	—	—

^{a} (N) Number of clams.
^{b} (N) Number of size measurements per age.

Table IV. Age–Height' Statistics for Samples of Clams from Sites 1, 3, 4, and 7

Site	Number of clams	Mean h' (mm)[a]	S.D.	Mean age (yr)	S.D.
1	101	43.12	12.81	3.92	1.70
3	74	46.82	12.57	4.25	1.72
4	64	49.74	9.65	5.03	1.31
7	38	46.82	17.05	4.03	1.79

[a] Height h' represents a dimension of the shell from the umbo to the ventral margin at a slight angle to the dorsal–ventral plane (see Plate IVi).

therefore, are in old age. The unimodal, negative skewness of the life assemblage distributions reveals a recent failure of annual recruitment in the bay.

Mass spawning and recruitment of M. mercenaria occur in the summer in New Jersey. If recruitment had been successful in the early and mid-1970's, modes corresponding to successive yearly broods should be found in the size-frequency distributions of the life assemblages. A theoretical model simulating these conditions is depicted in Fig. 5. This model delineates a polymodal size-frequency distribution of an ideal, stable clam population that experiences constant annual recruitment and no mortality. Modes in the model at 1.5, 10.9, 22.2, 33.0, and 42.4 mm correlate with the 1976, 1975, 1974, 1973, and 1972 year classes, respectively, and are derived from the average rate of growth per year for M. mercenaria in Barnegat Bay (Table VI). Troughs between modes represent periods of growth of each generation from one recruitment stage to the next. Beyond 50 mm, the modes merge and the age classes become obscured because of the spread of sizes of older clams and the reduction in growth rates, which causes an overlapping of the larger classes. This overlapping effect is also apparent in the observed size-frequency distributions of the life assemblages (see Fig. 4).

A comparison of the observed size-frequency distributions of the life assemblages in Barnegat Bay with the size-frequency distributions of the model reflects the conspicuous absence of single-year classes of clams in the estuary from 1973 to 1976. A sharp peak at 11.8 mm in the life-assemblage distribution for site 5 (Barnegat Inlet) corresponds to the 1975-year class, and it compares well to the population growth model. The absence of modes at smaller and larger sizes, however, is a consequence of poor recruitment at this site prior to, and following, 1975.

Young M. mercenaria are being added to populations in Barnegat Bay not on an annual basis, but sporadically. This recruitment pattern appears to be caused by the normal population dynamics of the species (Kennish, 1978). Recruitment may fail for as long as 5 or 10 continuous years, but

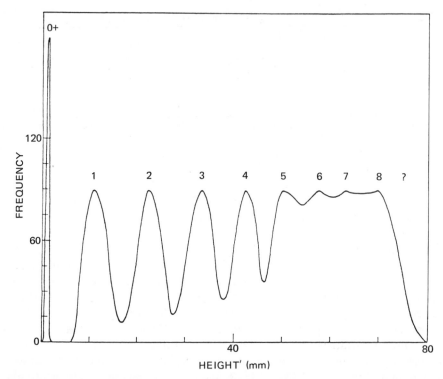

Figure 5. Theoretical population model for 8 years of growth of *Mercenaria mercenaria* in Barnegat Bay. The model assumes population stability with constant annual recruitment and zero mortality. Height' represents a dimension of the shell from the umbo to the ventral margin at a slight angle to the dorsal–ventral plane (see Plate IVi).

a population can be successfully maintained by a single year of good recruitment.

4.2. Mortality Patterns

Absolute mortality patterns in death assemblages of M. *mercenaria* at sites 1, 3, 4, and 7 were resolved by shell microgrowth analysis. Data on the season of death, as well as on the size and age of the individual specimen at the time of death, permitted the construction of size- and age-frequency distributions, death-frequency histograms, and life tables for each death assemblage. This information was used to specify seasons of peak stress in the species and to determine the relationship between the size and age of the clam and mortality in the population. In addition, the

Table VI. Mean Age–Height' Relationships
for the Pooled Death Assemblages of
Clams from Sites 1, 3, 4, and 7

Age	Number of size measurements per age	Number of clams: 277	
		Mean height' (mm)[a]	S.D.
1	277	10.92	4.44
2	256	22.17	5.51
3	216	33.05	5.67
4	167	42.35	5.47
5	87	49.82	6.11
6	41	57.35	6.65
7	22	62.69	4.37
8	7	69.97	3.73

[a] Height' represents a dimension of the shell from the umbo to the ventral margin at a slight angle to the dorsal–ventral plane (see Plate IVi).

information was utilized to reveal the effect of geographical location within the estuary on mortality of the organisms.

Table VII shows the frequency of death of M. mercenaria per season based on the analysis of 278 valves of specimens sampled at the four sites in 1974. Mortality reaches a maximum in the summer and winter and a minimum in the spring and fall. Figure 6 displays this trend and indicates that seasonal mortality is similar at all sites.

I attributed the high summer mortality to the physiological stress of spawning and to increased predator and parasitic activity during the warmer months of the year (Kennish, 1978). High winter mortality, however, I ascribed to physiological stresses related to harsh environmental conditions. During the winter, the bivalve lives on stored carbohydrate reserves because of low food supply and low water temperatures in the

Table VII. Death-Frequency Distributions per Season for Natural Populations of Mercenaria mercenaria at Sites 1, 3, 4, and 7

Site	Number of clams	Season of death[a]			
		Spring	Summer	Fall	Winter
1	101	12 (11.88%)	33 (32.67%)	25 (24.75%)	31 (30.69%)
3	75	13 (17.33%)	24 (32.00%)	13 (17.33%)	25 (33.33%)
4	64	13 (20.31%)	19 (29.69%)	12 (18.75%)	20 (31.25%)
7	38	9 (23.68%)	16 (42.11%)	7 (18.42%)	6 (15.79%)
All sites	278	47 (16.91%)	92 (33.09%)	57 (20.50%)	82 (29.50%)

[a] Percentage of mortality per season in parentheses.

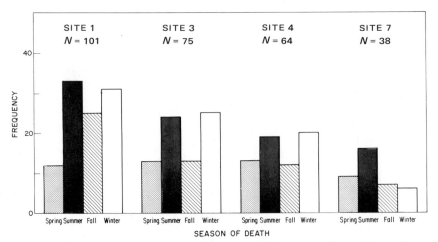

Figure 6. Death-frequency distributions per season for death assemblages of natural populations of *Mercenaria mercenaria* at sites 1, 3, 4, and 7.

bay. However, these reserves may be inadequate in some clams, especially during severe winters.

Absolute mortality of *M. mercenaria*, like seasonal mortality, is similar at sites 1, 3, 4, and 7. Evidence for this exists in the life tables constructed for the death assemblages at each location (Tables VIII–XI), which were formulated by using the methods of Deevey (1947), Barclay (1958), Reyment (1971), and Gani (1973). Larval mortality of *M. mercenaria* was not investigated; therefore, the life tables were constructed beginning at age 1 of a hypothetical cohort (radix) of 1000 clams.

All the life tables show low mortality for clams between the ages of 1 and 4 years. After the fourth year of life, mortality increases markedly, although less consistently, through the remaining age intervals. This pattern of mortality is graphically demonstrated by mortality curves drawn for the four assemblages (Fig. 7).

Mercenaria mercenaria larval mortality is known to be high (Thorson, 1950; Carriker, 1961); consequently, a survivorship curve projected for the entire ontogeny of this bivalve should be sigmoidal. A high probability of death exists for the meroplanktic stages because of disease, starvation, and predation, but the probability drops substantially after the organism settles onto a favorable substratum. The probability of death increases again in old age.

On the basis of the life table data, the risk of death of *M. mercenaria* begins to increase subsequent to the attainment of sexual maturity. To test this effect, I transplanted 1486 clams to the substratum at sites 1 and 3, and recorded the mortality patterns of the experimental populations 1

Table VIII. Life Table for Mercenaria mercenaria at Site 1

Age interval x to x + 1 years	Proportion dying in interval x to x + 1 ($1000q_X$)	Number living at age x (l_X)	Number dying in interval x to x + 1 (d_X)	Number of time spans lived in interval x to x + 1 (L_X)	Total number of time spans lived past age x (T_X)	Average life expectancy (yr) at age x (e_X)	Proportion surviving in interval x to x + 1 (P_X)
1–2	22.00	1000	22	989.0	4508.0	4.5080	0.9780
2–3	68.51	978	67	944.5	3519.0	3.5982	0.9315
3–4	83.42	911	76	873.0	2574.5	2.8260	0.9166
4–5	130.54	835	109	780.5	1701.5	2.0377	0.8695
5–6	544.08	726	395	528.5	921.0	1.2686	0.4559
6–7	314.20	331	104	279.0	392.5	1.1858	0.6868
7–8	1000.00	227	227	113.5	113.5	0.5000	0.0000

Table IX. Life Table for Mercenaria mercenaria at Site 3

Age interval x to x + 1 years	Proportion dying in interval x to x + 1 ($1000q_X$)	Number living at age x (l_X)	Number dying in interval x to x + 1 (d_X)	Number of time spans lived in interval x to x + 1 (L_X)	Total number of time spans lived past age x (T_X)	Average life expectancy (yr) at age x (e_X)	Proportion surviving in interval x to x + 1 (P_X)
1–2	7.00	1000	7	996.5	4844.0	4.8440	0.9930
2–3	27.19	993	27	979.5	3847.5	3.8746	0.9728
3–4	39.34	966	38	947.0	2868.0	2.9689	0.9607
4–5	113.15	928	105	875.5	1921.0	2.0700	0.8869
5–6	343.86	823	283	681.5	1045.5	1.2704	0.6561
6–7	829.63	540	448	316.0	364.0	0.6741	0.1704
7–8	978.26	92	90	47.0	48.0	0.5217	0.0217
8–9	1000.00	2	2	1.0	1.0	0.5000	0.0000

Table XI. Life Table for *Mercenaria mercenaria* at Site 7

Age interval x to x + 1 years	Proportion dying in interval x to x + 1 (1000q$_x$)	Number living at age x (l$_x$)	Number dying in interval x to x + 1 (d$_x$)	Number of time spans lived in interval x to x + 1 (L$_x$)	Total number of time spans lived past age x (T$_x$)	Average life expectancy (yr) at age x (e$_x$)	Proportion surviving in interval x to x + 1 (P$_x$)
1–2	19.00	1000	19	990.5	4951.0	4.9510	0.9810
2–3	34.66	981	34	964.0	3960.5	4.0372	0.9653
3–4	24.29	947	23	935.5	2996.5	3.1642	0.9757
4–5	124.46	924	115	866.5	2061.0	2.2305	0.8755
5–6	398.02	809	322	648.0	1194.5	1.4765	0.6020
6–7	377.82	487	184	395.0	546.5	1.1222	0.6222
7–8	1000.00	303	303	151.5	151.5	0.5000	0.0000

Table X. Life Table for *Mercenaria mercenaria* at Site 4

Age interval x to x + 1 years	Proportion dying in interval x to x + 1 (1000q$_x$)	Number living at age x (l$_x$)	Number dying in interval x to x + 1 (d$_x$)	Number of time spans lived in interval x to x + 1 (L$_x$)	Total number of time spans lived past age x (T$_x$)	Average life expectancy (yr) at age x (e$_x$)	Proportion surviving in interval x to x + 1 (P$_x$)
1–2	0.00	1000	0	1000.0	5074.0	5.0740	1.0000
2–3	0.00	1000	0	1000.0	4074.0	4.0740	1.0000
3–4	30.00	1000	30	985.0	3074.0	3.0740	0.9700
4–5	122.68	970	119	910.5	2089.0	2.1536	0.8773
5–6	324.32	851	276	713.0	1178.5	1.3848	0.6757
6–7	775.65	575	446	352.0	465.5	0.8096	0.2244
7–8	620.16	129	80	89.0	113.5	0.8798	0.3798
8–9	1000.00	49	49	24.5	24.5	0.5000	0.0000

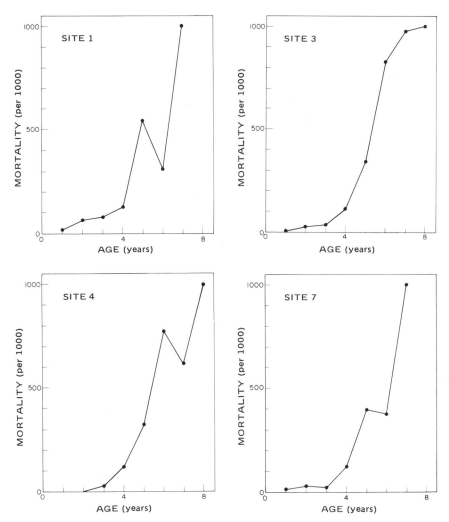

Figure 7. Mortality curves for natural populations of *Mercenaria mercenaria* at sites 1, 3, 4, and 7. Data are presented as mortality per 1000 for each age interval of 1 year (1000 q_x) plotted against the start of the interval.

year after transplantation (Kennish, 1978). Each site received 743 specimens of comparable size (Table XII).

The valves of the clams were sanded at the ventral margin prior to transplantation to produce reference notches for shell microgrowth analysis. The experimental populations were transplanted in June 1975 and retrieved in June 1976. Empty valves of dead specimens were studied microscopically to determine their absolute age and season of death.

Table XII. Size $(h')^a$-Frequency Distributions for Clams Transplanted to the Substratum at Sites 1 and 3 in June 1975

Size class (mm)	Frequency of clams transplanted to site 1	Frequency of clams transplanted to site 3
20–25	1	0
25–30	9	10
30–35	103	103
35–40	69	68
40–45	116	117
45–50	115	114
50–55	124	124
55–60	103	103
60–65	66	67
65–70	24	24
70–75	10	10
75–80	3	3
TOTAL:	743	743

^{a} Height h' represents a dimension of the shell from the umbo to the ventral margin at a slight angle to the dorsal–ventral plane (see Plate IVi).

A total of 1000 live and dead clams was recovered at the two experimental sites in June 1976, representing 67% of the specimens originally transplanted. Table XIII compares the size (h')-frequency distributions of these clams. Most of the dead M. mercenaria were larger than 50 mm in h' (Fig. 8) and greater than 5 years of age (Fig. 9); thus, they were reproductively mature at the time of death.

These distributions conform well to those of the death assemblages of natural populations, although the frequency of death in the smaller and younger classes is greater in the experimental populations. The differences in absolute mortality between the natural and experimental populations are not substantial, considering that mortality in the natural populations constitutes a cumulative effect over many years, whereas mortality in the experimental populations depicts only a single-year effect. The death assemblages of the natural populations give a more realistic picture of mortality trends for M. mercenaria.

As in the death assemblages of natural populations, the seasonal frequency of death of the experimental populations peaks in the summer (Fig. 10). The low frequency of winter death in the experimental populations, however, stands in contrast to the high frequency of winter death in the natural populations (see Fig. 6). This indicates the absence of a significant environmental stress on the experimental populations during the winter of 1975–1976.

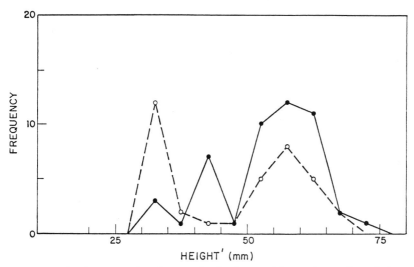

Figure 8. Size (h')-frequency distributions for death assemblages of transplanted clams recovered from the substratum at sites 1 and 3 in June 1976. (○----○) Site 1; (●——●) site 3. Height' represents a dimension of the shell from the umbo to the ventral margin at a slight angle to the dorsal–ventral plane (see Plate IVi).

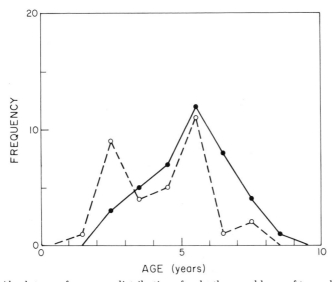

Figure 9. Absolute age-frequency distributions for death assemblages of transplanted clams retrieved from the substratum at sites 1 and 3 in June 1976. (○----○) Site 1; (●——●) site 3.

Table XIII. Size $(h')^a$-Frequency Distributions for Live and Dead Clams Recovered from the Substratum at Sites 1 and 3 in June 1976 (Clams Originally Transplanted in June 1975)

Size class (mm)	Frequency of live clams recovered from site 1	Frequency of dead clams recovered from site 1	Frequency of live clams recovered from site 3	Frequency of dead clams recovered from site 3
20–25	0	0	0	0
25–30	0	0	0	0
30–35	11	12	20	3
35–40	43	2	31	1
40–45	86	1	73	7
45–50	104	1	90	1
50–55	97	5	62	10
55–60	113	8	64	12
60–65	49	5	44	11
65–70	11	2	12	2
70–75	1	0	2	1
75–80	0	0	0	0
				+3 (broken)
TOTAL:	515	36	398	51

^{a} Height h' represents a dimension of the shell from the umbo to the ventral margin at a slight angle to the dorsal–ventral plane (see Plate IVi).

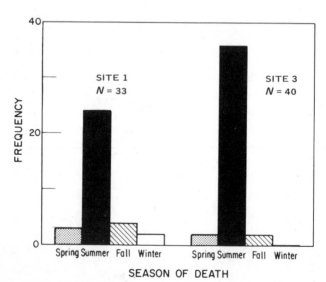

Figure 10. Death-frequency distributions per season for death assemblages of transplanted clams collected from the substratum at sites 1 and 3 in June 1976.

A number of inferences can be made from the population dynamics of M. mercenaria based on the size, age, and season-of-death data on the natural and experimental populations in Barnegat Bay. High mortality during both the larval and gerontic stages suggests that the bivalve is most susceptible to physiological and environmental stresses at these times. The life expectancy of the clam increases markedly after it successfully settles in an area and passes through the plantigrade stage ($\approx$10 mm in size). When the bivalve reaches sexual maturity at the age of 2 years, the probability of death is relatively low, which ensures the survival of the species. Because mortality is relatively low between the plantigrade and gerontic stages, a young population can persist reasonably intact for several years at a time without the need of large-scale recruitment.

The gerontic stage is a period of diminished metabolism (Hopkins, 1930), slowed growth, and increased mortality. High mortality at this time appears to be associated with physiological degeneration combined with environmental influences, although it is impossible to determine the relative contribution of either of these factors to observed mortality patterns by utilizing shell microgrowth analysis alone. Histological examination of soft parts in combination with the analysis of the shell microstructure may yield additional detailed information on the causes of mortality in many populations.

5. Conclusions and Future Research Efforts

Rhoads and Pannella (1970) discussed potential applications of molluscan shell microgrowth patterns in ecology and paleoecology. Many of these applications were formulated from observations on extant M. mercenaria. Since 1970, applications of shell microgrowth patterns to fossil M. mercenaria have lagged behind investigations on live specimens, but significant interdisciplinary advances in recent years have enabled applications to a number of paleoecological problems.

Perhaps the most useful application of shell microgrowth patterns in M. mercenaria is in the study of the population dynamics of the organism. For example, data on the season of death, age and size at the time of death, and absolute rate of growth of Recent populations of clams from various estuarine environments should be valuable to resource-management programs assessing this commercially important bivalve. Determining the season of peak environmental stress would be important to aquaculture and future stock-rebuilding programs.

Research on shell microgrowth patterns of live M. mercenaria in stressed environments may provide critical information on the biological consequences of man's activities on this species. For instance, data on shell microgrowth patterns of clams subjected to dredging and filling operations, domestic and industrial sewages, and oil spillages can reveal information of predictive value concerning population changes. By employing shells of death assemblages of M. mercenaria in these stressed environments, temporal and spatial changes in the populations can be related to anthropogenic factors.

Similar applications may be extended to other molluscan taxa. It is necessary, however, to establish the relationship of shell microgrowth patterns in these taxa to specific environmental parameters and physiological conditions. This requires a detailed analysis of live specimens from natural and experimental populations.

Once the analysis of shell microgrowth patterns in many extant molluscan taxa is accomplished, applications to fossil molluscan populations should proceed more rapidly. In the future, for example, it is conceivable that the analysis of microgrowth patterns in fossil molluscan populations from stratigraphic sequences of coastal plain formations along the Atlantic and Gulf coasts could be used to reconstruct paleolatitudes, paleoclimates, and paleobathymetry through the Tertiary period. It might also prove important in evolutionary studies of molluscs during this period of time.

Molluscan shell microgrowth patterns remain a viable tool for government, industry, and academic research. With continued interdisciplinary investigations into molluscan physiology, biochemistry, and shell structure, additional applications of the molluscan shell should be forthcoming. These applications should prove to be particularly helpful in the analysis of environmental and paleoenvironmental problems.

ACKNOWLEDGMENTS. Financial support for the author's research reported in this chapter was kindly provided by the Geological Society of America, the Society of Sigma Xi, the Marine Sciences Center of Rutgers University, and the Jersey Central Power & Light Company. Copeland MacClintock kindly supplied the photographs reproduced here as Fig. 1 and Plate Ii,j.

References

Ansell, A. D., 1968, The rate of growth of the hard clam Mercenaria mercenaria (L.) throughout the geographical range, J. Cons. Int. Explor. Mer. **31**:364–409.

Barclay, G. W., 1958, Techniques of Population Analysis, John Wiley, New York, 311 pp.

Barker, R. M., 1964, Microtextural variations in pelecypod shells, Malacologia **2**:69–86.

Carriker, M. R., 1961, Interrelation of functional morphology, behavior, and autecology in early stages of the bivalve Mercenaria mercenaria, J. Elisha Mitchell Sci. Soc. **77**:168–241.

Clark, G. R., 1974, Growth lines in invertebrate skeletons, Annu. Rev. Earth Planet. Sci. **2**:77–79.

Craig, G. Y., 1967, Size-frequency distributions of living and dead populations of pelecypods from Bimini, Bahamas, B.W.I., *J. Geol.* **75:**34–45.

Craig, G. Y., and Hallam, A., 1963, Size-frequency and growth-ring analyses of *Mytilus edulis* and *Cardium edule*, and their paleoecological significance, *Palaeontology* **6:**731–750.

Craig, G. Y., and Oertel, G., 1966, Deterministic models of living and fossil populations of animals, *Q. J. Geol. Soc. London* **122:**315–355.

Crenshaw, M. A., and Neff, J. M., 1969, Decalcification at the mantle–shell interface in molluscs, *Am. Zool.* **9:**881–885.

Cunliffe, J. E., 1974, Description, interpretation, and preservation of growth increment patterns in shells of Cenozoic bivalves, Ph.D. thesis, Rutgers University, New Brunswick, New Jersey, 171 pp.

Cunliffe, J. E., and Kennish, M. J., 1974, Shell growth patterns in the hard-shelled clam, *Underwater Nat,* **8:**20–24.

Deevey, E. S., 1947, Life tables for natural populations of animals, *Q. Rev. Biol.* **22:**283–314.

Gani, J., 1973, Stochastic formulations for life tables, age distributions and mortality curves, in: *The Mathematical Theory of the Dynamics of Biological Populations* (M. S. Bartlett and R. W. Hiorns, eds.), pp. 291–302, Academic Press, New York.

Gordon, J., and Carriker, M. R., 1978, Growth lines in a bivalve mollusk: Subdaily patterns and dissolution of the shell, *Science* **202:**519–521.

Hallam, A., 1972, Models involving population dynamics, in: *Models in Paleobiology* (T. J. M. Schopf, ed.), pp. 62–80, Freeman, Cooper, San Francisco.

Hopkins, H. S., 1930, Age difference and the respiration in muscle tissues of molluscs, *J. Exp. Zool.* **56:**209–239.

Kennish, M. J., 1976, Monitoring thermal discharges: A natural method, *Underwater Nat.* **9:**8–11.

Kennish, M. J., 1977a, Growth increment analysis of *Mercenaria mercenaria* from artificially heated coastal marine waters: A practical monitoring method, Proceedings of the XII International Society of Chronobiology Conference (Washington, D.C.), pp. 663–669, Casa Editrice "IL PONTE", Milano, Italy.

Kennish, M. J., 1977b, Effects of thermal discharges on mortality of *Mercenaria mercenaria* in Barnegat Bay, New Jersey, Ph.D. thesis, Rutgers University, New Brunswick, New Jersey, 161 pp.

Kennish, M. J., 1978, Effects of thermal discharges on mortality of *Mercenaria mercenaria* in Barnegat Bay, New Jersey, *Environ. Geol.* **2:**223–254.

Kennish, M. J., and Olsson, R. K., 1975, Effects of thermal discharges on the microstructural growth of *Mercenaria mercenaria*, *Environ. Geol.* **1:**41–64.

Levinton, J. S., and Bambach, R. K., 1970, Some ecological aspects of bivalve mortality patterns, *Am. J. Sci.* **268:**97–112.

Lutz, R. A., and Rhoads, D. C., 1977, Anaerobiosis and a theory of growth line formation, *Science* **198:**1222–1227.

MacClintock, C., 1967, Shell structure of patelloid and bellerophontoid gastropods (Mollusca), *Peabody Mus. Nat. Hist. Yale Univ. Bull.* **22:**1–140.

MacClintock, C., and Pannella, G., 1969, Time of calcification in the bivalve mollusk *Mercenaria mercenaria* during the 24 hour period (abstract), *Geol. Soc. Am. Annu. Mtg. Program* **1:**140.

McKerrow, W. S., Johnson, R. T., and Jakobson, M. E., 1969, Palaeoecological studies in the Great Oolite at Kirtlington, Oxfordshire, *Palaeontology* **12:**56–83.

Nelson, T. C., and Haskin, H. H., 1949, On the spawning behavior of the oyster and of *Venus mercenaria* with especial reference to the effects of spermatic hormones, *Anat. Rec.* **105:**484–485.

Pannella, G., and MacClintock, C., 1968, Biological and environmental rhythms reflected in molluscan shell growth, in: *Paleobiological Aspects of Growth and Development: A Symposium* (D. B. Macurda, ed.), *Paleontol. Soc. Mem.* **2** (*J. Paleontol.* **42** Suppl.):64–80.

Reyment, R. A., 1971, *Introduction to Quantitative Paleoecology*, Elsevier, Amsterdam, 226 pp.

Rhoads, D. C., and Pannella, G., 1970, The use of molluscan shell growth patterns in ecology and paleoecology, *Lethaia* **3**:143–161.

Saila, S. B., and Pratt, S. D., 1973, Mid-Atlantic Bight fisheries, in: *Coastal and Offshore Environmental Inventory, Cape Hatteras to Nantucket Shoals*, Marine Publ. Ser. No. 2, University of Rhode Island, Kingston, Rhode Island, 125 pp.

Schmidt, R. R., and Warme, J. E., 1969, Population characteristics of *Protothaca staminea* (Conrad) from Mugu Lagoon, California, *Veliger* **12**:193–199.

Sheldon, R. W., 1965, Fossil communities with multimodal size-frequency distributions, *Nature (London)* **206**:1336–1338.

Tevesz, M. J. S., 1972, Implications of absolute age and season of death data compiled for Recent *Gemma gemma*, *Lethaia* **5**:31–38.

Thompson, I., 1975, Biological clocks and shell growth in bivalves, in: *Growth Rhythms and the History of the Earth's Rotation* (G. D. Rosenberg and S. K. Runcorn, eds.), pp. 149–161, John Wiley & Sons, London.

Thompson, I. L., and Barnwell, F. H., 1970, Biological clock control and shell growth in the bivalve *Mercenaria mercenaria* (abstract), *Geol. Soc. Am. Annu. Mtg. Program* **2**:704.

Thorson, G., 1950, Reproductive and larval ecology of marine bottom invertebrates, *Biol. Rev.* **25**:1–45.

Chapter 8

Environmental Relationships of Shell Form and Structure of Unionacean Bivalves

MICHAEL J. S. TEVESZ and JOSEPH GAYLORD CARTER

1. Introduction	295
2. Unionacean Form	296
2.1. Form–Habitat Relationships	297
2.2. Other Factors	301
2.3. Sexual and Ontogenetic Variability	303
3. Growth Banding	305
3.1. Origin of Bands	305
3.2. Band Information	307
4. Ecological and Evolutionary Significance of Unionacean Shell Microstructure ..	309
4.1. Periostracum and Shell Protection	310
4.2. Adventitious Interior Conchiolin Layers	315
4.3. Evolutionary Limitation of Shell Form	316
4.4. Periostracal Color and Surface Texture	317
References	318

1. Introduction

Most fossil and Recent freshwater bivalves are members of the Unionoida and Corbiculacea, with the great concentration of genera (>150) and species (several hundred) occurring in the unionoid superfamily Unionacea [Permian (?), Triassic-Recent (Haas, 1969)]. Unionaceans are found worldwide in diverse freshwater habitats and show a remarkable amount of inter- and intraspecific shell form variability (see, for instance, Grier,

MICHAEL J. S. TEVESZ ● Department of Geological Sciences, The Cleveland State University, Cleveland, Ohio 44115.   JOSEPH GAYLORD CARTER ● Department of Geology, University of North Carolina, Chapel Hill, North Carolina 27514

295

1920a–c; Stratton, 1960; Haas, 1969; Clarke, A. H., 1973; Tadić,1975; Johnson, 1978). Much of the available freshwater bivalve literature relevant to the theme of this volume is concerned with analyzing the environmental relationships of this variability in the unionacean families Unionidae and Margaritiferidae. The balance of this literature pertains largely to obtaining information from distinctive "growth bands" and shell microstructure.

First, we will discuss the relationship between shell form and environment, paying particular attention to intraspecific variants that appear to be consistently correlated with physical ecological factors. Because form and environment are correlated in unionaceans, tracing certain morphological changes among Recent and fossil populations provides a way of inferring corresponding environmental changes. We will also investigate the general environmental factors that permit the maintenance of the great morphological variability of unionaceans. Understanding these factors may provide information that may allow accurate reconstructions of ancient aquatic ecosystems and thus increase the resolution by which we may infer environmental change through geological time.

Second, we will investigate the nature and origin of growth bands and try to identify what kinds of information they contain concerning the history of a bivalve and its environment. Growth bands appear as dark rings on the shell surface and as distinctive lines observable in shell cross sections.

Finally, we will provide an explanation of the environmental significance of unionacean shell microstructure and the periostracum. Because these are important features, about which there is little literature, we will present new data and interpretations to help complete the understanding of the shell structure of this group.

2. Unionacean Form

The literature on the ecological significance of shell form in unionaceans can be roughly divided into two categories. The first category is comprised of papers that describe correlations between shell form (usually size, outline, and proportions) and habitat (often generally designated as "lake," "large stream," "small stream," and the like). The second category consists of finer-resolution studies that attempt to explain the relationship of a specific aspect of the shell (say, the thickness of an individual valve) to a particular aspect of the environment (e.g., water hardness). We will discuss form–habitat relationships first.

2.1. Form–Habitat Relationships

Ball (1922) and Eagar (1948) provided, for their time, up-to-date reviews of an extensive, largely anecdotal literature on form–habitat relationships in unionaceans. Eagar's review concisely summarized previously published observations (including those of Ball) on the habitat relationships of shell outline and proportions. This summary included the following general observations (also see his references):

Obesity [measured as T/L (see Figs. 1 and 2A)]
1. Many species, when traced within a single stream, tend to increase in obesity downstream.
2. In a given species or considering a stream fauna as a whole, obesity is positively correlated with stream size and negatively correlated with water velocity.
3. In a given species, lake forms tend to be more obese than river forms.

Size [often measured as L (see Fig. 1)]
1. Small-stream or creek species are smaller than those characteristic of large rivers.
2. While within-individual species size may increase upstream, the entire fauna often seems to show a size increase in a downstream direction.

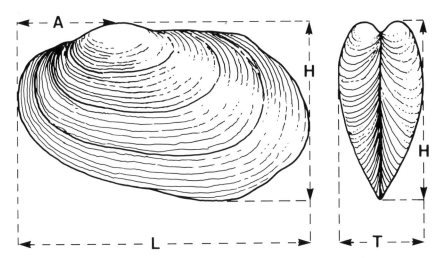

Figure 1. Morphometric definitions. (A) Anterior-to-umbo distance; (L) length; (H) height; (T) shell thickness. For lateral view outline: up = dorsal; down = ventral; left = anterior; right = posterior.

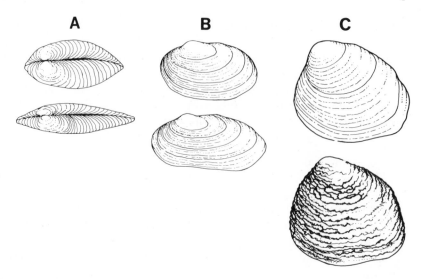

Figure 2. Intraspecific variations. (A) Obesity differences in *Elliptio complanata* (modified from Clarke, A. H., and Berg, 1959); (B) height differences in *Lampsilis radiata siliquoidea* (modified from Baker, 1928); (C) ornamentation differences in *Nyassunio nyassaensis* (modified from Haas, 1936).

    3. Individual members of species occupying exposed positions in lakes are generally smaller than individual members of the same species occupying other habitats.

Height [measured as H (See Figs. 1 and 2B)]
    1. Both inter- and intraspecifically, there tends to be an increase in relative height in species inhabiting lakes or slow-moving bodies of water compared to forms inhabiting more rapidly moving water.

Shell Outline
    1. Both inter- and intraspecifically, curved or well-rounded ventral borders, a long line, and an expanded or well-developed posterior end appear to be more typical of forms living in large lakes, rivers, or slow-flowing bodies of water.
    2. Alternatively, forms with a curved dorsal margin and straight or reflected ventral margin are generally found in relatively small streams or swift-flowing water.

These conclusions have generally been substantiated by subsequent studies. Some examples are given below.

A. H. Clarke (1973) found the largest *Lampsilis radiata siliquoidea,* *Ligumia recta,* and *Proptera alata* in the rivers of the Canadian interior basin. Lakes, by contrast, contained mainly smaller forms. Similarly, Harman (1970) mentioned that *Elliptio complanata, Anodonta grandis,* and

Lampsilis radiata were smaller in the deep, cold Finger Lakes of New York than in warmer, nutrient-rich streams. A. H. Clark (1973) further noted that such species as *Fusconaia flava*, *Quadrula quadrula*, and *Amblema plicata* were often more obese in larger streams and less obese in smaller streams in the southern part of their range. However, both large- and small-stream forms were relatively compressed when they occurred in northern regions. This, he suggested, may be a general trend among unionids.

More correlations between obesity (and other morphometric features) and latitude were observed by Cvancara (1963). *Lampsilis ventricosa*, *L. ovata*, and *L. excavata* from different latitudes were measured and the data represented on scatter plot diagrams (e.g., his Figs. 3 and 4). These diagrams showed that the values for height/total length and width/total length (= obesity as used here) increased toward lower latitudes. Although all three "species" did not reflect these trends individually, Cvancara suggested that further research could demonstrate that the three forms may represent only one species.

On the subject of length/height proportions, A. H. Clarke (1973) stated for certain *Anodonta* that values of the anterior-to-beak/length ratio were highest in specimens collected from large-lake habitats. Hendelberg (1960), after measuring the length/height ratio of 112 *Margaritifera margaritifera* from a Swedish river and then discussing literature relevant to the environmental "meaning" of the ratio, concluded that this meaning was still obscure. But he mentioned that pearl fishers along this river, Pärlälven, referred to anteroposteriorally elongate shells with concave ventral margins as "rapid shells." Similarly, Dell (1953) observed an increase in the height/length ratio from running-water to standing-water habitats in the genus *Hyridella* from New Zealand. Finally, Cvancara (1972) showed a trend of decreasing shell height with depth in *Anodontoides ferussacianus* and *Lampsilis radiata siliquoidea*, but not in *Anodonta grandis*, in Long Lake, Minnesota.

One aspect of the shell that influences form, but was not covered in the review by Eagar (1948), is shell sculpture. The most dramatic examples of habitat-correlated variability in this feature were reported by Haas (1936, 1969). Within single species of African unionaceans, he found lake-dwelling forms strongly sculptured and stream-dwelling forms smooth (Fig. 2C). However, Ortmann (1920) found no definite trends in this feature among species inhabiting the Ohio drainage region.

If a major change occurred in a particular environment, how quickly would the change be reflected in the shell form of the resident unionids? The damming of a river and subsequent formation of a reservoir provide, in the minds of some workers (Baker, 1928), a "natural experiment" whereby the response of a unionid fauna to rapid environmental change may be observed. Generally, such a change causes the local extinction of most of the original stream-dwelling unionids (Isom, 1971; Harman, 1974). Nevertheless, Baker (1928) showed for some man-made lakes in Wisconsin that *Anodonta grandis* and *Lampsilis radiata siliquoidea* apparently tolerated this kind of environmental change. Moreover, the surviving populations of both species showed shell-form differences compared to conspecific stream-dwelling forms. For example, compared to the riverine morphs, *A. grandis* in the new lakes were relatively short and wide; *L. siliquoidea* from the impoundments were relatively laterally compressed and high. Apparently, the populations surviving in the newly created lakes quickly assumed some of the shell proportions characteristic of typical lake-dwelling forms.

Evidence casting doubt on the generality (and, perhaps, validity) of Baker's findings was presented by van der Schalie (1936) when he described the riverine morph of *Lampsilis ventricosa* inhabiting Carpenter Lake, Michigan. He concluded that this unusual form–habitat association was caused by humans introducing into the lake fish parasitized by glochidia larvae derived from river-dwelling unionids.

Because human interference is thus a potential source of influence in these "natural experiments" (also see Kessler and Miller, 1978), the usefulness of unionacean shell form in documenting or inferring rapid environmental change is still unknown.

Knowledge of general form–habitat relationships of Recent unionaceans has been used extensively in paleoecological studies. Early studies employing this information include those of Davies and Trueman (1927) and Leitch (1936). The most prominent and prolific writer of late has been Eagar (e.g., 1948, 1953, 1974). These works all have a common focus, since they deal mainly with Carboniferous, nonmarine (fresh or brackish water or both) unionoid (but nonunionacean) bivalves of the family Anthracosiidae.

Eagar's work is the most detailed. Typically, he presents literature summaries of form–habitat information for modern unionaceans. Also, he describes the shell form in the fossil unionoids and places them in a stratigraphic setting. He then uses Recent form–habitat information along with stratigraphic information to help reconstruct the habitat and life position of the fossil forms. With this information, he can infer environmental changes among strata.

A common deficiency of most form–habitat studies of Recent and fossil unionoids is that the functional meaning of the form–habitat correlations is not well understood. The absence of precise ecological data has forced workers interested in discerning habitat controls on unionid proportions, size, and outline to make numerous assumptions concerning the nature of the habitat in which the bivalves were found. Large streams and lakes, for instance, are assumed to have soft, muddy substrata and slow currents. By comparison, small streams are thought to have coarser bottoms and swift currents. Through the same reasoning, individual streams are considered to take on "large-stream" or "lake" characteristics toward the mouth and "small-stream" characteristics toward the source.

Working with such premises, Wilson and Clark (1914) suggested that the typically compressed form of small-stream unionaceans was adaptive for crawling in gravel and withstanding strong currents. Also, they believed that the more obese forms of large rivers were adapted, through their low bulk density, to soft bottoms and weak currents. A. H. Clarke (1973) cautiously suggested that perhaps more obese specimens had a low surface area/volume ratio and thus were less likely than narrower forms to be dislodged from sediments by shifting currents that occur in lakes during storms. In rivers, compressed forms oriented parallel to the current were presumably less susceptible to disinterment than inflated forms.

It would be easy to propose several more hypotheses to explain form–habitat relationships of the unionacean shell, provided enough assumptions were made about the habitat. But as Ball (1922) pointed out, accurately generalizing the ecological conditions in a given freshwater body is difficult, not only because conditions vary within the environment, but also because the bivalves may inhabit an atypical portion of the environment. For example, large streams may have extensive areas of sand and gravel bottoms, and obese forms from these streams are likely to occur in such areas. Also, Fisher and Tevesz (1976) observed that while most of the bottom of Lake Pocotopaug, Connecticut, is soft mud, the great majority of unionid bivalves in that lake are restricted to a narrow band of coarse, hard substratum encircling the lake at depths of less than one meter.

We conclude that unionacean shell proportions, size, and outline are correlated with broad aspects of the environment. Thus, particularly in paleoecological studies, unionacean shell form may be a useful indicator of environmental change. Nevertheless, at present, functional morphological explanations of these form–habitat correlations are largely speculative.

2.2. Other Factors

Temperature is the most extensively studied environmental factor known to influence the shell. Its relationship to shell growth and size has been particularly well documented by Howard (1922) for North American unionids, by Negus (1966) for British unionids, and by Alimov (1974) for Russian freshwater bivalves (including unionids). In general, unionids from temperate climates grow only during the warmer months (e.g., April–October), and within the same general kind of habitat, specimens living in warm waters often grow faster and attain larger sizes than those living in cold waters (also see Harman, 1970).

The presence of "lime" or "hard water" is often correlated with valve thickness. A. H. Clarke (1973), for one, found that hard water often supported heavier specimens of certain species. Additionally, McMichael (1952, p. 351) reported for Australian unionaceans that "mussels from rivers whose waters are poor in lime are often quite thin and weak, while the same species from a river rich in lime will have a strong, thick shell." Similarly, Coker et al. (1921) explained the presence of thin, highly etched shells in eastern North American unionids as being caused by relatively acidic water (Fig. 3A).

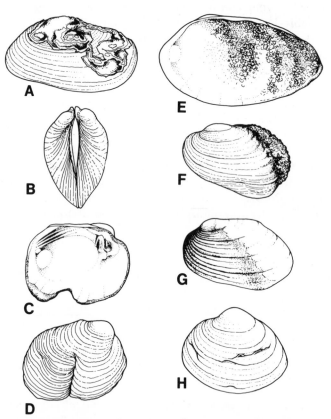

Figure 3. Shell aberrations. (A) Etching by acidic water; (B) gaping margins; (C) notched outline (after Coker et al., 1921); (D) furrowed shell surface (after Coker et al., 1921); (E) abscessed shell; (F) algal-related protuberances; (G) eroded shell posterior; (H) repaired shell.

Agrell (1949) believed "trophic degree of the environment" to be related to shell form. For Swedish unionids, he found that such features as anterior development, relative height, and obesity are positively correlated with trophic degree. In the same general geographical area, Björk (1962) found growth rate higher and "quantitative development" better in streams with a relatively high trophic degree and relatively hard water.

A particularly detailed study by Green (1972) showed a complex of environmental factors correlated with the condition of individual shell features. Through use of a multivariate statistical approach, he demonstrated that low-NaCl, high-alkalinity-relative-to-pH environments were inhibited by relatively thick-valved *Lampsilis radiata siliquoidea*. Conversely, specimens from high-NaCl low-alkalinity-relative-to-pH environments were characterized by thin valves. His statistical analysis also showed an inverse relationship between size and water turbulence. Moreover, he showed that such factors as shell height, overall shell thickness, and inside shell volume were conservative morphological variables with respect to measured environmental varaiables.

Certain shell-form aberrations can be directly linked to particular environmental causes. Gaping margins (Fig. 3B), a permanently notched outline (Fig. 3C), and a furrowed surface (Fig. 3D) were attributed by Coker *et al.* (1921) to an infestation of the mantle by parasitic mites.

These workers also explained that massive, asymmetrical protuberances on the external surface of mussels from lakes were formed as a result of accumulated lime and algae (Fig. 3F). Algae growing on the tip of the shell exposed above the substratum also produce an eroded shell surface (Fig. 3G).

Decksbach (1957) described inner shell surfaces that become abscessed when sedimentary particles were trapped between the mantle and shell. Extensive abscessing produced a shell with irregular outlines (Fig. 3E). Shell breakage and repair also produces aberrantly formed or marred shells [personal observation (Fig. 3H)]. One would intuitively feel that such breakage would be expected to be most prevalent in physically rigorous environments associated with hard substrata.

2.3. Sexual and Ontogenetic Variability

Unionaceans, especially those belonging to the subfamily Lampsilinae, are often markedly sexually dimorphic in shell features. For example, Ball (1922) showed that male or female forms of various species differed markedly in obesity. Also, Brander (1956) and Johnson (1978), among others, observed that such features as size and the development and shape of the posterior margin were often sexually controlled (Fig. 4).

Ontogeny has similarly marked effects on shell morphology. Both Ball (1922) and Ortmann (1920) found young unionids to be more obese than older specimens. Also, many of the environmentally distinctive aspects of form become apparent only after a bivalve has reached sexual maturity. Thus, spatial segregation of age classes or sexual morphs could produce a morphologically distinctive grouping that might be correlated with, but not necessarily causally related to, environmental factors.

As a final caveat, consider the remark by Eagar (1948) that changes in one shell dimension are often accompanied by perhaps compensating changes in others. Height, for example, generally is positively correlated with obesity (also see Tolstikova and Orlov, 1972; Hendelberg, 1960). This morphological covariance adds to the difficulty in discriminating between fortuitous, as opposed to causally related, correlations between form and environment.

While the genotypic and phenotypic controls on specific aspects of morphological variability in unionaceans are still incompletely known, it is nevertheless possible to provide an explanation of why this extensive variability can be maintained. Vermeij (1974) concluded from his analysis of an extensive literature that a decrease in competition often resulted in an increase in diversity within a higher taxon. At lower taxonomic levels, abnormalities and variants survived in species that either were under low levels of competition or predation or were relatively unspecialized.

Unionaceans live in environments in which competition and predation levels are low, and unionaceans may also be considered unspecialized with respect to the utilization of resources such as food and sub-

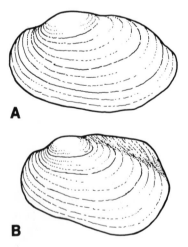

Figure 4. Sexual dimorphism in *Lampsilis radiata siliquoidea.* (A) Male; (B) female.

stratum or space (Coker *et al.*, 1921; Tevesz and McCall, 1978, 1979). Thus, it may be inferred that low biotic selective pressure, an inherent feature of their environment, is one of the factors that permits the maintenance of unionacean variability. In addition, the lack of specialization for particular resources in this group may also be related to low selective pressures. For example, numerous authors (see the reviews by Diamond, 1978; Tevesz and McCall, 1978, 1979) have shown that reduced levels of competition and predation often resulted in decreased specialization (niche-width expansion) of various species. This has been particularly well demonstrated in comparing mainland (high competition) vs. island (low competition) bird populations

This relationship among morphological variability, degree of specialization, and selection pressure has important paleoecological implications. The tendency among paleontologists in reconstructing ancient ecosystems has often been to gather life-habit information from Recent environments and apply this information directly to reconstruct the life habits and community dynamics of fossils. If, however, predation and competition levels of the modern environments differ significantly over geological time, then the "transferred-ecology" approach could lead to erroneous paleoecological inferences. It is therefore important to try to reconstruct these selective pressures directly from the fossils and their associated rock strata (Tevesz and McCall, 1978, 1979). In this respect, a thorough knowledge of the inter- and intraspecific variability of the fossil assemblage may prove informative.

3. Growth Banding

3.1. Origin of Bands

Examination of the outer surface of unionacean valves almost always reveals conspicuous dark-colored bands that extend circumferentially from the umbo and are similar in form to a view of the shell outline normal to the saggital plane. These bands may be divided into two varieties: wide, darker, macroscopically obvious bands occurring at fairly regular intervals and fainter, less dark, more irregularly spaced bands (Fig. 5). The same structures are also represented as dark lines in valve cross sections.

The significance and mode of formation of these bands were extensively discussed by Coker *et al.* (1921) (but also see, for example, von Hessling, 1859; Hazay, 1881; Israel, 1911; Ekman, 1905; Lafevre and Curtis, 1912; Isley, 1914; Altnöder, 1926; Crowley, 1957). Coker and his colleagues found that picking up a mussel and measuring it (i.e., disturbing it) invariably caused a band to be formed on the shell before shell growth

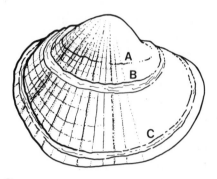

Figure 5. Annual (B, C) and pseudoannual (A) growth bands.

was resumed. On the basis of this observation and subsequent analysis of valve cross sections, they explained the formation of the bands as follows: The unionid shell grows by the deposition of successive laminae of $CaCO_3$ from the surface of the mantle, and the growing edges of the shell are deposited by the mantle margin. However, if the mantle margin withdraws within the shell, away from the growing edge, then when the new layers of prismatic shell and periostracum are deposited, they are not continuous but are overlapped by the old shell and periostracum (Fig. 6). This doubling-up of layers, a result of mantle retraction and reextension, produces the visual appearance of a dark "band" on the shell. The faint, irregularly spaced, "pseudoannual" type of band may be explained in this way.

Coker *et al.* (1921) further believed the darker, more regular bands to be formed by growth retardation due to repeated disturbance by low temperatures. They reasoned that if these bands are similarly formed and the discontinuity of the outer layers is caused by the withdrawal of the mantle because of cold weather, then one would expect several overlapping shell layers before a winter season, because the onset of seasonal cold weather is preceded by alternate warm and cold periods. Thus, these darker bands, or annual rings, are the result of frequent growth interruptions that produce multiple "doublings-up" of the shell along growth edges (Fig. 7).

Figure 6. Cross-sectional view of pseudoannual band. Vertical elements: prisms; horizontal elements: nacre. After Coker *et al.* (1921).

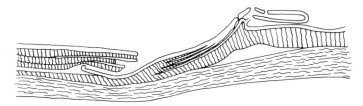

Figure 7. Cross-sectional view of annual band. Upper clear areas: periostracum; vertical elements: prisms; horizontal elements: nacre. After Coker *et al.* (1921).

3.2. Band Information

Several techniques make it easy to discriminate between the "annual" and "pseudoannual" bands of unionaceans (see Appendix 1.A.4). Thus, absolute age information is readily available for species from temperate latitudes. This information, in conjunction with information on population densities and shell weights and dimensions, has been used by many workers to infer life-history parameters such as growth rates, population age structure, annual production, recruitment and mortality rates, and season of death (see particularly Negus, 1966; but also see, among many others, Chamberlain, 1931; Tudorancea and Florescu, 1968a,b, 1969; Tudorancea and Gruia, 1968; Magnin and Stanczykowska, 1971; Harangy et al., 1964, 1965; Ray, 1977).

The life-history information that may be obtained because these bands yield absolute age information again exemplifies how unionaceans are potentially useful in monitoring environmental change. For example, because much information is already available on unionacean population age structure and mortality rates, new information on population dynamics that differs substantially from long-term averages may possibly be related to a change in the environment, for instance, the introduction of a pollutant. Also, once the typical season of death in certain unionaceans has been established, one could analyze the season of death from shells in a suspected "kill" and make inferences as to whether or not the mortality was natural or caused by an unusual modification of the environment.

An interesting way of relating environmental temperature change to band information also comes from the work of Negus (1966). In a very informative diagram (her Fig. 10), she plotted average percentage of departure from mean growth for a particular bivalve year class against the average departure from mean average temperature during growing season for each calendar year. This plot showed that slower growth during a particular year, as reflected by relatively closely spaced annual bands, was generally associated with a relatively cool growing season for that year. These findings invite speculation that unionaceans may be used as relative paleothermometers. However, Stansbery (1970) found that unionids inhabiting deep water and fine-grained sediments in lakes (i.e., according to him, a situation associated with reduced current) tended to grow slower, live longer, and show less ontogenetic change in growth rate than specimens from shallower depths. Thus, the spacing of annual bands may also be related in part to the physical rigor of the habitat.

Grier (1922) and Brown *et al.* (1938) noted the regular spacing and distinctness of all bands in lake-dwelling, as opposed to stream-dwelling, unionids. This condition resulted from a paucity of pseudoannuli on lake-dwelling forms. Stansbery (1967) explained that the stabilizing effects of a large lake ensured gradual uniform temperature changes, thereby largely eliminating the rapid temperature fluctuations that are an important cause of pseudoannual band formation. Thus, the number of pseudoannuli produced per year per unionid is inversely related to the temperature stability of the environment. This, in turn, is often related to the size, depth, and circulation patterns of a given body of water.

Because unionids concentrate trace elements, the shell, like that of other bivalves, is a repository of chemical information. If it were possible to analyze the concentration of various substances in an individual annual layer, then the shell might provide a chronological record of relative concentrations of those elements or compounds in the ambient environment. The presence of growth bands and their organic component help to facilitate such analyses. According to Sterrett and Saville (1974), several interruptions in the formation of the shell in the spring and fall produce an area of discontinuity visible as a dark line. The greater amount of periostracum at this border results in an area higher in protein than in other parts of the shell. Theoretically, if the shell is baked at high temperatures, the protein of the periostracum will ash, and the $CaCO_3$ will not, leaving a weak spot or crack that delineates part of the shell formed in a single growing season. A methodology for separating the annual layers based on this reasoning has been used by Nelson (1964) and Sterrett and Saville (1974) (see Appendix 1.A.4). The ease with which these layers may be separated makes the chemical concentrations in the shell for particular years readily analyzable and thus facilitates monitoring yearly changes in the chemical environment. The significance of these chemical concentrations in bivalves in general is discussed in Chapter 4.

Counting growth bands on the shell surface and in shell cross sections is not the only way of determining absolute age and thus the population dynamics of unionaceans. European workers have frequently employed the method of Wellmann (1938), whereby annuli are counted on the ligament (see, for example, Björk, 1962; Hendelberg, 1960). However, this method is limited in usefulness because annuli on the ligament are destroyed by erosion of the anterior portions of the ligament, which is a common phenomenon in unionaceans. This makes it necessary to esti-mate lost information. Over the long term, this method suffers because the entire ligament is much more easily destroyed than the shell.

Finally, the study of the significance of unionacean growth bands has been confined mainly to temperate latitudes, where seasonal cold stops growth for prolonged periods and is responsible for annual band forma-tion.

But what about warmer climates? To the best of our knowledge, the significance of growth banding in tropical unionaceans is little studied. Warm-climate unionids do form distinctive bands, but the causes behind the formation of these bands are different than for temperate species. For example, McMichael (1952, p. 351) mentions the following regarding Australian unionaceans: "During severe droughts, the rivers and creeks dry up, and the shells may become dormant. Fresh water mussels will then bury deeply into the mud and their shells will cease growing, be-coming thickened at the edge. When rain comes again, the shells may emerge and grow quickly, leaving a thick growth line on the shell which marks the dormant period."

Other seasonal effects in the tropics, such as extensive rain, could also conceivably cause growth interruptions. Stream-dwelling unionids, for instance, could be subjected to the effects of abnormally high amounts of suspended particulate matter during repeated floods caused by mon-soons. A prolonged stormy season could likewise cause lake-dwelling forms to be subject to abnormally long periods of agitated conditions. These and related phenomena could disturb the mussel enough to cause repeated and prolonged retractions of the mantle, thus producing a thick and distinctive band. Therefore, it is necessary to know the general cli-matic setting from independent evidence (e.g., temperate vs. tropical) before growth bands on unionaceans can be accurately interpreted.

4. Ecological and Evolutionary Significance of Unionacean Shell Microstructure

Unionaceans have a primitive shell microstructure consisting of outer prismatic and inner nacreous shell layers, with the former deposited on a nonreflected shell margin. They probably inherited this microstructure from the Trigoniacea, a superfamily inhabiting marine and estuarine en-vironments and now restricted to only a few Indo-Pacific species (Cox, 1960; Newell and Boyd, 1975). Unlike other major groups of freshwater bivalves (e.g., Corbiculacea, Dreissenacea), the Unionacea have appar-ently never evolved crossed lamellar microstructure, and have retained an outer simple prismatic to composite prismatic shell layer comparable to that in many Trigoniacea (Boggild, 1930; Beedham, 1965; Taylor et al., 1969, 1973; Cox, 1969; Tolstikova, 1972, 1973, 1974; Newell and Boyd, 1975). The Unionacea are likewise conservative in terms of the variability of their shell microstructure. Despite their considerable generic diversity, unionaceans show only minor interfamilial differences in this feature, represented largely by changes in the internal arrangement of the outer layer prisms and by the occurrence of adventitious thick organic (conchiolin) laminae within the nacreous layer of certain species (Tolstikova, 1974).

Like the shell of most other bivalves, the unionacean nacroprismatic shell provides protection of the soft parts, support of the ctenidia and viscera, an assist in burrowing, and a source of $CaCO_3$ for buffering metabolic acids during periodic anaerobic metabolism. In terms of protection, their nacreous microstructure may have been retained because of its relatively high mechanical resiliency in comparison with certain other common inner-layer microstructures [e.g., complex crossed lamellar and homogenous (see, for instance, Taylor and Layman, 1972)]. Many unionaceans inhabit streams, and their shallow burrowing habit exposes the shell posterior to physically rigorous conditions, especially during floods. Unionaceans may additionally have retained a primitive nacroprismatic shell microstructure because this offers maximum protection of their shell from dissolution in fresh water. Their vertical to reclined outer layer prisms show relatively thick interprism organic matrices, and the horizontal organic laminae separating nacreous laminae may likewise retard shell dissolution. Perhaps the most important adaptations of unionacean shells, with regard to preventing shell dissolution, are their persistent periostracum and occasional deposition of thick, horizontal laminae within the nacreous layer.

4.1. Periostracum and Shell Protection

The adaptive value of the unionacean periostracum for preventing exterior shell dissolution is clearly seen where this layer is abraded. In this case, dissolution of the underlying prismatic and nacreous layers occurs. This results in solution pits on the shell exterior, some of which penetrate through most of the shell thickness at the umbos (Figs. 3A and 8). Thick periostraca are rarely encountered in the Unionacea, but most species in this superfamily show a periostracum of intermediate thickness in the adult shell. By the present definition, intermediate-thickness periostracum is between 15 and 50 μm as measured near the margin of the adult shell. Thin periostraca (<15 μm) are equally rare in this superfamily.

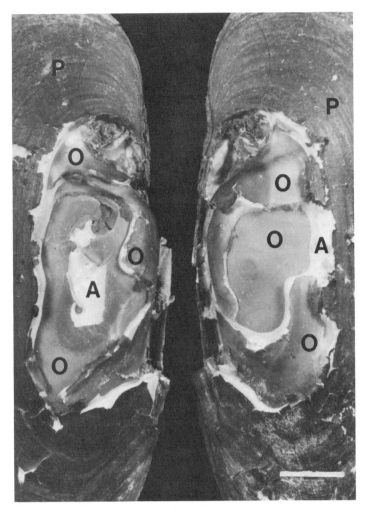

Figure 8. Natural dissolution of the shell exterior in the margaritiferid *Margaritifera falcata* (Gould) from the Ozette River, Washington. Yale University Peabody Museum No. 10158. (A) Aragonitic prismatic and nacreous shell layers; (O) prominent adventitious organic laminae sporadically distributed among the nacreous laminae; (P) periostracum. Scale bar: 1.0 cm.

This uniformity of periostracal thickness probably reflects the fact that thinner periostraca are not sufficiently protective, whereas thicker periostraca tend to peel away from the shell exterior after repeated desiccation.

Compared with most other bivalves, many unionaceans are remarkable for the resistance of their periostracum to peeling and cracking after

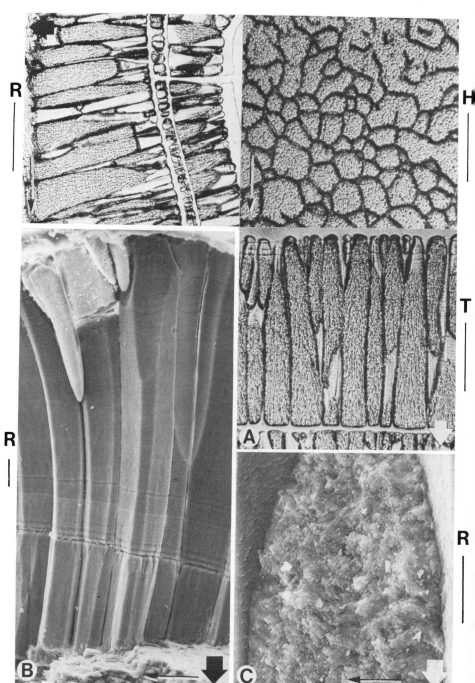

drying. Two other bivalve surperfamilies show this property, i. e., certain representatives of the Mytilacea and, to a lesser degree, the Trigoniacea. In mytilids such as *Mytilus edulis* Linnaeus, resistance to desiccation peeling is improved by the incorporation of fluid-filled vacuoles in the center of the periostracal layer. These vacuoles maintain the moisture of the periostracal layer with subaerial exposure, thereby preventing differential contraction of the outer and inner periostracal sublayers (Carter, 1976). In contrast, trigoniacean and unionacean periostraca are typically nonvacuolated, but they show a strong structural bond with the underlying prismatic layer. As described by Taylor *et al.* (1969) and illustrated in Figs. 9 and 10, the unionacean (and trigoniacean) periostracum conforms with the end of the prisms and extends deeply into the prismatic layer. This forms a strong bond between these two layers and retards peeling when the exterior of the periostracum contracts through drying. Consequently, even air-dried museum specimens of most unionacean species resist periostracal peeling and cracking, except for the species with unusually thick periostracal layers (e.g., certain *Cumberlandia* and *Amblema*). In nonunionacean and nontrigoniacean bivalves with nonreflected shell margins (e.g., most Mytilacea, Ostreacea, Pectinacea, and Myacea), the initial spherulites of the outer prismatic shell layer are deposited on a more or less planar inner periostracal surface. In these instances, the periostracum supports the initial spherulites and excludes environmental contaminants, but it does not envelop the ends of the prisms, nor does it extend into the prismatic layer to the extent that this occurs in the Unionacea and Trigoniacea (see, among others, Clarke, G. R., 1974).

The trigoniacean periostracum is generally thinner over the prism ends than in the Unionacea. Additionally, the trigoniacean periostracum does not extend as deeply into the outer prismatic shell layer. It therefore seems likely that the trigoniacean periostracal–prism relationship provided only a preadaptation for the evolution of a typical unionacean periostracum–prism layer bond. Interestingly, aside from thickening the periostracum and bonding this more strongly with the underlying prismatic shell layer, unionaceans have tended to reduce the prominence of apical

←
___

Figure 9. Outer prismatic shell layer in the unionid *Elliptio complanata* (Solander) from the Sangerfield River near Hamilton, New York. University of North Carolina No. 4923. The three photographs in the upper and middle right parts of the plate (A) comprise a three-dimensional view of the prismatic layer in the shell posterior, viewed in radial vertical (R), transverse vertical (T), and horizontal (H) sections (acetate peels of polished and etched sections). Scale bars: 50μm. (B, C) Two photographs representing radial vertical fractures through the outer prismatic shell layer viewed by scanning electron microscopy. The thick arrows indicate the direction toward the depositional surface; the thin arrows indicate the direction toward the posterior shell margin. Scale bars: 2 μm.

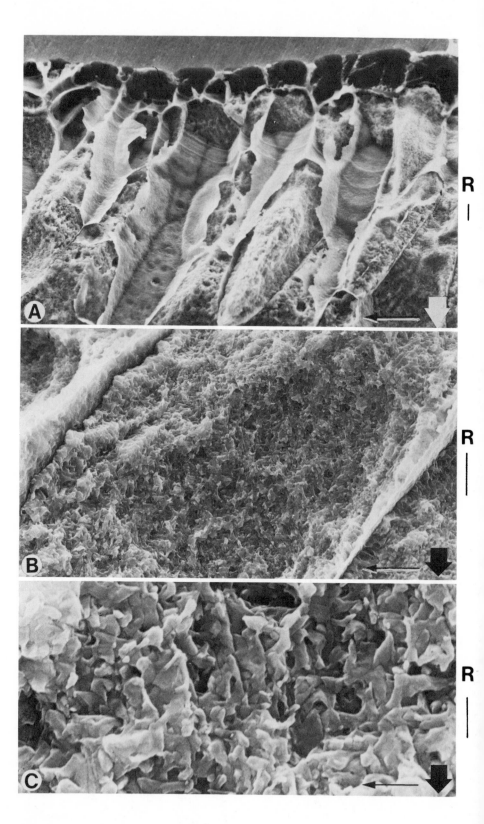

R

R

R

bosses that characteristically occur on the ends of trigoniacean prisms. These apical bosses are generally pyramidal and often structurally distinct from the underlying prism columns (see Taylor et al., 1969; Newell and Boyd, 1975). Loss or reduction of these structures is adaptive in freshwater bivalves because it reduces the likelihood of exposure of the shell $CaCO_3$ as a result of abrasion over the apical bosses. On the basis of this evolutionary trend, it may be possible to identify fossil trigoniaceans evolving microstructural adaptations for life in acidic fresh water. These transitional trigoniacean–unionacean species should show progressive reduction of their prism apical bosses and increased separation of their outer prism columns, i.e., indicative of a more deeply penetrating periostracal layer.

4.2. Adventitious Interior Conchiolin Layers

Thick, horizontal laminae of organic material (conchiolin) occur as adventitious sublayers within the nacreous layers of a large number of unionaceans. These sublayers can occur sporadically over the interior surface of the valve, but they are most commonly observed as irregular patches in or near the umbonal cavity. In addition, these organic deposits are most commonly found among unionaceans with relatively thin shells in the umbonal area (e.g., in thin-shelled species of the genera *Alasmidonta, Cristaria, Iridina, Elliptio,* and *Margaritifera,* but less commonly in thicker-shelled species of the genera *Pleurobema, Plethobasus,* and *Obovaria*). These organic sublayers can be deposited in response to shell penetration through chemical dissolution or as responses to contaminants entering between the shell and mantle (e.g., extraneous water, sediment, or other material entering through an abnormal separation between the mantle margin and the shell margin). Beedham (1965) suggested that inner-layer conchiolin patches in older *Anodonta* are secreted in response to outside water moving through cracks in the anterior part of the hinge. Beedham states that in *Anodonta,* secretion of the conchiolin patches is followed by the secretion of an underlying prismatic layer prior to resumption of normal nacre deposition. This sequence is identical to that which develops in the repair of damaged shell margins in this genus (Beedham, 1965). In the case of certain deeply eroded *Margaritifera falcata* (Gould), the distribution of the interior conchiolin patches clearly indicates that they are being deposited as a reaction to deep solution pitting

←

Figure 10. Scanning electron micrographs of radial vertical fractures through the outer prismatic shell layer of the unionid *Obliquaria reflexa* Rafinesque from the Meramec River, St. Louis, Missouri. Yale University Peabody Museum No. 9749. The periostracum appears in the upper part of (A). A higher magnification of the prism fracture in (B). The thick arrows indicate the direction toward the depositional surface; the thin arrows indicate the direction toward the posterior shell margin. Scale bars: 5 μm (A, B) and 1 μm (C).

over the umbos. But in other *M. falcata,* and in many other thinner-shelled unionaceans, similar but less abundant organic patches can be observed even in shells lacking appreciable umbonal dissolution. In some instances, these patches are superficially smooth and do not extend to the shell margins or hinge, suggesting that they have not been deposited in response to water leakage or a foreign body contaminant. Thus, it appears likely that unionaceans can deposit patches of adventitious conchiolin near their umbonal region as a safeguard against possible deep umbonal dissolution, in addition to utilizing the same strategy to seal over complete shell penetration and foreign contaminants. In some instances, these prophylaxis conchiolin layers can be distinguished from damage-response layers by their lack of an underlying prismatic layer prior to the resumption of normal nacre deposition.

The deposition of prophylaxis and damage-response conchiolin layers in unionacean shells can grossly alter the total shell conchiolin/CaCO$_3$ ratio. This fact must be considered when conchiolin/CaCO$_3$ ratios (or "scleroprotein"/CaCO$_3$ ratios) are utilized as indicators of paleosalinity, as attempted by Kolesnikov (1970). It may be that prophylaxis conchiolin layers are deposited by certain species with greater frequency and in thicker layers in more acidic freshwater environments. However, this aspect of their formation has yet to be investigated, and is at present entirely speculative.

4.3. Evolutionary Limitation of Shell Form

In addition to protecting the shell valves from exterior dissolution, the unionacean periostracum forms a water-tight seal at the shell margins on shell closure. This aids in water retention within the mantle cavity in species periodically exposed to subaerial conditions, and it effectively excludes unfavorable water conditions when these arise. The gasketlike role of the unionacean periostracum is especially useful because these species have tapering (i.e., nonreflected) shell margins. Consequently, without a periostracal seal, minor damage to their thin margins could seriously impair their ability to seal the mantle cavity and protect their mantle margins on shell closure. Bivalves with nontapering (i.e., reflected) shell margins can secrete a thick shell margin, thereby reducing the likelihood of shell-margin breakage (e.g., *Mercenaria* and many other Veneracea), and thus generally show only a thin periostracum.

It is possible that the Unionacea have been evolutionarily limited to a tapering shell margin because of their unique periostracum–prism adaptation for preventing periostracal peeling. With a fully reflected shell margin, a unionacean would necessarily deposit its outer prismatic shell layer directly on the curved surface of the previously deposited prismatic layer, rather than on the underside of the periostracum. Consequently, it would be impossible for such a unionacean to develop a deep interpenetrating bond between the outer ends of the simple prisms and the periostracal layer. Their periostracum, as in veneroid bivalves with fully reflected shell margins, would then be more prone to peeling and cracking on desiccation. Most living veneroid bivalves with fully reflected shell margins are seldom exposed to such conditions, but air-dried museum specimens are evidence of the transient bond between their periostracum and the underlying shell. Inasmuch as strongly crenulated shell margins and sharp ornamental features would increase the likelihood of local abrasion through the periostracum, the subdued ornament typical of unionacean shells may likewise be attributed, in part, to natural selection for resistance to shell dissolution in an acidic environment. Even in comparison with their relatives in the Trigoniacea, the modern and fossil Unionacea show only subdued and generally knobby or rounded ornament with the exception of rare spinose species such as *Canthyria spinosa* (Lea). Utilizing a similar shell microstructure and tapering shell margin, the Trigoniacea have evolved radial costae or plicae, concentric ridges, or even divaricate ornamentation (see, for instance, Newell and Boyd, 1975).

4.4. Periostracal Color and Surface Texture

Two features of the unionacean periostracum that have yet to be analyzed adequately for their ecological and evolutionary significance are color and exterior surface texture. After studying periostracal color in unionaceans from Lake Erie and the Upper Ohio River drainage system, Grier (1920c) concluded that color variation is often greater than standard species descriptions generally indicate. He suggested that this color variability might be related to aging, water acidity, and water silt content. He noted downstream and aging effects in which yellows and greens darken to deep browns and blacks. Other investigators have noted an aging effect in periostracal color among marine bivalves Ockelmann (1958) described an orange, yellow-brown, or red-brown periostracum in juvenile *Astarte*. In contrast, the periostracum secreted by the adult is darker brown to black. Holme (1959) mentioned a similar aging effect for *Lutraria* in which the periostracum changes from colorless in juveniles to brown when secreted by adults. With regard to salinity, Fischer (1887) noted that olive-green is especially common in freshwater bivalves. Carter (1976) noted that greens are especially common in periostraca of the two largest freshwater bivalve superfamilies (Unionacea and Corbiculacea), and that

greens are also common, but less ubiquitious, among two marine superfamilies (Nuculanacea and Solenacea). Inasmuch as nuculanacean shells
may be exposed to acidic sedimentary environments, and solenaceans
often inhabit estuarine environments, the possibility arises that green coloring is more common in all four goups as a result of an acidity or salinity
effect. A. H. Clarke (1973) has noted that green coloring in *Anodonta* is
associated with muddy substrata. Inasmuch as muddy substrata commonly contain abundant decomposing organic compounds, and a resulting acidic sedimentary environment, this would appear to reinforce the
idea of a pH effect on periostracal color.

Periostracal layers in the Unionacea are, by and large, generally superficially smooth and macroscopically featureless, except for occasional
wrinkles. However, a number of species, especially in the families Mu-
telidae and Etheriidae, show radial, concentric, or intersecting microscopic ridges on the surface of the periostracum (Simpson, 1900; Marshall,
1926; Bonetto and Ezcurra, 1965; Carter, 1976). According to Carter (1976),
these features are generally expressed only in the surface of the periostracum, and do not represent crenulations affecting the entire thickness
of the periostracum. The ecological and evolutionary significance of these
structures, like that of periostracal color, has yet be to be documented.

ACKNOWLEDGMENTS. We wish to thank Mr. Michael Ludwig for drafting
Figs. 1–7 and Dr. A. H. Clarke for reviewing the manuscript.

References

Agrell, I., 1949, The shell morphology of some Swedish unionides as affected by ecological
conditions, Ark. Zool. **41A**:1–30.

Alimov, A. F., 1974, Regularity of growth in freshwater bivalve molluscs, Zh. Obshch. Biol.
35:576–589 (in Russian, English summary).

Altnöder, K., 1926, Beobachtungen über die Biologie von *Margaritana margaritifera* and
über Ökologie ihres Wohnorts, Arch. Hydrobiol. **17**:423–491.

Baker, F. C., 1928, The influence of changed environment in the formation of new species
and varieties, Ecology **9**:271–283.

Ball, G. H., 1922, Variation in freshwater mussels, Ecology **3**:93–121.

Beedham, G. E., 1965, Repair of the shell in species of *Anodonta*, Proc. Zool. Soc. London
145:107–124.

Björk, S., 1962, Investigations on *Margaritifera margaritifera* and *Unio crassus*, Acta Limnol.
4:5–109.

Bøggild, O. B., 1930, The shell structure of the mollusks, K. Dan. Vidensk. Selsk. Skr. Naturvidensk. Math. Afd. 9 **2**:231–326.

Bonetto, A., and Ezcurra, I. D., 1965, Malacological notes. III: 5. The periostracum sculpture
in the *Anodontites* genus; 6. The lasmidium of *Anodontides trapezeus* (Spix); 7. The
lasmidium of *Mycetopoda siliquosia* (Spix), Physis (Buenos Aires) **25**:197–204.

Brander, T., 1956, Zur Frage der Geschlechtsdimorphismus bei den Unionazeen, *Arch. Molluskenkd.* **86**:177–181.

Brown, C. J., Clarke, C., and Gleissner, B., 1938, The size of certain naiades in western Lake Erie in relation to shoal exposure, *Am. Midl. Nat. Monogr.* **19**:682–701.

Carter, J. G., 1976, Ph.D. dissertation, Yale University, New Haven, Connecticut.

Chamberlain, T. K., 1931, Annual growth of freshwater mussels, *Bull. U.S. Bur. Fish.* **46**:713–739.

Clarke, A. H., 1973, The freshwater molluscs of the Canadian interior basin, *Malacologia* **13**:1–509.

Clarke, A. H., and Berg, C. O., The freshwater mussels of central New York, *Cornell Univ. Agric. Exp. Sta. Mem.* **367**:1–79.

Clarke, G. R., II, 1974, Calcification on an unstable subtrate: Marginal growth in the mollusk *Pecten diegensis, Science* **183**:968–970.

Coker, R. E., Shira, A. F., Clark, H. W., and Howard, A. D., 1921, Natural history and propagation of freshwater mussels, *Bull. Bur. Fish.* **37**:75–181.

Cox, L. R., 1960, Thoughts on the classification of the Bivalvia, *Proc. Malacol. Soc. London* **34**:60–88.

Cox, L. R., 1969, Family Pachycardiidae, in: *Treatise on Invertebrate Paleontology, Part N, Mollusca 6* (R. C. Moore, ed.), pp. N467–N468, Geological Society of America and University of Kansas Press, Lawrence.

Crowley, T. E., 1957, Age determination in *Anodonta, J. Conchol.* **24**:201–207.

Cvancara, A. M., 1963, Clines in three species of *Lampsilis* (Pelecypoda: Unionidae), *Malacologia* **1**:215–225.

Cvancara, A. M., 1972, Lake mussel distribution as determined with SCUBA, *Ecology* **53**:154–157.

Davies, J. H., and Trueman, A. E., 1927, A revision of the non-marine lamellibranchs of the coal measures and a discussion of their zonal sequence, *Q. J. Geol. Soc. London* **83**;210–259.

Decksbach, W. K., 1957, Abscess formation on the shell of *Anodonta anatina* L., *Zool. Zh.* **36**:787–788 (in Russian, English summary).

Dell, R. K., 1953, The fresh-water Mollusca of New Zealand. Part I. The genus *Hyridella, Trans. R. Soc. N. Z.* **81**:221–237.

Diamond, J. M., 1978, Niche shifts and the rediscovery of interspecific competition, *Am. Sci.* **66**:322–331.

Eagar, R. M. C., 1948, Variation in shape of shell with respect to ecological station: A review dealing with Recent Unionidae and certain species of the Anthracosiidae in Upper Carboniferous times, *Proc. R. Soc. Edinburgh Sect. B* **63**:130–148.

Eagar, R. M. C., 1953, Variation with respect to petrological differences in a thin band of Upper Carboniferous non-marine lamellibranchs, *Liverpool Manchester Geol. J.* **1**:163–190.

Eagar, R. M. C., 1974, Shape of shell of *Carbonicola* in relation to burrowing, *Lethaia* **7**:219–238.

Ekman, T., 1905, Undersökningar öfver flodpärlmusslans förekomst och lefnadsförhallanden i ljusnan och dess tillflöden inom Harjedalen (cited in Hendelberg, J., 1960, p. 157).

Fischer, P., 1887, *Manuel de Conchyliogie et de Paleontologie et Fossiles,* F. Savy, Paris, 1187 pp.

Fisher, J. B., and Tevesz, M. J. S., 1976, Distribution and population density of *Elliptio complanata* (Mollusca) in Lake Pocotopaug, Connecticut, *Veliger* **18**:332–338.

Green, R. H., 1972, Distribution and morphological variation of *Lampsilis radiata* (Pelecypoda, Unionidae) in some central Canadian lakes: A multivariate statistical approach, *J. Fish. Res. Board Canada* **29**:1565–1570.

Grier, N. M., 1920a, On the erosion and thickness of shells of freshwater mussels, *Nautilus* **34:**15–22.

Grier, N. M., 1920b, Variation in nacreous color of certain species of naiades inhabiting the upper Ohio drainage and their corresponding ones in Lake Erie, *Am. Midl. Nat.* **6:**211–243.

Grier, N. M., 1920c, Variation of epidermal color of certain species of naiades inhabiting the upper Ohio drainage and their corresponding ones in Lake Erie, *Am. Midl. Nat.* **6:**247–285.

Grier, N. M., 1922, Observations on the rate of growth of the shell of lake dwelling freshwater mussels, *Am. Midl. Nat.* **8:**129–148.

Haas, F., 1936, Binnen-Mollusken aus Inner-Africa, hauptsächlich gesammelt von Dr. F. Haas während der Schomburg-Expedition in den Jahren 1931–1932, *Senckenbergiana Naturforsch. Ges. Abh.* **431:**1–156.

Haas, F., 1969, Superfamily Unionacea, in: *Treatise on Invertebrate Paleontology, Part N, Mollusca 6* (R. C. Moore, ed.), pp. N411–N467, Geological Society of America and University of Kansas Press, Lawrence.

Haranghy, L., Balázs, A., and Burg, M., 1964, Phenomenon of aging in Unionidae, as example of aging in animals of telometric growth, *Acta Biol. Acad. Sci. Hung.* **14:**311–318.

Haranghy, L., Balázs, A., and Burg, M., 1965, Investigation on aging and duration of life in mussels, *Acta Biol. Acad. Sci. Hung.* **16:**57–67.

Harman, W. N., 1970, New distribution records and ecological notes on central New York Unionacea, *Am. Midl. Nat.* **84:**46–58.

Harman, W. N., 1974, The effects of reservoir construction and canalization on the molluscs of the upper Delaware watershed, *Bull. Am. Malacol. Union,* pp. 12–14.

Hazay, J., 1881, Die Molluskenfauna von Budapest. III. Biologische Mitteilungen, in den Malakozöolgischen Blättern, Kassel (cited in Rubbel, 1913).

Hendelberg, J., 1960, The fresh-water pearl mussel, *Margaritifera margaritifera* (L.), *Drottingholm: Statens Undersoknings-och forsoksonstalt for sotvattensfisket. Medd.* **41:**149–171.

Holme, N. A., 1959, The British species of *Lutraria* (Lamellibranchia), with a description of *L. angustior* Philippi, *J. Mar. Biol. Assoc. U. K.* **38:**557–568.

Howard, A. D., 1922, Experiments in the culture of fresh-water mussels, *Bull. U.S. Bur. Fish.* **38:**63-89.

Isley, F. B., 1914, Experimental study of growth and migration of fresh-water mussels, U.S. Commercial Fisheries Report, 1913, pp. 1–24, Bureau of Fisheries Document No. 792.

Isom, B. G., 1971, Mussel fauna found in Fort Loudoun Reservoir, Tennessee River, Knox County, Tennessee, in December 1970, *Malacol. Rev.* **4:**127–130.

Israel, F. B., 1911, Najadologische Miscellaneen, *Nachrichterbl. Dtsch. Malakal. Ges.* **1911:**10–17.

Johnson, R. I., 1978, Systematics and zoogeography of *Plagiola* (= *Dysnomia* = *Epioblasma*), an almost extinct genus of freshwater mussels (Bivalvia: Unionidae) from Middle North America, *Bull. Mus. Comp. Zool.* **148:**239–320.

Kessler, J., and Miller, A., 1978, Observations on *Anodonta grandis* (Unionidae) in Green River Lake, Kentucky, *Nautilus* **92:**125–129.

Kolesnikov, Ch. M., 1970, Paleobiochemical studies of fossils as applied to Mesozoic freshwater mollusks, *Paleontol. Zh.* **1970**(1):48–57 (in Russian).

Lefevre, G., and Curtis, W. C., 1912, Studies on reproduction and artificial propagation of fresh water mussels, *Bull. U.S. Bur. Fish.* **30:**105–201.

Leitch, D., 1936, The *Carbonicola* fauna of the Midlothian fifteen foot coal: A study in variation, *Trans. Geol. Soc. Glasgow* **19:**390–402.

Magnin, E., and Stanczykowska, A., 1971, Quelques donneés sur la croisance, la biomasse et la production annuelle de trois mollusques Unionidae de la région Montréal, *Can. J. Zool.* **49:**491–497.

Marshall, W. B., 1926, Microscopic sculpture of pearly freshwater mussel shells, *Proc. U.S. Natl. Mus.* **67**:1–14.

McMichael, D. F., 1952, The shells of rivers and lakes, I, *Aust. Mus. Mag.* **10**:348–352.

Negus, C. L., 1966, A quantitative study of the growth and production of unionid mussels in the River Thames at Reading, *J. Anim. Ecol.* **35**:513–532.

Nelson, D. J., 1964, Deposition of strontium in relation to morphology of clam (Unionidae) shells, *Verh. Int. Ver. Limnol.* **15**:893–902.

Newell, N. D., and Boyd, D. W., 1975, Parallel evolution in early trigoniacean bivalves, *Bull. Am. Mus. Nat. Hist.* **154**:53–162.

Ockelmann, K. W., 1958, The zoology of East Greenland: Marine Lamellibranchiata, *Medd. Grønland; Komm. Vidensk. Undersøg. I. Grønland* **122**:1–257.

Ortmann, A. E., 1920, Correlation of shape and station in freshwater mussels (naiades), *Proc. Am. Philos. Soc.* **59**:269–312.

Ray, R. H., 1977, Application of acetate peel technique to analysis of growth processes in bivalve unionid shells, *Bull. Am. Malacol. Union* **1977**:79–81.

Rubbel, A., 1913, Beobachtungen über das Wachstum von *Margaritana margaritifera*, *Zool. Anz.* **41**(4):156–162.

Simpson, C. T., 1900, Synopsis of the najades or pearly freshwater mussels, *Proc. U.S. Natl. Mus.* **22**:501–1044.

Stansbery, D. H., 1967, Growth and longevity of naiades from Fishery Bay in western Lake Erie, *Abstr. Cond. Pap. Am. Malacol. Union Inc.* **1967**:10–11.

Stansbery, D. H., 1970, A study of the growth rate and longevity of the naiad *Amblema plicata* (Say 1817) in Lake Erie (Bivalvia: Unionidae), *Am. Malacol. Union Inc. Bull.* **37**:78–79.

Sterrett, S. S., and Saville, L. D., 1974, A technique to separate the annual layers of a naiad shell (Mollusca, Bivalvia, Unionacea) for analysis by neutron activation, *Am. Malacol. Union Inc. Bull.* **1974**:55–57.

Stratton, L. W., 1960, Some variations in the Unionidae, *J. Conchol.* **24**:433–437.

Tadić, A, 1975, Some *Unio* and *Anodonta* species in various habitats, *Glas. Prir. Muz. Beogradu Ser. B* **30**:103–118 (in Serbian, German summary).

Taylor, J. D., and Layman, M., 1972, The mechanical properties of bivalve (Mollusca) shell structures, *Palaeontology* **15**:73–87.

Taylor, J. D., Kennedy, W. J., and Hall, A., 1969, The shell structure and mineralogy of the Bivalvia: Introduction: Nuculacea–Trigonacea, *Bull. Br. Mus. (Nat. Hist.) Zool. Suppl.* **3**:1–125.

Taylor, J. D., Kennedy, W. J., and Hall, A., The shell structure and mineralogy of the Bivalvia. II. Lucinacea–Clavagellacea: Conclusions, *Bull. Br. Mus. (Nat. Hist.) Zool.* **22**:255–294.

Tevesz, M. J. S., and McCall, P. L., 1978, Niche width in freshwater bivalves and its paleoecological implications, *Geol. Soc. Am. Annu. Mtg. Program* (abstract), p. 504.

Tevesz, M. J. S., and McCall, P. L., 1979, Evolution of substratum preference in bivalves (Mollusca), *J. Paleontol.* **53**:112–120.

Tolstikova, N. V., 1972, Microstructural variability of shells in *Unio tumidus* (Bivalvia), *Zool. Zh.* **51**:1565–1569 (in Russian, English summary).

Tolstikova, N. V., 1973, Systematic studies of the microstructure of the Unionids, *Vodoemy Sibiri i Perspectivy ikh Rybokhoz ispol'z*, Tomsk University, pp. 212–213 (in Russian).

Tolstikova, N. V., 1974, Microstructural characteristics of freshwater bivalves (Unionidae), *Paleontol. Zh.* **1974**(1):61–65 (in Russian).

Tolstikova, N. V., and Orlov, V. A., 1972, A study of variability of freshwater bivalve molluscs with computers, *Zool. Zh.* **51**:969–974 (in Russian, English summary).

Tudorancea, C. L., and Florescu, M., 1968a, Cu privire la fluxul energetic al populatiei de *Unio pictorum* din Balta Crapina, *An. Univ. Bucuresti Ser. Stiint. Nat.* **17**:233–243.

Tudorancea, C. L., and Florescu, M., 1968b, Considerations concerning the production and energetics of Unio tumidens Philipsson population from the Crapina marsh, Trav. Mus. Hist. Nat. "Grigore Antipa" **8**:395–409.

Tudorancea, C. L., and Florescu, M., 1969, Aspecte de productiei si energeticii populatiei de Anodonta piscinalis Nilsson disi Balta Crapina, Stud. Cercet. Biol. Ser. Zool. **21**:43–55.

Tudorancea, C. L., and Gruia, L. 1968, Observations on the Unio crassus Philipsson population from the Nera river, Trav. Mus. Hist. Nat. "Grigore Antipa" **8**:381–394.

Vermeij, G. J., 1974, Adaptation, versatility, and evolution, Syst. Zool. **22**:466–477.

van der Schalie H., 1936, An unusual naiad fauna of a southern Michigan lake, Am. Midl. Nat. **17**:626–628.

von Hessling, 1859, Die Perl Muschln und ihre Perlen, Leipzig (cited in Rubbel, 1913).

Wellmann, G., 1938, Untersuchungen über die Flussperlmuschel (Margaritana margaritifera L.) und ihren Lebensraum in Bachen der Lüneburger Heide, Z. Fisch. Deren Hilfswiss. Neudammund Berlin **36**:489–603.

Wilson, C. B., and Clark, H. W., 1914, The mussels of the Cumberland River and its tributaries, Bureau of Fisheries Document No. 781, 63 pp.

Chapter 9

Molluscan Larval Shell Morphology
Ecological and Paleontological Applications

DAVID JABLONSKI and RICHARD A. LUTZ

1. Introduction 323
2. Developmental Types in Marine Benthic Invertebrates 323
    2.1. Planktotrophic Larvae 324
    2.2. Nonplanktotrophic Larvae 327
3. Recognition of Molluscan Developmental Types from Larval Shell Morphology. 329
    3.1. Gastropoda 329
    3.2. Bivalvia 337
4. Ecological and Paleontological Applications 347
    4.1. Ecological Applications 347
    4.2. Paleontological Applications 355
    References 363

1. Introduction

Considerable information is contained within the molluscan shell on the ontogenetic history of the organism. In this chapter, we are concerned with the earliest calcified portion of the shell, that secreted by larvae and young juveniles. Careful interpretation of the morphological features of the shell apex or umbo region can provide insight into molluscan systematics, ecology, biogeography, and evolution (e.g., Thorson, 1950; Scheltema, 1971a,b, 1977, 1978, 1979a,b; Thiede, 1974; Kauffman, 1975; Lutz, 1977; Jablonski, 1978a, 1979; Hansen, 1978a,b: Jablonski and Lutz, 1978; Lutz and Jablonski, 1978a–c; 1979; Hansen and Jablonski, 1979).

2. Developmental Types in Marine Benthic Invertebrates

Many classifications have been proposed for the spectrum of developmental types found in marine benthic invertebrates (cf. Mileikovsky,

DAVID JABLONSKI ● Department of Geology and Geophysics, Yale University, New Haven, Connecticut 06520. *Present address:* Department of Geological Sciences, University of California, Santa Barbara, California 93106. RICHARD A. LUTZ ● Department of Oyster Culture, New Jersey Agricultural Experiment Station, Cook College, Rutgers University, New Brunswick, New Jersey 08903.

1974a). The classification employed depends largely on the perspective of the individual investigator. For example, a paleobiogeographer concerned with dispersal capability might emphasize the dichotomy between those larvae that feed during the pelagic state and those that do not (planktotrophy vs. nonplanktotrophy). Alternatively, such a worker might wish to classify larval forms according to their swimming capabilities (pelagic vs. nonpelagic). A population ecologist might finely subdivide developmental types according to amount of reproductive effort invested by the parent, while an embryologist might be more interested in the presence or absence of a distinct larval stage (direct vs. indirect development). In this study, we adopt an informal classification of general utility, a modified version of that devised by Thorson (1946, 1950) in the course of his monumental work on marine larval ecology. In such a classification scheme, the various types of larval development fall under two main categories: planktotrophic and nonplanktotrophic larvae. We stress this dichotomy in feeding type because it is readily recognizable using shell morphology and because planktotrophic larvae may have a prolonged pelagic stage and thus high dispersal capability. Nonplanktotrophic larvae may be pelagic or nonpelagic, and in either case will generally have lower dispersal capability than planktotrophs. This classification, which is briefly discussed in the following sections, is not all-encompassing; intergradations, exceptions, and combinations of two or more developmental types within a single life cycle render the scheme presented herein as artificial as any other simple pigeonholing of the complexity of the natural world.

2.1. Planktotrophic Larvae

2.1.1. Pelagic Larvae

Pelagic larvae spend a significant portion of their development time in surface waters as free-swimming veligers. Planktotrophic pelagic larvae depend on smaller planktic organisms (and possibly organic detritus) for nutrition during the pelagic stage. This involves development of complex structures that function in locomotion and feeding (Figs. 1 and 2), all of which are lost or resorbed during metamorphosis when the larva settles and assumes a benthic mode of life. Eggs are released virtually yolk-free, and thus there is a minimum expenditure per egg per parent. Species having planktotrophic larvae produce them in great numbers—up to 70 million eggs per individual in a single spawning of the oyster *Crassostrea virginica* (see Davis and Chanley, 1956) and 2.5–25 million eggs per individual per spawning in the abalone *Haliotis midae* (see Newman, 1967); most gastropods, however, produce fewer eggs (see Table IX in Webber,

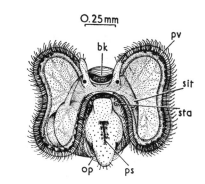

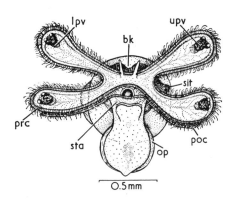

Figure 1. Gastropod veliger larvae. (A) *Nassarius reticulatus*, about 40 days old; (B) *Nassarius incrassatus*, swimming-crawling stage. Note interspecific differences such as pigmentation and velar lobe morpholgy. (bk) Beak of shell; (lpv) lower pigment patch of velum; (op) operculum; (poc) postoral cilia of velum; (prc) preoral ciliar of velum; (ps) pigment on sole of foot; (pv) pigment of velum; (sit) siphonal tube; (sta) statocyst; (upv) upper pigment patch of velum. Reprinted with permission from Fretter and Graham (1962).

1977). However, predation, starvation, and other mortality factors take a tremendous toll of planktotrophic larvae before and just after settlement [over 99% mortality (Thorson, 1950, 1966; Mileikovsky, 1971)].* Despite this extremely high larval mortality, the enormous numbers of pelagic larvae produced by populations of planktotrophic species are sufficient to ensure dispersal and survival. For example, Ayers (1956) calculated that only 0.0013% of the larvae need survive annually to maintain a population of *Mya arenaria* in Barnstable Harbor, Massachusetts.

Scheltema (1967) divided the pelagic stage of planktotrophic species into two phases: (1) growth and development, followed by (2) a "delay

---

* The actual sources of larval mortality are poorly known. Thorson (1950), Mileikovsky (1971, 1974b), and many others have emphasized the impact of predation. Laboratory studies (e.g., Loosanoff, 1954; Bayne, 1965; Millar and Scott, 1967) have led to suggestions concerning the role of larval starvation, but as Thorson (1950) points out, planktotrophic larvae found in nature rarely, if ever, exhibit the characteristics of starved laboratory specimens. However, because level of food availability [in addition to temperature and a number of other factors (e.g., Bayne, 1965)] may markedly affect growth rates, starvation may indirectly contribute to larval mortality by significantly prolonging the pelagic larval stage and thus the amount of time the larvae are exposed to predation (Vance, 1973a,b).

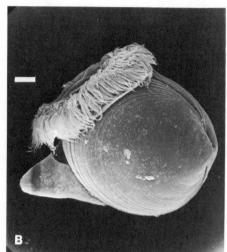

Figure 2. Critical-point dried bivalve larvae. (A) Veliger of *Ostrea edulis* (Ostreidae), 15 days after fertilization. Scale bar: 40 μm. Reprinted with permission from Waller (1980); scanning electron micrograph courtesy of Thomas R. Waller. (B) Pediveliger larva of *Lyrodus pedicellatus* (Teredinidae) showing extended foot and velum. Scale bar: 25 μm. Reprinted with permission from Turner and Boyle (1975); scanning electron micrograph courtesy of Ruth D. Turner.

period," in which development is essentially completed, but a final loss of pelagic adaptations is postposed until a substratum suitable for settlement is found. Scheltema (1971a, p. 306) further comments that "it is the delay period which gives to many species the greatest flexibility in the length of their larval life," since this stage may last from a few minutes to over 3 months, depending on the species.* The complex process of selection of settlement sites is reviewed by Meadows and Campbell (1972a,b), Scheltema (1974), and Crisp (1974), and has been modeled as an absorbing Markov chain by Doyle (1975, 1976).

The ability to subsist on planktic food not only obviates the need for deposition of yolk (in the egg) as an energy source for the developing larvae, but also allows a long pelagic larval duration for many species. The evolutionary success of this mode of development is indicated by the estimate of Thorson (1950) (also see Mileikovsky, 1971) that 70% of all benthic marine species undergo planktotrophic development. In temperate-water species, this pelagic stage generally lasts from 2 to 6 weeks

---

* If this delay period is extended beyond a certain limit—for example, by depriving larval cultures of a suitable settlement substratum—the larvae can lose their ability to metamorphose [e.g., in asteroids (Barker, 1977)].

(Thorson, 1961); Scheltema (1977) has pointed out that during this period, an ocean current of only 0.5 km/hr could carry a larva 150–500 km. Other species, notably among the tropical benthos, can remain in the plankton for 6 months or more, and thus may be carried great distances by ocean currents. Larvae of this type have been termed "teleplanic" [Gr. *teleplanos*, "far-wandering" (Scheltema, 1971b)] larvae. Scheltema (1971a, 1977) discussed several tropical gastropod species having larvae capable of traversing transoceanic distances, and thus maintaining the amphi-Atlantic distribution of the adults (also see Edmunds, 1977). The ecological and evolutionary implications of a long pelagic stage in essentially sessile benthic organisms will be considered in more detail later in this chapter.

2.1.2. Demersal Larvae

Demersal larvae undergo development while swimming or crawling, or both, in the near-bottom water layer, apparently feeding on organic detritus (or bacteria?) in the water close to or on the sediment. Some benthic feeding may occur, and inclusion of this larval type within the general planktotrophic category may be incorrect. Though poorly documented, this mode of development is probably an important one in high-latitude and deep-sea faunas (Pearse, 1969; Mileikovsky, 1971). Demersal development has been demonstrated in a number of polar echinoderms (Pearse, 1965, 1969; Pearse and Giese, 1966), but, apparently, in only a few molluscan species, notably the olivid gastropod *Olivella verreauxi* (Radwin and Chamberlin, 1973), and perhaps in *Hydrobia ulvae* (Fish and Fish, 1974, p. 696; also see Fish and Fish, 1977). In high latitudes, this dispersal mode appears to be an adaptive compromise, retaining some of the advantages of pelagic larvae, while reducing mortality by keeping larvae out of the polar surface waters (Pearse, 1969). Also, benthic dispersal may expose larvae to a more stable benthic food resource compared to the seasonally variable production at the sea surface (cf. Levinton, 1974; McCall, 1978).

2.2. Nonplanktotrophic Larvae

2.2.1. Lecithotrophic Larvae

Lecithotrophic larvae are nourished by the yolk of the eggs from which they develop (and thus are nonplanktotrophic). Such larvae are either entirely nonpelagic or remain in the plankton for little more than a few hours to a few days (Thorson, 1950, 1961).* Clearly, reproductive effort per offspring is much higher, and larval mortality much lower, than in planktotrophic species. Accordingly, far fewer eggs per parent are produced [4100 eggs per parent in the bivalve *Nucula proxima* and 1200 in the related species *N. annulata* (see Scheltema, 1972)].

In contrast to species having lecithotrophic larvae with a pelagic phase, oviparous species have lecithotrophic larvae that develop entirely within an egg mass or egg case (often attached to a hard substratum). Embryos pass through a lecithotrophic veliger stage, with metamorphosis taking place within the protection of the egg mass. The offspring emerge as juveniles (i.e., miniature adults), for example, *Nucula delphinodonta* (Drew, 1901). This mode of reproduction has been termed "direct development" (Thorson, 1946, 1950; Mileikovsky, 1971, 1974a; Webber 1977), but as Chia (1974) emphasized, this term should be applied only to species that do not pass through a distinct intermediate stage such as a veliger. Direct development should apply only to the embryos (which may or may not be free-living) that morphologically differentiate gradually into the adult. Certain opisthobranchs [see Bonar (1978), who uses the term "ametamorphic development"], most cephalopods (e.g., Arnold and Williams-Arnold, 1977; Wells and Wells, 1977; Boletsky, 1974), and some echinoderms (Chia, 1974) show true direct development.

Among the higher prosobranch gastropods (Neogastropoda), oviparous species may deposit, along with viable eggs, a supplementary food source in the form of nurse eggs. The ratio of nurse eggs to viable eggs may vary considerably within a species. For example, the gastropod *Buccinum undatum* deposits 50–2000 eggs per capsule, with only 10–30 of these eggs developing into juveniles (Portmann, 1925; for a tabulation of prosobranch data, see Webber, 1977). Spight (1976a) has shown that embryos feeding on nurse eggs show a greater range in hatching size than embryos in which the entire yolk supply is enclosed within their own egg membrane (also see Fioroni, 1966, 1967).

2.2.2. Brooded Larvae

Brooded larvae, characteristic of ovoviviparous species, are retained (sometimes encapsulated) within the parent throughout development, emerging as metamorphosed juveniles. Thus, ovoviviparous species can be regarded as investing the greatest amount of effort into the protection of their offspring during early ontogeny. This developmental type has

---

* A remarkable exception to this general observation is the lecithotrophic seastar *Mediaster aequalis*, which begins settling at approximately 30 days. Under laboratory conditions, though probably not natural conditions, this species is capable of delaying metamorphosis for up to 14 months, apparently without feeding (Birkeland *et al.*, 1971).

been termed "viviparity" (Thorson, 1946, 1950; Mileikovsky, 1971, 1974a), but Simpson (1977, p. 128) writes:

> . . . a viviparous condition requires that the developing embryo derives nutriment by close contact with maternal tissues without interposition of egg membranes, for example, as in placental mammals. Surely this must not be the case for the invertebrates frequently listed as viviparous in the literature. Ovoviviparous is the correct term in describing embryos that develop within the maternal organism from which they derive nutriment but are separated from the parent organism by egg membranes for most or all of their development, which is at the expense of yolk. It is an important distinction.

Purchon (1968, pp. 307–313) cites a number of examples that might be regarded as intermediate between ovoviviparity and oviparity, in which eggs are attached, and developing embryos are carried, on the surface of the shell or on the body of the parent. Such developing embryos might be placed, somewhat arbitrarily, in the general category of *brooded larvae*.

3. Recognition of Molluscan Developmental Types from Larval Shell Morphology

One of the fundamental dichotomies in larval adaptation, the presence or absence of a yolky food supply during development (nonplanktotrophy vs. planktotrophy), affects egg size, which is reflected in the morphology of the larval shell. Thus, the study of both larval and well-preserved early juvenile molluscan shells permits inferences to be made concerning developmental types of extant, or even long-extinct, species.

3.1. Gastropoda

3.1.1. Terminology

A number of different terminologies have been devised for the early stages of gastropod shell growth. In this section, we follow Thiriot-Quiévreux (1972) and Robertson (1974) in designating the *protoconch* as the entire shell formed prior to metamorphosis [following Robertson (1974), we consider this term synonymous with "larval shell"] (Fig. 3). *Protoconch I*, the "embryonic shell" of Fretter (1967) and Robertson (1971), is the initial shell, comprising less than two whorls, or even less than one whorl in some species, reportedly secreted by the shell gland while the embryo is still enclosed within the egg membrane (Raven, 1966; Kniprath, 1979). It is generally unornamented, except for granulations and punctae, which are apparently loci of shell deposition. *Protoconch II* is deposited by the mantle edge during the veliger stage and ranges from 1½ to 8 smooth to heavily ornamented whorls. The ornamentation pattern may not be the same on all the whorls and may not be consistent with the adult ornamentation (e.g., Fig. 3). The ultrastructure of the protoconch has not been described in detail; Togo (1977) reported simple prismatic structure and "irregularly complex" structure (disordered aragonite needles) in the archaeogastropod *Haliotis,* and a "protocrossed lamellar"structure in eight neogastropod species. The *teleoconch* (sometimes misspelled "teloconch") is the postlarval shell, laid down after metamorphosis (Fig. 3).

Because protoconchs are rarely planispiral, it is difficult to accurately define the number of whorls present. We prefer the method used by Taylor (1975). A line is drawn from the sutural point of the aperture perpendicular to and across the remaining apical sutures. The sutural point at the aperture is counted as one complete whorl; another whorl is added to the count each time the perpendicular line is crossed (Fig. 4).

3.1.2. Criteria

The key character in distingushing between planktotrophic and nonplanktotrophic larval shells was indicated in the "apex theory" of Thorson (1950, p. 33) [also well articulated by Powell (1942)], which states that "as a general rule, a clumsy, large apex points to a nonpelagic development, while a narrowly twisted apex, often with delicate sculpture, points to a pelagic development." That is, large rounded, often paucispiral protoconchs indicate larvae that spent little or no time in the plankton, while narrow, polygyrate protoconchs suggest planktotrophic larvae (also see Shuto, 1974). Thorson (1950) himself, as well as subsequent authors (e.g., Robertson, 1974; Taylor, 1975), suggested caution in utilizing this character, since numerous exceptions are known.*

With the increasing application of the scanning electron microscope (SEM) to larvel shell studies (e.g., Fretter and Pilkington, 1971; Robertson, 1971; Richter and Thorson, 1975; Bouchet, 1976), a more precise criterion, with a closer direct relationship to the underlying development process, can be used. A small Protoconch I (boundary readily discernible using the SEM) generally derives from a small (60–200 μm), yolk-poor egg; larval development is generally planktotrophic (Fig. 5A). In contrast, larval

---

* Thorson (1950) states: "As, however, the notion 'a clumsy or a narrowly twisted apex' is different in large and tiny adult shells and may even be different in termperate and tropical genera. . . a general rule valid for all apices cannot be given. Only when we know the type of apex derived from a pelagic or a non-pelagic development with the individual genus, can reliable results be obtained."

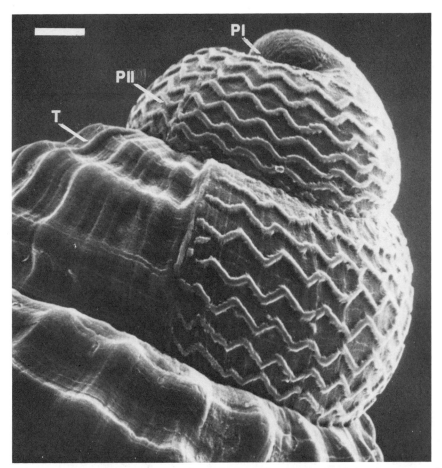

Figure 3. Scanning electron micrograph of *Alvania jeffreysi* (Rissoidae) showing gastropod larval shell terminology. (PI) Protoconch I; (PII) Protoconch II; (T) teleoconch (postlarval shell). Scale bar: 50 μm. After Rodriguez Babio and Thiriot-Quiévreux (1974); micrograph courtesy of Catherine Thiriot-Quiévreux.

shells in which the whorl has a large diameter and that lack a sharply demarcated Protoconch I grow from large, probably yolk-rich eggs; larval development is generally nonplanktotrophic (Robertson, 1971, 1974; Rodriguez Babio and Thiriot-Quiévreux, 1974; Shuto, 1974; Marshall, 1978) (Fig. 5B). The direct relationship between egg size and hatching size (and, thus, developmental type) has been investigated quantitatively by Amio (1963), Fioroni (1966), Shuto (1974), and Spight (1976b); for discussion of the more controversial relationship between egg size and developmental time, see Underwood (1974), Vance (1974), Strathmann (1977), and Steele (1977). Simpson (1977) has shown that the relationship be-

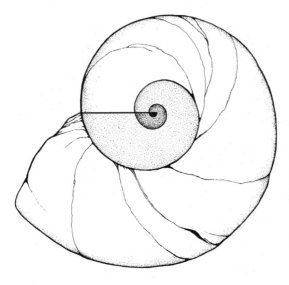

Figure 4. Method for determining number of protoconch whorls, following Taylor (1975). The sutural point at the aperture is counted as one whorl, with an additional whorl counted each time the reference line is crossed. The protoconch illustrated has 3½ whorls.

tween developmental type and egg size can be used to predict mode of development from examination of the ovaries.

Shuto (1974) has made some empirical observations concerning correlations between protoconch morphology and mode of larval life. The ratio of the maximum diameter (D, in mm) to the number of whorls or

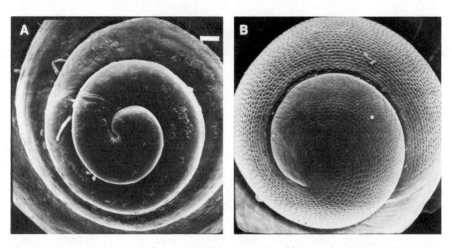

Figure 5. Protoconch morphology in rissoid gastropods of known developmental type. (A) *Rissoa guerini*, planktotrophic; (B) *Barleeia rubra*, nonplanktotrophic. Scanning electron micrographs at the same magnification. Scale bar: 50 μm. Reprinted with permission from Thiriot-Quiévreux and Rodriguez Babio (1975); micrographs courtesy of Catherine Thiriot-Quiévreux.

volutions (Vol) was found to be greater than 1.0 for species with lecithotrophic larvae (Fig. 6). Most protoconchs with D/Vol between 0.3 and
1.0 and fewer than 2.25 volutions also belong to species having lecithotrophic larvae, while larvae of species with D/Vol values less than 0.3 and
with more than 3 volutions are generally planktotrophic [also see Sohl
(1977) for a discussion of the relationship between Vol and length of pelagic life]. D/Vol values between 0.3 and 1.0 and with fewer than 3 volutions are found in species with planktotrophic larvae, as well as in those
with lecithotrophic larvae.

Shuto (1974) also considered certain types of protoconch ornamentation, such as close brephic axials and fine cancellate and reticulate ornamentation, to be indicative of planktotrophy (Fig. 7). Such ornamentation probably serves to strengthen the shell against the rigors of a more
or less prolonged free-swimming interval (also see Bandel, 1975a). A sinusigera lip on the larval aperture (evident in later life as the proto-

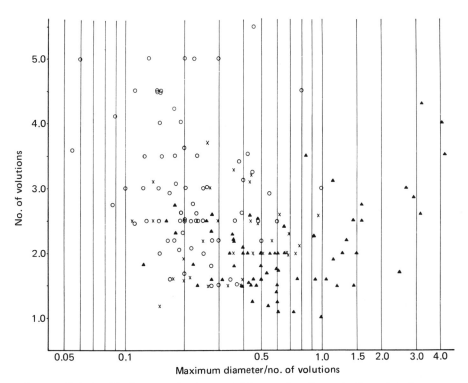

Figure 6. Relationship between developmental type and protoconch dimensions in prosobranch gastropods. (o) Planktotrophic larval type having moderate to long pelagic stage;
(×) planktotrophic larval type having short pelagic or demersal stage; (▲) directly developing
lecithotrophic larval type. After Shuto (1974).

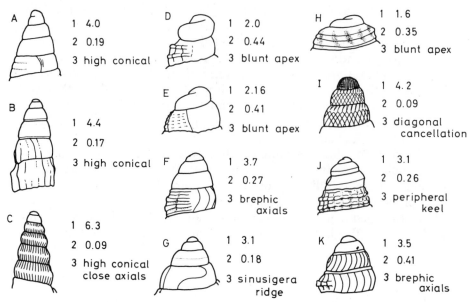

Figure 7. Relationship between morphology of protoconchs of living and fossil prosobranch gastropods and developmental type. (1) Number of volutions; (2) ratio of maximum diameter to number of volutions; (3) remarks. (A) *Tiara gerthi philippinensis*, Upper Miocene, inferred planktotrophic; (B) *Triplostephanus santosi*, Upper Miocene, inferred planktotrophic; (C) *Triphora (Notosinister) conaspersa*, Pleistocene, inferred planktotrophic; (D) *Ocinebrellus eurypteron*, Pleistocene, inferred lecithotrophic; (E) *Merica asperella varicosa*, Upper Miocene, inferred lecithotrophic; (F) *Latirus (Dolicholatirus) fusiformis*, Pliocene, inferred lecithotrophic; (G) *Zafra pumila*, Pleistocene, inferred planktotrophic; (H) *Iwakawatrochus urbanus*, Pleistocene, inferred lecithotrophic; (I) *Philbertina linearis*, Recent, planktotrophic; (J) *Nassarius (Nassarius) caelatus verbeeki*, Pliocene, inferred planktotrophic; (K) *Gemmula (Gemmula) speciosa*, Pliocene, inferred planktotrophic. Reprinted with permission from Shuto (1974).

conch–teleoconch boundary) accommodates the extended velar lobes of the larva and is best developed in teleplanic larvae (also see Robertson, 1974).

Finally, the most dependable method of distinguishing planktotrophic from nonplanktotrophic developmental types in gastropods is the approach advocated by Scheltema (1978, 1979a,b). To assess the mode of development, protoconchs of a given species should be compared to those of a congeneric or confamilial species the developmental characteristics of which are known. At these taxonomic levels, congruence of shell morphology probably indicates similarity of developmental history.* Even this is not an infallible technique; Robertson (1979, personal

---

* Note that the converse is often not the case; protoconch morphology can only occasionally form the sole basis for distinguishing supraspecific taxa (Dall, 1924; Smith, B., 1945, 1946;

communication) reports that planktotrophic and lecithotrophic *Boonea* (Pyramidellidae) have identical protoconchs.

Identification of larval gastropods (both living and fossil) is difficult because few species have been adequately described, and most published studies deal exclusively with newly hatched larvae. Some important papers that describe later pelagic veliger stages include Thorson (1946), Lebour (1936, 1937, 1945), Natarajan (1957), Fretter and Pilkington (1971), Pilkington (1974, 1976), and Taylor (1975).

Although some gastropods appear to have species-specific protoconch morphologies, identification of larval or early juvenile shells must generally be based on careful comparison with the apexes of metamorphosed juveniles that have attained an age at which diagnostic shell features are available. The protoconch is often abraded from adult specimens, so it is often necessary to assemble growth series. Excellent figures for comparative purposes have been presented in a number of SEM studies (e.g., Richter and Thorson, 1975; Bandel, 1975a,b; Rodriguez Babio and Thiriot-Quiévreux, 1974, 1975; Thiriot-Quiévreux and Rodriguez Babio, 1975; Jung, 1975).

3.1.3. Taxonomic Distribution

The taxonomic distribution of developmental types in prosobranch gastropods has been reviewed by Anderson (1960), Fretter and Graham (1962), Radwin and Chamberlin (1973), Taylor (1975), and Webber (1977). The following discussion is abstracted from these much more extensive surveys.

Archaeogastropods typically release eggs and sperm freely into the water, giving rise to a trochophore larval stage followed by a short (seldom longer than 2 or 3 days) planktotrophic veliger stage. Some acmaeids, trochids, fissurellids, and neritaceans deposit eggs in a gelatinous mass, from which pelagic veligers or benthic juveniles may emerge.

Mesogastropods and neogastropods exhibit internal fertilization, with oviparous or ovoviviparous development. Young may emerge as planktotrophic pelagic veligers, lecithotrophic pelagic veligers, or crawling juveniles. The mesogastropods are developmentally the most plastic order of prosobranchs, and larval types may vary within a single superfamily, family, genus, and in a few reported cases, a single species. The intriguing phenomenon of intraspecific variation in developmental types, reviewed by Robertson (1974) (also see Spight, 1975), is best documented in the

---

Robertson, 1974; Marshall, 1978). Despite this limitation, certain authors, such as Iredale (1910, 1911), Grabau (1912), and Finlay (1931), have considered protoconch morphologies to be highly reliable characters, and some genera and subgenera are still differentiated exclusively on the basis of protoconch differences.

Vermetidae (Hadfield *et al.*, 1972). On closer scrutiny, many supposed examples of intraspecific variation have proven to be due to unrecognized cryptic or sibling species [e.g., the polychaete *Capitella capitata* (see Grassle, J. P., and Grassle, J. F., 1976; Grassle, J. F., and Grassle, J. P., 1977), *Littorina saxatilis* (see Heller, 1975), *Rissoa membranacea* (see Rehfeldt, 1968; Robertson, 1974), and *Calyptraea dilatata* (see Gallardo, 1977a,b)].

The Littorinidae include nonpelagic as well as feeding and nonfeeding pelagic larvae (Mileikovsky, 1975) [pelagic egg capsules that give rise to free-swimming veligers are present in many species (e.g., Bandel, 1974)]. The Cerithiacea (Houbrick, 1973), Calyptraeacea (Hoagland, 1977), Cypraeacea (Ranson, 1967; Taylor, 1975), and Tonnacea (Taylor, 1975) include a variety of developmental types as well. Among the more conservative groups are the Strombacea, which are all reportedly oviparous with pelagic larvae, and the Rissoacea, which all follow one of two developmental pathways: those that hatch as crawling juveniles from relatively large eggs (140–320 μm) and those that hatch from relatively small eggs (60–130 μm) and spend 2–3 weeks as planktotrophic veligers (Taylor, 1975).

Some mesogastropods have an unusual larval form known as the echinospira larva, in which an outer transparent shell or membrane (scaphoconch) is developed in addition to the calcified shell, with the space between the two filled with sea water. Fretter and Graham (1962) have suggested that the outer "shell" is actually of proteinaceous, periostracal origin. This flotation aid, lost at metamorphosis, is reported in the Eratoidae, Lamellariidae, Capulidae (Lebour, 1935, 1937; Fretter and Graham, 1962), Trichotropidae (Pilkington, 1974), and Hipponicidae [(Cernohorsky, 1968; Robertson (1974, p.221) states that this observation requires confirmation].

The neogastropods, an ecologically more homogeneous group, show a greater uniformity in mode of development. Among the "stenoglossan"neogastropods (Muricacea, Volutacea, and Buccinacea), oviparous development predominates and culminates in a crawling juvenile stage.* This mode of development generally involves deposition of large eggs (400–600 μm) in capsules attached to hard substrata. Pelagic larvae are known, however, from all three superfamilies (Scheltema, 1971a; Radwin and Chamberlin, 1973; Taylor, 1975; Bandel, 1975b; Spight, 1976b, 1977). Marche-Marchad (1968) describes an unusual case of ovoviviparity in the volutid *Cymbium* ("*Cymba*"), which retains its egg capsules within an enlarged pedal gland. The toxoglossan neogastropods (Mitracea and Conacea) appear to have a higher proportion of pelagic larvae than do the stenoglossans.

---

* Of the 170 species listed by Radwin and Chamberlin (1973), 70% exhibited this mode of development.

Little information is available on the larval ecology of opisthobranchs. Here we will deal only with taxa most likely to have a good fossil record; for more extensive recent reviews, see Beeman (1977) and Bonar (1978). The most primitive group is the Order Heterogastropoda [Architectonicacea, including Mathildidae and Architectonicidae; Triphoroidea; Epitoniacea; Pyramidellacea (see Kosuge, 1966; Climo, 1975; Marshall, 1977a,b)].* Shelled opisthobranchs generally have heterostrophic protoconchs—the protoconch is sinistrally coiled, in the opposite direction to the dextral teleoconch. Architectonicaceans have planktotrophic larvae that can remain in the plankton for, in some cases, well over 3 months (Robertson, 1964; Robertson et al., 1970); eggs are as small as 63 μm, with settling specimens attaining sizes in excess of 3–6 Protoconch II whorls before metamorphosis (Taylor, 1975). Little is known of triphorid development; of the 27 species of *Triphora* reported from Hawaii, all species (with one possible exception) apparently have a pelagic veliger stage (Taylor, 1975). Marshall (1977a,b) provides scanning electron micrographs of triphorid protoconchs. For the Pyramidellacea, Rodriguez Babio and Thiriot-Quiévreux (1975) describe four types of pyramidellid protoconchs, two of which are apparently associated with planktotrophic development, and two in which the pelagic phase is brief or absent.

For the Cephalaspidea, various authors report planktotrophic development [*Philine* spp. (Brown, 1934; Horikoshi, 1967), *Acteocina canaliculata* (Franz, 1971)], lecithotrophic development [*Haloa japonica* (Usuki, 1966)], true direct development [*Runcina* spp. and *Acteocina senestra* (see references in Baba and Hamatani, 1959; Bonar, 1978)], and in capsulo veliger development, with crawl-away young [*Retusa obtusa* (Smith, S. T., 1967)].

The development of marine pulmonates (belonging to the lower Basommatophora) has been reviewed by Berry (1977). Most siphonariids hatch as planktotrophic veligers, while the remaining families (Ellobiidae, Otinidae, Amphinolidae, and Chilinidae), for the most part, exhibit nonpelagic development, emerging from egg capsules as crawling juveniles. As in the case of the opisthobranchs, there is a general tendency for protoconchs to be heterostrophic.

3.2. Bivalvia

3.2.1. Terminology

Bivalve larval shell terminology as employed herein is analogous to that for gastropods, and is generally compatible with the terminology of

---

* Some or all of these taxa may actually be prosobranchs, and the Heterogastropoda may in fact constitute an artificial group.

Werner (1939), Rees (1950), Ockelmann (1965), and Cox (1969). The *pro-dissoconch* is the entire shell formed prior to metamorphosis and is considered in this chapter to be synonymous with the term "larval shell." *Prodissoconch I* is the first shelled stage developing from the nonshelled trochophore, and is secreted by the shell gland and mantle epithelium (Carriker and Palmer, 1979). Prodissoconch I tends to have a granulated appearance under the optical microscope, which resolves into a coarse and irregular punctate surface texture under the SEM. This stage is equivalent to the special usage by Werner (1939) of the term *veliger*,* which has been followed by many European workers, and to the protostracum of Bernard (1898). On the basis of shape, this stage is also known as the "D-shaped" or "straight-hinged" veliger. The straight-hinged stage, however, does not exactly coincide with the Prodissoconch I stage and is considered to persist until total shell length is twice the length of the hinge line (Chanley and Andrews, 1971).

Prodissoconch II is deposited by the mantle edge along the shell margins and inside the Prodissoconch I valves (Carriker and Palmer, 1979). At this stage, the larva is referred to as a *veliconcha* by Werner (1939) and subsequent European workers. The thickened hinge region of the shell, which may support an interlocking tooth system, is termed the *provinculum* (Bernard, 1898). Prodissoconch II shows concentric growth lines, and sometimes exhibits other ornamentation as well (LaBarbera, 1974). However, the great diversity of surface features seen in gastropods at the protoconch stage is not present in the bivalve prodissoconch. The term *pediveliger* (Carriker, 1961) has been applied to a late larval stage marked by the coexistence of a velum and a "foot" that has undergone rapid growth "in preparation for" a benthic existence (see Fig. 2B).

The *dissoconch* constitutes the shell deposited after metamorphosis. Ornamentation, surface texture, and parameters of dissoconch shell growth may contrast markedly with the prodissoconch (Fig. 8). A sharp transitional line of demarcation, the *metamorphic line,* is often clearly evident at the Prodissoconch II–dissoconch boundary (Carriker and Palmer, 1979).

In bivalve larvae, the provinculum and larval shell commissure provide a datum for a standard plane of reference. The shell-dimension terminology followed in this chapter is illustrated diagrammatically in Fig. 9. *Length* is the greatest shell distance measured parallel to the hinge line. *Height* is the greatest shell distance in the dorsal–ventral (commissural)

---

* In this chapter, the term *veliger* refers to a shelled molluscan larva that possesses a velum (ciliated swimming organ), an adaptation for a free-swimming existence. This velum is lost or resorbed at the time of final settlement and metamorphosis. Certain species that undergo development through metamorphosis entirely within the confines of an egg capsule nevertheless pass through a veliger stage.

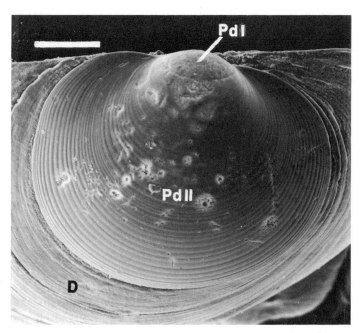

Figure 8. Bivalve larval and early juvenile shell terminology. (PdI) Prodissoconch I; (PdII) Prodissoconch II; (D) dissoconch. Scanning electron micrograph is of the umbonal region of a deep-sea planktotrophic mytilid from a hydrothermal vent on the Galapagos Rift. Scale bar: 100 μm.

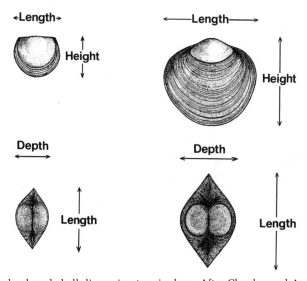

Figure 9. Bivalve larval shell-dimension terminology. After Chanley and Andrews (1971).

plane perpendicular to the length (= width of some authors). *Depth* is the maximum distance through the larva from right to left valves, perpendicular to the larval commissure; this dimension has variously been called thickness, convexity, and width by other authors. These and other morphological terms are defined in a glossary prepared by Chanley and Andrews (1971).

3.2.2. Criteria

As with gastropods, the key character in separating planktotrophs and nonplanktotrophs is the initial calcified growth stage, Prodissoconch I. The Prodissoconch I of species having planktotrophic larvae is relatively small (70–150 μm in length), reflecting the low yolk supply of the egg (Ockelmann, 1965). Prodissoconch II, deposited during the veliger stage, is large (200–600 μm in length) relative to Prodissoconch I. The Prodissoconch II stage of species with planktotrophic larvae is generally devoid of surface ornamentation except for growth lines (Fig. 10A). Species with lecithotrophic pelagic larvae generally have a relatively large Prodissoconch I [135–230 μm, depending on the species (Ockelmann, 1965)]. A distinct Prodissoconch II interposed between Prodissoconch I and the dissoconch is relatively small or absent (Fig. 10B). This probably reflects the short time lecithotrophs spend in the plankton, although a direct relationship between planktic duration and Prodissoconch II size has not

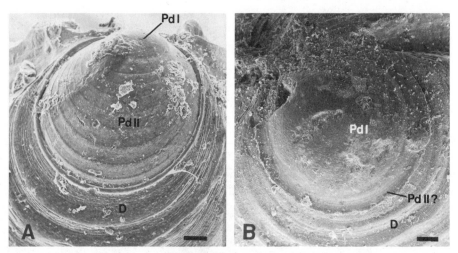

Figure 10. Scanning electron micrographs illustrating differences in larval shell morphology of two Late Cretaceous bivalves with different types of larval development. (A) *Uddenia texana* (planktotrophic); (B) *Vetericardiella crenalirata* (nonplanktotrophic). Scale bar: 20 μm.

been quantitatively demonstrated. Species with larvae that have no pelagic stage (e.g., certain brooded larvae) have the largest eggs of all, producing a Prodissoconch I of 230 to over 500 μm in length. The Prodissoconch I of such larvae is often inflated, lacking the distinctive D-shape (see Ockelmann, 1965; Chanley and Andrews, 1971; LaBarbera, 1974), and may show irregular folds and wrinkles. While many nonpelagic larvae are brooded within the parent's gill lamellae or pallial sinus, some nonpelagic species undergo development outside the parent. For example, *Musculus discors* and *M. niger* release egg strings into a byssal nest, from which young hatch as crawling juveniles (Thorson, 1935). Ockelmann (1965) suggests that the encapsulated eggs released by certain *Astarte* also hatch as benthic juveniles; a similar mode of development is reported by Blacknell and Ansell (1974) for the lucinacean *Thyasira gouldi*. The latter authors provide an excellent brief review of the distribution of lecithotrophic development in the Bivalvia.

While the criteria discussed above provide useful guidelines, it should be mentioned that interpretations based on egg and Prodissoconch I size are sometimes ambiguous. For example, Allen and Sanders (1973) report that while egg size in certain deep-sea Nuculanacea indicates a borderline planktotrophic or lecithotrophic mode of development, Prodissoconch I size indicates nonpelagic or lecithotrophic development. As with gastropods, the most reliable means of inferring developmental type from prodissoconch morphology is through comparison with confamilial or congeneric species for which the development is known (e.g., La-Barbera, 1974).

Identification of larval shells is facilitated by examination of the provinculum and associated lateral hinge system (Fig. 11). Denticulated structures associated with the provinculum are diagnostic at the familial or superfamilial level, although they have no direct relationship to the dentition that develops in the adult hinge [Rees (1950) and numerous subsequent authors]. Larval dentition is obscured shortly after metamorphosis by intraumbonal dissoconch growth. Rees (1950) provides diagrammatic figures (see Fig. 12) of the hinge apparatus of most of the North Sea bivalve superfamilies having pelagic larvae (for a useful discussion, see Guérin, 1973); the Pectinacea and Anomiacea were found to share a single larval hinge type, as were the Pteriacea and Ostreacea. Rees (1950) recognizes five major provinculum types (Table I):

1. Several or numerous rectangular teeth or crenulations in a row along a relatively thick provinculum, reminiscent of adult arcoid taxodont dentition (found in larval Mytilacea, Pectinacea, Anomiacea, Pteriacea, Ostreacea, some Veneracea, some Mactracea, and Tellinacea).

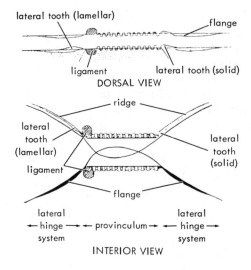

Figure 11. Terminology of bivalve larval hinge and associated structures. After Rees (1950); reprinted with permission from Cox (1969), courtesy of the Geological Society of America and the University of Kansas Press.

2. One, two, or three strong rectangular teeth in each valve arranged along a relatively thick provinculum (Astartacea, some Cardiacea, some Solenacea, Hiatellacea, Myidae and some Corbulidae, and Pholadacea).
3. A few teeth on a relatively thin provinculum (some Cardiacea, some Mactracea, Pandoracea, and Poromyacea).
4. Right valve has spiky teeth (on a thin strip) that insert into sockets on the left valve (some Cardiacea, some Solenacea, and some Corbulidae).
5. A thin provinculum from which teeth are entirely lacking (Lucinacea and Erycinacea).

In his studies, Rees (1950) relied on optical microscopy; however, structural details of the larval hinge apparatus are more readily discerned using SEM (Fig. 13).

Table I. Larval Hinge Types of Rees (1950)[a]

Provinculum	No teeth	Few teeth	Many teeth
Thick	—	Astartacea, some Cardiacea, some Solenacea, Hiatellacea, Myidae and some Corbulidae, Pholadacea	Mytilacea, Pectinacea, Anomiacea, Pteriacea, Ostreacea, Tellinacea, some Veneracea, some Mactracea
Thin	Lucinacea, Ercinacea	Some Cardiacea, some Mactracea, Pandoracea, Poromyacea	Some Cardiacea, some Solenacea, some Corbulidae

[a] Further subdivision of groups is possible using lateral hinge system, larval shell shape, and other characteristics.

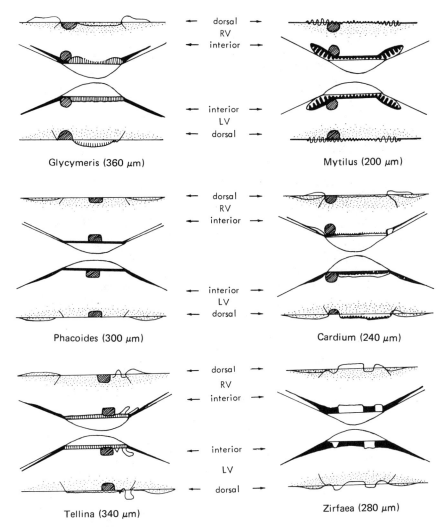

Figure 12. Prodissoconch hinges of six North Sea bivalve genera. After Rees (1950); reprinted with permission from Cox (1969), courtesy of the Geological Society of America and the University of Kansas Press.

A further aid to identification of larval shells is the configuration of the lateral hinge system (Rees, 1950; also see Cox, 1969), situated on the dorsal shell margin anterior or posterior, or both, to the provinculum (see Fig. 11). A projecting flange, when present, is generally on the left valve, interlocking with the right valve between the dorsal margin of the shell and an internal ridge. Near the junction of this ridge with the provinculum, there may be a "lamellar" or "solid" toothlike projection (Rees,

Figure 13. Scanning electron micrographs of bivalve larval hinge structures. Note distinctive hinge morphologies, diagnostic at superfamilial and lower taxonomic levels. (A) *Teredo navalis* (Pholadacea). Scale bar: 100 μm (B) *Ostrea edulis* (Ostreacea). Scale bar: 50 μm. From Calloway and Turner (1978). Micrographs courtesy of C. Bradford Calloway and Ruth D. Turner.

1950). Lamellar lateral teeth are thin and elongate, and interlock with a groove bordered by the rim of the opposite valve. Solid lateral teeth are more massive than lamellar lateral teeth and, unlike the latter, come into contact with the outside edge of the provinculum of the other valve when the shell is closed. There are also large special teeth that cannot be readily considered as part of either the provincular or the lateral hinge system (Rees, 1950). These larger projections are located at the juncture between the two hinge systems and exhibit various shapes; Rees reports them from the Astartacea, Mactracea, Tellinacea, and Poromyacea. Chanley (1969) has suggested that in the Tellinacea, the "special teeth" belong to the structure of the juvenile postsettlement hinge, and do not appear until late larval life or after metamorphosis. A much closer scrutiny of larval hinge features, as well as soft part morphology, is possible with the SEM (e.g., Scheltema, 1971b; Culliney and Turner, 1976; Turner and Boyle, 1975; LaBarbera, 1975; Lutz and Jablonski, 1978b), and may be used to distinguish larvae to the generic and species levels (e.g., Booth, 1977; Lutz and Jablonski, 1978c; Lutz, 1977, 1978; Lutz and Hidu, 1979).

Shell shape is also important in taxonomic placement of larval shells (Werner, 1939; Rees, 1950; Yoshida, 1953; Miyazaki, 1962). Relative prominence of the umbo (in the Prodissoconch II stage) and height/length and length/depth ratios are particularly useful. Chanley and Andrews (1971) provide a glossary and numerous examples of species-level separation on the basis of shape (also see Loosanoff et al., 1966).

The presence and position of ligaments or "ligament pits" ("fossettes ligamentaires") has been considered another diagnostic characteristic useful in superfamilial separation of larval specimens (Bernard, 1896; Rees, 1950; Ansell, 1962; Loosanoff et al., 1966; Chanley and Andrews, 1971; Le Pennec and Masson, 1976; Bayne, 1976). The designation of many of these structures as larval features may be misleading, if not entirely inappropriate (Lutz, 1977, 1978; Lutz and Jablonski, 1979; Lutz and Hidu, 1979). Lutz and Hidu (1979), in their study of hinge morphogenesis of larval and early postlarval mytilids, suggested that ligament attachment sites are not visible, using either optical or scanning electron microscopy, until larvae are at least "capable of metamorphosis," and concluded that "many of the morphological features heretofore interpreted in optical micrographs as larval ligaments are actually post-larval ligament pits." The aragonitic fibrillar structure of such ligament pits may be responsible for the optical-density differences observed in published micrographs.

3.2.3. Taxonomic Distribution

About 60–75% of shallow-water marine bivalves have planktotrophic larvae. Other developmental types are apparently more dependent on latitude and other ecological factors than on taxonomic relationships, with confamilial or congeneric species sometimes exhibiting very different developmental histories. References given in this section are intended only to provide recent information on larval types, and an entry into the literature; for older compendia, see Lebour (1938), Thorson (1946), Sullivan (1948), Ockelmann (1958), Loosanoff and Davis (1963), and Loosanoff et al. (1966).

The protobranchs, which include the Orders Nuculoida, Nuculanoida, and Solemyoida (Allen, 1978), exhibit an unusual [primitive? (Chanley, 1968)] mode of development (Fig. 14A and B). Larvae develop from large eggs and are lecithotrophic, with development often taking place within a free-swimming ciliated test that is cast off at metamorphosis after a short pelagic existence (Drew, 1899, 1901; Chanley, 1968) [nonpelagic species are also known (see Scheltema, 1972)]. Development within the arcoids (Chanley, 1966; Tanaka, 1971), mytiloids (Chanley, 1970; Ockelmann, 1965; Bayne, 1976; Booth, 1977), and pterioids (Tanaka, 1970;

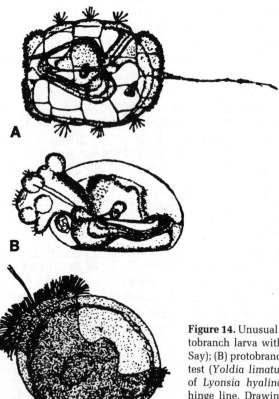

Figure 14. Unusual types of bivalve larvae. (A) Protobranch larva with ciliated test (*Yoldia limatula* Say); (B) protobranch larva in the process of casting test (*Yoldia limatula* Say); (C) pandoracean larva of *Lyonsia hyalina* (Conrad), showing indented hinge line. Drawings not to scale. From Chanley (1968).

Ota, 1961) is generally by means of the more familiar planktotrophic veliger larva, with some lecithotrophic and a few nonpelagic larval types known in each order. Among the ostreoids (Waller, 1978), the scallops also are dominantly planktotrophic (Sastry, 1965; LePennec, 1974; Culliney, 1974), while the oysters show a tendency toward mantle cavity brooding, generally accompanied by lecithotrophy. The brooding habit is present in *Ostrea*, *Alectryonella*, and *Lopha*, while *Crassostrea*, *Saccostrea*, and *Striostrea* are known to be nonbrooding (Stenzel, 1971). In the veneroids, the Lucinacea (Blacknell and Ansell, 1974), Galeommatacea (Chanley and Chanley, 1970; Franz, 1973), Cyamiacea, Carditacea (Jones, 1963; Yonge, 1969), and Crassatellacea (Harry, 1966) show a tendency toward lecithotrophy and ovoviviparity, while the Cardiacea (Kingston, 1974a,b), Tridacnacea (LaBarbera, 1975; Jameson, 1976), Mactracea (Chanley, 1965a), Solenacea, Tellinacea (Chanley, 1969), Arcticacea (Landers, 1976), and Veneracea (LaBarbera and Chanley, 1970; LePennec, 1973) are predominantly planktotrophic. The Myoida are also primarily planktotrophic in their development (Boyle and Turner, 1976; Chanley, 1965b; Culliney, 1975; Culliney and Turner, 1976). For the Pholadomyoida, Chanley (Chanley, 1968; Chanley and Castagna, 1966) states that the Pandoracea are lecithotrophic [although Ockelmann (1965) figures a planktotrophic species of *Thracia*]. The hinge line of pandoracean larvae is slightly indented; consequently, the straight-hinge stage is absent (Fig. 14C). The few species of Poromyacea that have been studied have lecithotrophic larvae (Ockelmann, 1965).

4. Ecological and Paleontological Applications

The distribution of larval types is clearly nonrandom in space and time, although we have yet to gain complete insight into the controlling factors. Furthermore, the evolutionary impact of different developmental histories is just beginning to be explored. In this section, we will briefly discuss a few areas of research in which studies of larval ecology and paleoecology could make a significant contribution.

4.1. Ecological Applications

4.1.1. Adaptive and Developmental Strategies

Phylogenetic relationships appear to be of secondary importance in determining developmental type in marine benthos. At first glance, the distribution of larval types appears to conform to the predictions of the r–K model for reproductive strategies (MacArthur and Wilson, 1967; Pianka, 1970; Gadgil and Bossert, 1970; Gadgil and Solbrig, 1972; Southwood, 1976; for a discussion and critique, see Stearns, 1976, 1977). In shallow marine waters, the more generalized, "opportunistic" species (supposedly "r-strategists") have a high rate of gamete production and produce small planktotrophic larvae that receive little or no parental care; populations may show wide annual fluctuations (Thorson, 1950, 1966; Coe, 1953, 1956). The more specialized, "equilibrium" species (supposed "K-strategists") produce fewer, larger eggs, invest more energy reserves per offspring, and have nonplanktotrophic larvae; Mileikovsky (1971) cites evidence that these species tend to have more stable populations. This mode could be invoked, for example, as an explanation for the presence of planktotrophic larvae in the opportunistic bivalve *Mulinia lateralis* (Calabrese, 1969, 1970; Calabrese and Rhodes, 1974) and pelagic lecithotrophic or nonpelagic larvae in *Nucula* and other protobranchs that are seemingly more K-selected (Rhoads *et al.*, 1978). Similar reasoning could be followed in explaining the prevalence of nonpelagic larvae in the carnivorous neogastropods, which are at a high trophic level and may be more specialized than many of the archeogastropod and mesogastropod groups that have pelagic larvae (cf. Radwin and Chamberlin, 1973; Shuto, 1974). When habitats or trophic resources are predictable or species are adapted to life within rather narrow environmental limits, survival of offspring is likely to be higher if larvae do not disperse any great distance from the parent [see Vermeij (1972) on tropical high intertidal gastropods and Hadfield (1963) on nudibranch development as related to prey population structure]. That is, a good indication of the suitability of a habitat is the survival of individual organisms to reproductive age. On the other hand, Strathmann (1974) suggests that when the parent cannot "predict" whether its present location is likely to provide a better setting for survival and reproduction of its offspring than some other location, high dispersal is the optimal strategy. When prey species exhibit relatively small population fluctuations, as in the tropics and other "favorable" environments, predator species can rely more heavily on fewer, more localized prey populations (MacArthur, 1972; Vermeij, 1978). For recent discussions of life-history strategies in nonmarine bivalves and gastropods, respectively, see Heard (1977) and Calow (1978).

Another interesting trend is seen in the latitudinal distribution of developmental types shown in Fig. 15. According to Thorson (1950, 1965), larvae of approximately 57–68% of the marine prosobranchs from Gibraltar to Trondheim on the Norwegian coast are pelagic (planktotrophic or lecithotrophic with a brief pelagic stage), while north of this point, the proportion of species with pelagic larvae declines until it reaches about 17% near Murmansk [also see Spight (1977) on the Muricacea]. Ockel-

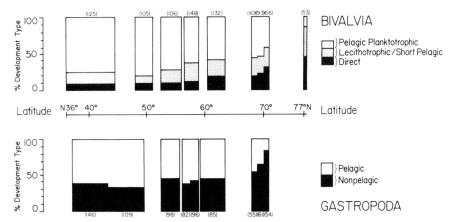

Figure 15. Molluscan developmental types as a function of latitude in the eastern North Atlantic. Bar widths delimit approximate latitudinal coverage of samples. Figures in parentheses indicate number of species. Subdivisions within bars indicate samples with overlapping latitudinal ranges. Bivalve data modified from Ockelmann (1965); gastropod data modified from Thorson (1965).

mann (1965) reports a similar trend [sometimes called "Thorson's rule" (Mileikovsky, 1971)] in the eastern Atlantic shelf Bivalvia, with about 75% of the Gibraltar bivalves having planktotrophic larvae (about 17% lecithotrophic with a short pelagic stage, and about 8% nonpelagic), while only 15% of the Spitsbergen bivalves have planktotrophic larvae (about 37% pelagic lecithotrophic, and 48% nonpelagic). Dell (1972) reports a similarly high incidence of viviparity and ovoviviparity in Antarctic benthos as well.*

The observations discussed above stand in contrast to our first-approximation assignment of species with planktotrophic larvae to the r-selected end of the adaptive strategy spectrum and species with nonplanktotrophic larvae to the K-selected end of the spectrum (also see Valentine and Ayala, 1978).† The latitudinal pattern may be more con-

---

* Dell (1972) points out that this short or nonexistent pelagic stage need not invariably lead to a limited geographic distribution. In Antarctic waters, at least, ovoviviparous species have often achieved the widest geographic distributions, and he attributes this to passive dispersal by attachment to floating algae (also see Arnaud, P. M., 1974; Arnaud, F. M., et al., 1976; Simpson, 1977). Dell (1972, p. 163) stresses that the "significance of such a bivalve arriving in a new area with several hundred well-developed young ready for release is obvious enough." However, it seems unlikely that this could serve as a general mechanism for widespread dispersal of subtidal benthos.

† This statement holds if one accepts that on the whole, density-independent physical factors are more important in temperate to polar waters, while density-dependent biological interactions predominate in tropical areas (Dobzhansky, 1950; Sanders, 1968; MacArthur, 1972).

sistent with the predictions of a stochastic model for the evolution of life history strategies [e.g., Murphy, G. I., 1968; Schaffer, 1974; Stearns, 1976 (the "bet-hedging model"), 1977]. This model suggests that where juvenile mortality fluctuates more than adult mortality, organisms with delayed reproduction, low reproductive efforts, and a few young are favored. Thus, for species or situations in which fluctuations in population densities result primarily from fluctuations in survival of early growth stages, evolution will lead to a complex of characteristics more typical of K-selection than of r-selection regimes. This may be the case in high-latitude molluscs, where harsh conditions in surface waters (e.g., salinity fluctuations; short, intense seasonal bursts of phytoplankton production; low temperatures* might lead to unusually high pelagic larvae mortality. Conversely, in situations where adult mortality fluctuates more than juvenile mortality, so-called r-selected traits would be expected. Thorson (1950) attributed this latitudinal pattern to the timing of primary production (for a critique, see Menge, 1975); phytoplankton blooms become progressively shorter in duration at high latitudes, and thus time available to planktotrophic larvae for completion of growth and metamorphosis is progressively restricted. This problem would be compounded by the slower metabolic and developmental rates brought about by the colder water temperatures of high latitudes (e.g., Vermeij, 1978).

The observation by Thorson (1950) of an increase in the proportion of pelagic to nonpelagic larvae from temperate to equatorial waters (Fig. 15) may be in part an artifact of the faunas studied to that date (Radwin and Chamberlin, 1973). Thorson's original data were obtained from the Persian Gulf (the waters of which are subject ot unusually high temperatures and salinities) and oceanic islands (Bermuda, and the Canary Islands, in which initial colonization would likely have been limited to planktotrophic species). Unless mainland species were fortuitously transported to the isolated islands, say by rafting or human activities, the only nonpelagic developmental histories present would have to evolve *in situ*. The fact that any species with nonpelagic development are found on oceanic islands probably testifies to the intensity of selection favoring this developmental type. Curtailment of the pelagic phase could be an adaptation to avoid mass loss of larvae that might otherwise be carried by prevailing currents away from the islands and into the open ocean. This may explain why young oceanic islands, such as Ascension Island, tend to have a molluscan fauna with predominantly pelagic larvae (Rosewater, 1975), while older islands have a higher proportion of species with nonpelagic larvae (cf. compilations of Thorson, 1965; Radwin and Chamber-

---

* Lower temperature might contribute indirectly to increased larval mortality by depressing developmental rates and thus prolonging the duration of exposure to predation while in the plankton (Vance, 1973a).

lin, 1973). The landward component of near-bottom coastal water movement may also play a role in preventing larvae from being lost to the open ocean (see Scheltema, 1975).

A similar trend in reproductive types is seen in intertidal–subtidal transects (Jackson, 1974), and from the subtidal shelf into the deep sea. Thorson (1950) predicted that deep-sea molluscs would dominantly show nonpelagic development with planktotrophic development being least common. More recent studies (Ockelmann, 1965; Knudsen, 1970; Scheltema, 1972; Grassle, J. F., and Sanders, 1973; Sanders and Allen, 1973, 1977) have shown that lecithotrophy (with a brief, nonfeeding pelagic stage) is the most common mode of development among deep-sea molluscs [78% of the 23 bivalve species examined by Knudsen (1970) had lecithotrophic development]. Planktotrophy appears to be rare among deep-sea molluscs, although it is apparently important in other deep-sea taxa [e.g., ophiuroids (Schoener, 1968, 1972)]. Despite this low dispersal capability, the relatively broad distribution of deep-sea molluscs (Scheltema, 1972) (also see Allen, 1978) suggests that this developmental strategy is a response to the long-term stability of the deep-sea habitat, and thus can again be linked to the K end of the r–K spectrum.* Dispersal could proceed in a stepwise fashion in a homogeneous environment over a long period of time. On the other hand, Bouchet (1976) has used the criteria discussed herein to infer planktotrophic development from shell morphology of a number of deep-sea gastropods (Fig. 16). He considers planktotrophy to be an important factor in maintenance of genetic continuity among populations in deep basins separated by major bathymetric features (e.g., the mid-Atlantic ridge). Bouchet believes that these species have pelagic larvae, which rise to feed in the photic zone, but it is not obvious how the larval shells of planktotrophic pelagic species and demersal species might be distinguished. Both types of larvae are capable of feeding during development, and thus each type would be expected to have a small Protoconch I. Pelagic or demersal larvae are also found in deep-sea molluscs that inhabit such rare, transient substrata as decomposing wood [pholadacean bivalves (Turner, 1973)].

Also underlying the distribution of larval types in marine organisms is the relationship between body size and developmental strategy (which of course need not be mutually exclusive with the relationship between life-history traits and development). Because of the great larval mortality associated with planktotrophy, species following this developmental mode must release immense numbers of larvae. As body size becomes small, so does the absolute amount of energy available for reproduction

---

* Note that in a certain sense, this reproductive pattern might be regarded as a taxonomic artifact, in that the protobranchs greatly dominate deep-sea bivalve collections (Allen, 1971, 1978).

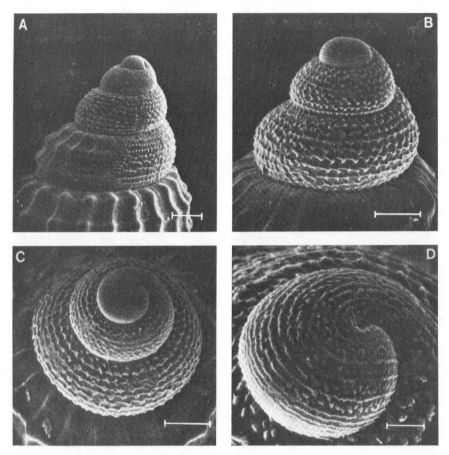

Figure 16. Apparent planktotrophy in deep-water gastropods. Scanning electron micrographs of *Alvania cimicoides* (Rissoidae) from a depth of 1050 m. (A–C) Scale bars: 100 μm; (D) scale bar: 10 μm. Reprinted with permission from Bouchet (1976).

(Giesel, 1976). Below a certain size, species will be unable to produce enough planktotrophic larvae to ensure survival of the population(s) because of the high mortality accompanying this mode of development. Here, the most efficient strategy is to invest a greater proportion of energy per offspring in a few eggs that have a relatively good chance of surviving to complete their development* (Mileikovsky, 1971; Arnaud, P. M., 1974; Chia, 1974; Menge, 1974, 1975; Simpson, 1977; McCall, 1977; Hoagland, 1977; Gould, 1977). For example, Sellmer (1967) presents a partial tabulation of known cases of brooding in bivalves, and, except for the oysters,

---

* It should be noted, however, that mortality of embryos in egg capsules is also not trivial; e.g., it is 42% in *Thais lamellosa* (Spight, 1975).

all species have a very small body size. Thus, in the case of the geographic and depth-related patterns discussed above, the diminutive size of most deep-sea (Sanders and Allen, 1973; Allen, 1978) and high-latitude (Arnaud, P. M., 1974; Nicol, 1964, 1966, 1978) molluscs, which is presumably related to limited nutrient supply (Thiel, 1975) or temperature-controlled growth rates (Gunter, 1957), or both, may govern the nature of their developmental histories. These species are "forced" to adopt a nonplanktotrophic strategy because they are unable to produce enough planktotrophic larvae per breeding episode to ensure adequate recruitment. Lack of a planktic food source has been evoked to explain the rarity of planktotrophic larvae in the deep sea [just as Thorson (1950) cited brief productivity episodes in high latitudes], but the ability of larvae to utilize detritus and dissolved nutrients cannot be ruled out (e.g., Pilkington and Fretter, 1970).

It appears that competition (Menge, 1974, 1975) and selective predation on larger-sized specimens (Brooks and Dodson, 1965; Dodson, S. I., 1970; Menge, 1973) can also lead to an optimum reproductive size that requires nonplanktotrophy. Spight (1976b) further notes that large hatching size may aid in reducing predation and physiological stress among juveniles (also see Woodin, 1976).

This adaptive link between body size and mode of reproduction is probably also a primary factor in the discordance between reproductive strategies in marine benthos and the r–K model. Many of the most strongly opportunistic species in the marine realm have nonpelagic larvae (Grassle, J. F., and Grassle, J. P., 1974, 1977). Small adult size, and attendant rapid onset of reproductive age, are characteristics of these species, which in many cases compensate for low adult yield per reproductive episode by reproducing several times, with only brief intervening pauses (e.g., Grassle, J. F., and Grassle, J. P., 1974, 1977). This is in contrast to the generalization of Pianka (1970) that r-selected species have a single reproductive episode per generation. Menge (1974, 1975) has suggested that this strategy occurs in the brooding starfish *Leptasterias hexactis*. Small body size is maintained by interactions with a superior competitor (*Pisaster ochraceus*, which is thus ostensibly the K-strategist of the pair) that releases large numbers of planktotrophic larvae. Grahame (1977) has attempted to reconcile these results with the r–K model by stressing the prediction (Gadgil and Solbrig, 1972) that r-selected species will allocate a greater proportion of their energy reserves to reproduction. *Leptasterias hexactis* does invest more energy in reproduction than *P. ochraceus*, and by this criterion could be regarded as more r-selected, despite its low gamete production per reproductive episode. In two species of the littorinid snail *Lacuna*, Grahame (1977) found that the species exhibiting greater reproductive effort, and showing other r-selected characteristics as well, had long-lived planktotrophic larvae. However, it is premature to generalize on the relationship between mode of development and energy expenditure of the parent in marine organisms; no universal correlation emerges from the sparse literature.

In an evolutionary context, small body size might be readily attained by progenesis (acceleration of gonad maturation relative to somatic growth), a mechanism Gould (1977) links to an "r-selected" regime. This, in turn, would rule out planktotrophic larvae in the life histories of these species if a critical threshold were crossed below which insufficient numbers of planktotrophic larvae could be produced. For example, Gould (1977, pp. 327–328) has suggested progenesis for the origin of the assemblage of minute bivalves described by Soot-Ryen (1960) from Tristan de Cunha; Gould envisions this as a response to superabundant resources by a number of normal-size species that drifted in via planktotrophic larvae. Seven or eight of the tiny, progenetic descendants of the initial colonizers are brooders. Gould (1977) presents a number of other marine examples, along with a discussion of the ecological situations most conducive to progenesis, such as unstable environments, colonization, parasitism, and a variety of habitats favoring small size *per se*.

Consideration of the selective forces and adaptive responses acting here can lead to a chicken–egg paradox. We must caution that the interrelationship among body size, adaptive strategy, dispersal capability, and environmental tolerance is a complex and fundamental one. It is difficult, and perhaps inappropriate, to single out any one factor in an attempt to explain the constellation of traits displayed by an assemblage of more or less disparate organisms. Clearly, a more precise understanding of the distribution of developmental types in all latitudes, depths, and colonization situations is desirable, and the link between development and larval shell morphology provides a relatively simple and direct way of accumulating the necessary data.

4.1.2. Environmental Monitoring

The recolonization of a sea floor or intertidal zone following a natural or anthropogenic perturbation is likely to reflect dispersal capabilities of the affected species. Two interrelated phenonema will underlie this response: (1) the differential ability of species to spread larvae into new areas or recolonize disturbed habitats and (2) the imprecise, but direct, correlation between dispersal capability and environmental tolerance. For example, following defaunation of a Florida intertidal zone by a red tide, two gastropod species that have nonpelagic larvae (*Nassarius vibex** and

---

* This is possibly a misidentification; Scheltema (1962) describes pelagic larvae for this species.

Prunum apicinum) had not returned by 24 months after repopulation had begun (Simon and Dauer, 1977). It appears, then, that examination of protoconch and prodissoconch morphology should allow environmental monitoring studies to accomplish the following aims:

1. "Factor out" the effects on slow colonizers following a disturbance. That is, species with nonplanktotrophic larvae, and especially those with nonpelagic larvae (with the exception of certain extreme opportunists), could be predicted *not* to return in the early stages of recolonization. Negligible recovery of populations of species having planktotrophic larvae would be a better indicator of extensive or lasting damage to environmental quality than would a gradual return of species having low-dispersal larvae, and these two groups should be weighted accordingly in assessing and predicting recolonization sequences.

2. Establish a cutoff point for recognizing "acceptable" pollution levels in regions in which there are temporal or spatial gradients in disturbance, such as point-source thermal or chemical effluents. A significant decline in the populations of species having nonplanktotrophic larvae will indicate that the community is becoming disturbed beyond its limits of resilience (see Rhoads *et al.*, 1978), since repopulation by these components of the ecosystem will be a long and haphazard process [e.g., greater than 2 years in the case of the gastropods mentioned by Simon and Dauer (1977)].

4.2. Paleontological Applications

Fossil shells sufficiently well preserved to be assigned with some confidence to the planktotrophic or nonplanktotrophic categories have been reported from Late Tertiary (e.g., Sorgenfrei, 1958; LaBarbera, 1974; Scheltema, 1978, 1979a,b), Early Tertiary (Hansen, 1978a,b), Late Cretaceous (Lutz and Jablonski, 1978a–c; Jablonski 1978a,b, 1979; Jablonski and Lutz, 1978), and even some Paleozoic deposits (Harrison, 1978; J. Kříž, personal communication). Possible protoconchs have been described for an extinct class of Paleozoic molluscs (?), the Hyolitha (Dzik, 1978). Application of the criteria outlined in this chapter to well-preserved fossil larval and early juvenile shells enables us to assess the effects of different developmental types from the unique time perspective of paleontology.

4.2.1. Evolutionary Patterns

Shuto (1974) and Scheltema (1977, 1978, 1979a,b) have developed models concerning the paleontological significance of differential dispersal capabilities. Planktotrophic species having a relatively long pelagic stage will have the ability to disperse over wide geographic areas in a single generation. Local catastrophes are unlikely to eliminate a species over its entire geographic range, and larvae from other, persistent populations will replenish populations reduced by local extinction. These effects will combine to produce a geologically long-lived species. Further, larval dispersal by ocean currents will maintain gene flow among disjunct populations of sedentary adults and thus suppress the genetic divergence required for allopatric speciation. There is some electrophoretic evidence for low levels of genetic differentiation among benthic species having pelagic larvae, relative to those having nonpelagic larvae (Berger, 1973; Gooch et al., 1972; Snyder and Gooch, 1973; reviewed in Gooch, 1975; also see Wilkins et al., 1978; Campbell, 1978). This pattern is, however, complicated by the effects of differential mortality in a heterogeneous environment (e.g., Struhsaker, 1968; Koehn, 1976; Koehn et al., 1976; Milkman and Koehn, 1977; Marcus, 1977; Levinton and Lassen, 1978) and recent genetic bottlenecks (Berger, 1977). Local intense natural selection may cause populations to diverge despite larval dispersal if propagules from distant populations do not survive to reproductive age.

Species having nonplanktotrophic larvae will tend to have smaller, necessarily more continuous geographic ranges. As a result of the more restricted geographic (and, often, ecological) range of species with nonplanktotrophic larvae, local catastrophes and random population fluctuations are more likely to result in the extinction of the species. Local populations will tend to remain isolated after initial colonization or separation from the parent population, and thus speciation will be more common among such groups.

The predictions discussed above, and the observations from which they were derived, were made for situations in which major oceanic barriers are present. However, there is some evidence that the effects of differential dispersal capabilities are significant even along a single continental shelf or within an epicontinental sea. Hansen (1978a,b) found, in the Lower Tertiary volutid gastropods of the Gulf Coast, that species having "planktonic" larvae had a mean species duration of 4.4 million years (m.y.), while those with "nonplanktonic" larvae had a mean duration of 2.2 m.y.; in addition, the geographic range of "planktonic" species averaged about twice that of "nonplanktonics" (Fig. 17). Similarly, in a survey of Gulf and Atlantic Coastal Plain gastropods from the Late Cretaceous, Jablonski (1978a, 1979) found that species having planktotrophic larvae had a mean duration of about 6 m.y., while nonplanktotrophic species had a mean duration of only about 3 m.y. In both studies, degree of environmental tolerance also played an important role in determining species longevity and geological range (cf. Jackson, 1974, 1977). This factor

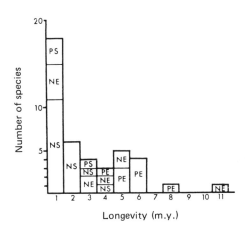

Figure 17. Developmental type and environmental tolerance of Lower Tertiary volutid gastropods and their relationship to species duration. (N) Nonplanktonic; (P) planktonic; (S) stenotopic; (E) eurytopic; (m.y.) million years. Reprinted with permission from Hansen (1978a). Copyright © by the American Association for the Advancement of Science.

may be responsible for the lack of a simple direct relationship between planktotrophy and species longevity in Late Cretaceous bivalves (Jablonski, 1978a).

Scheltema (1978) has elaborated on his earlier (Scheltema, 1977) conceptual model by suggesting that lineages or clades having planktotrophic larvae will produce new species via "punctuated equilibria" (Eldredge and Gould, 1972, 1977; Gould and Eldredge, 1977; Stanley, 1975a, 1978), by occasional isolation of small outlying populations that, once isolated, would tend to evolve rapidly into new species. Barriers that might disrupt the flow of larvae among marine populations are not easily envisioned. Sympatric and parapatric speciation processes (e.g., Endler, 1977; White, 1978; Rosenzweig, 1978) would of course not require physical barrier formation and may play an important role in the multiplication of marine species. Nonetheless, P. G. Murphy (1978) has suggested a climatic isolating mechanism for allopatric speciation in the acmaeid limpet *Collisella*. During warming trends, planktotrophic larvae of a warm-water species will successfully settle at higher latitudes, thus extending the species' range. Subsequent cooling intervals will cause geographic ranges to contract equatorward again, but some populations might become isolated in suitably warm, higher-latitude embayments. Should this isolation last long enough, speciation will occur (Murphy, P. G., 1978). Threshold effects may also allow for rapid speciation near the edges of species' ranges along environmental gradients (Dodson, M. M., and Hallam, 1977).

Scheltema (1978) further suggests that species with nonplanktotrophic larvae would be expected to undergo a gradual transformation into new species because populations are more localized and genetically more homogeneous; thus, they will be particularly susceptible to selection pressures generated by environmental changes. However, good cases of gradualism may be rare even among species with nonplanktotrophic larvae; species are more likely to become extinct or migrate along with their shifting habitat or temperature regime than they are to evolve in situ along with changing environmental conditions. Note also that if the parent population is small enough, evolutionary change in this lineage could be so rapid and localized as to be indistinguishable from punctuated equilibria in the fossil record. Schopf and Dutton (1976) present possible evidence in favor of phyletic gradualism in bryozoans having low dispersal capability, but the link between clinal geographic variation and speciation is still unclear (e.g., Gould and Johnston, 1972; Endler, 1977).

Evolutionary patterns in the fossil record will be affected by the bathymetric gradient in adaptive types discussed by Jackson (1974, 1977) (see Section 4.1.1). Jablonski (1978b, 1979) found evidence in Late Cretaceous coastal plain sediments that very nearshore fossil assemblages will tend to comprise species with greater geological and geographic ranges (the latter presumably a reflection of greater environmental tolerance), and will have a greater proportion of species with planktotrophic larvae, than deeper-water assemblages. Thus, although the more offshore assemblages inhabit a somewhat less heterogeneous habitat, low dispersal and relative stenotopy may combine to make speciation more likely in these molluscs than in those inhabiting very shallow water. This may be a more satisfying reconciliation between an allopatric model for speciation (Mayr, 1963, 1970; Bush, 1975; White 1978) and the observation that offshore assemblages are more diverse and less persistent than very nearshore ones (Bretsky and Lorenz, 1970; Bambach, 1977) than is the suggestion of Eldredge (1974) that new species arise onshore and migrate to the offshore areas.

With larval shell morphology giving an indication of dispersal capability, it will be possible to better separate the processes underlying extinction and speciation (Scheltema, 1978; Jablonski, 1979). Numerous authors (e.g., Moore, 1954, 1955; Kauffman, 1972, 1977a, 1978) have considered environmental stress to be responsible for both the extinction and origination components of species turnover rates. However, as Eldredge and Gould (1977, p. 36) (also see Eldredge, 1974; Bush, 1975) point out, "directions and intensities have never been directly correlated with speciation rates." If speciation proceeds primarily via the formation of peripheral isolates (allopatry), rates through time will depend on species' ability to maintain genetic continuity among populations, i.e., on their dispersal capability. Thus, lineages having planktotrophic larvae should have lower speciation rates than taxa having nonplanktotrophic larvae.* An assymmetrical pattern of speciation rates through a transgres-

---

* Again, this will not be completely exclusive of a eurytopy–stenotopy effect, since eurytopy and planktotrophy often co-occur in the same species.

sive–regressive cycle (e.g., Kauffman, 1972, 1977a) can be expected only if transgressive–regressive cycles are asymmetrical with regard to degree of spatial heterogeneity on the affected shelf, which may, in fact, be the case.

We agree with Kauffman (1977a) and others (e.g., Moore, 1954, 1955) that increasing environmental stress during regression will result in increasing extinction rates. Here, too, this effect will be ameliorated for species having planktotrophic larvae (see Jablonski, 1979). While extinctions within a single basin may be most closely correlated with degree of environmental tolerance as modeled by Kauffman (1977a), planktotrophic taxa will be more widespread and undergo a high proportion of local, rather than complete, exterminations. With the subsequent transgression, the same species (or a closely related descendant) can return to colonize the area. Nonplanktotrophic species will tend to be driven into extinction during regression (or any other major environmental perturbation), and lineages will often terminate. With the return of suitable habitats, recolonizing species will be derived from more distantly related stocks.

4.2.2. Biostratigraphy and Paleobiogeography

Kauffman (1975) has stressed that benthic species with pelagic larvae can spread over very wide geographic ranges instantaneously (in geological terms), and thus will make good biostratigraphic index fossils despite the sedentary habit of the adults. For example, the Cretaceous inoceramid bivalves are reportedly more useful for correlating sediments in different paleobiogeographic regions than such classic pelagic index fossils as the ammonites and planktic foraminifera (Kauffman, 1977b). The extensive geographic distribution of the inoceramids and certain other benthic species is probably due in large part to the presence of a pelagic larval stage (Kauffman, 1975, 1977b; Thiede and Dinkelmann, 1977), although passive dispersal of adults attached to floating material cannot be entirely ruled out.

Application of the criteria discussed above will enable the biostratigrapher to choose groups within a fauna that are likely to be the most useful in interprovincial and intercontinental correlations by recognizing those taxa with planktotrophic, and especially teleplanic, larvae. Thus, artificially differentiated faunas, saddled with provincial systematics, need not be entirely restudied before improved correlations can be made. Perhaps most significantly, larval ecological studies will shed some light on the processes underlying the different stratigraphic and geographic distributions already abundantly documented by biostratigraphers.

Herein lies a paradox, commented on by Scheltema (1977), Jackson

(1977), Kauffman (1977b), and others. Benthic species tolerant enough, and having larvae with a sufficient pelagic duration, to maintain a broad geographic distribution would not be expected to evolve very rapidly, in terms of either speciation or extinction rates. Yet valuable index species, including the inoceramids, must be sufficiently short-lived to enable the biostratigrapher to finely subdivide the stratigraphic succession. Some other aspect of the adaptive strategy [perhaps related to trophic level or life position or both (e.g., Levinton, 1974; Kauffman, 1977a, 1978)] must play a key role in keeping certain widespread lineages vulnerable to high extinction rates.

Recognition of differential dispersal capability can provide considerable insight into the processes underlying the geographic affinities of species assemblages as well as individual species. Partitioning of planktotrophic and nonplanktotrophic faunal components will allow paleogeographers to recognize more precisely the timing and nature of zoogeographic barrier formation between regions. For example, Shuto (1974) found that the number of nonplanktotrophic species shared among local faunas in the Philippine–Indonesian region declines markedly from the Late Miocene through the Pliocene and into the Early Pleistocene, while the percentage of shared planktotrophic species remains virtually constant. He infers that the Celebes Sea was forming during this interval, presenting a bathymetric barrier to the species with low dispersal capability, but was not wide enough to disrupt gene flow among populations of species with planktotrophic larvae.

4.2.3. Paleocommunity Studies

The population dynamics of ancient communities have only recently received much attention (e.g., Levinton and Bambach, 1970, 1975; Richards and Bambach, 1975; Walker and Alberstadt, 1975; Walker and Parker, 1976; Peterson, 1977). Study of larval shell morphology in fossil assemblages will provide insight into some of the long-term ecological and evolutionary processes underlying the establishment, maintenance, and alteration of marine biotic associations. Estimation of proportions of developmental types through time (in terms of numbers of both individuals and species) will allow the paleoecologists to make an assessment of processes of recruitment, dispersion, and adult reproductive effort. Faunal composition of small-size classes, as compared to adult-size classes, may be interesting from the viewpoint of failed recruitment episodes, particularly since the presence of a ligament pit may enable the investigator to differentiate metamorphosed juveniles from pelagic larval shells (Lutz, 1977, 1978; Lutz and Hidu, 1979). By analogy with present-day forms, the presence of species with nonplanktotrophic larvae within more than one fossil assemblage may indicate relative spatial homogeneity on the sea floor, while species with planktotrophic larvae, as a result of both high dispersal and accompanying eurytopy, could be expected to be present in a wider spectrum of assemblages. Similarly, examination of larval types within richly fossiliferous horizons or shell concentrations should give insights into the mode of recruitment of the constituent populations. Among the molluscs (if not the polychaetes), species with nonplanktotrophic larvae will have small population sizes and low turnover rates, and will tend to be stenotopic. They are more likely to constitute a significant portion of fossil assemblages when the deposits have been accumulating over ecologically long periods of time under relatively constant conditions; mass mortalities involving single spatfalls will infrequently involve large populations of species with planktotrophic larvae. Stratigraphic condensation could be responsible for compressing this time span into an apparant single event-horizon (e.g., Fürsich, 1978).

4.2.4. Paleoclimatology

Lutz and Jablonski (1978b) recently suggested that the morphometry of the bivalve Prodissoconch II could be used as a relative, and in some cases absolute, paleotemperature indicator. Of all the factors influencing the duration of delay of metamorphosis in modern bivalves, temperature appears to play the single most important role (e.g., Bayne, 1965, 1971). For the few species that have been examined in detail, strong negative correlations have been found between temperature and maximum larval shell size at metamorphosis (Bayne, 1965; Sullivan, 1948; Savage and Goldberg, 1976). Figure 18 illustrates this relationship for living *Mya arenaria*. Maximum reported size at metamorphosis of *M. arenaria* cultured at 19–24°C was 228 μm (Loosanoff and Davis, 1963), while a maximum of 310 μm was found at lower culture temperatures fluctuating between 16 and 19°C (Savage and Goldberg, 1976). Similar temperature–settlement size relationships have been found in several mytilids (Bayne, 1965; Siddall, 1978) and in the oyster *Crassostrea virginica* (Carriker, 1951, p. 24). At high temperatures, the high rate of growth may be insufficient to compensate for the short time available for feeding (due to temperature-accelerated developmental rates), and this results in a smaller maximum size at higher temperatures (Bayne, 1965).*

Data concerning maximum size at metamorphosis are available in fossil material in the form of the prodissoconch–dissoconch boundary (Lutz and Jablonski, 1978b). Variations in the maximum size (and perhaps mean

---

* This apparent temperature-mediated decoupling of rates of somatic growth and ontogenetic change (cf. Gould, 1977) is of considerable interest in a group of organisms in which paedomorphosis has played a significant evolutionary role (Stanley, 1975b).

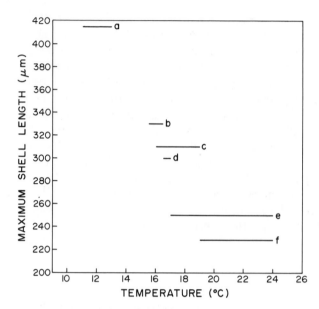

Figure 18. Maximum shell length of larval *Mya arenaria* as a function of culture or environmental temperature. Environmental termparature ranges represent reported values for those periods of time when mature larval specimens were sampled and measured. Letters designate studies from which shell length and temperature data were obtained. Authors, localities, and additional pertinent information are as follows: (a) Stafford (1912) and Davidson (1933); St. Andrews, Canada. Temperature range represents average surface water values for July and August, 1924 to 1931, as reported by Davidson (1933) for waters adjacent to St. Andrews (station 6). (b) Savage and Goldberg (1976); Hampton Beach, New Hampshire. Temperature range represents surface water extremes of weekly measurements made in August 1975 off Isles of Shoals, New Hampshire (R. A. Lutz, unpublished data). (c) Savage and Goldberg (1976); laboratory culture. (d) Thorson (1946); Øresund, Denmark. Temperature range is that reported by Thorson (1946) for surface waters of the sound during August. (e) Sullivan (1948); Malpeque Bay, Canada. Temperature range is that reported by Sullivan (1948) for July and August. (f) Loosanoff and Davis (1963); laboratory culture. From Lutz and Jablonski (1978b). Copyright © by the American Association for the Advancement of Science.

size, although adequate Recent data are not yet available) of the prodissoconch within a species should be largely a function of temperature. Empirical observations reviewed by Lutz and Jablonski (1978b) indicate that while such factors as individual genetic differences, food availability, and salinity will contribute to variation in larval size at metamorphosis, they will not mask the temperature effect.

Prodissoconch–dissoconch boundaries of specimens in sediments as old as the Late Cretaceous are readily distinguished (e.g., Fig. 10) and, when combined with size distribution data for larval specimens, can be used to reconstruct relative and absolute temperatures.

4.2.4.1. Relative Temperatures. Detailed examination of changes in prodissoconch dimensions in a species through time at a single locality, or in a series of localities along a single horizon, should provide an indication of temporal or spatial gradients for any interval containing adequately preserved faunas. Such paleotemperature-gradient reconstructions should serve as an independent test of data derived from other shelled microfossils or stable isotope analyses. They will also be useful in nearshore deposits, where holoplanktic paleoclimatic indicators may be rare.

4.2.4.2. Absolute Temperatures. The long geological duration of most bivalve species (Stanley, 1977) suggests that absolute paleotemperature estimates might be achieved for Holocene, Pleistocene, and Pliocene environments using regressions of prodissoconch length on temperature for a variety of extant species. However, as with other quantitative paleoclimatic techniques (Imbrie and Kipp, 1971; CLIMAP Project Members, 1976; Sachs *et al.*, 1977), the assumption of evolutionary stasis within species with regard to physiological processes may not be entirely warranted, and should be used with caution.

ACKNOWLEDGMENTS. We thank the following for critical review of the manuscript and helpful discussions: Robert Robertson, Ruth D. Turner, Stan Rachootin, Richard R. Strathmann, Earl D. McCoy, Thomas R. Waller, Karl M. Waage, Donald C. Rhoads, and Susan M. Kidwell. We are especially grateful to Rudolf and Amelie Scheltema for their hospitality and thorough review and comments, which significantly improved this chapter. T. R. Waller, C. Thiriot-Quiévreux, and R. D. Turner generously provided some of their excellent scanning electron micrographs. Maxine Lane, Susan Shiminski, and Betsy Dabakis kindly typed the manuscript. The research reported herein was supported in part by NSF grant EAR78-15536, EPA grant R804-909-010, and NOAA Sea Grant 04-7-158-44034. This is paper No. 5307 of the Journal Series, New Jersey Agricultural Experiment Station, Cook College, Rutgers University, New Brunswick, New Jersey 08903.

References

Allen, J. A., 1971, Evolution and functional morphology of the deep water protobranch bivalves of the Atlantic, *Proceedings of the Joint Oceanography Assembly* (Tokyo, 1970), pp. 251–253.

Allen, J. A., 1978, Evolution of the deep sea protobranch bivalves, *Philos. Trans. R. Soc. London Ser. B* **284**:387–401.

Allen, J. A., and Sanders, H. L., 1973, Studies on deep-sea Protobranchia (Bivalvia): The families Siliculidae and Lametilidae, *Mus. Comp. Zool. Bull.* **145**:263–310.

Amio, M., 1963, A comparative embryology of marine gastropods, with ecological considerations, *J. Shimonoseki Coll. Fish.* **12**:229–358.

Anderson, D. T., 1960, The life histories of marine prosobranch gastropods, *J. Malacol. Soc. Aust.* **1**:16–29.

Ansell, A. D., 1962, The functional morphology of the larva, and the post-larval development of *Venus striatula* (da Costa), *J. Mar. Biol. Assoc. U. K.* **42**:419–443.

Arnaud, F. M., Arnaud, P. M., Intès, A., and LeLeouff, P., 1976, Transport d'invertébrés benthiques entre l'Afrique du Sud et Sainte Hélène par les laminaires (Phaeophyceae), *Bull. Mus. Natl. Hist. Nat. Ecol. Gen.* **30**(384):49–55.

Arnaud, P. M., 1974, Contribution à la bionomie marine benthique des régions antarctiques et subantarctiques, *Tethys* **6**:465–656.

Arnold, J. M., and Williams-Arnold, L. D., 1977, Cephalopoda: Decapoda, in: *Reproduction of Marine Invertebrates*, Vol. IV, *Molluscs: Gastropods and Cephalopods* (A. C. Giese and J. S. Pearse, eds.), pp. 243–290, Academic Press, New York.

Ayers, J. C., 1956, Population dynamics of the marine clam, *Mya arenaria*, *Limnol. Oceanogr.* **1**:26–34.

Baba, K., and Hamatani, I., 1959, The direct development in *Runcina setoensis* Baba (Opisthobranchia—Cephalaspidea), *Publ. Seto Mar. Biol. Lab.* **7**:281–290.

Bambach, R. K., 1977, Species richness in marine benthic habitats through the Phanerozoic, *Paleobiology* **3**:152–167.

Bandel, K., 1974, Studies on Littorinidae from the Atlantic, *Veliger* **17**:92–114.

Bandel, K., 1975a, Embryonale und larvale Schale einiger Prosobranchier (Gastropoda, Mollusca) der Oosterschelde (Nordsee), *Hydrobiol. Bull.* **9**:3–22.

Bandel, K., 1975b, Embryonalgehäuse karibischer Meso- und Neogastropoden (Mollusca), *Akad. Wiss. Lit., Mainz, Abh. Math.-Naturwiss. Kl.* **1975**(1):1–133.

Barker, M. F., 1977, Observations on the settlement of the brachiolaria larvae of *Stichaster australis* (Verrill) and *Cescinasteria calameria* (Gray) (Echinodermata: Asteroidea) in the laboratory and on the shore, *J. Exp. Mar. Biol. Ecol.* **30**:95–108.

Bayne, B. L., 1965, Growth and the delay of metamorphosis of the larvae of *Mytilus edulis* (L.), *Ophelia* **2**:1–47.

Bayne, B. L., 1971, Some morphological changes that occur at the metamorphosis of the larvae of *Mytilus edulis*. in: *Proceedings of the Fourth European Marine Biology Symposium*, Bangor, 1969 (D. J. Crisp, ed.), pp. 259–280, Cambridge University, Press, Cambridge.

Bayne, B. L., 1976, The biology of mussel larvae, in: *Marine Mussels: Their Ecology and Physiology* (B. L. Bayne, ed.), pp. 81–120, Cambridge University Press, Cambridge.

Beeman, R. D., 1977, Gastropoda, Opisthobranchiata, in: *Reproduction of Marine Invertebrates*, Vol. IV, *Molluscs: Gastropods and Cephalopods* (A. C. Giese and J. S. Pearse, eds.), pp. 115–179, Academic Press, New York.

Berger, E. M., 1973, Gene–enzyme variation in three sympatric species of *Littorina*, *Biol. Bull.* **145**:83–90.

Berger, E. M., 1977, Gene–enzyme variation in three sympatric species of *Littorina*. II. The Roscoff population, with a note on the origin of North American *L. littorea*, *Biol. Bull.* **153**:255–264.

Bernard, F., 1896, Deuxième note sur le developpment et la morphologie de la coquille chez les lamellibranchs, *Bull. Soc. Geol. Fr.* **24**(3):54–82.

Bernard, F., 1898, Recherches ontogenetiques et morphologiques sur la coquille des lamellibranches. I. Taxodontes et anisomyaires, *Ann. Sci. Nat.(a) (Zoologie)* **8**:1–208.

Berry, A. J., 1977, Gastropoda: Pulmonata, in: *Reproduction of Marine Invertebrates*, Vol. IV, *Molluscs: Gastropods and Cephalopods* (A. C. Giese and J. S. Pearse, eds.), pp. 181–226, Academic Press, New York.

Birkeland, C., Chia, F.-S., and Strathmann, R. R., 1971, Development, substratum selection, delay of metamorphosis and growth in the seastar, *Mediaster aequalis* Stimpson, *Biol. Bull.* **144:**99–108.

Blacknell, W. M., and Ansell, A. D., 1974 (1976), The direct development of the bivalve *Thyasira gouldi* (Philippi), *Thalassia Jugosl.* **10:**23–43.

Boletsky, S. v., 1974 (1976), The "larvae" of Cephalopoda: A review, *Thalassia Jugosl.* **10:**45–76.

Boñar, D. B., 1978, Morphogenesis at metamorphosis in opisthobranch molluscs, in: *Settlement and Metamorphosis of Marine Invertebrate Larvae* (F.-S. Chia and M. E. Rice, eds.), pp. 177–196, Elsevier, New York.

Booth, J. D., 1977, Common bivalve larvae from New Zealand: Mytilacea, *N. Z. J. Mar. Freshwater Res.* **11:**407–440.

Bouchet, P., 1976, Mise en évidence de stades larvaires planctoniques chez des Gastéropodes Prosobranches de étages bathyal et abyssal, *Bull. Mus. Nat. Hist. Nat. Zool.* **277:**947–972.

Boyle, P. J., and Turner, R. D., 1976, The larval development of the wood boring piddock *Martesia striata* (L.) (Mollusca: Bivalvia: Pholadidae), *J. Exp. Mar. Biol. Ecol.* **22:**55–68.

Bretsky, P. W., and Lorenz, D. W., 1970, Adaptive response to environmental stability: A unifying concept in paleoecology, *Proceedings of the North American Paleontological Convention,* E, pp. 522–550.

Brooks, J. L., and Dodson, S. I., 1965, Predation, body size, and composition of plankton, *Science* **150:**28–35.

Brown, H. H., 1934, A study of a tectibranch gasteropod mollusc, *Philine aspera* (L.), *Trans. R. Soc. Edinburgh* **58:**179–210.

Bush, G. L., 1975, Modes of animal speciation, *Annu. Rev. Ecol. Syst.* **6:**339–364.

Calabrese, A, 1969, *Mulinia lateralis:* Molluscan fruit fly?, *Proc. Natl. Shellfish. Assoc.* **59:**65–66.

Calabrese, A., 1970, Reproductive cycle of the coot clam, *Mulinia lateralis* (Say), in Long Island Sound, *Veliger* **12:**265–269.

Calabrese, A., and Rhodes, E. W., 1974 (1976), Culture of *Mulinia lateralis* and *Crepidula fornicata* embryos and larvae for studies of pollution effects, *Thalassia Jugosl.* **10:**89–102.

Calloway, C. B., and Turner, R. D., 1978, New techniques for preparing shells of bivalve larvae for examination with the scanning electron microscope, *Am. Malacol. Union Inc. Bull.* **1978:**17–24.

Calow, P., 1978, The evolution of life-cycle strategies in fresh-water gastropods, *Malacologia* **17:**351–364.

Campbell, C. A., 1978, Genetic divergence of *Thais lamellosa,* in: *Marine Organisms: Genetics, Ecology, and Evolution* (B. Battaglia and J. A. Beardmore, eds.), pp. 157–170, Plenum Press, New York.

Carriker, M. R., 1951, Ecological observations on the distribution of oyster larvae in New Jersey estuaries, *Ecol. Monogr.* **21:**19–38.

Carriker, M. R., 1961, Interrelation of functional morphology, behavior, and autecology in early stages of the bivalve *Mercenaria mercenaria, J. Elisha Mitchell Sci. Soc.* **77:**168–241.

Carriker, M. R., and Palmer, R. E., 1979, Ultrastructural morphogenesis of prodissoconch and early dissoconch valves of the oyster *Crassostrea virginica,* *Proc. Natl. Shellfish. Assoc.* **69:**103–128.

Cernohorsky, W. O., 1968, Observations on *Hipponix conicus* (Schumacher, 1817), *Veliger* **10:**275–280.

Chanley, P. E., 1965a, Larval development of the brackish water mactrid clam, *Rangia cuneata, Chesapeake Sci.* **6:**209–213.

Chanley, P. E., 1965b, Larval development of a boring clam, *Barnea truncata, Chesapeake Sci.* **6:**162–166.

Chanley, P. E., 1966, Larval development of the large blood clam, *Noetia ponderosa* (Say), *Proc. Natl. Shellfish. Assoc.* **56:**53–58.

Chanley, P. E., 1968, Larval development in the Class Bivalvia, *Mar. Biol. Assoc. India Symp. Mollusca* **II**:475–481.

Chanley, P. E., 1969, Larval development of the Coquina clam, *Donax variabilis* Say, with a discussion of the structure of the larval hinge in the Tellinacea, *Bull. Mar. Sci.* **19**:214–224.

Chanley, P. E., 1970, Larval development of the hooked mussel, *Brachidontes recurvus* Rafinesque (Bivalvia: Mytilidae), including a literature review of larval characteristics of the Mytilidae, *Proc. Natl. Shellfish. Assoc.* **60**:86–94.

Chanley, P. E., and Andrews, J. D., 1971, Aids for identification of bivalve larvae of Virginia *Malacologia* **11**:45–119.

Chanley, P. E., and Castagna, M., 1966, Larval development of the pelecypod *Lyonsia hyalina*, *Nautilus* **79**:123–128.

Chanley, P., and Chanley, M. H., 1970, Larval development of the commensal clam, *Montacuta percompressa* Dall, *Proc. Malacol. Soc. London* **39**:59–67.

Chia, F.-S., 1974 (1976), Classification and adaptive significance of developmental patterns in marine invertebrates, *Thalassia Jugosl.* **10**:121–130.

CLIMAP Project Members, 1976, The surface of the ice-age Earth, *Science* **191**:1131–1137.

Climo, F. M., 1975, The anatomy of *Gegania valkyrie* Powell (Mollusca: Heterogastropoda: Mathildidae) with notes on other heterogastropods, *J. R. Soc. N. Z.* **5**:275–288.

Coe, W. R., 1953, Resurgent populations of littoral marine invertebrates and their dependence on ocean currents, *Ecology* **34**:225–229.

Coe, W. R., 1956, Fluctuations in populations of littoral marine invertebrates, *J. Mar. Res.* **15**:212–232.

Cox, L. R., 1969, Ontogeny, in: *Treatise on Invertebrate Paleontology, Part N, Mollusca 6, Bivalvia* (R. C. Moore, ed.), Vol. 1, pp. 91–102, Geological Society of America and University of Kansas Press, Lawrence.

Crisp, D. J., 1974, Factors influencing the settlement of marine invertebrate larvae, in: *Chemoreception in Marine Organisms* (P. T. Grant and A. N. Mackie, eds.), pp. 177–265, Academic Press, New York.

Culliney, J. L., 1974, Larval development of the giant scallop *Placopecten magellanicus* (Gmelin), *Biol. Bull.* **147**:321–332.

Culliney, J. L., 1975. Comparative larval development of the shipworms *Bankia gouldi* and *Teredo navalis*, *Mar. Biol.* **29**:245–252.

Culliney, J. L., and Turner, R. D., 1976, Larval development of the deep-water wood boring bivalve, *Xylophaga atlantica* Richards (Mollusca, Bivalvia, Pholadidae), *Ophelia* **15**:149–161.

Dall, W. H., 1924, On the value of nuclear characters in the classification of marine gastropods, *J. Washington Acad. Sci.* **14**:177–180.

Davidson, V. M., 1933, Fluctuations in the abundance of planktonic diatoms in the Passamaquoddy region, New Brunswick, from 1924 to 1931, *Contrib. Can. Biol. Fish.* **8**:357–407.

Davis, H. C., and Chanley, P. E., 1956, Spawning and egg production of oysters and clams, *Biol. Bull.* **110**:117–128.

Dell, R. K., 1972, Antarctic benthos, *Adv. Mar. Sci.* **10**:1–216.

Dobzhansky, T. H., 1950, Evolution in the tropics, *Am. Sci.* **38**:209–221.

Dodson, M. M., and Hallam, A., 1977, Allopatric speciation and the fold catastrophe, *Am. Nat.* **111**:415–433.

Dodson, S. I., 1970, Complementary feeding niches sustained by size-selection predation, *Limnol. Oceanogr.* **15**:131–137.

Doyle, R. W., 1975, Settlement of planktonic larvae: A theory of habitat selection in varying environments, *Am. Nat.* **109**:113–126.

Doyle, R. W., 1976, Analysis of habitat loyalty and habitat preference in the settlement behavior of planktonic marine larvae, *Am. Nat.* **110**:719–730.

Drew, G. A, 1899, Some observations on the habits, anatomy, and embryology of members of the Protobranchia, *Anat. Anz.* **15**:493–518.

Drew, G. A., 1901, The life history of *Nucula delphinodonta*, *Q. J. Microsc. Sci.* **44**:313–392.

Dzik, J., 1978, Larval development of hyolithids, *Lethaia* **11**:293–299.

Edmunds, M., 1977, Larval development, oceanic currents, and origins of the opisthobranch fauna of Ghana, *J. Moll. Stud.* **43**:301–308.

Eldredge, N., 1974, Stability, diversity, and speciation in Paleozoic epeiric seas, *J. Paleontol.* **48**:541–548.

Eldredge, N., and Gould, S. J., 1972, Punctuated equilibria: An alternative to phyletic gradualism, in: *Models in Paleobiology* (T. J. M. Schopf, ed.), pp. 82–115, Freeman, Cooper, San Francisco.

Eldredge, N., and Gould, S. J., 1977, Evolutionary models and biostratigraphic strategies, in: *Concepts and Methods of Biostratigraphy* (E. G. Kauffman and J. E. Hazel, eds.), pp. 25–40, Dowden, Hutchinson & Ross, Stroudsburg, Pennsylvania.

Endler, J. A., 1977, *Geographic Variation, Speciation, and Clines*, Princeton University Press, Princeton, New Jersey, 246 pp.

Finlay, H. J., 1931, On *Austrosassia, Austropharpa* and *Austrolithes*, with some remarks on the gastropod protoconch, *N. Z. Inst. Trans.* **62**:1–10.

Fioroni, P., 1966, Zur Morphologie und Embryogenese des Darmtraktes und der transitorischen Organe bei Prosobranchiern (Mollusca, Gastropoda), *Rev. Suisse Zool.* **73**:621–876.

Fioroni, P., 1967, Quelques aspects de l'embryonogènse des prosobranches (Mollusca, Gastropoda), *Vie Milieu Ser. A* **18**:153–174.

Fish, J. D., and Fish, S., 1974, The breeding cycle and growth of *Hydrobia ulvae* in the Dovey estuary, *J. Mar. Biol. Assoc. U. K.* **54**:685–697.

Fish, J. D., and Fish, S., 1977, The veliger larva of *Hydrobia ulvae* with observations on the veliger of *Littorina littorea* (Mollusca: Prosobranchia), *J. Zool.* **182**:495–503.

Franz, D. R., 1971, Development and metamorphosis of the gastropod *Acteocina canaliculata* (Say), *Trans. Am. Microsc. Soc.* **90**:174–182.

Franz, D. R., 1973, The ecology and reproduction of a marine bivalve, *Mysella planulata* (Erycinacea), *Biol. Bull.* **144**:93–106.

Fretter, V., 1967, The prosobranch veliger, *Proc. Malacol. Soc. London* **37**:357–366.

Fretter, V., and Graham, A., 1962, *British Prosobranch Mollusks*, The Ray Society, London, 755 pp.

Fretter, V., and Pilkington, M. C., 1971, The larval shell of some prosobranch gastropods, *J. Mar. Biol. Assoc. U. K.* **51**:49–62.

Fürsich, F. T., 1978, The influence of faunal condensation and mixing on the preservation of fossil benthic communities, *Lethaia* **11**:243–250.

Gadgil, M., and Bossert, M., 1970, Life history consequences of natural selection, *Am. Nat.* **104**:1–24.

Gadgil, M., and Solbrig, O. T., 1972, The concept of "r" and "K" selection: Evidence from wild flowers and some theoretical considerations, *Am. Nat.* **106**:14–31.

Gallardo, C. S., 1977a, Two modes of development in the morphospecies *Crepidula dilatata* (Gastropoda: Calyptraeidae) from southern Chile, *Mar. Biol.* **39**:241–251.

Gallardo, C., 1977b, *Crepidula philippiana* n.sp., nuevo gastropodo Calyptraeidae de Chile con especial referencia cal patron de desairollo, *Stud. Neotrop. Fauna Environ.* **12**:177–185.

Giesel, J. T., 1976, Reproductive strategies as adaptations to life in temporally heterogeneous environments, *Annu. Rev. Ecol. Syst.* **7**:57–79.

Gooch, J. L., 1975, Mechanisms of evolution and population genetics, in: *Marine Biology*, Vol. II, *Physiological Mechanisms, Part I.* (O. Kinne, ed.), pp. 349–409, Wiley, New York.

Gooch, J. L., Smith, B. S., and Knupp, D., 1972, Regional survey of gene frequencies in the mud snail *Nassarius obsoletus*, *Biol. Bull.* **142**:36–48.

Gould, S. J., 1977, *Ontogeny and Phylogeny*, Harvard University Press, Cambridge, Massachusetts. 501 pp.

Gould, S. J., and Eldredge, N., 1977, Punctuated equilibria: The tempo and mode of evolution reconsidered, *Paleobiology* **3**:115–151.

Gould, S. J., and Johnston, R. F., 1972, Geographic variation, *Annu. Rev. Ecol. Syst.* **3**:457–498.

Grabau, A. W., 1912, Studies of Gastropoda. IV. Value of the protoconch and early conch stages in the classification of Gastropoda, *Proceedings of the 7th International Zoological Congress*, pp. 753–766.

Grahame, J., 1977, Reproductive effort and r- and K-selection in two species of *Lacuna* (Gastropoda: Prosobranchia), *Mar. Biol.* **40**:217–224.

Grassle, J. F., and Grassle, J. P., 1974, Opportunistic life histories and genetic systems in marine benthic polychaetes, *J. Mar. Res.* **32**:253–284.

Grassle, J. F., and Grassle, J. P., 1977, Temporal adaptations in sibling species of *Capitella*. in: *Ecology of Marine Benthos* (B. C. Coull, ed.), pp. 177–189, University of South Carolina Press, Columbia.

Grassle, J. F., and Sanders, H. L., 1973, Life histories and the role of disturbance, *Deep-Sea Res.* **20**:643–649.

Grassle, J. P., and Grassle, J. F., 1976, Sibling species in the marine pollution indicator *Capitella capitata* (Polychaeta), *Science* **192**:567–569.

Guérin, J. P., 1973, Contribution à l'étude systématique, biologique et écologique des larves méroplanctoniques de polychètes et de mollusques du Golfe de Marseille. 2. Le cycle des larves de lamellibranches, *Tethys* **5**:55–70.

Gunter, G., 1957, Temperature, *Geol. Soc. Am. Mem.* **67**(1):159–184.

Hadfield, M. G., 1963, The biology of nudibranch larvae, *Oikos* **14**:85–95.

Hadfield, M. G., Kay, E. A., Gilette, M. U., and Lloyd, M. C., 1972, The Vermetidae (Mollusca: Gastropoda) of the Hawaiian Islands, *Mar. Biol.* **12**:81–98.

Hansen, T. A., 1978a, Larval dispersal and species longevity in Lower Tertiary gastropods, *Science* **199**:885–887.

Hansen, T. A., 1978b, Ecological control of evolutionary rates in Paleocene–Eocene molluscs, Unpublished Ph.D. dissertation, Yale University, New Haven, Connecticut, 310 pp.

Hansen, T. A., and Jablonski, D., 1979, Larvae of marine invertebrates: Paleontological significance, in: *The Encyclopedia of Paleontology* (R. W. Fairbridge and D. Jablonski, eds.), pp. 416–419, Dowden, Hutchinson & Ross, Stroudsburg, Pennsylvania.

Harrison, W. B., 1978, The occurrence of larval and young post-larval juvenile Mollusca in the Upper Ordovician Cincinnatian Series, *Geol. Soc. Am. Abstr.* **10**:256 (abstract).

Harry, H. W., 1966, Studies on bivalve molluscs of the genus *Crassinella* in the northwestern Gulf of Mexico: Anatomy, ecology and systematics, *Publ. Inst. Mar. Sci.* **11**:65–89.

Heard, W. H., 1977, Reproduction of fingernail clams (Sphaeriidae: *Sphaerium* and *Musculum*), *Malacologia* **16**:421–455.

Heller, J., 1975, The taxonomy of some British *Littorina* species, with notes on their reproduction (Mollusca: Prosobranchia), *Zool. J. Linn. Soc.* **56**:131–151.

Hoagland, K. E., 1977, Systematic review of fossil and recent *Crepidula* and discussion of evolution of the Calyptraeidae, *Malacologia* **16**:353–420.

Horikoshi, M., 1967, Reproduction, larval features and life history of *Philine denticulata* (J. Adams) (Mollusca—Tectibranchia), *Ophelia* **4**:43–84.

Houbrick, R. S., 1973 (1974), Studies on the reproductive biology of the genus *Cerithium* (Gastropoda: Prosobranchia) in the western Atlantic, *Bull. Mar. Sci.* **23**:875–904.

Imbrie, J., and Kipp, N. G., 1971, A new micropaleontological method for quantitative paleoclimatology: Application to a Late Pleistocene Caribbean core, in: *The Late Cenozoic*

Glacial Ages (K. K. Turekian, ed.), pp. 71–183, Yale University Press, New Haven, Connecticut.

Iredale, T., 1910, On marine Mollusca from the Kermadoc Islands, and on the "*Sinusigera* apex," *Proc. Malacol. Soc. London* **9**:68–79.

Iredale, T., 1911, On the value of the gastropod apex in classification, *Proc. Malacol. Soc. London* **9**:319–323.

Jablonski, D., 1978a, Late Cretaceous gastropod protoconchs, *Geol. Soc. Am. Abstr.* **10**:49 (abstract).

Jablonski, D., 1978b, Transgressions, regressions, and endemism in Gulf Coast Cretaceous mollusks, *Geol. Soc. Am. Abstr.* **10**:427–428 (abstract).

Jablonski, D., 1979, Paleoecology, paleobiogeography, and evolutionary patterns of Late Cretaceous Gulf and Atlantic Coastal Plain mollusks, Unpublished Ph.D. dissertation, Yale University, New Haven, Connecticut, 604 pp.

Jablonski, D., and Lutz, R. A., 1978, Larval and juvenile bivalves from Late Cretaceous sediments, *Geol. Soc. Am. Abstr.* **10**:49 (abstract).

Jackson, J. B. C., 1974, Biogeographic consequences of eurytopy and stenotopy among marine bivalves and their evolutionary significance, *Am. Nat.* **108**:541–560.

Jackson, J. B. C., 1977, Some relationships between habitat and biostratigraphic potential of marine benthos, in: *Concepts and Methods of Biostratigraphy* (E. G. Kauffman and J. E. Hazel, eds.), pp. 65–72, Dowden, Hutchinson & Ross, Stroudsburg, Pennsylvania.

Jameson, S. C., 1976 (1977), Early life history of the giant clams *Tridacna crocea* Lamarck, *Tridacna maxima* (Röding), and *Hippopus hippopus* (Linneaeus), *Pac. Sci.* **30**:219–233.

Jones, G. F., 1963, Brood protection in three southern California species of the pelecypod *Cardita, Wassman J. Biol.* **21**:141–148.

Jung, P., 1975, Quaternary larval gastropods from Leg 15, Site 147, Deep Sea Drilling Project, preliminary report, *Veliger* **18**:109–126.

Kauffman, E. G., 1972, Evolutionary rates and patterns of North American Cretaceous Mollusca, *Proc. 24th Int. Geol. Congr.* **7**:174–189.

Kauffman, E. G., 1975, Dispersal and biostratigraphic potential of Cretaceous benthonic Bivalvia in the Western Interior, *Geol. Assoc. Can. Spec. Pap.* **13**:163–194.

Kauffman, E. G., 1977a, Evolutionary rates and biostratigraphy, in: *Concepts and Methods of Biostratigraphy* (E. G. Kauffman and E. Hazel, eds.), pp. 109–141, Dowden, Hutchinson & Ross, Stroudsburg, Pennsylvania.

Kauffman, E. G., 1977b, Systematic, biostratigraphic, and biogeographic relationships between Middle Cretaceous Euramerican and North Pacific Inoceramidae, *Palaeontol. Soc. Jpn. Spc. Pap.* **21**:169–212.

Kauffman, E. G., 1978, Evolutionary rates and patterns in Cretaceous Bivalvia, *Philos. Trans. R. Soc. London Ser. B* **284**:277–304.

Kingston, P. F., 1974a, Studies on the reproductive cycles of *Cardium edule* and *C. glaucum*, *Mar. Biol.* **28**:317–324.

Kingston, P. F., 1974b, Some observations on the effects of temperature and salinity upon the growth of *Cardium edule* and *Cardium glaucum* larvae in the laboratory, *J. Mar. Biol. Assoc. U. K.* **54**:309–318.

Kniprath, E., 1979, The functional morphology of the embryonic shell-gland in the conchiferous molluscs, *Malacologia* **18**:549–552.

Knudsen, J., 1970, The systematics and biology of abyssal and hadal Bivalvia, *Galathaea Rep.* **11**:1–241.

Koehn, R. K., 1976, Migration and population structure in the pelagically dispersing invertebrate, *Mytilus edulis*, in: *Isoenzymes*, Vol. IV, *Genetics and Evolution* (C. L. Markert, ed.), pp. 945–959, Academic Press, New York.

Koehn, R. K., Milkman, R., and Mitton, J. B., 1976, Population genetics of marine pelecypods.

IV. Selection, migration and genetic differentiation in the blue mussel *Mytilus edulis*, *Evolution* **31:**2–32.

Kosuge, S., 1966, The family Triphoridae and its systematic position, *Malacologia* **4:**297–324.

LaBarbera, M., 1974, Larval and post-larval development of five species of Miocene bivalves (Mollusca), *J. Paleontol.* **48:**256–277.

LaBarbera, M., 1975, Larval and post-larval development of the giant clams *Tridacna maxima* and *Tridacna squamosa* (Bivalvia, Tridacnidae), *Malacologia* **15:**69–79.

LaBarbera, M., and Chanley, P., 1970, Larval development of *Chione cancellata* Linne (Veneridae, Bivalvia), *Chesapeake Sci.* **11:**42–49.

Landers, W. S., 1976, The reproduction and early development of the ocean quahog, *Arctica islandica*, in the laboratory, *Nautilus* **90:**88–92.

Lebour, M. V., 1935, The echinospira larvae (Mollusca) of Plymouth, *Proc. Zool. Soc. London* **1935:**163–174.

Lebour, M. V., 1936, Notes on the eggs and larvae of some British prosobranchs, *J. Mar. Biol. Assoc. U. K.* **20:**547–566.

Lebour, M. V., 1937, The eggs and larvae of the British prosobranchs with special reference to those living in the plankton, *J. Mar. Biol. Assoc. U. K.* **22:**105–166.

Lebour, M. V., 1938, Notes on the breeding of some lamellibranchs from Plymouth and their larvae, *J. Mar. Biol. Assoc. U. K.* **23:**119–144.

Lebour, M. V., 1945, The eggs and larvae of some prosobranchs from Bermuda, *Proc. Zool. Soc. London* **114:**462–489.

LePennec, M., 1973, Morphogenèse de la charnière chez 5 especès de Veneridae, *Malacologia* **12:**225–245.

LePennec, M., 1974, Morphogenèse de la coquille de *Pecten maximus* (L.) élevé au laboratoire, *Cah. Biol. Mar.* **15:**475–482.

LePennec, M., and Masson, M., 1976, Morphogenèse de la coquille de *Mytilus galloprovincialis* (Lmk.) élevé au laboratoire, *Cah. Biol. Mar.* **17:**113–118.

Levinton, J. S., 1974, Trophic group and evolution in bivalve molluscs, *Palaeontology* **17:**579–585.

Levinton, J. S., and Bambach, R. K., 1970, Some ecological aspects of bivalve mortality patterns, *Am. J. Sci.* **268:**97–112.

Levinton, J. S., and Bambach, R. K., 1975, A comparative study of Silurian and Recent deposit-feeding bivalve communities, *Paleobiology* **1:**97–124.

Levinton, J. S., and Lassen, H. H., 1978, Experimental mortality studies and adaptation at the Lap locus in *Mytilus edulis*, in: *Marine Organisms: Genetics, Ecology, and Evolution* (B. Battaglia and J. A. Beardmore, eds.), pp. 229–254, Plenum Press, New York.

Loosanoff, V. L., 1954, Advances in the study of bivalve larvae, *Am. Sci.* **42:**607–624.

Loosanoff, V. L., and Davis, H. C., 1963, Rearing of bivalve mollusks, in: *Advances in Marine Biology*, Vol. I (F. S. Russell, ed.), pp. 1–136, Academic Press, London.

Loosanoff, V. L., Davis, H. C., and Chanley, P. E., 1966, Dimensions and shapes of larvae of some marine bivalve mollusks, *Malacologia* **4:**351–435.

Lutz, R. A., 1977, Shell morphology of larval bivalves and its use in ecological and paleo-ecological studies, *Geol. Soc. Am. Abstr.* **9:**1079 (abstract).

Lutz, R. A., 1978, A comparison of hinge line morphogenesis in larval shells of *Mytilus edulis* and *Modiolus modiolus*, *Proc. Natl. Shellfish. Assoc.* **68:**83 (abstract).

Lutz, R. A., and Hidu, H., 1979, Hinge morphogenesis in the shells of larval and early post-larval mussels [*Mytilus edulis* L. and *Modiolus modiolus* (L.)], *J. Mar. Biol. Assoc. U. K.* **59:**111–121.

Lutz, R. A., and Jablonski, D., 1978a, Cretaceous bivalve larvae, *Science* **199:**439–440.

Lutz, R. A., and Jablonski, D., 1978b, Larval bivalve shell morphometry: A new paleoclimatic tool?, *Science* **202:**51–53.

Lutz, R. A., and Jablonski, D., 1978c, Classification of bivalve larvae and early post-larvae using scanning electron microscopy, *Am. Zool.* **18**:647 (abstract).

Lutz, R. A., and Jablonski, D., 1979, Micro- and ultramorphology of larval bivalve shells: Ecological, paleoecological, and paleoclimatic implications, *Proc. Natl. Shellfish. Assoc.* **69**:197–198 (abstract).

MacArthur, R. A., 1972, *Geographical Ecology*, Harper & Row, New York, 269 pp.

MacArthur, R. A., and Wilson, E. O., 1967, *The Theory of Island Biogeography*, Princeton University Press, Princeton, New Jersey, 203 pp.

Marche-Marchad, I., 1968, Un nouveau mode de developpement intercapsulaire chez les mollusques prosobranches néogastropodes: l'Incubation intrapedieuse des *Cymba* (Volutidae), *C. R. Acad. Sci. Ser. D* **266**:706–709.

Marcus, N. H., 1977, Genetic variation within and between geographically separated populations of the sea urchin, *Arbacia punctulata, Biol. Bull.* **153**:560–576.

Marshall, B. A., 1977a, The Recent New Zealand species of *Triforis* (Gastropoda: Triforidae), *N. Z. J. Zool.* **4**:101–110.

Marshall, B. A., 1977b, The dextral triforid genus *Metaxia* (Mollusca: Gastropoda) in the south-west Pacific, *N. Z. J. Zool.* **4**:111–117.

Marshall, B. A., 1978, Cerithiopsidae (Mollusca: Gastropoda) of New Zealand, and a provisional classification of the family, *N. Z. J. Zool.* **5**:47–120.

Mayr, E., 1963, *Animal Species and Evolution*, Harvard University Press, Cambridge, Massachusetts, 797 pp.

Mayr, E., 1970, *Populations, Species, and Evolution*, Harvard University Press, Cambridge, Massachusetts, 453 pp.

McCall, P. L., 1977, Community patterns and adaptive strategies of the infaunal benthos of Long Island Sound, *J. Mar. Res.* **35**:221–266.

McCall, P. L., 1978, Spatial–temporal distributions of Long Island Sound infauna: The role of bottom disturbance in a nearshore habitat, in: *Estuarine Interactions* (M. L. Wiley, ed.), pp. 191–219, Academic Press, New York.

Meadows, P. S., and Campbell, J. I., 1972a, Habitat selection and animal distribution in the sea: The evolution of a concept, *Proc. R. Soc. Edinburgh Sect. B.* **73**:145–157.

Meadows, P. S., and Campbell, J. I., 1972b, Habitat selection by aquatic invertebrates, *Adv. Mar. Biol.* **10**:271–382, 493.

Menge, B. A., 1973, Effect of predation and environmental patchiness on the body size of a tropical pulmonate limpet, *Veliger* **16**:87–92.

Menge, B. A., 1974, Effect of wave action and competition on brooding and reproductive effort in the seastar, *Leptasterias hexactis, Ecology* **55**:84–93.

Menge, B. A., 1975, Brood or broadcast? The adaptive significance of different reproductive strategies in the two intertidal sea stars *Leptasterias hexactis* and *Pisaster ochraceus, Mar. Biol.* **31**:87–100.

Mileikovsky, S. A., 1971, Types of larval development in marine bottom invertebrates, their distribution and ecological significance: A re-evaluation, *Mar. Biol.* **10**:193–213.

Mileikovsky, S. A., 1974a (1976), Types of larval development in marine bottom invertebrates: An integrated ecological scheme, *Thalassia Jugosl.* **10**:171–179.

Mileikovsky, S. A., 1974b, On predation of pelagic larvae and early juveniles of marine bottom invertebrates by adult benthic invertebrates and their passing alive through their predators, *Mar. Biol.* **26**:303–311.

Mileikovsky, S. A., 1975, Types of larval development in Littorinidae (Gastropoda: Prosobranchia) of the World Ocean, and ecological patterns of their distribution, *Mar. Biol.* **30**:129–135.

Milkman, R., and Koehn, R. K., 1977, Temporal variation in the relationship between size, numbers, and an allele-frequency in a population of *Mytilus edulis, Evolution* **31**:103–115.

Millar, R. H., and Scott, J. M., 1967, The larva of the oyster *Ostrea edulis* during starvation, *J. Mar. Biol. Assoc. U. K.* **47:**475–484.

Miyazaki, I., 1962, On the identification of lamellibranch larvae, *Bull. Jpn. Soc. Sci. Fish.* **28:**955–966.

Moore, R. C., 1954, Evolution of Late Paleozoic invertebrates in response to major oscillations of shallow seas, *Bull. Mus. Comp. Zool.* **112:**259–286.

Moore, R. C., 1955, Expansion and contraction of shallow seas as a causal factor in evolution, *Evolution* **9:**482–483.

Murphy, G. I., 1968, Pattern in life history and the environment, *Am. Nat.* **102:**390–404.

Murphy, P. G., 1978, *Collisella austrodigitalis* sp. nov.: A sibling species of limpet (Acmaeidae) discovered by electrophoresis, *Biol. Bull.* **155:**193–206.

Natarajan, A. V., 1957, Studies on the egg masses and larval development of some prosobranchs from the Gulf of Mannar and the Palk Bay, *Proc. Indian Acad. Sci. Sect. B* **46:**170–228.

Newman, G. G., 1967, Reproduction of the South Africa abalone *Haliotis midae*, *Invest. Rep. Div. Sea Fish. S. Afr.* **64:**1–24.

Nicol, D., 1964, An essay on size of marine pelecypods, *J. Paleontol.* **38:**968–974.

Nicol. D., 1966, Size of pelecypods in Recent marine faunas, *Nautilus* **79:**109–113.

Nicol, D., 1978, Size trends in living pelecypods and gastropods with calcareous shells, *Nautilus* **92:**70–79.

Ockelmann, K. W., 1958, Marine Lamellibranchiata, *Medd. Grønland* **122:**1–256.

Ockelmann, K. W., 1965, Developmental types in marine bivalves and their distribution along the Atlantic coast of Europe, in: *Proceedings of the First European Malacological Congress,* London, 1962 (L. R. Cox and J. F. Peake, eds.), pp. 25–35, Conchological Society of Great Britain and Ireland and the Malacological Society of London, London.

Ota, S., 1961, Identification of the larva of *Pinna atrina japonica* (Reeve), *Bull. Jpn. Soc. Sci. Fish.* **27:**107–111.

Pearse, J. S., 1965, Reproductive periodicities in several contrasting populations of *Odontaster validus* Koehler, a common Antarctic asteroid, *Biology of the Antarctic Seas.* II, *Antarct. Res. Ser.* **5:**39–85.

Pearse, J. S., 1969, Slow developing demersal embryos and larvae of the Antarctic sea star *Odontaster validus, Mar. Biol.* **3:**110–116.

Pearse, J. S., and Giese, A. C., 1966, Food, reproduction and organic constituents of the common Antarctic echinoid *Sterechinius neumayeri* (Meisner), *Biol. Bull.* **130:**387–401.

Peterson, C. H., 1977, The paleoecological significance of undetected short-term temporal variability, *J. Paleontol.* **51:**976–981.

Pianka, E. R., 1970, On "r" and "K" selection, *Am. Nat.* **104;**592–597.

Pilkington, M. C., 1974, The eggs and hatching stages of some New Zealand prosobranch molluscs, *J. R. Soc. N. Z.* **4:**411–431.

Pilkington, M. C., 1976, Description of veliger larvae of monotocardian gastropods occurring in Otago plankton hauls, *J. Moll. Stud.* **42:**337–360.

Pilkington, M. C., and Fretter, V., 1970, Some factors affecting the growth of prosobranch veligers, *Helgol. Wiss. Meeresunteres.* **20:**576–593.

Portmann, A., 1925, Der Einfluss der Nähreier auf die Larven-Entwicklung von *Buccinum* and *Purpura, Z. Morphol. Oekol. Tiere* **3:**526–541.

Powell, A. W. B., 1942, The New Zealand Recent and fossil Mollusca of the family Turridae, *Bull. Auckland Inst. Mus.* **2:**1–188.

Purchon, R. D., 1968, *The Biology of the Mollusca,* Pergamon, Oxford, 560 pp.

Radwin, G. E., and Chamberlin, J. L., 1973, Patterns of larval development in stenoglossan gastropods, *Trans. San Diego Soc. Nat. Hist.* **17:**107–117.

Ranson, G., 1967, Les protoconques ou coquilles larvaires des Cyprées, *Mem. Mus. Natl. Hist. Nat. Ser. A Zool.* **47**:93–126.

Raven, C. P., 1966, *Morphogenesis: The Analysis of Molluscan Development,* rev. ed., Pergamon, Oxford, 365 pp.

Rees, C. B., 1950, The identification and classification of lamellibranch larvae, *Hull Bull. Mar. Ecol.* **3**:73–104.

Rehfeldt, N., 1968, Reproductive and morphological variations in the prosobranch "*Rissoa membranacea,*" *Ophelia* **5**:157–173.

Rhoads, D. C., McCall, P. L., and Yingst, J. Y. 1978, Disturbance and production on the estuarine sea floor, *Am. Sci.* **66**:577–586.

Richards, R. P., and Bambach, R. K., 1975, Population dynamics of some Paleozoic brachiopods and their paleoecological significance, *J. Paleontol.* **49**:775–798.

Richter, G., and Thorson, G., 1975, Pelagische Prosobranchier-larven des Golfes von Neapel, *Ophelia* **13**:109–185.

Robertson, R., 1971, Scanning electron microscopy of planktonic larval marine gastropod shells, *Veliger* **14**:1–12.

Robertson, R., 1974 (1976), Marine prosobranch gastropods: Larval studies and systematics, *Thalassia Jugosl.* **10**:213–236.

Robertson, R., Scheltema, R. S., and Adams, F. W., 1970, The feeding, larval dispersal, and metamorphosis of *Philippia* (Gastropoda: Architectonicidae), *Pac. Sci.* **24**:55–65.

Rodriguez Babio, C., and Thiriot-Quiévreux, C., 1974, Gastéropodes de la région de Roscoff: Étude particulière de la protoconque, *Cah. Biol. Mar.* **15**:531–549.

Rodriguez Babio, C., and Thiriot-Quiévreux, C., 1975, Trochidae, Skeneidae et Skeneopsidae (Mollusca, Prosobranchia) de la région de Roscoff: Observations au microscope électronique à balayage, *Cah. Biol. Mar.* **16**:521–530.

Rosenzweig, M. L., 1978, Competitive speciation, *Biol. J. Linn. Soc.* **10**:275–289.

Rosewater, J., 1975, An annotated list of the marine mollusks of Ascension Island, South Atlantic Ocean, *Smithsonian Contrib. Zool.* **189**:1–41.

Sachs, H. M., Webb, T., III, and Clark, D. R., 1977, Paleoecological transfer functions, *Annu. Rev. Earth Planet. Sci.* **5**:159–178.

Sanders, H. L., 1968, Marine benthic diversity: A comparative study, *Am. Nat.* **102**:243–282.

Sanders, H. L., and Allen, J. A., 1973, Studies on deep-sea Protobranchia (Bivalvia): Prologue and Pristiglomidae, *Mus. Comp. Zool. Bull.* **145**:237–262.

Sanders, H. L., and Allen, J. A., 1977, Studies on the deep-sea Protobranchia: The family Tindariidae and the genus *Pseudotindaria, Mus. Comp. Zool. Bull.* **148**:23–59.

Sastry, A. N., 1965, The development and external morphology of pelagic larval and postlarval stages of the bay scallop, *Aequipecten irradians concentricus* Say, reared in the laboratory, *Bull. Mar. Sci.* **15**:417–435.

Savage, N. B., and Goldberg, R., 1976, Investigation of practical means of distinguishing *Mya arenaria* and *Hiatella* sp. larvae in plankton samples, *Proc. Natl. Shellfish. Assoc.* **66**:42–53.

Schaffer, W. M., 1974, Optimal reproductive effort in fluctuating environments, *Am. Nat.* **108**:783–790.

Scheltema, R. S., 1962, Pelagic larvae of New England intertidal gastropods. I. *Nassarius obsoletus* Say and *Nassarius vibex* Say, *Trans. Am. Microsc. Soc.* **81**:1–11.

Scheltema, R. S., 1967, The relationship of temperature to the larval development of *Nassarius obsoletus* (Gastropoda), *Biol. Bull.* **132**:253–265.

Scheltema, R. S., 1971a, Larval dispersal as a means of genetic exchange between geographically separated populations of shallow-water benthic marine gastropods, *Biol. Bull.* **140**:284–322.

Scheltema, R. S., 1971b, The dispersal of the larvae of the shoal-water benthic invertebrate species over long distances by ocean currents, in: *Fourth European Marine Biology Symposium* (D. J. Crisp, ed.), pp. 7–28, Cambridge University Press, Cambridge.

Scheltema, R. S., 1972, Reproduction and dispersal of bottom dwelling deep-sea invertebrates: A speculative summary, in: *Barobiology and the Experimental Biology of the Deep Sea* (R. W. Brauer, ed.), pp. 58–66, University of North Carolina, Chapel Hill.

Scheltema, R. S., 1974 (1976), Biological interactions determining larval settlement of marine invertebrates, *Thalassia Jugosl.* **10**:263–296.

Scheltema, R. S., 1975, The relationship of larval dispersal, gene flow and natural selection to geographic variation of benthic invertebrates in estuaries and along coastal regions, in: *Estuarine Research*, Vol. I: *Chemistry, Biology and the Estuarine System* (L. E. Cronin, ed.), pp. 372–391, Academic Press, New York.

Scheltema, R. S., 1977, Dispersal of marine invertebrate organisms: Paleobiogeographic and biostratigraphic implications, in: *Concepts and Methods of Biostratigraphy* (E. G. Kauffman and J. E. Hazel, eds.), pp. 73–108, Dowden, Hutchinson & Ross, Stroudsburg, Pennsylvania.

Scheltema, R. S., 1978, On the relationship between dispersal of pelagic veliger larvae and the evolution of marine prosobranch gastropods, in: *Marine Organisms: Genetics, Ecology and Evolution* (B. Battaglia and J. A. Beardmore, eds.), pp. 303–322, Plenum Press, New York.

Scheltema, R. S., 1979a, Dispersal of pelagic larvae and the zoogeography of Tertiary benthic gastropods, in: *Historical Biogeography, Plate Tectonics and the Changing Environment* (A. J. Boucot and J. Gray, eds.), pp. 391–397, Oregon State University Press, Corvallis.

Scheltema, R. S., 1979b, Mode of reproduction and inferred dispersal of prosobranch gastropods in the geologic past: Consequences for biogeography and species evolution, *Geol. Soc. Am. Abstr.* **11**:126 (abstract).

Schoener, A., 1968, Evidence for reproductive periodicity in the deep sea, *Ecology* **49**:81–87.

Schoener, A., 1972, Fecundity and possible mode of development of some deep-sea ophiuroids, *Limnol. Oceanogr.* **17**:193–199.

Schopf, T. J. M., and Dutton, A. R., 1976, Parallel clines in morphologic and genetic differentiation in a coastal zone marine invertebrate: The bryozoan *Schizoporella errata*, *Paleobiology* **2**:255–264.

Sellmer, G. P., 1967, Functional morphology and ecological life history of the gem clam, *Gemma gemma* (Eulamellibranchia: Veneridae), *Malacologia* **5**:137–223.

Shuto, T., 1974, Larval ecology of prosobranch gastropods and its bearing on biogeography and paleontology, *Lethaia* **7**:239–256.

Siddall, S. E., 1978, The development of the hinge line in tropical mussel larvae of the genus *Perna*, *Proc. Natl. Shellfish. Assoc.* **68**:86 (abstract).

Simon, J. L., and Dauer, D. M., 1977, Reestablishment of a benthic community following natural defaunation, in: *Ecology of Marine Benthos* (B. C. Coull, ed.), pp. 139–154, University of South Carolina Press, Columbia.

Simpson, R. D., 1977, The reproduction of some littoral molluscs from Macquarie Island (sub-Antarctic), *Mar Biol.* **44**:125–142.

Smith, B., 1945, Observations on gastropod protoconchs, *Palaeontogr. Am.* **3**(19):220–270.

Smith, B., 1946, Observations on gastropod protoconchs, *Palaeontogr. Am.* **3**(21):285–304.

Smith, S. T., 1967, The development of *Retusa obtusa* (Montagu) (Gastropoda, Opisthobranchia), *Can. J. Zool.* **45**:737–746.

Snyder, T. P., and Gooch, J. L., 1973, Genetic differentiation in *Littorina saxatilis* (Gastropoda), *Mar Biol.* **22**:177–182.

Sohl, N. F., 1977, Utility of gastropods in biostratigraphy, in: *Concepts and Methods of*

Biostratigraphy (E. G. Kauffman and J. E. Hazel, eds.), pp. 519–539, Dowden, Hutchinson, & Ross, Stroudsburg, Pennsylvania.

Soot-Ryen, T., 1960, Pelecypods from Tristan da Cunha, *Res. Norw. Sci. Exped. Tristan da Cunha* **49**:1–47.

Sorgenfrei, T., 1958, Molluscan assemblages from the marine Middle Miocene of South Jutland and their environments, *Dan. Geol. Unders.* [*Afh.*] *Raekke 2* **79** (2 vols.).

Southwood, T. R. E., 1976, Bionomic strategies and population parameters, in: *Theoretical Ecology: Principles and Applications* (R. M. May, ed.), pp. 24–48, Saunders, Philadelphia.

Spight, T. M., 1975, Factors extending gastropod embryonic development and their selective cost, *Oecologia* **21**:1–16.

Spight, T. M., 1976a, Hatching size and the distribution of nurse eggs among prosobranch embryos, *Biol. Bull.* **150**:491–499.

Spight, T. M., 1976b, Ecology of hatching size for marine snails, *Oecologia* **24**:283–294.

Spight, T. M., 1977, Latitude, habitat, and hatching type for muricacean gastropods, *Nautilus* **91**:67–71.

Stafford, J., 1912, On the recognition of bivalve larvae in plankton collections, *Contrib. Can. Biol.* **1906–1910**:221–242.

Stanley, S. M., 1975a, A theory of evolution above the species level, *Proc. Natl. Acad. Sci. U.S.A.* **72**:646–650.

Stanley, S. M., 1975b, Adaptive themes in the evolution of the Bivalvia, *Ann. Rev. Earth Planet, Sci.,* **3**:361–385.

Stanley, S. M., 1977, Trends, rates, and patterns of evolution in the Bivalvia, in: *Patterns of Evolution* (A. Hallam, ed.), pp. 209–250, Elsevier, Amsterdam.

Stanley, S. M., 1978, Chronospecies' longevities, the origin of genera, and the punctuational model of evolution, *Paleobiology* **4**:26–40.

Stearns, S. C., 1976, Life-history tactics: A review of the ideas, *Q. Rev. Biol.* **51**:3–47.

Stearns, S. C., 1977, The evolution of life history traits: A critique of the theory and a review of the data, *Annu. Rev. Ecol. Syst.* **8**:145–171.

Steele, D. H., 1977, Correlation between egg size and developmental period, *Am. Nat.* **111**:371–372.

Stenzel, H. B., 1971, Oysters, in: *Treatise on Invertebrate Paleontology, Mollusca 6, Part N, Bivalvia* (R. C. Moore, ed.), Vol. 3, pp. N953–N1224, Geological Society of America and University of Kansas Press, Lawrence.

Strathmann, R. R., 1974, The spread of sibling larvae of sedentary marine invertebrates, *Am. Nat.* **108**:29–44.

Strathmann, R. R., 1977, Egg size, larval development, and juvenile size in benthic marine invertebrates, *Am. Nat.* **111**:373–376.

Struhsaker, J. W., 1968, Selection mechanisms associated with intraspecific shell variation in *Littorina picta* (Prosobranchia; Mesogastropoda), *Evolution* **22**:459–480.

Sullivan, C. M., 1948, Bivalve larvae of Malpeque Bay, P. E. I., *Bull. Fish. Res. Board Can.* **77**:1–36.

Tanaka, Y., 1970, Studies on molluscan larvae. 2. *Pinctada margaritifera, Venus* **29**:117–122.

Tanaka, Y., 1971, Studies on molluscan larvae. 3. *Anadara (Scapharca) broughtonii, Venus* **30**:29–34.

Taylor, J. B., 1975, Planktonic prosobranch veligers of Kaneohe Bay, Unpublished Ph.D. dissertation, University of Hawaii, Honolulu, 606 pp.; *Diss. Abstr.* **36B**:2110–2111.

Thiede, J., 1974, Marine bivalves: Distribution of mero-planktonic shell-bearing larvae in eastern North Atlantic surface waters, *Palaeogeogr. Palaeoclimatol. Palaeoecol.* **15**:267–290.

Thiede, J., and Dinkelmann, M. G., 1977, Occurrence of *Inoceramus* remains in Late Mesozoic pelagic and hemipelagic sediments, *Initial Reports, Deep Sea Drilling Project*, Vol. 39, pp. 899–910, U.S. Government Printing Office, Washington, D.C.

Thiel, H., 1975, The size structure of the deep-sea benthos, *Int. Rev. Ges. Hydrobiol.* **60:**575–606.

Thiriot-Quiévreux, C., 1972, Microstructures de coquilles larvaries de prosobranches au microscope électronique à balayage, *Arch. Zool. Exp. Gen.* **113:**553–564.

Thiriot-Quiévreux, C., and Rodriguez Babio, C., 1975, Étude des protoconques de quelques prosobranches de la région de Roscoff, *Cah. Biol. Mar.* **16:**135–148.

Thorson, G., 1935, Studies on the egg-capsules of Arctic marine prosobranchs, *Medd. Grønland* **100:**1–71.

Thorson, G., 1946, Reproduction and larval development of Danish marine bottom invertebrates, *Medd. Dan. Fisk. Havunders.* **4:**1–523.

Thorson, G., 1950, Reproductive and larval ecology of marine bottom invertebrates, *Biol. Rev.* **25:**1–45.

Thorson, G., 1961, Length of pelagic life in marine bottom invertebrates as related to larval transport by ocean currents, in: *Oceanography* (M. Sears, ed.), *Am. Assoc. Adv. Sci. Publ.* **67:**455–474.

Thorson, G., 1965, The distribution of benthic marine Mollusca along the N.E. Atlantic shelf from Gibraltar to Murmansk, in: *Proceedings of the First European Malacological Congress*, London, 1962 (L. R. Cox and J. F. Peake, eds.), pp. 5–23, Conchological Society of Great Britain and Ireland and the Malacological Society of London.

Thorson, G., 1966, Some factors influencing the recruitment and establishment of marine benthic communities, *Neth. J. Sea Res.* **3:**267–293.

Togo, Y., 1977, The shell structure of the protoconch and innermost shell layer of the teleoconch in marine prosobranch gastropods, *J. Geol. Soc. Jpn.* **83:**567–573.

Turner, R. D., 1973, Wood-boring bivalves, opportunistic species in the deep sea, *Science* **180:**1377–1379.

Turner, R. D., and Boyle, P. J., 1975, Studies of bivalve larvae using the scanning electron microscope and critical point drying, *Am. Malacol. Inc. Bull. Union* **1974:**59–65.

Underwood, A. J., 1974, On models for reproductive strategy in marine benthic invertebrates, *Am. Nat.* **108:**874–878.

Usuki, I., 1966, The life cycle of *Haloa japonica* (Pilsbry). I. The larval development *Sci. Rep. Niigata Univ. Ser. D* **3:**87–105.

Valentine, J. W., and Ayala, F. J., 1978, Adaptive strategies in the sea, in: *Marine Organisms: Genetics, Ecology, and Evolution* (B. Battaglia and J. A. Beardmore, eds.), pp. 323–345, Plenum Press, New York.

Vance, R. R., 1973a, On reproductive strategies in marine benthic invertebrates, *Am. Nat.* **107:**339–352.

Vance, R. R., 1973b, More on reproductive strategies in marine benthic invertebrates, *Am. Nat.* **107:**353–361.

Vance, R. R., 1974, Reply to Underwood, *Am. Nat.* **108:**879–880.

Vermeij, G. J., 1972, Endemism and environment: Some shore molluscs of the tropical Atlantic, *Am. Nat.* **106:**89–101.

Vermeij, G. J., 1978, *Biogeography and Adaptation: Patterns of Marine Life*, Harvard University Press, Cambridge, Massachusetts, 332 pp.

Walker, K. R. and Alberstadt, L. P., 1975, Ecological succession as an aspect of structure in fossil communities, *Paleobiology* **1:**238–257.

Walker, K. R., and Parker, W. C., 1976, Population structure of a pioneer and a later stage species in an Ordovician ecological succession, *Paleobiology* **2:**191–201.

Waller, T. R., 1978, Morphology, morphoclines and a new classification of the Pteriomorphia (Mollusca: Bivalvia), *Philos. Trans. R. Soc. London Ser. B* **284:**345–365.

Waller, T. R., 1980, Morphology and development of the veliger larvae of the European oyster, *Ostrea edulis* Linné, observed by scanning electron microscopy, *Smithsonian Contrib. Zool.* (in press).

Webber, H. H., 1977, Gastropoda: Prosobranchia, in: *Reproduction of Marine Invertebrates*, Vol. IV, *Molluscs: Gastropods and Cephalopods* (A. C. Giese and J. S. Pearse, eds.), pp. 1–97, Academic Press, New York.

Wells, M. J., and Wells, J., 1977, Cephalopods: Octopoda, in: *Reproduction of Marine Invertebrates*, Vol. IV, *Molluscs: Gastropods and Cephalopods* (A. C. Giese and J. S. Pearse, eds.), pp. 291–336, Academic Press, New York.

Werner, B., 1939, Über die Entwicklung und Artunterscheidung von Muschellarven des Nordseeplanktons, unter besonderer Berüchsichtigung der Schalenentwicklung, *Zool. Jahrb. Abt. Anat. Ontog.* **66:**1–54.

White, M. J. D., 1978, *Modes of Speciation*, Freeman, San Francisco, 455 pp.

Wilkins, N. P., O'Reagan, D., and Maynihan, E., 1978, Electrophoretic variability and temperature sensitivity of phosphoglucose isomerase and phosphoglucomutase in littorinids and other marine molluscs, in: *Marine Organisms: Genetics Ecology, and Evolution* (B. Battaglia and J. A. Beardmore, eds.), pp. 141–155, Plenum Press, New York.

Woodin, S. A., 1976, Adult–larval interactions in dense infaunal assemblages: Patterns of abundance, *J. Mar. Res.* **34:**25–41.

Yoshida, H., 1953, Studies on larvae and young shells of industrial bivalves in Japan, *J. Shimonoseki Coll. Fish.* **3:**1–106.

Yonge, C. M., 1969, Functional morphology and evolution within the Carditacea, *Proc. Malacol. Soc. London* **38:**493–528.

Chapter 10

Gastropod Shell Growth Rate, Allometry, and Adult Size
Environmental Implications

GEERAT J. VERMEIJ

1. Introduction	379
2. Geometry of Shell Growth	379
3. Allometry and Adaptation	384
4. Growth Rate and Adult Size	387
5. Application to Environmental Reconstructions	390
References	391

1. Introduction

Gastropod shells increase in size primarily at the outer lip of the aperture. The shell is usually external to the soft parts, and a more or less complete record of postlarval ontogeny is preserved in the shell. This makes gastropods suitable for the study of environmental and genetic controls of growth, and for monitoring changes in the environment.

This chapter is not so much a review of existing data as it is a statement of some general principles and an enumeration of promising lines of research. I shall emphasize the interdependence of growth rate, shape, and adult size, and I shall draw attention to the striking variability of growth rate among members of the same species. It must be pointed out from the beginning that many of the issues raised here apply equally to other animals.

2. Geometry of Shell Growth

The gastropod shell is essentially a narrow cone of calcium carbonate that grows at the wide end. In limpetlike gastropods, the growth axis of

GEERAT J. VERMEIJ ● Department of Zoology, University of Maryland, College Park, Maryland 20742.

the cone is nearly a straight line, but in coiled shells, the axis of the cone traces a curve closely approximating a right-handed (rarely a left-handed or planispiral) logarithmic conispiral. This curve is technically the geodesic on a right circular cone, that is, the shortest distance between two points on the surface of the cone. The geodesic coils around a straight axis running from the apex to the base of the shell. In *Balcis*, and a few other living gastropods, the shell axis is itself slightly curved. Numerous land snails and a number of Paleozoic marine genera display a marked change in direction of the shell axis during ontogeny.

Marginal growth in coiled shells takes place at the outer (abaxial) rim of the aperture. In an idealized shell, shape remains constant throughout postlarval growth (Thompson, 1942); that is, each point along the growing edge traces a geodesic of a right circular cone as the shell grows. Because points on different parts of the growing edge lie at different distances from the axis of coiling, the rate and direction of growth are unique at each point. For a shell of a given coiling direction, growth along a specified geodesic can be described in terms of three quantities: (1) length (L) of the geodesic deposited per unit of time; (2) expansion rate (W), a measure of the rate at which a point on the geodesic moves radially away from the axis of coiling; and (3) translation rate (T), a measure of the rate at which a point on the geodesic moves parallel to the axis of coiling in a direction away from the apex. Expansion rate is defined as

$$W^{\theta/2\pi} = r_2/r_1 \tag{1}$$

where r_1 and r_2 are distances between the axis of coiling and two points lying θ radians apart on the geodesic; r_2 is defined to be greater than r_1 so that W is greater than unity. Translation rate is defined as

$$T = dz/dr \tag{2}$$

where z and r are distances parallel and perpendicular to the axis of coiling, respectively (Fig. 1). The quantities T and W vary independently among species, but within individuals, one is usually an inverse function of the other (Raup, 1961, 1966, 1967) (for other aspects of shell geometry, see Vermeij, 1971a,b, 1973b; Graus, 1974; Linsley, 1977).

The geometry of shell growth has implications for the measurement of marginal growth in snails. Marginal growth is most accurately measured by following a specified geodesic over the course of time. For convenience, one might use the geodesic at the suture or one corresponding to a distinctive color band or sculptural feature. This method, which requires marking the shell and periodically measuring the rate and direction of

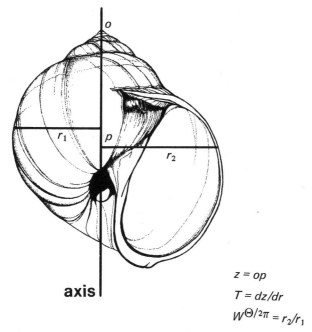

$$z = op$$
$$T = dz/dr$$
$$W^{\Theta/2\pi} = r_2/r_1$$

Figure 1. Geometry of the gastropod shell. op is the axis of the shell. $T = dz/dr$, where T is translation rate, z is the distance along the axis, and r is the distance perpendicular to the axis of the shell. $W^{\theta/2\pi} = r_2/r_1$ is the expansion rate of the whorls, where r_1 and r_2 are distances from the axis to points on the geodesic separated by an angular distance of θ radians.

growth, is preferable in many ways to the conventional approach of measuring increments in overall shell length. Increase in length is only one component of marginal growth, and its measurement in many snails is complicated by erosion near the apex (Spight, 1973; Spight et al., 1974).

Moreover, most gastropods do not conform to the idealized condition of constancy of shape during growth. Typically, they display some degree of allometry (change in shape with increasing size). This means that the geometrical properties T and W of a given geodesic change with increasing size (and age) of the shell; furthermore, if translation rate T decreases allometrically while W increases, marginal growth that is inferred from measuring shell length will be underestimated.

Allometrical change in T and W is frequently caused or accompanied by a progressive or sudden change in the shape of the growing edge or generating curve. The aperture of many littorinids and zonitid land snails becomes gradually less angular in outline (Gould, 1968; Andrews, 1974). Many gastropods in which growth is determinate (that is, marginal growth ceases at maturity) have the adult aperture modified greatly from that of the actively growing juvenile stage; obvious examples among Recent snails include Strombidae, Cypraeidae, Cerithiidae, Cymatiidae, Muricidae, and a large number of land snails with a reflected outer lip.

Gould (1968) suggested that doming (allometry in which relative spire height increases during ontogeny) is very common among land snails. Similar allometry has been described in later growth stages of littorinids, intertidal trochids, a Paleocene *Turritella*, and *Thais lapillus* (Moore, 1936, 1937; Regis, 1969; Vermeij, 1973a; Andrews, 1974). Shells that display doming can be recognized by the convex lateral profile of the spire. (Fig. 2). By this criterion, doming also occurs in most marine Modulidae, Turbinidae, Cerithiidae, and Terebridae, among other families. Allometry in limpets usually involves an increase in shell height relative to basal diameter, and is thus also a kind of doming. Doming is especially well marked in high-intertidal temperate species such as *Acmaea digitalis* and *A. persona* in the northeastern Pacific, *A. orbignyi* in western South America, and *Helcion pectunculus* and *Patella granularis* in southern Africa (Hamai, 1937; Giesel, 1969; Vermeij, 1973a; Branch, 1975a).

The opposite type of allometry (flaring) is comparatively rare in the later growth stages of gastropods, and it has not been studied quantitatively in any species. Flaring can be recognized superficially by the concave lateral profile of the spire (Fig. 2). Using this criterion, I suspect that flaring is widespread in Cypraeidae, Olividae, Conidae, and several other short-spired marine groups, as well as in *Clausilia*, *Papuina*, and a few other land snails. It is also a characteristic gerontic feature of *Trochus niloticus* (Gail and Devambez, 1958) and *Telescopium telescopium* (Butot, 1954), two large marine species with a protruding spire. Some shells in which external spines increase disproportionately in size during ontogeny (*Strombus*, *Tympanotonus*, *Vasum*, *Melongena*, *Drupa*, and some populations of *Thais haemastoma* in West Africa) may also have a concave lateral profile of the spire. Flaring may be very common in the early stages of postlarval growth. This is strongly hinted at in the work of Moore (1936, 1937) on *Thais lapillus* and *Littorina littorea*. It may also occur in species of the marine genera *Tectarius*, *Nodilittorina*, *Cerithium*, and *Terebra*, among others.

Allometric changes in shape may be accompanied by changes in color pattern and shell sculpture (also see Gould, 1968, 1971a,b). The western tropical American pulmonate limpet *Siphonaria gigas* has very strong fluted radial ribs in the early stages of shell growth when the shell is relatively low-conic. As growth proceeds, shell height increases allometrically with basal diameter, and the ribs become much weaker. The intensity of ribbing in this limpet, and in other species, appears to be dictated by the direction of growth at the shell edge; the more the growth

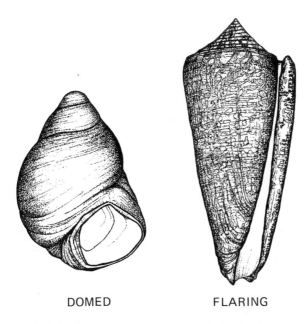

DOMED   FLARING

Figure 2. Gastropod shell allometry. Doming is recognizable by the convex lateral profile of the spire, whereas flaring is recognizable by a concave profile.

direction departs from the horizontal plane (the plane of the base of the shell), the weaker the ribs become (Vermeij, 1971b). A similar relationship between sculpture and shape exists in *S. lessoni* in Argentina (Bastida *et al.*, 1971).

In addition to allometric relationships involving spire height, many gastropods display qualitative changes in shell sculpture. In most cases, spiral or axial (collabral) ornamentation that is present on early postnuclear whorls is reduced or entirely lost on later whorls. Examples may be found in the marine families Trochidae (*Trochus niloticus, Tectus pyramis, Calliostoma* spp., and many others), Potamididae (*Cerithidea*), Strombidae (*Strombus, Tibia*), Cerithiidae (some *Rhinoclavis*), Planaxidae (*Planaxis nucleus*), Phasianellidae (*Tricolia*), Thaididae (*Thais melones*), Buccinidae (*Northia*), Fasciolariidae (*Fusinus*), and Terebridae (*Terebra*). A much smaller number of taxa shows the reverse pattern, in which new types of sculpture are introduced during postlarval ontogeny. This latter pattern is found principally in species that develop varices or spines; examples include *Tympanotonus radula* (Potamididae), some *Pachymelania* (Melaniidae), and *Terebra crenulata* (Terebridae). Excluded from these cases of sculptural allometry are numerous examples in which the number of axial or spiral elements is increased or reduced during growth.

3. Allometry and Adaptation

Several attempts have been made to explain gastropod allometry in adaptational terms. Gould (1968) suggested that rounding of the aperture and doming in land snails reduces the rate at which the adhesive area of the foot decreases relative to foot volume as the snail grows; that is, allometry ameliorates the otherwise sharp, steady decline in the surface/volume ratio during ontogeny and allows large snails to cling to substrata almost as effectively as small ones. I suggested, but did not prove, that limpets and littorinids in the intertidal zone show a positive allometry of shell height or spire height relative to apertural size as an adaptive response to desiccation stress (Vermeij, 1973a). Allometry accentuates the rate at which foot surface area decreases with volume; thus, it could improve protection against area-dependent desiccation that is already implicit in large animals that do not display allometry (see Segal, 1956; Davies, 1969). Andrews (1974) thought that rounding of the aperture and increase in translation rate with increasing size in *Turritella* permit the relative expansion of filter-feeding organs. These explanations are plausible, and I have little doubt that substantial apertural modifications at later growth stages are consistent with adaptational demands; however, none of the adaptational explanations has been subjected to experimental analysis.

Various lines of evidence have led me to believe that many instances of gradual changes in shell allometry (especially doming) are geometrically tied to growth rate. Growth rate accelerates to a peak early in post-larval life and then declines gradually toward maturity (Moore, 1936, 1937; Thompson, 1942; von Bertalanffy, 1957). This pattern corresponds to the ontogeny of spire height as it is qualitatively understood in many snails. An episode of slight flaring is followed by isometry (constancy of shape with increasing size) and eventually by doming. As a first approximation, translation rate ($T = dz/dr$) may be proportional to the rate of change of growth rate (dL/dt):

$$T \propto d^2L/dt^2 \tag{3}$$

As growth slows down, the growth rates of the geodesics farthest from the axis of coiling are affected most (R. C. Highsmith, personal communication). This is because more skeletal material is needed to traverse an angular increment far from the axis of coiling than to traverse the same increment closer to the axis. To reduce the large linear growth increment of abaxial geodesics, and to maintain a more or less planar generating curve at the growing edge, expansion rate (W) can be reduced. This often results in a slight increase in translation rate (T) and in a gradual abapical shift in the point of contact between the growing edge and the previous whorl. This geometry describes doming.

Flaring in later growth stages may similarly have a geometric explanation. In their study of the growth of *Ceratostoma foliatum* (a muricid), Spight and Lyons (1974) noted that this snail alternates a spiral growth phase with a radial growth phase. In spiral growth, a roughly constant angular increment of shell is added to the growing edge during ontogeny. In radial growth, shell material is laid down in the form of a varix or thickening of the outer lip. The spiral and radial phases increase in duration as the animal grows. If the radial phase begins to predominate over the spiral, the varices, or other collabral sculptural elements formed during the radial phase, will increase disproportionately in size. Thus, a concave profile of the spire would be created. Variations on the growth theme observed in *Ceratostoma* are known in the Cymatiidae, Muricidae, and other related families (Laxton, 1970; Spight *et al.*, 1974).

The best evidence that there is a geometric connection between growth rate and intensity of allometry comes from work on cap-shaped limpets. Many high-intertidal species of *Acmaea* and *Patella* display pronounced allometry in which shell height increases at a faster rate than basal diameter. Limpets living among barnacles in the high-intertidal zone have a slower rate of growth, and a relatively higher-peaked shell, than do animals in lower zones and animals living near a richer supply of algal food (Lewis and Bowman, 1975). When Haven (1973) removed all *Acmaea digitalis* from a mixed assemblage of *A. digitalis* and *A. scabra* in Central California, the remaining *A. scabra* began to grow faster than undisturbed controls, and new increments of shell were flared relative to the older portion. Removal of *A. scabra* from sites where both species lived together had a similar but less pronounced effect on *A. digitalis*. Unfortunately, growth vectors of the shells were not explicitly measured by Haven; hence, the correlation between allometry and growth rate remains to be established quantitatively.

In some preliminary studies of *Thais lamellosa*, Spight (1973) showed that snails growing in the laboratory or in the field in the presence of densely packed barnacles (their principal food) acquired fatter shells with a lower spire than did snails in environments where food was scarce. The effect of temperature on growth rate and shell shape was not considered, but it is plausible that snails grown at low temperatures grow more slowly and have higher spires than snails in warmer surroundings. Phillips *et al.* (1973) performed laboratory experiments showing that temperature and food have a profound effect on shell form, growth rate, and sculpture in Australian thaidids of the genus *Dicathais*. When a spirally ribbed form (*D. orbita*) from Victoria and New South Wales was maintained on a diet of mussels at temperatures normally experienced by the nodulose western

Australian form (*D. aegrota*), the shell of *D. orbita* became smooth; control *D. orbita* did not change in form or sculpture. It was concluded that these two forms, as well as a smooth to weakly striate form from southern and eastern Australia (*D. textilosa*), are variants of a single highly polymorphic species, and that the variants represent environmental responses to differences in temperature and food supply, factors that control growth rate and therefore shell architecture.

The small European limpet *Patina pellucida* consists of two distinct growth forms that were formerly regarded as separate species. Graham and Fretter (1947) showed that the shell of the form typically found on kelp fronds is oval (mean breadth/length ratio 0.68), low-conic (height/length ratio 0.27), and nearly transparent, and has the apex near the anterior margin. The form that lives on kelp holdfasts has a rounded shell (breadth/length ratio 0.82) that is more high-conic (height/length ratio 0.34) and opaque, and the apex of which is more central in position. Careful examination reveals that the form on holdfasts has an early growth stage that is identical in its features to the frond form.

Freshwater snails are notorious for their plasticity of form (Fretter and Graham, 1963). Although no carefully controlled work seems to have been done on environmentally determined form in freshwater species, Schwetz (1954) found that bulinids that were transferred from a lake or river to an aquarium acquired relatively higher spires and larger apertures. Frömming (1956) noted, from a lifetime of culturing snails, that maximum adult size depends on the size of the culture vessel in which the snails are kept. In experiments with the polymorphic European freshwater neritid *Theodoxus fluviatilis*, Neumann (1959) discovered that the color pattern on the shell was influenced by the ratio of calcium to magnesium ions.

Ino (1949) noted that *Turbo cornutus* grown under conditions of reduced temperature and salinity in laboratory aquaria loses the spines characteristic of shells collected in the field in Japan. When the abalone *Haliotis kamchatkana* is transplanted from northern Japan to warmer southern waters, its form becomes round and flattened (Ino, 1952). Beu (1970) has pointed out that deep-water specimens of *Charonia* and other cymatiids have a thinner, more delicate, and often higher-spired shell than specimens from shallow water. This difference, which is frequently believed to distinguish species, may be little more than an expression of different environments on animals that are genetically similar. The same may be true of many other gastropods (e.g., many Muricidae and Conidae), but experiments are badly needed on this topic.

In archaeogastropods, and perhaps some other snails, shell pigment is evidently influenced by diet. Japanese *Turbo cornutus* raised on the brown alga *Eisenia bicyclis* become white in color, whereas those fed on the red algae *Cheilosporum maximum* and *Corallina pilulifera* retain the greenish-brown color typical of specimens collected in the field. The red alga *Gelidium* induces a greenish-black color (Ino, 1949, 1953). Leighton (1961) noted that the abalone *Haliotis rufescens* is white when it feeds on the holdfasts of the brown kelp *Macrocystis pyrifera* in California, whereas the shell is dark brick-red when the snail feeds on red algae.

The effects of the environment on growth rate, shape, and color of snail shells, and the postulated correspondence between growth rate and allometry, have obvious implications for gastropod systematics. Shell features may be so profoundly altered by the environment that the unwary taxonomist may regard variants as separate species when in fact the morphs are merely different phenotypes determined by different environmental regimes. The taxonomic question can be resolved only if the forms are grown reciprocally in each other's environments. This point has been made forcefully by Gould (1968, 1971a,b, 1977). He examined Pleistocene populations of the Bermudian land snail genus *Poecilozonites*. Shell thickness, intensity of doming, and the tendency for flames of color to coalesce into continuous color bands on the shell all increase as the rate of marginal growth declines. Paedomorphic *Poecilozonites* may attain the same size as "normal" specimens, but the former retain such juvenile features as a thin shell, blotched color pattern, a wide-open umbilicus, and a low intensity of doming at adult sizes. Gould believes that retention of these attributes reflects a faster than normal growth rate or, perhaps more accurately, a slower rate of decline of growth rate with increasing size and age. Similar relationships have been postulated in Jurassic oysters of the genus *Gryphaea* (Gould, 1972). The important point is that many seemingly independent traits are directly controlled by growth rates; that is, a morphologically distinct shell is produced when only growth rate is altered.

4. Growth Rate and Adult Size

One of the most striking features of snails with determinate growth is their great variability in adult size, both within and among populations. Adult *Cerithium nodulosum* from the inner and middle portions of the reef flat at Pago Bay, Guam (Mariana Islands), range in shell length from 64 to 98 mm, whereas those on the outer flat range from 69 to 102 mm (Yamaguchi, 1977). At St. John in the Virgin Islands, adult queen conchs (*Strombus gigas*) range in length from 151 to 261 mm; a similar size range (143–264 mm) is found in the Bahamas (Randall, 1964). Analysis of a large population of the cowry *Cypraea lynx* killed during a spell of unusually low sea level in Pago Bay during October 1972 revealed that adults of this species range from about 25 to over 50 mm in length (Yamaguchi, 1975). Adult *Morula granulata* collected from various sites show enormous variation in shell size among populations; some representative adult mean lengths are 10.4 mm (N = 11) for specimens from a lava shore near Makapu on the north shore of Oahu, Hawaii; 15.8 mm (N = 20) for a limestone bench at Kahuku, Oahu; 20.1 mm (N = 6) for Coconut Island, again on the north coast of Oahu; and 22.5 mm (N = 20) for the outer reef flat at Pago Bay, Guam. Similar differences in mean adult sizes are evident in *M. uva, Drupa ricinus, Clypeomorus morus, Cerithium columna, Cypraea moneta, Columbella mercatoria, Cantharus ringens,* and many other species (also see Taylor, 1978). Foin (1972) gives some preliminary evidence for a geographic pattern in adult size of the tiger cowry, *Cypraea tigris;* the largest specimens are found in the central Pacific, and the smallest in areas in or near the Indonesian archipelago and western Pacific arc.

There is some evidence that adult size is determined to an important extent by marginal growth rate. Tagged specimens of *Cerithium nodulosum* in Guam grow (in shell length) at rates varying from 2.3 to 4.6 mm per month; this variation is reflected in the wide range of adult size referred to earlier (Yamaguchi, 1977). Studies on *Ceratostoma foliatum* convinced Spight *et al.* (1974) that adult size in this snail and in other muricids can be predicted from juvenile growth rates. Laxton (1970) found that some specimens of the New Zealand cymatiid *Cabestana spengleri* added four varices to their shell in 1 year, whereas other specimens of similar age and from the same locality (Mill Bay) added none during 2 years. The average length of *C. spengleri* from Mill Bay after 1 year was about 55 mm, whereas the average length of animals of the same age from Ahipara was only 28 mm. These differences in growth rate were reflected in adult size, and were correlated by Laxton with food supply; ascidians, which appear to comprise the major part of the diet of *C. spengleri*, were far more abundant at Mill Bay than at Ahipara. Different specimens of *Thais lamellosa* from a single population have growth rates varying by a factor of more than 2 even when food, population density, and temperature are kept constant (A. R. Palmer, personal communication).

There is also great variation in growth rates of snails with indeterminate growth (Bretos, 1978). The South African limpet *Patella granularis* grows at a slower rate in the presence of densely packed barnacles (which restrict the limpet's movement) than on surfaces where limpets are less confined and where food is more available (Branch, 1976). Lewis and Bowman (1975) found that *P. vulgata* in Yorkshire has a similar response to barnacle cover. Limpets above a critical population density (400 individuals per square meter for *P. vulgata*) grow more slowly than those living under less crowded conditions. This fact has been established experimentally for limpets in various parts of the world (Breen, 1972; Haven, 1973; Branch, 1975b, 1976; Lewis and Bowman, 1975).

Numerous studies have established that growth rate declines in an upshore direction in most intertidal organisms, apparently because time available for feeding is reduced. A good demonstration of this fact has been provided by Lewis and Bowman (1975). In Yorkshire, *Patella vulgata* with an initial length of 10 mm grew by an increment of 10 mm over three growing seasons at midtide level, and by an increment of 8.25 mm in the high-intertidal zone. Corresponding increments for specimens with an initial length of 20 mm were 7.5 and 3.5 mm, respectively (also see Branch, 1974).

Population density and intertidal height occasionally counteract each other in their effect on growth rate. Sutherland (1970), for example, has shown that *Acmaea scabra* in Central California grows faster in Zone 1 (high intertidal) than in Zone 2 (high middle intertidal) because the limpets' density in the higher zone is sufficiently low to compensate for the retarding effects on growth imposed by greater exposure to air.

Another local factor that influences growth rate is quantity or quality of food. Frömming (1956) noted that the freshwater snail *Planorbarius corneus* grows more rapidly on a diet of *Potamogeton* than do snails at similar population densities on a diet of *Mentha aquatica*. The difference in growth rate does not apply, however, to *Lymnaea stagnalis*. Unpublished data (J. Lubchenco) on the growth of *Littorina littorea* in New England show that growth is negligible when the snail is fed the brown rockweed *Ascophyllum nodosum* in the laboratory, whereas growth is quite rapid if ephemeral green algae (e.g., *Ulva*, *Enteromorpha*) are offered as food. One of the most thorough studies of the effect of food quality on the growth of herbivores is that of Vadas (1977) on the sea urchins of the genus *Strongylocentrotus* in the temperate northeastern Pacific. *Strongylocentrotus droebachiensis* prefers the large brown alga *Nereocystis leutkeana* over all other potential foods, and is least prone to consume two species of *Agarum*. Growth in body weight and test diameter is highest when the urchins' diet consists of *Nereocystis*, and least when only *Agarum* is eaten. Gonad weight and reproductive value are also highest on a diet of *Nereocystis*.

Vadas's results underscore the importance of the environment in the determination of growth rate and reproductive capacity of the individual animal. The ability of one animal to compete effectively against another, and the rate at which an animal can achieve a size at which it has become immune from predation, or other causes of mortality that affect primarily the young, are influenced to an important extent by environmental conditions.

Despite the emphasis on environmental controls on shell growth and size, I do not wish to imply that these two characteristics are determined solely by external influences. There is substantial evidence that shell sculpture, color, growth rate, and even adult size are under partial genetic control (see, for example, Struhsaker, 1968; Clarke and Murray, 1969; Largen, 1971; Cole, 1975).

5. Application to Environmental Reconstructions

Although more experiments are needed to establish causal links among growth rate, shell shape, and adult size in Recent gastropods, snail shells may prove to be helpful in the interpretation of ecological conditions in fossil communities. For example, comparison of tropical western Atlantic gastropods with closely related eastern Pacific species shows that adult size of the latter is typically 20–50% greater than that of the former (for example, see Radwin and D'Attilio, 1976; Hoagland, 1977; Vermeij, 1978). This difference in size is correlated with food supply; waters of the tropical eastern Pacific are more productive than those in the western Atlantic. In fact, productive areas in the western Atlantic (e.g., western Venezuela, the west coast of Florida) support larger snails than do regions with clear, nutrient-poor water such as Jamaica (Vermeij, 1978). When Recent shells are compared with close relatives in the Caloosahatchee and Pine Crest beds of Pliocene age in southern Florida, it is seen that the fossil shells have sizes similar to, or even larger than, those of present-day eastern Pacific species (for examples, see Olsson and Harbison, 1953; Hoagland, 1977). These generalizations apply to species in many unrelated genera and families; therefore, it is plausible that the waters of the Pliocene seas of Florida were richer in nutrients for shallow-water benthic animals than are most waters in the present-day tropical western Atlantic.

Gould (1969) found that the land snail *Cerion uva* from barren seaside limestone outcrops in Curaçao is only about two thirds as long in the adult phase as a snail living on well-vegetated substrates 5 m away. This difference in adult size is probably due to the greater availability of food on the vegetated substratum. Fossil (Pleistocene) *C. uva* are generally much larger than Recent examples, and it is possible that conditions at the time of deposition were more favorable to the production of dense vegetation than they are today on this desert island.

In another study, Gould (1970) used the morphology of land snails in his interpretations of Pleistocene climatic cycles in Bermuda. Thin shells of paedomorphic *Poecilozonites* and other species dominated red-soil sediments poor in calcium carbonate, whereas thicker-shelled "normal" forms were typical of calcium-rich aeolinite dune sands. The calcium-poor red soils were apparently deposited during the cooler, wetter glacial stages, whereas the aeolinites accumulated as dune sands during the drier interglacial episodes.

Many more applications will be possible when more is known about daily or other periodic growth increments in gastropods. Struhsaker (1968) observed macroscopic growth lines in Hawaiian *Nodilittorina picta*; these increments appear to have been laid down daily during early postlarval life. No one has attempted to study snail-shell growth by means of acetate-peel or other high-resolution techniques that have been so successful in bivalves (see Chapter 7). Gastropods with crossed lamellar shell microstructure should be well suited for such studies. Interpretation of data on the widths of growth increments will be complicated somewhat by the tight conispiral coiling of gastropods. Because a shell must be cut by a plane section for acetate-peel analysis, observed increment widths must be corrected for by taking into account the changing angle between the increment and the plane of the section cutting through it to arrive at the true increment width. This problem arises in bivalves as well; however, there is usually a planispiral geodesic on the shells of bivalves along which the section can be cut. High-resolution techniques in gastropods will be most helpful in estimating growth rates over specific intervals of angular growth; however, because of the geometric constraints, determinations of age will usually not be possible.* It will also be instructive to study the development of varices and other evenly spaced collabral thickenings on a microscopic level, since features such as these, which are geometrically like the concentric ridges of many bivalves and brachiopods, are very common in gastropods.

References

Andrews, H. E., 1974, Morphometrics and functional morphology of *Turritella mortoni*, *J. Paleontol.* **48:**1126–1140.

Bastida, R., Capezzani, A., and Torti, M. R., 1971, Fouling organisms in the port of Mar del Plata (Argentina). I. *Siphonaria lessoni*: Ecological and biometric aspects, *Mar. Biol.* **10:**297–307.

Beu, A. G., 1970, The Mollusca of the genus *Charonia* (family Cymatiidae), *Trans. R. Soc. N. Z. Biol. Sci.* **11:**205–223.

Branch, G. M., 1974, The ecology of *Patella* Linnaeus from the Cape Peninsula, South Africa. III. Growth-rates, *Trans. R. Soc. S. Afr.* **41:**161–193.

Branch, G. M., 1975a, Ecology of *Patella* Linnaeus from the Cape Peninsula, South Africa. IV. Desiccation, *Mar. Biol.* **32:**179–188.

---

* Estimating the age of untorted gastropods (archeogastropods) may be possible using acetate-peel or thin-sectioning techniques, since a single medial section will include the entire ontogenetic growth record.

392

Branch, G. M., 1975b, Intraspecific competition in *Patella cochlear* Born, *J. Anim. Ecol.* **44**:263–282.

Branch, G. M., 1976, Interspecific competition experienced by South African *Patella* species, *J. Anim. Ecol.* **45**:507–529.

Breen, P. A., 1972, Seasonal migration and population regulation in the limpet *Acmaea digitalis*, *Veliger* **15**:133–141.

Bretos, M., 1978, Growth in the keyhole limpet *Fissurella crassa* Lamarck (Mollusca: Archaeogastropoda) in northern Chile, *Veliger* **21**:268–273.

Butot, L. J. M., 1954, On *Telescopium telescopium* (Linné) and the description of a new species from P. Panaitan (Prinsen Island), Straits of Sunda, *Basteria* **18**:1–18.

Clarke, B., and Murray, J., 1969, Ecological genetics and speciation in land snails of the genus *Partula, Biol. J. Linn. Soc.* **1**:31–42.

Cole, T. J., 1975, Inheritance of juvenile shell colour of the oyster drill *Urosalpinx cinerea*, *Nature (London)* **257**:794–795.

Davies, P. S., 1969, Physiological ecology of *Patella*. III. Desiccation, *J. Mar. Biol. Assoc. U. K.* **49**:291–304.

Foin, T. C., 1972, Ecological observations on the size of *Cypraea tigris* L., 1758, in the Pacific, *Proc. Malacol. Soc. London* **40**:211–218.

Fretter, V., and Graham, A., 1963, The origin of species in littoral populations. in: *Speciation in the Sea* (J. P. Harding and N. Tebble, eds.), *Syst. Assoc. London Publ.* **5**:99–107.

Frömming, E., 1956, *Biologie der Mitteleuropäischen Süsswasserschnecken*, Duner und Humpert, Berlin, 311 pp.

Gail, R., and Devambez, L., 1958, A selected annotated bibliography of *Trochus, South Pac. Comm. Tech. Pap.* **111**:1–17.

Giesel, J. T., 1969, Factors influencing the growth and relative growth of *Acmaea digitalis*, a limpet, *Ecology* **50**:1086–1087.

Gould, S. J., 1968, Ontogeny and the explanation of form: An allometric analysis, *J. Paleontol.* **42**, Paleontol. Soc. Mem. (Part II) **2**:81–98.

Gould, S. J., 1969, Character variation in two land snails from the Dutch Leeward Islands: Geography, environment, and evolution, *Syst. Zool.* **18**:185–200.

Gould, S. J., 1970, Coincidence of climatic and faunal fluctuations in Pleistocene Bermuda, *Science* **168**:572–573.

Gould, S. J., 1971a, Environmental control of form in land snails: A case of unusual precision, *Nautilus* **84**:86–93.

Gould, S. J., 1971b, Precise but fortuitous convergence in Pleistocene land snails from Bermuda, *J. Paleontol.* **45**:409–418.

Gould, S. J., 1972, Allometric fallacies and the evolution of *Gryphaea*: A new interpretation based on White's criterion of geometric similarity, *Evol. Biol.* **6**:91–118.

Gould, S. J., 1977, *Ontogeny and Phylogeny*, Belknap Press of Harvard University, Cambridge, Massachusetts, 501 pp.

Graham, A., and Fretter, V., 1947, The life history of *Patina pellucida* (L.), *J. Mar. Biol. Assoc. U. K.* **26**:590–601.

Graus, R. R., 1974, Latitudinal trends in the shell characteristics of marine gastropods, *Lethaia* **7**:303–314.

Hamai, I., 1937, Some notes on relative growth with special reference to the growth of limpets, *Sci. Rep. Tohoku Imp. Univ. Biol.* **12**(4):71–95.

Haven, S. B., 1973, Competition for food between the intertidal gastropods *Acmaea scabra* and *Acmaea digitalis*, *Ecology* **54**:143–151.

Hoagland, K. E., 1977, Systematic review of fossil and Recent *Crepidula* and discussion of evolution of the Calyptraeidae, *Malacologia* **16**:353–420.

Ino, T., 1949, Ecological studies of *Turbo cornutus* Solander, *J. Mar. Res.* **8**:1–5.

Ino, T., 1952, Biological studies on the propagation of Japanese abalone (genus *Haliotis*), *Bull. Tokai Regional Fish. Res. Lab.*, No. 3, pp. 1–102.

Ino, T., 1953, Ecological studies of *Turbo cornutus* Solander. I. Changes of the spines on the shell due to environments, *Bull. Jpn. Soc. Sci. Fish.* **19**:410–414.

Largen, M. J., 1971, Genetic and environmental influences upon the expression of shell sculpture in the dog whelk (*Nucella lapillus*), *Proc. Malacol. Soc. London* **39**:383–388.

Laxton, J. H., 1970, Shell growth in some New Zealand Cymatiidae (Gastropoda: Prosobranchia). *J. Exp. Mar. Biol. Ecol.* **4**:250–260.

Leighton, D. L., 1961, Observations of the effect of diet on shell coloration in the red abalone, *Haliotis rufescens* Swainson, *Veliger* **4**:29–32.

Lewis, J. R., and Bowman, R. S., 1975, Local habitat-induced variations in the population dynamics of *Patella vulgata* L., *J. Exp. Mar. Biol. Ecol.* **17**:165–203.

Linsley, R. M., 1977, Some "laws" of gastropod shell form, *Paleobiology* **3**:196–206.

Moore, H. B., 1936, The biology of *Purpura lapillus*. I. Shell variation in relation to environment, *J. Mar. Biol. Assoc. U. K.* **21**:61–89.

Moore, H. B., 1937, The biology of *Littorina littorea*. I. Growth of the shell and tissues, spawning, length of life and mortality, *J. Mar. Biol. Assoc. U. K.* **21**:721–742.

Neumann, D., 1959, Morphologische und experimentelle Untersuchungen über die Variabilität der Farbmuster auf der Schale von *Theodoxus fluviatilis* L., *Z. Morphol. Oekol. Tiere* **48**:349–411.

Olsson, A. A., and Harbison, A., 1953, Pliocene Mollusca of southern Florida with special reference to those from North Saint Petersburg, *Acad. Nat. Sci. Philadelphia Monogr.* **8**:1–457.

Phillips, B. F., Campbell, N. A., and Wilson, B. R., 1973, A multivariate study of geographic variation in the whelk *Dicathais*, *J. Exp. Mar. Biol. Ecol.* **11**:27–69.

Radwin, G. E., and D'Attilio, A., 1976, *Murex Shells of the World*, Stanford University Press, Palo Alto, California, 284 pp.

Randall, J. E., 1964, Contribution to the biology of the queen conch, *Strombus gigas*, *Bull. Mar. Sci. Gulf Caribb.* **14**:246–295.

Raup, D. M., 1961, The geometry of coiling in gastropods, *Proc. Natl. Acad. Sci. U.S.A.* **47**:602–609.

Raup, D. M., 1966, Geometric analysis of shell coiling: General problems, *J. Paleontol.* **40**:1178–1190.

Raup, D. M., 1967, Geometric analysis of shell coiling: Coiling in ammonoids, *J. Paleontol.* **41**:43–65.

Regis, M. B., 1969, Écologie et aspects quantitatifs de la croissance de quelques monodontes et gibbules de la Mediterranée, *Rec. Trav. Sta. Mar. Endoume Bull.* **45**:199–304.

Schwetz, J., 1954, L'influence du milieu sur la taile et la forme du même planorbe ou du même *Bulinus*, *Ann. Soc. Roi. Zool. Belg.* **85**:23–34.

Segal, E., 1956, Adaptive differences in water-holding capacity in an intertidal gastropod, *Ecology* **37**:174–178.

Spight, T. M., 1973, Ontogeny, environment, and shape of a marine snail *Thais lamellosa* Gmelin, *J. Exp. Mar. Biol. Ecol.* **13**:215–228.

Spight, T. M., and Lyons, A., 1974, Development and functions of the shell sculpture of the marine snail *Ceratostoma foliatum*, *Mar. Biol.* **24**:77–83.

Spight, T. M., Birkeland, C. E., and Lyons, A., 1974, Life histories of large and small murexes (Prosobranchia: Muricidae), *Mar. Biol.* **24**:229–242.

Struhsaker, J. W., 1968, Selection mechanisms associated with intraspecific shell variation in *Littorina picta* (Prosobranchia: Mesogastropoda), *Evolution* **22**:459–480.

Sutherland, J. P., 1970, Dynamics of high and low populations of the limpet, *Acmaea scabra* (Gould), *Ecol. Monogr.* **40**:169–188.

Taylor, J. D., 1978, Habitats and diet of predatory gastropods at Addu Atoll, Maldives, *J. Exp. Mar. Biol. Ecol.* **31**:83–103.

Thompson, D. W., 1942, *On Growth and Form*. 2nd ed., Cambridge University Press, London, 1116 pp.

Vadas, R. L., 1977, Preferential feeding: An optimization strategy in sea urchins, *Ecol. Monogr.* **47**:337–369.

Vermeij, G. J., 1971a, Gastropod evolution and morphological diversity in relation to shell geometry, *J. Zool. London* **163**:15–23.

Vermeij, G. J., 1971b, The geometry of shell sculpture, *Forma Functio* **4**:319–325.

Vermeij, G. J., 1973a, Morphological patterns in high intertidal gastropods: Adaptive strategies and their limitations, *Mar. Biol.* **20**:319–346.

Vermeij, G. J., 1973b, Adaptation, versatility, and evolution, *Syst. Zool.* **22**:466–477.

Vermeij, G. J., 1978, *Biogeography and Adaptation: Patterns of Marine Life*, Harvard University Press, Cambridge, Massachusetts, 332 pp.

von Bertalanffy, L., Quantitative laws in metabolism and growth, *Q. Rev. Biol.* **32**:217–231.

Yamaguchi, M., 1975, Sea level fluctuations and mass mortalities of reef animals in Guam, Mariana Islands, *Micronesica* (*J. Coll. Guam*) **11**:227–243.

Yamaguchi, M., 1977, Shell growth and mortality rates in the coral reef gastropod *Cerithium nodulosum* in Pago Bay, Gaum, Mariana Islands, *Mar. Biol.* **44**:249–263.

Chapter 11

Growth-Line Analysis as a Test for Contemporaneity in Populations

JOHN F. DILLON and GEORGE R. CLARK II

1. Introduction 395
    1.1. Growth Lines 395
    1.2. Environmental Relationships 398
    1.3. Population Studies 399
2. Analysis of a Recent Population: *Pecten maximus* 401
    2.1. Suitability for Study 401
    2.2. Analytical Considerations 404
    2.3. Comparisons between Pairs of Specimens 406
    2.4. Identification of Contemporaneous Growth 411
3. Guidelines for Future Research 413
    References 414

1. Introduction

1.1. Growth Lines

It is a fundamental premise in ecology that an organism is both sensitive and responsive to its environment. If the organism is one of the many that accrete, or add to their shells during growth, such variations may be preserved within, or on, the shells as growth lines.

There is much confusion about the terminology of growth lines. A large part of this confusion is related to the natural tendency of specialists working with different taxa to settle on definitions that fit their individual needs. Much of the remainder seems to be a legacy of generations of fishermen and lumbermen, who instigated the first research on "annual rings"

JOHN F. DILLON ● Continental Oil Company, Albuquerque, New Mexico 87112. GEORGE R. CLARK II ● Department of Geology, Kansas State University, Manhattan, Kansas 66506.

and "tree rings" and naturally wanted the reports written in a language they could understand.

Even the definition of "growth lines" is the subject of controversy. Although most growth lines are like tree rings in that they are both well defined and repetitive, some, such as the undulations on the epithecae of corals (Wells, 1963; Scrutton, 1965; Barnes, 1972), are poorly defined and can be recognized only by their repetition. Others, such as the "annual ring" of some short-lived bivalves (Belding, 1910), are well defined but occur only once on any shell. Some authors have suggested that "growth increment" should be used in preference to "growth line," but this would work only where the lines, or increments, are repetitive. Any comprehensive definition of growth lines must apply to each of these situations.

It is also important to note that growth lines can be expressed in a wide variety of ways. They may involve changes in surface morphology, shell structure, organic content, shell transparency, direction of growth, or even trace element or isotopic composition. A good definition will include all these features. The general definition suggested by Clark (1974b, p. 77) seems to satisfy these requirements. He defines growth lines as "abrupt or repetitive changes in the character of an accreting tissue."

It may help to consider the geometric relationships involved in the formation of an accretive skeleton. Unlike the skeletons of echinoderms and vertebrates, in which the skeletal tissue is highly porous and infused with living tissue, accretive skeletons are in contact with the living tissues only along a single surface called the *surface of deposition* (although dissolution may also take place at this surface). Any interruptions or variations in growth will cause changes along this surface, but as new material continues to cover the old, we cannot observe these changes directly without cutting a section. Any distinctive variations in shell deposition are visible on such sections as *internal growth lines*.

In some cases, as in trees and the molluscan pearl, the surface of deposition extends completely around the skeleton and internal growth lines are the only kind that can be observed. In more common situations, such as mollusc shells and solitary corals, the living tissues are in contact with only a part of the surface of the skeleton, so that the surface of deposition is limited in extent. The remainder of the skeletal surface can be thought of as the accumulation of the margins of the surfaces of deposition, and is called the *surface of accretion*. Any variations in shell deposition can be observed directly on the surface of accretion as *external growth lines*. Figure 1 illustrates these concepts.

Although there is often a direct relationship between internal and external growth lines, either one can occur independently of the other. In corals, for example, the daily external growth lines reflect a variation in the direction of growth rather than growth rate (Barnes, 1972), and

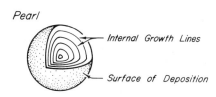

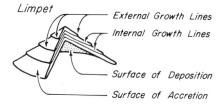

Figure 1. Accretive skeletons and their major features. Reprinted with permission from Clark (1974b). Copyright © 1974 by Annual Reviews Inc., all rights reserved.

internal growth lines remain obscure. In the bivalve Mercenaria mercenaria, internal growth lines seem to reflect environmental variations far better than external lines (see Chapter 7), possibly because the most dramatic events, such as shell dissolution (Lutz and Rhoads, 1977), occur while the shell valves are closed and the tissues are not in contact with the shell margin. On the other hand, the bivalve Pecten diegensis seems to grow continuously, with internal growth lines being rare or absent; the well-developed external growth lines reflect a daily variation in direction, rather than degree, of marginal growth (Clark, 1975).

A final consideration is that of practical application. A line, by definition, can have no thickness; yet many features, such as tree rings and the "annual rings" on shells, have an obvious breadth. It is good practice to attempt to define the position of the growth "line" as precisely, and as narrowly, as possible. In the two examples given, it is fairly simple to mark the sudden transition from fast growth to slow growth as the growth line. The precise position of the growth line is of great importance only if the interval between lines—the growth increment—is to be measured. If the main interest in the growth lines is their number, they can be readily counted with little concern for their precise position.

It is common practice to subdivide one type of increment into two or more increments of shorter periodicity, especially where a major increment is formed of two or more well-defined, different kinds of material. Thus, a daily increment might be divided into a dark-colored, organic-rich subdaily increment and a light-colored, carbonate-rich subdaily increment. When the major increment is annual, the subdivisions are often referred to as "bands," so that an annual increment might consist of a narrow winter band and a wide summer band. Growth lines can be positioned both at the transitions from dark to light and at the transitions from light to dark, but one or the other must be selected to serve as the reference boundaries of the major increment. The position of the annual line is usually defined as the end of the summer band. Additional terminology and explanatory figures can be found in Clark (1979).

1.2. Environmental Relationships

To return to the thought expressed at the beginning of this chapter, growth lines frequently form in direct response to variations in the environment. Thus, in a sense, they preserve a record of environmental variation.

The interpretation of such a record depends on an understanding of the different kinds of relationships that can exist between environmental stimuli and the resulting growth lines. We are probably not yet aware of every possible relationship, but the most common ones seem well established.

Environmental stimuli range from traumatic events such as severe storms or attacks by predators to very subtle events such as gradual changes in temperature in moderately deep water. Physiological responses to traumatic events are usually sudden, and the resulting growth lines are well defined. Physiological responses to subtle events may themselves be subtle, forming ill-defined growth lines that are apparent only if they are repetitive. Alternatively, a physiological response may depend on a "threshold" factor and form an abrupt growth line when the stress reaches a certain limit. In this latter case, it should be noted that the growth lines will probably not correspond to the extreme in the environmental stimuli, so that, for example, the growth lines separating winter growth bands from summer growth bands may form at various times during the spring and fall. In contrast, the subdaily growth lines separating the "organic-rich" increments from the "carbonate-rich" increments within the bivalve *Mercenaria* may directly follow the closing or opening of the shell (Whyte, 1975; Lutz and Rhoads, 1977); in such cases, the lines not only mark true extremes of environmental stimuli, but also more accurately reflect natural periodicities in the environment.

A number of observations point to other needs for caution. One is that an annual increment will often consist of a very wide summer band and an extremely narrow winter band; any estimates of short-term growth rates, or of the number of daily growth lines in a year, must compensate for this effect.

A second problem is the difficulty of distinguishing growth lines due to one stimulus, such as winter temperature stress, from growth lines due to another stimulus, such as a severe storm. This problem can be resolved, but often requires detailed examination (see Chapter 7).

A third problem is that organisms may vary their growth for morphological or structural reasons, producing external "growth lines" unrelated to environmental variation. Clark (1969) found this to be the case for the "concentric" spinose ornamentation on some species of the bivalve *Chlamys*, and similar situations are suspected to be represented in other molluscs.

One other observation is truly remarkable. There is good evidence that at least some physiological variations and growth lines need not be directly tied to environmental stimuli, but can continue as endogenous rhythms (Millar, 1968; Clark, 1969, 1975; Thompson and Barnwell, 1970; Thompson, 1975). This, together with the previous observation, means that a series of growth lines is not necessarily an indication of life in a variable environment. Well-developed growth lines on living specimens of a bivalve species dredged by one of us (G.R.C.) from depths exceeding 1000 m support this conclusion.

Any attempt at the interpretation of growth-line patterns should begin with consideration of whether the environmental stimuli appear to be random, suggesting events such as storms or attacks by predators, or periodic, suggesting events such as semidaily tidal exposures or annual temperature variations. Although both random and periodic growth lines can provide useful information about the environment, the periodic lines offer an additional advantage. Because they mark off equal time units in an accreting skeleton, they permit direct measurements of rates of growth during the lifetime of the organism. This is the case even if the periodicity is in response to an endogenous rhythm. Such information can be a powerful tool for the study of both species and population paleobiology and paleoecology.

1.3. Population Studies

A remarkable amount of information can be obtained from growth-line patterns under the right circumstances. A combination of daily (or tidal) and annual lines, for example, could provide information such as average life span, maximum life span, average rate of growth, variations in growth rate with age, relationships between mortality and environmental stress (such as winter), seasonality of spatfall and recruitment, and, if spawning is traumatic enough, the season of spawning and even the average duration of the planktonic larval stage.

If the environment, and thus the growth rate, is variable enough, then day-by-day or year-by-year variations in spacing between growth lines can be used to determine which specimens lived at the same place and time; given a sufficiently large population of fairly long-lived organisms, overlapping life spans and their growth records can be combined into a chronology of environmental variability, much as has long been accomplished with tree-ring data in the American Southwest. With such a chronology, population studies could be made with reasonable estimates of annual recruitment, mortality, and year-by-year population size.

Fossil populations could be carefully examined to reject transported specimens brought in after death, to determine the approximate size of the standing crop, to determine whether mortality was continuous or catastrophic, and perhaps even to judge variations in burrowing depth with age in extinct bivalve species, or to estimate rates of sedimentation around sessile brachiopods.

Complex tidal growth-line patterns might serve to pinpoint the position and extent of ancient tidal flats, and the frequency and seasonality of growth lines attributable to storms could provide useful data for paleoclimatology.

Some steps have been taken in this direction. Posgay (1950), working with the living scallop *Placopecten magellanicus*, noted that specimens from the same "bed" could be recognized by the patterns of disturbance lines, and Craig and Hallam (1963) noted that Recent cockles (*Cerastoderma edule*) collected in one locality had a distinct disturbance line not present on specimens from a locality about a mile distant. Fairbridge (1953) mentioned that another commercial scallop, *Notovola meridionalis*, displayed enough variability in annual growth to permit specimens from different fishing grounds to be recognized by their growth-line patterns.

Clark (1968, 1969, 1975) carried this approach even further by constructing growth-rate curves (Fig. 2) to graphically compare the daily growth-line patterns of scallops maintained in a single aquarium. The striking similarities in these patterns were easily correlated, but the environmental variations in the aquarium, which included occasional water-flow blockages and consequent temperature increases, were probably more extreme than would be expected in all but the most marginal marine environments.

The potential value of growth-line analysis in paleoecological studies seemed beyond doubt, but direct application of this technique to a fossil assemblage could be premature. We decided to attempt a growth-line analysis of a living population to develop a practical working approach. To test the sensitivity of growth-line analysis, as well as its application to the widest variety of paleoecological situations, we decided to avoid

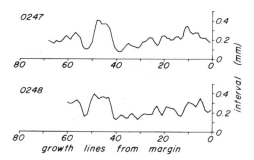

Figure 2. Growth-rate curves constructed from variations in spacing ("interval") between successive growth lines in two experimentally grown specimens of *Pecten diegensis* (after Clark, 1968, 1969). Direction of growth is to the right; growth lines are numbered from the margin for convenience.

intertidal populations in favor of organisms living in a relatively low-variability, open marine environment.

2. Analysis of a Recent Population: *Pecten maximus*

2.1. Suitability for Study

Pecten maximus, a common bivalve in the eastern North Atlantic, is nearly ideal for a population-based growth-line analysis. It is a commercially important shellfish, occurring in large numbers over a wide geographic range. Like most commercial bivalves, the species has been intensively studied, and its growth rate, reproductive habits, and general ecology are well documented (Tang, 1941; Mason, 1953, 1957, 1958; Gibson, 1956).

Pecten maximus (Fig. 3; also see Fig. 4) has two distinct types of external growth lines. The "annual ring," reflecting a period of very slow growth or a growth halt during the winter months, consists of a well-defined narrow band of light-colored shell; much broader, darker bands represent fast growth during the remainder of the year. Both these bands include large numbers of fine growth lines, expressed as narrow ridges. These ridges are spaced relatively far apart in the fast-growth band and very close together in the winter slow-growth band. Although the annual periodicity of the winter band has been well established (Mason, 1953, 1957), the periodicity of the fine growth lines (the narrow ridges) is not documented; preliminary observations and reports on related species (Clark, 1968, 1975) suggest that they may be daily.

For the purposes of this study, the exact periodicity of the fine growth lines is not critical as long as they mark off approximately the same time intervals on all the shells in a population. In this way, each shell will preserve a record of variations in growth rate reflecting corresponding variations in environmental conditions.

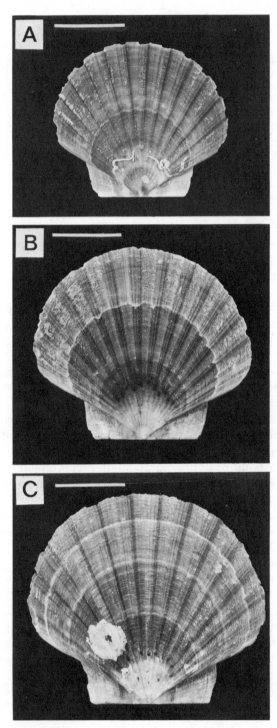

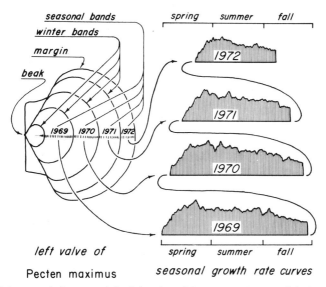

Figure 4. Major growth features of the left valve of *Pecten maximus* and their relationships to the seasonal growth-rate curves used in this study.

The presence of the annual lines is a disadvantage in the sense that variations in growth rate cannot be determined for significant parts of the year because of the extreme crowding of the fine ridges. This is more than compensated for, however, by the manner in which the winter bands separate the successive fast-growth bands (herein after called "seasonal bands" or "seasonal growth records") (Fig. 4). This is a tremendous advantage for studies of contemporaneous growth, for it is much easier to compare well-defined seasonal growth records than entire lifetime growth records. For one thing, seasonal growth records are fairly short, extending across 120–200 fine growth lines compared to the thousands of fine growth lines formed during the lifetime of the scallop. Even more important, each seasonal growth record can be assumed to begin in the spring and end

---

Figure 3. Left valves of specimens of *Pecten maximus*. Narrow, light-colored bands mark slow winter growth, and wide, dark-colored bands mark rapid growth in the spring, summer, and fall. The two types of bands are termed "winter" and "seasonal" bands. The annual lines are defined as the boundaries between seasonal bands and the following winter bands. The first seasonal band is not complete, since it includes growth from spatfall to the first winter. Also see Fig. 4. Scale bars: 30 mm. (A) Specimen 305. The margin represents the third winter. Other winter bands separate the partial 1970 seasonal band, the complete 1971 seasonal band, and the 1972 seasonal band. (B) Specimen 306. The margin represents the fourth winter. Other winter bands separate the partial 1969 seasonal band and the complete seasonal bands for 1970, 1971, and 1972. (C) Specimen 318. The margin represents the fifth winter. Other winter bands separate the partial 1968 seasonal band and the complete seasonal bands for 1969, 1970, 1971, and 1972. Note the decline in width of the last two bands; this is a typical ontogenetic effect.

in the fall (Fig. 4); no such assumptions are permissible for the lifetime growth record (spatfall to death) of an individual specimen. As an additional advantage, the three or four seasonal growth records on each shell (corresponding to 1970, 1971, and 1972, for example) can be treated as independent groups of data. This means that seasonal growth records of known contemporaneity (all those formed in 1972, for example) or of noncontemporaneous origin (1970, 1971, and 1972) can both be studied in a single population.

A final advantage of using *Pecten maximus* is that the pectinids are well suited to growth-line analysis in fossil populations. The family ranges from the Ordovician to the Recent, and, although best known from the large and varied assemblages so common in Tertiary formations, pectinids are found as gregarious populations in the Paleozoic as well (Clark, 1978). The calcite shells of pectinids are much more resistant to alteration or dissolution than the aragonitic shells of most bivalves, and the use of external growth lines permits one to utilize recrystallized shells, or even good external molds of dissolved shells to derive all the information essential to a comprehensive growth-line analysis.

2.2. Analytical Considerations

A sample population of *Pecten maximus* was dredged from the Firth of Clyde, Scotland, at a depth of 16–20 m. More than a hundred specimens were collected on March 9, 1973, from an area of less than three square kilometers. At the time of collection, the population was in the winter growth phase, so that each seasonal (fast-growth) band was complete, with the outermost representing 1972 (Figs. 3 and 4).

Only the left valves were used in this study. The fine growth lines are generally not well preserved on the right valves, since these rest on the substratum and experience abrasion when the organism shifts position. The left valves are also relatively flat and thus suitable for detailed examination by photographic methods.

Many of the specimens proved unsuitable for this study. Shells with more than five seasonal bands had too little growth in the outer (most recent) bands for meaningful comparisons, and many others had chipped or badly abraded edges. About 25 shells were eventually selected for measurement.

These shells were rinsed and treated with a commercial solution of 5.25% sodium hypochlorite to remove organic matter from the shell surface. Because the ridges forming the fine growth lines stand high above the shell surface and frequently curl over, the precise position of the growth line is not always visible (see illustrations in Clark, 1974a, 1975).

This problem was resolved by lightly brushing the shell surface to break off the upper parts of the ridges and expose the contact between the outer ridge surface and the shell surface (this was selected as the precise position of the growth line). The increments between successive growth lines were then measured from photographic enlargements of known scale, and compiled into growth-rate curves (Fig. 4).

On *Pecten maximus*, as on many other pectinids, the fine growth lines are abraded or obscure near the juvenile portion of the shell, so it is easiest to begin the measurements at the shell margin and work toward the beak. The growth record nearest the margin is also the most useful for determining whether mortality is castastrophic or continuous, so if time is limited, the outermost seasonal bands should be processed first. For both these reasons, the fine growth lines and their corresponding growth increments are numbered from the margin. The direction of numbering has no effect on the data processing that follows, but visual plots of growth-rate curves, shown in Figs. 5–9, must be numbered from right to left in order to show the direction of growth, and the passage of time, in the conventional left to right.

Although it is intuitively satisfying to make visual comparisons of growth-rate curves, a more objective approach is required if large numbers of growth records are to be compared. Digital-computer processing permits a mathematical comparison to be made using a standard correlation coefficient; this method has enjoyed considerable success in tree-ring research (Fritts, 1963) and varve correlation (Anderson and Kirkland, 1966). In this approach, a data set corresponding to one growth-rate curve is compared mathematically with a data set corresponding to a second growth-rate curve. As one data set is "moved" past the other, a correlation coefficient is computed for each possible point of overlap. The highest correlation coefficient is assumed to mark the position at which the two sequences match. Because such a comparison requires the computing of about twice the number of coefficients as the number of points in each data set, and because each specimen's growth-rate curve must be compared with every other curve (unlike tree-ring studies, in which each data set is compared with a master chronology), a large amount of computer time and memory is required for even a relatively small number of specimens. We found that we could conveniently compare about 20 growth-rate curves, each corresponding to a seasonal band, at one time.

A potential problem in the comparison of the growth-rate curves was the possibility of missing data points.* Clark (1968, 1969, 1975) demonstrated that although fine growth lines in some species of *Pecten* are

---

* Data point or "point" refers to both the abscissa, or growth-line number, and the ordinate, or measured growth increment, on the growth-rate curve.

formed with a daily periodicity, some days passed with no corresponding line being formed in some specimens.

To minimize the potential problem of incomplete data and to reduce the effects of errors in the measuring process, it was decided to perform the initial correlations by breaking the growth-rate curves into segments. Each segment would then be compared to every possible corresponding segment on the other curve. This not only improved the computed correlation for different parts of curves with missing data points, but also permitted a check on the validity of the correlation process; if the best fits for two successive segments of one curve were, for example, in a reversed order on the other curve, it would suggest that neither fit was very good.

As a check on the validity and sensitivity of the method, two separate traverses were sometimes measured on the same shell; these data were then compared, and the correlation coefficients obtained were utilized as standards for evaluating comparisons between different shells. These "sister" growth-rate curves were also used to establish the most suitable length of segments. Correlation coefficients were computed several times using segments of different lengths, until the 30-point segment was selected as the best compromise between sensitivity and significance. An example of such a comparison is shown graphically in Fig. 5, together with the computed correlation coefficients for each segment.

Another problem encountered in working with these data was the seasonal variation in growth rate. As can be seen in Figs. 4 and 5, the growth-rate curve for each seasonal band begins with a sharp rise, representing rapid spring growth, continues with a slow decline in growth rate during the summer, and ends with a more rapid decline in the fall. Because this general growth pattern is found in every specimen, a certain degree of correlation, especially in the spring and fall, would be expected even between noncontemporaneous growth-rate curves.

To minimize this problem, the data were filtered. This involved constructing a low-frequency growth-rate curve (Fig. 6A) by computing a 30-point running mean of the original data (Fig. 6B) and then subtracting the low-frequency data from the original data. This produces a high-frequency, or residual, growth-rate curve (Fig. 6C), which accentuates the day-to-day variations in growth rate. As expected, correlation coefficients based on the high-frequency data (Fig. 7) are lower.

2.3. Comparisons between Pairs of Specimens

Correlation coefficients were computed for a number of combinations of 1972 growth records. In each case, both original and high-frequency

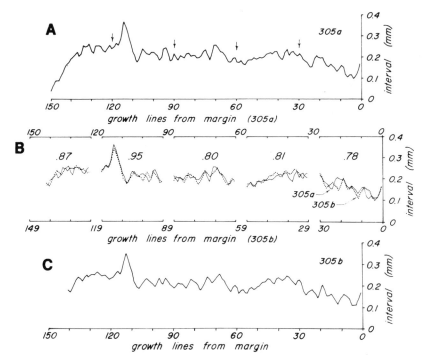

Figure 5. Growth-rate curves and correlations of "sister" transects (a) and (b) on specimen 305. Direction of growth is to the right. (A) Seasonal growth-rate curve (1972) for transect 305a, with limits of 30-point segments indicated (arrows). (B) Comparison of transects 305a and 305b. Segments of transect 305a are superimposed on the best-matching segments of transect 305b. Numbers along the upper x-axis indicate the position of the arbitrarily selected 30-point segment of transect 305a. Numbers along the lower x-axis indicate the position of the computer's choice of the best-fitting 30-point segment of transect 305b. Any overlaps in these segments suggest possible missing lines in transect 305b. Gaps between segments suggest missing lines in transect 305a. Computed correlation coefficients for the best fits appear above each segment. (C) Seasonal growth-rate curve for (1972) transect 305b.

data were compared separately. An example of such a comparison is shown graphically in Fig. 8, with the computed correlation coefficient shown for each line segment.

The results from this initial sample permitted some generalized observations. First, the correlations between contemporaneous (in this case, 1972) growth records were not much less than the correlations computed for "sister" growth records (see Figs. 5 and 7); some correlations were actually better than these standards.

Second, the best correlations usually involved some compensation for missing lines. This is the case even for "sister" growth records, as seen in Figs. 5 and 7; here, a missing line in traverse 305a is indicated by the

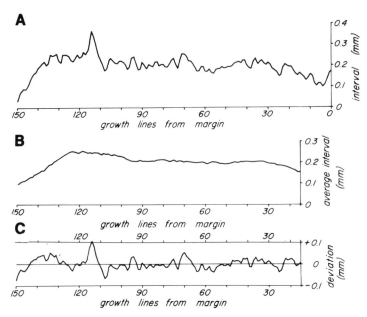

Figure 6. Growth-rate curves for transect 305a. Direction of growth is to the right. (A) Original 1972 seasonal growth-rate curve. (B) Low-frequency growth-rate curve obtained by taking a 30-point running mean of the data points in (A). This shows the strong seasonal bias (fastest growth in spring) in the original growth-rate curve. (C) High-frequency growth-rate curve obtained by subtracting each point in (B) from its equivalent in (A). This accentuates the high-frequency, i.e., day-to-day, variations in growth rate.

overlap of the two right-hand segments of traverse 305b. Correlations of both the original data and the high-frequency data locate the missing line in the same region, so it seems likely to be a real phenomenon.

In contrast, many of the "missing lines" indicated by other comparisons may represent cases in which several good correlations between segments of growth-rate curves exist at slightly different overlaps; the computer was programmed to select the best correlation without considering how well this fits the adjoining line segments. In Fig. 8, for example, it appears that at least 4 lines are missing from the margin of specimen 306, and that at least 3 more lines are missing along the traverse, but a comparison of best matches along the original-data curves and the high-frequency data curves shows little agreement on the positions of these missing lines.

Third, the early part (spring) of the seasonal growth record, as seen on the left side of Figs. 5, 7, 8, and 9, tends to yield better correlations than the later part (the right side). This is especially true for the original data (Figs. 5, 8A, and 9A), since rapid growth following the winter growth

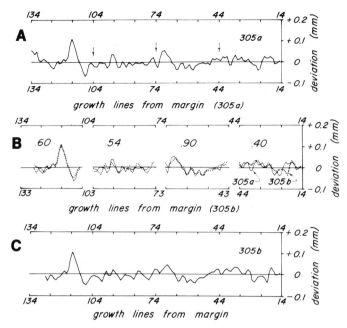

Figure 7. High-frequency growth-rate curves and correlations for "sister" transects 305a and 305b. Compare with Fig. 5.

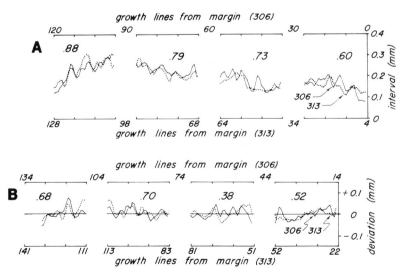

Figure 8. Comparison of contemporaneous (1972) seasonal growth-rate curves for specimens 306 and 313. Best matches of 30-point segments are shown together with computed correlation coefficients. Direction of growth is to the right. (A) Original growth-rate curves; (B) high-frequency growth-rate curves.

halts is common to all growth records. The high-frequency growth-rate curves (Figs. 7, 8B, and 9B) have compensated for this particular effect, but usually still show the poorest correlations on the right, late in the season; this accompanies a general decline in growth rate that may limit the ability of individual specimens to respond to external stimuli.

Correlation coefficients were also computed for contemporaneous growth records from seasons prior to 1972. In general, these results were not greatly different from the 1972 growth records discussed above.

The next step was the computation of correlation coefficients for non-contemporaneous growth records. Figure 9 shows a graphic representation of such a comparison, together with the computed correlation coefficients. Examination of several noncontemporaneous correlations, and comparison with contemporaneous correlations, permit some additional observations.

First, the correlations between noncontemporaneous growth records are, on the average, much poorer than correlations between contemporaneous growth records. Unfortunately, the highest correlation coefficients from noncontemporaneous growth records are better than the lowest correlation coefficients from contemporaneous growth records. This is particularly apparent for original-data growth-rate curves, where the overall seasonal trends are as heavily weighted in the computation of correlation coefficients as the day-by-day variations, but some overlap was noted even for the high-frequency growth-rate curves.

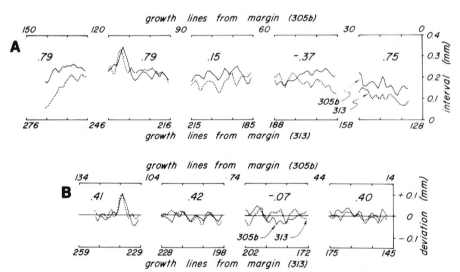

Figure 9. Comparison of noncontemporaneous growth-rate curves; the 1972 seasonal growth-rate curve for specimen 305 is compared with the 1971 seasonal growth-rate curve for specimen 313. (A) Original growth-rate curves; (B) high-frequency growth-rate curves.

Second, individual 30-point segments of noncontemporaneous growth-rate curves showed a great range of correlation coefficients, with many higher than the average for 30-point segments in contemporaneous growth-rate curves. This was especially true for the early (spring) portions of the growth-rate curves, and most pronounced in the original unfiltered data. Such high correlations often overshadowed low or negative correlations in midseason growth, especially if the total growth record was short, as in older shells. For *Pecten maximus*, growth records of at least 120 data points (yielding four 30-point segments on the original-data growth-rate curve, or three 30-point segments on the high-frequency growth-rate curve*) are required for dependable correlations; growth records of fewer than 90 data points are essentially useless.

For practical purposes, this means that growth records from third- and fourth-year growth are the most useful for determining correlations between individual specimens of *Pecten maximus*. Fifth-year growth generally adds only about 90–100 recognizable growth ridges, and later growth is even more abbreviated (see Fig. 4). To compound this problem, these shorter growth records show less day-by-day variation in growth rate, so the differences between contemporaneous and noncontemporaneous records are less obvious.

A fundamental question, of course, is whether any means could be found that could determine infallibly whether two growth records were contemporaneous or not. Attempts to limit the data to third- and fourth-year growth, to mid- and late-season data, and to high-frequency data alone were helpful but ultimately unsatisfactory. No method could be found by which all contemporaneous comparisons yielded higher correlation coefficients than all noncontemporaneous comparisons.

2.4. Identification of Contemporaneous Growth

Although it was not found possible to compare two individual growth records and determine unequivocally whether they are contemporaneous, it was observed that in general, individual growth records correlated better with contemporaneous records than with noncontemporaneous records (Fig. 10).

This led to a number of attempts to use standard weighted pair-group cluster analysis to group contemporaneous growth records. Twelve specimens, some with as many as four seasons of growth, were measured to form a data set of 22 growth records. Correlation coefficients were computed for all possible combinations of these growth records, first for the

---

* Construction of the high-frequency growth-rate curves involves a filtering process that cuts off 15 data points from each end of the growth record.

		Specimen 319			
		1972	1971	1970	1969
Specimen 318	1972	(0.64)	0.35	0.45	0.45
	1971	0.34	(0.51)	0.36	0.31
	1970	0.28	0.48	(0.67)	0.41
	1969	0.35	0.42	0.22	(0.54)

Figure 10. Comparison of correlation coefficients for contemporaneous and noncontemporaneous growth seasons in specimens 318 and 319. Coefficients for contemporaneous growth seasons are in parentheses.

original data and then for the high-frequency data. Each group of coefficients was then set up as a matrix for cluster analysis, but neither produced a completely satisfactory grouping of growth seasons.

Careful examination of the relationships between correlation coefficients and known contemporaneity of the corresponding seasonal bands suggested that some sort of average of the two sets of correlation coefficients (original data and high-frequency data) might be better than either set alone. Because the high-frequency correlation coefficients are generally much lower than the original-data correlation coefficients, the two sets had to be normalized before meaningful averages could be calculated. This was accomplished by multiplying each high-frequency correlation

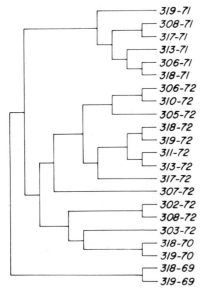

Figure 11. Dendrogram constructed by cluster analysis of 22 individual growth seasons of 12 specimens. The last two digits of each number identify the season. The dendrogram clearly shows that the greatest similarities between seasonal growth records are for contemporaneous years, demonstrating the potential of growth-line analysis for determining contemporaneity.

coefficient by the ratio of the mean of all original-data correlation coefficients to the mean of all high-frequency data correlation coefficients. Each pair of coefficients (original data and normalized high-frequency data) was then averaged and a new matrix constructed.

A cluster analysis based on these combined coefficients proved far more sensitive than previous attempts. Figure 11 shows a dendrogram based on such an analysis. Twenty-two seasonal growth records are clearly grouped into three sets, with all the 1971 growth records in one group, the two 1969 growth records in another group, and the two 1970 growth records forming a tight subgroup within the last group, which otherwise consists entirely of the 1972 growth records. This remarkable separation of contemporaneous from noncontemporaneous growth records is convincing evidence that comparisons of growth-line records can be used to test for contemporaneity.

3. Guidelines for Future Research

It is more than ten years since Clark (1968) suggested that growth-line analysis could provide a powerful tool in determining the contemporaneity of fossil assemblages. The results of this study carry us a step further toward this goal, but the final test of its capabilities awaits an analysis of a fossil population.

Next to the results themselves, the most important observations to come from this study are those relating to the design of future research. Experienced growth-line workers will have noted such points in the earlier discussion, but in the hope that others will be stimulated into considering growth-line analysis as a means of testing fossil assemblages for contemporaneity, we would like to offer a few suggestions.

First, the growth lines to be used must be distinct and easily defined throughout complete traverses. The worker must be able not only to count the lines but also to determine their precise positions so that growth increments can be measured accurately and consistently. It is good practice to measure a few sample traverses more than once to test this consistency.

Second, the number of data points needed for satisfactory correlations will depend on the variability in the size of the growth increments. The *Pecten maximus* in this study required traverses across more than 90 growth lines, but much shorter sequences might prove adequate for species exhibiting greater variability.

Third, because one of the most difficult steps will be the determination of the level of correlation (or clustering) corresponding to contemporaneity, it would be sound practice to make "sister" traverses on every specimen and treat them as independent growth records. The distribution of these in the resulting dendrograms could then be of great help in making that determination.

Fourth, every effort should be made to emphasize those aspects of the growth-rate curves that are directly related to random variations in the environment. In working with daily increments, seasonal variations in growth rate should be suppressed in favor of short-term variations. In working with annual increments [see observations by Posgay (1950) and Fairbridge (1953) above], variations in growth rate due to physiological changes with age should be suppressed. The possibility of subtle interference by such periodic effects as fortnightly tidal cycles could be explored by Fourier analysis (see Appendix 3.A), and the effects, if any, suppressed. Completely random growth lines can be of help if they can be shown to be related to events, such as storms, that affect the entire population. Correlation of such lines between individual specimens of a population is simultaneously proof of contemporaneity and of the widespread influence of the environmental stimuli; a lack of correlation, however, might be due either to a lack of contemporaneity or to the lines being caused by local events.

ACKNOWLEDGMENTS. We thank Dr. James Mason and the Marine Laboratory, Aberdeen, Scotland, for their interest and generosity in providing the specimens of *Pecten maximus* used in this study. This chapter is based on a Master's thesis (Dillon, 1974), and was supported by National Science Foundation Grant GA-33494, G. R. Clark, principal investigator. This is contribution No. 22 of the Palaeobiology Laboratory, Kansas State University.

References

Anderson, R. Y., and Kirkland, D. W., 1966, Intrabasin varve correlation, *Geol. Soc. Am. Bull.* **77**:241–256 (5 plates).

Barnes, D. J., 1972, The structure and formation of growth-ridges in scleractinian coral skeletons, *Proc. R. Soc. London Ser. B* **182**:331–350 (4 plates).

Belding, D. L., 1910, *A Report Upon The Scallop Fishery of Massachusetts*, Commonwealth of Massachusetts, Boston, 150 pp., 41 plates.

Clark, G. R., II, 1968, Mollusk shell: Daily growth lines, *Science* **161**:800–802.

Clark, G. R., II, 1969, Shell characteristics of the family Pectinidae as environmental indicators, Ph.D. thesis, California Institute of Technology, Pasadena, 101 pp.

Clark, G. R., II, 1974a. Calcification on an unstable substrate: Marginal growth in the mollusk *Pecten diegensis*, *Science* **183**:968–970.

Clark, G. R., II, 1974b, Growth lines in invertebrate skeletons, *Annu. Rev. Earth Planet. Sci.* **2**:77–99.

Clark, G. R., II, 1975, Periodic growth and biological rhythms in experimentally grown bivalves, in: *Growth Rhythms and the History of the Earth's Rotation* (G. D. Rosenberg and S. K. Runcorn, eds.), pp. 103–117, John Wiley, London.

Clark, G. R., II, 1978, Byssate scallops in a Late Pennsylvanian lagoon, *Geol. Soc. Am. Abstr. Program* **10**:380.

Clark, G. R., II, 1979, Growth lines, in: *Encyclopedia of Paleontology* (R. W. Fairbridge and D. Jablonski, eds.), pp. 359–364, Dowden, Hutchinson & Ross, Inc., Stroudsburg, Pennsylvania.

Craig, G. Y., and Hallam, A., 1963, Size-frequency and growth-ring analyses of *Mytilus edulis* and *Cardium edule*, and their palaeoecological significance, *Palaeontology* **6**:731–750.

Dillon, J. F., 1974, Growth line analysis of a Recent scallop population and its potential for paleoecology, M. S. thesis, University of New Mexico, Albuquerque, 68 pp.

Fairbridge, W. S., 1953, A population study of the Tasmanian "commercial" scallop, *Notovola meridionalis* (Tate) (Lamellibranchiata, Pectinidae), *Aust. J. Mar. Freshwater Res.* **4**:1–41.

Fritts, H. C., 1963, Computer programs for tree-ring research, *Tree-Ring Bull.* **25**:(3–4):2–7.

Gibson, F. A., 1956, Escallops (*Pecten maximus* L.) in Irish waters, *Sci. Proc. R. Dublin Soc.* **27**:253–271.

Lutz, R. A., and Rhoads, D. C., 1977, Anaerobiosis and a theory of growth line formation, *Science* **198**:1222–1227.

Mason, J., 1953, Investigations on the scallop [*Pecten maximus* (L.)] in Manx waters, Ph.D. thesis, University of Liverpool.

Mason, J., 1957, The age and growth of the scallop, *Pecten maximus* (L.), in Manx waters, *J. Mar. Biol. Assoc. U.K.* **36**:473–492.

Mason, J., 1958, The breeding of the scallop, *Pecten maximus* (L.), in Manx waters, *J. Mar. Biol. Assoc. U.K.* **37**:653–671 (2 plates).

Millar, R. H., 1968, Growth lines in the larvae and adults of bivalve molluscs, *Nature (London)* **217**:683.

Posgay, J. A., 1950, Investigations of the sea scallop, *Pecten grandis*, in: *Third Report on Investigations of Methods of Improving the Shellfish Resources of Massachusetts*, Chapt. 4, pp. 24–30, Commonwealth of Massachusetts, Boston.

Scrutton, C. T., 1965, Periodicity in Devonian coral growth, *Palaeontology* **7**:552–558 (2 plates).

Tang, S.-F., 1941, The breeding of the escallop [*Pecten maximus* (L.)] with a note on the growth rate, *Proc. Trans. Liverpool Biol. Soc.* **54**:9–28.

Thompson, I. L., 1975, Biological clocks and shell growth in bivalves, in: *Growth Rhythms and the History of the Earth's Rotation* (G. D. Rosenberg and S. K. Runcorn, eds.), pp. 149–162, John Wiley, London.

Thompson, I. L., and Barnwell, F. H., 1970, Biological clock control and shell growth in the bivalve *Mercenaria mercenaria*, *Geol. Soc. Am. Abstr. Program* **2**:704.

Wells, J. W., 1963, Coral growth and geochronometry, *Nature (London)* **197**:948–950.

Whyte, M. A., 1975, Time, tide and the cockle, in: *Growth Rhythms and the History of the Earth's Rotation* (G. D. Rosenberg and S. K. Runcorn, eds.), pp. 177–189, John Wiley, London.

Chapter 12

Demographic Analysis of Bivalve Populations

ROBERT M. CERRATO

1. Introduction 417
2. Construction of Class-Frequency Histograms 418
3. Interpretation of Class-Frequency Histograms 420
   3.1. Recruitment, Growth, and Age Structure 420
   3.2. Recognition of Age Classes 426
   3.3. Quantitative Estimates of Age Classes 427
   3.4. Mortality and Life Tables 440
   3.5. Survivorship 443
   3.6. Further Uses of Death Assemblages 451
4. Adaptiveness 453
5. Population Responses to Environmental Stress 455
   References 463

1. Introduction

A species possesses characteristics that cannot be expressed at the individual level and that exist only in a statistical sense (Allee *et al.*, 1949; Odum, 1971). Such properties are termed "group" or "population" attributes (Allee *et al.*, 1949). These quantities can be divided into two categories: (1) demographic characteristics associated with population structure and its maintenance (e.g., abundance, age structure, birth rates, age-specific growth rates, and mortality) and (2) genetic properties (i.e., adaptiveness, reproductive fitness, and persistence) (Odum, 1971). Population attributes reflect the ecology of a species, and they can therefore provide information relevant to the study of the spatial and temporal extent of environmental stress. Accurate determination of one or more group attributes in natural populations is not a simple task. However, the potential

ROBERT M. CERRATO • Department of Geology and Geophysics, Yale University, New Haven, Connecticut 06520.

for extracting these characteristics is particularly high if the organisms studied possess a calcareous skeleton.

In this chapter, I will use molluscan populations to illustrate how population attributes can be deduced from an analysis of shell parameters. The statistical unit of study will be the class-frequency histogram. As we shall see, many group attributes can be derived either directly or indirectly from an evaluation of this fundamental structure. This chapter will review, at an elementary level, methods of constructing and interpreting class-frequency histograms, present examples of specific techniques, and indicate how the derived information can be used for the analysis of environmental stress. Bivalve populations are given preferential treatment, since their demographic properties have been extensively studied and this class is distributed throughout the marine ecosystem. In addition, I consider only shallow-water populations living in a temperate or boreal climate.

2. Construction of Class-Frequency Histograms

The simplest population characteristic to determine is species abundance. In many situations, a temporal or spatial analysis of abundance patterns yields useful information about an organism and its ecology. However, individual organisms within a population are not developmentally, physiologically, or genetically identical, and may respond in different ways to various environmental stimuli. It is meaningful, therefore, to analyze a population at a higher level of resolution.

Population samples are most conveniently arranged for study by dividing the life span of the collected organisms into a number of nonoverlapping groups, called "classes." The class frequencies for the sample are found by determining the number of individual organisms belonging to each class. A frequency table is usually constructed giving the class arrangement, i.e., the way in which the classes were defined, and the corresponding class frequencies (Table I). The graphic equivalent of the frequency table is the frequency histogram (Fig. 1). Terms commonly used to describe the qualitative shape of class-frequency histograms are shown in Fig. 2.

For molluscan populations in which shell microstructural growth patterns are clearly defined (see Chapters 6, 7, 8, and 11), the class arrangement may be based on age. Age-class intervals may be chosen to distinguish year classes or recruitment events, or they may represent even smaller periods of time. When reliable age estimates are not available, class-frequency histograms may be based on other shell parameters such

Table I. An Example of a
Frequency Table

Length (mm)	Frequency (number of individuals)
28–30	9
31–33	17
34–36	32
37–39	24
40–42	19
43–45	6

as anterior–posterior length, shell width, or height. Size-class intervals are normally quite small and generally range from 1 to 5% of the size of the largest specimen in the population. Size-frequency histograms are useful even when ages can be determined. By constructing histograms with small size-class intervals, variations in growth within a year or recruitment class can be resolved. Size-frequency histograms may therefore provide additional details about age-specific growth dynamics.

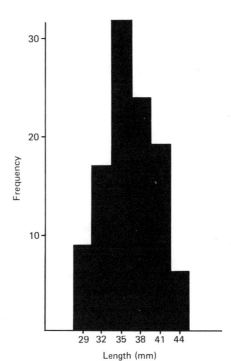

Figure 1. An example of a class-frequency histogram. Data from Table I.

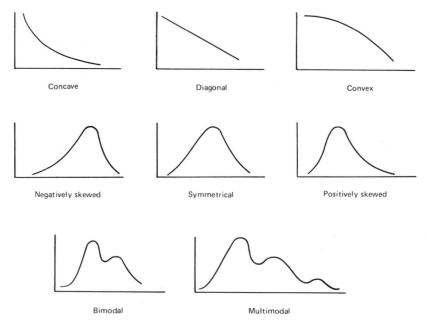

Concave

Diagonal

Convex

Negatively skewed

Symmetrical

Positively skewed

Bimodal

Multimodal

Figure 2. Terms commonly used to describe the shape of class-frequency distributions.

3. Interpretation of Class-Frequency Histograms

In this section, I will show how population attributes may be derived from class-frequency histograms. Many of these attributes are time-spe-cific (e.g., seasonal recruitment, age-specific growth and mortality). It is important, therefore, to relate class assignments to real time. As we shall see, many of the analytical techniques discussed are devoted to estab-lishing this relationship.

3.1. Recruitment, Growth, and Age Structure

For populations of molluscs living in a temperate or boreal climate, recruitment is often seasonal with one or more distinct spawning periods during the year. However, unless one can sample a population frequently during the spawning season(s), estimates of recruitment commonly de-pend on a knowledge of other population attributes. For example, consider a hypothetical sedentary population in which the spawning season is restricted to a 4-week period beginning in June. Let us also assume that recruitment and mortality do not vary from year to year. For such a population, a representative age-frequency histogram obtained from a census collected in July—that is, just after the cessation of spawning—is given in Fig. 3A. The decreasing pattern of abundance with age in this example is due only to mortality. If this same population were sampled 6 months later, say in January, it might appear as shown in Fig. 3B. No recruitment has taken place during the intervening period, and the size of each cohort is further reduced by mortality.

Variations in both recruitment and mortality will complicate this simple picture. A quantitative assessment of spatial and temporal recruitment patterns is possible only if losses due to mortality can be accurately estimated. In addition, older age classes usually comprise only a small percentage of the total sample, and they are especially prone to collection bias. Thus, only the more recent recruitment events can ordinarily be used in an analysis.

Exceptional departures from normal recruitment patterns are the easiest to recognize. Such an event can often be identified for a considerable period of time after it occurs. For example, the poor numerical representation of the 2 + year class of *Spisula solida* (Fig. 4) suggests weak recruitment 2 years prior to the census date. As a second example, consider the age structure of the bivalve *Venerupis pullastra* in Fig. 5. If one or more unusually large and successful spawning events take place, recruitment in subsequent years may be suppressed (Odum, 1971). Thus, it is possible for a population to be numerically dominated by one or a few large, older age classes. The age structure of *Venerupis pullastra* (Fig. 5) may reflect this dominant-age-class phenomenon.

When age determinations are not available, shell-size parameters form the basis for class-frequency histograms. Size-frequency histograms are common in the literature on molluscan populations. They are easy to construct and require less subjective judgment than counting and interpreting shell-growth increments.

The shape of a size-frequency histogram is influenced not only by recruitment and mortality, as I have illustrated with age-frequency distributions, but also by ontogenetic growth rates. The relationship between a size parameter such as anterior–posterior length and the age of a mollusc is nonlinear and species-specific. The growth curve is often sigmoidal in form (Fig. 6), but there is no general model that can account satisfactorily for all the different types of growth curves that are observed. The size–age relationship also depends on many environmental factors. Temporal and spatial variations in average rates of shell growth are often evident. For example, average growth curves, reported by Seed (1976) for several populations of *Mytilus edulis*, are shown in Fig. 7. For a complete review of those environmental factors that may influence shell-growth rates, the reader is referred to Hallam (1965).

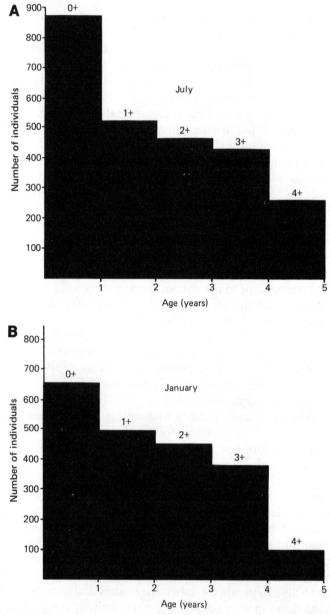

Figure 3. A hypothetical sedentary population showing the effects of mortality on age structure. In this example, it is assumed that recruitment and mortality do not vary from year to year. Note mortality effects both within and between age classes. (A) Population as it might appear in July, just after annual spatfall; (B) Population as it might appear in January. No recruitment has taken place during the intervening period.

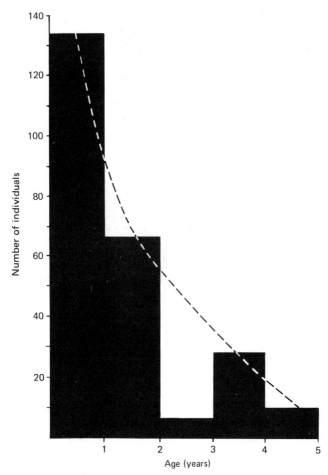

Figure 4. Age-frequency histogram of *Spisula solida* collected in 1972. Notice the weak representation of the 2+ year class. This is probably due to poor recruitment during 1970. Data estimated from Smith (1974).

Within a population, individual organisms of the same age will not be the same size. When a cohort is plotted as a size-frequency histogram, it will generally appear as a normal or slightly positively skewed distribution (Craig and Hallam, 1963). The degree of variation in size among organisms in the same age class (cohort) will usually increase in proportion to the mean or average length (Simpson and Roe, 1939; Craig and Hallam, 1963).

Because recruitment is seasonal, a size-frequency histogram will be polymodal, with each peak representing one or more spawning events. The general form of size-frequency histograms is illustrated in Fig. 8. In

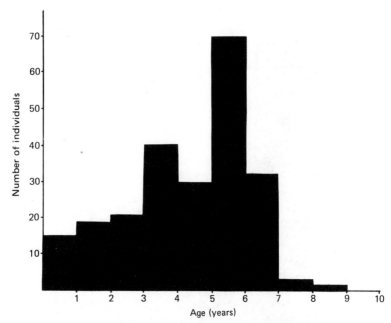

Figure 5. Age-frequency histogram of *Venerupis pullastra*. Note the numerical dominance of the older age classes (especially the 5 + year class). This particular age structure may be an example of the dominant age-class phenomenon described in the text. After Johannessen (1973).

this hypothetical example, the individual recruitment classes making up the histogram are included for clarity. The trend of increased dispersion in older recruitment classes is qualitatively characteristic of most distributions. The size-frequency histograms shown in Fig. 8 have the same age structure as the hypothetical population in Fig. 3. Figure 8A shows the population as it would appear in July, just after annual recruitment. The most recent cohort has been labeled the 0 + year class. Figure 8B shows the same population as it might appear in January. The size modes are all displaced to the right as a result of 6 months' growth. Once the position of each peak is determined, the relationship between size and absolute time can be established. One can construct an estimate of the average ontogenetic growth curve from this information (Fig. 9).

For molluscan populations, much information about recruitment, growth, and age structure can be obtained from size-frequency histograms. For example, Fig. 10 is a temporal sequence of size-frequency histograms for the bivalve *Donax incarnatus*. Two year classes are present in the samples. These are easily identified and followed in time. By plotting mean shell length for both year classes vs. time, an average growth curve for this population is derived (Fig. 11).

As a second example, I have chosen a population of *Mytilus edulis* (Fig. 12). Age-class assignments given in this figure were determined by Craig and Hallam (1963) on the basis of mode identification supplemented by growth data collected in this same general locality by Savage (1956). Recruitment classes are not as clearly defined as in the previous example. Based on Savage's growth data (Fig. 13), the peak of the 1 + year class in the April 1961 sample should have been located at about 12 mm. Absence of a mode here suggests inadequate sampling and poor annual recruitment during 1960 (Craig and Hallam, 1963). The 1960 cohort appears to be represented as a small minor mode in the November 1961 sample (Craig and Hallam, 1963), but no clear interpretation can be made.

In this example, only the more recent spawning events are sharply defined. Because of size dispersion within a recruitment class, and because of the marked decrease in growth rate with age in *M. edulis*, older age groups tend to merge together. Because of decreasing ontogenetic rates

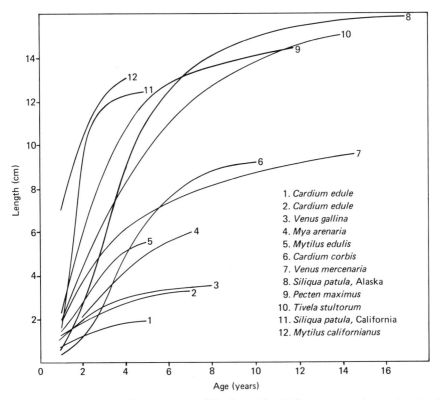

1. *Cardium edule*
2. *Cardium edule*
3. *Venus gallina*
4. *Mya arenaria*
5. *Mytilus edulis*
6. *Cardium corbis*
7. *Venus mercenaria*
8. *Siliqua patula*, Alaska
9. *Pecten maximus*
10. *Tivela stultorum*
11. *Siliqua patula*, California
12. *Mytilus californianus*

Figure 6. Growth curves for 11 species of bivalves. After Hallam (1967), who cites original sources.

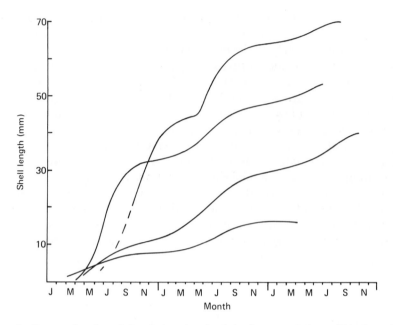

Figure 7. Temporal and spatial variations in growth for four populations of *Mytilus edulis.* After Seed (1976).

of growth with age, Craig and Hallam (1963) suggest that typical size-frequency histograms will have up to three clearly separated peaks at any season. This is an ideal upper limit, since these workers also point out that the number of clearly defined modes may be reduced by poor recruitment in some years, by mortality, or if sampling takes place at a time long after the most recent spawning event. The recognition of peaks in size-frequency histograms represents, therefore, a very important aspect in determining information about recruitment, growth, and age structure in populations.

3.2. Recognition of Age Classes

A useful technique for distinguishing among separate age classes in a size-frequency histogram is due to Sheldon (1965). This method is based on earlier work by Walford (1946). For many shellfish species, Walford determined that a plot of the size of an individual specimen at any given age (L_t) against its size 1 year later (L_{t+1}) is approximately linear (Fig. 14). In addition, he found that the line produced intercepts a point of no growth, that is, a point at which $L_t = L_{t+1}$.

Consider a size-frequency histogram in which only the modes of the first two recruitment classes are clearly defined. Sheldon's method for recognizing other modes in the distribution is as follows: The average sizes of the two initial modes are plotted as a point on a graph of L_{t+1} vs. L_t. Using an estimate of the mean size of the oldest age class in the population, a second point located at $L_t = L_{t+1}$ is also plotted. For this point, Sheldon proposes taking that size class that is larger than 98% of the individual specimens in the population. By drawing a line between these two points, the position of successive modes can be estimated. Starting on the abscissa with the average size of the last discernible mode, the approximate location of the next year class may be read off the line. This process is generally repeated several times using the current estimate to predict the average size of the next older age class.

An example of this method applied to the fossil brachiopod *Chonetes laguessiana* is shown in Fig. 15. The average sizes of the first two modes were estimated by Sheldon (1965) to be 0.25 and 0.69 cm, respectively. These modes are assumed to represent two separate year classes. The mean maximum size was taken by Sheldon to be 1.22 cm. Figure 16 shows a straight line joining the two points: (1) L_2 plotted against L_1 and (2) the mean maximum size, i.e., $L_t = L_{t+1}$. From this line, one can determine that the peak of the third year class should be located at 0.93 cm. By referring to Fig. 15, we see that a mode is located at approximately the same position as predicted. Peaks corresponding to a fourth or fifth year class cannot be recognized in the size-frequency histogram. Their presence may have been obscured by the merging effect common in older age groups. In many cases, it will be important to distinguish the numerical contribution of individual recruitment classes when they overlap as they do in this example. The problem of overlapping age classes in size-frequency histograms is discussed in the next section.

To demonstrate the usefulness of the technique in evaluating recruitment, consider the size-frequency histogram of the bivalve *Spisula solida* in Fig. 17. The location of the first two modes, and an estimate of the mean maximum size, can be used to form a Walford plot (Fig. 18). This plot predicts a mode for the 1970 year class at about 38 mm. The absence of this mode supports the conclusion by Smith (1974), based on growth-ring analysis, that recruitment was poor in 1970.

3.3. Quantitative Estimates of Age Classes

Because the shape of most age classes approximates a normal distribution, one can make effective use of probability paper in distinguishing separate modes in a polymodal size-frequency histogram. This technique

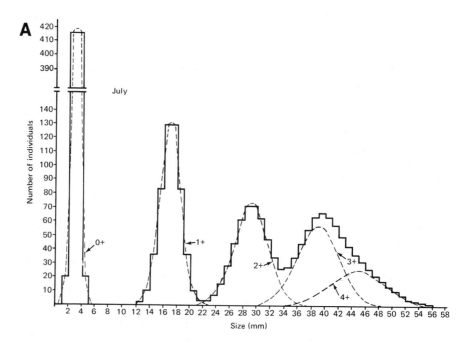

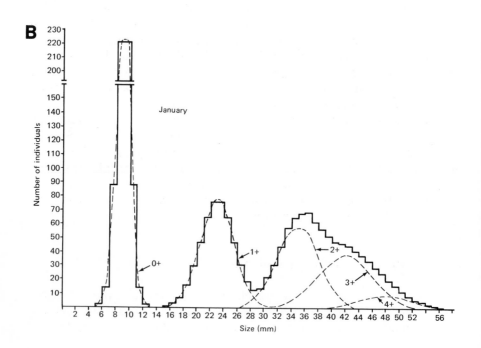

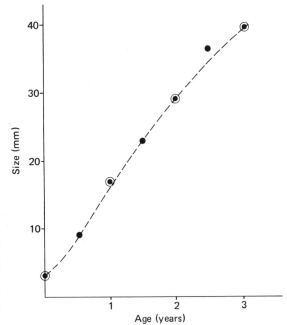

Figure 9. Growth curve constructed from the example in Fig. 8 by plotting estimates of size for each discernible mode. (◉) Points taken from the July histogram; (●) points taken from the January histogram. Note that some of the older age classes are not represented.

is, unfortunately, rarely used in population studies. I will therefore consider this method in detail.

First, we will consider how a normal or gaussian distribution plots on standard probability paper. A size-frequency distribution made up of 1572 measurements is shown in Fig. 19. Data from this histogram are plotted on probability paper as cumulative percentages. For example, the first size class consists of one specimen. Therefore, $(1/1572) \times 100$, or 0.06% of the sample, falls below 19.25 mm in length. This point is plotted as P_1. The second class consists of three specimens, so $[(3 + 1)/1572] \times 100$, or 0.25% of the population, lies below 19.75 mm in length (P_2). The third class has 30 specimens, so $[(30 + 3 + 1)/1572] \times 100$, or 2.2%, lies below 20.25 mm (P_3). The fourth class has 39 specimens, hence $[(39 + 30 + 3 + 1)/1572] \times 100$, or 4.7%, lies below 20.75 mm (P_4). This process is continued for all the classes in the distribution. Thus, class

←

Figure 8. A hypothetical example of the general form of size-frequency histograms for a molluscan population. (– – –) Individual recruitment classes. Notice the increased dispersion in size with age. Note also the overlap of the older age classes. The histograms in this figure have the same age structure as the hypothetical population in Fig. 3. (A) Population as it might appear in July; (B) population as it might appear in January. Note that the modes in (B) are displaced to the right of those in (A) as a result of 6 months' growth.

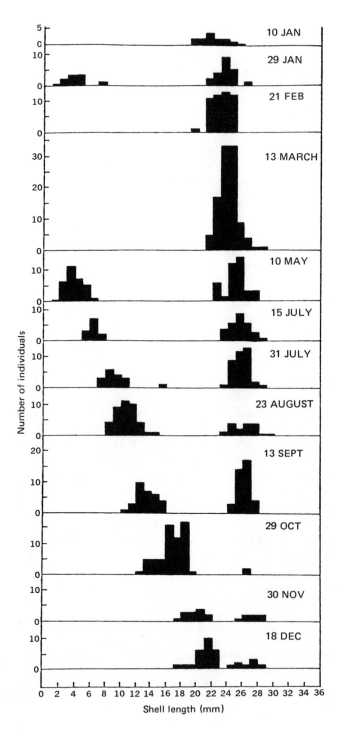

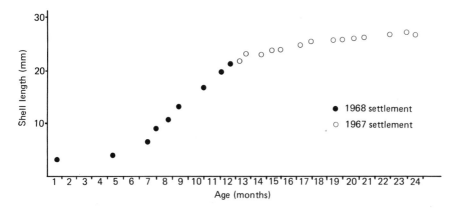

Figure 11. Growth curve for *Donax incarnatus* derived by plotting mean shell lengths in Fig. 10 as a function of time. After Ansell *et al.* (1972).

length vs. cumulative frequency plots as a straight line for a normal distribution.

Population parameters may be derived from this line. The mean of the sample is estimated as the size at the 50% probability point (M). The standard deviation is found by subtracting the size at the 84.13% probability point (S_2) from the 15.87% probability point (S_1) and dividing by 2. In this example

$$\bar{x} \equiv \text{mean} = L_{50\%} = 22.35 \text{ mm}$$

$$s \equiv \text{standard deviation} \quad = (L_{84.13\%} - L_{15.87\%})/2$$

$$\doteq 23.4 - 21.35/2$$

$$= 1.025 \text{ mm}$$

The percentages $(L_{50\%}, L_{84.13\%}, \text{ and } L_{15.87\%})$ are the same for any normal distribution.

A class-frequency distribution that is multimodal, but made up of a number of overlapping normal distributions, will not plot as a straight line on probability paper. The characteristic shape of a bimodal normal distribution is sigmoidal (Figs. 20 and 21), with the curve having a single point of inflection (located at 50% in Fig. 20 and at 97% in Fig. 21). In general, the number of points of inflection will be one less than the number

←

Figure 10. Size-frequency histograms of *Donax incarnatus* from Shertallai beach during 1968. Two year classes representing the 1967 and 1968 spatfalls are present in these samples. Note the displacement of modes to the right in this series due to growth. After Ansell *et al.* (1972).

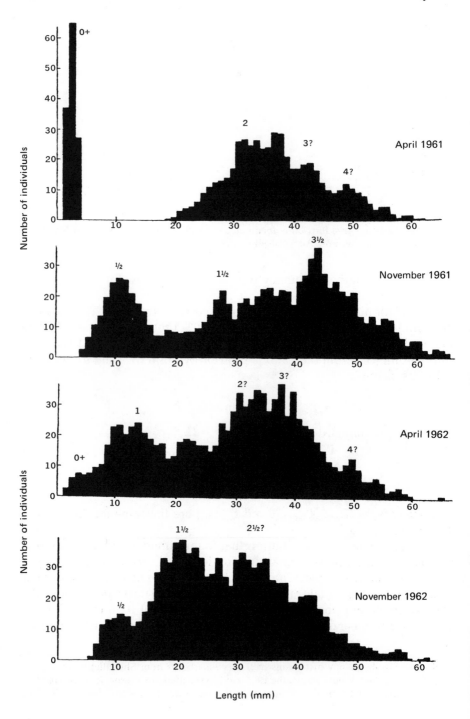

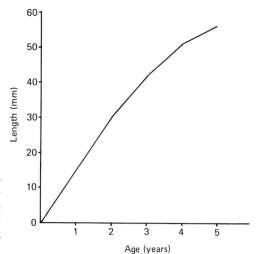

Figure 13. Growth curve derived for *Mytilus edulis* from data provided by Savage (1956). These growth data were used by Craig and Hallam (1963) to assign age classes in Fig. 12. After Craig and Hallam (1963).

of modes making up the distribution. In Fig. 22, there are four points of inflection, located at 10, 66, 91, and 98.6%. We therefore expect the population in this figure to be composed of five overlapping normal distributions.

The process of separating modes from one another will be illustrated using Fig. 22. The first mode in the original distribution is extracted by multiplying the original percentages of all the points up to the first point of inflection by 100 divided by the percentage of the first inflection point. In this case, the location of the first inflection point is 10%, and so one multiplies by 100/10. Thus, for the first point (P_1), the original percentage is 2.3%; therefore

$$P_1' = 2.3\% \times 100/10 = 23\%$$

For the second point (P_2), the original percentage is 5.1%; therefore

$$P_2' = 5.1\% \times 100/10 = 51\%$$

For the third point (P_3), the original percentage is 8.6%; therefore

$$P_3' = 8.6\% \times 100/10 = 86\%$$

Figure 12. Size-frequency histograms of *Mytilus edulis* from Ferny Ness, East Lothian. Sampling dates are given. Age-class assignments shown were determined by Craig and Hallam (1963) on the basis of mode identification and growth information (Fig. 13) reported by Savage (1956). After Craig and Hallam (1963).

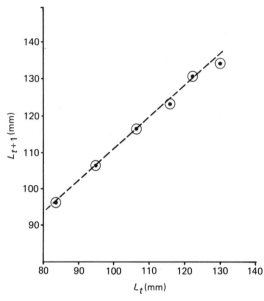

Figure 14. Walford plot for *Placopecten magellanicus* derived from length-at-age data collected by Merrill *et al.* (1965). Note that the data plot as a straight line.

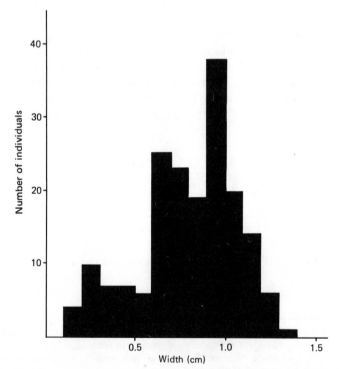

Figure 15. Size-frequency histogram of the fossil brachiopod *Chonetes laguessiana*. After Sheldon (1965).

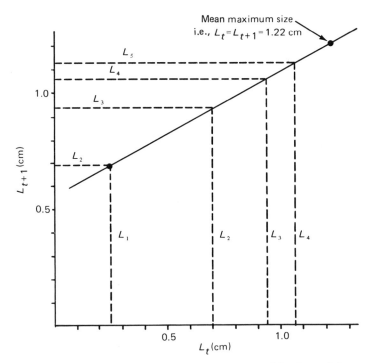

Figure 16. Walford plot for *Chonetes laguessiana*. Estimates of L_1, L_2, and the mean maximum size are from Fig. 15. The Walford plot suggests that a mode representing the third year class should be located at 0.93 cm. In Figure 15, note that a mode is located at approximately the same position as predicted. After Sheldon (1965).

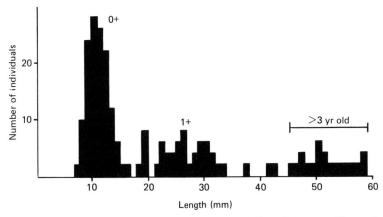

Figure 17. Size-frequency histogram of *Spisula solida* collected in 1972. The peaks at 12 and 27 mm represent the 0+ and 1+ year classes, respectively. From the Walford plot in Fig. 18, a peak representing the 2+ year class should have been located at 38 mm. The absence of this peak suggests poor recruitment during 1970. After Smith (1974).

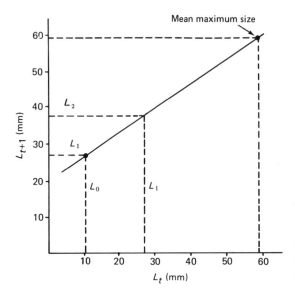

Figure 18. Walford plot of *Spisula solida*. Data from Fig. 17. The Walford plot predicts a peak in the size-frequency histogram at about 38 mm.

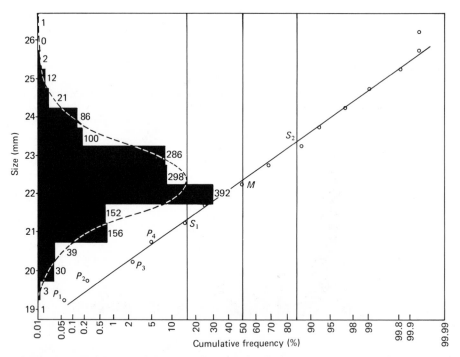

Figure 19. Size-frequency histogram of 1572 individual size measurements plotted as cumulative percentages on probability paper. The horizontal scale represents the cumulative percentage. The vertical scale is size in millimeters. Note that a normal distribution plots as a straight line on probability paper. After Harding (1949).

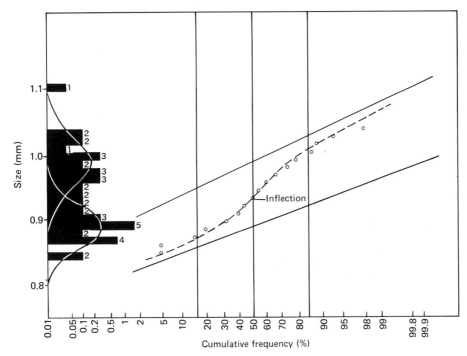

Figure 20. Plot on probability paper of two overlapping normal distributions. Note the single point of inflectin at 50%. After Harding (1949).

Finally, for the last point (P_4), the original percentage is 9.8%; therefore

$$P_4' = 9.8\% \times 100/10 = 98\%$$

These newly calculated percentages are then plotted with their appropriate lengths, and the resulting line, denoted by A in Fig. 22, is the distribution of the first mode. It has a mean of about 3.22 mm and a standard deviation of 0.27 mm. It makes up a total of about 10% of the original population as determined from the location of the inflection point.

The second mode is extracted in a similar way, except that the location of the first inflection point must be subtracted both from the original percentages and from the location of the second inflection. All points located between the first (10%) and second (66%) inflection points are used. For example, the first point (P_5) beyond the first inflection is located at 10.9%; therefore

$$P_5' = \frac{(\text{original percentage} - \text{location of first inflection})}{(\text{location of second inflection} - \text{location of first inflection})} \times 100$$

$$= [(10.9 - 10)/(66 - 10)] \times 100 = 1.6\%$$

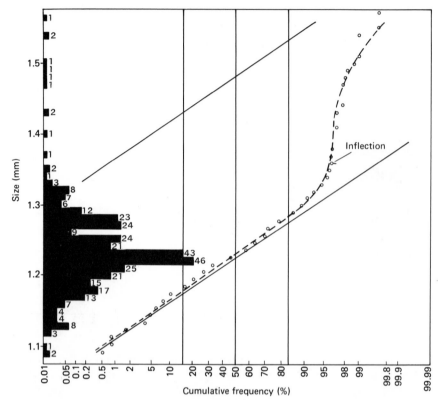

Figure 21. Plot on probability paper of two overlapping normal distributions. Note the single point of inflection at 97%. After Harding (1949).

For the second point (P_6) beyond the first inflection:

$$P_6' = [(12.5 - 10)/(66 - 10)] \times 100 = 4.46\%$$

When all these points are plotted, they yield the straight line labeled B in Fig. 22. The second mode has a mean of 5.33 mm and a standard deviation of 0.62 mm. It comprises (66 − 10) or 56% of the original population.

The third mode is extracted using points located between the second and third inflection points, and percentages determined from

$$P_i' = \frac{\text{(original percentage − location of second inflection)}}{\text{(location of third inflection − location of second inflection)}} \times 100$$

The resulting line is labeled C in Fig. 22.

The fourth mode is determined in a similar way (line D). The fifth mode is too small to resolve.

For small samples (i.e., fewer than 25 measurements), it may not be possible to define meaningful class assignments. However, the data set may still be plotted on normal probability paper. With small samples, individual specimens are arranged according to increasing size. The cumulative percentage of the ith point in this ordered sample is then simply $i/n \times 100$, where n is the number of specimens in the sample. Plotting these cumulative percentages vs. size will generate an acceptable probability plot. Note, however, that when a sample is plotted in this way, the last point will have a cumulative value of 100%. This point cannot, there-

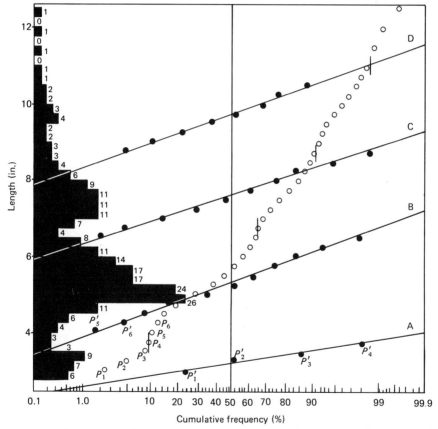

Figure 22. Size-frequency histogram of 256 measurements plotted as cumulative percentages on probability paper. Notice the four points of inflection (at 10, 66, 91, and 98.6%). They imply that the population is made up of at least five overlapping normal distributions. Modified from Cassie (1954).

fore, be included in the analysis. To correct for this problem, Guttman and Wilks (1965) suggest that $100/2n$ be subtracted from the cumulative percentage of all points.

Size-frequency histograms are sometimes truncated at one or both ends. Truncation may occur, for example, when benthic samples are sieved, and the mesh size used is larger than the smallest specimens in the population. A truncated mode will not plot as a straight line on probability paper. To correct for truncation in a given mode, the following equations may be applied (Cassie, 1954). For truncation on the small size end:

$$p' = p(1 - p''/100) + p''$$

where p is the uncorrected cumulative percentage, p' is the corrected cumulative percentage, and p'' is an estimate of the percentage truncation. For truncation on the large size end:

$$p' = p(1 - p''/100)$$

3.4. Mortality and Life Tables

Information on mortality is generally summarized in a life table. A life table is simply a columnar tabulation of statistical quantities relating to survivorship, rates of mortality, and the mean expectation of life (Table II). The information appearing in a life table is stated as a function of the age of the organism, where age itself may be expressed in absolute terms (e.g., in months or years), or in relative terms by comparing one part of the life cycle with another (Kurtén, 1964; Levinton and Bambach, 1969). The statistical quantities appearing in a life table are formally defined as:

x = age
x' = age as a percentage deviation from the mean length of life
l_x = the number of survivors at the start of age interval x
d_x = the number dying during the age interval x to $x + 1$
q_x = the rate of mortality during the age interval x to $x + 1$
L_x = the average number of individuals alive during the age interval x to $x + 1$
T_x = the number of individual time units left in the population at the start of age interval x
e_x = the mean expectation of life for an organism at the start of age interval x

The standard practice is to normalize the population at age zero to 1000

Table II. Life Table for the California Pismo Clam (*Tivela stultorum*)[a]

Age (yr) (x)	Age as % deviation from mean length of life (x')	Number surviving at beginning of age interval out of 1000 born (l_x)	Number dying in age interval out of 1000 born (d_x)	Mortality rate per 1000 alive at beginning of interval ($1000q_x$)	Average number of individuals alive during age interval (L_x)	Number of individual time units left at beginning of interval (T_x)	Expectation of life, or mean lifetime remaining to those attaining age intervals (yr) (e_x)
0–1	−100	1000	550	550	725	1569	1.57
1–2	−36	450	202	449	349	844	1.87
2–3	+27	248	72	290	212	495	1.99
3–4	+91	176	60	341	146	283	1.60
4–5	+155	116	60	517	86	137	1.18
5–6	+219	56	38	678	37	51	0.90
6–7	+282	18	13	722	11.5	14	0.75
7–8	+346	5	5	1000	2.5	2.5	0.50

[a] Recomputed from Hallam (1967). Mortality data from Fitch (1950).

individuals (Table II). Not all of the columns shown in this example appear in a given life table. T_x and L_x are often omitted.

The statistical quantities defined above are computed from observations of either l_x or d_x (Table III). Information about mortality may be obtained from three sources. These sources reflect different sampling strategies, and they may be classified as follows (Deevey, 1947):

1. Direct observations of the age at death (d_x). This strategy generally requires a census of the death assemblage.
2. Direct observation of survival (l_x) in the life assemblage. Individuals of a specific age class (cohort) are identified and followed in time by periodic sampling.
3. Indirect observation of the age at death (d_x). A census of the life assemblage is collected at a given time. The quantity d_x is inferred from the numerical differences between successively older age classes.

Certain problems associated with mortality data are important to mention at this point. Factors affecting the death assemblage such as transport or differential shell destruction will tend to bias the results in case (1). Furthermore, to use cases (1) and (3) safely, one must assume that the living population is stable in time (Deevey, 1947). To ensure this, it may necessary to average a time sequence of class-frequency histograms (Levinton and Bambach, 1969; Deevey, 1947). Finally, the application of case (2) will avoid most of the problems discussed above, but it will give mortality information for only one age class. While this may be desirable in

Table III. Relationships for Computing the Statistical Quantities Appearing in a Life Table

$$x' = 100(x - e_0)/e_0$$

$$d_x = l_x - l_{x+1}$$

$$q_x = d_x/l_x$$

$$L_x = \int_x^{x+1} l_x\, dx \Big/ \int_x^{x+1} dx$$

$$\doteq (l_x + l_{x+1})/2 \quad \text{if the age interval } \Delta x \text{ is small}$$

$$T_x = \int_x^{\infty} l_x\, dx$$

$$\doteq \sum_{i=x}^{\infty} L_x\, \Delta x \quad \text{if the age interval } \Delta x \text{ is small}$$

$$e_x = T_x/l_x$$

many specific situations (such as following the mortality effects of a pollution event), the resulting life table may or may not represent the mortality patterns of the entire population. However, averaging the life tables of several cohorts is usually acceptable in this situation.

Because one cannot screen out all these problems, life tables constructed for the same population, but using different sampling strategies, will yield somewhat different results.

3.5. Survivorship

It has become traditional, in an analysis of mortality, to concentrate on one particular statistical quantity, l_x, the survivorship. Survivorship is conveniently displayed by plotting l_x on a logarithmic scale as a function of age (x), or age as a percentage deviation from the mean length of life (x').

Various types of theoretical survivorship curves are shown in Fig. 23. They have been discussed in general by many authors, including Pearl (1940), Deevey (1947), Craig and Hallam (1963), Kurtén (1964), and Craig and Oertel (1966). Curve I is the negatively skewed rectangular type, and represents a population with low mortality early in life, but with high mortality in old age. Curve II is diagonal and characteristic of constant mortality throughout life, i.e., age-independent mortality. Curve III, the positively skewed rectangle, is found in populations with high mortality in the youngest growth stages. Curves Ia and IIIa are less extreme versions of I and III.

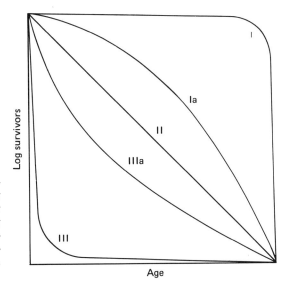

Figure 23. Theoretical survivorship curves. (I) Low mortality early in life, but high mortality in old age; (II) constant mortality throughout life; (III) high mortality early in life; (Ia, IIIa) less extreme versions of (I) and (III). From Deevey (1947).

Kurtén (1964) suggests that the survivorship patterns shown in Fig. 23 are approximated in nature only during a portion of the entire life span of an organism. When the complete record of a species population is considered, including all larval, juvenile, and adult stages, the shape of the survivorship curve is sigmoidal (Fig. 24), with high mortality in both young and old age classes (Kurtén, 1964).

I have chosen several examples from the literature to illustrate survivorship. The first example is the California Pismo clam, *Tivela stultorum*, studied by Fitch (1950). Censuses of living populations collected over several years allowed Fitch to estimate mortality. From this information, Hallam (1967) constructed a life table (see Table II). The survivorship curve (Fig. 25) taken from this life table has a sigmoidal form, even though mortality was estimated only after spat settling. As noted by Hallam, mortality is high during the first year, begins to decline, and then increases sharply again after the fourth year of life. Man has had a major effect on the shape of this survivorship curve. At about 4 years of age, the clams reach a marketable size (Hallam, 1967). Without human predation, a clam may live for 35 years or more (Hallam, 1967).

In another example, Craig and Hallam (1963) studied mortality in the mussel *Mytilus edulis* from the coast of Great Britain. Mortality data were obtained from size-frequency histograms of dead shell assemblages (Fig. 26). Age was estimated from growth data provided by Savage (1956) (see Fig. 13). Only postlarval specimens were considered in the mortality data;

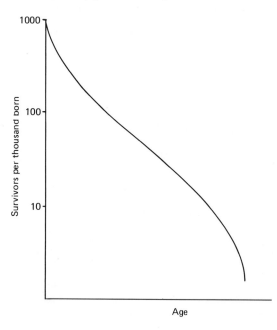

Figure 24. Hypothetical shape of a survivorship curve representative of a population in which a complete record is available, including all larval, juvenile, and adult stages. Suggested by Kurtén (1964).

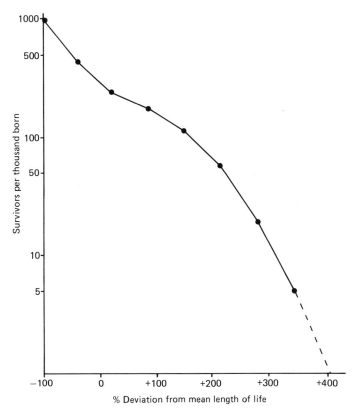

Figure 25. Survivorship curve for the California Pismo clam,*Tivela stultorum*. Survivorship data from Table II. Note the sigmoidal shape even though only postlarval animals were used in the analysis. Dashed portion of curve has been extrapolated.

hence, the survivorship curves from the two localities are not sigmoidal (Fig. 26). The curves indicate that the mortality factors of Ferny Ness and Craigelaw Bay act on the populations in similar ways.

The final example to be considered is from Levinton and Bambach (1969). When constructing life tables or survivorship curves, one must know the age of individual members in a population. An erroneous estimate of age will make the tabulation of mortality inaccurate. When other methods of absolute aging fail, Levinton and Bambach (1969) propose application of a technique developed specifically for bivalves that may be used to assign an approximate relative age scale to individual animals in a population. Levinton and Bambach (1969) have observed that a relative growth model given by the logarithmic relationship

$$D = s \ln (T + 1)$$

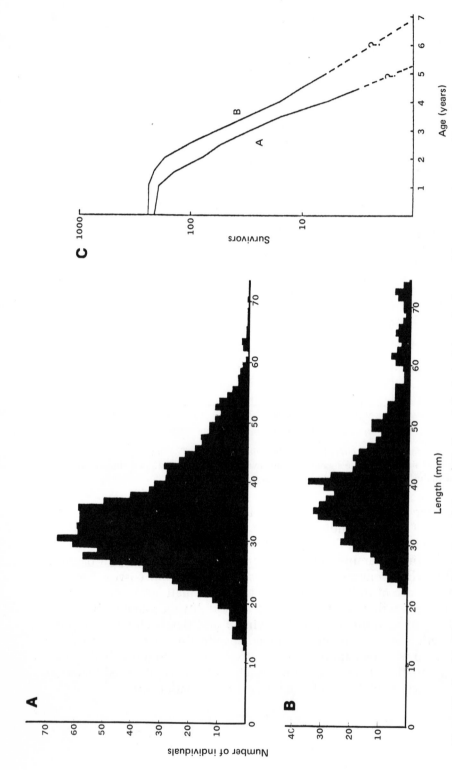

Figure 26. Survivorship curves for two populations of *Mytilus edulis* derived from death assemblages. (A) Valve assemblage from Ferny Ness; (B) valve assemblage from Craigelaw Bay; (C) survivorship curves computed from (A) and (B). Note the close similarity in survivorship. After Craig and Hallam (1963).

where D is the size, T is the relative age, and s is a constant, seems to account adequately for a number of growth curves found in Hallam (1967) (Fig. 27). Their technique is simple to apply. If a relative age at a particular size is chosen, s may be determined. The relative age of any other animal may then be estimated from the inverse of the preceding equation, i.e.,

$$T = e^{D/s} - 1$$

For example, suppose we wish to give the largest animal in a population a relative age of 10. If the size of this animal is 54.30 mm, then s becomes

$$s = D/\ln (T + 1) = 54.3 \text{ mm}/\ln (10 + 1) = 54.30/2.398 = 22.64$$

From the inverse equation, the relative age of an animal 40.00 mm long is then

$$T_{D=40} = e^{D/s} - 1 = e^{40/22.64} - 1 = 4.85$$

In Levinton and Bambach (1969), information on mortality for two bivalve species was derived from size-frequency histograms of the death assemblage. Relative age was determined by the method described above. The smallest size classes in all sampled death assemblages were lost by sieving. The two species considered were *Yoldia limatula*, an infaunal deposit-feeding nuculanid, and *Mulinia lateralis*, an infaunal filter-feeding mactrid. All populations were collected from shallow-water, muddy bottoms in central Long Island Sound.

Two survivorship curves for *Yoldia* are shown in Fig. 28. Higher juvenile mortality in the second *Yoldia* population was attributed to sub-

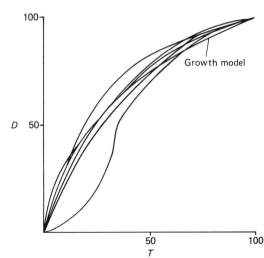

Figure 27. Fit of the growth model $D = s \ln (T + 1)$ to five growth curves chosen at random from Hallam (1967). The growth curves may be found in Fig. 6. In the model, D is the size, T is the relative age, and s is a constant. Both D and T in the figure are expressed as percentages. After Levinton and Bambach (1969).

stratum stability. This sample was collected in an area that had a thin (1-cm), noncohesive, mobile layer at the sediment–water interface (Levinton and Bambach, 1969). This zone provided poor physical support for juveniles (Levinton and Bambach, 1969). Adults, because they have the capacity to live in the more cohesive substratum below this zone, were not affected (Levinton and Bambach, 1969).

A survivorship curve obtained for *Mulinia* by Levinton and Bambach

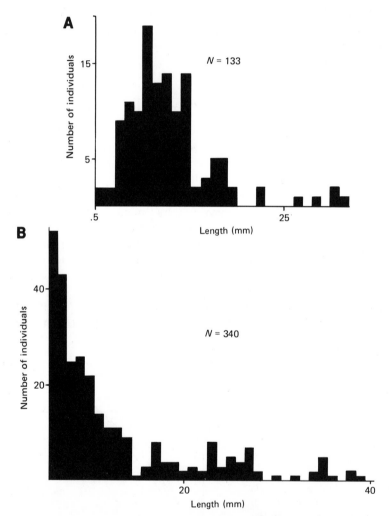

Figure 28. Survivorship curves for two populations of *Yoldia limatula* derived from death assemblages. Both populations are from a shallow-water mud habitat in Long Island Sound. (A) Size-frequency histogram of a death assemblage taken at a water depth of 7 m. (B) Size-frequency histogram of a death assemblage collected at a water depth of 20 m. This second

(1969) is given in Fig. 29. *Mulinia* appears to show higher juvenile mortality than *Yoldia*. Levinton and Bambach ascribe this to the observation that juvenile filter-feeders are especially sensitive to high levels of suspended sediment.

The examples from Craig and Hallam (1963) and from Levinton and Bambach (1969) are interesting because they explicitly use the death assemblage to estimate mortality. The survivorship curves are, however, inaccurate in each case. The growth curves used to estimate age are one source of concern. There are no assurances that the size–age relationship is accurate for any specific population in these examples. In addition, variations in size among animals of the same age were not considered in either paper.

In both papers, the raw size data are transformed so that no reinterpretation of the survivorship method is needed for the analysis. No at-

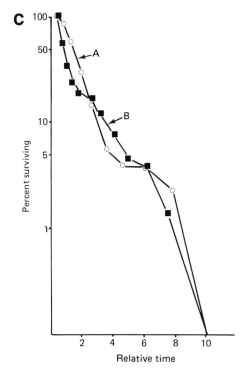

area was characterized by a thin, noncohesive, mobile layer at the sediment–water interface. (C) Survivorship curves computed from (A) (○) and (B) (■). Note the relative time scale. Slightly higher juvenile mortality in (B) is attributed to poor physical support provided by the substrate. Lower size classes in (A) and (B) excluded by sieving. After Levinton and Bambach (1969).

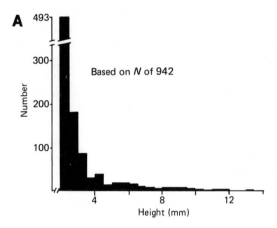

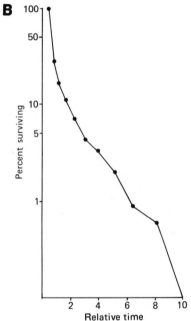

Figure 29. Survivorship curve for a population of *Mulinia lateralis* derived from a death assemblage. (A) Size-frequency histogram of a death assemblage collected in Long Island Sound; (B) survivorship curve computed from (A). Note the relative time scale. Lower size classes excluded by sieving. After Levinton and Bambach (1969).

tempt should be made to compare the resulting curves to ordinary survivorship curves. The methods used in these examples produce only an analogue of "true" survivorship. I believe, however, that the analogy is a potentially useful one. The alternative would be to develop a reliable stochastic model of shell growth. This is not easily done.

For many species, and for molluscs in particular, mortality is dependent more on size than on absolute age (Emlen, 1973). Thus, it is entirely appropriate for comparative purposes to construct "size" survivorship curves directly from the raw data. The nonlinear transformations used by Craig and Hallam (1963) and by Levinton and Bambach (1969) appear to be consistently applied, but they are not necessary. In either the transformed or the untransformed cases, the conclusions reached can be safely interpreted only in terms of "size" survivorship. Either case does, however, provide useful information about mortality.

3.6. Further Uses of Death Assemblages

Death assemblages can be used to obtain other information on population attributes under one of two conditions. First, if mass mortality occurs, the dead shell assemblage created by the event represents a quantitative census of the living population (Levinton and Bambach, 1969; Kurtén, 1964). Population attributes prior to the event can be evaluated just as though the population were still alive. These estimates can be used later to assess the extent of population recovery by newly settled animals. Care must be taken to exclude shells not associated with the single mortality event, because the mixture would destroy the features of the census. Growth-increment patterns may be used to reject all shells that do not have the same season of death (see Chapters 6, 7, and 11).

Second, if relatively high mortality occurs during a season of slow or little growth, then the death assemblage will tend to resemble the size-frequency distribution of the living population (Sheldon, 1965; Craig and Oertel, 1966). The season of slow growth for molluscan populations is during the winter. When mortality is high enough, and when there is little "contamination" of the sample by shells of specimens that did not die during this period, the death assemblage becomes a biased census of the population. The census is biased because mortality is age- (or size-)specific. If a correction for this bias is made, the data can be used just as though they were a winter census of the living population.

To justify this assertion, consider in Fig. 30 the size-frequency histograms of live Cerastoderma (= Cardium) edule and articulated and disarticulated death assemblages of C. edule collected as a census by Craig and Hallam (1963). Cerastoderma has clearly defined winter growth rings; a histogram of growth-ring size frequencies from the living population is also included in Fig. 30. The living population at Peffer Burn is unimodal and represents essentially one cohort. Two peaks (14.5 and 21.5 mm) in the growth-ring size-frequency histograms coincide with peaks in the articulated shell death assemblage (Fig. 30B) and one peak (14.5 mm) with a peak in the disarticulated shell death assemblage (Fig. 30C).

Disarticulation may be attributed to the gradual decomposition and

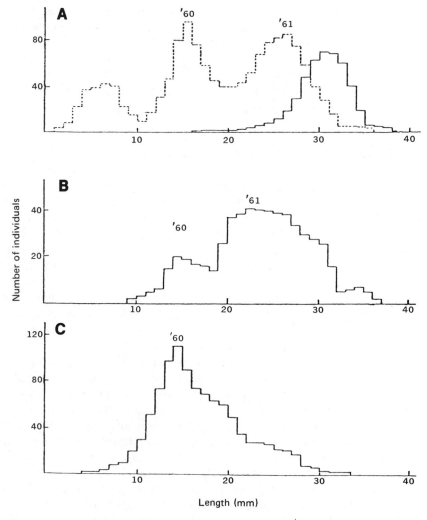

Figure 30. Size-frequency histograms of *Cerastoderma* (= *Cardium*) *edule* from Peffer Burn, East Lothian. This census was taken in May 1962. (A) Living population (——) and growth-ring size frequencies (– – –). The living population is unimodal and probably represents a single year-class. Two peaks in the growth-ring size-frequency histogram correspond to the winters of 1960 and 1961. The smaller peak at 6.5 mm represents a growth ring that apparently developed shortly after spatfall in summer 1960. Because of shell abrasion in the umbonal region, this growth ring was discernible on only part of the living population. (B) Articulated dead shells. Note the correlation between peaks in the growth-ring size-frequency distribution and peaks in this death assemblage. Specimens dying in the most recent winter are represented to a large extent by articulated shells. (C) Disarticulated death assemblage (right valves only). Note the correlation between the peak here and the 1960 winter peak in the growth-ring size-frequency histogram (A, – – –). After Craig and Hallam (1963).

disintegration of the ligament joining the two valves together (Craig and Hallam, 1963). In this example, specimens dying in the more recent winter (1961) are represented to a large extent by articulated shells. Those dying in the previous winter (1960) are mostly disarticulated. A 2-year record of the population during the winter is clearly preserved, therefore, in this death assemblage.

4. Adaptiveness

The genetic properties of a population (adaptiveness, reproductive fitness, and persistence) cannot be determined from a short-term analysis of class-frequency histograms. However, because of an interesting correlation, one genetic property, adaptiveness, can be qualitatively inferred from population attributes already discussed. If one studies the biological attributes of species populations colonizing "new" or recently disturbed habitats, it is possible to recognize patterns in survival adaptations (Odum, 1969; Rhoads et al., 1978). These patterns of adaptation are remarkably similar in taxonomically distinct organisms, and can be observed in terrestrial as well as marine species (Hutchinson, 1953; Margalef, 1958). One aspect of adaptation is the relationship between population attributes and habitat stability.

An organism colonizing a new or short-lived unpredictable environment must be able to: (1) discover a suitable habitat quickly because of a high dispersal capacity; (2) grow, mature, and reproduce quickly before other organisms enter the habitat and compete for resources; and (3) either disperse quickly or suffer high mortality as resources decline or are used up (Wilson and Bossert, 1971). Such a species is an opportunist. Opportunists are adapted for life in a biologically unsaturated community (Pianka, 1970). They have followed an evolutionary pathway that favors a high intrinsic rate of increase at the expense of competitive ability. An opportunist is often called an "r-strategist," a reference derived from the logistic equation of population growth. In the logistic equation, the symbol r represents the intrinsic rate of increase of the population.

An organism adapted to a stable environment has less need for rapid dispersal, growth, or reproductive capabilities (Wilson and Bossert, 1971). In stable habitats, population abundances are close to the carrying capacity of the environment (Pianka, 1970). To maintain itself in such an environment, an organism is dependent primarily on its ability to compete for resources. Here, natural selection favors increased specialization at the expense of a high reproductive capacity. An organism adapted to a stable environment will therefore be a specialist. A specialist is often called an "equilibrium species" or a "K-strategist," a term that again refers to the logistic equation of population growth. In the logistic equation, K is a symbol that represents the carrying capacity of the environment.

Correlates of r- and K-selection are presented in Table IV. These strategies represent opposite ends of a continuous spectrum that runs from unstable or unpredictably changing environments dominated by opportunists to stable or relatively constant environments dominated by equilibrium species (e.g., Sanders and Hessler, 1969; Slobodkin and Sanders, 1969; Pianka, 1966; Klopfer and MacArthur, 1961). Most species will lie somewhere between these two extremes.

The r and K correlates in Table IV may be used to place species in a relative order of opportunism. For example, consider two species commonly encountered in Long Island Sound: *Mulinia lateralis* and *Nucula*

Table IV. Some of the Correlates of r-Selection and K-Selection[a]

	r-Selection	K-Selection
Climate	Variable and/or unpredictable; uncertain	Fairly constant and/or predictable; more certain
Mortality	Often catastrophic, nondirected, density-independent	More directed, density-dependent
Survivorship	Often Type III (Deevey, 1947)	Usually Types I and II (Deevey, 1947)
Population size	Variable in time, nonequilibrium; usually well below carrying capacity of environment; unsaturated communities or portions thereof; ecological vacuums; recolonization each year	Fairly constant in time, equilibrium; at or near carrying capacity of the environment; saturated communities; no recolonization necessary
Intra- and interspecific competition	Variable, often lax	Usually keen
Relative abundance	Often does not fit MacArthur's broken-stick model (King, 1964)	Frequently fits the MacArthur model (King, 1964)
Selection favors:	1. Rapid development	1. Slower development, greater competitive ability
	2. High r_{max}	2. Lower resource thresholds
	3. Early reproduction	3. Delayed reproduction
	4. Small body size	4. Larger body size
	5. Semelparity: single reproduction	5. Iteroparity: repeated reproductions
Length of life	Short, usually less than 1 year	Longer, usually more than 1 year
Leads to:	Productivity	Efficiency

[a] Reprinted from Pianka (1970) by permission of the University of Chicago Press. Copyright © 1970 by the University of Chicago.

annulata. Mulinia lateralis is characterized by rapid development, a short generation time, and a life span rarely exceeding 3 years. *Nucula annulata* develops more slowly, has delayed reproduction, and has a life span sometimes exceeding 8 years in Long Island Sound. One would expect, therefore, that *Mulinia* is more opportunistic than *Nucula*. This is observed in Long Island Sound. *Mulinia* is a rather transient species, appearing sporadically and colonizing parts of the seafloor recently disturbed by bottom current scour or dredge spoiling. In only a year or two, a population may appear in an area, reach very high densities, and then experience catastrophic mortality (Levinton, 1970; Levinton and Bambach, 1969). Large populations of *Nucula*, however, tend to be more stable in time. Using such indicator species, it is possible to identify areas of local disturbance (Johnson, 1972; McCall, 1975).

5. Population Responses to Environmental Stress

In general, the population attributes of shallow-water molluscan species can be related to many environmental factors, including food supply, salinity, temperature, oxygen tension, water movement and turbidity, substratum composition and physical stability of the bottom, population den-

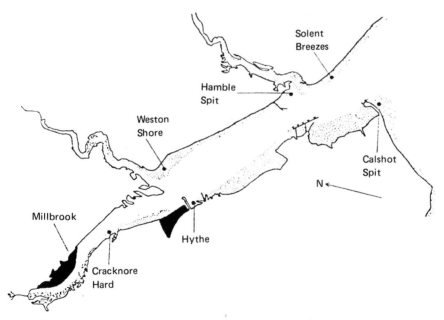

Figure 31. Sketch map of Southampton Water. Population samples in Fig. 32 were collected from the stations shown. The littoral zone is stippled, and the areas in the process of reclamation are shown blackened. After Barnes (1973).

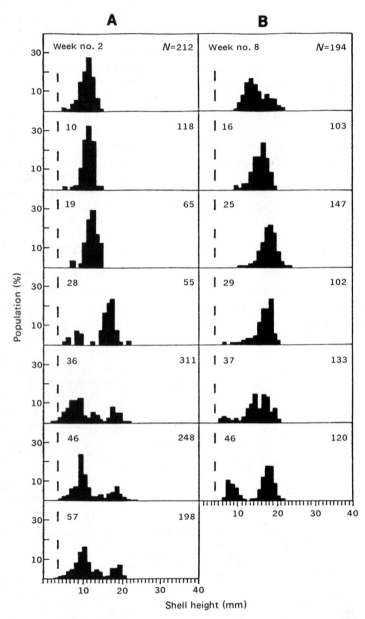

Figure 32. Size-frequency histograms of *Cerastoderma edule* from Southampton Water during 1971. Collecting stations (see Fig. 31): (A) Hythe; (B) Millbrook; (C) Weston Shore; (D) Hamble Spit; (E) Solent Breezes; (F) Calshot Spit; (G) Cracknore Hard. Note variations in recruitment and age structure among localities. After Barnes (1973).

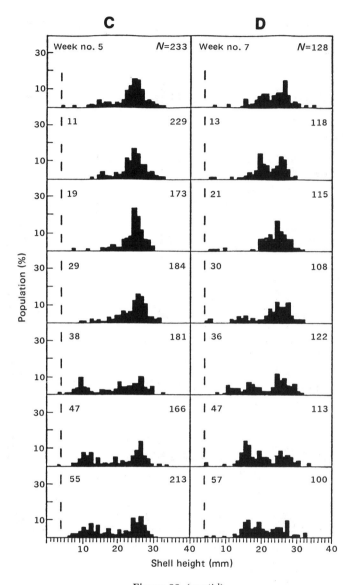

Figure 32. (cont'd)

sity (both intra- and interspecific interactions), and the presence and concentration of chemical pollutants (Hallam, 1965; Kinne, 1970; Moore, 1958; Hedgpeth, 1963). The utility of population attributes in deducing environmental change is related to the degree of resolution afforded by class-frequency histograms, and how specific the environmental effect is

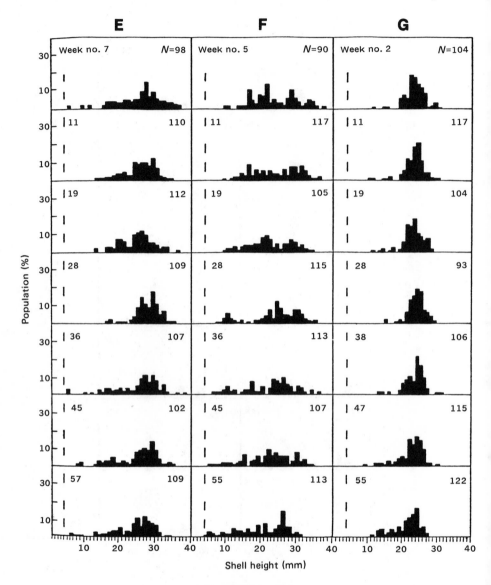

Figure 32. (cont'd)

on a population. For example, reconsider the age structure of the bivalve *Spisula solida* (see Fig. 4). The age structure suggests an environmental effect of probably short duration that inhibited recruitment 2 years prior to the sampling date. The stress effect was probably of short duration, because subsequent recruitment classes do not appear to have been affected.

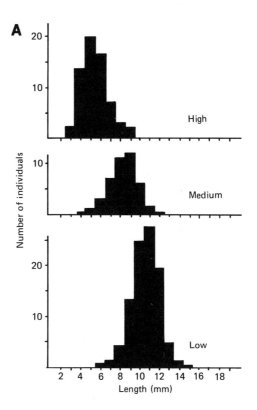

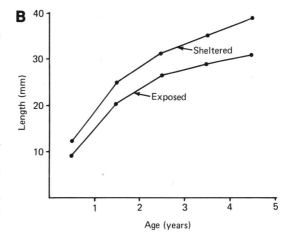

Figure 33. Differences in population attributes along a stress gradient. (A) Size-frequency histogram for the 0 + year-classes of *Cerastoderma* (= *Cardium*) *edule* from Llanrhidian Sands, Burry Inlet, South Wales. The stations shown differ in height in relation to low-water level. Average shell lengths reflect rates of growth. Higher stations are more exposed and individuals have a shorter feeding period. After Cole (1956). (B) Average shell growth for two populations of *Cerastodermaedule* showing effect of exposure to wave action. Both localities are in Essex. Data from Cole (1956).

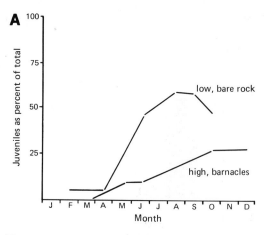

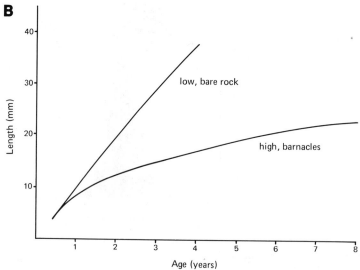

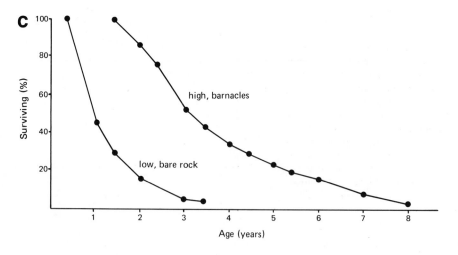

In many instances, temporal and spatial trends in population attributes can be identified rather easily. For example, Barnes (1973) studied intertidal populations of the bivalve *Cerastoderma edule* from Southampton Water during 1971 (Fig. 31). The observed trends (Fig. 32) ranged from populations near the head of the water markedly influenced by annual recruitment, and comprised mostly of young specimens, to populations at the mouth little affected by year-to-year recruitment, and dominated by older specimens (Barnes, 1973). The Cracknore Hard location (Fig. 32G) was an exception to the described pattern. At this site, the population was stable and was probably composed of one or two older age classes showing little growth during the year; recruitment was insignificant (Barnes, 1973). The population at this location is probably an example of the "dominant-age-class" phenomenon discussed in Section 3.1.

In many cases, trends in population attributes can be related to environmental gradients, as shown, for example, in the study of *Cerastoderma* (= *Cardium*) *edule* (Fig. 33). In other instances, the source of environmental stress may not be easily recognized. Complex interactions among environmental factors are sometimes difficult to resolve without additional information on the biology of the species being studied. For example, Orton and Lewis (1931) correlated increased mortality in oyster drills (*Murex erinaceus* and *Thais lapillus*) with temperatures 2–6°C below normal. Temperature, however, was not the direct cause of mortality. Increased mortality at low temperatures was due to the inability of the organisms to tolerate salinities that, at normal temperatures, were within their tolerance range (Moore, 1958).

In a second example illustrating complex interactions among environmental factors, Lewis and Bowman (1975) studied variations in recruitment, growth, and mortality in intertidal populations of the limpet *Patella vulgata* (Fig. 34). Trends ranged from populations at low tidal level on bare rock that were greatly influenced by annual recruitment, and showed high rates of growth and mortality, to populations in high-tidal-level, barnacle-encrusted areas in which growth and mortality were low, and where annual spawning represented a small percentage of the standing crop. Many populations studied correlated well with this tidal-level pattern. However, it was also found that the presence or absence of *Balanus balanoides* or *Mytilis edulis* or both could locally reverse pop-

←———————————————————————————————

Figure 34. Differences in population attributes in intertidal populations of the limpet *Patella vulgata*. (A) Recruitment; (B) growth; (C) mortality. General trends are illustrated. The population structure at low tidal level on bare rock was markedly influenced by annual recruitment, and showed high rates of growth and mortality. Populations at high tidal level living in barnacle-encrusted areas showed lower rates of growth and mortality, and annual spawning represented a smaller percentage of the standing crop. After Lewis and Bowman (1975).

ulation attributes based solely on the position of a population with respect to tidal level (Lewis and Bowman, 1975).

As a final example, Brush (1974) derived population attributes for *Mulinia lateralis*, a filter-feeding mactrid, and *Nucula annulata*, a deposit-feeding nuculid, for several areas in central Long Island Sound. One area considered was a dump site for dredge spoils from New Haven Harbor. A second location several kilometers from this dump site, and thought to be relatively undisturbed for a long period of time, was also studied. In comparing samples from these two localities, Brush concluded that the opportunistic bivalve *Mulinia* showed higher abundances, growth rates, and survivorship at the dump site than at the undisturbed station. The *K*-selected species, *Nucula*, showed the opposite trend. In view of the earlier discussion of the adaptive strategies of these two species, the spatial patterns in population attributes are exactly what one would expect.

To further interpret these patterns, one needs, however, additional information. The dump site was similar to the undisturbed station in sediment type. Both areas had a silt–clay substratum (Rhoads *et al.*, 1975). The undisturbed area was dominated by a deposit-feeding community (Rhoads and Michael, 1974). The sediment was highly bioturbated and uncompacted, and the bottom was frequently resuspended by tidal scour (Rhoads *et al.*, 1975). Bottom resuspension would result in burial of juvenile *M. lateralis* and interfere with filter-feeding mechanisms (Rhoads and Young, 1970; Levinton and Bambach, 1969). The poor growth and survivorship of *Mulinia* at this station may therefore have been due to the presence of a well-developed deposit-feeding community. This phenomenon is known as "trophic group amensalism" (Rhoads and Young, 1970). On the other hand, the dump site had a dense population of opportunistic tubicolous polychaetes. These dense tube aggregations may have bound the sediment surface and prevented erosion or resuspension (Brush, 1974). *Mulinia*, as a filter-feeder, would therefore be expected to do better on such a firm bottom where it was not competing with deposit-feeders.

In conclusion, trends in population attributes can be used to infer spatial and temporal gradients in ecological stress. Relating these trends to specific environmental factors may, however, require supplementary biological information. If the species chosen for study are carefully selected, most problems with interpretation could, I believe, be resolved. Molluscs are an ecologically diverse group ranging in adaptive types from opportunists to equilibrium species. They are found in epifaunal, infaunal, and pelagic habitats and represent a wide range of reproductive, dispersal, and feeding types. The potential for assessing environmental change through class-frequency analysis of molluscan populations is therefore very high.

ACKNOWLEDGMENTS. The preparation of this chapter was made possible through support of the Environmental Protection Agency, Environmental Research Laboratory, Narragansett, Rhode Island (Grant R804-909-010).

References

Allee, W. C., Emerson, A. E., Park, O., Park, T., and Thomas, K. P., 1949, *Principles of Animal Ecology*, W. B. Saunders, Philadelphia.

Ansell, A. D., Sivadas, P., Narayanan, B., and Trevallion, A., 1972, The ecology of two sandy beaches in southwest India. III. Observations on the population of *Donax incarnatus* and *D. spiculum*, *Mar. Biol. (Berlin)* **17**(4):318–332.

Barnes, R. S. K., 1973, The intertidal Lamellibranchs of Southampton water, with particular reference to *Cerastoderma edule* and *C. glaucom*, *Proc. Malacol. Soc. London* **40**:413–433.

Brush, A., 1974, The effect of environmental perturbation on three species of bivalve mollusc in central Long Island Sound, Unpublished senior essay, Department of Geology and Geophysics, Yale University, New Haven, Connecticut.

Cassie, R. M., 1954, Some uses of probability paper in the analysis of size-frequency distributions, *Aust. J. Mar. Freshwater Res.* **5**:513–522.

Cole, H. A., 1956, A preliminary study of growth-rate in cockles (*Cardium edule* L.) in relation to commercial exploitation, *J. Cons. Cons. Int. Explor. Mer.* **22**:77–90.

Craig, G. Y., and Hallam, A., 1963, Size-frequency and growth ring analyses of *Mytilus edulis* and *Cardium edule* and their paleoecological significance, *Paleontology* **6**:731–750.

Craig, G. Y., and Oertel, G., 1966, Deterministic models of living and fossil populations of animals, *Q. J. Geol. Soc. London* **122**:315–355.

Deevey, E. S., Jr., 1947, Life tables for natural populations of animals, *Q. Rev. Biol.* **22**:283–314.

Emlen, J. M., 1973, *Ecology: An Evolutionary Approach*, Addison-Wesley, Reading, Mass.

Fitch, J. E., 1950, The Pismo clam, *Calif. Fish Game* **36**:285–312.

Guttman, I., and Wilks, S. S., 1965, *Introductory Engineering Statistics*, John Wiley, New York.

Hallam, A., 1965, Environmental causes of stunting in living and fossil marine benthonic invertebrates, *Palaeontology* **8**:132–155.

Hallam, A., 1967, The interpretation of size-frequency distributions in molluscan death assemblages, *Palaeontology* **10**:25–42.

Harding, J. P., 1949, The use of probability paper for the graphical analysis of polymodal frequency distributions, *J. Mar. Biol. Assoc. U. K.* **28**:141–153.

Hedgpeth, J. W., 1963, Treatise on marine ecology and paleoecology, *Geol. Soc. Am. Mem.* **67**.

Hutchinson, G. E., 1953, The concept of pattern in ecology, *Proc. Acad. Nat. Sci. Philadelphia* **105**:1–12.

Johannessen, O. H., 1973, Population structure and individual growth of *Venerupis pullastra* (Mortagu) (*Lamellibranchia*), *Sarsia* **52**:97–116.

Johnson, R. G., 1972, Conceptual models of benthic marine communities, in: *Models in Paleobiology* (T. J. M. Schöpf, ed.), pp. 148–159, Freeman, Cooper, San Francisco.

King, C. E., 1964, Relative abundance of species and MacArthur's model, *Ecology* **45**:716–727.

Kinne, O., 1970, *Marine Ecology*, Vol 1, Wiley Interscience, London.

Klopfer, P. H., and MacArthur, R. H., 1961, On the causes of tropical species diversity: Niche overlap, *Am. Nat.* **95**:223–226.

Kurtén, B., 1964, Population structure in paleoecology, in: *Approaches to Paleoecology* (J. Imbrie and N. Newell, eds.), pp. 91–106, John Wiley, New York.

Levinton, J. S., 1970, The paleoecological significance of opportunistic species, *Lethaia* **3**:69–78.

Levinton, J. S., and Bambach, R. K., 1969, Some ecological aspects of bivalve mortality patterns, *Am. J. Sci.* **268**:97–112.

Lewis, J. R., and Bowman, R. S., 1975, Local habitat-induced variations in the population dynamics of *Patella vulgata* L., *J. Exp. Mar. Biol. Ecol.* **17**:165–203.

Margelef, R., 1958, Mode of evolution of species in relation to their places in ecological succession, *XVth International Congress on Zoology*, Section 10, Paper 17, pp. 787–789.

McCall, P. L., 1975, Disturbance and adaptive strategies of Long Island Sound infauna, Ph.D. dissertation, Department of Geology and Geophysics, Yale University, New Haven, Connecticut.

Merrill, A. S., Posgay, J. A., and Nichy, F. E., 1965, Annual marks on shell and ligament of sea scallop (*Placopecten magellanicus*), *U.S. Fish Wildl. Serv. Fish. Bull.* **65**:299–311.

Moore, H. B., 1958, *Marine Ecology*, John Wiley, New York, 493 pp.

Odum, E. P., 1969, The strategy of ecosystem development, *Science* **16**:262–270.

Odum, E. P., 1971, *Fundamentals of Ecology*, W. B. Saunders, Philadelphia.

Orton, J. H., and Lewis, H. M., 1931, On the effect of the severe winter of 1928–1929 on the oyster drills (with a record of five year's observations on sea temperature on the oyster beds) of the Blackwater estuary, *J. Mar. Biol. Assoc. U. K.* **17**(2):301–313.

Pearl, R., 1940, *Introduction to medical biometry and statistics*, 3rd ed., W. B. Saunders, Philadelphia and London, xv + 537 pp.

Pianka, E. R., 1966, Latitudinal gradients in species diversity: A review of concepts, *Am. Nat.* **100**:33–46.

Pianka, E. R., 1970, On r- and K-selection, *Am. Nat.* **104**:592–597.

Rhoads, D. C., and Michael, A., 1974, Summary of benthic biology sampling in central Long Island Sound and New Haven Harbor (prior to dredging and dumping) July 1972–August 1973, Unpublished report of the U.S. Army Corps of Engineers, New England Division, Waltham, Massachusetts.

Rhoads, D. C., and Young, D. K., 1970, The influence of deposit-feeding benthos on bottom sediment stability and community trophic structure, *J. Mar. Res.* **28**:150–178.

Rhoads, D. C., Tenore, K., and Browne, M., 1975, The role of resuspended bottom mud in nutrient cycles of shallow embayments, *Estuarine Res.* **1**:563–579.

Rhoads, D. C., McCall, P. L., and Yingst, J., 1978, Disturbance and production on the estuarine seafloor, *Am. Sci.* **66**:577–586.

Sanders, H. L., and Hessler, R. R., 1969, Ecology of the deep-sea benthos, *Science* **163**:1419–1424.

Savage, R. E., 1956, The great spatfall of mussels (*Mytilus edulis* L.) in the River Conway estuary in spring 1940, *Fish Invest. Ser. II* **20**:1–21.

Seed, R., 1976, Ecology, in: *Marine Mussels: Their Ecology and Physiology* (B. L. Bayne, ed.), International Biological Programme **10**:13–65, Cambridge University Press, Cambridge.

Sheldon, R. W., 1965, Fossil communities with multi-modal size-frequency distributions, *Nature (London)* **206**:1336–1338.

Simpson, G. G., and Roe, A., 1939, *Quantitative Zoology*, Harcourt, Brace, New York and London.

Slobodkin, L. B., and Sanders, H. L., 1969, On the contribution of environmental predictability to species diversity, in: *Diversity and Stability in Ecological Systems, Brookhaven Symp. Biol.*, No. 22.

Smith, S. M., 1974, Mollusca dredged off Musselburgh, Firth of Forth, Scotland, in 1972, with particular reference to the population of *Spisula solida* (L.), *J. Conchol.* **28:**217–224.

Walford, L. V., 1946, A new graphic method of describing the growth of animals, *Biol. Bull.* **30**(4):453–467.

Wilson, E. O., and Bossert, W. H., 1971, *A Primer of Population Biology*, Sinauer Associates, Stamford, Connecticut, 192 pp.

II

Other Taxa

Chapter 13

Barnacle Shell Growth and Its Relationship to Environmental Factors

EDWIN BOURGET

1. Introduction 469
2. Shell Macrostructure 470
3. Shell Formation and Growth 471
4. Shell Growth Patterns 473
    4.1. Internal Growth Bands 473
    4.2. External Growth Ridges 482
5. Ecological and Paleoecological Applications 485
    References 489

1. Introduction

Ridges on the outer surface of the barnacle shell reflect patterns of growth. Darwin (1854) proposed that this irregular growth was related to the molting cycle, and this explanation has been widely accepted. However, in recent years, microscopic examination of barnacle shells has uncovered the presence of growth marks on or in the test that permit the identification of the influences of molting, feeding, tidal immersion, tidal level, and temperature on skeletal growth. Thus, many factors, such as age, age and season of death, and growth rate, may be determined from growth patterns found within shells. Therefore, detailed examination of both internal and external shell structures may provide important information for both ecological and paleoecological studies.

This chapter reviews experiments that have been carried out to identify the effects of various environmental and physiological factors. I have

EDWIN BOURGET • Groupe Interuniversitaire de Recherches Océanographiques du Québec (GIROQ), Département de Biologie, Faculté des Sciences et de Génie, Université Laval, Québec, Canada G1K 7P4.

deliberately excluded from this review aspects dealing with the crystalline structure and barnacle shell figures,* since these aspects have not yet been related to specific environmental factors. For further information on these structures, the reader is referred to Cornwall (1962), Newman *et al.* (1967), and Bourget (1977). The experimental approach used so far for evaluating the influence of any factor is to control the environment so that the influence of only that factor predominates. The experimentalist then proceeds to evaluate the influence of this factor on either the frequency or of skeletal growth increments. The limitations on interpreting this kind of experimental data are related to the complex and dynamic environment to which barnacles are exposed in the intertidal zone.

The influence of environmental variables on growth patterns can best be understood if I first describe the structure of a barnacle shell and review briefly the mechanism of growth and formation of the shell.

2. Shell Macrostructure

The macrostructure of a sessile barnacle shell is illustrated in Fig. 1. Typically, the shell has the shape of a truncated cone enclosing a mantle cavity in which the animal is suspended. There are up to eight overlapping calcareous wall plates and two bipartite opercular valves that are capable of closing the aperture formed by the ring of plates. The cone of plates rests on a membranous basal disk that may or may not be calcified. As shown in Fig. 1, the cuticle forming the base itself, or contained in the calcareous basal disk, extends over the external surface of the shell (Bocquet-Védrine, 1963, 1964, 1965, 1966a; Bourget and Crisp, 1975a; Bubel, 1975; Klepal and Barnes, 1975a; Bourget, 1977). In adults, the cuticular membrane covering the plates is usually eroded off in the older top portions of the shell.

Each paries (singular of *parietes*) is reinforced internally in the dorsal part by the addition of a downward-growing layer of shell to which is attached the opercular membrane. In some species, this element may be apposed to the parietes and called a "sheath" (Darwin, 1854) (see Figs. 1 and 4); when it is not clearly separate from the paries, it has been referred to as a "pseudosheath" (Bourget, 1977). Winglike secondary structures (alae and radii) that develop at the margins of the parietes also increase the mechanical strength at the junction of the plates (Darwin, 1854; Barnes *et al.*, 1970). Barnacle wall plates may be further complicated by the pres-

---

* "Shell figures" are lines of discontinuity within the shell matrix, which reflect the presence of projections near the basal margin of the shell plates of most barnacles having a complex interlocking relationship between the calcareous base and the shell wall plates.

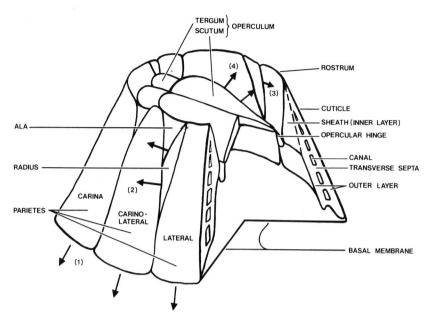

Figure 1. Surface and cutaway view of a typical barnacle shell. Arrows indicate directions of growth: (1) downward growth at the basal margin; (2) growth at the margin of the radii; (3) growth at the margin of the alae; (4) growth at the margin of the operculum.

ence of canals running from the base to the apex (Fig. 1). In many species, these canals may be partitioned by transverse septa, while in others they may become secondarily filled with columns of shell material (see Bourget, 1977).

3. Shell Formation and Growth

The barnacle shell grows in height by the addition of shell material at the basal edge of the parietes and in girth by adding shell material at the margins of the radii and alae (Fig. 1). However, in species in which complete concrescence of the parietes has occurred (e.g. *Tetraclita* and *Pyrgoma*), growth proceeds entirely through addition of increments at the basal margins (Darwin, 1854). These two modes of growth for concrescent and nonconcrescent shells are discussed in some detail by Crisp and Patel (1967) and Bourget and Crisp (1975b).

Growth of the barnacle shell proceeds by deposition of shell material by the epithelium of the mantle underlying the plates and base (Bocquet-Védrine, 1963, 1964, 1965, 1966a; Bourget and Crisp, 1975a,c; Bubel, 1975; Klepal and Barnes, 1975a). Barnacles have the ability to build up a calcareous exoskeleton while, like all crustaceans, still retaining the capacity to molt the chitinous portions of the body.

There are two distinct processes involved in the growth of the barnacle shell, both of which are associated with the activity of the epithelium and are important in the formation of growth patterns. The first concerns the formation of cuticular tissue by the epithelium at the basal margin of the shell, while the second concerns the secretion of calcium onto this newly formed cuticle. Bocquet-Védrine (1963, 1964, 1965, 1966a), Klepal and Barnes (1975a), and Bubel (1975) showed that a cuticular slip is produced by specialized cells underlying the base near the edge of the shell at each molting cycle (Fig. 2). The secretory activity of that part of the epithelium is synchronous with the premolt period of the molting cycle. During the molting cycle, this newly formed slip of cuticle is stretched out at the margin and later becomes progressively filled with calcite secreted by the epithelium underlying the inner region of the parietal plates. In contrast to the secretory activity of the epithelium underlying the base, that of the epithelium of the mantle lining the shell is little influenced

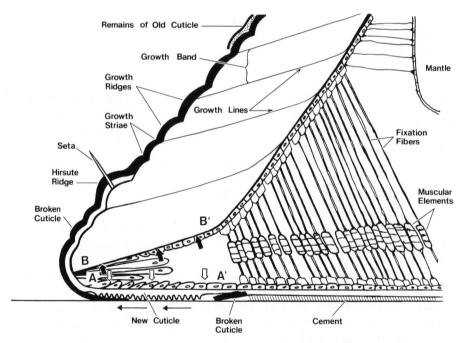

Figure 2. Growing edge at the basal margin of the barnacle shell plate. The cuticular covering does not show the different constituent layers. Each cuticular scale is produced during one molting cycle. Scales overlap one another on the outer surface of the shell plate. (→) Direction of movement of the newly formed cuticular slip; (⇨) region of formation of the cuticle (A–A'); (➡) region of CaCO₃ deposition (B–B'). After Bourget and Crisp (1975a).

by the molting cycle (Costlow and Bookhout, 1953, 1956; Costlow, 1956; Bourget and Crisp, 1975a,c; Bubel, 1975) (also see Section 4). Bourget and Crisp (1975a) suggested the possible involvement of coelomic pressures, controlled by the muscles of the prosoma and the small muscular elements and fixation fibers joining the shell plates to the base, to account for the capacity of the animal to lay down shell material outside the perimeter of the original calcified layers. In short, pressures developed in the body fluids would push part of the new cuticle out from the calcified edge of the shell (Fig. 2, region A). Progressively, the old cuticle, produced during the preceding intermolt period, becomes stretched and eventually breaks. The small muscular elements joining the base to the shell plate (Fig. 2) would counterbalance these pressures. The stretched part of the new cuticle, which is presumably sufficiently elastic to hold until the next intermolt period, then constitutes the template on which $CaCO_3$ is subsequently deposited.

4. Shell Growth Patterns

There is some confusion related to growth terminology in the literature, and I will therefore define the terms that will be used in this chapter. The term "growth band" refers to thin (up to 20 μm) internal incremental layers (similar to microgrowth increments of bivalve shells) of shell material seen in thin sections of shell plates that are separated by a dark line, the "growth line" (Fig. 3). The term "growth ridge" refers to the small surface crests separated by fine linear depressions (striae) on the outer surface of the shell plates (Fig. 4). The distinction between bands and ridges is necessary, since these structures are produced somewhat independently of each other (Bourget and Crisp, 1975a).

4.1. Internal Growth Bands

Little mention is made of the growth bands in the literature. This is not surprising since they are visible only in thin sections of shell and must be acid-etched to be seen and counted accurately. Darwin (1854) stated that very thin layers of shell material could be seen in shell sections. However, it is not clear whether he was referring to the fine growth bands defined here or the thicker layers of shell material that can sometimes be seen between organic sheets. Gutmann (1960) observed growth bands in the outer lamina (see Fig. 1) of *Balanus balanoides* L., but his representation of the banding in the sheath is imprecise and clearly shows that he associated the bands with the thin slip of cuticle remaining after ecdysis.

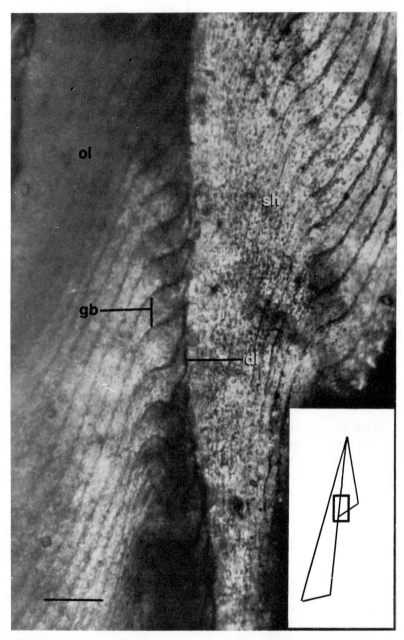

Figure 3. Growth bands (gb) as seen in thin section of *Balanus balanoides*. The figure shows the discontinuity (di) at the junction of the outer layer (ol) and sheath (sh). Scale bar: 50 μm The inset shows the location of the photograph on the shell section.

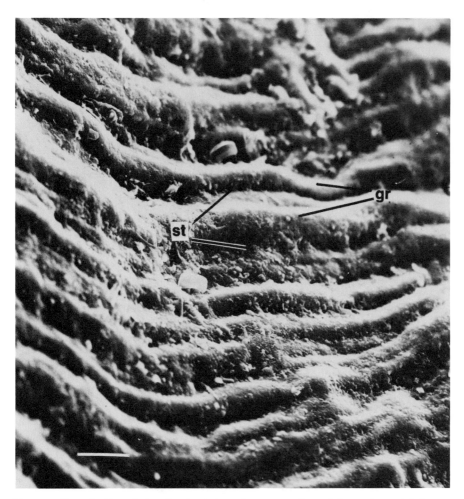

Figure 4. Growth ridges (gr) and striae (st) on the outer surface of the parietes of *Balanus balanoides*. Scale bar: 20 μm.

Two major arrangements of the bands, leading to different banding patterns, have been observed in barnacle shells (Bourget, 1977). In some species (i.e., chthamalids), the bands extend without interruption along the entire inner surface of the shell plate (Fig. 5A). In other species, those within the Balanidae where a shell sheath is present, the secretory activity of the epithelium underlying the shell is not continuous along the entire inner surface of the shell, but is either interrupted, or exceedingly thin, in the lower inner portion of the shell (Fig. 5B). In the latter case, shell growth proceeds at the basal margin of the parietes and in the upper half

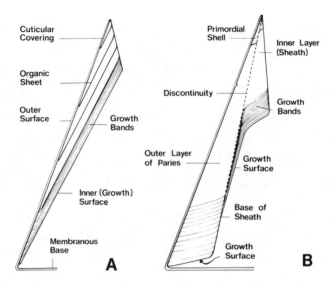

Figure 5. Distribution of the growth bands as seen in radial sections of a barnacle shell. (A) Bands continuous along the inner surface of the shell in *Chthamalus*; (B) bands discontinuous near the basal margin of the shell plate in *Balanus*. Note the absence of sheath in *Chthamalus*. After Bourget (1977).

of the shell plate where the sheath is present. This type of deposition leads to two series of growth bands, those forming the lower part of the paries and those forming the sheath (Fig. 5B).

A comparison of Fig. 5A and B shows that the band widths resulting from these two modes of shell deposition differ markedly. Those shells growing uniformly along the inner surface have much thinner bands than shells growing in two separate regions. Indeed, in practice, uniformly deposited bands of chthamalids are too thin (< 1 µm) to be precisely measured.

The formation of intra- and interlaminate figures (see Cornwall, 1962; Newman *et al.*, 1967, 1969; Bourget, 1977) and canals within the parietes complicates the interpretation of growth bands. However, this problem can be avoided by studying the growth bands within the sheath, which correspond to those of the parietes, but are more easily observed (Bourget and Crisp, 1975c; Crisp and Richardson, 1975).

4.1.1. Relationship to Environmental Factors

Experiments to determine the effects of environmental factors on band production in *B. balanoides* and *Elminius modestus* have been carried out under both laboratory and field conditions (Bourget and Crisp,

1975c; Crisp and Richardson, 1975). Two bands per day were observed to be deposited in these taxa. Research efforts, which are discussed below, focused on environmental factors that have a pronounced periodicity, such as photoperiod, immersion, feeding, and temperature.

4.1.1a. Photoperiod. The question of the effect of light on the growth rate of barnacles was first raised by Klugh and Newcombe (1935), who claimed that growth was negatively affected by direct sunlight. This observation was later questioned by Barnes (1953) and Crisp and Patel (1960), who suggested that reduced growth in illuminated conditions in the field was probably caused by the detrimental effect of enhanced algal growth; sessile algae reduce water flow over the barnacles and thus hinder feeding behavior. This whole question was reexamined experimentally in the absence of sessile algae by Bourget and Crisp (1975c). A group of newly metamorphosed B. balanoides was kept in continuous light for 20 days, while a second group was exposed to alternating conditions of illumination (12 hr) and darkness (12 hr) for the same period. High-resolution growth measurements of the basal diameter of the shell made at 6-hr intervals confirmed the absence of any significant short-term effect of illumination on growth. However, some of the animals kept in the field by Bourget and Crisp (1975a) showed greater rates during daytime than during nighttime. It is thought that increased surface water temperature in the intertidal zone during daytime might be responsible for increased metabolic activity. This could explain the differences between the laboratory and field experiments (see Appendix 3.B).

Two other experiments, one in the laboratory and one in the field, were specifically designed to measure the effect of light on the production of growth bands (Bourget and Crisp, 1975a). Two groups of newly metamorphosed barnacles were held for 30 days either under continuous illumination or with alternating light and dark periods of 12 hr. In the field, animals that had settled on plates were attached inside and outside a light-tight box that was submerged under a raft continuously just below the surface for nearly a month.

By counting the number of growth bands on shell sections, we found that all animals with countable bands, whether kept under laboratory or field conditions, produced about two bands per day regardless of light regime. Furthermore, there was no greater definition of the bands in the discontinuously illuminated animals than in those kept in constant light. These experiments show conclusively that the elimination of diurnal changes in light intensity has no effect on the frequency of shell bands.

4.1.1b. Immersion and Feeding. Immersion period indirectly affects growth, since it limits feeding, which may also be affected by the tidal levels, currents, and wave exposure (Hatton and Fischer-Piette, 1932; Moore, 1934, 1936; Hatton, 1938; Pyefinch, 1950; Barnes and Powell,

1953; Savilov, 1957; Crisp, 1960). In the laboratory, Bourget and Crisp (1975c) examined the effects of immersion, emersion, and feeding on short-term growth. In this study, the growth rate of three groups of newly metamorphosed barnacles was measured every 6 hr for 30 days for animals: (1) immersed and fed continuously; (2) continuously immersed but fed for 6 out of 12 hr; and (3) immersed and fed for 6 hr every 12 hr (normal tidal regime). The results clearly indicated that shell growth continues during nonfeeding periods but at a diminished rate, that during emersion growth stopped entirely, and that the immersion–emersion sequence was the major factor controlling shell growth. A separate experiment showed that shell growth stopped immediately following emersion and resumed on immersion (Bourget and Crisp, 1975c) (Fig. 6).

Having proved that the immersion–emersion cycle affected short-term growth, Bourget and Crisp (1975a) investigated the effects of this cycle on the number of bands produced. Barnacles were held under continuous illumination and subjected to the following types of immersion–emersion cycles in the laboratory: (1) total immersion; (2) 12-hr immersion/12-hr emersion; (3) 24-hr immersion/24-hr emersion; (4) 48-hr immersion/48-hr emersion; and (5) 96-hr immersion/96-hr emersion. In the field, some barnacles were continuously immersed, while others were subjected to normal tidal immersion. These experiments conclusively demonstrated that the clarity of the growth bands is related to immersions; animals contin-

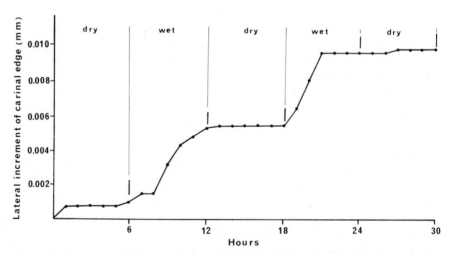

Figure 6. Influence of immersion and emersion on shell growth of *Balanus balanoides*. Measurements of the lateral growth of the carinal edge of the shell were taken at hourly intervals during three dry and two wet periods. Little growth occurs while the animal is emersed, but shell increments are observed when the animal is immersed and calcium becomes available. After Bourget and Crisp (1975c).

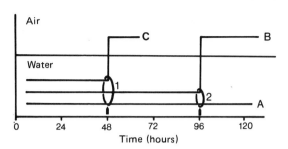

Figure 7. Diagram illustrating the immersion–emersion cycles to which some newly settled *Balanus balanoides* were submitted in the laboratory. (A) Continuously immersed animals; (B) 96 hr immersed/96 hr emersed; (C) 48 hr immersed/48 hr emersed.

uously immersed have poorly defined bands, while those subjected to immersion–emersion cycles have distinct growth bands. Animals that were continuously immersed deposited two growth bands per day, as did animals subjected to normal semidiurnal immersion–emersion cycles. However, Bourget and Crisp (1975a) stated that barnacles experiencing 24-, 48-, and 96-hr immersion–emersion cycles in the laboratory deposited one band per immersion. As can be seen in Fig. 7, all experiments started with the immersion phase of the cycle using test animals of identical pretreatment backgrounds, and it is therefore difficult to explain why barnacles that had received identical treatment after 48 or 96 hr (Fig. 7, points 1 and 2) would show any differences in the number of bands deposited. As has been previously stated (Bourget and Crisp, 1975a), the definition of growth bands is very poor in immersed animals compared with those subjected to immersion–emersion cycles. It is therefore possible that the growth line formed at the time of emersion was so distinct that those ill-defined bands deposited during prolonged immersion were overlooked or were not visible. Such a difference in band contrast would not occur between the growth bands in continuously immersed animals and thus, for some animals, rings would be counted. One must therefore be careful in analyzing shell sections and not be biased by marked contrasts between successive growth bands.

4.1.1c. Tidal Level and Temperature. The location of the barnacle relative to the high and low tide level will influence the duration of each immersion and emersion period. Bourget and Crisp (1975a) carried out experiments to determine whether the duration of the immersion period affected the formation of growth bands. Two groups of newly settled *B. balanoides* were placed on a pier, one at the mean high water level (MHW) and the other at the mean low water level (MLW). A third group attached to the undersurface of a raft was continually immersed. Again, all barnacles deposited approximately two bands per day. All bands for the intertidal barnacles were distinct, but animals placed near the MHW had clearer bands than those animals at MLW. Those immersed continuously formed diffuse bands. However, the widths of the growth bands were largest for those animal continuously immersed and smallest for those placed near MHW.

A further experiment was designed by Bourget and Crisp (1975c) to correlate the width of the growth bands with the duration of the immersion period. For this experiment, the need for a marker growth band was essential. However, efforts using vital stains (Hidu and Hanks, 1968) and radioactive-labeled ^{45}Ca were unsuccessful. It was found that if $CaCl_2$ was added to experimental tanks (resulting in twice the Ca ion concentration of normal seawater), barnacles deposited an easily recognizable "marker" band. Animals thus marked were placed for 12 days at the two stations at mean high and low tidal levels. Analysis of the shells showed that the banding pattern was similar in all specimens, with a good correlation among the widths of successive bands in different animals. Band width and the immersion period were also positively correlated.

Analyses revealed that animals located near MHW did not deposit growth bands on days 7 and 8. On those days, the expected bands were replaced by a prominent dark line, referred to as a "stress band" (see Fig. 8). Analysis of meteorological data showed that a stress band occurs immediately following long periods of emersion associated with high ambient air temperature at low tide and long periods of insolation (see Fig. 8). Thus, it would appear that shell deposition is temporarily halted during periods of heat stress. *Balanus balanoides* is capable of using atmospheric oxygen (Barnes *et al.*, 1963b), yet desiccation at high temperatures causes the complete closure of the valves, resulting in anaerobic respiration and the accumulation of acidic end products (Borsuk, 1929; Barnes *et al.*, 1963b). Thus, it is possible that the darker stress band is the result of a process homologous to that put forward by Lutz and Rhoads (1977) to explain growth line formation in molluscs. It was also noted that the correlation between band width and immersion period was significantly reduced following a period of heat stress. Considering that the effects of heat stress were distinct in May in North Wales (United Kingdom) where the experiments were carried out, it is probable that the effects would be even more pronounced in summer months and in more southerly latitudes. It is evident that our knowledge of band deposition as it relates to heat-stress periods is cursory, and further research in this area is essential.

4.1.1d. Other Factors. It is evident from the preceding discussion that immersion is the only environmental factor that is recorded with a certain regularity in the shell. However, it will be readily recognized that factors such as food, wave action, barnacle orientation, water flow, competition, season, breeding events, age, molting, tide, and temperature (see Appendix 3.B), which are known to affect barnacle growth rate (see Barnes, 1955, 1961, 1962; Barnes and Barnes, 1959; Crisp, 1960; Bourget and Crisp, 1975c), are also likely to influence band width. For instance,

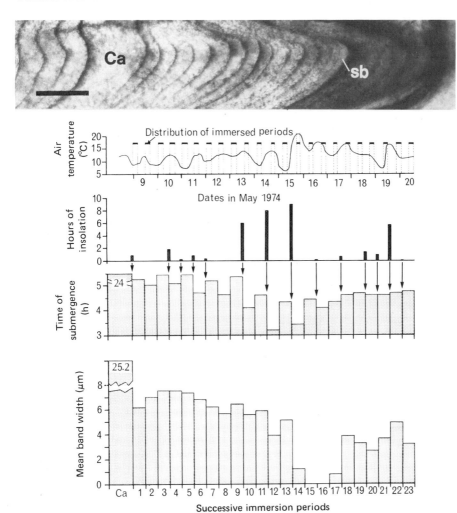

Figure 8. Mean band width in 4 specimens of *Balanus balanoides* from MLW. The time of submergence, the hours of insolation during the emersion periods, and the air temperature are plotted against the mean band width. The photograph is of a region of the sheath showing the marker band (Ca) and darker stress band (sb). Scale bar: 25 μm. After Bourget and Crisp (1975c).

significant variations in barnacle shell growth occur from one season to another. The seasonal cycle of growth in *B. balanoides* is characterized by a maximum in the spring, a reduction in the summer, and a further reduction during winter (Hatton and Fischer-Piette, 1932; Moore, 1934; Hatton, 1938; Barnes and Barnes, 1959; Barnes, 1961, 1962). As expected, winter bands are reported to be much thinner than summer bands (Bourget and Crisp, 1975a).

The influence of age is of importance in *B. balanoides, B. crenatus,* and *E. modestus.* Growth rate accelerates soon after metamorphosis, stabilizes in maturing animals, and then declines in older animals (Crisp, 1960).

Breeding also significantly influences the seasonal pattern of growth. Barnes (1962) showed that growth rate is reduced during the breeding season. It was also shown to increase when breeding was inhibited (Crisp and Patel, 1961).

Bourget and Crisp (1975a) demonstrated that shell deposition increased significantly (between 5 and 10%) shortly after ecdysis and slowed down just prior to the next molt. Considering that the normal intermolt period for *B. balanoides* can be relatively short (3–5 days) (Crisp and Patel, 1960; Barnes *et al.,* 1963a; Bourget and Crisp, 1975c), such variations in growth rates cannot be neglected when analyzing the relationship between immersion period and band width.

4.1.2. Exogenous vs. Endogenous Control

One of the problems in trying to elucidate the formation of the growth bands is the production of bands in continuously immersed animals, whether kept in the laboratory or in the field. This problem raises the following question: is the formation of bands under endogenous or exogenous control in barnacles? The following observations may be considered: (1) semidiurnal bands are observed subtidally in newly metamorphosed animals; (2) the production of semidiurnal bands is immediate after metamorphosis; (3) the production of semidiurnal bands cannot be arrested in uniform conditions in the laboratory; (4) bands were observed in subtidal species (Bourget, 1977); and, finally, (5) bands are also observed in the deep-sea *Bathylasma corraliforme* sampled at a depth of 7000 m in Ross Sea, Antarctic (see Bourget and Crisp, 1975a). This evidence supports the hypothesis of endogenous control. According to Bourget and Crisp (1975a), the bands would merely be kept in phase and reinforced by the semidiurnal cycle, at least in *B. balanoides.*

4.2. External Growth Ridges

4.2.1. Distribution

In the literature, the term *growth ridges* usually refers to the large prominent ridges often present on the four valves of the opercula (Fig. 9A), which together with those of the surface of the rim above the op-

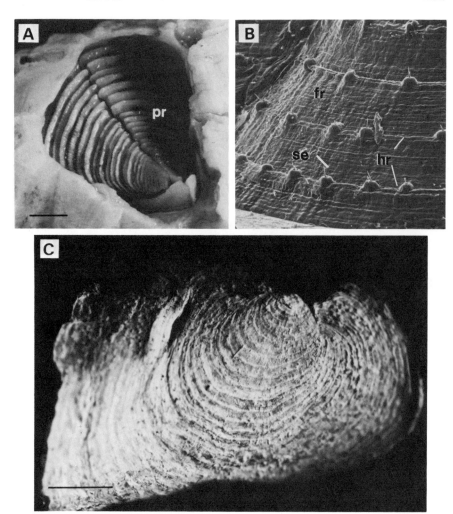

Figure 9. Growth ridges on the outer surface of the shell plates. (A) Scanning electron micrograph of prominent ridges (pr) on the opercula of *Balanus balanus*. Scale bar: 2 mm. (B) Scanning electron micrograph of fine surface ridges (fr) interspaced with more prominent hirsute ridges (hr), bearing setae (se), on the lateral paries of *B. balanoides*. From Bourget (1977). (C) Photomicrograph of the undersurface of the calcareous base of *Balanus tintinnabulum*. Scale bar: 5 mm. Supplied by C. Muir, Department of Zoology, The University, St. Andrews, Scotland.

ercular hinge have for a long time been associated with the molting cycle (Darwin, 1854; Bocquet-Védrine, 1965, 1967). Fine surface ridges, scarcely visible to the naked eye, also occur as small crests between the fine striae on the outer surface of all plates, including the opercula (Fig. 9B) (Petersen, 1966; Bourget and Crisp, 1975a; Klepal and Barnes, 1975b). On the parietal plates, these ridges are interspaced by slightly more prominent ridges, the "hirsute ridges," which bear a series of minute setae projecting outward from the shell plates (Fig. 9B) (Darwin, 1854; Bocquet-Védrine, 1966b; Bourget and Crisp, 1975a,c; Bourget, 1977). A further type of ridge has also been observed as concentric striations on the external surface of the calcareous base of barnacles (Fig. 9C) (Saroyan et al., 1970). Thus, growth ridges can be observed on all parts of the shell surface running parallel to the growing edge of the shell plate. Each ridge is formed by the crystallization of shell material on the inner surface of the chitinous sheets adhering to and covering the shell plates and thus reflects the position of the cuticular slip near the edge at the time of shell deposition.

4.2.2. Relationship to Environmental and Physiological Factors

4.2.2a. Fine Surface Ridges. At first sight, it would seem probable that the finer ridges ($\approx$ 10 μm) observed on the outer surface of the shell correspond to the growth bands observed in the internal shell structure of the plates. However, counts of external ridges on acetate peels made from the outer surface of the shell of B. balanoides later sectioned for the study of growth bands indicated that substantial differences exist between the rate of formation of the finer outer surface ridges and that of growth bands (Bourget and Crisp, 1975a). For instance, the fastest growing, continuously immersed B. balanoides produced four to six ridges a day; moderately fast-growing specimens from the mid- and lower-intertidal zones produced about three ridges per day, while slow-growing specimens from the higher portion of the shore produced about two ridges per day. Thus, the number of ridges laid down in B. balanoides increases with increasing growth rate, although only one growth band per immersion is formed.

4.2.2b. Hirsute Ridges. Bourget and Crisp (1975a,c) showed that hirsute ridges are formed 6–12 hr before the casting off of the exuvia during the molting cycle. It should be noted that the formation of hirsute ridges always occurs in conjunction with a molt, but molting may occur without the formation of a ridge.

4.2.2c. Prominent Ridges. Darwin (1854) associated the formation of the most prominent ridges observed on the opercula and inner rim of the shell with the periods of molting. Indeed, crests on the inner rim of the sheath, and probably those of the opercula, are associated with a thin slip of cuticle, the opercular hinge (see Fig. 1), which is cast off at the time of ecdysis (Darwin, 1854; Gutmann, 1960; Bocquet-Védrine, 1963, 1964, 1965, 1966a; Petersen, 1966). Bourget and Crisp (1975a) explained the extension of the cuticle outside the original perimeter of shell by pressure changes in body fluid. High coelomic pressure was assumed to push the cuticle outside the outer edge of the base, forming a convex cuticular ridge. The prominence of the hirsute ridge, which was shown to be directly associated with molting, supports this view, since the water intake, and consequently the coelomic pressure, is highest in most Crustacea at the time of ecdysis (Lockwood, 1968). As for the hirsute ridges on the outer surface of the parietes, some opercular hinges are apparently not cast off at all; ecdysis and molting of this tissue may be delayed until the following ecdysis (Bocquet-Védrine, 1965). Thus, one probably cannot accurately relate the number of ridges on the rim of the sheath and on the operculum to the molting cycle, since molting may take place without leaving a record on the shell.

4.2.2d. Winter Rings. The outer surface of the shell of barnacles living in subarctic and arctic intertidal zones shows clear disturbance rings seen as distinct overhangs or notches where reduction of growth and retreat of the growing edge has occurred (Fig. 10).

There are at least four examples of the utilization of winter rings to age animals (Kuznetsov and Matveeva, 1949; Crisp, 1954; Petersen, 1966; Klepal and Barnes, 1975b). Distinct winter rings are apparently not always formed in more temperate climates, although reduction of growth is observed during the winter period (see Barnes and Barnes, 1959).

5. Ecological and Paleoecological Applications

The different growth patterns observed on or in the barnacle shell are summarized in Table I together with information pertaining to their formation. Let us first examine the use of external growth patterns: growth ridges and winter rings. To date, the information available indicates that fine growth ridges cannot be related to any environmental factor, nor can they be considered reliable indicators of age, since they seem to result from the buckling of the epicuticular scale produced at the basal margin of the shell plate at each molt. The production of "hirsute ridges" has been directly associated with the molting cycle, since they are produced at the basal margin about 6–12 hr before ecdysis. Unfortunately, they are unreliable indicators of the molting frequency, since they are not always formed at each molt. Thus, the frequency of the ridges of the external shell surface provides a minimum estimate of the number of ecdyses. However, even if some species were to produce "hirsute ridges" regularly in the laboratory, they might not be observed on animals grown in the field, due to shell abrasion.

Winter rings formed on the outer surface of the shells may be used for age determination. The specific growth rate can be estimated from the distance between successive rings providing that one has previously es-

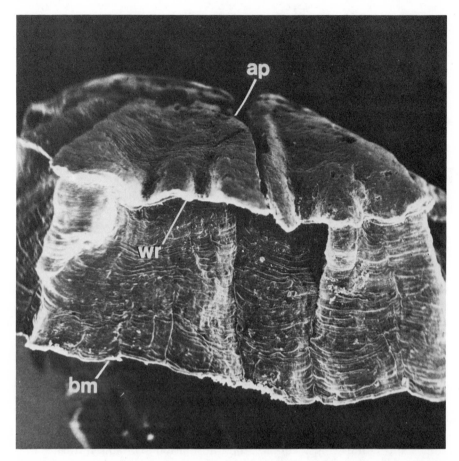

Figure 10. Scanning electron micrograph of parietal shell surface of *Balanus balanoides* showing the winter ring (wr) as a distinct overhang on a lateral shell plate. (ap) Aperture; (bm) basal margin.

tablished that these rings are indeed winter rings and not disturbance rings. Winter rings have been used to age *B. balanoides* from the east Murman coast (U.S.S.R.) (Kuznetsov and Matveeva, 1949) and from Greenland (Petersen, 1966) and to estimate the age of *Balanus balanus* (*B. porcatus*) from the Irish Sea (Crisp, 1954) and *Chthamalus depressus* from the Adriatic (Klepal and Barnes, 1975b). However, as Petersen (1966) pointed out, this technique has a number of drawbacks. Old animals produce smaller winter rings, and the rings on the older portions of the plates tend to be worn away. Thus, prior to age determination, it is necessary to establish that the oldest winter ring on the shell is indeed the first

Table I. Summary of Growth Patterns Observed on or in Barnacle Shells

Growth pattern	Characteristics	Location on or in the shell	Frequency	Cause
External growth patterns				
Fine surface ridges	Fine crests (8–20 μm wide) scarcely visible to the naked eye and running parallel to the growing edges of shell plates	Surface of all parietes, opercular valves, and calcareous base	No known periodicity; 1.5–6 ridges/day produced	Buckling of the cuticle near the edge of the shell plate
Prominent ridges	Ridges wider than 100 μm usually comprising a number of fine surface ridges	Opercular valves and inner rim of the sheath	Usually one per ecdysis, but sometimes probably missing	Produced by shell deposition near the opercular hinge
Hirsute ridges	Prominent ridges bearing a series of minute setae	Outer surface of parietes	Usually one per ecdysis, but sometimes missing	Increased coelomic pressures at time of ecdysis (?)
Disturbance rings	Distinct overhang or notch on the shell surface	Outer surface of parietes and opercular valves	Associated with the frequency of the event (i.e., winter ring)	Events causing reduction of growth and retreat of the growing edge
Internal growth patterns				
Growth bands	Thin layers of shell interspaced by distinct discontinuities	Within the structure of the shell parietes parallel to the surface of shell deposition	One per immersion period of two per day in continuously immersed animals	Possibility of an endogenous rhythm that would be kept in phase and reinforced by the semidiurnal tidal cycle
Stress band	Opaque growth band preceded by a pronounced growth line	Within the structure of the shell parietes	Associated with the frequency of the event (i.e., strong period of heat)	Disturbance events. So far, only high temperature has been shown to cause stress bands.

winter ring produced. This step is not easy in arctic and subarctic regions because of heavy abrasion of the shells by ice.

It should be possible to obtain some rough estimate of the season of death of a barnacle by evaluating the proportion of estimated annual growth accomplished since the deposition of the most recent winter ring. However, the following facts should be kept in mind: (1) barnacle growth rate declines with age (Crisp, 1960); (2) there are marked seasonal variations in growth rates throughout the year, growth being rapid in the spring and considerably reduced in summer, autumn, and winter (Hatton and Fischer-Piette, 1932; Moore, 1934; Hatton, 1938; Barnes and Barnes, 1959; Barnes, 1961, 1962); and (3) growth rates may vary greatly among animals in nonuniform habitats (Crisp, 1960). Hence, prior to estimating the period of mortality, one should obtain growth curves for different age classes, and have some idea, from previous yearly growth increments, whether the animal is a fast or a slow grower. Estimates are likely to be more reliable if growth is appreciable. It follows that spring mortality should be easily separable from summer, autumn, and winter mortality, but death occurring during the latter three seasons might not be so easily distinguishable.

Tidal immersion and periods of marked stress are the only events recorded as internal growth patterns. Other factors may be recorded indirectly by causing variation in growth band width; however, their importance is secondary to duration of immersion. Since one band is laid down per immersion period and the band width is correlated with the duration of immersion, the band record may possibly be used to: (1) determine the number of days in the lunar cycle or in the year or (2) make inferences about the type of tides (diurnal vs. semidiurnal). The use of banding patterns is limited because band deposition may be temporarily halted, as seen with the formation of stress bands. However, the number of missing bands can be determined by: (1) backdating the bands if the series is short and the time of death is known (Bourget and Crisp, 1975a) or (2) comparing banding patterns of individual specimens occurring along an environmental gradient (i.e., tidal level) and assuming the longest series to be the most complete, if relatively reproducible. The effects on growth banding of cyclic factors such as feeding, breeding, climate, and molting are incompletely known and therefore may also present a major obstacle to the interpretation of growth-band patterns. Their influences are complex, frequently interacting, and may be strongly influenced by factors such as day length and temperature (Barnes, 1958, 1962; Barnes et al., 1963a,b; Barnes and Stones, 1974; Crisp, 1957, 1959; Crisp and Patel, 1960; Crisp and Clegg, 1960) (also see Appendix 3.B). Band series could be analyzed mathematically for determining the respective influences of various factors. In Appendix 3.B, the procedures for analyzing the influence of various periodic factors are described using time-series analysis and growth functions. However, regardless of the techniques utilized to analyze banding patterns, the absence of sufficient data concerning the impact of extreme environmental conditions on banding patterns remains a serious problem. Information presently available is too sparse, and pertains to too few barnacle species, to make valid generalizations. However, the potential for using barnacle shells for reconstructing environmental conditions is considerable and may well be realized after required experimentation has been carried out.

ACKNOWLEDGMENTS. I would like to express my thanks to Drs. P. Morisset and J. N. McNeil for critically reading the manuscript and C. Muir for kindly supplying the photograph of the calcareous base shown in Fig. 9.

References

Barnes, H., 1953, The effect of light on the growth rate of two barnacles *Balanus balanoides* (L.) and *B. crenatus* Brug. under conditions of total submergence, *Oikos* **4**:104–111.

Barnes, H., 1955, The growth rate of *Balanus balanoides* (L.), *Oikos* **6**:109–113.

Barnes, H., 1958, Regarding the southern limits of *Balanus balanoides* (L.), *Oikos* **9**:139–157.

Barnes, H., 1961, Variation of the seasonal growth rate of *Balanus balanoides* with special reference to the presence of endogenous factors, *Int. Rev. Gesamte Hydrobiol. Hydrogr.* **46**:427–528.

Barnes, H., 1962, So-called anecdysis in *Balanus balanoides* and the effect of breeding upon growth of the calcareous shell of some common barnacles, *Limnol. Oceanogr.* **7**:462–473.

Barnes, H., and Barnes, M., 1959, A comparison of the annual growth patterns of *Balanus balanoides* (L.) with special reference to the effect of food and temperature, *Oikos* **10**:1–18.

Barnes, H., and Powell, H. T., 1953, The growth of *Balanus balanoides* (L.) and *B. crenatus* Brug. under varying conditions of submersion, *J. Mar. Biol. Assoc. U. K.* **32**:107–128.

Barnes, H., and Stones, R. L., 1974, The effect of food, temperature and light period daylength on molting frequency in *Balanus balanoides*, *J. Exp. Mar. Biol. Ecol.* **15**:275–284.

Barnes, H., Barnes, M., and Finlayson, D. M., 1963a, The metabolism during starvation of *Balanus balanoides*, *J. Mar. Biol. Assoc. U. K.* **43**:184–213.

Barnes, H., Finlayson, D. M., and Platigorsky, J., 1963b, The effect of desiccation and anaerobic conditions on the behavior, survival and general metabolism of three common cirripedes, *J. Anim. Ecol.* **32**:233–254.

Barnes, H., Read, R., and Topinka, J. A., 1970, The behavior on impaction by solids of some common cirripedes in relation to their normal habitat, *J. Exp. Mar. Biol. Ecol.* **5**:70–87.

Bocquet-Védrine, J., 1963, Structure du test calcaire chez *Chthamalus stellatus* (Poli), *C. R. Acad. Sci.* **257**:1350–1352.

Bocquet-Védrine, J., 1964, Relation entre la croissance basilaire du test du cirripède operculé *Elminius modestus* Darwin et le cycle d'intermue de la masse viscérale, *C. R. Acad. Sci.* **258**:5060–5062.

Bocquet-Védrine, J., 1965, Étude du tégument de la mue chez le cirripède operculé *Eliminus modestus* Darwin, *Arch. Zool. Exp. Gen.* **105**:30–76.

Bocquet-Védrine, J., 1966a, Structure et croissance du test chez le cirripède operculé *Acasta spongites* (Poli), *Arch. Zool. Exp. Gen.* **107**:693–702.

Bocquet-Védrine, J., 1966b, Les soies et les expansions épineuses du test calcaire chez le cirripède operculé *Acasta spongites* (Poli), *Arch. Zool. Exp. Gen.* **107**:337–348.

Bocquet-Védrine, J., 1967, La mue de la charnière reliant l'opercule à la muraille chez *Balanus crenatus* Brugière (cirripède operculé), *Archs. Zool. Exp. Gen.* **108**:447–459.

Borsuk, V. N., 1929, O vleyanii exeneniya obshcheĭ colenosti morskoĭ vodȳ na gazoobmen u *Balanus balanoides* (Investigations on the respiratory gas exchange in *Balanus balanoides* in air medium), *Rab. Murman Biol. St.* **2**:25–44.

Bourget, E., 1977, Shell structure in sessile barnacles, *Nat. Can.* **104**:281–323.

Bourget, E., and Crisp, D. J., 1975a, An analysis of the growth bands and ridges of barnacle shell plates, *J. Mar. Biol. Assoc. U. K.* **55**:439–461.

Bourget, E., and Crisp, D. J., 1975b, Early changes in the shell form of *Balanus balanoides* (L.), *J. Exp. Mar. Biol. Ecol.* **17**:221–237.

Bourget, E., and Crisp, D. J., 1975c, Factors affecting deposition of the shell in *Balanus balanoides* (L.), *J. Mar. Biol. Assoc. U. K.* **55**:231–249.

Bubel, A., 1975, An ultrastructural study of the mantle of the barnacle *Elminius modestus* Darwin in relation to shell formation, *J. Exp. Mar. Biol. Ecol.* **20**:287–334.

Cornwall, I. E., 1962, The identification of barnacles with further figures and notes, *Can. J. Zool.* **40**:621–629.

Costlow, J. D., 1956, Shell development in *Balanus improvisus* Darwin, *J. Morphol.* **99**:359–415.

Costlow, J. D., Jr., and Bookhout, C. G., 1953, Molting and growth in *Balanus improvisus*, *Biol. Bull.* **105**:420–433.

Costlow, J. D., Jr., and Bookhout, C. G., 1956, Molting and shell growth in *Balanus amphitrite*, *Biol. Bull.* **110**:107–117.

Crisp, D. J., 1954, The breeding of *Balanus porcatus* (DaCosta) in the Irish sea, *J. Mar. Biol. Assoc. U. K.* **33**:473–496.

Crisp, D. J., 1957, Effect of low temperature on the breeding of marine animals, *Nature (London)* **179**:1138–1139.

Crisp, D. J., 1959, Factors influencing the time of breeding of *Balanus balanoids*, *Oikos* **10**:275–289.

Crisp, D. J., 1960, Factors influencing growth-rate in *Balanus balanoides*, *J. Anim. Ecol.* **29**:95–116.

Crisp, D. J., and Clegg, D. J., 1960, The induction of the breeding conditions in *Balanus balanoides* (L.), *Oikos* **11**:265–275.

Crisp, D. J., and Patel, B. S., 1960, The moulting cycle in *Balanus balanoides* (L.), *Biol. Bull.* **118**:31–47.

Crisp, D. J., and Patel, B., 1961, The interaction between breeding and growth rate in the barnacle *Elminius modestus* Darwin, *Limnol. Oceanogr.* **6**:105–115.

Crisp, D. J., and Patel, B., 1967, The influence of the contour of the substratum on the shapes of barnacles, in: *Symposium on Crustacea, Proc. Mar. Biol. Assoc. India (Part II)*, pp. 612–629.

Crisp, D. J., and Richardson, C. A., 1975, Tidally produced internal bands in the shell of *Elminius modestus*, *Mar. Biol.* **33**:155–160.

Darwin, C., 1854, *A Monograph of the Sub-class Cirripedia*, Vol. 2, Ray. Society, London, 684 pp.

Gutmann, W. F., 1960, Funktionelle Morphologie von *Balanus balanoides*, *Abh. Senckenb. Naturforsch. Ges.* **500**:1–43.

Hatton, H., 1938, Essais de bionomie explicative sur quelques espèces intercotidales d'algues et d'animaux, *Ann. Inst. Oceanogr. Monaco (N.S.)* **17**:241–348.

Hatton, H., and Fischer-Piette, E., 1932, Observations and expériences sur le peuplement des côtes rocheuses par les cirripèdes, *Bull. Inst. Oceanogr. Monaco* **592**:1–15.

Hidu, H., and Hanks, J. E., 1968, Vital staining of bivalve mollusk shells with alizarin sodium monosulfonate, *Proc. Natl. Shellfish. Assoc.* **58**:37–41.

Klepal, W., and Barnes, H., 1975a, A histological and scanning electron microscope study of the formation of the wall plates in *Chthamalus depressus* (Poli), *J. Exp. Mar. Biol. Ecol.* **20**:183–198.

Klepal, W., and Barnes, H., 1975b, The structure of the wall plate in *Chthamalus depressus* (Poli), *J. Exp. Mar. Biol. Ecol.* **20**:265–285.

Klugh, A. B., and Newcombe, C. L., 1935, Light as a controlling factor in the growth of *Balanus balanoides*, *Can. J. Res. Sect. D* **13**:39–44.

Kuznetsov, W. W., and Matveeva, T. A., 1949, The influence of the density of the population on certain biological processes in *Balanus balanoides* (L.) from eastern Murman, *Dokl. Adad. Nauk SSSR* **64**:413–415.

Lockwood, A. P. M., 1968, *Aspects of the Physiology of Crustacea*, Oliver and Boyd, Edinburgh and London, 328 pp.

Lutz, R. A., and Rhoads, D. C., 1977, Anaerobiosis and a theory of growth line formation, *Science* **198**:1222–1227.

Moore, H. B., 1934, The biology of *Balanus balanoides*. I. Growth rate and its relation to size, season and tidal level, *J. Mar. Biol. Assoc. U. K.* **19**:851–868.

Moore, H. B., 1936, The biology of *Balanus balanoides*. V. Distribution in the Plymouth area, *J. Mar. Biol. Assoc. U. K.* **20**:701–716.

Newman, W. A., Zullo, V. A., and Wainwright, A., 1967, A critique on recent concepts of growth in Balanomorpha (Cirripedia, Thoracica), *Crustaceana* **12**:167–178.

Newman, W. A., Zullo, V. A., and Withers, T. H., 1969. Cirripedia, in: *Treatise on Invertebrate Paleontology*, Part R, *Arthropoda* (R. C. Moore, ed.), pp. R206–R295, University of Kansas Press, Lawrence.

Petersen, G. H., 1966, *Balanus balanoides* (L.) (Cirripedia) life cycle and growth in Greenland, *Medd. Groenl.* **159**:1–114.

Pyefinch, K. A., 1950, Notes on the ecology of ship-fouling organisms, *J. Anim. Ecol.* **19**:29–35.

Saroyan, J. R., Lindner, E., and Dooley, C. A., 1970, Repair and reattachment in the Balanidae as related to their cementing mechanism, *Biol. Bull.* **139**:333–350.

Savilov, A. I., 1957, Growth and its variability in the invertebrates of the White Sea: *Mytilus edulis*, *Mya arenaria* and *Balanus balanoides*. II. *Balanus balanoides* in the White Sea, *Trudy Inst. Okeanol. Akad. Nauk SSSR* **23**:216–236.

Chapter 14

Skeletal Growth Chronologies of Recent and Fossil Corals

RICHARD E. DODGE and J. RIMAS VAIŠNYS

1. Introduction 493
2. General Features of Coral Growth 494
3. Applications of Coral Growth Features 498
   3.1. Ecological Applications 498
   3.2. Chemical and Paleoecological Applications 509
   3.3. Geophysical and Paleontological Applications 511
4. Suggestions for Future Research 513
   References 514

1. Introduction

Most examples of skeletal growth in this volume are based on the onto-genetic records of individual organisms. Colonial organisms that secrete mineralized tissues are also capable of recording growth events in the communal skeleton. In this chapter, we consider in detail growth patterns within one such group of organisms—corals.

The phylum Coelenterata (Cnidaria) contains many colonial taxa with calcareous skeletons, and it is well known that certain physical and/or chemical variables, capable of affecting a coral's physiology, are also capable of being preserved in the growth record of the colony. Both short-term (solar day) and longer-term (annual) growth periodicities may be recorded in corals. The Scleractinia, and the extinct Rugosa and Tabulata, are of special interest because the former are the dominant metazoan in modern reefs and the latter were widespread and common in the Paleo-zoic. The Scleractinia contain well-preserved internal growth banding, while many Paleozoic corals have external epithecal ridges that reflect growth periodicities.

RICHARD E. DODGE ● Nova University Ocean Sciences Center, Dania, Florida 33004.
J. RIMAS VAIŠNYS ● Department of Geology and Geophysics, Yale University, New Haven, Connecticut 06520.

Through the study of environmental events recorded in extant and extinct corals, we have learned much about the earth's rotational history (e.g., Rosenberg and Runcorn, 1975; Scrutton and Hipkin, 1973), and we are beginning to understand relationships between a coral's physiology and present and past environmental change (e.g., Buddemeier and Kinzie, 1976; Vaišnys and Dodge, 1978). It is important to note that application of coral growth records to problems of environmental reconstruction is possible only after the coupling between a coral's physiology and its ambient environment is understood in modern corals. The emphasis of this chapter will be on this physiological–environmental coupling, which is the focus of much current research, not only for reconstructing past events, but also for understanding the growth of extant species in relation to environmental perturbations. An attempt is made to review the available literature in some detail. Pertinent examples for more lengthy discussion are drawn primarily from the current and prior research of the authors.

2. General Features of Coral Growth

Good reviews of the subject of coral growth are provided by Wells (1957), Yonge (1963), Stoddart (1969), and Buddemeier and Kinzie (1976). In general, reef-building (hermatypic) corals are mutualistically associated with photosynthetic dinoflagellate algae, called "zooxanthellae," which reside within the living coral tissue. These small algal cells are protected from grazing and, in turn, provide the coral with nutrients, and may aid in removal of the coral's metabolic wates. The coral–algal relationship is thought to be very important in determining the abundance, high growth rate, and reef-building abilities of these corals.

The algal association of many coral species requires that their distribution be within the photic zone. There appear, however, to be differences in light-level requirements among certain of the hermatypic coral species. The temperature regime for reef-building species must be tropical to subtropical. Extremely high temperature may be lethal, and coral reefs are rare or absent where the mean annual temperature falls below approximately 18°C. Another important requirement for hermatypic coral growth is that turbidity in the ambient water be low. Particulate material in the water increases light attenuation and adversely affects coral health through reduced light and sedimentation. Also, for optimum development, corals require sufficient quantities of nutrients in the form of zooplankton or dissolved organics.

Throughout their growth, certain species of corals form internal or external skeletal growth increments or bands. Whitfield (1898) (cited in Wells, 1966) was probably the first to explicitly consider external growth features. He noted regular undulations on the upper surface of the frond-like branches of the common Caribbean coral species *Acropora palmata*, which he felt were formed over a period of a year in response to annual water temperature variations [more recent research on this topic is reported by Barnwell (1977)]. Wells (1970) discusses an earlier, more oblique, reference to external time-dependent skeletal growth features.

Internal, time-dependent, incremental banding on a greater than daily scale was first suggested by Ma (1933), who observed cycles of thinning and thickening in elements of the skeletal architecture of both Recent and fossil corals. He suggested that each cycle was annual and that the width of an annual cycle was directly proportional to the ambient water temperature. Ma used his results to calculate paleotemperatures from fossil collections representing a wide geographic and temporal range (Ma, 1934, 1937a,b, 1938). In his later work, he became primarily interested in applications of coral growth data for determining paleolatitudes and establishing evidence for continental drift (Ma, 1953, 1960). Faul (1943) used Ma's ideas of skeletal "growth rings" to measure the annual increase of a Devonian tetracoral.

Fine external growth increments, probably formed over a daily cycle, were first considered by Wells (1963). He counted clusters of approximately 360 fine growth increments on the epitheca of the modern West Indian scleractinian *Manicina areolata* and assumed that each cluster represented the growth in one year. Numbers of increments per cluster were found to be higher for certain Paleozoic corals, and Wells postulated a greater number of days per year in the past. Barnes (1972) has presented observational and experimental evidence for modern *M. areolata* to confirm Wells's (1963) premise of daily increment formation. Barnes reports that the growth ridges in the epithecae of this species are produced from daily changes in the shape of the epithecal tissues secreting the skeleton. Because of the coral–algal association, it is probable that the solar light–dark cycle induces the formation of daily increments.

Interest in annual coral growth features has recently undergone a resurgence, initiated by the work of Knutson et al. (1972). These workers compared nuclear-bomb-test-induced autoradiographic bands in coral skeletons from Eniwetok to density banding that could be observed inside the coral with X-radiography. The comparison indicated that the density banding was annual: a high- and a low-density band forms each year. The results of Knutson et al. (1972) thus suggested confirmation of Ma's (1933) original idea of internal annual growth increments.

An experimental method for testing the temporal significance of coral growth banding became available when Moore (1969) suggested that ^{228}Ra might be useful for estimating skeletal growth rate of corals. ^{228}Ra is present in surface ocean waters of many areas at levels of activity comparable to ^{226}Ra, a nuclide known to be incorporated in the coral skeleton during deposition. Moore and Krishnaswami (1972) and Moore et al. (1973) were able to confirm that ^{228}Ra does indeed decrease with increasing depth into the coral skeleton. On the basis of this isotopic evidence, they established mean growth rates for some Caribbean and Pacific corals.

Dodge and Thomson (1974) and Dodge et al. (1974), to establish conclusively that bands were annual, sampled the internal growth bands of *Diploria labyrinthiformis* from Bermuda and *Montastrea annularis* from Jamaica and analyzed these bands for ^{228}Ra and ^{210}Pb. The data, when plotted sequentially for each band, or group of bands, obeyed laws of radioactive systematics. Figure 1 shows an X-radiograph positive print of *D. labyrinthiformis* from Bermuda similar to the one analyzed. Researchers have also confirmed the existence of annual banding in other corals. Buddemeier et al. (1974), in a follow-up to their initial work (Knutson et al., 1972), discussed details of their results of autoradiography and X-radiography comparisons for various Pacific corals. Macintyre and Smith (1974) compared growth rates as revealed by density banding in *Stephanocoenia* sp. from Jamaica, *Solenastrea hyades* from North Carolina, and *Favia speciosa* from the Gulf of Kutch, India, with radiometric growth determinations of Moore and Krishnaswami (1972) and Moore et al. (1973) and found good results. Moore and Krishnaswami (1974) also reported similar results with their own X-ray and radiochemical comparisons. In addition, Macintyre and Smith (1974) compared X-radiographic growth bands with direct growth measurements of *Montastrea annularis* (from Florida) by Vaughan (1915) and by Shinn (1972). Results suggested that density-band couplets are annual. Preliminary data on staining experiments suggested that density bands in *Pavona gigantea* and *Agariciella planulata* from the Gulf of Panama were annual (Macintyre and Smith, 1974). Noshkin et al. (1975) demonstrated annual bands in *Favites virens* from Bikini lagoon by comparing bomb-nuclide autoradiographs and X-radiographs of internal density bands. Hudson et al. (1976) suggested that annual bands existed in Florida *M. annularis* by observing the seasonal formation of bands through sequential sampling and X-radiography. Stearn et al. (1977) has confirmed the presence of internal annual density banding in *M. annularis* from Barbados by staining experiments in the field and later examination by X-radiography.

At present, we know very little about the physical characteristics and the internal microstructure of coral density bands. X-radiographs show skeletal density variations averaged over a number of cubic millimeters of thinly sliced skeletal material. Results depend on X-ray output, film type, sample types, and skeletal geometries. Over a year's growth, the amount of open space within the skeletal architecture changes. The an-

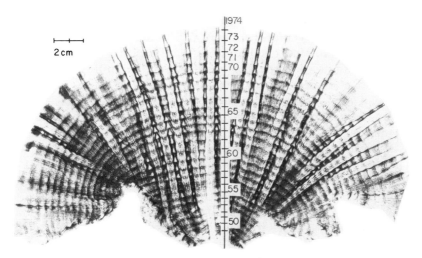

Figure 1. X-radiograph positive of medial section of *Diploria labyrinthiformis* from Castle Harbor, Bermuda, collected live on May 17, 1974. The annual density-band couplets (dark areas are of greater density on the X-ray positive) are circumferential, while the axial structures radiating from the coral base are the traces of corallite (polyp) walls in the skeleton. A typical band-width transect is shown where annual growth is measured from the top edge of a high-density band portion to the top edge of the next older, lower, high-density band portion. Bands have been assigned to their respective years of formation.

nual skeletal cycle consists of two parts, the high-density (HD) portion and the low-density (LD) portion.

Dodge and Thomson (1974) have shown for *Diploria labyrinthiformis* from Bermuda that average annual band width and average annual band density are inversely related, while mass deposition per year is positively related to average band width. In terms of width–density–mass characteristics of density-band portions, Dodge (1978) demonstrated for *Diploria* sp. from Bermuda and *M. annularis* from Barbados that HD bands generally are narrower and have less mass than LD band portions. Buddemeier *et al.* (1974) reported quantitative measurements of skeletal density and density variations for various coral species from Eniwetok, and Buddemeier and Kinzie (1975) suggested that fine density structure within the annual bands of *Porites lobata* from Hawaii may represent lunar cycles of skeletal deposition. Weber *et al.* (1975) and Weber and White (1974, 1977) studied that dimensions of density-band portions of a variety of coral species at many locations in relation to season and cause of formation.

Macintyre and Smith (1974) have demonstrated for *Pavona gigantea* from the Gulf of Panama that spacing of skeletal elements (dissepiments) does not differ between HD and LD bands. Buddemeier and Kinzie (1975) have found that trabecular spacing in *Porites lobata* is relatively constant along corallite growth axes regardless of band position. They report that density variations arise from the thickening and thinning of the trabecular walls. From qualitative observation of specimens of well-banded *Diploria* spp. and *M. annularis* from Bermuda, Barbados, and Jamaica, we conclude that skeletal structural elements (septa and dissepiments) become thicker in higher-density bands, while spacing between the midpoints of septa and dissepiments remains relatively constant over the annual cycle.

Although the overall cycle of high and low density is known to be annual, the precise times of deposition and causal environmental factors that initiate changes in skeletal density remain elusive. For *M. annularis*, we have found that the HD bands form during months with highest water temperature in specimens from Bermuda, Barbados, and Jamaica (Dodge, 1978; Fairbanks and Dodge, 1979). Weber *et al.* (1975), Dodge (1978), Fairbanks and Dodge (1979), Goreau (1977a), Emiliani *et al.* (1978), and Highsmith (1979) consider this problem of the season of band-portion formation for *M. annularis* and other coral species in detail.

3. Applications of Coral Growth Features

3.1. Ecological Applications

There are as yet no direct studies relating the dimensions of coral growth increments to amounts of solar radiation. Stearn *et al.* (1977) have studied the position of density-band portions in relation to sunlight hours per month and other environmental parameters. Bak (1974) followed the growth of corals by measuring changes in colony weight and suggested that there appeared to be a positive correlation between skeletal mass deposited and number of hours of direct sunlight per month. Dodge *et al.* (1977) (also see Dodge, 1978) found a positive correlation between the yearly band widths of a single *Diploria strigosa*, collected at approximately 10 m depth on a Barbados reef, and incident solar radiation (Fig. 2). However, specimens of *M. annularis* at depths of 1–2 m in the same location did not give similar results. Support for the positive relationship between coral growth and solar radiation is presented in studies conducted by Buddemeier *et al.* (1974) and Baker and Weber (1975). Their results showed decreased linear growth rate with increasing water depth. Because food, sedimentation, and temperature are presumably relatively constant over the depths studied, light intensity as a causal factor is indicated. It should be noted that a strong positive correlation between sunlight exposure and coral annual band width (or annual band average den-

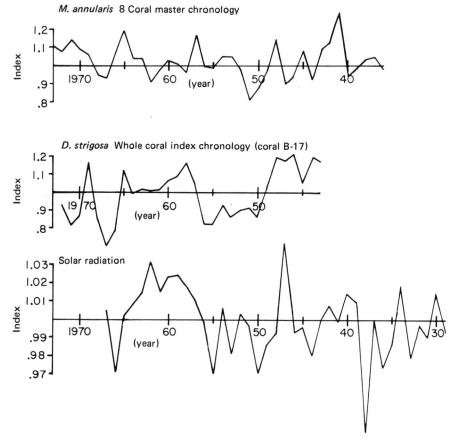

Figure 2. (A) Master index chronology based on 8 *Montastrea annularis* specimens collected at approximately 2 m depth from the Recent reef off Bellairs Institution, Barbados; (B) single coral chronology of one *Diploria strigosa* collected at a depth of approximately 10 m in Barbados; (C) average solar radiation for Barbados. Solar radiation data were calculated from sunlight-hours data of C. C. Skeet (personal communication). On each graph, the horizontal axis denotes the year and the vertical axis shows the percentage of the particular series mean divided by 100 (termed the "index" value).

sity or mass) may be present only within a certain depth range on the reef. Barnes and Taylor (1973) have noted that shallow-water specimens are in the zone of saturating light intensity for photosynthesis and are thus probably less light-limited during the day than deeper-living specimens.

High sedimentation rates are deleterious to coral growth. Marshall and Orr (1931) pioneered work on this subject. Hubbard and Pocock (1972) have ranked certain coral species according to their sediment-rejection ability (capability of coral tissues to remove sediment by various means), as have Bak and Elgershuizen (1976). Dodge *et al.* (1974) and Aller and Dodge (1974) studied growth rates of *Montastrea annularis* in Discovery Bay, Jamaica, and found that average annual band widths of specimens decreased in bottom regions of high resuspension of bottom sediments. The work of Loya (1976) on Puerto Rican reefs that were subject to high sedimentation rates supports these aforementioned results.

Dodge and Vaišnys (1977) examined the ecological effects of dredging on corals in Castle Harbor, Bermuda, by means of growth-band analysis. These workers found catastrophic coral mortality in the harbor coincident with dredging for airport construction during the period of 1941–1943. The conclusion was supported by a number of independent lines of evidence. Inside the harbor, collections were made of both living and dead corals. For growth reference, other living specimens were collected from reefs located outside the harbor. All corals were subsequently aged by counting growth bands within each coral head. Knowledge of the collection efficiency allowed construction of age-frequency diagrams of living and dead harbor coral populations and living populations external to the harbor. This evidence coupled with information on maximum longevity, species composition, and the known date of the dredging event indicated that a mass mortality of Castle Harbor corals had taken place. For example, Fig. 3 shows the age-frequency distributions of living coral populations measured along various transect areas both inside and outside the harbor. It is clear that harbor coral density and longevity are greatly reduced in comparison to those on more open ocean reefs.

Further evidence for mortality of populations affected by dredging was reflected in coral growth patterns over time. Living corals in the harbor had similar yearly patterns of growth. Figure 4 shows representative growth chronologies of several living corals collected at one station in Castle Harbor. The figure also shows the formation of a station master chronology by sequential averaging, by year, all growth bands of all corals at that station. Station masters were also well correlated and are convenient for summarizing the data. Figure 5 shows the results for a selection of Castle Harbor stations. Figures 4 and 5 indicate that living corals in Castle Harbor have similar patterns of yearly growth. Banding patterns of the collection of dead corals from the harbor were investigated under the following premise: if the Castle Harbor dead corals died at approximately the same time in response to a lethal harborwide disturbance, then the expectation would be that their growth patterns should also be similar. Figure 6 gives typical examples of dead coral chronologies at a single station. The plot of band widths for each coral commences at the surface (time of death) and extends to the right (predeath chronology). It should be noted that because of coral surface relief and bioerosion and/or encrustations on the dead coral surfaces, there will be several years' un-

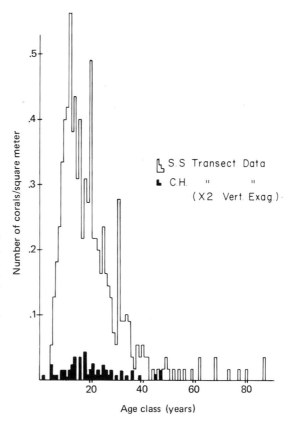

Figure 3. Comparison of the age-frequency characteristics of corals measured along various transect areas on reefs in Castle Harbor and on the south shore of Bermuda. The vertical axis represents the number of corals per square meter along the transect area; the horizontal axis depicts age classes of the corals in 1 year intervals. (□) South shore data; (■) Castle Harbor data. The Castle Harbor data are vertically exaggerated by a factor of 2 for clarity.

certainty in the location of the death surface. Although the absolute time of band formation for the corals of Fig. 6 is not precisely known, the growth patterns do appear to contain common features. Also note the decline in growth rate (band width) prior to death. This was a common feature of the Castle Harbor dead corals; a sample X-radiograph is shown in Fig. 7.

It was originally expected that an absolute time chronology could be assigned to the growth bands of the Castle Harbor dead corals by comparing these chronologies with old living corals in the harbor that lived through the dredging operation. Unfortunately, we were unable to find sufficient extant specimens to fulfill these requirements. Dendrochronologists typically require a minimum of 50–100 years to be able to reliably date wood samples of unknown age (Fritts, 1972). Further reconnaissance, however, may locate several old large "rosetta" corals that could provide a precise date for the coral mortality event. It would be especially inter-

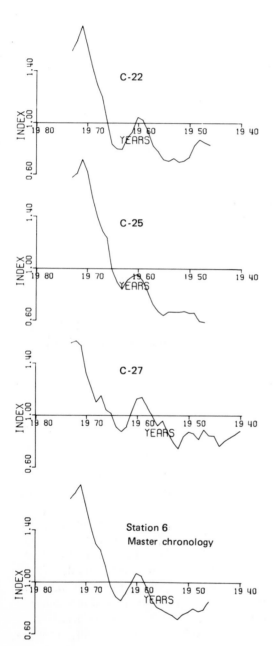

Figure 4. Band-width chronologies. (A–C) Time series of band widths plotted as index values (percentage of the average coral growth rate divided by 100) for three typical live coral colonies at station 6 in Castle Harbor, Bermuda. Each band-width chronology has been smoothed by a 3-year moving average. (D) Station summary or master chronology of the three corals. This is the yearly average (ensemble average) of the coral chronologies shown in (A–C).

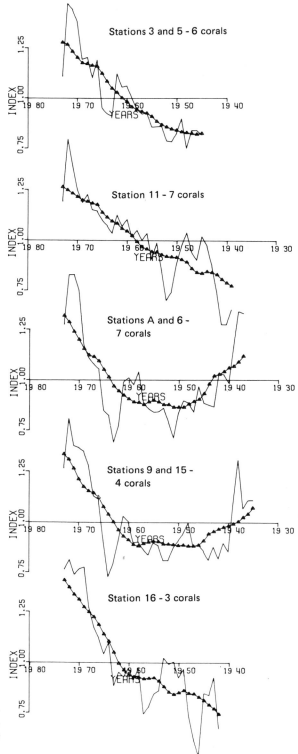

Figure 5. Examples of station chronologies of live corals from Castle Harbor, Bermuda. (▲) Smoothed (13-year moving average) chronology.

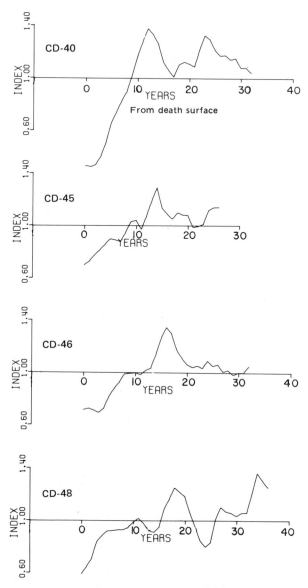

Figure 6. Chronologies of several representative dead corals from station 13 in Castle Harbor, Bermuda, each smoothed by a 3-year moving average. Growth rates (normalized to the coral mean: index values) are plotted on the vertical axis against time on the horizontal axis beginning at the band approximately coincident with the time of death (x = 0).

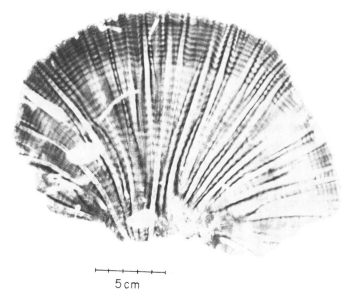

5 cm

Figure 7. X-radiograph positive print of a specimen of *Diploria labyrinthiformis* collected dead from Castle Harbor, Bermuda, showing an abruptly decreasing growth rate toward the surface of death. Note the bioerosion at the surface of the coral.

esting to know whether corals died relatively quickly during, or immediately after, the dredging, or whether they lived several years beyond the cessation of dredging, only to succumb eventually to the increased turbidity or changed water-circulation patterns, or both. The curves of Fig. 6 suggest that the latter is the case; however, it is possible that the early 1940's were poor years for coral growth and that the decline in growth rate prior to death merely reflects such a natural occurrence.

Other studies of annual growth bands have attempted to determine the response of coral growth to seawater temperature differences over wide latitudinal ranges. Weber and White (1974, 1977) indicate that growth-band thickness increases in regions of higher water temperature for *M. annularis* in the Caribbean and *Platygyra* sp. in the Pacific. These results need to be tested rigorously for the effect of other factors that vary latitudinally, such as incident solar radiation, nutrient supply, and wind-wave stress. In terms of coral growth response to environmental parameters at a single location, Glynn and Wellington (1980) have found through analysis of growth-band dimensions and climatic records that linear growth in massive *Pavona* spp. from the Galapagos Islands is accelerated during periodic events of relatively warm, less saline, and nutrient-poor tropical water (El Niño). The findings may well be useful in determining the historical frequency of such El Niño events over long time periods.

A relationship between nutrient supply and coral growth rate has been presented by Dodge and Vaišnys (1975a,b) and Dodge (1978). Similarities in the year-to-year growth of individual Bermuda corals outside Castle Harbor allowed construction of a multiple (27)-coral chronology representing a span of approximately 70 years. Growth was found to be inversely related to annual water temperature (as deduced from air-temperature records) over the long-term period (Fig. 8). Menzel and Ryther (1961) have found that in Bermuda, nutrient supply is related to upwelling of deeper, colder water and thus inversely related to annual water temperature. Apparently Bermuda corals react over the long-term by showing slightly higher growth in years of slightly lower water temperature associated with relatively high nutrient levels. Glynn (1977) has examined upwelling as an effect on the growth (by physical measurements) of *Pocillopora damicornis* in Panama. He reports that reef areas that experience upwelling are lower in both coral cover and abundance and that corals in those areas have lower growth rates compared to those in non-upwelling areas. Interestingly, coral growth rate and zooplankton abundance were found to be positively correlated in the upwelling area.

Banding in corals has also been utilized to estimate bioerosion rates of the skeletons of dead corals (Hudson, 1977) and the contribution of the

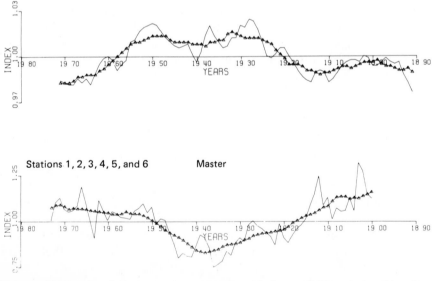

Figure 8. Comparison of maximum air temperature with the grand master index chronology of all 27 corals for all six south shore reef stations. Horizontal and vertical axes are as in Fig. 5–7. (△) 13-Year moving average.

component of coral calcification in reef $CaCO_3$-budget studies (Stearn et al., 1977). Glynn (1974) used X-radiography and other techniques to age and study mobile coralliths in the Gulf of Panama. Isdale (1977) measured growth rates of coral specimens on Fairy Reef (Great Barrier Reef) and found high variability from one coral to another. In terms of spatial distributions of coral growth at a reef locality, Dodge (1978, 1981) determined the growth rates by density-band measurement of many *Diploria* spp. corals at a number of stations over the Bermuda platform. Figure 9 illustrates the Bermuda results. Highest mean growth rates were found in Castle Harbor (CH) as a whole and on the central lagoon reefs (N3, WBC). Reef margin localities (NR, S1, S2, S3, S4, S5, S6) show lower growth-rate values. For the southeast coast stations, growth rate increases eastward toward the mouth of Castle Harbor. Growth variability (as measured by percent standard deviation) is highest in Castle Harbor and the easternmost southeast coast stations. Lowest variability occurs in the central lagoon (N3, WBC). Reef margins have, in general, low to moderate variability.

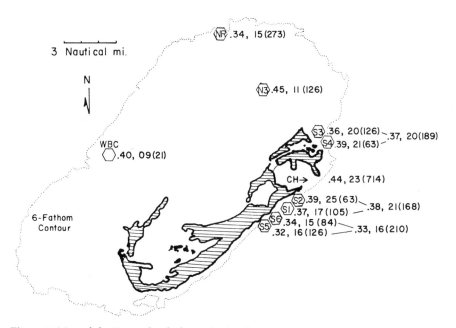

Figure 9. Map of the Bermuda platform. Station locations are designated by hexagons. To the right of each station is provided the average coral growth rate in cm/yr (over the period 1951–1971) for all corals at that station, followed by the percent standard deviation, and in parentheses the number of band widths used in the calculation. Also indicated are the values for the pooled groupings of stations S5 and S6, S1 and S2, and S3 and S4. The pooled growth value for all Castle Harbor (CH) stations is also shown.

Of all the environmental variables such as spatial patterns of water temperature, nutrient supply, and solar energy influx over the Bermuda platform, the distribution of wind energy offers the best explanation of coral growth rate and variability. Wind energy is presumably correlated with wave activity (Upchurch, 1970), which can negatively affect coral growth by causing (1) high resuspension of sediment; (2) high water motion around coral tissue, impeding food capture (Hubbard, 1974); and (3) reduced light levels through turbidity effects. Figure 10 shows that in winter, wind is predominatly from the northwest and southwest in Bermuda, while in summer it is predominantly from the southwest and south. The areas most protected from the wind fetch are Castle Harbor and the central lagoon, and these show highest mean growth rates. The most windward exposed reef margin stations show lowest mean growth rates. The trend of increasing growth rate toward the mouth of Castle Harbor on the southeast coast is apparently related to a decreasing wind–wave energy

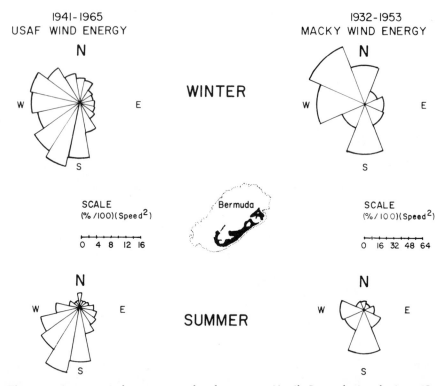

Figure 10. Average wind-energy rose plots for summer (April–September) and winter (October–March) from the data of Macky (1956) and the U.S. Air Force station, Kindley Field, Bermuda. The data are presented as percentage frequency of wind from a given direction divided by 100 and multiplied by the wind speed squared.

gradient. Growth variance patterns also appear to be explained by wind-energy patterns. Lowest variability is found in areas of highest wind–wave activity on the reef margins, while more protected stations near the mouth of Castle Harbor have higher variability. The lowest variability found in the central lagoon is thought to be a result of low-stress conditions persisting year after year. The higher variability in growth in Castle Harbor may be correlated with its shallowness and greater turbidity levels.

3.2. Chemical and Paleoecological Applications

Living coral tissue deposits $CaCO_3$ from dissolved ions present in seawater. Other trace materials are often included in minute amounts. As previously discussed, radionuclide uptake and inclusion into the coral skeleton has been utilized for establishing the annual nature of banding. Other studies have used mainly ^{14}C and ^{90}Sr from nuclear fallout to measure nuclide incorporation and hence skeletal growth rate (Knutson and Buddemeier, 1972; Moore and Krishnaswami, 1972). Coral bands have also been studied for detritus inclusions (Barnard et al., 1974; Macintyre and Smith, 1974), as well as for the skeletal distribution of stable isotopes (Goreau, 1977a,b; Fairbanks and Dodge, 1978, 1979; Nozaki et al., 1978; Emiliani et al., 1978; Druffel and Linick, 1978).

The information contained in density features of the coral skeleton is useful in paleoecological and paleoclimatological investigations. Well-preserved fossil corals also display annual density banding in the skeleton (Weber et al., 1975) (also see Fig. 11). Dodge et al. (1977) found significantly lower growth rate and growth variability in M. annularis samples from a raised Barbados reef dated at 105,000 B.P. compared with Recent M. annularis from the island. These observations suggest that the variability of past climates may be assessed from coral growth records.

Several current lines of research utilizing chemical studies of the coral skeleton may prove valuable for reconstructing paleoclimatological and paleooceanographic events in the recent past. Weber and Woodhead (1972) established that the oxygen isotopic composition of hermatypic coral skeletons reflects the temperature of the surrounding water in which the coral was growing. Values of ^{18}O are, however, displaced from equilibrium values by a "vital" effect. Both Goreau (1977a,b) and Emiliani et al. (1978) have found yearly carbon and oxygen isotopic variations in the skeletons of the coral M. annularis from Jamaica and the Florida Keys, respectively. Druffel and Linick (1978) report and discuss the implications of radiocarbon measurements on a 175-year growth of M. annularis from the Florida Keys. Fairbanks and Dodge (1978, 1979) have analyzed skeletal samples of M. annularis from Bermuda, Barbados, and Jamaica for oxygen

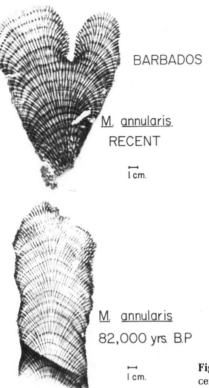

BARBADOS

<u>M. annularis</u>
RECENT

⊢—⊣
1 cm.

<u>M. annularis</u>
82,000 yrs. B.P

⊢—⊣
1 cm.

Figure 11. X-radiograph positive prints of a Recent and a fossil *Montastrea annularis* from Barbados showing annual density bands.

and carbon isotopes. The ratios $^{18}O/^{16}O$ and $^{13}C/^{12}C$ show annual periodicities. In addition, *M. annularis* was found to have a constant displacement from oxygen isotopic equilibrium and to accurately record seasonal temperature variations via the temperature-dependent aragonite–water
fractionation factor. Light intensity, mediated through the photosynthesis
and respiration of the coral's endosymbiotic algae, appears to determine
seasonal variations in the carbon isotopic composition of the skeleton.
These results have marked implications for assessing paleoseasonality of
the waters surrounding ancient coral reefs (Fairbanks and Dodge, 1979).
In another study dealing with the chemistry of coral bands, Nozaki *et al.*
(1978) sequentially sampled a 200-year-old coral (*Diploria labyrinthiformis*) from Bermuda by density band. Determination of carbon and oxygen
isotopic composition revealed that the coral reflects atmospheric variations in carbon isotopes over the last 80 years caused by human activities
(e.g., bomb ^{14}C, dilution of ^{14}C and ^{13}C by fossil fuel burning). Before 1900,
differences between coral and tree-ring carbon isotopes may be related

to changes in worldwide plant production and possibly oceanic upwelling rates. These results are intriguing and appear well-suited for paleo-ocean-ography studies dealing with ocean–atmosphere exchange of carbon and upwelling histories of various water masses.

The usefulness of coral skeletons (and other carbonates) in carbon and/or uranium dating of fossil deposits in the Quaternary is well known. A full discussion is beyond the scope of this chapter. Veeh and Green (1977) provide a comprehensive review of this subject.

3.3. Geophysical and Paleontological Applications

Wells (1963) made counts of fine epithecal banding on well-preserved Devonian fossil rugose corals and found approximately 400 fine incre-ments within major rugosities or ridges on the outer portion of the roughly conical skeleton. The major ridges were assumed to represent one year's growth and the fine increments to be daily periods of skeleton deposition. Surprisingly, Wells's (1963) estimate of a 400-day Devonian year is in good agreement with astronomical estimates that the rotational rate of the earth is decreasing 0.002 second per century. Other workers, using similar approaches with growth increments, have estimated the number of days per month and, with the data of Wells (1963), the months per year in the Devonian and Carboniferous. Geochronometry is associated with some problems: specimens must be exceptionally well preserved, band counts are often made between subjectively determined skeletal features, and inferences must be made about the time relationships of the major growth constrictions. The proven presence of daily and annual growth period icities in the skeletons of modern corals provides an analogy for the interpretation of fossil corals. Figure 12 illustrates a Devonian coral with presumed daily and annual bands. Scrutton and Hipkin (1973) and Ro-senberg and Runcorn (1975) offer the most complete and comprehensive reviews on the growth increments of corals and other organisms and pa-leogeophysics.

Another interesting and potentially productive line of research is aimed at establishing the geological history of the coral–algal association. Sorauf (1974) has reported daily growth lines on the structural elements of the skeleton of a few species of Silurian favositid tabulate corals. Be-cause the deposition of solar growth bands in living Scleractinians may be strongly influenced by symbiotic zooxanthellae, Sorauf suggests that ancient corals had analogous growth bands for the same reason. An in-dependent line of evidence supporting this hypothesis is related to growth-rate estimates. By measuring the thickness of the daily increments

Figure 12. (A) Devonian cup coral *Cyathophyllum helianthoides* showing presumed annual (major constrictions) and daily increments; (B) enlargement of portion of the epitheca of (A) showing details of presumed daily increments. Scale bars: (A) 1 cm; (B) 1 mm.

of fossil corals described by Wells (1963) and other workers, one is able to establish that fossil corals had growth rates comparable to those of extant hermatypic corals (see, for example, Faul, 1943). Modern corals without algal symbionts almost invariably have lower growth rates when compared with their hermatypic relatives. If one accepts this algal–coral growth-rate hypothesis, these observations suggest that the coral–algal symbiosis has existed since at least the Silurian.

4. Suggestions for Future Research

Coral growth-band research is still in its infancy, and many opportunities exist for productive research. For coral growth banding to be useful for environmental reconstruction, it is necessary to have a firm knowledge of the physiological processes of recent corals, how those processes are affected by environmental factors, and how, in turn, the environmental changes are recorded, if at all, in the coral skeleton.

More information is needed about the coral skeleton itself to couple the skeletal deposition rate with physiological–environmental variables and to establish relationships between skeletal microstructure and time of deposition. These kinds of studies will help to provide an understanding of skeletal deposition on a subannual time scale and will be of great interest in resolving the fate of various materials enclosed within pores and cavities of the skeleton (such as endolithic algae, their associated pigments, and trapped seawater), the mechanics of trace-element incorporation, and the processes of skeletal isotope fractionation.

At present, there are a variety of techniques that can be employed for productive investigations. For example, long-period *in situ* and controlled laboratory growth experiments can be coupled with monitoring and manipulation of the natural environment. Skeletal staining can give a datum for studying growth changes in relation to density features, and *in situ* recording of temperature, light, wave energy, and other factors, can provide invaluable information on the coral growth response. Further analysis should include chemical aspects of the corals and their environment over the staining-growth period. Once subannual growth responses of corals are better understood, attention can be focused on long-term growth records of corals. For example, time-series analysis of growth vs. environmental variables should become more productive when short-term coral growth periodicities are better known.

When firm relationships of the growth response of extant corals are established, it should be possible to use long series of annual bands for the reconstruction of climatic and environmental parameters. Reconstruction may be keyed to the recent past for historical information retrieval or may be focused on certain periods of the fossil record for paleoenvironmental reconstruction. Certain corals are extremely long-lived (e.g., 200 years is not an uncommon age for some species), and thus data available in no other way may be supplied through coral research for very precise and lengthy periods in the past.

ACKNOWLEDGMENTS. In the course of our research, funding has been provided by NSF Grant No. OCE 75-17618, The Geological Society of America, The Woman's Seaman's Friend Society of Conn., Inc., The Schuchert Fund of the Yale Peabody Museum, and The Yale Department of Geology and Geophysics. We thank Prof. K. K. Turekian and Prof. D. C. Rhoads for valuable discussions and the editors for their assistance with the manuscript.

References

Aller, R. C., and Dodge, R. E., 1974, Animal–sediment relations in a tropical lagoon—Discovery Bay, Jamaica, *J. Mar. Res.* **32**:209–232.

Bak, R. P. M., 1974, Available light and other factors influencing growth of stony corals through the year in Curacao, *Proc. 2nd Int. Coral Reef Symp.* **2**:229–233.

Bak, R. P. M., and Elgershuizen, J. H. B. W., 1976, Patterns of oil-sediment rejection in corals, *Mar. Biol.* **37**:105–113.

Baker, R. A., and Weber, T. N., 1975, Coral growth rate: Variation with depth, *Earth Planet. Sci. Lett.* **27**:57–66.

Barnard, L. A., Macintyre, I. G., and Pierce, J. W., 1974, Possible environmental index in tropical reef corals, *Nature (London)* **252**:219–220.

Barnes, D. J., 1972, The structure and formation of growth-ridges in scleractinian coral skeletons, *Proc. R. Soc. London Ser. B* **182**:331–350.

Barnes, D. J., and Taylor, D. L., 1973, *In situ* studies of calcification and photosynthetic carbon fixation in the coral *Montastrea annularis*, *Helgol. Wiss. Meersunters.* **24**:284–291.

Barnwell, F. H., 1977, Phase synchrony of skeletal rhythms within populations of corals (*Millepora* and *Acropora*) at Boca Chica, Dominican Republic, *Abstr. Am. Zool.* **17**:869.

Buddemeier, R. W., and Kinzie, R. A., III, 1975, The chronometric reliability of contemporary corals, in: *Growth Rhythm and the History of the Earth's Rotation* (G. D. Rosenberg and S. K. Runcorn eds.), pp. 135–147, John Wiley, London.

Buddemeier, R. W., and Kinzie, R. A., III, 1976, Coral growth, *Oceanogr. Mar. Biol. Annu. Rev.* **14**:183–225.

Buddemeier, R. W., Maragos, J. E., and Knutson, D. W., 1974, Radiographic studies of reef coral exoskeleton: Rates and patterns of coral growth, *J. Exp. Mar. Biol. Ecol.* **14**:179–200.

Dodge, R. E., 1978, The natural growth records of reef building corals, Ph.D. thesis, Yale University, New Haven, Connecticut, 237 pp.

Dodge, R. E., 1981, Coral growth in Bermuda: Spatial distribution in relation to the environment in: *Bermuda—Anatomy of an Oceanic Island* (W. Sterrer, ed; A. Logan, reef section ed.), Wiley Interscience, New York (in prep.).

Dodge, R. E., and Thomson, J., 1974, The natural radiochemical and growth records in contemporary hermatypic corals from the Atlantic and Caribbean, *Earth Planet. Sci. Lett.* **23**:313–322.

Dodge, R. E., and Vaišnys, J. R., 1975a, Corals as environmental indicators, *Geol. Soc. Am. Abstr. Program* **7**:1055.

Dodge, R. E., and Vaišnys, J. R., 1975b, Hermatypic coral growth banding as environmental recorder, *Nature (London)* **258**:706–708.

Dodge, R. E., and Vaišnys, J. R., 1977, Coral populations and growth patterns: Responses to sedimentation and turbidity associated with dredging, *J. Mar. Res.* **35**:715–730.

Dodge, R. E., Aller, R. C., and Thomson, J., 1974, Coral growth related to resuspension of bottom sediments, *Nature (London)* **247**:574–577.

Dodge, R. E., Turekian, K. K., and Vaišnys, J. R., 1977, Climatic implications of Barbados coral growth, *Proc. 3rd Int. Coral Reef Symp.* **2**:361–365.

Druffel, E. M., and Linick, T. W., 1978, Radiocarbon in annual coral rings of Florida, Geophys. *Res. Lett.* **5**:913–916.

Emiliani, C., Hudson, J. H., Shinn, E. A., and George, R. Y., 1978, Oxygen and carbon isotopic growth record in a reef coral from the Florida Keys and a deep-sea coral from Blake Plateau, *Science* **202**:627–629.

Fairbanks, R. G., and Dodge, R. E., 1978, Coral thermometry: A method for determining past surface water temperature variations, Presented at 1978 spring American Geological Union meeting, Miami.

Fairbanks, R. G., and Dodge, R. E., 1979, Annual periodicity of the $^{18}O/^{16}O$ and $^{13}C/^{12}C$ ratios in the coral *Montastrea annularis*, *Geochim. Cosmochim. Acta* **43**:1009–1020.

Faul, H., 1943, Growth-rate of a Devonian reef-coral (*Prismatophyllum*), *Am. J. Sci.* **241**:579–582.

Fritts H. C., 1972, Tree-rings and climate, *Sci. Am.* **226**:93–100.

Glynn, P. W., 1974, Rolling stones among the Scleractinia: Mobile coralliths in the Gulf of Panama. *Proc. 2nd Int. Coral Reef Symp.* **2**:183-198.

Glynn, P. W., 1977, Coral growth in upwelling and non-upwelling area of the Pacific coast of Panama, *J. Mar. Res.* **35**:567–585.

Glynn, P. W., and Wellington, G. M., 1980, An ecological study of corals and coral reefs in the Galapagos Archipelago, *Smithsonian Contrib. Zool.* (in prep.).

Goreau, T. J., 1977a, Coral skeletal chemistry: Physiological and environmental regulation of stable isotopes and trace metals in *Montastrea annularis*, *Proc. R. Soc. London Ser. B* **196**:291–315.

Goreau, T. J., 1977b, Seasonal variations of trace metals and stable isotopes in coral skeletons: Physiological and environmental controls, *Proc. 3rd Int. Coral Reef Symp.* **2**:425–430.

Highsmith, R. C., 1979, Coral growth rates and environmental control of density banding, *J. Exp. Mar. Biol. Ecol.* **37**:105–125.

Hubbard, J. A. E. B, 1974, Scleractinian coral behavior in calibrated current experiments: An index to their distribution patterns, *Proc. 2nd Int. Coral Reef Symp.* **2**:107–125.

Hubbard, J. A. E. B., and Pocock, Y. P., 1972, Sediment rejection by recent scleractinian corals: A key to palaeo-environmental reconstruction, *Geol. Rundsch.* **61**:598–626.

Hudson, J. H., 1977, Long-term bioerosion rates on a Florida reef: A new method, *Proc. 3rd Int. Coral Reef Symp.* **2**:491–497.

Hudson, J. H., Shinn, E. A., Halley, R. B., and Lidz, B., 1976, Sclerochronology: A tool for interpreting past environments, *Geology* **4**:361–364.

Isdale, R., 1977, Variations in growth rate of hermatypic corals in a uniform environment, *Proc. 3rd. Int. Coral Reef Symp.* **2**:403–408.

Knutson, D. W., and Buddemeier, R. W., 1972, Distribution of radionuclides in reef corals: Opportunity for data retrieval and study of effects, in: *Radioactive Contamination of the Marine Environment*, pp. 735–746, International Atomic Energy Agency, Vienna.

Knutson, D. W., Buddemeier, R. W., and Smith, S. V., 1972, Coral chronometers: Seasonal growth bands in reef corals, *Science* **177**:270–272.

Loya, Y., 1976, Effects of water turbidity and sedimentation on the community structure of Puerto Rico corals, *Bull. Mar. Sci.* **26**:450–466.

Ma. T. Y. H., 1933, On the seasonal change of growth in some Paleozoic corals, *Proc. Imp. Acad. (Tokyo)* **9**:407–408.

Ma, T. Y. H., 1934, On the seasonal change of growth in a reef coral, *Favia speciosa* (Dana), and the water-temperature of the Japanese seas during the latest geological times, *Proc. Imp. Acad. (Tokyo)* **10**:353–356.

Ma, T. Y. H., 1937a, On the growth rate of reef corals and its relation to sea water temperature, *Paleontol. Sinica. Ser. B* **16**:1–426.

Ma, T. Y. H., 1937b, On the growth rate of *Calapoecia canadensis* Billings and the climate of the arctic regions during the Ordovician period, *Geol. Soc. China Bull.* **17**:177–182.

Ma, T. Y. H., 1938, On the water temperature of the western Pacific during early and late Pleistocene as deduced from the growth rate of fossil corals, *Geol. Soc. China Bull.* **18**:349–418.

Ma, T. Y. H., 1953, The sudden total displacement of the outer solid earth shell by sliding relative to the fixed rotating core of the earth, *Res. Past Climate Cont. Drift* **6**:1–15.

Ma, T. Y. H., 1960, Climate and the relative positions of continents during the upper carboniferous as deduced from the growth values of reef corals, *Res. Past Climate Cont. Drift* **16**:1–2.

Macintyre, I. G., and Smith, S. V., 1974, X-radiograph studies of skeletal development in coral colonies, *Proc. 2nd Int. Coral Reef Symp.* **2**:277–287.

Macky, W. A., 1956, The surface wind in Bermuda, Bermuda Meteorological Office Technical Note, No. 7, 31 pp.

Marshall, S. M. and Orr, A. P., 1931, Sedimentation on Low Isles Reef and its relation to coral growth, *Br. Mus. (Nat. Hist.) Great Barrier Reef Exped. Sci. Rep.* **1**:93–133.

Menzel, D. W., and Ryther, J. H., 1961, Annual variation in primary productivity of the Sargasso Sea off Bermuda, *Deep-Sea Res.* **7**:282–288.

Moore, W. S., 1969, Oceanic concentrations of radium-228 and a model for its supply, Ph.D. thesis, State University of New York at Stony Brook.

Moore, W. S., and Krishnaswami, S., 1972, Coral growth rates using ^{228}Ra and ^{210}Pb, *Earth Planet. Sci. Lett.* **15**:187–190.

Moore, W. S., and Krishnaswami, S. W., 1974, Correlation of X-radiography revealed banding in corals with radiometric growth rate, *Proc. 2nd Int. Coral Reef Symp.* **2**:269–276.

Moore, W. S., Krishnaswami, S., and Bhat, S. G., 1973, Radioactive determination of coral growth rate, *Bull. Mar. Sci.* **23**:157–176.

Noshkin, V. E., Wong, K. M., Eagle, R. J., and Gatrousis, G., 1975, Transuranics and other radionuclides in Bikini lagoon: Concentration data retrieved from aged coral sections, *Limnol. Oceanogr.* **20**:729–742.

Nozaki, Y., Rye, D. M., Turekian, K. K., and Dodge, R. E., 1978, A 200 year record of carbon-13 and carbon-14 variations in a Bermuda coral, *Geophys. Res. Lett.* **5**:825–828.

Rosenberg, G. D., and Runcorn, S. K. (eds.), 1975, *Growth Rhythms and the History of the Earth's Rotation,* John Wiley, London, 538 pp.

Scrutton, C. T., and Hipkin, R. G., 1973, Long-term changes in the rotation rate of the earth, *Earth Sci. Rev.* **9**:259–274.

Shinn, E. A., 1972, Coral reef recovery in Florida and in the Persian Gulf, Shell Oil Co., Houston, Texas, 9 pp.

Sorauf, J. E., 1974, Growth lines on Tabular of *Favosites* (Silurian, Iowa), *J. Paleontol.* **48**:553–555.

Stearn, C. W., Scoffin, T. P., and Martindale, W., 1977, Calcium carbonate budget of a fringing reef on the west coast of Barbados. Part I. Zonation and productivity, *Bull. Mar. Sci.* **27**:479–510.

Stoddart, D. R., 1969, Ecology and morphology of Recent coral reefs, *Biol. Rev.* **44**:433–498.

Upchurch, S., 1970, Sedimentation on the Bermuda platform, *U.S. Lake Surv. Res. Rep. 2-2*, 192 pp.

Vaišnys, J. R., and Dodge, R. E., 1978, Coelenterata, *McGraw Hill Yearbook of Science and Technology, 1978,* pp. 121–123.

Vaughan, T. W., 1915, The geological significance of the growth rate of the Florida and Bahama shoal-water corals, *J. Wash. Acad. Sci.* **5**:591–600.

Veeh, H. H., and Green, D. C., 1977, Radiometric geochronology of coral reefs, in: *Biology and Geology of Coral Reefs*, Vol. 4, *Geology 2* (O. A. Jones and R. Endeam, eds.), pp. 183–200, Academic Press, New York.

Weber, J. N., and White, E. W., 1974, Activation energy for skeletal aragonite deposited by the hermatypic coral *Platygyra* spp., *Mar. Biol.* **26:**353–359.

Weber, J. N., and White, E. W., 1977, Caribbean reef corals *Montastrea annularis* and *Montastrea cavernosa*—Long term growth data as determined by skeletal X-radiograph, in: *Reefs and Related Carbonates—Ecology and Sedimentology* (S. H. Frost, M. P. Weiss, and J. B. Saunder, eds.), pp. 171–179, A.A.P.G. Studies in Geology, No. 4.

Weber, J. N., and Woodhead, P. M. J., 1972, Temperature dependence of oxygen-18 concentration in reef coral carbonates, *J. Geophys. Res.* **77:**463–473.

Weber, J. N., White, E. W., and Weber, P. H., 1975, Correlation of density banding in reef coral skeletons with environmental parameters; the basis for interpretations of chronological records preserved in the coralla of corals, *Paleobiology* **1:**137–149.

Wells, J. W., 1957, Coral reefs, *Mem. Geol. Soc. Am.* **67:**609–631.

Wells, J. W., 1963, Coral growth and geochronometry, *Nature (London)* **197:**948–950.

Wells, J. W., 1966, Paleontological evidence of the rate of the Earth's rotation, in: *The Earth–Moon System* (B. F. Marsden and A. G. W. Camaron, eds.), pp. 70–81, Plenum Press, New York.

Wells, J. W., 1970, Problems of annual and daily growth-rings in corals, in: *Palaeogeophysics* (S. K. Runcorn, ed.), pp. 3–9, Academic Press, London and New York.

Whitfield, R. P., 1898, Notice of a remarkable specimen of the West India coral *Madrepra (Acropora) palmeta*, *Bull. Am. Mus. Nat. Hist.* **10:**463–464.

Yonge, C. M., 1963, The biology of coral reefs, *Adv. Mar. Biol.* **1:**209–260.

Chapter 15
Growth Patterns in Fish Sagittae

GIORGIO PANNELLA

1. Introduction	519
2. Morphology and Function of Sagittae	521
3. Calcification	524
4. Cyclical Growth Patterns	529
4.1. Subdaily Patterns	530
4.2. Daily Increments	531
4.3. Lunar Patterns	544
4.4. Seasonal Patterns	546
5. Growth Discontinuities	549
6. Growth Patterns and Ecology	552
6.1. Daily Increments	554
6.2. Tidal Patterns	554
6.3. Seasonal Patterns	555
6.4. Discontinuities	555
7. Fossil Otoliths and Paleoecology	555
References	556

1. Introduction

The peculiarly shaped carbonate bodies in the auditory system of fishes have been noted since at least Aristotle's time (Adams, 1940), but it is only in the zoological literature of the mid-19th century that they began to be methodically studied and to be known as *otoliths* (Adams, 1940). The inorganic bodies are not unique to fishes; bodies with similar auditory or balance functions are found in other vertebrates, where they are generally called *statoconia*, and in invertebrates, where they are called *statoliths* (Carlstrom, 1963).

All fishes have inorganic bodies in their auditory systems; cartilaginous fishes, for instance, have rudimentary bodies consisting of quartz sand grains bound together by an organic matrix or they have calcitic

GIORGIO PANNELLA • Department of Geology, University of Puerto Rico, Mayaguez, Puerto Rico 00708.

statoconia, but it is in teleosts that otoliths have developed into a large variety of forms and sizes. This chapter will deal only with sagittae of teleosts.

Commonly, the term *otolith* is incorrectly used in place of the term *sagitta*. The term *otoliths* should be used only in referring to all three bodies, located on each side of the head cavities at the base of the semicircular canals, called *utriculus*, *sacculus*, and *lagena*. The three mineralized structures are named *lapillus*, *sagitta*, and *astericus*, respectively. Attempts to change the names of the latter structures into *utriculith*, *sacculith*, and *lagenalith* have not found many followers (Adams, 1940).

Of the three otoliths, the largest* is the sagitta (pl. sagittae), which, possibly because of its size, has been more intensely studied than the others.

The three pairs of otoliths are morphologically different. The sagittae, ranging in size from millimeters to centimeters, are generally concavo-convex, subelliptical, scutiform bodies; the asteriscuses are small (millimeter-size), flat, semicircular bodies with crenulated outlines; and lapilli are small (millimeter-size), circular to irregularly shaped flat bodies. Because of their small size, the last two pairs of otoliths are often lost during dissection. The asteriscus is slightly larger than the lapillus.

Mineralogically, otoliths are composed of calcium carbonate crystals and organic material. In teleosts, the mineral part is aragonite, whereas the organic material is a protein, rich in aspartic-glutamic acid and low in aromatic and basic amino acids, called *otolin* (Degens *et al.*, 1969) (Table I).

The microstructures and chemistry of sagittae have been compared to those of molluscan shells, but the resemblance is only superficial (Irie, 1955; Mugiya, 1964), since growth and calcification are different in these two taxa. Sagittae have one feature in common with mollusc shells as well as other calcified taxa; i.e., inorganic deposition always follows the formation of the organic matrix.

Since Reibisch's (1899) suggestion of using sagittae to estimate the age of fish, many studies have been published on the method of aging temperate or cold-water fish by the growth zones found in the sagittae [see Bagenal (1974) and Williams and Bedford (1974) for general reviews of methods of aging fish]. Sagittae have also been used in taxonomic, phylogenetic, and evolutionary studies of fish (Bassoli, 1906, 1909; Bauza-Rullan, 1958; Chaine, 1956; Frizzell and Lamber, 1961; Frizzell, 1965; Weiler, 1968).

Recent studies (Pannella, 1971, 1974; Brothers *et al.*, 1976; Ralston, 1976; Struhsaker and Uchiyama, 1976; Taubert and Coble, 1977) have

---

* Exceptions are found in the Ostariophysi.

Table I. Comparison of Sagittae Protein with Proteins of Other Calcified Tissues[a]

Protein	Otolin Pisces[b]	Enamel[c]	Dentine collagen[d]	Keratin[b]	Mercenaria[d]	Snail shell[c]
Hydroxyproline	30	0	99	—	—	3
Aspartic acid	162	29	46	65	148	183
Threonine	57	37	17	58	47	22
Serine	44	46	33	103	75	91
Glutamic acid	170	185	74	111	71	81
Proline	50	271	116	75	128	47
Glycine	125	49	329	84	146	265
Alanine	96	24	112	54	70	83
Valine	70	37	25	68	32	32
Methionine	1	47	5	1	14	1
Isoleucine	30	32	9	34	24	15
Leucine	73	94	24	74	35	34
Tyrosine	1	22	6	25	42	35
Phenylalanine	12	26	16	26	33	21
Hydroxylysine	—	2	10	—	—	6
Lysine	20	11	22	23	68	52
Histidine	7	72	5	6	5	4
Arginine	14	6	52	59	41	15
Cysteic acid (half)	10	0	0	114	21	10

[a] Values are in parts per thousand protein (by weight).
[b] From Degens et al. (1969).
[c] From Piez (1961).
[d] From Degens et al. (1967).

suggested the possibility of using the growth patterns present in sagittae as a daily record of fish ontogeny. The potential of this application in ecology and paleoecology is the focus of this chapter.

2. Morphology and Function of Sagittae

A short description of the geometry of growth and of the structural features of sagittae is necessary before calcification and growth patterns can be discussed.

A sagitta pair consists of two elliptical, laterally compressed, distally concave bodies with their long axis oriented in an anterior–posterior direction. Each sagitta is the mirror image of the other. The morphological features of sagittae are shown in Fig. 1. The convex or proximal side, oriented toward the fish central axis, is divided into two areas by a deep groove, the *sulcus acusticus*. Nerve fibers extending from the sacculus walls are inserted in this groove. The sulcus terminates anteriorly in a deep sinus in the marginal outline, the *excisura*, and posteriorly in the *postcaudal trough*. The *collum* is a smooth ridge that separates the sulcus

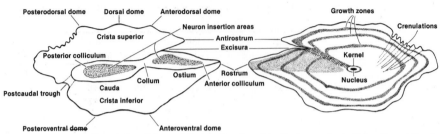

Figure 1. Proximal and distal face of an idealized sagitta. Redrawn from Frizzell and Dante (1965).

into a posterior and anterior *colliculum*. The ventral margin is generally rounded and smooth, while the dorsal margin often shows crenulations or pointed extensions that correspond to internal structural elements consisting of bundles of aragonitic needles (Plates Ia and b and IIIa and c).

In relatively young fishes, the central area of sagittae generally stands out because it is translucent and surrounded by an opaque ring, the diameter of which varies in different specimens. The tiny central area, known as the *kernel*, represents the earliest phase of otolith growth and, together with the opaque ring, constitutes the core or nucleus. In temperate fishes, the core or nucleus represents that part of the sagitta deposited up to the first winter. It has been found that the central or nuclear area of the otolith of temperate and cold-water fishes can provide a basis for racial grouping (Einarsson, 1951; Parrish and Sharman, 1958; Postuma, 1974). For instance, herring spawning early in the year have a nucleus of larger average size than herring spawning later in the same year. The difference is related to the growth rate after hatching. Similar size differences in the central area are present in tropical fishes, but their significance is still obscure and needs to be studied. The importance of nuclear growth for the reconstruction of early life history cannot be overemphasized (Blaxter, 1974). In many species, thin sections or acetate-peel replicas of the spherical or slightly elliptical kernel do not show growth increments. The kernel thus represents the prerhythmic phase of growth around which incremental layers are periodically added concentrically (Plate Ic, e, and f). The earliest phase of rhythmic growth consists of sequences of nearly spherical concentric layers of even thickness. Bundles of aragonitic needles develop radially and perpendicularly to these concentric growth surfaces (Plates Ia and b and IIIa and c). During a second growth phase, concentric spherical layers become compressed in some regions, and growth gradients develop in different directions, with the highest growth rate being along the anterior–posterior axis, intermediate growth in the dorsal–ventral direction, and minimal growth in the proximal–distal direction. Because of these growth gradients, the shape of sagittae is compressed, and the best sections for study of growth patterns are those transecting the otolith in the central plane along the maximum growth axis. It is along the sagittal anterior–posterior direction that the sequences of growth increments are the thickest and have the least number of growth interruptions.

In most sagittae, three stages of postkernel growth are recognizable. The first is characterized by a fast growth rate, rare or no interruptions in the continuity of growth sequences, and nearly isotropic growth in all directions. In the second stage, the average thickness (and thus the growth per unit of time) of increments is reduced, and interruptions in the deposition of daily increments become frequent. In the third stage, the growth rate is further reduced, and discontinuities and unconformities in growth sequences become evident (Fig. 2).

These three stages are expressed differently in different areas of the sagitta and also in different taxonomic groups. Along the excisura groove of the Lutjanidae, for instance, growth patterns have a characteristic scalloped pattern (Plate Id). The scalloped pattern is also visible in horizontal sections cutting through the sagitta center. Because of the lateral variations of patterns, the precise position of the section should always be kept in mind when interpreting growth patterns, and the section orientation should always be provided in illustrations and photographs. Werner (1928) has suggested a set of symbols to represent graphically the orientation of the section. His symbols are shown in Fig. 3.

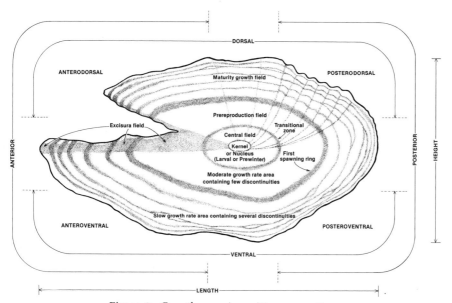

Figure 2. Growth areas in sagitta cross section.

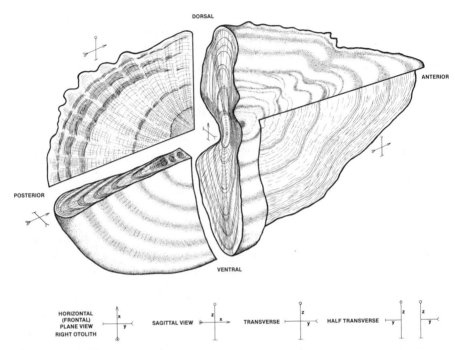

DORSAL

ANTERIOR

POSTERIOR

VENTRAL

HORIZONTAL (FRONTAL) PLANE VIEW RIGHT OTOLITH   SAGITTAL VIEW   TRANSVERSE   HALF TRANSVERSE

Figure 3. Symbolic representation of oriented sections of sagittae. After Werner (1928).

Sagittae are part of the ear of fishes and are thought to fulfill, together with the lagena, the function of sound perception. It has been shown that damaging the sagitta makes the fish unable to hear. The Ostariophysi, which have the widest range of sound detection among teleosts, show quite complicated structures in both the asteriscus and the sagitta. Although the mechanism of sound perception is still obscure, it has been suggested that sagittae, with their piezoelectric properties, act to transmit mechanical stimuli to neuron fibers inserted in the sulcus acusticus (Morris and Kittleman, 1967). It is not known whether ontogenetic morphological changes of the sagittae affect the frequency range and discrimination of sounds, nor is it known whether the carbonate concretionary deposits found in sagittae of old fish may hamper hearing.

3. Calcification

Calcium deposition in otoliths can be understood only in the framework of calcium metabolism in fish. Unfortunately, calcium metabolism in fish is highly complex and still poorly understood, although it has been

525

the topic of many studies (i.e., Moss, 1961a,b, 1963; Norris et al., 1963; Fleming, 1967; Berg, 1968; Simmons, 1971; Johnson, 1973; Pang, 1973; Simkiss, 1974). It appears that calcium levels in fish are not in isotonic equilibrium with environmental levels, with the exception of some primitive fish (Simkiss, 1974). Teleosts show a plasma calcium content on the order of 300–350 mmol/liter in marine species and 250–275 mmol/liter in freshwater species (Simkiss, 1974, p. 2). Fish, like all other vertebrates, contain calcium plasma in two forms: one free and diffusible and the other protein-bound and nondiffusible.

Female osteichthyes increase the protein-bound calcium level during reproduction, while ionic calcium remains at the same level (Urist and Schjeide, 1961; Woodhead, 1968a,b). During the reproductive period, yolk proteins, synthesized by the liver, are transported to the ovary as calcium complexes. Because sagittae in females show spawning rings that are less marked than in male sagittae (Pannella, unpublished data), it seems logical to infer that it is the ionic calcium that contributes to the calcification of the otoliths.

Without getting involved in the complexities of fish osteogenesis, it will be sufficient here to point out some basic chemical and mineralogical differences between bone and otoliths as calcified tissues. "Bone consists of an essentially collagenous fiber organic matrix with associated mucoproteins and is calcified by hydroxyapatite salts" (Moss, 1961b; p. 102), whereas sagittae consist of aragonitic needles and otolin, a mesh of organic fibrous matrix composed of mucopolysaccharides (Degens et al., 1969). Collagen and otolin are quite different in their amino acid composition; the latter is closer to the organic matrix of bivalve shells (Mercenaria mercenaria) than the calcified organic matrix of vertebrate bone (Table I). The similarity in the inorganic and organic composition between fish sagittae and the molluscan shell has stimulated comparative studies (Irie, 1955; Mugiya, 1964, 1966a,b). Infrared spectra obtained from powders of sagitta of Merluccius bilinearis (Mitchill) and homogeneous shell layer of Mercenaria mercenaria (L.) are almost exactly the same (Fig. 4), with typical aragonitic bands of $v_1 = 1080$, $v_2 = 865$, $v_4 = 712–699$ (Farmer, 1974).

Another basic difference between bones and sagittae lies in the cellular structures that are thought to be responsible for calcium deposition. Fish skeletal bones are either cellular or acellular. In the case of acellular bone, which is common in teleosts with a few exceptions for species belonging to the primitive Clupeiformes and Scopeliformes, Moss (1961a,b) distinguished two processes of osteogenesis, one related to the periosteum and the other to the chondroid tissue. "Both modes participate in the formation of both cellular and acellular bone" (Moss, 1961b, p. 100). In the first process, osteoblastic cells of the periosteum withdraw from the

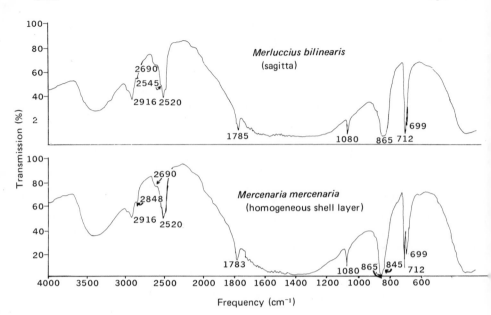

Figure 4. Infrared spectra of aragonitic material of a bivalve shell and a fish sagitta obtained from powders.

site of calcification as the matrix material is deposited. This is similar to what is observed in mammals. Chondroidal osteogenesis involves intra-cytoplasmic calcification, which obliterates osteoblastic sites and appears to be a unique feature of piscine osteogenesis (Moss, 1961b). Scale mineralization of the bony hyalodentine layer is also attributed to the action of osteoblasts or scleroblasts present in the dermal papillae. No osteoblasts have been reported in the area of active calcification of sagittae, but on the other hand, only a few studies have been published on otolith mineralization.

It has been suggested that maculae, the sensory areas on the saccular wall, are involved in the calcification process. Radiocalcium experiments have shown that in the rainbow trout, isotope deposition in the otoliths was greatest on the medial surface, on dorsal and ventral margins, and on anterior and posterior areas (Mugiya, 1974). These are areas where the thickness of increments is relatively great and indicates sites of active and rapid deposition. Also, these are areas where macular cells are concentrated. Although the relationships and the mechanisms involved are still obscure, Mugiya (1974) suggested that calcium is transported via blood capillaries to the macular cells, which then secrete calcium carbonate into the otolith fluid. The greatest concentration of calcium is around the sites of secretion. Macular activity also appears to be involved in the calcification of otoconia of other vertebrates (Lyon, 1955). In the same study, Mugiya (1974) found that calcium transport into the endolymphatic fluid was very rapid and its incorporation into the sagitta continued for 72 hr.

Two stages appear to be involved in otolith mineralization. A preliminary stage involves secretion of an organic matrix from particular sites associated with the endolymphatic sac. This is then followed by deposition of $CaCO_3$ around the organic fibers. Dunkelberger et al. (1980) found two layers of organic material present in the sacculus: (1) a tightly arranged fibrous material (gelatinous layer) extending from the tips of the sensory hairs to the sagitta and (2) a nonstructured zone (subcupular meshwork) of loose fibers that possibly incorporate into the otolith as an organic matrix. Noting the relationships between the precursor organic matrix and the following inorganic deposition, Mugiya (1964) sampled monthly the prealbumin fraction and the protein concentration in saccular endolymph and concluded that both substances increase during the season of rapid growth of the sagitta. In another study (Mugiya, 1966a), the endolymph composition of rainbow trout and flatfish was found to contain a hyaluronatelike substance and a PAS-positive material that, "together with proteins, are considered to be components essential to the formation of organic matrix of otolith itself [sic]" (Mugiya, 1966a, p. 123). It was also found that the concentration of total calcium in endolymph of the same two species was lower than that of their respective sera and "nearly equal to the diffusible calcium in the area" (Mugiya, 1966b, pp. 555–556). In the endolymph, the diffusible calcium was 72.7% (of the total calcium) in flatfish and 73.7% in rainbow trout. Seasonal variations in the flatfish ranged from 65.4% during fast otolith growth to 79.1% during formation of the hyaline zone.

Notwithstanding the basic differences in the calcification of bones, scales, and otoliths, calcium metabolism is integrated in such a way as to affect bone, scales, and otoliths at the same time; the growth of one tissue is paralleled by the growth of the others. Evidence for calcifying episodes being coordinated in all tissues is provided by comparison of growth patterns in different tissues of the same specimen. Several skeletal parts, sagittae, and scales of specimens of *Megalops atlanticus* Valenciennes were studied, and growth bands were counted and measured with a binocular microscope. The results reveal a consistent relationship between major episodes of rhythmic calcification in different structures (Table II). The number of growth bands and biochecks (interruptions of growth sequences) and the relative growth rate are generally consistent among most skeletal parts (Weiler and Pannella, 1980). Because of structural differences, however, some tissues show the patterns better than others.

Environmental variables such as temperature, salinity, pH, and ionic

Table II. Average Measurements of Growth Zones in Different Calcified
Tissues of a 5-Year-Old Tarpon (*Megalops atlanticus* Valenciennes)[a]

Tissue	Number of bands per zone					Total number of bands	Length or radius (mm) of zones					Total length or radius (mm)
	R_1	R_2	R_3	R_4	R_5		R_1	R_2	R_3	R_4	R_5	
Vertebrae	5	7	7	8	7	34	3.1	5.2	7.3	8.3	9.0	9.0
Scales	5	7	6	7	8	33	13	18	22.7	26.2	29	29
Sagittae	11	13	12	12	13	61	8.2	12.1	15.4	17.0	18.5	18.5
Cleithrum	10	11	12	11	12	56	35	48.2	59	66	68.9	68.9
Opercular bone	10	11	12	11	12	56	29	39	48.5	55.5	58.5	58.5

[a] The tarpon was captured on September 13, 1977, in the Mona passage (Puerto Rico). A zone is defined
by periodic variations in color or structural details and consists of one light, fast-growth (opaque in
transmitted light) and one dark, slow-growth (hyaline in transmitted light) sequence of bands. Zones
are progressively numbered from the center, and are not necessarily annual.

ratios seem to play important roles in regulating internal water and salt
balance in fish (Schlieper, 1930; Krogh, 1939; Shaw, 1960; Parry, 1966;
Rao, 1969) and thus are, directly or indirectly, important in regulating
calcification rates. There is no doubt that winter slowdowns in calcification rate, reflected in the winter bands of temperate sagittae, are controlled in part by temperature changes. The relationships between other
parameters and otolith calcification need to be investigated. Although the
data are still scanty, another relationship seems to exist between calcium
deposition and fish feeding activity. Bhatia (1932) noted the relationship
between growth of scales and feeding, and Bilton and Robins (1971) have
shown that the number of circuli in scales of *Oncorhyncus nerka* is a
function of feeding levels. Bilton (1974) found that circuli in scales of
salmonids formed during periods of starvation and suggested the possibility of scale resorption during periods of stress and starvation. Growth
rate of scales and overall body biomass has been shown to be affected by
variations in growth hormones; hormone levels are, in turn, related to
food consumption (Swift and Pickford, 1965). Some evidence relating
feeding activity to otolith growth has been presented in the literature. For
example, Liew (1974) noted that in the eel (*Anguilla rostrata*), sagittae
stop growing during periods of starvation. Taubert (1975) has shown experimentally that the amount of food consumed by *Lepomis cyanellus*
Refinesque directly controls the thickness of growth increments. In *Tilapia* sp. grown in experimental ponds and regularly fed once a day, there
is a close relationship between the number of days and the number of
sagittal increments. In fishes fed twice a day, the number of increments
is generally greater than the number of days (personal observation). Irie
(1960) found that calcium deposition was influenced by water tempera-

ture and food intake, but suggested that the organic matrix is produced at a constant rate throughout the year. He also noted seasonal variations in fish feeding activity and related fast sagitta growth to peaks of such activity.

All the aforementioned works point strongly to a close relationship between feeding and otolith deposition. This phenomenon is undoubtedly more complex than this review may suggest, but it provides in its simplification a model to be checked experimentally, and it leaves open the possibility of the existence of increments representing time intervals other than 24 hr. Most species found to have daily increments in the sagitta are known to feed with daily rhythms.

Before concluding my remarks on calcification, I must add a final consideration. Osteolysis, or resorption of bones and scales, is a common phenomenon in fish and has been widely reported in the literature. So far, however, similar phenomena have not been noted in fish otoliths, and the idea that these structures are not resorbed is common in the literature. This notion, however, should be discarded; there are obvious episodes of resorption recorded in sagittae (Pannella and Weiler, 1980). There is further evidence that otoliths participate in the total systemic calcium metabolism. From literature sources and the considerations discussed above, the following conclusions may be drawn: (1) calcification of fish sagittae can be understood only in the framework of total (whole-body) calcium metabolism; (2) although mineralogically and chemically different from bones or scales, otoliths appear to be affected by periodic variation in diffusible calcium in a manner similar to that affecting bones and scales; (3) an organic precursor is necessary for calcium deposition in fish skeletal tissues including otoliths; (4) endolymph physicochemical changes control the deposition of organic and inorganic components of otoliths*; and (5) feeding activities appear to have some effect on endolymph composition and therefore on the growth of sagittae.

4. Cyclical Growth Patterns

For many years, only the gross morphology or the seasonal zonation of fish sagittae were studied from a taxonomic point of view or for the purposes of making age determinations. Finer structures such as lamellae or subseasonal rings did not attract attention, although some references to them can be found in the literature. Hickling (1931), after decalcifying a sagitta of hake, noted lamellae a few microns in thickness within sea-

---

* In temperate regions, calcium content of fish endolymph varies with season and with reproductive activity; hormones also control the calcium content of endolymph.

sonal zones. Subseasonal rings were also mentioned by Irie (1955, 1960). However, references to subseasonal growth patterns were scarce and fragmentary until the acetate-peel replication technique was applied to sagittae studies (Pannella, 1971).

The technique allows the detection of fine structural and growth details without requiring the time-consuming preparation of ultrathin sections (see Appendix 1.C). The application of replicas opened up the field of growth patterns, which, though still in its infancy, promises to be revolutionary in terms of ontogenetic studies of fish. Much work remains to be done to adequately interpret otolith growth patterns. Nevertheless, over the relatively few years since the development of the acetate-peel technique, it has become apparent that there is an enormous amount of life-history information stored in sagittae. Because each sagitta represents a unique ontogenetic history, the variety of growth patterns is practically infinite. However, there are basic recurrent patterns that are common to many sagittae. In attempting to recognize and classify growth patterns, quantitative analysis of growth patterns can be used to distinguish among periodic, aperiodic, and random growth records. In this way, causative agents can be looked for to explain the physiological rhythms as they are recorded in sagittae.

The causative factors responsible for the deposition of rhythmic patterns are more easily identified than are those for randomly located growth bands. Rhythmic patterns are common or perhaps even universal, and can be reduced to only a few categories. According to Pannella and MacClintock (1968), growth patterns may be categorized into three types: (1) *private* patterns recorded in only one member of a population; (2) *universal* patterns common to all members of a population; and (3) *semiuniversal* patterns common to only a fraction of a population. This classification is adopted herein. More elaborate systems can also be used (Rosenberg, 1974), but for the purposes of this chapter, the simple classification of growth patterns presented in Table III is sufficient.

4.1. Subdaily Patterns

Growth patterns representing frequencies of more than one per 24-hr interval are infrequently encountered, but have been detected within sequences of thick daily increments and in those parts of the sagittae that are deposited rapidly, that is, in the early stages of growth (Ralston, 1975, 1976; Taubert and Coble, 1977). In sagittae of tropical fish, which generally grow from 3 to 10 times faster than sagittae of temperate or cold-water fishes, subdaily patterns are common, suggesting that they may appear to be rare in slow-growing sagittae only because they are impossible to

Table III. Classification of Growth Patterns in Sagittae Based on Recorded Periodicities

Periodic	*Subdaily*: Includes all patterns with a periodicity of less than 24 hr. In fish otoliths, they are rare and are best observed in relatively thick increments. *Daily*: Patterns with a circadian periodicity. *Multidaily*: In this category are included all patterns formed by more than one daily increment. The periodicity ranges from a cluster of a few daily increments, to fortnightly, monthly, multimonthly (seasonal), and annual clusters. *Interruptions*: Patterns formed when growth sequences are interrupted and interruptions are periodic. Interruptions may be both nondepositional and resorptive.
Aperiodic and random	A broad category that includes all growth phenomena that do not have an apparent periodicity.

detect in increments 2–3 μm thick. In the first 15–25 days of growth, subdaily patterns are visible in the 20- to 40-μm-thick daily increments of *Lutjanus vivanus* and *Tilapia* sp. (Plate IIIb). The patterns are made up of very thin and faintly expressed layers composed of aragonite and an organic matrix. Their number in a 24-hr interval ranges from 3 to 20 in *Lutjanus vivanus* and from 3 to 8 in *Tilapia* sp. The environmental or physiological significance of the patterns is unknown.

"Subdaily rings" have been observed in the early growth record of *Lepomis gibbosus* (Taubert and Coble, 1977). Subdaily patterns are present in sagittae of larval fishes and often are not distinguishable from daily increments (Brothers *et al.*, 1976). In general, subdaily patterns etch in acid less well than the organic-rich matrix regions that bound daily increments (Plate IIIb). The separation of subdaily from daily patterns is particularly difficult in planktic larval fish. Growth patterns of planktic forms, whether they be invertebrates or vertebrates, are characterized by an extremely high number of increments in relation to their age as measured in days. Pteropod shells, for instance, show an average ratio of 12 subdaily lines to 1 daily increment. The presence of numerous subdaily lines in the sagitta nucleus can be used to recognize the planktic larval growth stage and to determine its length in terms of daily increments.

4.2. Daily Increments

Acetate replicas or ultrathin sections of medial sections of sagittae reveal discrete increments located within seasonal zones or rings. Neglecting for a moment specific variations and details, an increment is characterized by a structurally homogeneous layer bounded by two structurally different surfaces. The homogeneous layer consists of a calcified mesh of organic material and has a lighter-colored appearance (in peels) than the boundary surfaces. The thin darker surfaces are "formed by a much denser mesh of organic fibers which, in cross section, appear as black dots" (Pannella, 1971, p. 1124). These regions are not calcified. The fibrous matrix is "oriented in a kind of corrugated pattern with well-defined lineages and the individual chains are organized in a helical fashion. At certain intervals the chains are twisted to such an extent that lumps or knots of apparently tangled fibers appear about 0.1 to 0.3 microns apart" (Degens et al., 1969, p. 109). Polished surfaces of sagittae, when etched with a 1% aqueous solution of HCl, show that contrary to what one would expect, the calcium-carbonate-rich layer resists dissolution better than the organic-rich layer (Plate IIf). When observed under the scanning electron microscope, polished surfaces show that etching is deeper in areas of concentrated organic matrix than in highly calcified areas. Thus, a daily increment is etched differentially into a relatively deep and narrow groove gradually merging into a thicker calcified prominent layer (Plate IId and e). Because hydrochloric acid does not affect the organic matrix, it must be the water that dissolves the organic fibers. Similar etching behavior has been described in bivalve shells (Dolman, 1974).

An increment represents a cycle of deposition that starts on an organic surface (growth line). This is followed by relatively rapid deposition of calcium carbonate and organic matrix. An increment is completed when calcification ceases and organic material concentrates in a thin layer. A similar cyclic structure is characteristic of molluscan growth increments. It has been suggested that anaerobic and aerobic phases in metabolism are at the base of the cyclicity in molluscan shells and that the concentration of organic material may actually result from dissolution of the shell by metabolic acids (Lutz and Rhoads, 1977). A similar mechanism appears improbable in the case of sagitta deposition, but may not be totally inapplicable in all cases. The sharpness of incremental boundaries varies from species to species, from population to population, from habitat to habitat, and, within the same specimen, from the early growth stage to the mature stage. It appears that a close relationship exists between the sharpness of the increment boundary and the nature of the physiological transition between cycles of activity and rest. Larval stages of fishes are much more active than later stages (Blaxter, 1974). Those species that are known to be almost constantly active show indistinct or faint separations between increments, while those species experiencing rest periods show marked incremental boundaries (Plate IIc). Often, the same correlation is found when one compares growth patterns of a fish that is diurnally active, but has nocturnal rest periods (Plate IIe and f), with a deep-water non-migratory species (Plate IIa and b). Ionic exchanges and respiration are

Text continues on p. 542.

Plate I. Microstructural features and nuclear growth of sagitta. Scanning electron micrographs: (a) of an acid-etched transverse section; (b–f) of acid-etched sagittal sections. Scale bars: (a,b,d) 100 μm; (c,e,f) 2 μm. (a) *Ocyurus chryurus* (Bloch). Radial arrangement of aragonitic needles from the nuclear area (lower left). Distal surface is in the direction of the lower left. (b) *Lutjanus vivanus* (Cuvier). Aragonitic needles arranged in a fan-shaped fashion in a dorso-posterior field. Note the low density of aragonitic needles at the bundle contact. (c) Kernel of *Lutjanus buccanella* (Cuvier). The longest axis of the eliptically shaped kernel (K) is about 200 μm. A structurally dense layer marks the boundary between the kernel and the beginning of the increment sequence. (d) Lower magnification of the same area pictured in (c) showing scalloped patterns radiating away from the kernel (K) anteriorly in the excisura field. (e) Kernel (K) of the surgeon fish, *Acanthurus chirurgus* (Bloch). The longest axis is about 50 μm. (f) Kernel of *Ocyurus chrysurus* (Bloch). Note the sharp boundary (arrows) between the kernel (K) and the larval growth sequence. The lighter surface is a sagittal section; the darker, on the lower right, is a transverse cut.

Plate II. Daily growth lines. All figures are scanning electron micrographs of acid-etched sagittal sections. Scale bars: (a–d) 20 μm; (e, f) 10 μm. Growth direction upward. (a) *Hilsa kelee*. Most increments are indistinct, in part because they are very thin. (b) *Hilsa kelee*. Another example of indistinct increments. Note the deeply incised discontinuity surfaces; the discontinuities record three different events: the continuous discontinuities in the right-hand corner represent a spawning (S) and a winter (W) episode; the others record minor stresses (P). (c) *Lutjanus buccanella* (Cuvier). Increments in the early growth stage. The transition between increments is represented by a gradual decrease in calcification rate resulting in a slight increase in the proportion of organic matrix. (d) Surgeon fish *Acanthurus chirurgus* (Bloch). Daily increments are defined by cyclic increases in otolin calcification followed by a sudden decrease in mineralization. Differential etching enhances these cyclic compositional changes. Subdaily patterns are visible within some of the thickest increments. (e) *Anchosargus rhomboidalis*. Mineralization cycles reversed relative to (d). A peak in calcification is followed by a gradually decreasing rate of mineralization. (f) *Archosargus rhomboidalis*. Sharply defined increments. Note the organic layers in which acid dissolution has produced a characteristic vacuolar structure.

Plate III. Daily growth increments and disturbances in *Tilapia* sp. raised in ponds and aquaria. All figures were taken of thin sections with a transmission optical microscope. All sections are sagittal. All scale bars equal 10 μm, with the exception of (c) (scale = 100 μm). The growth direction is upward. (a) Undisturbed sequences of daily increments characterize the early growth stage before sexual maturity. Each increment consists of a relatively thick calcium-carbonate-rich layer and in most cases a relatively thin organic-rich layer. Note the fan-shaped arrangement of the aragonitic bundles, which are separated by thick organic sheaths. (b) Detail of (a) showing subdaily organic lines representing changing rates of calcification. Calcification increases toward the top of the figure in each increment; darker color in thin section characterizes the calcified layers. (c) Traumatic disturbance and resorption of sagitta due to capture and storage of the fish in a very crowded anaerobic and high-temperature water tank for 6 hr. The animal almost died, but recovered in a pond, and the sagitta began to grow very fast after the deposition of a thick organic layer around the surface of disturbance (arrows). (d) Growth disturbances due to transplantation of a fish to one aquarium for 5 days (T_1) and then to another (T_2). After the second transplantation, the fish lived for over a month before being sacrificed. (e) Detail of same sequence of events as in (d) showing daily increments.

Plate IV. Daily growth increments and tidal patterns. All photographs are scanning electron micrographs of sagittal sections. The growth direction is upward. All scale bars: 10 μm. (a) Sharply defined thick increments in *Lutjanus vivanus* (Cuvier). No tidal patterns are apparent in this daily sequence. (b) Sharply defined thin increments in *Arcosargus rhomboidalis*. Tidal periodic patterns are marked by recurrent clusters (C). (c, d) Sharply defined thick increments of surgeon fish, *Acanthurus chirurgus* (Bloch). (c) Tidal patterns are marked by thickness variations and periodic interruptions (arrows). Note the progressive thinning of increments upward toward a major discontinuity related to reproduction (R) and the progressive thickening of increments afterward. (d) Higher magnification of sequence in (c). (e, f) Sharply defined thick increments of *Arcosargus rhomboidalis*. The organic-rich layers (O) are thicker than the calcium-rich (C). Tidal patterns (T) are marked by thickening and thinning of increments.

Plate V. Hyaline and opaque zones in tropical fishes. All figures are scanning electron micrographs of acid-etched sections. All are sagittal sections with the exception of (e). All scale bars: 10 μm. The growth direction is toward the upper right. (a) *Ocyurus chrysurus* (Bloch). The abrupt change in structure (arrow) in the early stage of growth marks the migration of this fish from an eelgrass habitat to a reef habitat. The decrease in thickness of increments is expressed macroscopically as a hyaline band (h). (b) *Rhomboplites aurorubens* (Cuvier). The change from the first fast-growth stage is marked by a cluster of interruptions (n) that form a hyaline band. (c) *Lutjanus buccanella* (Cuvier). Hyaline bands correspond to the concentration of growth discontinuities (d); opaque zones correspond to continuous fast-growth calcium-rich sequences. (d) *Coelorinchus chilensis*. Hyaline (n) and opaque (o) zones reflect concentrations of growth discontinuities (n) or continuous depositional sequences (o). (e) *Archosargus rhomboidalis*. The hyaline zone (n) represents two spawning epidodes (s) separated by a recovery period (v). Transverse section. (f) *Hilsa kelee*. The thick organic layer represents a major growth interruption, followed by slow recovery period. This interval (arrows) forms a hyaline band.

Plate VI. Growth discontinuities in sagittae. All photographs are scanning electron micrographs of acid-etched sagittal sections with the exception of the transverse section shown in (c). The growth direction is toward the upper right. All scale bars: 20 μm. (a) Winter (W), spawning (S) and minor discontinuities in a dorso-posterior area of *Hilsa kelee*. Note the change in the depth of the spawning break from right to left and the pinch out of the growth record. The depth and thickness of the discontinuity surface are proportional to the magnitude of the temporal hiatus. The spawning hiatus is estimated to be 15–20 days (on the upper left). The winter hiatus represents about one month. (b) Growth discontinuities in a dorso-central area of *Rhomboplites aurorubens* (Cuvier). The magnitude of the discontinuities is marked with numbers indicating the number of days missing at that position. (c) Growth discontinuity (d') caused by capture and a 3-day starvation period in an aquarium before the fish, *Ocyurus chrysurus* (Bloch), was transferred to a submerged cage in the sea. The second discontinuity (d'') was caused by an unknown event that occurred at neap tide. (d) Minor growth discontinuities (m_1, m_2) in *Lutjanus vivanus* (Cuvier) related to the lateral pinching-out of one daily increment. (e, f) Minor growth discontinuities in *Ocyurus chrysurus* (Bloch) similar to those shown in (d).

Plate VII. Spawning breaks, periodic discontinuities, and tidal patterns. All photographs are scanning electron micrographs of acid-etched sagittal sections. All scale bars: 20 μm. Arrows indicate the growth direction. (a) Surgeon fish, *Acanthurus chirurgus* (Bloch), showing tidal growth patterns and the first spawning break (s') in the antirostrum. A second spawning break (s'') occurs 1 lunar month later. (b) Same sagitta as shown in (a). Double spawning breaks (s) are separated by an interval of a few days (a week?) in the posterior area. Note the number, and degree of development, of the growth interruptions of the 1-year interval represented between annual hyaline bands h_1 and h_2. (m) Growth margin. (c–f) Growth patterns in *Lachnolaimus maximus* (Walbaum). (c) Anterior part of the rostrum. The fish was captured on October 29, 1977. The last major growth interruption prior to capture occurred on October 1 (I'). The next break occurred 28 days before (I''), and a third 56 days before (I'''). Minor interruptions affect only the dorsal and ventral areas of the rostrum. (d) Structural details of minor interruptions. Note how increments pinch out on the dorsal side (toward the bottom of the micrograph). (e) Two double spawning breaks. (s_1, s_2). The breaks interrupt a regular fast-growth sequence and are followed by a 6- to 7-day recovery period. (f) Structural details of a growth sequence before a spawning break (s_1). Note the fortnightly tidal patterns marked by subtle changes in structural features of daily increments (t).

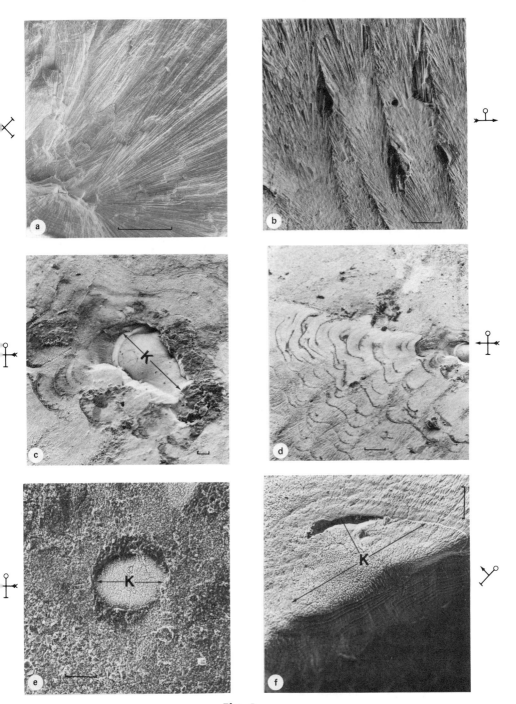

Plate I

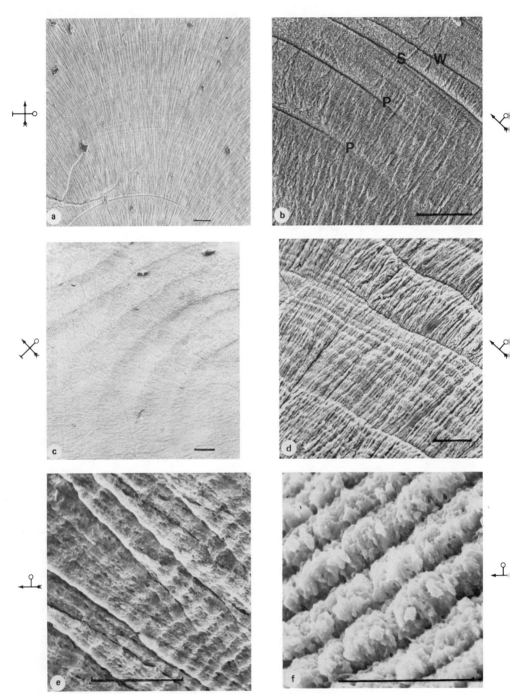

Plate II

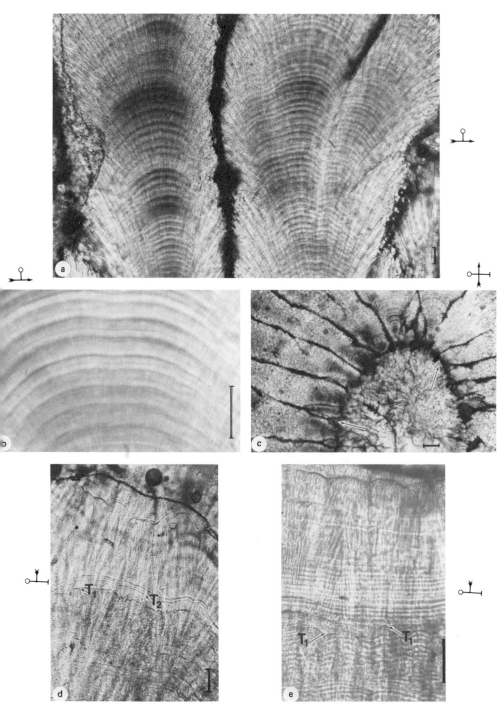

Plate III

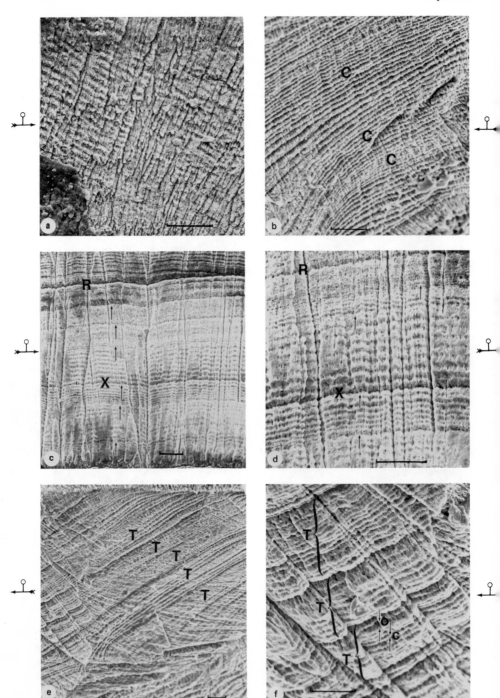

Plate IV

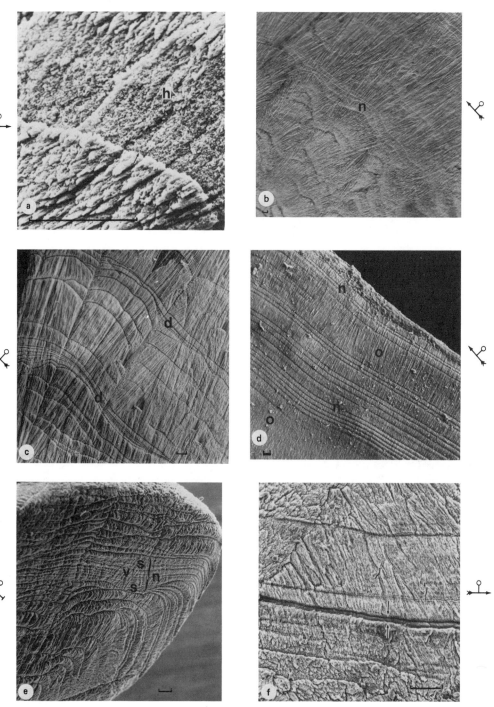

Plate V

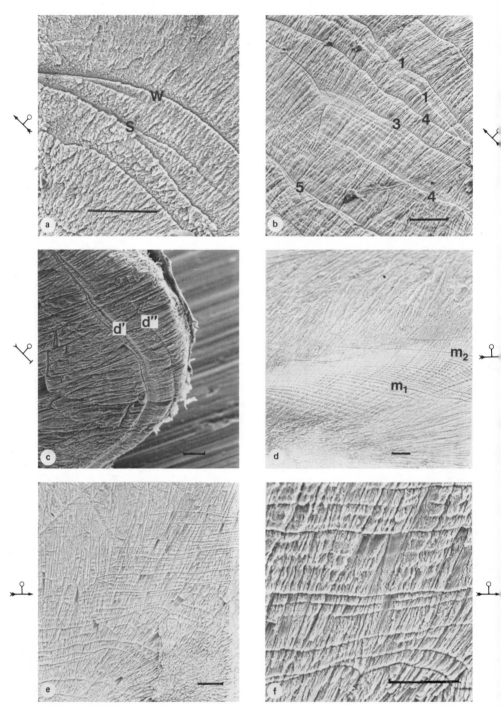

Plate VI

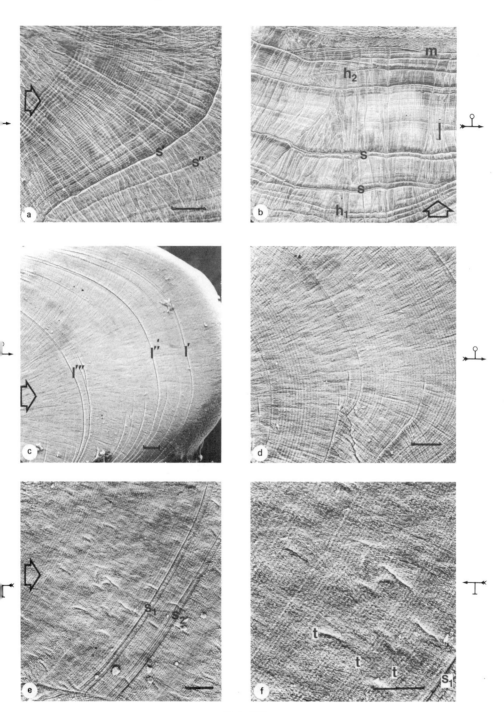

Plate VII

enhanced during fish activity, whereas during rest periods all metabolic functions are reduced and the respiratory process is kept at a minimum. It is not far-fetched to speculate that anaerobic conditions are created in the endolymph during rest. The problem is interesting in terms of the Lutz–Rhoads hypothesis and deserves attention.

Increment boundaries seem to have an important physiological–ecological significance, and a classification of increments based on sharpness of growth lines may become a useful tool in the interpretation of fish life history and habits. In Table IV, increments are divided into four categories based on boundary features and on the observation of over 150 different species of tropical, temperate, demersal, reef, and deep-water fish. There is no doubt that this classification will have to be improved and will become more accurate and detailed as more observations become available. In fact, it is probable that an *ecological* classification will be constructed in the near future.

Several factors appear to control increment features. One such factor is genetic; members belonging to the same family have similar increment

Table IV. Classification of Growth Increments in Sagittae According to Boundary Features

Increment type	Description	Comments
Indistinct Plates IIa,b; Vb,c	The transition from one increment to the next is not marked by clear structural or chemical changes.	Frequent in early growth stages, in deep-water species, in growth patterns formed under laboratory conditions and stress—in winter bands of temperate fish.
Gradual or diffuse Plate IIc	Subtle and gradual changes in texture and composition mark the boundary between increments.	Frequent in the larval stage of some tropical and temperate fishes, in fishes that are active during most of the 24-hr period, and in relatively deep-water fishes.
Distinct thick Plates IId,e; IIIa,b; VIe,f; VIIa–f	Increments are made of one thick calcified layer and one slightly thinner organic-rich layer. The organic-rich layer etches in HCl aqueous solutions more deeply than the mineralized layer.	In fishes with strong diurnal and nocturnal habits, characteristically developed in herbivorous fishes and in many reef fishes.
Distinct thin Plates IIf; IVe,f; VId	The boundary between increments is marked by a thin otolin layer that, in places, interrupts the perpendicular structural needles.	In many mature-growth sequences, in many temperate and fewer tropical fishes. Possibly the most common type in demersal fishes.

types even if they live in different habitats. A second factor is related to water depth and environmental factors. Deep-water fish have different increment types from shallow-water fish. Feeding habits, rhythmic behavior, migration, and latitude are also factors affecting the type of increments and increment patterns. Characteristics and frequencies of growth increments also change in the same fish during ontogeny; thickness decreases with age and the thickest increments are always in the early growth stage (in the nuclear part of the sagitta). Relatively thick increments follow the kernel up to sexual maturity, after which thin increments, interruptions, and resorption become common. Only faintly developed and thin growth lines or subtle textural changes mark the boundaries of early increments, which are, in fact, so characteristic that they can be set apart and called *larval increments*. Larval increments appear to be present only in fishes with pelagic larval stages. Sharply bounded increments appear in fast-growing zones after the larval stage. These are associated with frequent growth interruptions and unconformities.

In some temperate species known to have annual bands in their sagittae, it was found that up to a certain age, each "*annulus*" consists of an average of 360 increments (with a small variance), and it was concluded that they must be daily. Later in ontogeny, when the number of increments per annulus decreases, the increments are still deposited with a daily periodicity, but some days are not recorded, so that growth sequences become discontinuous (Pannella, 1971).

Recent work has confirmed experimentally the presence of daily increments in several species (Ralston, 1975; Taubert, 1975; Brothers *et al.*, 1976; Struhsaker and Uchiyama, 1976). Experiments under different dark and light and feeding cycles with *Tilapia mossambica* (Peters) suggest that a 24-hr light–dark cycle entrains an internal, diurnal clock that is required for daily ring production (Taubert and Coble, 1977, p. 332). Experiments with different tropical marine species did not produce as clear-cut results as in the case of *Tilapia*. Transplants from natural conditions to artificial aquarium conditions created stress patterns in which daily increments were indistinct or absent (Miranda-Brandt, 1978). Miranda-Brandt's experiments showed that the history of feeding or starvation was recorded by the thickness of increments and the formation of interruption lines, respectively, and that diurnal deposition continued under constant light conditions.

The results of experiments and observations do not prove that distinct increments of all fish sagittae are daily. In fact, it has been suggested that some fish do not deposit them diurnally. However, these suggestions may be based on the misinterpretation of the patterns. The general tendency is for sagittae to grow in response to diurnal activity cycles. However, one must keep open the possibility that other growth periodicities might be recorded in sagittae.

One source of misinterpretation about increment periodicity can be related to the fact that more than one increment (generally two) may be deposited in 24 hr, or, alternatively, increments may not be deposited for one or more days. Moreover, in the later stages of growth, increments already deposited may be eliminated by resorption (Pannella and Weiler, 1980). These three phenomena produce variance in the growth record that results in a number of increments different from the actual number of days of growth.

Although the reasons for double increments are still unknown, one may speculate that they could be related to twice-a-day feeding activity. Many fish become active at dusk and dawn, and the activity of other fish follows the flow and ebb of the tide (Briggs and Caldwell, 1957; Randall and Brock, 1960; Randall, 1961; Jones, 1968; Hobson, 1972). The surgeon fish, *Acanthurus chirurgus* (Bloch), known to feed continuously during daylight, shows very definite one-per-day increments (Plate VIIa). Predators, such as snappers, which depend on the availability of prey, show occasionally double increments and more frequently interrupted sequences (Plate Vb and c). If feeding is a major factor controlling the growth of sagitta, rhythmically feeding fish should have rhythmic growth patterns, and disturbance and other features of feeding must be reflected in the increments as interruptions, change in thickness, and other characteristics. This relationship provides an extremely useful tool for inferring feeding habits and rhythmicity of behavior. In conclusion, increment types (Table IV) appear to be a source of data on events affecting growth on a daily basis.

4.3. Lunar Patterns

Groupings of increments in regularly repeated clusters are often found within seasonal zones. Periodic clustering was first reported in temperate fish (Pannella, 1971). In *Merluccius bilinearis* (Mitchill), for instance, a cluster of 6–7 thin daily increments is regularly followed by a group of 6–8 thick increments, and these two groups make up fortnightly patterns of 13–15 increment bands. A periodic pattern is also shown in the succession of bands: one band with distinct and thick increments alternating with another with less distinct and thinner increments.

Because of their periodicity, the factors producing these patterns have been suggested to be related to, or controlled by, tidal fluctuations. Not all fish show these patterns, but those that are known to synchronize their activities with the tides always show strongly defined patterns. In *Pseudopleuronectes americanus* (Walbaum), for instance, tidal bands are marked, in the summer zones, by growth sequences of 6–14 increments and thick interruption surfaces, each sequence representing a cycle of growth, interrupted (or even resorbed) with a tidal periodicity (Pannella, 1974).

Tidal patterns are also present in some tropical or even deep-water fishes. *Lachnolaimus maximus* (Walbaum), collected on the reef of southwest Puerto Rico, shows monthly and fortnightly patterns. These are recognized by subtle changes in textural features of increments and by discontinuity surfaces that occur with a periodicity of 7, 14, or 28 increments (Plate VIIc–f). Spawning marks also appear to occur with a tidal periodicity (Plate VIIe). Similar patterns have also been found in the surgeon fish, *Acanthurus chirurgus* (Bloch), collected from the same area. Periodic changes in increment thickness, growth interruptions, and spawning marks show rhythms of about 7–14 and 27 increments (Plate VIIa and b).

Tidal periodicities suggest behavioral and physiological synchronization with subtle environmental changes accompanying the less-than-one-foot-amplitude tidal fluctuations. Tidal influence in growth patterns is also recognizable in demersal fish (Weiler and Pannella, 1980). The knowledge that nektic animals are affected by tides or phases of the moon is utilized by fishermen, who know that their catches are larger around full and new moon and when tidal currents flow in certain directions during the day or the night. That feeding activity could be directly or indirectly regulated by tidal currents and by the direction and velocity of these currents is understandable because tides are a ubiquitous rhythmic factor in the sea. Examples of tidally controlled behavior of marine organisms are numerous (for a review, see Palmer, 1973, 1974) and have been reported in certain fish (Gibson, 1967), and evidence from skeletal growth patterns seems to indicate that the phenomenon is widespread. There seems to be an obvious selective advantage in synchronizing feeding and reproductive activity in littoral organisms, and it may also be advantageous to organisms living in deeper, tidally influenced water.

The evidence from growth patterns opens up a series of questions. At this point, one may ask, for instance, whether "daily" increments are, in fact, "lunar." If feeding is tidally regulated and is the major factor controlling otolith deposition, then the time represented in one increment could be either 12.4 or 24.8 hr instead of the sidereal 24-hr interval. Observed patterns cannot be explained without admitting that two periodicities are operating in otolith calcification, one lunar and the other solar. The combination of the two periodicities results in patterns of 14 and 28 increments.

I am now in the process of studying the effect of different tidal regimes on growth patterns. Irregular semidaily tides are present on the northern coast of Puerto Rico, and irregular daily tides are present on the southern coast. Puerto Rico is therefore an ideal area in which to test the hypothesis that a close relationship may exist between tidal periodicity and otolith growth patterns. Parenthetically, it is interesting to note that *Tilapia* spp. living in isolated (nontidal) ponds show no tidal patterns in their otoliths (Plate IIIa and b), while those collected from tidal lagoons do show tidal patterns.

Another problem requiring further work concerns the relationship between lunar phases and reproductive activities. For example, the case of the California grunion, which come ashore to deposit and fertilize the eggs at a precise tide, is well known. In most temperate species, reproduction time is controlled mainly by seasonal changes, while in tropical species there is some evidence that spawning may be cued to tides (Randall and Brock, 1960; Starck and Schroeder, 1970). Reshetnikov and Claro (1976a,b) noted that the spawning of *Lutjanus synagris* (L.) coincided with the full moon in the Batabano area of Cuba. Randall (1961) found that adults of surgeon fishes spawn more frequently at the full moon when spring ebb-tidal currents carry the fertilized eggs to the open sea. The question of cyclic fish reproduction deserves more attention, and otolith growth patterns can contribute to solution of this problem.

4.4. Seasonal Patterns

The existence of seasonal zones in sagittae of temperate fish has been known for a long time and has been used for age determination* (Chugunova, 1959; Blacker, 1974). The zones can be seen in unsectioned sagittae with a binocular microscope (20×) under reflected light; the *hyaline zone*, deposited during periods of slow (winter in temperate fishes) growth, appears as a translucent thin band, whereas the *opaque zone*, formed during periods of fast growth, appears as an optically dense and thick band. Some confusion exists in the use of the terms opaque and hyaline in the literature, and several papers have addressed this problem (Mina, 1968; Blacker, 1969; Pannella, 1974). One of the reasons for this

---

* The term *zone* is used here to denote growth rings resulting from seasonal changes in growth and calcification rates. Two zones make up an annual cycle of growth (in nontropical fishes) and together should be called an *annulus*. However, this term is used improperly to denote the markings of winter or spawning, and it will not be used here because it implies annual periodicity and cannot be used in tropical growth patterns. The term *band* denotes growth rings within a zone related to minor changes in growth and calcification rates. Many bands could be found in a zone. Each band consists of several growth increments.

confusion is that opaque zones in reflected light appear to be hyalinelike in transmitted light. In thin-section or acetate-peel replicas, the difference is based on the otolin/aragonite ratio, which is relatively easy to determine visually.

Microscopically and in sectioned sagittae, the two zones show different structural features. In the hyaline winter band, aragonitic needles are short and thin, often indistinguishable from the mesh of organic fibers surrounding and penetrating them. The structural changes reflect the fact that the amount of calcium deposited in the cold months becomes insignificant, and the otolin appears to dominate. It has been shown that the amount of organic matrix decreases in the endolymph during the winter and is at a maximum during the periods of fast calcification (Mugiya, 1966b). In cold months, it appears that the deposition of organic matrix decreases, but much less than calcium carbonate; therefore, otolin becomes the dominant structural material in winter zones (Blacker, 1969; Degens et al., 1969). In summer opaque bands, aragonitic needles are long and relatively thick, and organic fibers are neatly organized in layers. Growth increments, which develop at right angles to the c axis of the needles, are distinct and relatively thick. Winter increments are often so thin as to be almost indistinguishable. In Merluccius bilinearis, their thickness drops to one tenth of that of summer increments.

Structural and chemical variations in the bands are controlled by changes that take place in fish physiology due to seasonal environmental fluctuations. We still do not know all the external factors influencing these changes. Temperature is possibly the most important factor and food supply possibly of second-order importance. In some instances, feeding rate, locomotor activity, and migration may be reduced in colder months and may therefore account for the decreased rate in otolith growth. No winter zones are found in tropical fish, but "annuli" have been reported in many species (Menon, 1953; McErlean, 1963; De Bont, 1967; Moe, 1969; Fagade, 1974). Various hypotheses have been offered to explain the formation of annual zones in tropical species. Chevey (1933) concluded that an annual variation of only 4–5°C was apparently sufficient to slow growth and leave a mark on scales. De Sylva (1963) assumed that an annual temperature change of 6.5°C was sufficient to cause annuli on scales of the barracuda, Sphyraena barracuda (Walbaum), in the waters of southern Florida. Menon (1953) speculated that "an inherent physiological rhythm was a more possible causative factor [than temperature] in the formation of growth checks." Moe (1969) suggested that annulus formation in sagittae of the red grouper, Epinephelus morio (Valenciennes), from the Gulf of Mexico was the result of physiological changes associated with spawning time, increased photoperiod, and elevated bottom water temperature in the spring. However, he goes on to state that temperature may not significantly influence annulus formation because fluctuations are only 5–6°C annually.

Beaumariage (1973) studied king mackerel, *Scomberomorus cavalla* (Cuvier and Valenciennes), and concluded that there appeared to be a relationship between annuli and gonadal maturation. Bruger (1974), discussing age determination of *Albula vulpes* (L.), concluded that no "single time can presently be established for annulus formation in bonefish." However, he uses the growth marks in scales and sagittae to determine the age and to reconstruct growth curves of *Albula*. Fagade (1974) attributed the growth zones in the opercular bones of *Tilapia melanotheron* (Ruppel) to changes in salinity related to dry and wet seasons. Several other opinions could be added to this list, but the aforementioned studies clearly indicate that the formation of growth zones probably has more than one cause, that it does not occur at a fixed time of the year, and that the zones are not as clearly developed and obvious as in temperate fishes. In none of the works mentioned have the microstructural features of the zones been studied. Pannella (1974) published a preliminary study attributing zones in tropical sagittae to reproductive events, based on microstructural details. As described in detail in the next section, a spawning ring in some species represents a slowing down or an interruption of calcification and often a resorption of the otolith. The effect is an increase in the concentration of organic material and the appearance of a hyaline band. Any event, not only spawning, that has the effect of decreasing the mineral/organic ratio in the sagittae creates a hyaline or less opaque band. At high latitudes, winter is by far the most important annual event that changes the inorganic/organic ratio. In the tropics, many events can produce the change, and the changes are not as marked; hence, there is a disparity of opinion on the validity and meaning of the bands. Careful observation, with the dissecting microscope, of a tropical sagitta reveals not one but many hyaline and opaque zones that are obviously not annual. Their counts give numbers that are too high to represent years. Counting only the major ones produces values that could be acceptable as years of growth. However, often the "major" zones are not that different from minor zones, and their interpretation becomes very subjective (Weiler and Pannella, 1980). When observed with a transmission or scanning electron microscope, acid-etched cross sections of sagittae reveal that opaque zones correspond to continuous sequences of fast-growing increments and hyaline zones correspond to slow-growth, discontinuous series (Plate Vb–f). A concentration of discontinuities creates a concentration of organic lines and thus a hyaline zone. Discontinuities are related to different physiological stress conditions ranging from storm events, starvation, and spawning to migrations and other physiologically induced changes (Pannella and Weiler, 1980). In *Ocyurus chrysurus* (Bloch), for instance, the juvenile migration from a turtlegrass environment to open water near reefs is accompanied by a drastic change in sagitta growth-pattern features; growth rate is reduced to 10% for 2–4 weeks, the diameter of aragonitic needles is drastically reduced, and calcification of organic matrix drops considerably (Plate Va). Macroscopically, these changes are reflected in a hyaline band around the nucleus of the sagitta. Similar, but less drastic, changes in the patterns of *Rhomboplites aurorubens* (Cuvier) occur about 20 days after hatching (Weiler and Pannella, 1980). Growth rate decreases and the increments change from the indistinct larval type to the distinct type (see Table IV and Plate Vb).

Spawning breaks form a marked hyaline band in those species in which reproduction is a physiologically stressful event affecting the animal for a length of time. In *Lutjanus buccanella* (Cuvier), spawning breaks are marked by resorption lines followed by a series of interruption lines (Plate Vc). A combination of spawning breaks and seasonally repeated stress interruptions produces hyaline bands in *Coelorhynchus cilensis* (Plate Vd).

Periodic growth disturbances related to tides also produce faint, but macroscopically distinguishable, minor hyaline and opaque bands within the zones of several tropical sagittae. The periodicity and interpretation of hyaline and opaque zones in tropical sagittae can be discovered only through microscopic examination of acetate-peel replicas or thin sections of sagittae. Because each species, stock, or population has its own characteristic patterns, any generalization about the origin and meaning of the zones is bound to be incorrect. This does not eliminate the possibility of using the zones for age determination; it only suggests that a full understanding of how specific patterns are formed is necessary before such an application can yield unequivocal data.

5. Growth Discontinuities

Terms such as interruptions, discontinuities, and resorption have been frequently used in the preceding pages during the description of growth patterns. Gaps and breaks in the regularity of incremental sequences are the rule rather than the exception in growth patterns. Discontinuities yield interesting information because they record unusual events in the rhythmic growth of fish otoliths. Several types of discontinuities have been noted; they range from small gaps that grade into uninterrupted sequences (Place VIIc) to major unconformities that can be traced continuously along the entire sagitta (Plate VIa and c). The magnitude of the hiatus reflects the importance of the event in the life history of the fish that causes it.

Various criteria may be used to recognize the magnitude of growth discontinuities. A simple one is to trace the interruption laterally and observe whether it continues (or merges) into a normal sequence of increments. If so, it is often possible to determine the number of days missing in the hiatus. Structural criteria provide clues to the magnitude. Minor interruptions do not disturb the structural continuity of aragonitic needles or change their diameter (Plate IVd). More important gaps abruptly interrupt structural continuity of the needles and may be followed by thinner aragonitic needles (Plates IVc and VIb).

Major discontinuities are marked by: (1) interruptions of structural features (Plate Vf); (2) a paraunconformity in which the surface of discontinuity is made of a relatively thick organic layer (Plate Vc); and (3) an angular unconformity that cuts across growth sequences at a low angle and indicates not only an interruption of growth but also a resorption of calcium carbonate (Plate VIIb). Obvious structural changes take place across major discontinuities.

Growth-rate changes manifested in the changing thicknesses of increments before and after a discontinuity also provide a clue to the magnitude and the cause of the disturbance. Several combinations are possible (Table V)*: (1) no growth change before or after a discontinuity (FF); (2) growth slowdown before and a gradual resumption of the prebreak growth rate or discontinuity (SS); (3) growth slowdown before but not after a discontinuity (SF); and (4) no change in growth rate before a discontinuity followed by slow recovery after the disturbance (FS). Because a slowdown in growth rate indicates a gradual deterioration in environmental and physiological conditions, these combinations give a clue to the abruptness and cause of the interruption. Thus, a winter disturbance is preceded and followed by a gradual thinning of increments, whereas a spawning break may be preceded, or followed, by a slowing down in growth depending on the species and sex of the fish (Pannella, 1971, 1974) (Plates Vc and f, and VIIe).

Gradual changes in growth rate preceding a disturbance are characteristic of tidal breaks, whereas no anticipated change in growth prior to a discontinuity indicates an abrupt event. The rate of recovery to predisturbance increment thickness provides an indication of the magnitude of the stress causing the growth interruption.

Additional information about the nature of disturbances is provided by the time of their occurrence. Using the sagitta margin as a reference point, one can date quite accurately the time of the stress event and attempt to correlate this with environmental events. When discontinuities

---

* Because of the variety of causative agents and of morphological expressions, only a simplified description of growth discontinuities is presented in Table V.

Table V. Classification of Discontinuities in Growth Sequences of Fish
Sagittae

Type[a]		Description	Possible causes
FS	U	Fast growth cut by erosional unconformity, followed by slow recovery. Structural continuity is interrupted.	Extreme physiological stresses, disease, reproduction, starvation, migrations (e.g., from pelagic shallow water to deep water)
SS	U	Gradual slowing down; increment sequence is cut by an erosional unconformity, followed by slow recovery. Structural continuity is interrupted.	Seasonal or continuous physiological deterioration prior to peak stress
DD	PU PU	Double organic lines delimiting thin indistinct sequences (paraunconformity). Marked structural changes.	Repeated stresses of reproduction or migrations
SF	D	Gradual slowing down followed by fast recovery after nonerosional interruption. Structural continuity disturbed.	Continuous stresses associated with reproduction
FF	D	Fast growth interrupted by a nonerosional discontinuity followed by resorption at prebreak growth rate. Structural continuity maintained.	Moderate short-term stresses such as tides or storms

[a] Abbreviations: (FS) fast growth—discontinuity—slow growth; (SS) slow growth—discontinuity—slow growth; (DD) double nonerosional discontinuity; (SF) slow growth—discontinuity—fast growth; (FF) fast growth—discontinuity—fast growth; (U) unconformity (erosional discontinuity); (PU) paraunconformity; (D) nonerosional discontinuity.

appear at a certain ontogenetic stage of several specimens, this suggests a developmental change. In sagittae of *Ocyurus chrysurus* (Bloch), for instance, the nuclear field is bounded by a strong discontinuity marked by major structural changes that represent the time of migration of the young from an eelgrass habitat to a reef environment (Plate Va). Some morphologically similar discontinuites occur at different times in different members of a population and suggest unsynchronized physiological events. An environmental origin is indicated when a disturbance is recorded synchronously by a multispecies community.

Experiments have shown that the magnitude and features of growth disturbances are directly related to the magnitude of the stress and to the time span during which the stress is experienced. One day of intense thermal and anaerobic stress, which brought the animal near death, left a profound disturbance, FS-type dicontinuity, accompanied by resorption in the sagitta of *Tilapia* (Plate IIIc). A less intense disturbance (FS) in the same animal was caused by the transport of the fish from a pond to an aquarium; the stress period lasted 4 hr (Plate IIId and e). Similar disturbances were found in transplant experiments using marine fishes. The difference between *Tilapia* and marine fishes seems to be in the recovery time; marine species never seem to recover fully from the transplantation stress and to resume normal growth in aquarium conditions.

Discontinuities also appear to be related to ontogeny and habitat. They become more frequent with age and are particularly numerous in fishes that live in reef and other shallow-water environments. Sagittae of the herbivorous *Acanthurus chirurgus* (Bloch) from southwestern Puerto Rican reefs have very reularly spaced, continuous sequences of growth increments up to the first year of growth. Then growth interruptions begin to appear and become very frequent and intense during the winter months (January–March) (Plate VIIa and b). This is a time of reproduction and also of choppy seas and relatively high tides. Another diurnal herbivorous inhabitant of the same reefs, *Lachnolaimus maximus* (Walbaum), shows similar patterns and discontinuties (Plate VIId and e). Another example of the relationships between habitat and discontinuities is the difference in the frequencies of growth interruptions between shallow- and deep-water Lutjanidae, the former showing higher frequency than the latter (Weiler and Pannella, 1980).

Continuity of growth patterns characterized the first 2- to 3-year growth sequences in *Merluccius bilinearis* (Mitchill) and *Urophycis chuss* (Walbaum), in contrast to the many discontinuities found even in the early stages of *Pseudopleuronectes americanus* (Walbaum) (Pannella, 1971, 1974).

A positive correlation exists between habitats and growth breaks. Together with other growth patterns, growth breaks promise to provide important imformation on ecology, habits, and life history of fish.

6. Growth Patterns and Ecology

Comparative studies of growth patterns in sagittae from different species living in the same or in different habitats provide a way to separate features that are environmentally controlled and common to the whole community from those that are genetically or physiologically controlled.

Several growth-pattern features common to all fish of a reef community are considered to be habitat-controlled, whereas features shared

Table VI. Working Hypothesis on the Origin of Information Stored in Periodic Growth Patterns of Fish Sagittae

Pattern	Diagnostic features	Possible causative factors	Ecological–biological meaning
Subdaily	Number per day	Internal physiological rhythms; Intermittent activities	Activity and feeding cycles; Subdaily rhythms in calcium turnover
Daily (increments)	Boundary sharpness	Rest–activity cycles, circadian rhythms, magnitude of daily changes in environmental parameters	Feeding and activity rhythms and types; Daily migratory patterns
	Thickness	Food supply, feeding activity, temperature	Food availability and types; Temperature
	Rhythmic changes	Changes in environmental parameters (e.g., food, temperature, currents, salinity); Biological clocks	Environmental and biological rhythms
Tidal (bands)	Periodical changes in thickness, structures, and continuity of increments	Tidally controlled food availability; Tidal migratory rhythms; Tidally controlled changes in environmental parameters	Reproduction, migratory, and activity rhythms; Tidal types and effects on environmental parameters
Seasonal (zones, annuli)	Periodicity, length	Seasonal changes in environmental and physiological parameters; Seasonal migrations, reproduction	Length of season, latitudinal position; Season of reproduction and death; Age, growth rate, population dynamics
Discontinuities	Periodicity	Environmental and physiological stress; Starvation, reproduction, storms, predation, sickness	External and internal stresses, their periodicity, type, and duration

by only one taxonomic group, say the Lutjanidae, living in different habitats, are considered to be genetically or physiologically controlled. Some features such as the presence of discrete increments arranged in periodic patterns, the progressive thinning of individual increments throughout ontogeny, and the increase in the number of disturbances with age appear to be present in sagittae of all fish and point to universal mechanisms controlling the formation of sagitta.

Relationships between fish ecology and growth patterns do exist, but they are now recognized only on an intuitive basis. The ecology and life history of most fishes are still unknown. These data could be reconstructed from growth patterns, but before this is feasible, studies relating known life histories to growth patterns are necessary (Hobson, 1975). At this early stage in the study of sagittae, one can offer only preliminary considerations and generalizations on the relationships between growth patterns and habitat, ecology, and life habits. Table VI presents a tentative synopsis of the potential ecobiological information to be found in growth patterns.

6.1. Daily Increments

The distinctiveness of increments seems to be a function of fish rhythmic activity, feeding habits, latitude, and bathymetry. Sharply defined increments are found in fish with known diurnal or nocturnal behavior (Plates IId–f and IVb–f). Continuously feeding diurnal herbivores, such as surgeon fishes, have distinct regularly spaced daily increments with rather thick organic layers showing a diurnal peak in matrix calcification (Plate IVc and d). Shallow-water predators show irregularities in spacing, more abrupt changes in increment thickness, and sharp interruptions. Diurnal predators of benthos appear to have more distinct increments than nocturnal ones.

Reef fishes generally have continuous and long sequences of uniformly spaced increments, in contrast to littoral and demersal fishes, which show discontinuities and irregularly spaced sequences. Deep-water nonmigratory fish have poorly defined increments. Tropical fishes have increments thicker than those found in temperate and cold-water sagittae.

6.2. Tidal Patterns

Tidal patterns, manifested by changes in incremental thickness, are present in those fishes that synchronize their feeding activity with tidal fluctuations. These fishes include not only littoral and reef inhabitants but also demersal fishes, which prey on organisms that migrate with the tides. Interruptions in growth sequences recurring with tidal periodicities are present in forms such as Blennidae living in tidal pools. Deep-water and demersal forms, in general, do not show obvious tidal patterns.

6.3. Seasonal Patterns

Seasonal growth patterns are typically present in fishes that live where marked seasonal changes in environmental parameters occur. The distinctiveness of the patterns is proportional to the magnitude of those changes. Studies relating the distinctiveness of seasonal patterns to latitude have not been attempted. It is obvious from the literature, however, that aging by means of sagitta annuli becomes rather difficult at latitudes lower than 28–30°. The relationships between seasonal growth zones in sagittae and latitude are complicated by migratory behavior.

6.4. Discontinuities

Although the relationships between discontinuities and environment are still obscure, the type and frequency of growth interruptions appear to have some relationship to the organism's habitat. Littoral forms typically have more growth disturbances than deep-water and demersal fishes. Reef and shallow-water species experience and record more disturbances than pelagic forms. A fertile field of research will be the application of Fourier and time-series analysis to cyclic growth patterns and the establishment of a quantitative basis for ecological interpretation (see Appendices 3.A and 3.B).

7. Fossil Otoliths and Paleoecology

Fossil otoliths are not uncommon in Cenozoic and late Mesozoic sediments. They are usually found in samples prepared for microfossil examination. They have been the object of several taxonomic and phylogenetic studies and have proved extremely useful in tracing the evolutionary history of certain teleost groups (for a complete bibliography, see Weiler, 1968). Because it is possible to determine the systematic position of a fish from sagittae morphology, and to infer its habitat from modern relatives, otoliths have been used in bathymetric, climatic, and environmental reconstructions (Weiler, 1968).

No attempts have been made to study growth patterns of fossil sagittae. An attempt has been made to check the preservation of patterns.

The conclusion was that patterns were not visible in Cretaceous and Miocene fossils (Pannella, 1971). Further attempts to document the preservation of patterns in fossils have led to the conclusion that because of diagenesis, they are rarely preserved. In one fossil assemblage, otoliths and bivalve shells were found together, but only the shells showed daily increments, whereas the sagittae showed only seasonal zones and what would be interpreted as tidal bands. Some Pliocene and Miocene otoliths that show better preservation than this may provide an opportunity to reconstruct environmental and paleobiological information. If nothing else, age and season-of-death determinations appear possible even in the poorly preserved Cretaceous forms.

As in the case of fossil bivalves, the study of growth patterns can be applied only to well-preserved material that, in the instance of aragonitic sagitta, appears to be even rarer than in molluscs. However, I expect that otoliths with preserved daily patterns will be found and that their usefulness as paleoecological indicators will be thus enhanced.

ACKNOWLEDGMENTS. The investigation that led to most of the ideas and conclusions expressed in this work was supported by National Science Foundation Grant No. OCE 77-09167. The author wishes to express his appreciation to Dr. D. S. Erdman of the Commercial Fishery Laboratory in Mayaguez, Puerto Rico, for providing ideas and materials, to Mrs. D. W. Arneson for the help in the preparation and study of the material, to Dr. P. Colin and Ms. I. Clavijo for providing many otoliths, and to O. Miranda-Brandt for his experimental work and preparation of the material.

References

Adams, L. A., 1940, Some characteristic otoliths of American Ostariophysi, *J. Morphol.* **66**:497–519.

Bagenal, T. B. (ed.), 1974, *The Ageing of Fish*, Old Working, Surrey, England, 234 pp.

Bassoli, G., 1906, Otoliti fossili terziari dell'Emilia, *Riv. Ital. Paleontol.* **12**:36–58.

Bassoli, G., 1909, Otoliti di pesci, *Att. Soc. Nat. Matem. Modena* **12**(4):39–44.

Bauza-Rullan, J., 1958, Otolitos de peces actuales, *Bol. R. Soc. Esp. Hist. Nat.* **56**:111–126.

Beaumariage, D. S., 1973, Age, growth and reproduction of king mackerel, *Scomberomorus cavalla*, in Florida, *Fla. Dept. Nat. Resour. Mar. Res. Rep.*, No. 1, 45 pp.

Berg, A., 1968, Studies on the metabolism of calcium and strontium in freshwater fish. I. Relative contribution of direct and intestinal absorption, *Mem. Ist. Ital. Idrobiol.* **23**:161–96.

Bhatia, D., 1932, Factors involved in the production of annual zones on the scale of the rainbow trout (*Salmo irideus*). II, *J. Exp. Biol.* **9**:6–14.

Bilton, H. T., 1974, Effect of starvation and feeding on circulus formation on scales of young sockeye salmon, in: *The Ageing of Fish* (T. B. Bagenal, ed.), pp. 40–70, Old Working, Surrey, England.

Bilton, H. T., and Robins, G. L., 1971, Effects of feeding level on circulus formation on scales of young sockeye salmon (*Oncorhynchus nerka*), *J. Fish. Res. Board Can.* **28**:861–868.

Blacker, R. W., 1969, Chemical composition of the zones in cod (*Gadus morhua* L.) otoliths, *J. Cons. Cons. Int. Explor. Mer* **33**(1):107–108.

Blacker, R. W., 1974, Recent advances in otolith studies, in: *Sea Fisheries Research* (F. R. H. Jones, ed.), pp. 67–90, Wiley New York.

Blaxter, J. H. S., 1974, *The Early Life History of Fish*, Springer-Verlag, New York, 765 pp.

Briggs, J. C., and Caldwell, D. K., 1957, *Acanthurus randalli*, a new surgeon fish from the Gulf of Mexico, *Bull. Fla. State Mus. Biol. Ser.* **2**(4):43–51.

Brothers, E. B., Mathews, C. F., and Lasker, R., 1976, Daily growth increments in otoliths from larval and adult fishes, *Mar. Fish. Serv. Fish. Bull.* **74**(1):1–8.

Bruger, G. E., 1974, Age, growth, food habits and reproduction of bonefish, *Albula Vulpes*, in South Florida waters, *Fla. Dept. Nat. Resour. Mar. Rep.*, No. 3, 20 pp.

Carlstrom, D., 1963, A crystallographic study of vertebrate otoliths, *Biol. Bull. (Woods Hole, Mass.)*, **125**:441–463.

Chaine, J., 1956, Recherches sur les otolithes des poissons: Étude descriptive et comparative de la sagitta des Téléosteens, *Bull. Cent. Int. Rech. Sci.* **1**:159–275.

Chevy, P., 1933, The method of reading scales and the fish of the intertropical zone, *Proc. 5th Pacif. Sci. Congr.* **5**:3817–3829.

Chugunova, N. I., 1959, *Age and Growth Studies in Fish*, Academy of Sciences of the U.S.S.R., Department of Biological Sciences, Board of Ichthyology, A. N. Severtsov Institute of Animal Morphology; English translation, National Science Foundation, Washington and Jerusalem, 1963, 132 pp.

De Bont, A. F., 1967, Some aspects of age and growth of fish in temperate and tropical waters, in: *The Biological Basis of Freshwater Fish Production* (S. D. Perking, ed.), pp. 67–88, Blackwell, Oxford.

Degens, E. T., Spencer, D. W., and Parker, R. H., 1967, Paleobiochemistry of molluscan shell proteins, *Comp. Biochem. Physiol.* **20**:533–579.

Degens, E. T., Deuser, W. G., and Haedrich, R. L., 1969, Molecular structure and composition of fish otoliths, *Int. J. Life Oceans Coastal Waters* **2**(2):105–113.

De Sylva, D. P., 1963, Systematics and life history of the great barracuda, *Sphyraena barracuda* (Walbaum), *Stud. Trop. Oceanogr.* **1**:1–179.

Dolman, J., 1974, An investigation of growth in bivalves, Ph.D. thesis, University of Newcastle Upon Tyne, 740 pp.

Dunkelberger, D. G., Dean, J. M., and Watabe, N., 1980, The ultrastructure of otolithic membrane and otolith in the Teleost, *Fundulus heteroclitus*, *J. Morphol.* (in press).

Einarsson, H., 1951, Racial analysis of Icelandic herrings by means of otoliths, *Rapp. P.-V. Reun. Cons. Int. Explor. Mer* **128**(1):55–74.

Fagade, S. O., 1974, Age determination in *Tilapia melanotheron* (Ruppel) in the Lagos lagoon, Lagos, Nigeria, in: *The Ageing of Fish* (T. B. Bagenal, ed.), pp. 71–77, Old Working, Surrey, England.

Farmer, V. C. (ed.), 1974, The infrared spectra of minerals, *Mineral. Soc. Monogr.* No. 4, 539 pp.

Fleming, W. R., 1967, Calcium metabolism in teleosts, *Am. Zool* **7**:835–842.

Frizzell, D., 1965, Otolith-based genera and lineages of fossil bonefishes (Clupeiformes, Albulidae), *Senckenbergiana Lethapa* **46a**:85–110.

Frizzell, D., and Dante, J., 1965, Otoliths of some early Cenozoic fishes of the Gulf Coast, *J. Paleontol.* **39**:687–718.

Frizzell, D., and Lamber, C. K., 1961, New genera and species of myripristid fishes, in the Gulf Coast Cenozoic known from otoliths (Pisces Beryciformes), *Bull. Univ. Missouri School Mineral. Metallurg. Tech. Ser.* **100**:1–25.

Gibson, R. N., 1967, Experiments on the tidal rhythm of *Blennius pholis, J. Mar. Biol. Assoc. U. K.* **47**:97–111.

Hickling, C. F., 1931, The structure of the otolith of the hake, *Q. J. Microsc. Sci.* **74**:547–563.

Hobson, E. S., 1972, Activity of Hawaian reef fishes during the evening and morning transitions between daylight and darkness, *U.S. Fish. Wildl. Serv. Fish. Bull.* **70**:715–740.

Hobson, E. S., 1975, Feeding patterns among tropical reef fishes, *Am. Sci.* **63**:382–392.

Irie, T., 1955, The crystal texture of the otolith in a marine teleost, *Pseudosciaena, J. Fac. Fish. Anim. Husb. Hiroshima Univ.* **1**:1–8.

Irie, T., 1960, The growth of the fish otolith, *J. Fac. Fish. Anim. Husb. Hiroshima Univ.* **3**:203–221.

Johnson, D. W., 1973, Endocrine control of hydromineral balance of teleosts, *Am. Zool.* **13**:799–818.

Jones, R. S., 1968, Ecological relationships in Hawaian and Johnston Island Acanthuridae (surgeonfishes), *Micronesica* **4**:309–361.

Krogh, A., 1939, *Osmotic Regulation in Aquatic Animals*, Cambridge University Press, Cambridge, 223 pp.

Liew, P. K. L., 1974, Age determination of American eel based on the structure of their otoliths, in: *The Ageing of Fish* (T. B. Bagenal, ed.), pp. 124–136, Old Working, Surrey, England.

Lutz, R. A., and Rhoads, D. C., 1977, Anaerobiosis and a theory of growth line formation, *Science* **198**:1222–1227.

Lyon, M. F., 1955, The development of the otoliths of the mouse, *J. Embryol. Exp. Morphol.* **3**:213–229.

McErlean, A. J., 1963, A study of the age and growth of the gag, *Mycteroperca microlepis* Goode and Bean (Pisces: Serranidae), on the west coast of Florida, *Fla. State Board Conserv. Tech. Ser.*, No. 21, 29 pp.

Menon, M. D., 1953, The determination of age and growth of fishes in tropical and subtropical waters, *Bombay Nat. Hist. Soc.* **51**:623–635.

Mina, M. V., 1968, A note on a problem in the visual qualitative evaluation of otolith zones, *J. Cons. Cons. Int. Explor. Mer* **32**(1):93–97.

Miranda-Brandt, O., 1978, Patrones de lineas de crescimiento en organismos tropicales, Ph.D. thesis, Department of Marine Sciences, University of Puerto Rico, Mayaguez, 214 pp.

Moe, M. A., 1969, Biology of the red grouper, *Epinephelus morio* (Valenciennes), from the Eastern Gulf of Mexico, *Fla. Dept. Nat. Resour. Mar. Res. Lab. Prof. Pap. Ser.*, No. 10, 95 pp.

Morris, K. W., and Kittleman, L. R., 1967, Piezoelectric property of otoliths, *Science* **158**:368–370.

Moss, M. L., 1961a, Studies of acellular fish bone. I. Morphological and systematic variations, *Acta Anat.* **46**:343–426.

Moss, M. L., 1961b, Osteogenesis of acellular fish, *Am. J. Anat.* **108**:99–110.

Moss, M. L., 1963, The biology of acellular teleost bone, *Ann. N. Y. Acad. Sci.* **109**:337–350.

Mugiya, Y., 1964, Calcification in fish and shell-fish. III. Seasonal occurrence of a prealbumin fraction in the otolith fluid of some fish corresponding to the period of opaque zone formation in the otolith, *Bull. Jpn. Soc. Sci. Fish.* **30**:955–967.

Mugiya, Y., 1966a, Calcification in fish and shell-fish. V. A study on paper electrophoretic patterns of the acid mucopolysaccharides and PAS-positive materials in the otolith fluid of some fish, *Bull. Jpn. Soc. Sci. Fish.* **32**(2):117–123.

Mugiya, Y., 1966b, Calcification in fish and shell-fish. VI. Seasonal changes in calcium and magnesium concentration of the otolith fluid in some fish, with special reference to the zone of formation of their otolith, Bull. Jpn. Soc. Sci. Fish. **32**(7):549–557.

Mugiya, Y., 1974, Calcium-45 behavior at the level of the otolith organs of rainbow trout, Bull. Jpn. Soc. Sci. Fish. **40**(5):457–463.

Norris, W. P., Chavin, W., and Lombard, L. S., 1963, Studies of calcification in a marine teleost, Ann. N. Y. Acad. Sci. **109**:312–336.

Palmer, J. D.,1973, Tidal rhythms: The clock control of the rhythmic physiology of marine organisms, Biol. Rev. **48**:377–418.

Palmer, J. D., 1974, Biological Clocks in Marine Organisms, Wiley, New York, 173 pp.

Pang, P. K. T., 1973, Endocrine control of calcium metabolism in Teleosts, Am. Zool. **13**:775–792.

Pannella, G., 1971, Fish otoliths: Daily growth layers and periodical patterns, Science **173**:1124–1126.

Pannella, G., 1974, Otolith growth patterns: An aid in age determination in temperate and tropical fishes, in: The Ageing of Fish (T. B. Bagenal, ed.), pp. 28–39, Old Working, Surrey, England.

Pannella, G., and Weiler, D. W., 1980, Resorption and discontinuities in growth sequences of fish sagittae, Science (in press).

Pannella, G., and MacClintock, C., 1968, Biological and environmental rhythms reflected in molluscan shell growth, J. Paleontol. Mem. **42**:64–80.

Parrish, B. B., and Sharman, D. P., 1958, Some remarks on methods used in herring racial investigation with special reference to otolith studies, Rapp. P.-V. Reun. Cons. Int. Explor. Mer **143**:67–80.

Parry, G., 1966, Osmotic adaptation in fishes, Biol. Rev. **41**:392–444.

Piez, K. A., 1961, Amino acid composition of some calcified proteins, Science **134**:841–842.

Postuma K. H., 1974, The nucleus of the herring otolith as a racial character, J. Cons. Cons. Int. Explor. Mer **35**(2):121–129.

Ralston, S. V., 1975, Aspects of the age and growth reproduction and diet of millet-seed butterflyfish Chaetodon miliaris (Pisces: Chaetodontidae), a Hawaian endemic, M.S. Thesis, University of Hawaii, Honolulu.

Ralston, S. V., 1976, Age determination of a tropical reef butterflyfish utilizing daily growth rings of otoliths, Fish. Bull. **74**:990–994.

Randall, J. E., 1961, A contribution to the biology of the convict surgeon-fish of the Hawaiian Islands, Acanthurus triostegua sandvicensis, Pac. Sci. **15**:215–272.

Randall, J. E., and Brock, V. E., 1960, Observations on the ecology of epinepheline and lutjanid fishes of the Society Islands, with emphasis on food habits, Trans. Am. Fish. Soc. **89**:9–16.

Rao, G. M. M., 1969, Effect of activity, salinity and temperature on plasma concentration of rainbow trout, Can. J. Zool. **47**:131–134.

Reibisch, J., 1899, Über die Eizahl bei Pleuronectes platessa und die Alterbestimmung dieser Form aus den Otolithen, Wissensch. Meeresuntersuch. Ab. Kiel (N.F.) **4**:231–248.

Reshetnikov, Yu. S., and Claro, R. M., 1976a, Cycles of biological processes in tropical fishes with reference to Lutjanus synagris, J. Ichthyol. **16**(5):711–723.

Reshetnikov, Yu. S., and Claro, R. M., 1976b, On time of formation of the annulus in Lutjanidae, Gidrobiol. Zh. **12**(3):30–35.

Rosenberg, G. D., 1974, A comment on terminology: The increment and the series, in: Growth Rhythms and the History of the Earth's Rotation (G. D. Rosenberg and S. K. Runcorn, eds.), pp. 1–8, Wiley, New York.

Schlieper, C., 1930, Die Osmoregulation wasserlebender Tiere, Biol. Rev. **5**:309–356.

Shaw, J., 1960, The mechanisms of osmoregulation, in: *Comparative Biochemistry*, Vol. 2 (M. Florkin and H. S. Mason, eds.), pp. 471–513, Academic Press, New York.

Simkiss, K., 1974, Calcium metabolism of fish in relation to ageing, in: *The Ageing of Fish* (T. B. Bagenal, ed.), pp. 1–12, Old Working, Surrey, England.

Simmons, D. J., 1971, Calcium and skeletal tissue physiology in teleost fishes, *Clin. Orthop. Relat. Res.* **76**:244–280.

Starck, W. A., and Schroeder, R. E., 1970, Investigations of the gray snapper *Lutjanus griseus*, *Stud. Trop. Oceanogr.* No. 10, 224 pp.

Struhsaker, P., and Uchiyama, J. H., 1976, Age and growth of the nehu, *Stolephorus purpureus* (Pisces: Engraulidae), from the Hawaiian Islands as indicated by daily growth increments of sagittae, *U.S. Fish. Wildl. Serv. Fish. Bull.* **74**(1):9–17.

Swift, D. R., and Pickford, G. E., 1965, Seasonal variations in the hormone content of the pituitary gland of the perch (*Perca fluviatilis*), *Gen. Comp. Endocrinol.* **5**:354–365.

Taubert, B. D., 1975, Daily growth rings in the otoliths of *Lepomis* sp. and *Tilapia mossambica* (Peters), 37th Midwest Fish & Wildlife Conference, Abstracts.

Taubert, B. D., and Coble, D. W., 1977, Daily rings in otoliths of three species of *Lepomis* and *Tilapia mossambica*, *J. Fish. Res. Board Can.* **34**(3):332–340.

Urist, M. R., and Schjeide, O. A., 1961, The partition of calcium and protein in the blood of oviparous vertebrates during estrus, *J. Gen. Physiol.* **44**:743–756.

Weiler, W., 1968, *Fossilium Catalogus I: Animalia, Otolithi piscium* **117**, 196 pp.

Weiler, D. W., and Pannella, G., 1980, Biology and age of some Lutjanidae from otolith readings (in prep.).

Werner, C. F., 1928, Studien über die Otolithen der Knochenfische, *Z. Wiss. Zool.* **131**:502–587.

Williams, T., and Bedford, B. C., 1974, The use of otoliths for age determination, in: *The Ageing of Fish* (T. B. Bagenal, ed.), pp. 114–123, Old Working, Surrey, England.

Woodhead, P. M. J., 1968a, Seasonal changes in the calcium content of the blood of Arctic cod, *J. Mar. Biol. Assoc. U. K.* **48**:81–91.

Woodhead, P. M. J., 1968b, On levels of calcium and of vitamin A aldehyde in the blood of Arctic cod, *J. Mar. Biol. Assoc. U. K.* **48**:93–96.

Chapter 16

Growth Lines in Polychaete Jaws (Teeth)

P. J. W. OLIVE

1. Introduction 561
2. Evidence of Growth Lines in Recent Polychaete Jaws 563
   2.1. Phyllodocemorpha 563
   2.2. Eunicemorpha 571
3. Chemical Composition and Preservation Potential of Polychaete Jaws 572
4. Ecological and Paleoecological Applications 581
   4.1. Analysis of Polychaete Population Age Structure 581
   4.2. Ecological and Paleoecological Interpretations 586
5. Conclusions 589
   References 590

1. Introduction

The discovery of annual growth lines in the jaws of *Nephtys hombergii* (Kirkegaard, 1970) was one of very great importance, not only to polychaete workers but also to benthic biologists, for whom the estimation of the population age structure and productivity of large, long-lived polychaetes has been a very difficult problem. The discovery also raises the possibility that growth records might be preserved in the numerous fossilized Paleozoic and Mesozoic polychaete jaws known as *scolecodonts*. In view of the considerable biological and ecological importance that growth lines could have in polychaete jaws, it is necessary to establish the validity and general applicability of this method for determining the age and growth rates of the Polychaeta.

Polychaete jaws are proboscoidal armaments that are widespread in the errant polychaete families; they are unknown in sedentary families except in the aberrant ampharetid *Gnathampharete paradoxa* (Des-

P. J. W. OLIVE ● Department of Zoology, University of Newcastle Upon Tyne, Newcastle, England NE1 7RU.

bruyères, 1977). Two fundamentally different types of proboscis are found in the Polychaeta (Dales, 1962). There has been some controversy regarding the taxonomic value of this dichotomy (Dales, 1962, 1977; Orrhage, 1964, 1973), but this feature does permit the separation of two major groupings of errant families: the Phyllodocemorpha and the Eunicemorpha, which have quite different types of jaws.

In the Phyllodocemorpha, the proboscis is axial. The buccal tube covers the proboscis externally when it is everted (Michel and DeVillez, 1978). The length of this sheath determines the length of the everted proboscis. In several families of phyllodocemorph polychaetes, a small number of relatively large (up to about 4 mm) jaws (*Kiefer*, German; *mâchoires*, French) are inserted into the epidermis at, or shortly behind, the junction of the pharyngeal or proboscoidal sheath and the pharynx. Thus, when the proboscis is everted, the jaws are deployed at the opening that forms the functional mouth.

In the Nereidae and Nephtyidae, there is a single pair of jaws, and in the Nereidae, the proboscoidal sheath is also armed with smaller teeth (<1 mm) that are referred to as "paragnaths" (Fig. 1). In the Polynoidae

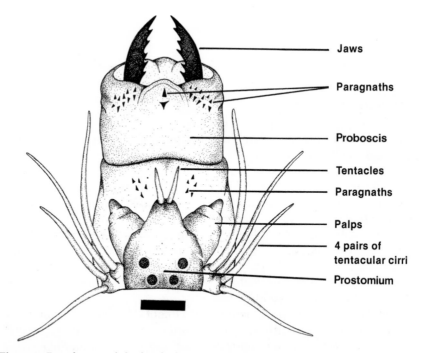

Figure 1. Dorsal view of the head of a generalized polychaete, *Nereis pelagica*, with the proboscis everted to show the jaws and some of the groups of paragnaths. This is an example of a phyllodocemorph jaw structure. Scale bar: 1 mm.

and Sigalionidae, there are four jaws arranged in two pairs, of which the dorsal and ventral pairs are hinged together. The jaws of the Polyodontidae resemble those of the Polynoidae, though they are more complex. In the Glyceridae, there are four diagonally arranged jaws, each of which is a hollow-pointed fang. In the Goniadidae, there are smaller and larger teeth at the tip of the everted proboscis and often smaller armaments arranged in a chevron pattern on the sides of the proboscoidal sheath. The Syllidae and Hesionidae may have small, often single jaws inserted into the wall of the proboscis. The morphology and arrangement of jaws in these and other families have been reviewed by Wolf (1976). The proboscis of the Phyllodocidae and some other families is unarmed.

The jaws of eunicemorph polychaetes (e.g., Eunicidae, Lumbrinereidae, Onuphidae, and Dorvilleidae) are quite different from those of the phyllodocemorph polychaetes. The buccal tube leads dorsally into the esophagus and ventrally into the saclike muscular evagination that forms the eunicid ventral proboscis (Desière, 1967; Hartmann-Schroeder, 1967). The proboscis is armed with a complex array of jaws arranged to form a single jaw apparatus. There is considerable variation in the arrangement of the individual elements in the jaw apparatus (for a review, see Kielan-Jaworowska, 1966), which have been named by analogy with the insect and crustacean mouth parts, though they are in no way homologous.

In Recent eunicids other than the Dorvilleidae, the jaw apparatus consists of a pair of elongated ventral jaws referred to as "mandibles," a series of dorsal jaws referred to as "maxillae," and a carrier piece usually paired but fused in the midline. There may be as many as six pairs of maxillae, though the arrangement is often asymmetrical. The maxillae are numbered from back to front as MI–MVI. The maxillae MI are curved hooks— the "forceps"—and the second pair, MII, which articulate with the ventral surfaces of the first, are a pair of powerful crushing jaws with numerous denticles in the opposing surfaces. Desière (1967), Hartmann-Schroeder (1967), and Kielan-Jaworowska (1966) have described in detail the jaws of recent eunicids; a typical eunicid jaw apparatus is illustrated in Fig. 2.

2. Evidence of Growth Lines in Recent Polychaete Jaws

2.1. Phyllodocemorpha

2.1.1. Nephtyidae

Kirkegaard (1970) first drew attention to apparent growth lines in the jaws of *Nephtys hombergii*, which he interpreted as annual growth lines.

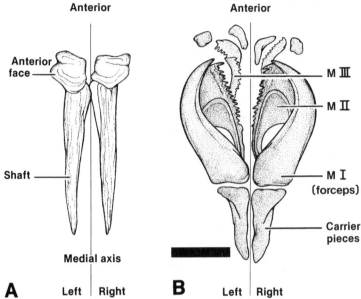

Figure 2. (A) Dorsal view of the ventral jaws (mandibles) of *Marphysa bellii*. The posteriorly directed shaft serves for muscle attachments; the anterior face may serve for chiseling hard substrates. (B) Dorsal view of the dorsal jaw apparatus (maxillae) of *Marphysa bellii*, articulated in the retracted position. Disassociated jaws of this species are illustrated in Plate V. (MI) First maxillae (forceps); (MII) second maxillae; (MIII) third maxillae. The more anterior jaws are not strictly paired. For a review of alternative forms of terminology of eunicid jaw apparatus, see Desière (1967). This is an example of a eunicemorph jaw structure. Scale bar: 1 mm.

Subsequent studies (Estcourt, 1975; Retière, 1976; Olive, 1977) have firmed that the growth lines can indeed be used to age several species of Nephtyidae. The original descriptions based on light microscopy have been supplemented by scanning electron microscopy (SEM) (Retière, 1976; Olive, 1977).

The jaws of Nephtyidae are hollow pyramids inserted at their base into the junction of the pharyngeal sheath and the pharynx. Several species have now been examined in our laboratory using both light and electron microscopy. In every case, growth lines occur on the jaws, but the recognition and resolution of annual growth lines requires investigations of both the jaw structure and the population age structure and growth, which has been attempted only for *N. hombergii* (Retière, 1976; Olive, 1977), *N. caeca* (Olive, 1977), and *Aglaophamus verillii* (Estcourt, 1975). The ecological implications of the results of these investigations will be discussed below.

The shape of the jaws in the Nephtyidae varies among species and is roughly species-specific. In some species, e.g., N. hombergii and N. longosetosa, the jaw is an irregularly shaped pyramid in which the anterior face is the largest. Growth lines are most easily seen on the larger anterior face and so can be detected in jaws with this shape more easily than in those of N. caeca, N. caecoides, N. punctata, N. bucera, and N. ciliata, for instance, which have regular pyramid-shaped jaws.

In all nephtyid species, growth lines can be seen in transmitted or incident light with a light microscope (magnification $\times 10$). Plate Ia shows a light micrograph of a young specimen of N. hombergii shortly after the appearance of the first growth line, while Plate Ib shows a jaw of an older specimen of N. caeca with five growth lines. Note that the growth lines can be seen through the anterior lamella of the jaw here viewed from the inner (posterior) surface. SEM reveals that the inner (posterior) surface of the jaw is quite smooth (see Plate Ig) or, in the case of N. caeca, has a honeycomb appearance (Olive, 1977). The jaws are secreted by the glandular cells of the pharyngeal epithelium (Michel, 1971). It appears that newly formed tooth material is deposited inside and projecting beyond the older, more distal pyramids. This can be seen more clearly in scanning electron micrographs. In some specimens, such as the jaw of N. hombergii illustrated in Plate Ic, there are a number of very prominent growth lines that give the jaw the appearance of a stack of superimposed pyramids, the smallest and oldest being at the apex. There is little difficulty in determining the number of rings in such a jaw. In other specimens, however, there are minor growth lines between major year rings, making the temporal interpretation of these increments more difficult to assess (Olive, 1977). An electron micrograph of such an N. hombergii jaw is shown in Plate If. Similar variations in the prominence of what are thought to be annual growth lines have been encountered in the jaws of all the species we have examined. Usually, the annual growth lines can be resolved more easily when the teeth are viewed with a simple dissecting microscope (magnification $\times 10$–20), which will not usually resolve minor growth lines with a periodicity less than 1 year. The jaw of N. caeca shown in Plate Id was originally identified as having two major growth lines with a light microscope. These two growth lines can still be observed with the electron microscope (see arrows in Plate Id), but many less prominent growth lines can also be seen between them.

In an extreme case, such as the jaw of a large and presumably old specimen of N. bucera (Plate Ie), annual lines cannot readily be distinguished from the numerous growth lines present. In the original light-microscopic examination, this jaw was regarded as having six lines. This resolution problem is usually more difficult in jaws from the largest and presumably oldest specimens of long-lived species such as *N. caeca, N. bucera,* and *N. ciliata.* The significance of variation in jaw morphology and growth patterns is not yet understood. In some cases, the growth of the jaws appears to have been more or less continuous between annual interruptions of growth (Fig. 3A), but in other cases, growth in the intervening period is also discontinuous as in Plate If. There are undoubtedly differences among species in this regard, but also marked differences in the jaws of different specimens of the same species even when they have been collected at the same time in the same vicinity. Figure 3 presents a diagrammatic representation of how the jaws of *Nephtys* could grow with single annual growth lines, multiple annual growth lines, or multiple growth lines that cannot be readily resolved as annual lines. The jaw of *N. hombergii* shown in Plate Ig appears to have been displaced by injury at some stage from the buccal epithelium, resulting in the formation of several growth lines at the margin of the jaw. These lines, though prominent, do not represent annual growth lines, since the size of the specimen would indicate that it was between 2 and 4 years old.

The careful preparation of the jaws of nephtyids for ring analysis is important. They must be freed from the attached muscles, and the buccal membrane must be removed from their outer surface. This is especially important in *N. caeca,* in which the buccal epithelium covering the jaws is black and opaque. We routinely clean the jaws in sodium hypochlorite at room termperature for between 30 sec and 5 min according to the size of the jaw, prior to ultrasonic cleaning and viewing (see Appendix 1.D). The best illumination of the rings is obtained with the jaws mounted in glycerin and viewed with both incident and transmitted illumination, at a magnification of about ×10–20.

2.1.2. Polynoidae and Sigalionidae

The horny teeth or jaws of the scale worms in the families Polynoidae and Sigalionidae bear prominent growth lines, which may provide a skeletal record of the changing growth rates of the jaw caused directly by periodic fluctuations in environmental conditions or indirectly through the influences of the environment on the annual cycles of growth and reproduction. Growth lines in the jaws of these families have not been described previously, and their biological significance has yet to be fully substantiated by biometric studies.

Both families of scale worms have a dorsal and ventral pair of horny beaklike jaws that project through the wall of the pharynx immediately behind the papillae at the anterior tip of the everted proboscis (Hoop, 1941) (Plate IIa and b). The jaws of *Harmothoe imbricata, Halosydna brevisetosa, Lepidonotus clava, Eunoe oerstedii* (Polynoidae), and *Sthenelais*

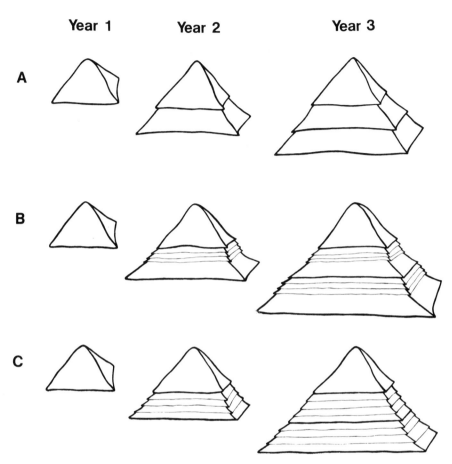

Year 1   **Year 2**   **Year 3**

A

B

C

Figure 3. Diagrammatic representation of the growth of Nephtys jaws during three successive growth periods. (A) Growth in each phase is continuous, with the result that a jaw develops in which "annual" growth lines are particularly prominent. (B) Growth in successive growth phases is at first discontinuous or interrupted, but an extended period of continuous growth eventually ensues. Such jaws are more difficult to read; each annual growth line appears as a multiple. (C) Growth in successive growth phases is discontinuous, and a series of growth lines are interpolated between annual growth lines. Aging of polychaete jaws with this kind of growth pattern is difficult unless the "annual" growth lines are more distinctly developed than the subordinate lines (see Plate If). This is not always so in the Nephtyidae.

boa (Sigalionidae) have been examined with reflected and transmitted light and with SEM. In each method of observation, two categories of growth lines can be recognized. The first description of these is given below. Opening the buccal cavity (by dissection) reveals the triangular buccal plate of the jaws (Plate IIc). Note that the anterior part of the jaw appears darker than the posterior part, separated by a prominent line where the jaw is inserted into the epithelium (Plate IIc, single arrow). This feature is not a growth line, but the line marked with the double arrow in Plate IIc was revealed in subsequent examination to be a prominent growth line (see Plate IIIf).

The two dorsal and the two ventral jaws are hinged together as shown in Plate IId. The general features of Polynoidae (*Lepidonotus clava*) jaws are shown in Plate IIf and Ig. Each jaw consists of the triangular buccal plate and fang, the hinge, the labial plate, and the central shaft or root (Hoop, 1941), which is fused with the buccal and labial plates at its anterior end. The inner surface of the jaws is usually smooth, but there may be an area for muscle attachment as in *Halosydna brevisetosa* (Plate IIe). In this species, a row of small holes can be seen on the inner surface of the fang. The inner surface of the jaws does not show growth lines (see Plate IId); the shaft appears smooth with faintly developed longitudinal striations, though toward the posterior end of the specimen illustrated in Plate IId, there are two structural discontinuities that may mark successive growth positions

Examination of the jaws in reflected and transmitted light and with SEM reveals two distinct types of growth-record patterns in the jaws.

Plates IIIa and b show the shaft of a jaw of *Harmothoe imbricata*. There are numerous parabolic growth lines on this structure, some of which are more pronounced than others. Similar lines can be seen very easily in the jaws of all the Polynoidae we have examined. In the shaft illustrated, there are several hundred growth lines along its length. They are sufficiently numerous to be consistent with their interpretation as daily growth lines, but this hypothesis has not yet been rigorously tested. Some of the lines are particularly prominent and may mark major growth discontinuities. The shaft of the jaw is secreted during its growth, and each line seems to represent a previous edge of the jaw. These growth lines are not readily seen with SEM; compare, for example, Plate IIIa and b with Plates IIIe and IIf and g. The SEM shows the surface of the shaft to be smooth or nearly so; presumably the growth lines observed with a light microscope are visible because of internal refraction and adsorption within the matrix of the jaw.

The growth lines in the shaft are similar to those in the jaws of the Nereidae (Plate IVc and d), which also mark successive positions of the edge of the jaw during growth, and the periodicity of which is obviously very much greater than annual. Further investigations of the short-term growth lines are urgently required. One approach that may be possible would be the labeling of a growth increment with radioactive amino acids followed by autoradiography of the jaws after a known time interval. Such studies, though feasible, have not yet been attempted.

The jaws of the Sigalionidae are in many ways similar to those of the

Polynoidae, to which they are closely allied, but the shaft is tubular (Plates IVa and b) and does not reveal growth lines as readily as that of the Polynoidae.

Growth lines occur on the buccal and labial plates of the jaws of both the Polynoidae and the Sigalionidae. They can be resolved easily with light microscopy and even more clearly with SEM. The number of lines is such that they are likely to be annual growth lines, but the necessary confirmation of this has yet to be obtained. Some Polynoidae live for only 2 years [e.g., *Harmothoe imbricata* (Daly, 1972) and *H. sarsi* (Sarvala, 1971)], but larger species such as *Eunoe oerstedii* probably live longer. I have examined the jaws from a number of specimens of *E. oerstedii*, but have not had access to a population of this species. Plate IIIc shows a light micrograph of the posterior margin of the buccal plate of a jaw that has three prominent growth lines. SEM supports this observation (Plate IIId). The complete jaw illustrated in Plate IIIe from the posterior–buccal aspect has two prominent growth lines; a third one is also discernible very close to the edge of the jaw, suggesting that it has recently entered the 4th growth phase.

The distance from the tip of the fang to the first prominent growth line on the jaw suggests either that this represents more than 1 year's growth or that the growth rate of the jaw declines sharply after the first growth line has been produced. A similar situation was described in *N. caeca* and *N. hombergii* in the Tyne estuary (Olive, 1977), and it was clearly established that there were two year groups in these populations with a ring count of zero. For animals with annual growth lines, therefore, the number of rings plus 1 represents the age of the animal in years. By analogy, it would not be unreasonable to suppose that these specimens of *Eunoe oerstedii* are 4 + years old.

The life history and reproductive biology of the larger Polynoidae such as *E. oerstedii* are unknown, and observations on their population structure, growth, and breeding will be necessary to test the validity of my interpretation of the structure of their jaws. Valuable evidence could be obtained from measurements of the distance of the last growth line from the edge of the jaw during a reproductive cycle. The life history of *Harmothoe imbricata* has been investigated in detail (Daly, 1972). The specimens illustrated in Plate IIa–c were 2 years old, having produced one generation of gametes. The scanning electron micrograph of one of the jaws (Plate IIIf) shows the presence of one very prominent growth line (arrow) among numerous minor ones.

The jaws of *Sthenelais boa* (Sigalionidae) also show growth records in the buccal and labial plates, which suggest an annual periodicity. Plates IVa and b illustrate jaws from a large specimen that could be interpreted as having three growth lines. If these lines are annual growth lines, the specimen from which the jaws were taken may have been 3 + or 4 + years old according to the status of the first line. We have also examined the jaws of the species *Phloe minuta*, which has a maximum length of about 10 mm and is usually much smaller (1–3 mm). The larger specimens often have a single growth line, which if interpreted as an annual growth line, correlates well with estimates of their age from population studies (Olive and Christie, unpublished observations).

2.1.3. Nereidae

The paired jaws of Nereidae are situated laterally in the anterior margins of the pharynx. The jaws are hollow, curved, forceplike structures with a series of denticles along the buccal edge. They continue to grow during adult life, and the area of deposition is at the posterior edge of the tooth around the central fossa where this edge is inserted into the secretory epithelium (Schwab, 1966). The pattern of growth is such that each of the numerous growth lines marks a previous position of the edge of the jaw (Plate IVc and d). In the future, growth lines may provide useful information about the age of individual worms, but as yet the necessary studies have not been made. At this point, it is not clear whether the number of lines is strictly age-dependent or is size-dependent. It is not possible to identify growth discontinuities with an annual periodicity in this family. For example, the jaw illustrated in Plate IVc was taken from a sexually mature epitokous specimen of *Perinereis cultrifera*. All nereid species spawn only once at or near the end of the total life span, which in *P. cultrifera* is thought to be 3 years. Plate IVd shows part of one of a large number of jaws of *Nereis diversicolor* that we have examined. An analysis of growth and mortality in a population of this species enables us to estimate the age of specimens with reasonable accuracy (Olive and Garwood, unpublished data). The specimen illustrated was approaching 2 years of age and was becoming sexually mature, yet there was no evidence of an annual growth line. Plate IVd shows clearly the numerous growth lines in the jaw, and we hope to obtain more information from this growth record. The question of the age at death of *Nereis* species has not yet been answered satisfactorily (Mettam, 1979), and in the populations we are investigating, there is evidence that some individuals breed at 2 years old, whereas others do not breed until they are 3 years old. Because *Nereis* species are easily maintained and fed under controlled conditions, it should be possible to determine the temporal and biological significance of growth lines in the jaws experimentally.

2.1.4. Glyceridae

Glycerid jaws consist of four relatively massive curved fangs that are hollow and characteristically have two pulp cavities; one serves for the hypodermic injection of a digestive venom (Michel, 1966, 1970a,b; Michel and Robin, 1972; Michel et al., 1973), while the larger one serves for muscle attachments (Michel et al., 1973). Each jaw has a characteristic wing (ailette, French), which may be attached to the right or left side, and the four jaws are arranged as a pair of right-handed and a pair of left-handed fangs. The venom duct is usually illustrated as ending at the tip of the jaw (Michel, 1966; Michel et al., 1973), but in an SEM study of the congeneric Cretaceous scolecodont Glycera glaucopsammensis, Charletta and Boyer (1974) have described a series of pores on the ventral side of the jaw that communicate with the venom duct. A similar series of pores is also present in the jaws of Glycera convoluta (Plate IVe and f), and they have been found in the jaws of all Recent Glyceridae (Wolf, 1977). It is thought that the pores are added at the anterior margin of the venom duct at the point where the venom duct enters the jaw (Plate IVf, arrow). Such a system would permit an increase in the venom-injecting capacity of the jaw during growth, which would otherwise be limited by the size of the terminal pore. The series of pores and associated growth lines provides a record of a rhythmic growth process, though this may well be endogenously controlled. The inner surface of the jaw of G. glaucopsammensis has a ridged appearance, and the curved dorsal face has a faintly developed striation that may represent growth lines. Similar lines have been described on the jaws of G. dibranchiata (Schwab, 1966) and many other species (Wolf, 1977). The morphology of the jaws and the ailette is sufficiently distinctive to be a valuable taxonomic character (Wolf, 1977). The growth lines are relatively prominent in G. convoluta (Plate IVe and f).

There are no marked growth discontinuities in the Glyceridae that might have an annual periodicity. It is of interest that the families that show such marked annuli (Nephtyidae and Polynoidae) have an iteroparous annual cycle of reproduction, whereas in the Nereidae and Glyceridae, reproduction occurs only once per lifetime (see Olive and Clark, 1978) and major growth annuli are absent.

2.2. Eunicemorpha

Because of the extraordinary predominance of "eunicid" jaws among the fossil scolecodonts (Kielan-Jaworowska, 1966), considerable importance would be attached to any evidence of growth lines in the jaws of Recent Eunicidae. Unfortunately, no growth lines seem to occur (see detailed descriptions of Kielan-Jaworowska, 1966; Desière, 1967; Hartmann-Schroeder, 1967). This may indicate that the pattern of jaw growth in eunicids is fundamentally different from that of Nephtyidae, Glyceridae, and Nereidae, in which the jaws continue to grow at the posterior margin during adult life. Detailed studies of jaw development in eunicids have not been made, but Kielan-Jaworowska (1966) has described specimens of *Halla parthenopeia* in which the jaws were incomplete and not fully differentiated; these observations, and earlier ones of Ehlers (1864–1868), led Kielan-Jaworowska (1966) to suggest that the jaws of some eunicids may be replaced during growth by a process of molting. Mierzejewski (1978) has described the occurrence of "pharate" jaws, which are assemblages of morphologically identical jaws arranged as a jaw-within-a-jaw, in which the outermost jaw, which he calls the "primary" jaw (after Schwab, 1966), is the oldest and functional one. The secondary jaws of, which there may be more than one set, are younger replacement jaws. According to Mierzejewski (1978), it is very likely that molting of jaws was a typical phenomenon in fossil Eunicida having a jaw apparatus of the placognath type. There is no proof that molting of jaws could occur in other fossil Eunicida besides the placognath forms, but there is some evidence that it occurred in some labidognath forms.

Similarly, it is known that the morphology of the jaws of *Ophryotrocha puerilis* changes during development (see Pfannenstiel, 1977), suggesting that the earlier jaws are replaced. The development of the jaws is not well known, but larval jaws of *Eunice kobiensis* are secreted by a single cell (Åkesson, 1967). If further studies confirm that the jaws of eunicids are deciduous, this could explain the apparent lack of growth lines in most fossil scolecodonts. Kielan-Jaworowska (1966) has demonstrated that there was a considerable radiation of eunicid orders in Paleozoic times and that most of these became extinct.

The large disassociated jaws of the Recent eunicid *Marphysa belli* are illustrated in Plate V. The jaws are hard and calcareous, unlike those of most phyllodocimorphs, which are composed of tanned protein. They are quite opaque to light, so that they can be viewed only by surface illumination. The surface of each jaw is smooth, though after drying it may show a scaly appearance due to the cracking of a calcareous white covering on the surface. None of the jaws shows any indications of growth lines. The specimen of *M. belli* from which the jaws illustrated in Plate V were taken was large and must have been several years old, but there is no evidence of periodic growth of the jaws.

3. Chemical Composition and Preservation Potential of Polychaete Jaws

Fossil scolecodonts do not represent the full range of Recent polychaete families; many of the Recent polychaetes that have prominent jaws are unknown as scolecodonts. Scolecodonts congeneric with Recent forms

Text continues on p. 580.

Plate I. Growth increment records in the jaws of Nephtyidae. (a) Light micrograph of the left-hand jaw of a specimen of Nephtys hombergii shortly after the formation of the first growth line (gl 1). The jaw is seen from the posterior showing the inside surface of the anterior lamella. (b) Light micrograph of the right-hand jaw of a large and old specimen of Nephtys caeca. Five major growth lines are present, and they can be seen through the anterior lamella (arrows). Perspective as in (a). The first growth line does not appear until the end of the second year after reproduction for the first time (Olive, 1977); therefore, the suggested age for this specimen is 6 + years. From Olive (1977) by permission of Cambridge University Press. (c) Scanning electron micrograph of the right-hand jaw of Nephtys hombergii showing three growth lines (arrows). The growth pattern of this jaw is of the type illustrated in Fig. 3A. Such a jaw is ideal for age determination, but not all jaws of Nephtys hombergii have this form. The jaw is viewed from the labial surface, with the large anterior face to the right. (d) Scanning electron micrograph of a jaw of Nephtys caeca showing weakly developed growth lines, which make age determination more difficult. The arrows indicate two growth lines that were visible with a light microscope. The jaw is viewed from the anterior, with the buccal edge to the right; the labial edge and spur are to the left. (e) Scanning electron micrograph of a large specimen of Nephtys bucera from an apical, anterior view; the labial spur is at the bottom left. There are numerous growth lines, of which six (1–6) were visible with the light microscope ($\times 10$). (f) Detailed view of the margin of a jaw of Nephtys hombergii. The growth lines on this jaw suggest that the growth pattern was of the form represented in Fig. 3C. Nevertheless, it is possible to distinguish major growth lines, which are thought to indicate annual growth interruptions, from the more frequent minor growth lines. The arrow indicates a major growth line. From Olive (1977) by permission of Cambridge University Press. (g) Scanning electron micrograph of an unusual jaw of Nephtys hombergii in which a large number of growth lines have been formed at the edge. This may have been caused by the partial displacement of the jaw from the epidermis, which has resulted in a repair process during which the multiple lines were formed. Note the smooth inner surface of the fossa. The jaw is viewed from the posterior [as in (a)], with the apex to the left. Plate reproduced at 48%.

Plate II. Jaw structure in Polynoidae. (a) Dorsal view of the everted proboscis of Harmothoe imbricata showing the papillae surrounding the functional mouth. (b) Anterior view of the proboscis showing the paired jaws within the circle of papillae. (c) Dissection showing the ventral pair of jaws in situ. A dark line marks the region where the jaw emerges from buccal epithelium (bel) (single arrow); a major growth line (gl) can also be seen on the buccal plate (double arrow). (d) Scanning electron micrograph of a pair of isolated jaws of Halosydna brevisetosa viewed from the labial aspect. The dorsal and ventral jaws are mirror images; a labial view is therefore a ventral view of the dorsal jaws or a dorsal view of the ventral jaws. A buccal view is a ventral view of the dorsal jaws and a dorsal view of the ventral jaws. (e) Detail from (d) showing the smooth shaft and the inner surface of the buccal plate with a prominent roughened area for muscle attachment. The function of the small row of holes (arrow) is not known. (f) Scanning electron micrograph of a buccal view of a jaw of Lepidonotus clava. The shaft appears smooth, but growth lines are visible on the buccal plate. Also see (g). (g) Scanning electron micrograph of a labial view of a jaw of Lepidonotus clava. Two main growth lines on the jaw are marked by arrows. Plate reproduced at 48%.

Plate III. Growth increment records in the jaws of Polynoidae. (a) Light micrograph with transmitted light of the shaft of a jaw of *Harmothoe imbricata* from the labial surface. Note the very numerous growth lines (gl). These are visible only faintly on the shaft surface using SEM [see (e)]. (b) Light micrograph of the same jaw viewed in reflected light from the same (labial) aspect. (c) Light micrograph of the medial posterior corner of the buccal plate of a jaw of *Eunoë oestedii* showing prominent growth lines, which are interpreted as possible annual growth rings (arrows). (d) Scanning electron micrograph of the same part of a jaw of *Eunoë oestedii* as illustrated in (c). The jaw appears to have recently entered the 4th growth phase. (e) Scanning electron micrograph of an apical/buccal view of a jaw of *Eunoë oestedii*. There are two prominent growth lines on the buccal plate and a third very close to the edge. The age of the worm could be determined accurately only if the age when the first major growth line (gl 1) appeared were known (see text for discussion), but is likely to be 4 + years. Note the faint growth lines and central longitudinal ridge on the shaft of this jaw. (f) Scanning electron micrograph of a jaw of *Harmothoe imbricata* illustrated in (a–c) from the buccal aspect. Numerous fine growth lines are present, but only one prominent line (arrow). This line was also visible with light microscopy and is indicated by a double arrow in Plate IIc. Plate reproduced at 48%.

Plate IV. Growth records in the jaws of Sigalionidae, Nereidae, and Glyceridae. (a) Scanning electron micrograph of the jaw of the sigalionid *Sthenelais boa* from the labial aspect. Note the smooth, tubular shaft of the jaw and the poorly developed labial plate. (b) Scanning electron micrograph of a buccal view of the jaw of *Sthenelais boa*. Light microscopy suggests the presence of three growth lines (arrows). (c) Light micrograph of part of the jaw of *Perinereis cultrifera* viewed from the dorsal surface. The medial cutting edge of the jaw, which has a series of denticles, appears below; the apex is to the left. The jaw lacks any major growth lines that could be thought to have an annual periodicity. A region of the jaw equivalent to that marked "a" is illustrated in (d). (d) Light micrograph, using the Zeiss Nomarski interference technique, which reveals the numerous growth lines in the jaws of Nereidae. The specimen illustrated is a fragment of a jaw of a *Nereis diversicolor* 2 years old; it was taken from a region of the jaw similar to that marked "a" in (c). (e) Scanning electron micrograph of a jaw of the glycerid *Glycera convoluta*. The venom duct is partly opened distally; the row of holes is formed as the jaw grows at the posterior edge. The jaw is viewed from the medial surface. (f) Detail showing the growing posterior edge of the jaw where the venom duct opens. A new pore will be formed at the region marked with an arrow. Numerous growth lines can be seen. Plate reproduced at 48%.

Plate V. Surface structure and absence of growth lines in the jaws of *Marphysa bellii* (Eunicidae). (a) Side view of a mandible of *Marphysa bellii*. The chisel-like anterior face is to the left. The scaly covering of the shaft has been partly removed at one point (arrow), but no surface growth lines are apparent. (b) Detail of the mandible of *Marphysa bellii* from the same aspect. Some authors have indicated that growth lines can be seen on the anterior face of the mandible. However, they are not visible on the surface as seen using SEM. (c, d) Labial (c) and buccal (d) aspects of the maxilla jaws (MI) of *Marphysa bellii*. (e, f) Dorsal (e) and ventral (f) aspects of the maxilla jaw (MII) of *Marphysa bellii*. The jaws of *Marphysa bellii* illustrate the general absence of growth records on the surface of the jaws of Eunicidae. Plate reproduced at 48%.

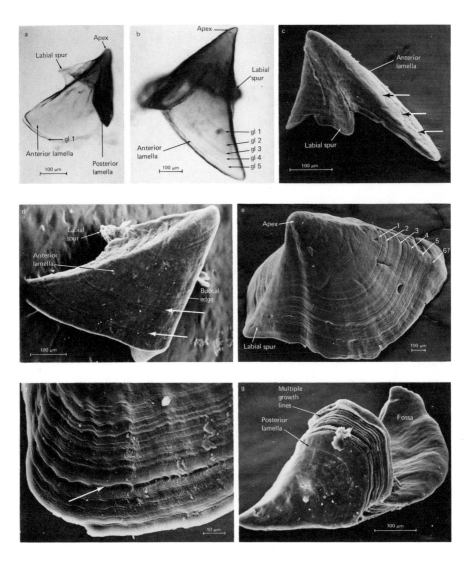

Plate I

Plate II

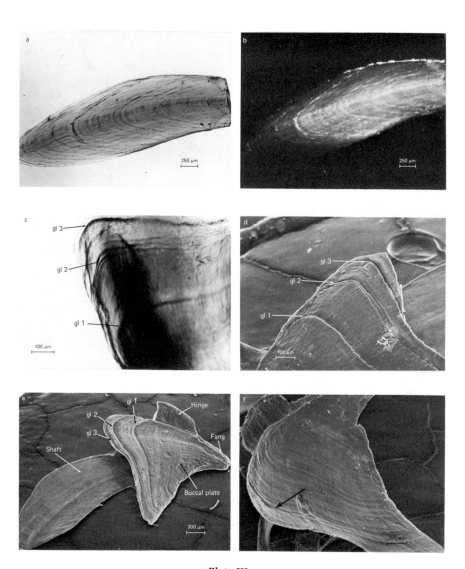

Plate III

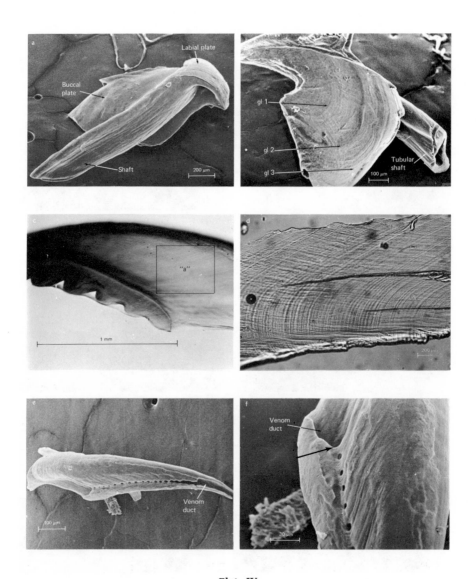

Plate IV

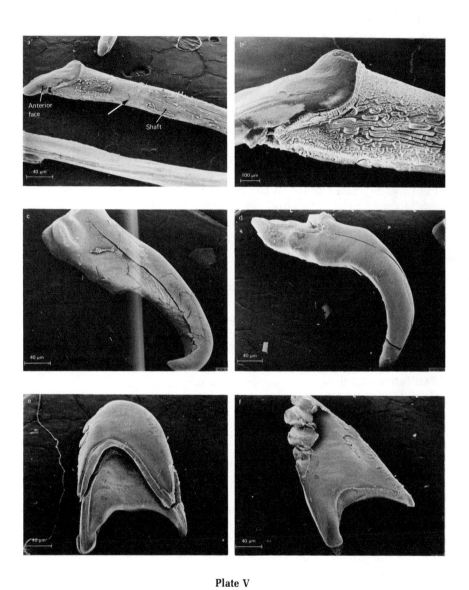

Plate V

are restricted to the Eunicidae and allies, the Glyceridae, and the Goniadidae (Szaniawski, 1974; Charletta and Boyer, 1974).

Indeed, it is most unfortunate that the jaws of the Nephtyidae and Polynoidae do not appear to have been preserved as fossils. This is almost certainly due to basic differences in chemical composition (Szaniawski, 1974).

The jaws of polychaetes do not contain chitin, despite frequent casual references to their chitinous nature in the literature (Desière and Jeuniaux, 1968; Voss-Foucart et al., 1973). The jaws of phyllodocemorph polychaetes consist of hardened protein (Michel and DeVillez, 1978). Those of Nephtys hombergii are composed of proteins hardened by quinone tanning involving the reaction of polyphenol oxidase with aromatic proteins (Michel, 1971). The jaw consists of inner and outer layers; tanning probably takes place in the outer layer. In chemical structure, the jaws of the Nereidae, Nephtyidae, and Sigalionidae are very similar; all consist of tanned protein with a high proportion of aromatic amino acids and glycine and very low amounts of sulfur-containing amino acids (Voss-Foucart et al., 1973). The jaws of the Nereidae contain large concentrations of zinc (Bryan and Gibbs, 1979). The harder jaws of the Glyceridae also lack chitin and are composed of tanned protein, but in addition, heavy metals, especially iron and copper, are concentrated in these jaws, so that only approximately 50% by weight of the jaw is protein (Michel et al., 1973). The amino acid component of the jaws of Glycera convoluta contrasts with that of the other phyllodocimorph polychaetes in that the percentage of aromatic amino acids is very low, and glycine and hystidine constitute about 86% of the total amino acid residues.

The jaws of Glycera convoluta and other phyllodocemorph polychaetes are not heavily impregnated with carbonate. In this respect, they differ fundamentally from the jaws of eunicids, which have a high concentration of calcium carbonate. The jaws of Marphysa sanguinea and Eunice torquata contain 80% by weight calcium carbonate and 3% by weight magnesium carbonate, whereas in Sthenelais boa and Perinereis cultrifera, carbonates account for less than 1% of jaw weight.

These major differences in chemical composition may explain the lack of fossil Nereidae, Nephtyidae, and Polynoidae scolecodonts. The jaws of eunicids and the Glyceridae, for instance, are not soluble in KOH, whereas the jaws of the Nephtyidae and Nereidae are readily soluble (Szaniawski, 1974).

The harder jaws of the Glyceridae and Eunicidae are opaque, so that they cannot be viewed with transmitted light. It is possible that growth features in some of these jaws could be detected using techniques developed for other invertebrate skeletal material (see Appendix 1.D), such as surface etching and the preparation of thin sections.

4. Ecological and Paleoecological Applications

4.1. Analysis of Polychaete Population Age Structure

Polychaete worms grow by the addition of new segments at a pre-pygidial growth zone (Olive, 1974) and by the enlargement of existing segments. In most species, the rate of addition of new segments decreases with size, and in a few (especially the Polynoidae and Maldanidae), a definitive segment number is reached during early development. Polychaetes are soft-bodied animals that can regenerate lost posterior segments, and a high proportion of animals in a population may show signs of caudal regeneration. The loss of posterior segments and their subsequent regeneration will compound difficulties in estimating their age from their length-frequency distribution. For the most part, worms continue to grow as they age, so that large animals can usually be regarded as being older than smaller ones in a given species. There are, however, important differences in growth rate among individual specimens. These will be due in part to differences of local habitat and effective feeding conditions, and in part will express inherited differences in growth potentiality. In a detailed study of the growth of individual specimens of *Spirorbis spirorbis*, Daly (1978) revealed gross differences in growth rates among specimens of the same age even when growing on a single frond of *Fucus serratus*. Similarly, studies of the growth of *Pectinaria hyperborea* (Peer, 1970) and *P. australis* (Estcourt, 1974), which breed and die when 2 years old, show how variation in growth rates among individual specimens causes a rapid increase in the variance of the mean size for a year class as it ages (Fig. 4A).

Similar variations in growth rates are also likely to occur in species that live for several years. Figure 4B shows the size-frequency histograms for a population of *N. hombergii* (Warwick and Price, 1975), and an even more complex picture emerges for long-lived species such as *N. ciliata* (Alheit, 1979). It is therefore extremely difficult to resolve all the component year groups in a population of long-lived polychaetes, even when the breeding season is sharply defined; when the breeding season is extended, it becomes virtually impossible by size-frequency analysis (Olive, 1970; Warren, 1978).

Under favorable circumstances, it is usually possible to recognize the first year class; its growth and productivity can be measured with considerable accuracy (Buchanan and Warwick, 1974; Warwick and Price, 1975). It is also often possible to recognize the second year class, especially at the beginning of the second year, but where older year classes are present in the population, the second year class gradually merges with them so that they cannot easily be separated at 3 years old. The size modalities

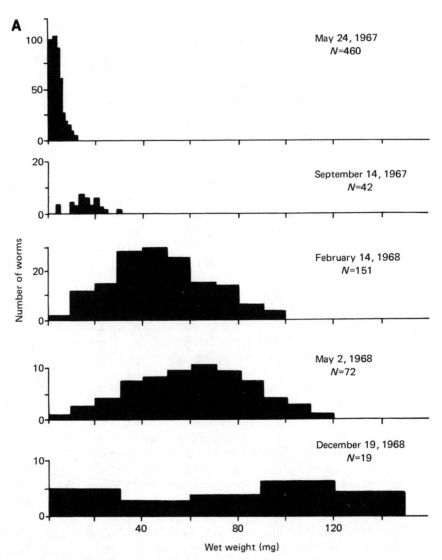

Figure 4. (A) Temporal change in the population structure of *Pectinaria hyperborea*. Note how the variance of wet weight values increases dramatically during the 19-month growth period. The population consists of a single generation of worms. After Peer (1970) by permission of the Canadian Fisheries Board. **(B)** Population structure of *Nephtys hombergii* in

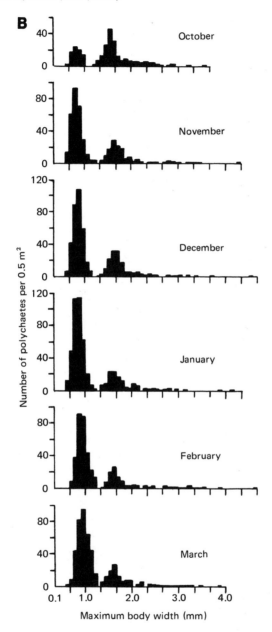

the Lynher estuary. Note that only the 0 and 1 + groups can be identified from size-frequency analysis, yet independent evidence suggests that the animals live for up to 7 years and do not breed until 2 years old. After Warwick and Price (1975) by permission of Cambridge University Press.

of subsequent year classes are much more difficult to detect. Size-frequency distributions for long-lived species usually show a long tail to the right, which indicates the presence of a smaller number of large, old animals in the population.

Several authors have attempted to analyze the age structure of complex polymodal size-frequency distributions with the aid of statistical methods based on the probability-paper methods of Harding (1949) and Cassie (1950) (see, for instance, George, 1964; Gibbs, 1968; Olive, 1970; Hutchings, 1973; Warren, 1978) (also see Chapter 11). Ideally, however, these methods require an independent estimate of the age of the component classes, and, at best, they give only an estimated age for a given animal. Furthermore, the application of different mathematical treatments may yield different results. Snow and Marsden (1974) were unable to resolve component classes in polymodal size-frequency histograms of the large nereid *Nereis virens* using the methods of Harding (1949) and Taylor (1965), but they were able to do so using the method of Bhattacharya (1967). Their study emphasizes the difficulties inherent in the interpretation of size-frequency data for long-lived species. They arrive at markedly different conclusions about the age at maturity and death of *N. virens* from those of Brafield and Chapman (1967), although their interpretations are based on similar size-distributions.

Tagging experiments, which would enable one to test hypotheses about the growth rates and aging of polychaete populations, are likely to be impracticable; therefore, the suggestion that at least some polychaetes may have permanent and cumulative growth records in the jaws is an important one that must be evaluated. This process of evaluation has begun for the family Nephtyidae, but similar detailed studies of polynoid growth in relation to jaw records are also likely to be fruitful. It may even be possible to determine whether the multiple growth lines in nereid and glycerid jaws are age- or size-related.

Recent investigations of the biology of the Nephtyidae have confirmed that *Nephtys hombergii* (Olive, 1977; Retière, 1976; Warwick and Price, 1975), *N. caeca* (Olive, 1977), and *Aglaophamus verillii* (Estcourt, 1975) are potentially long-lived species, living for at least 5 years, and in the case of *N. caeca*, as long as 7–9 years. In all these species, the jaws permit a size-independent estimate of age. Indeed, the older year groups cannot be resolved without recourse to the jaws in the populations investigated by Olive (1977), and the only alternative would be to group all post-2-year old animals into a single size grouping of indeterminate age.

Even when some of the jaws' growth records prove difficult to interpret (as is invariably the case with some of the older animals), the rings in specimens with clear annual growth lines permit an estimate of the approximate growth rate, longevity, and death rate of the species.

Kirkegaard (1970), Retière (1976), and Estcourt (1975) have all re-garded animals with one ring in the jaws as 1-year-old animals, and Retière in particular provides evidence that this is valid for the population of N. *hombergii* in the Rance estuary. The 0 group animals are shown as forming the first ring during the spring, which is immediately before the breeding season and the period of recruitment. On the other hand, in populations of N. *hombergii* and N. *caeca* from the Tyne estuary, the first ring is not formed until the second year (Olive, 1977). In October 1975, there were two groups of animals with zero rings in the jaws: the newly recruited 0 group settling from the plankton during July and August, and the 1 + year group, which had been recruited during 1974. It appeared from this study that the growth arrest that resulted in the formation of a prominent growth line was not caused simply by winter cold, but rather by com-pletion of a period of gametogenesis and sexual maturation. Further ob-servations suggested that some animals may not become sexually mature until the third year and that such animals would have only one ring in the jaw when 3 years old (Olive, 1978). For the Tyne estuary populations, the ring number plus 1 year represents the *minimum* age of individual worms. The reasons for the apparent differences in the age at which in-dividual animals become sexually mature for the first time in the popu-lations investigated by Retière and Olive are unknown. These studies em-phasize that the status of the growth lines can be established only through careful population analysis. In C. *ciliata*, the jaws have numerous growth lines, which supports the observations of Alheit (1979) that this is a long-lived species. However, the jaws are difficult to interpret, partly because the apex of the jaw erodes, and partly because there are numerous sub-ordinate growth lines located between the major ones. Several workers have found difficulty in utilizing information from the jaws because of the problem of finding an objective criterion for separating supposed an-nual growth lines from the others. Much more work is necessary before the method can be claimed to have general validity for all populations of *Nephtys* species.

Annual growth annuli are not present in the jaws of the Nereidae, but the jaws are nevertheless valuable in establishing the age structure of populations of species within this family. Chambers and Milne (1975) and Mettam (1979) have found that the most useful criterion of size in *Nereis diversicolor* is jaw length. Biometric studies of *Nereis diversicolor* (Cham-bers and Milne, 1975) suggested that the jaw length was related to body length as a function:

$$\log_{10}(\text{body length in mm}) = a + b \log_{10}(\text{jaw length in mm})$$

$$a = 1.68, b = 1.57$$

We found in a Blyth River Northumberland population of the same species that jaw length and body length were related arithmetically (Olive and Garwood, unpublished data), but our estimate of jaw length included the whole jaw length, whereas Chambers and Milne (1975) and Mettam (1979) measured only the length of the denticulate part of the jaw. There is a logarithmic relationship between jaw length and body weight in both populations.

There are several advantages in using jaw length as a measure of size. The jaw is a hardened part that is not affected by the degree of contraction of the worm, which reduces the accuracy of body-length measurements. Since jaw length is not affected by the loss of posterior segments, all members of a population (not only intact, nonregenerating specimens) can be used in population studies.

4.2. Ecological and Paleoecological Interpretations

When jaws are present, it is possible to define a "standard animal" by reference to jaw length or jaw weight; this greatly facilitates estimation of changes in somatic and germinal tissues during the reproductive cycle. It is therefore possible to measure the caloric values of somatic and germinal tissues, and to compare the values obtained in different animals by reference to the jaw as an independent parameter of size. Similarly, estimates of the changing values of biochemical components such as lipid, protein, or carbohydrate could be expressed in absolute terms and not simply as relative proportions. It may be possible to define a standard animal when jaws are absent,* but it is likely to be much more difficult.

When jaws are present, it will be possible to compare fecundities, reproductive effort, and relative conversion of energy to germ cells in specimens living in a variety of different ecological conditions, since each parameter measured can be related to a standard index of size. The question of reproductive effort and efficiency is an interesting one in polychaetes. The r–K selection theory does not at present provide an explanation of the different reproductive strategies exhibited by the Polychaeta. For instance, many of the larger polychaetes, which, by most of the criteria of Pianka (1970), could be regarded as K-selected, are also the most highly fecund, which would indicate that they are r-selected. It will be necessary to obtain data on the relative values of reproductive effort in polychaetes before progress can be made in the analysis of the selective forces controlling their reproductive strategies.

It is possible to recognize three different reproductive strategies, one of which is a form of semelparity, and the two others of which are different

---

* The weight of a muscular proboscis, for instance, may be a suitable index of size.

forms of iteroparity (see Bell, 1976). They have been designated by Olive and Clark (1978) as follows:

Monotely ≡ semelparity
Polytely ≡ iteroparity with discrete, usually single, annual spawning crises
Semicontinuous ≡ iteroparity with the release of small broods repeatedly during an extended breeding season

It is a striking feature of the morphology of the jaws of phyllodoce-morph polchaetes that in families with polytelic reproduction (Polynoi-dae, Nephtyidae, and Sigalionidae), there are marked annual annuli in the jaws, whereas in species with monotelic reproduction (Nereidae and Glyceridae), such annuli are absent. If such a correspondence of jaw morphology and reproductive strategy is confirmed by further studies, the reproductive strategies of some fossil polychaetes could be inferred from their scolecodonts.

Species with semicontinuous reproduction are usually small, but it is not known whether there are any permanent records in the form of jaw micromorphology that reflect the adoption of this form of reproduction. However, since the reproductive strategy in contemporary polychaetes is influenced by absolute size, it may be possible to infer from the size of scolecodonts their likely reproductive strategies. To do so, it will be necessary, first, to document the body size/jaw size relationship for a wide range of polychaetes and, second, within the individual families that are represented as scolecodont fossils, to document the relationship between maximum body size (jaw size) and reproductive strategy. For effective paleoecological work based on scolecodonts, there are two basic requirements: (1) the widespread occurrence of well-preserved fossils and (2) an adequate knowledge of the biological functions and growth of the jaws in contemporary species. The bulk of Paleozoic and Mesozoic scolecodonts are eunicid jaws or jaw apparatuses. Scolecodonts that can reliably be assigned to the families Nereidae, Polynoidae, and Nephtyidae have not yet been found. All the phyllodocemorph scolecodonts discovered so far belong to the Glyceridae or Goniadidae (Charletta and Boyer, 1974; Szaniawski, 1974). The discovery of fossilized jaws of the other phyllo-docimorph families would obviously be of great importance. The jaws of the Glyceridae are promising material for future work, since glycerid jaws have a cumulative permanent growth record, and scolecodonts that closely resemble the jaws of modern glycerids range from the Cretaceous to the Holocene, or, if *Paranereites* Eisenack is properly regarded as a glycerid, as suggested by Charletta and Boyer (1974) and Szaniawski (1974), from the Jurrassic to the Holocene. Interpretation of the fossil material is hampered by incomplete knowledge of the growth of the jaws of contemporary glycerids, but their morphology is now well known (Wolf, 1977). It should be possible in the future to estimate the size of Mesozoic glycerids from the scolecodont record, and from this it may eventually be possible to estimate other size-related parameters such as longevity and fecundity. It would be of value to determine the mortality of modern glycerid species in relation to age, as well as at different times during the year. If mortality is not constant, then size-frequency analysis of scolecodont assemblages might be expected to yield information that would permit the reconstruction of size-frequency distributions for the fossil populations.

All Paleozoic and most Mesozoic scolecodonts known so far are of eunicemorph species. Interpretation of this material is hampered by the complexity of the jaws and our lack of knowledge of their function in many of the living forms.

The jaws can be assigned to one of several morphological types, and the jaws are always in the form of a complex jaw apparatus. Contemporary eunicids were assigned by Ehlers (1864–1868) to two types: the labidognathan, which includes the Recent families Lumbrinereidae, Eunicidae, and Onuphidae, and the prionognathan, which includes the Recent families Arabellidae and Lysaretidae. Each of these types of jaw apparatus was abundant in Paleozoic times and is known from numerous Paleozoic scolecodont assemblages. A third type known as the placognathan and a fourth type known as the ctenognathan were defined by Kielan-Jaworowska (1966). Both were abundant in Paleozoic times, but the placognathan type is not represented in the Recent species, and the ctenognathan type is now represented only by the small specialized family Dorvilleidae. More recently, Mierzejewski and Mierzejewska (1975) have defined a fifth type known as the xenognathan that is known from a single Ordovician genus, *Archaeoprion*.

The majority of Paleozoic eunicid lines have disappeared without any descendants after a period of abundance in Ordovician and Devonian times. Interpretation of the possible functions of the different kinds of Paleozoic jaw apparatuses is hampered by an almost complete lack of comparative studies in Recent forms. The most detailed studies, which are those of Desière (1967) and Hartmann-Schroeder (1967), refer only to Recent Eunicidae.

There are difficulties in drawing inferences from isolated jaws; homologous elements from quite different apparatuses may be very similar or identical. According to Szaniawski (1974), the identification of detached scolecodonts on the basis of their similarity to single elements of a known apparatus is open to error.

There is some evidence, as explained above, that some eunicid jaws may be deciduous, but much more rigorous investigations of living eunicids are necessary to establish whether this is true in all species, and if it is true, to establish the periodicity of jaw replacement and the relationship to growth and reproduction. If eunicid jaws are deciduous, this may explain the apparent lack of growth records in them. It may also explain why isolated jaws are so much more numerous than intact jaw apparatuses. Biometric studies of isolated jaws and intact jaw apparatuses in the same deposits could yield data about the population structure, growth, and size at death in Paleozoic eunicids. The intact apparatuses were presumably deposited at death, whereas the isolated jaws may have been shed by younger growing worms. It may be possible to obtain estimates of population structure from biometric studies of isolated jaws, especially if molting was a seasonal phenomenon.

5. Conclusions

The jaws in several families of phyllodocemorph polychaetes grow during adult life, and a record of their periodic growth is preserved in growth annuli with a probable periodicity of 1 year. Annual growth lines, when present, offer a valuable tool in the studies of the longevity, productivity, and mortality of polychaetes, but much further work is needed for the proper evaluation of the temporal and biological significance of these growth patterns.

In the families Glyceridae and Nereidae, the growth lines in the jaws have a periodicity of much less than 1 year, possibly about 1 day; annual growth lines appear to be absent. In the Polynoidae, both types of growth record are present in the jaws.

The absence of annual growth lines in the jaws of the Glyceridae and Nereidae is thought to reflect their reproductive strategies, in which reproduction is delayed until the last year of life. The size of the jaws of polychaetes is a convenient index of body size that can be the basis for a definition of a "standard animal" and that is not affected by seasonal variations in condition or sexual maturity.

The jaws of the Glyceridae occur as fossilized scolecodonts in Mesozoic deposits, but the jaws of the Nereidae, Nephtyidae, Polynoidae, and Sigalionidae have not yet been discovered as fossils.

The jaws of the eunicemorph polychaetes differ fundamentally from those of phyllodocemorph polychaetes. They are highly calcified, whereas those of the phyllodocemorph species are composed of tanned proteins, sometimes with the deposition of iron and copper (Glyceridae). Eunicemorph jaws are arranged to form a complex jaw apparatus that has a pair of mandibles and up to six pairs of maxillae. Surface growth lines are absent in these jaws. They are thought to be deciduous, but much further work is needed to clarify how they grow and are replaced.

The majority of Paleozoic and Mesozoic scolecodonts are the fossilized jaws of eunicid polychaetes, most lines of which are now extinct. It has not yet been possible to find growth lines in the fossil eunicid scolecodonts, but biometric studies of scolecodont assemblages may yield information on possible population structure.

ACKNOWLEDGMENTS. Original observations have been supported with a grant from the Natural Environment Research Council (United Kingdom). I would like to thank Dr. P. R. Garwood for valuable assistance, and Professor R. B. Clark, who kindly allowed me to examine the jaws of poly chaetes in his collection. My wife Ann has been of great help in the preparation of the manuscript.

References

Åkesson, B., 1967, The embryology of *Eunice kobiensis*, *Acta Zool.* **48:**141–192.

Alheit, J., 1979, The annual cycles of three species of the polychaete genus *Nephtys*, in *Proceedings of the 13th European Marine Biology Symposium* (E. Naylor and R. G. Hartnoll, eds.), pp. 49–56, Pergamon Press, London.

Bell, G., 1976, On breeding more than once, *Am. Nat.* **110:**57–77.

Bhattacharya, C. G., 1967, A simple method of resolution of a distribution of Gaussian components, *Biometrics* **23:**115–135.

Brafield, A. E., and Chapman, G., 1967, Gametogenesis and breeding in a natural population of *Nereis virens*, *J. Mar. Biol. Assoc. U. K.* **47:**619–627.

Bryan, G. W., and Gibbs, P. E., 1979, Zinc—A major inorganic component of nereid jaws, *J. Mar. Biol. Assoc. U.K.* **59:**969–973.

Buchanan, J. B., and Warwick, R. M., 1974, An estimate of benthic macrofaunal production in the offshore mud of the Northumberland coast, *J. Mar. Biol. Assoc. U. K.* **54:**197–222.

Cassie, R. M., 1950, The analysis of polymodal frequency distributions by the probability paper method, *N. Z. Sci. Rev.* **8:**89–91.

Chambers, M. R., and Milne, H., 1975, Life cycle and production of *Nereis diversicolor* O. F. Muller in the Ythan Estuary, Scotland, *Estuarine Coastal Mar. Sci.* **3:**133–144.

Charletta, A., and Boyer, P. S., 1974, Scolecodonts from the Cretaceous greensand of the New Jersey coastal plain, *Micropaleontology* **20:**354–366.

Dales, R. P., 1962, The polychaete stomodeum and the inter-relationships of the families of Polychaeta, *Proc. Zool. Soc. London* **139:**389–428.

Dales, R., 1977, The polychaete stomodaeum and phylogeny, in: *Essays on Polychaetous Annelids* (J. Reish and K. Fauchald, eds.), pp. 525–545, Alan Hancock Foundation, Los Angeles.

Daly, J. M., 1972, The maturation and breeding biology of *Harmothoe imbricata* (Polychaeta: Polynoidae), *Mar. Biol.* **12:**53–66.

Daly, J. M., 1978, Growth and fecundity in a Northumberland population of *Spirorbis spirorbis* (Polychaeta: Serpulidae), *J. Mar. Biol. Assoc. U. K.* **58:**177–190.

Desbruyères, D., 1977, Un Ampharetidae (Annélide Polychètes sedentaires) à structure buccale aberrante, *Gnathampharete paradoxa* gen sp. n., *C. R. Acad. Sci. Ser. D* **286**:281–284.

Desière, M., 1967, Morphologie de l'organe buccal ventral de *Marphysa bellii* (Audouin & Edwards) (Polychaeta, Eunicidae), *Ann. Soc. R. Zool. Belg.* **97**:65–90.

Desière, M. and Jeuniaux, C., 1968, Observations préliminaires sur la constitution chimique des pièces mandibulaires de quelques Annelides Polychètes, *Ann. Soc. R. Zool. Belg.* **98**:43–48.

Ehlers, E., 1864–1868, *Die Borstenwürmer (Annelida, Chaetopoda)*, pp. 1–748, Engelmann, Leipzig.

Estcourt, I. N., 1974, Population study of *Pectinaria australis* (Polychaeta) in Tasman Bay, *N. Z. J. Mar. Freshwater Res.* **8**:283–290.

Estcourt, I. N., 1975, Population structure of *Aglaophamus verilli* (Polychaeta: Nephtyidae) from Tasman Bay, *N. Z. Oceanogr. Inst. Rec.* **2**:149–154.

George, J. D., 1964, The life history of *Cirriformia* (= *Audouinia*) *tentaculata*, *J. Mar. Biol. Assoc. U. K.* **44**:47–65.

Gibbs, P. E., 1968, Observations on the population of *Scoloplos armiger* at Whitsable, *J. Mar. Biol. Assoc. U. K.* **48**:225–254.

Harding, J. F., 1949, The use of probability paper for the graphic analysis of polymodal frequency distributions, *J. Mar. Biol. Assoc. U. K.* **28**:141–153.

Hartmann-Schroeder, G., 1967, Feinbau und Funktion des Kieferapparates der Euniciden am Beispiel von *Eunice* (Palola) *Siciliensis* Grube (Polychaeta), *Mitt. Hamb. Zool. Mus. Inst.* **64**:5–27.

Hoop, M., 1941, Der Kieferapparat von *Harmothoë imbricata* L., *Zool. Anz.* **135**:171–175.

Hutchings, P. A., 1973, Age structure and spawning of a Northumberland population of *Melinna cristata* (Polychaeta: Ampharetidae), *Mar. Biol.* **18**:218–227.

Kielan-Jaworowska, Z., 1966, Polychaete jaw apparatuses from the Ordovician and Silurian of Poland and a comparison with modern forms, *Palaeontol. Pol.* **16**:1–52 (plates 1–36).

Kirkegaard, J. B., 1970, Age determination of *Nephtys* (Polychaeta: Nephtyidae), *Ophelia* **7**:277–281.

Mettam, C., 1979, Seasonal cycles in populations of *Nereis diversicolor* (O. F. Muller) from the Severn Estuary, U. K., in: *Proceedings of the 13th European Marine Biological Symposium* (E. Naylor and R. G. Hartnoll, eds.), pp. 123–130, Pergamon Press, London.

Michel, C., 1966, Mâchoires et glandes annexes de *Glycera convoluta* (Keferstein) Annélide Polychète Glyceridae, *Cah. Biol. Mar.* **7**:367–373.

Michel, C., 1970a, Role physiologique de la trompe chez quatre Annélides Polychètes appartenant aux genres: *Eulalia, Phyllodoce, Glycera* et *Notomastus*, *Cah. Biol. Mar.* **11**:209–228.

Michel, C., 1970b, Étude histophysiologique de la trompe d'Annélides Polychètes appartenant aux genres: *Eulalia, Phyllodoce, Glycera* et *Notomastus*, Thèse Docterat d'État, University of Paris, CNRS No. A0 4190, 123 pp.

Michel, C., 1971, Mise en évidence d'un système de tannage quinonique au niveau des mâchoires de *Nephtys hombergii* (Annélide, Polychète), *Ann. Histochim.* **16**:273–282.

Michel, C., and DeVillez, E. J., 1978, Digestion, in: *Physiology of Annelids* (P. J. Mill, ed.), pp. 509–554, Academic Press, New York.

Michel, C., and Robin, Y., 1972, Premières données biochimiques sur les glandes à venin de *Glycera convoluta* Keferstein (Annélide Polychète), *C. R. Soc. Biol. Fr.* **166**:853–857.

Michel, C., Fonze-Vignaux, M. T., and Voss-Foucart, M. F., 1973, Données nouvelles sur la morphologie, l'histochimie et la composition chimique des mâchoires de *Glycera convoluta* Keferstein (Annélide Polychète), *Bull. Biol.* **107**:301–321.

Mierzejewski, P., 1978, Moulting of the jaws of the early Palaeozoic Eunicida (Annelida, Polychaeta), *Acta Palaeontol. Pol.* **23**:73–88.

Mierzejewski, P., and Mierzejewska, G., 1975, Xenognath type of polychaete jaw apparatuses, *Acta Palaeontol. Pol.* **20:**437–443.

Olive, P. J. W., 1970, Reproduction of a Northumberland population of the polychaete *Cirratulus cirratus*, *Mar. Biol.* **5:**259–273.

Olive, P. J. W., 1974, Cellular aspects of regeneration hormone influence in *Nereis diversicolor*, *J. Embryol. Exp. Morphol.* **32:**111–131.

Olive, P. J. W., 1977, The life history and population structure of the polychaetes *Nephtys caeca* and *Nephtys hombergii* with special reference to the growth rings in the teeth, *J. Mar. Biol. Assoc. U. K.* **57:**133–150.

Olive, P. J. W., 1978, Reproduction and annual gametogenic cycles in *Nephtys hombergii* and *N. caeca* (Polychaeta: Nephtyidae), *Mar. Biol.* **46:**83–90.

Olive, P. J. W., and Clark, R. B., 1978, Physiology of reproduction, in: *Physiology of Annelids* (P. J. Mill, ed.), pp. 271–368, Academic Press, New York.

Orrhage, L., 1964, Beiträge zur Kenntnisder spiomorphen Polychaeta, *Acta Univ. Ups.* **50:**1–16.

Orrhage, L., 1973, Two fundamental requirements for phylogenetic scientific work as a background for an analysis of Dales' (1962) and Webb's (1969) theses, *J. Zool. Syst. Evol.* **11:**161–173.

Peer, D. L., 1970, Relationship between biomass, productivity and loss to predators in a population of the marine benthic polychaete *Pectinaria hyperborea*, *J. Fish. Res. Board Can.* **27:**2143–2153.

Pfannenstiel, H. D., 1977, Experimental analysis of the "paarkultureffekt" in the protandric polychaete *Ophryotrocha puerilis* Clap. Mecz, *J. Exp. Mar. Biol. Ecol.* **28:**31–40.

Pianka, E. R., 1970, On r- and K-selection, *Am. Nat.* **104:**592–597.

Retière, C., 1976, Détermination des classes d'âge des populations de *Nephtys hombergii* (Annelida polychète) par lecture des mâchoires en microscopie électronique à balayage: Structure des populations en Rance maritime, *C. R. Acad. Sci. Ser. D.* **282:**1553–1556.

Sarvala, J., 1971, Ecology of *Harmothoe sarsi* (Malmgren) (Polychaeta, Polynoidae) in the northern Baltic area, *Ann. Zool. Fenn.* **8:**231–309.

Schwab, K. W., 1966, Microstructure of some fossil and Recent scolecodonts, *J. Paleontol.* **40:**416–423.

Snow, D. R., and Marsden, J. R., 1974, Life cycle weight and possible age distribution in a population of *Nereis virens* (Sars) from New Brunswick, *J. Nat. Hist.* **8:**513–527.

Szaniawski, J., 1974, Some Mesozoic scolecodonts congeneric with Recent forms, *Acta Palaeontol. Pol.* **19:**179–197.

Taylor, B. J. R., 1965, The analysis of polymodal frequency distributions, *J. Anim. Ecol.* **34:**445–452.

Voss-Foucart, M. F., Fonze-Vignaux, M. T., and Jeuniaux, C., 1973, Systematic characters of some annelid Polychaetes at the level of the chemical composition of the jaws, *Biochem. Syst.* **1:**119–122.

Warren, L., 1978, A population study of the polychaete *Capitella capitata* at Plymouth, *Mar. Biol.* **38:**209–216.

Warwick, R. M., and Price, R., 1975, Macrofauna production in an estuarine mud-flat, *J. Mar. Biol. Assoc. U. K.* **55:**1–18.

Wolf, G., 1976, Bau und Funktion der Kieferorgane von Polychaeten, Private publication, Hamburg, pp. 1–70.

Wolf, G., 1977, Kieferorgane von Glyceriden (Polychaeta)—ihrer Funktion und ihr taxonomischer Wert, *Senckenbergiana Marit.* **9:**261–283.

Appendices

Introduction

Anyone interested in doing skeletal growth research may find the following appendices useful. Much of the practical information about preparation techniques and methods of analysis is scattered throughout a diverse literature. We have attempted to draw some of this information together.

Appendix 1 deals with the "nuts-and-bolts" aspects of preparing thin sections, acetate peels, fractured sections, and X-radiography sections of skeletal materials. The range of techniques outlined in Appendix 1 can be used to study most skeletal structures. Our readers should understand that some of these techniques require skillful preparation. To gain finesse with the techniques requires some initial patience and experience. If problems arise, it is often most efficient to call or write a specialist to resolve difficulties.

In some cases, the authors of Appendix 1 mention specific products or vendors for supplies and instruments. These suppliers and products are listed only as examples and do not represent the sole sources, nor do the listings represent our individual or collective endorsement of products or suppliers.

Appendix 2 is supplementary to Chapter 2, although it has broader implications for unraveling the complexities of molluscan mineralogy and microstructure in general. Carter's microstructure guide represents one perspective on bivalve shell microstructure and differs somewhat from other classificatory schemes that are referenced in his guide. For this reason, some of Carter's terms used to describe shell microstructural types may differ from those used by other workers.

Three examples of quantitative analyses of skeletal growth data comprise Appendix 3. Appendices 3.A and 3.B are examples of how chemical or structural ontogenetic growth records can be quantitatively compared. Appendix 3.C is an interesting demonstration of how formal hypotheses concerning the relationships between observed skeletal growth phenomena and possible environmental causes can be formulated and tested. The innovative feature of this probabilistic approach is that it allows one to include in hypotheses qualitative, as well as quantitative, information.

Preparation and Examination
of Skeletal Materials for Growth Studies

1. Preparation of Acetate Peels and Fractured Sections for Observation of Growth Patterns within the Bivalve Shell

MICHAEL J. KENNISH, RICHARD A. LUTZ, and
DONALD C. RHOADS

1. Acetate Peels

Throughout various chapters of this book (especially Chapters 6 and 7), reference has been made to acetate-peel replicas of polished and etched bivalve shell sections. Such peel replicas are easily and rapidly prepared; seven procedural steps are involved: (1) embedding, (2) sectioning, (3) grinding, (4) polishing, (5) acid-etching, (6) washing and drying, and (7) application of acetone and acetate. We will describe these steps and give examples of specific products various workers have used.

1.1. Embedding

As an initial step, individual shell valves are embedded in an epoxy resin to prevent fracturing during sectioning. Rhoads and Pannella (1970) have reported success using Epon 815 resin and DTA curing agent from

MICHAEL J. KENNISH • Environmental Affairs Department, Jersey Central Power & Light Company, Morristown, New Jersey 07960.   RICHARD A. LUTZ • Department of Oyster Culture, New Jersey Agricultural Experiment Station, Cook College, Rutgers University, New Brunswick, New Jersey 08903.   DONALD C. RHOADS • Department of Geology and Geophysics, Yale University, New Haven, Connecticut 06520.

the Miller Stephenson Chemical Co., Inc., Danbury, Connecticut 06810. A modification of their method follows.

The epoxy solution should be prepared by thoroughly mixing the reagents in a ratio of 1 part DTA curing agent to 10 parts Epon 815. A greater proportion of curing agent should be avoided due to the exothermic reaction of the curing agent—resin mixture. Each pouring of epoxy should be restricted to thicknesses no greater than 2 cm; pouring thicknesses greater than this may result in a violent exothermic reaction. Embedding of large shells, which require greater thicknesses of epoxy, should be accomplished by building up 2-cm-thick layers one at a time. Disposable plastic containers (e.g., PEEL-A-WAY tissue embedding molds, Lipshaw Manufacturing Co., 7446 Central Avenue, Detroit, Michigan 48210) can be used as nonsticking molds for the epoxy. After a shell valve is placed in the liquid epoxy mixture, the entire container should be placed under vacuum. After approximately 5 min, the vacuum is released, and the evacuation is then repeated for another 5 min. This procedure should be repeated until no air bubbles remain in the epoxy. The mixture may then be left at ambient pressure to cure. Curing time for the mixture is approximately 6–8 hr at room temperature.

1.2. Sectioning

After the epoxy has hardened, the embedded shell may be sectioned along the desired axis, using a diamond rock saw (e.g., various models of saws from Buehler Ltd., 2120 Greenwood Street, P.O. Box 830, Evanston, Illinois 60204; solid rim blades with No. 1 concentration of diamonds from Felker Operations, Dresser Industries, Inc., 1900 South Crenshaw Boulevard, Torrance, California 90509). Because the epoxy is transparent, it is possible to examine the embedded shell and to define a transect for this section; this transect may be scratched directly on the epoxy surface to serve as a guide during sectioning. In the majority of studies of internal bivalve growth patterns, the valve or valves are sectioned along the axis of maximum growth (Fig. 1, a–b) (also see Chapter 1). This cut is oriented so that growth increments intersect the plane of section at right angles (Fig. 1).

1.3. Grinding

Once cut, valve cross sections are ground sequentially on cast iron lapidary wheels using carborundum powders in the following order of grit sizes: 120, 240, 320, 400, and 600 (e.g., Buehler carborundum grits). If the saw cut is sufficiently smooth, it may not be necessary to use the

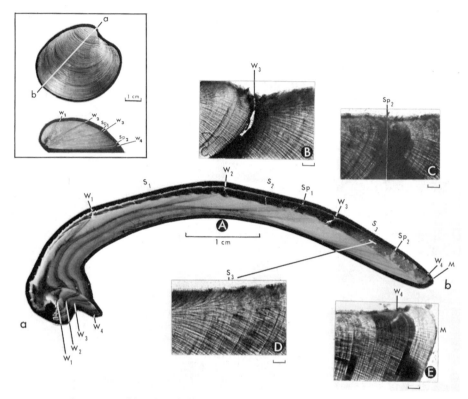

Figure 1. Illustration of bivalve shell sectioning technique. Box inset at upper left depicts axis of maximum growth (a–b) and external shell growth patterns of *Mercenaria mercenaria*. (A) Polished section along axis of maximum growth (a–b) of a specimen (depicted in inset) collected during the fourth winter (W_4). (S_1, S_2, S_3) Summer growth; (Sp_1, Sp_2) spawning breaks (see Chapter 7); (M) shell margin. (B–E) Microgrowth increments in specific shell regions of (A). Unlabeled scale bars: 100 μm.

coarsest grits early in this grinding sequence. Shell surfaces and all lapidary surfaces must be thoroughly cleaned after each grinding step to avoid contamination of the successively finer grits by the earlier coarse grits. As an alternative grinding procedure, one may utilize pressure-sensitive grinding discs (e.g., Buehler Ltd., AB carbimet special silicon carbide wet grinding papers) that are mounted on the lapidary wheels. These discs are available with the same size grits as the carborundum powders described above and should be used in the same sequential order (i.e., 120, 240, 320, etc.). If lapidary equipment is not available, grinding may be done directly on glass plates. For the final two grinding steps on glass, one should use 2000-grade, followed by 3000-grade, carborundum grits. No polishing (next step) is necessary if the surfaces are finely ground on glass in this manner.

1.4. Polishing

After sequential grinding (down to a 600-grade carborundum grit), surfaces are polished with 2000, followed by 3000, alumina powders [e.g., LINDE (Union Carbide, Crystal Products, Electronics Division) alumina powder abrasive] on a high-speed lapidary wheel covered with a polishing cloth (e.g., Buehler Ltd., AB microcloth No. 40-7208). A separate cloth must be used for each grit, and the lapidary wheel must be thoroughly rinsed after each polishing step.

1.5. Acid-Etching

After being polished, sections are etched by immersing them in a dilute solution of hydrochloric acid [1 part concentrated (38%) HCl to 100 parts distilled H_2O] for periods ranging from a few seconds to a few minutes. Optimal etching time will vary with the species examined and is related to shell structure, mineralogy, organic content, and state of preservation. The acid-etching process is a critical step in the technique, and it is recommended that a series of etching times be carried out to determine optimum etching periods for a particular specimen.

1.6. Washing and Drying

Immediately after the specimen is etched, the acid must be thoroughly rinsed from the sectioned surface. The section should be set aside to air-dry. Drying time can be reduced by spraying acetone on the etched surface.

1.7. Application of Acetone and Acetate

The etched shell surface is flooded with acetone, and a piece of sheet acetate (≈ 3 mm thick) is firmly applied to this surface. To eliminate air bubbles, one end of the piece of acetate should be placed at the edge of the cross section and then slowly lowered onto the meniscus of acetone, applying firm pressure. This procedure is similar to placing a cover slip onto a glass slide. A small weight should then be placed on top of the acetate to insure a firm contact of the epoxy block with the acetate. After the acetone has evaporated (20–30 min), the peel is removed from the cross section and, if desired, mounted onto a glass slide with Canada balsam or one of a number of synthetic resins prior to examination under the microscope.

Acetate-peel replicas may alternatively be prepared using extremely thin acetate. If this is done, the acetate should be placed on a *clean* piece of glass. The shell cross section is then flooded with acetone and firmly applied onto the acetate. Once again, air-bubble formation may be minimized by placing one end of the cross section at the edge of the acetate and slowly lowering it onto the acetate, applying firm pressure throughout this process. A weight should be placed on top of the epoxy block. After 20–30 min, the acetate can be removed and should be immediately mounted between two glass slides to avoid curling and wrinkling.

2. Fractured Sections

Fractured sections of bivalve shells are extremely easy to prepare for scanning electron microscopic (SEM) examination. Shell valves should be thoroughly dried prior to fracturing; placement of the valves in a desiccator is advisable for shells from which soft tissues have recently been removed. Once dry, the shells may be fractured (i.e., broken) along the desired axis. For fracturing, it is generally advisable to place the shell on the edge of a solid object (e.g., a firm table) with the desired transect superimposed directly over the edge (of the table) itself. Firm pressure may then be applied to the unsupported portion of the shell extending beyond the table edge, until the shell breaks. It is important to break the shell away from any shell surface that is of interest. For example, if one is interested in examining the periostracum of a given shell valve, the shell should be placed on the edge of the object (table), with the periostracum side of the valve up; pressure is then applied downward so that the break itself is away from the periostracum. After being fractured, the shell fragment may be mounted on a standard SEM stub. Care should be taken to ensure electrical conductivity between the specimen and the stub; the use of silver paint or graphite-containing solutions (e.g., DAG 154, Ted Pella, Inc., Cat. No. 1603-16) as mounting media generally adequately provides such conductivity. Clean compressed air should be used to remove all contaminants from the surface of the fractured section. Finally, the specimen should be coated (under vacuum) with gold–palladium or a combination of gold and carbon.

ACKNOWLEDGMENTS. This is Paper No. 5308 of the Journal Series, New Jersey Agricultural Experiment Station, Cook College, Rutgers University, New Brunswick, New Jersey 08903.

Reference

Rhoads, D. C., and Pannella, G., 1970, The use of molluscan shell growth patterns in ecology and paleoecology, *Lethaia* 3:143–161.

Preparation and Examination
of Skeletal Materials for Growth Studies

2. Study of Molluscan Shell Structure and Growth Lines Using Thin Sections

GEORGE R. CLARK II

Early studies of molluscan shell structure, culminating in the classic work of Bøggild (1930), were based on examination of petrographic thin sections of shells. The grinding of such sections required a considerable investment of time and labor, and because of the tendency of many shells to split along growth surfaces or crystal boundaries, modern high-speed thin-sectioning equipment has not appreciably lessened this problem.

For this reason, acetate peels (see Appendix 1.A.1) have largely replaced thin sections in recent research on shell structure and growth lines. Although they provide much less information than thin sections, acetate peels can be prepared in a fraction of the time with little training.

New equipment, however, may again reverse the trend. The introduction of low-speed diamond saws, such as Buehler's Isomet 11-1180, makes it possible to prepare high-quality thin sections nearly as easily as acetate peels.

The micrographs in Plate I, which are all from a recent report on seasonal growth variations in *Mercenaria* (Clark, 1979; also see Clark, 1977), illustrate the type of information available in thin sections at relatively low magnifications. In addition to those features visible on acetate peels, these sections show variations in transparency, pigmentation, and crystallographic orientation. Higher-magnification studies can make use of the finite thickness of the thin section to make limited observations in three dimensions. Thin sections are also much easier than peels to store and study, having no surface relief to protect and no tendency to curl up.

GEORGE R. CLARK II ● Department of Geology, Kansas State University, Manhattan, Kansas 66506.

603

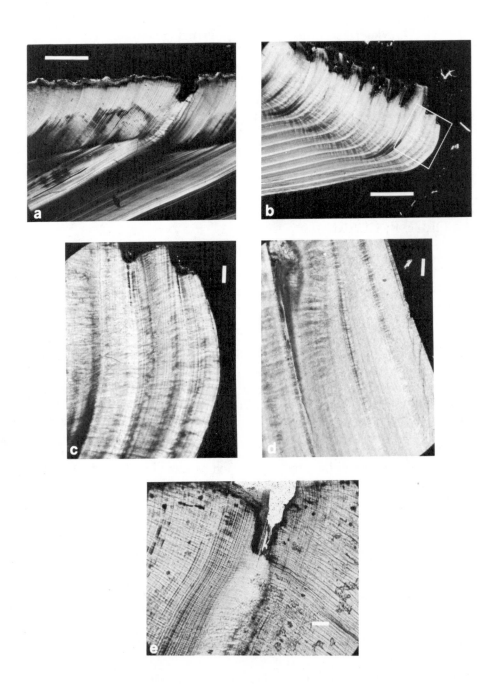

Since thin sections can now be prepared with little more effort than peels, they should become a useful supplement if not a replacement.

The first step, as in making a peel, is to cut the shell along the desired plane of section. This can be accomplished with no preliminary embedding on a low-speed saw. When the cut is completed, the sectioned surface is lightly sanded or ground using 600 or 1000 grit. Some care should be taken to ensure that only a single planar surface is present after this operation, since it is easy to develop two facets by a back-and-forth grinding motion. This surface should be thoroughly cleaned to remove grit, cutting oils, or other contaminants; ultrasonic cleaning may help if the shell is not too delicate.

After the section is dried, the sectioned surface must be cemented to a petrographic slide. I find that the bonds are improved when the slides are "frosted" beforehand with 600 grit. Both the section and the slide must be dry and free from oils or fingerprints. A variety of cements may be used, but thermoplastic cements such as Lakeside and high-temperature epoxies such as "Petropoxy" will often fail due to thermal stress on cooling. I find Ward's "Bioplastic," which cures at 70°C, satisfactory for most purposes. Only a few drops of cement should be used, for a large buildup of excess cement at the base of the section will prolong the cutting time and may warp the saw blade.

When the bond between section and slide is well developed, the slide should be mounted parallel to the saw blade and a second cut made about 200 μm (0.008 inch) from the surface of the slide. The section should be lowered carefully onto the blade to make sure the blade is not deflected by the cement.

On completion of the second cut, the slide can be ground by hand on a glass plate with 600 or 1000 grit in less than 10 min. The goal here is transparency, not a specific thickness, so progress can be readily judged and a tendency to grind too fast on one side can be compensated for before it is too late.

After being cleaned, the thin section can have a cover slip permanently attached with Canada balsam or epoxy, or it can be studied with a temporary cover slip using immersion oil as a bond. The latter method permits staining or additional grinding at a later time.

Plate I. (a) Thin section of mature region of a specimen of Mercenaria mercenaria. Scale bar: 1 mm. (b) Thin section of margin and senile region of a specimen of Mercenaria mercenaria. Rectangle indicates position of micrograph (c). Scale bar: 1 mm. (c) Thin section of the margin and outer shell layer of the senile region of a specimen of Mercenaria mercenaria. Scale bar: 0.1 mm. (d) Thin section of the middle shell layer and inner surface in the senile region of a specimen of Mercenaria mercenaria. Scale bar: 0.1 mm. (e) Thin section of the outer shell layer of the mature region of a specimen of Mercenaria mercenaria. Scale bar: 0.1 mm. All micrographs from Clark (1979).

References

Bøggild, O. B., 1930, The shell structure of the mollusks, *K. Dan. Vidensk. Selsk. Skr. Naturvidensk. Mathem. Afd. Ser. 9* **2**:231–326 (15 plates).

Clark, G. R., II, 1977, Seasonal growth variations in bivalve shells and some applications in archeology, *J. Paleontol.* **51**(Suppl. to No. 2):7 (abstract).

Clark, G. R., II, 1979, Seasonal growth variations in the shells of recent and prehistoric specimens of *Mercenaria mercenaria* from St. Catherines Island, Georgia, *Am. Mus. Nat. Hist. Anthropol. Pap.* **56**(1):161–179.

3. Techniques for Observing the Organic Matrix of Molluscan Shells

GEORGE R. CLARK II

Although mollusc shells are built primarily of various structural arrangements of calcite and aragonite, these minerals are invariably accompanied by organic constituents. Some of the organic matter forms obvious functional structures such as the ligament and the periostracum, but much of it is distributed throughout the shell as the organic matrix.

The organic matrix appears to be an essential participant in the calcification process, and for this reason has attracted the attention of many investigators [see literature summaries in Wilbur (1972, 1976), for example]. Most of the older work is directed toward questions of composition and ultrastructure of the matrix, but an increasing amount of attention is being paid to the morphological relationships between the organic matrix and carbonate crystals. Numerous workers (Bevelander and Nakahara, 1969; Erben, 1972, 1974; Erben and Watabe, 1974; Mutvei, 1970, 1972a,b; Taylor and Kennedy, 1969; Towe, 1972; Towe and Hamilton, 1968; Towe and Thompson, 1972; Travis, 1968; Wada, 1961, 1972; Watabe, 1963, 1965; Wise, 1970; Wise and Hay, 1968) have noted that the most prominent organic matrix forms sheets that separate layers of crystals (interlamellar matrix) or isolate individual crystals (intercrystalline matrix). The precise physical relationships are of considerable interest, for it appears that the nucleation of individual crystals takes place on the organic surfaces and that these influence the mineralogy and crystallographic orientation of the crystals. This "template" hypothesis has received wide acceptance, but a further interpretation of the matrix–crystal

GEORGE R. CLARK II • Department of Geology, Kansas State University, Manhattan, Kansas 66506.

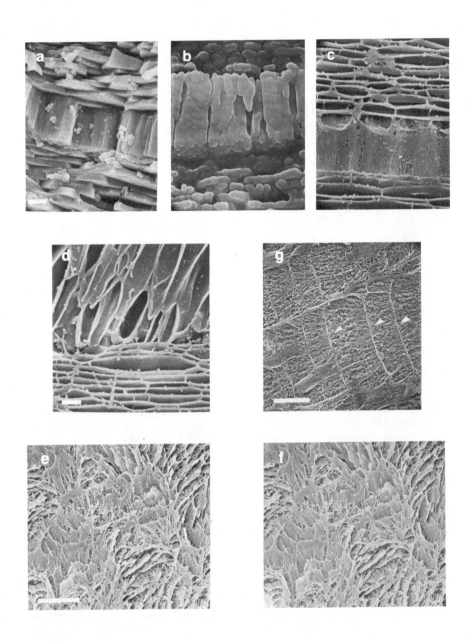

relationship, by which the sheets of matrix are thought to confine and eventually terminate the growth of individual crystals (the "compartment" hypothesis), is the subject of considerable controversy (Bevelander and Nakahara, 1969; Erben, 1972, 1974; Lutz and Rhoads, 1977; Mutvei, 1970, 1972b; Towe, 1972; Towe and Hamilton, 1968; Wada, 1972).

Our understanding of the organic matrix has been limited by difficulties in preparation. Light microscopy of thin sections cannot reach the magnification required to distinguish most matrix elements, and scanning electron microscopy (SEM) cannot readily discriminate between organic and crystalline elements in fractured or polished sections. Decalcification can eliminate the crystalline elements, but the matrix is then prone to collapse or shrink when dried. Transmission electron microscopy can make such distinctions, but the preparation of ultrathin sections showing undistorted relationships between matrix and crystals is extremely difficult, and even when achieved will exhibit only two dimensions of a complex three-dimensional structure.

I have found that the application of critical-point drying to oriented sections that have been polished and then very slightly etched results in organic matrix preparations highly suitable for SEM (Clark, 1978). Limiting the decalcification to a depth of only a few microns ensures that the exposed organic matrix will maintain dimensional stability and retain its proper orientation with respect to the dissolved portions of the carbonate crystals. Using critical-point drying eliminates the crushing effect of surface tension experienced during drying in air. Polishing the sectioned surface before etching is necessary to ensure that the very thin zone of matrix exposed by etching is free from damage (scratches) and mechanical distortion. This also serves as a check on the success of the final preparation, for if no distortion has occurred during etching or drying, the exposed matrix should appear truncated along a planar surface.

Plate Ia–c illustrates the basic advantages of this technique as applied to an examination of the organic matrix relationships between the pallial

← _____

Plate I. (a–c) Pallial myostratum (thick layer in middle of micrographs) and surrounding nacreous layers in Mytilus californianus. (a) Fractured section; the organic matrix is present but cannot be readily distinguished from the aragonite crystals. (b) Polished and etched section dried in air. The organic matrix exposed by etching has collapsed over the etched surface. (c) Polished and etched section prepared by critical-point drying. The organic matrix stands in high relief above the etched surface. Scale bar for a–c (depicted in a): 1 μm. (d) Transition from prismatic layer (top) to nacreous layer (bottom) in Mytilus californianus. Scale bar: 1 μm. (e,f) Transition zone between outer prismatic layer and middle homogeneous layer in Mercenaria mercenaria. Stereo viewing shows this to be a variety of complex crossed lamellar structure. Scale bar: 10 μm. (g) Prismatic layer of Mercenaria mercenaria showing prominent growth lines (arrows) formed by concentration of organic matter. Scale bar: 10 μm.

myostracum (center band) and the nacreous layer in a specimen of *Mytilus californianus*. The organic matrix is present in the fractured section (Plate Ia), but cannot be readily distinguished. In Plate Ib, the organic matrix has been collapsed onto the etched surface in an air-dried preparation. Plate Ic, prepared by critical-point drying after etching, shows the organic matrix clearly, apparently undistorted. Note that this section, unlike the fractured section, could be made at any angle to these crystals. Other organic matrices prepared in this manner are shown in Plate Id–g. The use of stereo pairs (Plate Ie and f) permits the direct visualization of three-dimensional structures.

In addition to improving studies of the organic matrix itself, preparations of this type should be useful in more general investigations. Studies of molluscan shell structure, for example, have long been hampered by the small size and complexity of many arrangements of crystals, and a variety of distinct structures classified as "complex" or "homogeneous" since the classic work of Bøggild (1930) remain to be adequately described. Stereo pairs of the matrix (Plate Ie and f) can be used to determine the physical orientation of crystals and crystallites. Also, because matrix preparations can be oriented in any direction and, in a sense, look into the shell, they should provide information not available from fractured or standard (air-dried) etched sections.

Studies on growth lines, particularly those involving variations in shell structure with stress (Kennish and Olsson, 1975; Clark, 1979) and variations in the accumulation of organic matter (Lutz and Rhoads, 1977), can also make good use of this sort of preparation. Various shell structures can be readily interpreted (Plate Id–g), and accumulations of organic matter along growth lines (Plate Ig) are very prominent.

Although the details of my method of preparation are of much less importance than the principles, I list them here. Sections of mollusc shells are cut at desired orientations using a low-speed diamond saw (e.g., Buehler Isomet 11-1180) to minimize stress and fractures. The sections are embedded in epoxy to facilitate the polishing operations, ground on a glass plate with 600 grit until flat, and then polished with a series of alumina compounds to a final polish with 0.05 μm alumina. After being washed with distilled water, the sections are etched for 15–30 sec in 0.1 N HCl, and then, keeping a miniscus of liquid on the specimen surface during every transfer, passed through a series of distilled water rinses and a series of alcohols until they are in 100% ethanol. The specimens are then dried in CO_2 using a critical-point drying apparatus; some care is required to devise holders that protect the etched surfaces during this process, since some turbulence is involved. Finally, the specimens are coated with a combination of carbon and gold–palladium to a total thickness of about 200 Å. Micrographs presented in this appendix were taken using an ETEC Autoscan scanning electron microscope.

References

Bevelander, G., and Nakahara, H., 1969, An electron microscope study of the formation of the nacreous layer in the shell of certain bivalve molluscs, *Calcif. Tissue Res.* **3**:84–92.

Bøggild, O. B., 1930, The shell structure of the mollusks, *K. Dan. Vidensk. Selsk. Skr. Naturvidensk. Math. Afd. Ser. 9* **2**:231–326 (15 plates).

Clark, G. R., II, 1978, Organic matrix morphologies in fossil and Recent mollusk shells, *Geol. Soc. Am. Abstr. Program* **10**:380.

Clark, G. R., II, 1979, Seasonal growth variations in the shells of recent and prehistoric specimens of *Mercenaria mercenaria* from St. Catherines Island, Georgia, *Am. Mus. Nat. Hist. Anthropol. Pap.* **56**(1):161–179.

Erben, H. K., 1972, Über die Bildung und das Wachstum von Perlmutt, *Biomineralisation* **4**:15–46.

Erben, H. K., 1974, On the structure and growth of the nacreous tablets in gastropods, *Biomineralisation* **7**:14–27.

Erben, H. K., and Watabe, N., 1974, Crystal formation and growth in bivalve nacre, *Nature (London)* **248**:128–130.

Kennish, M. J., and Olsson, R. K., 1975, Effects of thermal discharges on the microstructural growth of *Mercenaria mercenaria*, *Environ. Geol.* **1**:41–64.

Lutz, R. A., and Rhoads, D. C., 1977, Anaerobiosis and a theory of growth line formation, *Science* **198**:1222–1227.

Mutvei, H., 1970, Ultrastructure of the mineral and organic components of molluscan nacreous layers, *Biomineralisation* **2**:48–61.

Mutvei, H., 1972a, Ultrastructural relationships between the prismatic and nacreous layers in *Nautilus* (Cephalopoda), *Biomineralisation* **4**:80–86.

Mutvei, H., 1972b, Formation of nacreous and prismatic layers in *Mytilus edulis* L. (Lamellibranchiata), *Biomineralisation* **6**:96–100.

Taylor, J. D., and Kennedy, W. J., 1969, The influence of the periostracum on the shell structure of bivalve molluscs, *Calcif. Tissue Res.* **3**:274–283.

Towe, K. M., 1972, Invertebrate shell structure and the organic matrix concept, *Biomineralisation* **4**:1–14.

Towe, K. M., and Hamilton, G. H., 1968, Ultrastructure and inferred calcification of the mature and developing nacre in bivalve mollusks, *Calcif. Tissue Res.* **1**:306–318.

Towe, K. M., and Thompson, G. R., 1972, The structure of some bivalve shell carbonates prepared by ion beam thinning, *Calcif. Tissue Res.* **10**:38–48.

Travis, D. F., 1968, The structure and organization of, and the relationship between the inorganic crystals and the organic matrix of the prismatic region of *Mytilus edulis*, *J. Ultrastruct. Res.* **23**:183–215.

Wada, K., 1961, Crystal growth of molluscan shells, *Bull. Natl. Pearl Res. Lab.* **7**:703–828.

Wada, K., 1972, Nucleation and growth of aragonite crystals in the nacre of some bivalve mollusks, *Biomineralisation* **6**:141–159.

Watabe, N., 1963, Decalcification of thin sections for electron microscope studies of crystal–matrix relationships in mollusc shells, *J. Cell Biol.* **18**:701–703.

Watabe, N., 1965, Studies on shell formation. XI. Crystal–matrix relationships in the inner layers of mollusk shells, *J. Ultrastruct. Res.* **12**:351–370.

Wilbur, K. M., 1972, Shell formation in mollusks, in: *Chemical Zoology*, Vol. VII, *Molluscs* (M. Florkin and B. T. Scheer, eds.), pp. 103–145, Academic Press, New York.

Wilbur, K. M., 1976, Recent studies of invertebrate mineralization, in: *The Mechanisms of Mineralization in the Invertebrates and Plants* (N. Watabe and K. M. Wilbur, eds.), pp. 79–108, University of South Carolina Press, Columbia.

Wise, S. W., Jr., 1970, Microarchitecture and deposition of gastropod nacre, *Science* **167**:1486–1488.

Wise, S. W., Jr., and Hay, W. W., 1968, Scanning electron microscopy of molluscan shell ultrastructures. II. Observation of growth surfaces, *Trans. Am. Microsc. Soc.* **87**:419–430.

Preparation and Examination
of Skeletal Materials for Growth Studies

4. Study of Annual Growth Bands in Unionacean Bivalves

MICHAEL J. S. TEVESZ and JOSEPH GAYLORD CARTER

1. Distinguishing Annuli

Chamberlain (1931) suggested a simple method for identifying the two types of growth bands of unionaceans. He found the more conspicuous, annual bands to be clearly distinguishable by holding thin-shelled forms up to monochromatic, yellow light from an incandescent lamp. For thicker-shelled forms, he held the shells up to an ultraviolet light. This caused the $CaCO_3$, but not the periostracum, to fluoresce, and the annual bands again showed up as the most prominent lines.

In an article that concerned both marine and freshwater bivalves, Haranghy et al. (1965) suggested the use of fluorescence-microscopic analysis in distinguishing annual rings. Moreover, they also advocated the following (p. 61): "Polarization optical examinations showed that, while the annual rings cannot be delimited in the polished preparations, the polarizational optical analyses display a pale stripe of double refraction in the region of the annual rings."

Stansbery (1961) listed several differences between the two kinds of bands based on his observations of Lake Erie unionids. False annuli, he said, were relatively thin and incomplete (i.e., did not extend around the shell in a uniform, unbroken manner). Also, the two kinds of bands consistently differed in the same shell in such properties as color and texture. Moreover, while true annuli passed between interruptions in color rays on the shell surface, false annuli passed through the uninterrupted por-

MICHAEL J. S. TEVESZ • Department of Geological Sciences, The Cleveland State University, Cleveland, Ohio 44115. JOSEPH GAYLORD CARTER • Department of Geology, University of North Carolina, Chapel Hill, North Carolina 27514.

613

tions of the rays. Finally, false annuli lacked the regularity of spacing or periodicity of the other bands.

2. Separating Shell Layers

To separate annual shell layers for chemical analysis, Nelson (1964) first cut the shells in cross sections with a hacksaw. The cut, he said, was along the longest radius connecting the umbo and shell margin. After the cut surface was polished with No. 320 grit emery cloth, annual layers became distinct. The thin lines separating the layers were scored with a scalpel. The sections were then placed in a muffle furnace at 400°C for 4 hr. During cooling in the furnace, CO_2 was put into the furnace to reconstitute carbonate that might have been oxidized. The sections were then easily pried apart along the growth lines by aid of a scalpel.

Sterrett and Saville (1974) modified the method by making two radial cuts on the valve. This, they said, facilitated layer separation after cooling. Moreover, they found that an automatic saw with a carborundum-charged blade made a smoother cut and was easier to use than a hacksaw. Finally, they said that using a shorter baking method (i.e., 500°C for 10 min or even 600°C for 5 min) facilitated separating the layers.

References

Chamberlain, T. K., 1931, Annual growth of freshwater mussels, *Bull. U.S. Bur. Fish.* **46:**713–739.

Haranghy, L., Balázs, A., and Burg, M., 1965, Investigation on aging and duration of life in mussels, *Acta. Biol. Acad. Sci. Hung.* **16:**57–67.

Nelson, D. J., 1964, Deposition of strontium in relation to morphology of clam (Unionidae) shells, *Verh. Int. Ver. Limnol.* **15:**893–902.

Stansbery, D. H., 1961, The naiades (Mollusca, Pelecypoda, Unionacea) of Fishery Bay, South Bass Island, Lake Erie, *Sterkiana* **5:**1–37.

Sterrett, S. S., and Saville, L. D., 1974, A technique to separate the annual layers of a naiad shell (Mollusca, Bivalvia, Unionacea) for analysis by neutron activation, *Am. Malacol. Union Inc. Bull.* **1974:**55–57.

Preparation and Examination
of Skeletal Materials for Growth Studies

Preparation of Coral Skeletons for Growth Studies

RICHARD E. DODGE

To observe density differences in hermatypic coral skeletons, X-radiography has proven to be a most useful technique; the method is summarized below. Also briefly discussed are coral staining techniques and methods for taking samples from the coral skeleton.

Roughly hemispherical coral heads are best suited for X-ray study. Cores from the top to the base of individual coral columns or heads are also suitable [e.g., see Hudson *et al.* (1976) and Stearn and Colassin (1979) for coring techniques]. For entire corals (and cores with appropriate modifications) the specimen is sectioned with a diamond or carborundum saw along a plane that includes the point of origin of the colony and points of maximum relief on the growth surface. A second parallel cut is made to obtain a thin slab that is several corallite diameters in thickness. We have had best results with slab thicknesses of about 0.5 cm. In addition, we find it practical to make several such slabs to ensure obtaining a section that includes the entire growth history of the coral. In the field, it is often relatively easy to rent a masonry-type rock saw (used primarily for cement-block cutting). Building contractors, even on small islands (e.g., Bermuda and Barbados), often have such equipment available. The machine consists of a saw blade mounted on a pivoting arm that can be moved up and down by a foot pedal and arm handle. The specimen is placed on a movable tray with rollers above a water trough. Water is pumped continuously over the blade for lubrication and cooling. This type of saw is excellent for obtaining thick sections that can later be resectioned in the laboratory. With practice, it is possible to obtain reasonably thin and parallel-sided

RICHARD E. DODGE • Nova University Ocean Sciences Center, Dania, Florida 33004.

slabs. If more precision is desired, thick slabs (approximately 2–4 cm) can be embedded in plaster of paris and, after hardening, mounted in a laboratory diamond-blade rock saw.

Once the desired slab is obtained, the specimen is ready for X-radiography. We use a dental X-ray unit with the collimator removed. Machine settings are approximately 10 mA and 100 kV. Exposure time varies with cathode output, specimen thickness, and distance of the X-ray source to the subject. For a 50-inch distance and 0.5-cm slabs, times of 10-50 sec are usually sufficient to produce good negatives. Trial-and-error methods are often necessary to obtain successful exposures.

In practice, a variety of films can be used. We have found best results with Kodak AA2 Industrial Ready-Pac X-ray film. The film can be obtained in 8 × 10 and 10 × 12 sheets, each sheet individually wrapped in paper. It seems impossible to buy this film directly through Kodak, and one must rely on retail outlets. It is best to order well in advance of planned work. The specimen is placed over the paper-covered film and X-rays are directed through the coral and onto the film, thus recording density differences through differential absorption of X-rays in the specimen. We have also found sheet Polaroid film to be handy for preliminary inspection or quick viewing of a specimen. As an alternative, hospitals with X-ray facilities are often willing to X-ray a few specimens at no cost, if the investigator is willing to accommodate their schedule.

Once the negative is suitably developed, a positive print can be made by contact printing. We have found it easier to work with a positive for visual assessment of band dimensions, although other workers (e.g., Weber and White, 1974, 1975) have used the negative exclusively. Once the print or negative is obtained, measurements of band width or other dimensions can be made either directly on the positive or on acetate overlays on the negative. Some distortion may occur during the developing and printing process, and it is often useful to enclose markers (opaque reference points) on the specimen.

A scanning densitometer has been shown to be useful for quantifying density differences over the slab. First, the optic density of the X-radiograph must be calibrated to the skeletal density of the coral. This can be done by including standards of known density in the X-radiograph. Dodge and Thomson (1974) used powdered coral standards, Buddemeier et al. (1974) used an aluminum step wedge, and Dodge (1978) used a calcite wedge of known dimensions. After calibration, it becomes possible to measure skeletal density changes. Figure 4 from Dodge and Thomson (1974) illustrates the kind of information that is obtainable. It is possible to determine density, width, and mass characteristics of desired areas.

There are several methods of imaging skeletal differences that have not yet been tried but that may prove useful for studying coral growth.

These include transmitted infrared photography (Rhoads and Stanley, 1966) or transmitted tungsten light through sectioned specimens. In addition, medical techniques of employing a variety of X-ray sources and positions coupled with computer image analysis [computerized axial tomography (CAT) scans] may be directly applicable to coral growth studies. Using this medical diagnostic method, it should be possible to eliminate coral sectioning altogether.

A method for checking time of growth-band formation and growth-band dimensions (as well as general coral growth rates from a reference marker) involves staining living corals with Alizarin red (Barnes, 1972) while they grow in the field or laboratory. The coral is then allowed to grow for an interval of time and finally it is collected, sectioned, and X-radiographed to enable one to relate skeletal density banding to the stained growth reference datum. However, there are some problems with the technique. Buddemeier and Kinzie (1975) were not able to resolve all their staining events in corals from Hawaii they worked with. Many workers find Alizarin useful (e.g., see Buddemeier and Kinzie, 1976), and more research with this type and different types of $CaCO_3$ stain would be helpful to advance the methodology.

Sampling the coral skeleton for material for chemical or other analysis can be done in a variety of ways. We have used a coarse-toothed metal file to sequentially scrape layers from a coral head for radiochemical analysis (Dodge and Thomson, 1974). In addition, a small chisel and hammer are useful for sampling skeletal material from prepared coral sections (e.g., Nozaki et al., 1978). Smaller samples can be obtained by using an appropriately sized masonry drill bit and catching the cuttings (e.g., Fairbanks and Dodge, 1979). A very thin-blade gem-cutting saw can also be used for microsampling (e.g., Goreau, 1977a,b). Coral skeletons are very brittle and require care when any small-scale sampling is required.

References

Barnes, D. J., 1972, The structure and formation of growth-ridges in scleractinian coral skeletons, *Proc. R. Soc. London Ser. B* **182**:331–350.

Buddemeir, R. W., and Kinzie, R. A., III, 1975, The chronometric reliability of contemporary corals, in: *Growth Rhythms and the History of the Earth's Rotation* (G. D. Rosenberg and S. K. Runcorn, eds.), pp. 135–147, John Wiley, London.

Buddemeir, R. W., and Kinzie, R. A., III, 1976, Coral growth, *Oceanogr. Mar. Biol. Annu. Rev.* **14**:183–225.

Buddemeir, R. W., Maragos, J. E., and Knutson, D. W., 1974, Radiographic studies of reef coral exoskeleton: Rates and patterns of coral growth, *J. Exp. Mar. Bio. Ecol.* **14**:179–200.

Dodge, R. E., 1978, The natural growth records of reef-building corals, Ph.D. thesis, Yale University, 237 pp.

Dodge, R. E., and Thomson, J., 1974, The natural radiochemical and growth records in contemporary hermatypic corals from the Atlantic and Caribbean, *Earth Planet. Sci. Lett.* **23**:313–322.

Fairbanks, R. G., and Dodge, R. E., 1979, Annual periodicity of the $^{18}O/^{16}O$ and $^{13}C/^{12}C$ ratios in the coral *Montastrea annularis, Geochim. Cosmochim. Acta* **43**:1009–1020.

Goreau, T. J., 1977a, Coral skeletal chemistry: Physiological and environmental regulation of stable isotopes and trace metals in *Montastrea annularis, Proc. R. Soc. London Ser. B* **196**:291–315.

Goreau, T. J., 1977b, Seasonal variations of trace metals and stable isotopes in coral skeletons: Physiological and environmental controls, *Proc. 3rd Int. Coral Reef Symp.* **2**:425–430.

Hudson, J. H., Shinn, E. A., Halley, R. B., and Lidz, B., 1976, Sclerochronology: A tool for interpreting past environments, *Geology* **4**:361–364.

Nozaki, Y., Rye, D. M., Turekian, K. K., and Dodge, R. E., 1978, A 200 year record of carbon-13 and carbon-14 variations in a Bermuda coral, *Geophys. Res. Lett.* **5**:825–828.

Rhoads, D. C., and Stanley, D. J., 1966, Transmitted infrared radiation, *J. Sediment. Petrol.* **36**:1144–1149.

Stearn, C. W., and Colassin, C., 1979, A simple underwater pneumatic hand drill, *J. Paleoontol.* **53**:1257–1259.

Weber, J. N., and White, E. W., 1974, Activation energy for skeletal aragonite deposited by the hermatypic coral *Platygyra spp., Mar. Biol.* **26**:353–359.

Weber, J. N., and White, E. S., 1977, Caribbean reef corals *Montastrea annularis* and *Montastrea cavernosa*—Long term growth data as determined by skeletal X-radiography, in: *Reefs and Related Carbonates—Ecology and Sedimentology* (S. H. Frost, M. P. Weiss, and J. B. Saunder, eds.), *AAPG Studies in Geology* No. 4, pp. 171–179, AAPG, Washington, D.C.

Preparation and Examination of Skeletal Materials for Growth Studies

Methods of Preparing Fish Sagittae for the Study of Growth Patterns

GIORGIO PANNELLA

1. Extraction

Two methods are commonly used to extract otoliths from their cranial position:

1. An incision is made along the medial ventral line, the branchial isthmus, and the branchial apparatus, freeing the area at the base of the brain case where otolith capsules are located. In certain groups (i.e., the Cyanidae), the capsules are quite large and evident and easily opened with an incision of the medial wall. In others, the capsules are not visible but can be located by careful incision in the medial area where the vertebral column joins the cranium. I personally prefer this method to the one described below.
2. An incision is made in the dorsal area of the cranium extending posteriorly from the eyes to the posteroventral part of the brain case into the otolith capsules (Fitch, 1951; Bas, 1960). This method requires cutting through cranial bones, which, in some fish, can be quite hard. After extraction, the otoliths are washed in distilled water or alcohol, freed of tissue fragments, and dried.

2. Preparation of Sagittae for Observation with a Dissecting Microscope

Several methods have been suggested to emphasize the contrast between hyaline and opaque zones in unsectioned (whole) otoliths. Contrast

GIORGIO PANNELLA • Department of Geology, University of Puerto Rico, Mayaguez, Puerto Rico 00708.

between these zones increases when sagittae are immersed in a liquid and observed in reflected light against a dark or black background. Different types of immersion liquids have been used. Water increases the contrast only slightly. Immersion of otoliths in glycerine or 50% glycerine and 50% water for at least a week improves contrast. In the method of Lawler and McRae (1961), the otoliths are kept in 50% water and 50% glycerine for 4 months and then heated at 170–195°C for 5–20 min. This method is time-consuming and does not seem to improve the contrast in sagittae of tropical fishes.

Another method requires the immersion of otoliths for 48 hr in creosote. Observations are then made in polarized, transmitted light (Pino, 1961).

Immersion in xylol also improves the contrast between zones, but is unpleasant and dangerous to the observer breathing the fumes.

A dyeing technique has been suggested to increase the contrast between zones (Albrechtsen, 1968). A dye is prepared by adding 0.05 g "Methyl violet B" to 30 ml distilled water, and 1 ml concentrated (38%) HCl to the solution after the dye has been completely dissolved. The solution must be used within a few hours. The solution is applied to sectioned otoliths with a soft brush; within 20–40 sec, the solution is neutralized by calcium carbonate from the otolith and changes in color to a light violet. The solution is absorbed by the organic matrix and winter zones in temperate fish. The method works less satisfactorily in tropical sagittae.

The otolith-burning technique consists of carbonizing the organic matrix of the otolith by keeping it in an alcohol or gas flame for 10–30 sec [according to the size of the otolith and the temperature of the flame (Christensen, 1964, p. 73)]. The otolith is then broken with a needle pressed in the center of the convex side. The method works well in temperate and boreal sagittae, but not in tropical otoliths. Variations of this method have been proposed (Chugunova, 1963).

Metallographic methods of examining otolith surface structures have been applied to the study of annual zones (Wiedeman Smith, 1968). The sagittae must be embedded as described below, cut, polished, and etched with acid, and examined under a metallographic microscope. The difference in reflectivity of hyaline and opaque zones is highly enhanced by this technique.

3. Embedding

Whenever sagittae are to be sectioned, embedding techniques are suggested. Many embedding media have been used, ranging from gypsum

(Taning, 1938) to epoxy (Pannella, 1971). The embedding process protects the otoliths from breaking when sectioned or ground and permits better control of the surface to be polished so that it can be made perfectly planar. Depending on size, sagittae are either covered with the embedding media in a small container or attached to a glass slide with epoxy (e.g., Epon 812 or 815, Miller Septhenson Chemical Co., P.O. Box 628, Danbury, Connecticut 08610) or with Canada balsam. Before being embedded, the otolith should be firmly positioned in such a way as to facilitate the cut along the desired section after the embedding medium has cured.

4. Thin-Sectioning and Preparation of Acetate-Peel Replicas

The techniques described to enhance the contrast between density-different zones in sagittae do not work satisfactorily with tropical specimens and when detailed subseasonal growth patterns have to be studied. To observe microstructural features, either thin sections or acetate-peel replicas must be prepared.

Thin sections of otoliths offer an advantage over peel replicas by preserving some of the original color of the mineralized tissue, but the sections must be ultrathin (5–20 μm) to resolve the thin daily increments. Thin sections also require more preparation time and care than replicas.

To make acetate-peel replicas, the technique used is the same one described for molluscan shells (Pannella and MacClintock, 1968; Pannella, 1971) (also see Appendix 1.A.2). A major difficulty is encountered in attempting to obtain a section that cuts growth surfaces at right angles. Because of the mode of sagitta growth (see Chapter 15), the sagitta plane, perpendicular to the growth surfaces, is curved, and no planar section can follow it along the total length of the sagitta. Dorsoventral transverse sections perpendicular to growth surfaces are easier to prepare, but are less desirable because they show discontinuous sequences of increments. The most satisfactory solution is to break the sagitta at the center in a dorsoventral direction and polish the two parts separately. The longitudinal curvature of sagittae is thus decreased, and each part can be ground down to the central sagittal plane with loss of only a small area.

Procedures to prepare a sagittal section of a medium-sized (6–15 mm) convex sagitta are as follows:

1. Break the sagitta at the center along a dorsoventral direction, using thumb and index finger of both hands or by pressing a needle downward in the central nuclear area. After a few attempts, the ability of breaking by hand is easily acquired. For large sagittae,

cutting with a diamond radial saw may be better. Even if the break does not produce bilaterally equal parts, the two parts, if not too small, are still usable.

2. Grind the two parts using 1100 or 2600 grade carborundum powder (American Optical Co.), checking frequently under the binocular microscope to see whether growth lines are visible. When the central plane is reached, many growth lines will be apparent. This is the most delicate part of the preparation. A few too many lapidary strokes can "overshoot" the central plane.

3. Replicas of the cut and polished surface can be made immediately after etching the polished surface for 25–60 sec with 1% HCl solution. The etched surface is then washed and dried. Thin acetate sheets (1/10,000 or 1/20,000 of an inch) are preferable to thick acetate sheets. The details of applying and peeling acetate sheets from sections are covered in Appendix 1.A.1.

4. Alternatively, after being acid-etched, the polished surface may be cemented to a microscope slide and the other side ground down until a thin section (10–20 μm thick) is obtained. If the thin section is not covered with glass, peel replicas can be made directly from the thin section after etching.

5. Acetate-peel replicas must be trimmed as closely as possible and mounted between a microscope glass slide and a cover glass with transparent tape immediately after preparation.

Taubert and Coble (1977) have followed the methods above for small (≈6 mm) otoliths and a variation of these methods for larger sagittae that were ground as horizontal sections parallel to the anteroposterior axis. Before the nucleus was reached in the grinding process, the polished surface was attached to a slide with Canada balsam and the other surface ground until the nucleus was reached. The preparation was then etched (15–45 sec with 1% HCl) and viewed under an optical microscope after placing a drop of glycerine on the slide surface or immersing it in water or ethanol for a few hours to enhance increment contrast.

5. Marking Sagittae for Growth-Pattern Studies

The reconstruction of the ontogeny of a fish from its growth record begins from the last event recorded by the increment at the margin of sagittae before death. This event is generally known and is an essential reference point in time. In many instances, a second reference point during the life of the fish is useful to check the effect of physiological and environmental events on growth patterns.

The problem is to induce the fish to produce a stress notch that is different from other stress marks but not so stressful as to impair or kill the fish. The stress should interrupt otolith growth only temporarily, but for a long enough interval to leave a permanent mark.

The time of transplantation from natural marine conditions to experimental conditions in an aquarium is a traumatic event that leaves a stress mark in the sagitta. In postjuveniles, however, this mark is difficult to separate from other kinds of interruption marks. Starvation for 2 or 3 days will also produce an interruption of otolith growth. In *Tilapia*, this technique produces well-defined marks in the sagitta. Creating a weak (2–6 ohms) electrical field for a few hours in an aquarium affects the fish, and the trauma is marked in the sagitta. Chemical notching has also been used. Mugiya (1974) used intraperitoneal injections of tetracycline in a dose of 80 mg/kg body weight. Calcium-45 has also been used, and it was found to be incorporated in the sagittae of rainbow trout within 72 hr after the intraperitoneal injection (Mugiya, 1974).

In summary, any event that traumatizes the fish is recorded in the growth patterns. The selection of a notching method depends on many parameters (e.g., age, habitat, species) and should be done after a series of preliminary experiments.

References

Albrechtsen, K., 1968, A dyeing technique for otolith age reading, *J. Cons. Coms. Int. Explor. Mer* **32**:278–280.

Bas, C., 1960, Consideraciones acerca del crecimiento de la Caballa (*Sgomber sgomberus*) en el Mediterraneo espanol, *Inv. Pesq. Patron. J. Cievra* **16**:13–38.

Christensen, J. M., 1964, Burning of otoliths, a technique for age determination of soles and other fish, *J. Cons. Coms. Int. Explor. Mer* **29**:78–81.

Chugunova, M. I., 1959, *Age and Growth Studies in Fish* (English translation by A. N. Severtsov), National Science Foundation, Washington, D.C., 132 pp.

Fitch, J., 1951, Age and composition of the southern California catch of Pacific mackerel 1939–40 through 1950–1951, *Calif. Dept. Fish Game Bur. Mar. Fish. Fish. Bull.* **83**:1–290.

Lawler, G. H., and McRae, G. P., 1961, A method for preparing glycerin-stored otoliths for age determination, *J. Fish. Res. Board Can.* **18**(1):47–50.

Mugiya, Y., 1974, Calcium-45 at the level of the otolithic organs of rainbow trout, *Bull. Jpn. Soc. Sci. Fish.* **40**(5):457–463.

Pannella, G., 1971, Fish otoliths: Daily growth layers and periodical patterns, *Science* **173**:1124–1127.

Pannella, G., and MacClintock, C., 1968, Biological and environmental rhythms reflected in molluscan shell growth, *J. Paleontol. Mem.* **42**:64–80.

Pino, Z. R., 1962, Estudios estadisticos y biologicos sobre la biajaiba (*Lutjanus synagris*), *Notas Invest. Cent. Inv. Pesq. (Habana)* **4**:1–91.

Taning, A. V., 1938, Method for cutting sections of otolith of cod and other fish, *J. Cons. Coms. Int. Explor. Mer* **13**(2):238–240.

Taubert, B. D., and Coble, D. W., 1977, Daily rings in otoliths of three species of *Lepomis* and *Tilapia mossambica, J. Fish. Res. Board Can.* **34**(3):332–340.

Wiedemann Smith, S., 1968, Otolith age reading by means of surface structure examination, *J. Cons. Coms. Int. Explor. Mer* **32**:270–277.

Preparation and Examination of Skeletal Materials for Growth Studies

Preparation of Polychaete Jaws for Growth-Line Examination

P. J. W. OLIVE

Polychaete jaws have not received a great deal of attention; the methodology for determination of their micromorphology is certainly underdeveloped. It is to be hoped that techniques developed primarily for the analysis of skeletal material of other organisms (see other sections of this appendix and Appendix 3) will in the future be modified for use with polychaete jaws.

1. Preparation of clean jaws for light- and electron-microscopic examination

    a. Dissect out the jaws with as little adhering musculature as possible.
    b. Transfer the jaws to a watch glass containing 1 N sodium hypochlorite. Leave the jaws in this solution for one to several minutes, depending on the size of the jaws. Since the jaws are soluble in 1 N sodium hypochlorite, they should be watched almost continuously while in this solution; special care is necessary to ensure that the edge of the jaw is not dissolved, particularly when the outer growth line is near the edge.
    c. Remove the sodium hypochlorite and wash the jaws with water.
    d. Clean the jaws ultrasonically by exposure to ultrasonic vibration at 20 kcycles/sec while the jaws are suspended in water.
    e. Mount the jaws in glycerin and view with transmitted light or mixed transmitted and incident light, or use Zeiss Nomarski interference microscopy to enhance the contrast of the growth lines.

P. J. W. OLIVE • Department of Zoology, University of Newcastle Upon Tyne, Newcastle, England NE1 7RU.

f. For scanning electron microscopy (SEM), the jaws should be air-dried after ultrasonication, mounted on appropriate SEM stubs, and coated (under vacuum) with gold, gold–palladium, or a combination of gold and carbon.

This method yields jaw specimens that are free of adhering epithelial and muscle tissue. Jaws of the Glyceridae and Eunicidae can also be prepared by this method, but they are not rendered translucent by it, and can be viewed only with reflected light. After treatment with sodium hypochlorite, the jaws of eunicids usually become disassociated.

2. Preparation of bleached scolecodonts for examination in transmitted light

Fossilized polychaete jaws, known as *scolecodonts*, are opaque; they can be rendered translucent by prolonged exposure (up to about 24 hr) to a bleaching solution.

a. Schulze's treatment [method based on Schwabb (1966)]
   (1) Bleach the scolecodonts with a mixture of 3 parts nitric acid and 2 parts potassium chlorate. Treatment may continue for up to 24 hr.
   (2) Nitric acid and potassium chlorate release humic acids; these may be removed by treatment with NH_4OH or KOH. When humic acids are not present, the basic reagent may be omitted.

b. Sodium hypochlorite [method based on Tasch and Shaffer (1961)]
   Treatment for one to several hours with a solution of sodium hypochlorite may also render scolecodonts translucent. The commercial products Clorox, a 5.3% solution of sodium hypochlorite, and Domestos can be used. The Clorox should be replenished during bleaching and not allowed to dry out. The reaction can be slowed by the addition of water; some scolecodonts, or parts of scolecodonts, may be completely digested in undiluted Clorox.

References

Schwabb, K. W., 1966, Microstructure of some fossil and recent scolecodonts, *J. Paleontol.* **40**:416–423.

Tasch, P., and Shaffer, B. L., 1961, Study of scolecodonts by transmitted light, *Micropaleontology* **7**:369–371.

Bivalve Shell Mineralogy and Microstructure

Part A

Selected Mineralogical Data for the Bivalvia

JOSEPH GAYLORD CARTER

Introduction

In the years following the work of Bøggild (1930), numerous investigators have documented the distribution of aragonite and calcite in bivalve shells. Although several of these works are discussed in Chapter 2, data from many additional references are summarized only diagrammatically in Chapter 2, Fig. 1. Table I of this appendix documents the more important mineralogical data used in the construction of Fig. 1 in Chapter 2, while Tables II and III provide new information for mytilacean shells and for a variety of calcified accessory structures, respectively. These tables are introduced below with notes and references (numerical superscripts in the tables) provided at the end of this appendix. The reader is referred to Chapter 2 for discussion of these data.

Table I. Taxonomic Summary of Original Mineralogy for Selected Bivalves

Table I is not intended to represent an exhaustive summary of mineralogical data for fossil bivalves. Rather, this table is selective in that it includes only (1) the geologically oldest representatives of each bivalve superfamily for which mineralogical data have been published and (2) a few more recent species that differ mineralogically from most representatives of their superfamily (e.g., some Lucinacea, Veneracea, and Chamacea). Except for the latter three superfamilies, the earliest known shell

JOSEPH GAYLORD CARTER • Department of Geology, University of North Carolina, Chapel Hill, North Carolina 27514.

627

Table I. Taxonomic Summary of Original Mineralogy for Selected Bivalves
Superscript numbers refer to Notes and References

Superfamily	Taxon	Geological age	Original mineralogy
Nuculacea	*Nuculoidea* cf. *N. opima* (Hall)[22]	Devonian	Arag.[1,27]
	"*Nucula*" sp. A	Carboniferous	Arag.[2,52]
	Nucula stantoni Stephenson	Cretaceous	Arag.[2,7]
Nuculanacea	"*Leda*" sp.	Carboniferous	Arag.[48]
	Nuculites oblongatus Conrad	Devonian	Arag.[1,27]
	"*Yoldia*" sp.	Carboniferous	Arag. and calc.(?)[6,52]
Solemyacea	*Solemya parallela* Beede and Rogers	Carboniferous	Arag.[1,7]
Cyrtodontacea	*Ptychodesma knappianum* (Hall and Whitfield)	Devonian	Arag.[1,3,28]
Arcacea	*Cucullaea obesa* (Pictet and Roux)	Cretaceous	Arag.[2,32]
	Striarca perovalis (Conrad)	Cretaceous	Arag.[2,7]
	Trigonarca elongata Stephenson	Cretaceous	Arag.[2,7]
	Parallelodon sp.	Carboniferous	Arag. and calc.(?)[2,52]
Limopsacea	*Glycymeris* sp.	Cretaceous	Arag.[1,7]
	Limopsis scalacor (Sowerby)	Eocene	Arag.[2,50]
Ambonychiacea	*Myalina recurvirostris* Meek and Worthen	Carboniferous	Arag. and calc.[3,25]
	"*Byssonychia*" (= junior synonym of *Ambonychia*)[42]	Ordovician	Arag. and calc.[3,41]
	Gosseletia triquetra (Conrad)	Devonian	Arag. and calc.[1,27]
Pteriacea	*Actinodesma erectum* (Conrad)	Devonian	Arag. and calc.[1,27]
	Ptychopteria (*Cornellites*) *fasciculata* (Goldfuss)	Devonian	Arag. and calc.[1,3,27]
Pinnacea	*Pinna flexistria*[8]	Carboniferous	Arag. and calc.[3,25]
Pectinacea	*Limipecten morsei* Newell	Carboniferous	Arag. and calc.[2,7,38]
	Chaenocardia ovata Meek and Worthen	Carboniferous	Arag. and calc.[48]
	Euchondria	Carboniferous	Arag. and calc.[39]
	Oxytoma inequivalvis (J. Sowerby)	Jurassic	Calc.[2,32]
	Pecten aequivalvis [= *Pseudopecten equivalvis* (J. Sowerby)]	Jurassic	Arag. and calc.[25]
	?*Placunopsis undulata*[8]	Cretaceous	Calc.[25]
Limacea	Several species	Jurassic and Cretaceous	Arag. and calc.[25]
Anomiacea	*Anomia olmstedi* Stephenson	Cretaceous	Arag. and calc.[2,7]

Table I. (*Continued*)

Superfamily	Taxon	Geological age	Original mineralogy
	Anomia argentaria Morton	Cretaceous	Arag. and calc.[2,7]
Ostreacea	*Enantiostreon spondyloides* (Goldfuss)	Triassic	Calc.[50]
	Ostrea monotiscaprilis Klipstein	Triassic	Arag. and calc.[50]
Modiomorphacea	*Modiomorpha concentrica* (Conrad)	Devonian	Arag.[1,3,7]
Mytilacea	*Volsellina* sp.	Carboniferous	Arag.[2,52]
	"*Lithodomus*" *lingualis* [= *Lithophaga lingualis* (Phillips)]	Carboniferous	Outer layer calc.[25]
	Mytilus mirabilis[8]	Jurassic	Arag. and calc.[5,25]
Trigoniacea	*Myophoria kejersteini* (Münster)	Triassic	Arag.[4,32]
	Trigonia bronni[8]	Jurassic	Arag.[25]
Unionacea	*Unio andersoni* Hudson	Jurassic	Arag.[50]
Megalodontacea	*Pachyrisma* sp.	Triassic	Largely or entirely arag.[3,35]
	Pachymegalodon crassa (Böhm)	Jurassic	Arag.[2,35]
Hippuritacea	*Eodiceras eximium* (Bayle)	Jurassic	Arag. and calc.[3,45]
	Apricardia toucasi[8]	Cretaceous	Arag. and calc.[1,45]
	Plagioptychus partschii[8,9]	Cretaceous	Arag. and calc.[3,25]
	Hippurites socialis Douvillé	Cretaceous	Arag. and calc.[46]
	Hippurites boehmi Douvillé	Cretaceous	Arag. and calc.[35]
	Biradiolites angulosissimus Toucas	Cretaceous	Arag. and calc.[45]
Lucinacea	*Lucina parva* Stephenson	Cretaceous	Arag.[2,7]
	Paracyclas rugosa (Goldfuss)	Devonian	Arag.[3,7]
	Lucina concinna (Damon)	Jurassic	Arag.[4,32]
	Anodontia bialata Pilsbry	Recent	Arag. and calc.[2,7]
Chamacea	*Chama pellucida* Broderip	Recent	Arag. and calc.[2,36,49]
	Chama haueri Zittel	Cretaceous	Shows traces of calcite, but there is no evidence that this

(*Continued*)

Table I. (Continued)

Superfamily	Taxon	Geological age	Original mineralogy
			was an original feature of the shell.[36,51]
Leptonacea	*Erycina* sp.	Eocene	Arag.[25]
Carditacea	*Myoconcha stampensis*[8]	Jurassic	Arag.[5,25]
	Cardita austriacea (Hauer)	Triassic	Arag.[4,32]
Crassatellacea	*Astartella* sp.	Carboniferous	Arag.[2,48,52]
	Crassatellites conradi (Whitfield)	Cretaceous	Arag.[2,7]
Cardiacea	*Protocardia parahillana* Wade	Cretaceous	Arag.[2,7]
	Loxocardium obliquum (Lamarck)	Eocene	Arag.[2,51]
Tridacnacea	Several species	Recent (probably evolved from the Cardiacea[47])	Arag.[51]
Mactracea	*Cymbophora formosa* (Meek and Haydn)	Cretaceous	Arag.[2,32]
Solenacea	*Solena* (*Eosolen*) *plagiaulax* (Cossmann)	Eocene	Arag.[2,32]
Tellinacea	*Tancredia americana* (Meek and Haydn)	Cretaceous	Arag.[2,32]
	Tancredia axiniformis and *T. jarneri*[8]	Jurassic	Arag.[25]
Dreissenacea	*Dreissena brandii*[8]	Eocene	Arag.[2,51]
Arcticacea	"*Cyprina*" *syssollae* and "*C.*" *mosquensis*[8]	Jurassic	Arag.[25]
	Arctica cordiformis (Sowerby)	Cretaceous	Arag.[2,32]
Glossacea	"*Isocardia*" *forchhammeri*[8]	Miocene	Arag.[25]
Corbiculacea	*Corbicula cordata* (Morris)	Eocene	Arag.[2,51]
Veneracea	*Protothaca staminea* (Conrad)	Recent	Arag. and calc.[2,26]
	Cyclorisma cf. *C. vectensis* (Forbes)	Cretaceous	Arag.[2,32]
	Cyprimeria depressa Conrad	Cretaceous	Arag.[2,7]
	Aphrodina regia Conrad	Cretaceous	Arag.[2,7]
Gastrochaenacea	*Gastrochaena linsleyi* Carter	Cretaceous	Arag.[1,7]
Myacea	*Corbula* sp.	Cretaceous	Arag.[2,32]
	Corbula oxynema Conrad	Cretaceous	Arag.[2,7]
Hiatellacea	*Panopea mandibulata* Sowerby	Cretaceous	Arag.[2,32]
Pholadacea	*Martesia procurva* Wade	Cretaceous	Arag.[2,7]
Pholadomyacea	*Gresslya seebachi*[8]	Jurassic	Arag.[25]
Pandoracea	*Thracia oblata* (Sowerby)	Eocene	Arag.[2,32]
Poromyacea	Several species	Recent	Arag.[51]
Clavagellacea	Several species	Recent	Arag.[51]

mineralogies for a superfamily generally do not differ from their more recent descendants. Exceptions to this rule may occur in the Nuculanacea and Arcacea, in which calcite has been reported in Upper Paleozoic species (Yochelson *et al.*, 1967). The latter observations are tentatively regarded as artifacts of diagenesis or sample contamination until they can be verified by microstructural analysis.

The reader may note that despite the occurrence of numerous Upper Paleozoic bivalves with preserved aragonite (Newell, 1938; Stehli, 1956; Yochelson *et al.*, 1967; Hall and Kennedy, 1967), these species are still largely unknown with respect to the distribution of aragonite and calcite in their shell layers. Preserved original aragonite is unknown among Lower and Middle Paleozoic faunas. However, Bøggild (1930) and other workers have successfully made inferences about original shell mineralogy for several Lower and Middle Paleozoic bivalves on the basis of differential preservation of aragonitic and calcitic shell layers. Inference of mineralogy from differential preservation is most reliable where the aragonitic and calcitic shell layers are equally prominent, so that dissolution of one of the layers becomes obvious. Caution must be exercised, however, in applying this criterion to fossils with an extremely thin middle·aragonitic layer (e.g., certain Upper Paleozoic Pectinacea) or with a thin outer calcitic layer (e.g., many Lower Paleozoic Ambonychiacea).

In some instances, diagenetically altered shells may retain relics of their original microstructure, thereby allowing inferences about their original mineralogy based on comparative microstructure. It is not uncommon to find microstructures preserved in calcitic shell layers among Lower Paleozoic bivalves, and relict microstructure is also preserved in (originally) aragonitic shell layers in a few Middle Paleozoic species (see Carter and Tevesz, 1978a,b). In the latter case, it is important to differentiate between diagenetic replacement textures and relict shell microstructures; in certain fossil bivalves, this distinction is not immediately obvious. The reader is referred to Bathurst (1964, 1975), James (1974), and Sandberg (1975) for discussion of replacement textures, and to Appendix 2.B for illustrations of shell microstructures. Neomorphic calcite replacing aragonite may show only faint traces of original accretion banding, or it may preserve the finest details of shell microstructure observable with an optical microscope. It is possible to infer something about original mineralogy from relict nacreous, regularly foliated, and fine complex crossed lamellar microstructures (see mineralogical data in Appendix 2.B). Certain other microstructures occur as both aragonite and calcite in modern bivalves, but their mineralogical varieties may be structurally distinct and therefore diagnostic of one mineralogy or another [e.g., certain crossed lamellar and cone complex crossed lamellar microstructures (see Bøggild, 1930; MacClintock, 1967)]. A few other microstructures (e.g., fibrous prismatic, spherulitic prismatic, and irregular simple prismatic) are not diagnostic of aragonite or calcite. In such cases, differential preservation may provide the only reliable criterion for making inferences about original shell mineralogy.

Table II. New Observations on Mytilacean Shell Mineralogy

Inasmuch as the Mytilacea are the most commonly studied bivalves in connection with environmental controls on shell mineralogy, it is especially important to document their species-level mineralogical variability. Toward this end, several new observations of mytilacean shell mineralogy are presented in Table II and are incorporated into the discussion of mytilacean mineralogy in Chapter 2.

Mytilacean mineralogy can be relatively complex in its ontogenetic and other spatial distributions in the shell. Consequently, complete mineralogical analysis of a mytilacean shell may require a combination of several analytical approaches, including:

1. Microstructural analysis to determine the distribution of the major shell layers and to determine whether there is calcification within the periostracum.
2. X-ray diffraction of scrapings from individual shell layers.
3. Removal of the periostracum by dissolving it in Clorox (5% sodium hypochlorite), followed by thorough rinsing of the shell in tap water, and then Feigl staining (see below) of the air-dried exterior shell surfaces. This procedure identifies spatial variations in outer-layer mineralogy.
4. Feigl staining of polished vertical sections through the shell, in conjunction with microstructural diagnosis, to determine possible mineralogical variations between the outer and inner parts of the outer shell layer.
5. Isolation of the periostracum, dissolution of its organic portion in Clorox, and concentration of any contained calcified periostracal spikes or granules by gravity settling from the Clorox solution. The periostracal origin of the spikes must then be confirmed by histological section of the mantle margins with the periostracum intact.

The use of Feigl solution (Feigl, 1937) to discriminate between calcite and aragonite has been described as a rapid, sure way of differentiating these two calcium carbonate polymorphs in skeletal materials (Schneidermann and Sandberg, 1971, p. 350). I have found that the modified

Table II. New Observations on Mytilacean Shell Mineralogy
Superscript numbers refer to Notes

Taxon[19]	Outer layer	Middle and inner layers	Remarks	Species range
Subfamily Mytilinae				
Mytilus edulis Linnaeus				
YPM 9542 Long Island Sound	Thick violet to white calcitic prismatic	Aragonitic nacreous and prismatic		Arctic Ocean to South Carolina[17]
Perumytilus purpuratus (Lamarck)				
YPM 9120 Bahía de Pucasana, south central Peru	Thick aragonitic granular to prismatic	Aragonitic nacreous and prismatic		Gulf of Guayaquil to Chile[18]
Brachidontes exustus (Linnaeus)				
YPM 8751 Fernando de Noronha	Intermediate thickness, fibrous to spherulitic prismatic, possibly entirely aragonitic	Aragonitic nacreous and prismatic	Feigl staining of the shell exterior suggests the presence of calcite, but X-ray diffraction of scrapings from the outer shell layer shows only aragonite.	North Carolina to Brazil, Uruguay[17]
Ischadium recurvum (Rafinesque)				
YPM 930 West Florida	Calcitic prismatic[10] underlain by thicker aragonitic prismatic	Aragonitic nacreous and prismatic	Larval shell (0.5 mm in length) is aragonitic.	Cape Cod to West Indies[17]
Mytella guyanensis (Lamarck)				
YPM 731 Panama	Thin aragonitic prismatic	Aragonitic nacreous and prismatic		Baja California to northern Peru[18]
Semimytilus algosus (Gould)				
YPM 9624 Callao-LaPunta, Peru	Aragonitic finely prismatic	Aragonitic nacreous and prismatic		Southwest South America[18]
Septifer bilocularis (Linnaeus)				
YPM 9994 Palau	Thin aragonitic irregularly prismatic	Aragonitic nacreous and prismatic		Tropical, but entire range is unknown.

(Continued)

Table II. (*Continued*)

Taxon[19]	Outer layer	Middle and inner layers	Remarks	Species range
	underlain by thicker aragonitic homogeneous			
Trichomya hirsuta (Lamarck)				
YPM 9667 Japan	Calcitic fibrous to irregularly prismatic underlain by aragonitic prismatic[11]	Aragonitic nacreous and prismatic	Larval shell (0.9 mm in length) is aragonitic; periostracum contains aragonitic spikes with hexagonal cross sections.	Tropical and temperate, but entire range is unknown.
Subfamily Crenellinae				
Crenella decussata (Montagu)				
YPM 9995 Frenchman's Bay, Maine	Thick calcitic prismatic	Aragonitic nacreous, prismatic, and irregular complex crossed lamellar	Larval shell (0.9 mm diameter) is aragonitic.	Greenland to North Carolina[17]
Crenella glandula Totten				
YPM 9194 Cape Cod Bay, Massa- chusetts	Thick calcitic prismatic	Aragonitic nacreous, prismatic, irregular complex crossed lamellar, and cone complex crossed lamellar	Larval shell (0.8 mm diameter) is aragonitic.	Labrador to North Carolina[17]
Crenella serica Conrad				
UNC 4920 Upper Cretaceous Ripley Formation, Georgia	Thick calcitic prismatic	Aragonitic nacreous and prismatic	Larval shell (0.6 mm diameter) is aragonitic.	Entire range is unknown.
Gregariella coralliophaga (Gmelin)				
YPM 10132 Bermuda	Thick calcitic homogeneous to prismatic[12]	Aragonitic nacreous and prismatic	Larval shell (0.7 mm in length) is	Bermuda to Brazil[17]

Table II. (Continued)

Taxon[19]	Outer layer	Middle and inner layers	Remarks	Species range
			aragonitic; periostracum contains aragonitic granules with hexagonal cross sections.	
Lioberus castaneus (Say)				
YPM 9996 Sanibel, Florida	Thin calcitic prismatic[10]	Aragonitic nacreous, prismatic, and irregular complex crossed lamellar		Florida to Brazil[17]
Musculus discors (Linnaeus)				
YPM 38 Eastport, Maine	Thin aragonitic homogeneous	Aragonitic nacreous, prismatic, and fine complex crossed lamellar		Arctic Sea to Long Island Sound[17]
Subfamily Lithophaginae				
Lithophaga nigra (Orbigny)				
UNC 4915 Windley Key, Florida	Calcitic prismatic[13]	Aragonitic nacreous and prismatic	Larval shell (1.0 mm in length) is aragonitic.	Bermuda to Brazil[17]
Lithophaga antillarum (Orbigny)				
YPM 9521 Florida	Calcitic prismatic[10]	Aragonitic nacreous and prismatic		Southeast Florida to Brazil[17]
Lithophaga aristata (Dillwyn)				
YPM 9529 Lakeworth, Florida	Calcitic prismatic[14]	Aragonitic nacreous and prismatic		North Carolina to West Indies
Adula falcata (Gould)				
YPM 10000 San Diego, California	Thin aragonitic (?) prismatic underlain by thicker calcitic fibrous prismatic	Aragonitic nacreous and prismatic		Oregon to Baja, California[17]

(Continued)

Table II. (*Continued*)

Taxon[19]	Outer layer	Middle and inner layers	Remarks	Species range
Subfamily Modiolinae				
Modiolus americanus (Leach)				
YPM 10249 Western Atlantic	Thin calcitic prismatic	Aragonitic nacreous and prismatic	Larval shell (1.1 mm in length) is aragonitic.	Bermuda to Brazil[17]
Modiolus capax (Conrad)				
YPM 2329 Pearl Islands, Panama	Intermediate thickness calcitic homogeneous to finely prismatic[10]	Aragonitic nacreous, prismatic, and complex crossed lamellar	Larval shell (0.9 mm in length) is aragonitic.	California to Peru[17]
Modiolus cf. *M. agripetus* (Iredale)				
UNC 9990 Japan	Intermediate thickness calcitic irregular to fibrous prismatic, pink to white, underlain by thinner calcitic(?) irregular to fibrous prismatic to homogeneous	Aragonitic nacreous and prismatic[15]	Larval shell (0.9 mm in length) is aragonitic.	Temperate?
Amygdalum dendriticum (Megerle von Mühlfeld)				
YPM 9997 West Indies	Thin calcitic (?) crust underlain by aragonitic prismatic	Aragonitic nacreous and prismatic		Tropical, but entire range is unknown.
Botula fusca (Gmelin)				
YPM 9998 Bermuda	Intermediate thickness calcitic prismatic, locally absent,[16] underlain by aragonitic homogeneous	Aragonitic nacreous, prismatic, and irregular complex crossed lamellar	Larval shell is aragonitic.	North Carolina to Brazil[17]
Dacrydium vitreum (Holböll)				
YPM 9999 Off Martha's Vineyard,	Thin to absent aragonitic prismatic	Aragonitic crossed lamellar and		Greenland to Gulf of Mexico[17]

Table II. (*Continued*)

Taxon[19]	Outer layer	Middle and inner layers	Remarks	Species range
Massa-chusetts		complex crossed lamellar		
Geukensia demissa (Dillwyn)				
YPM 131 Long Island Sound	Thin calcitic homogeneous crust (violet to white) underlain by thicker aragonitic prismatic	Aragonitic nacreous and prismatic	Larval shell (0.7 mm in length) is aragonitic.	Gulf of St. Lawrence to northeast Florida[17]
Idasola argentea (Jeffreys)				
YPM 9596 Off Martha's Vineyard, Massa-chusetts	Thick calcitic fibrous prismatic	Middle layer alternating nacreous and fine complex crossed lamellar; inner layer irregular complex crossed lamellar	Larval shell (0.4 mm in diameter) is aragonitic.	Deep water[17]

formula for Feigl solution presented by Cheetham *et al.* (1969), and the original formula described by Feigl (1937), can in some instances provide misleading results because of the influence of shell microstructure on stain uptake.

The discriminating property of Feigl solution is its more rapid reaction with aragonite than with calcite, which results from the greater solubility of the former. As described by Huegi (1945), dissolution of the aragonite is followed by a precipitation of MnO_2 and Ag^0 on the crystal surface, thereby staining it black. The less-soluble calcite resists dissolution and staining for a longer period of time, although it will eventually acquire a black stain in Feigl solution. However, extremely small crystal sizes of calcite are characterized by increased solubility (Chave and Schmalz, 1966), so finer-grained microstructures will naturally acquire a Feigl stain more readily than coarser microstructures of the same mineralogy. In addition, the black coating may simply become visible sooner in layers with more exposed crystal surface area, as in a finely crystalline homogeneous shell layer. These microstructure effects are especially striking in *Modiolus* cf. *M. agripetus* (Iredale), in which the coarsely textured inner aragonitic prismatic layer resists Feigl staining much longer than the adjacent aragonitic nacreous layer (Table II). In this case, the nacreous layer acquires a black stain even before the aragonitic prismatic layer begins to darken, and as the latter becomes darker, the outer calcitic prismatic shell layer has already begun to discolor. Philippon (1975) noted that crystal orientation in gastropod aragonitic crossed lamellar microstructure can likewise contribute to a misleading negative reaction with Feigl stain.

Fortunately, misleading *positive* Feigl staining of calcitic shell layers is uncommon in the Bivalvia because calcitic microstructures are generally rather coarsely textured (e.g., most simple prismatic and foliated microstructures). A possible exception occurs in mytilaceans showing an extremely thin, fine-grained calcium carbonate crust beneath the periostracum. This exterior crust may be too thin to be observed by optical microscopy of vertical sections through the shell. However, its presence is apparent using scanning electron microscopy of the shell exterior cleaned of periostracum by dissolution in Clorox. Different parts of this exterior crust can give negative or positive reactions with Feigl stain, even in the same shell [e.g., in Bermuda specimens of *Botula fusca* (Gmelin) (see Table II)]. At present, the differential reaction of portions of this exterior crust with Feigl stain cannot be separated from possible effects of small crystal size and accompanying increased solubility. The exterior crusts in *B. fusca* are unfortunately too thin to be separated from the underlying calcitic layer for independent X-ray analysis.

Aside from these thin, exterior crusts, X-ray diffraction comprises the basis for the present mineralogical determinations, with Feigl staining used as an accessory technique to map the distribution of mineralogy over the shell exterior and in vertical sections.

Table III. Mineralogy of Accessory Calcification in the Bivalvia

Accessory calcification includes any calcification other than that comprising the two shell valves and the periostracum. This generally occurs in the form of tubes or burrow linings, or both, among coral-, rock-, or wood-boring species, e.g., in the Gastrochaenacea, Clavagellacea, Pholadacea, and Mytilacea. Accessory calcification also includes posterior encrustations on the shells of certain *Lithophaga*, and adventitious shell plates and pallets in certain Pholadacea. Mineralogical data for accessory calcification are compiled here to provide a preliminary indication of environmental vs. biological controls on its variability, as discussed in Chapter 2. The mineralogical data in Table III are based on X-ray diffraction, unless indicated otherwise.

Table III. Mineralogy of Accessory Calcification in the Bivalvia
Superscript numbers refer to Notes

Taxon[19]	Accessory structure	Mineralogy
Superfamily Mytilacea		
Lithophaga bisulcata (Orbigny)		
UNC 5204	Middle burrow lining	Arag. and trace of calc.
Castle Harbor, Bermuda	Posterior encrustation	Arag. and calc.
Lithophaga bisulcata (Orbigny)		
YPM 9528	Posterior encrustation	Arag. and trace of
Sanibel Island, Florida		calc.
Lithophaga bisulcata (Orbigny)		
YPM 9456	Posterior burrow lining	Arag.
Discovery Bay, Jamaica	Middle burrow lining	Arag.
	Posterior encrustation	Arag.
Lithophaga plumula (Hanley)		
YPM 2035	Posterior encrustation	Arag.
Panama		
Lithophaga aristata (Dillwyn)		
YPM 658	Posterior encrustation	Calc. and minor arag.
Bay of Panama		
Lithophaga sp.		
UNC 5193	Entire tube	Calc.
Shallow pit near Nocatee, Florida (Pliocene?)		
Lithophaga sp. cf. *L. bisulcata* (Hanley)		
UNC 5201	Posterior burrow lining	Arag.
Myrtle Beach, South	Middle burrow lining	Arag.
Carolina, Pleistocene,	Posterior encrustation	Arag.
Sangamon Age, Canepatch Formation		
Fungiacava eilatensis Soot-Ryen		
YPM 9545	Anterior burrow lining	Arag.
Gulf of Eilat, Red Sea		
Superfamily Gastrochaenacea		
Gastrochaena hians (Gmelin)		
YPM 9416	Anterior "igloo"	Arag.
Whalebone Bay, Bermuda		
Gastrochaena sp. cf. *G. stimpsonii* (Tryon)		
UNC 5202	Entire "igloo"	Arag.
Beaufort, North Carolina		
Gastrochaena sp. cf. *G. stimpsonii* (Tryon)		
UNC 4979	Entire burrow lining	Arag.
Calabash, North Carolina, Pliocene (?), Waccamaw Formation		
Gastrochaena hians (Gmelin)		
YPM 9488	Posterior burrow lining	Arag. and Mg calc.[20]
Discovery Bay, Jamaica	Middle burrow lining	Arag. and possibly

(Continued)

Table III. (*Continued*)

Taxon[19]	Accessory structure	Mineralogy
		minor Mg calc.[20]
	Anterior burrow lining	Arag.
Gastrochaena hians (Gmelin)		
YPM 9484	Posterior burrow lining	Arag. and minor Mg
Discovery Bay, Jamaica		calc.[20]
	Middle burrow lining	Arag.
	Anterior burrow lining	Arag.
Gastrochaena (*Cucurbitula*) *cymbium* Spengler		
YPM 10218a	Anterior "igloo"	Arag.
Mindoro, Philippines		
Spengleria rostrata (Spengler)		
YPM 9480	Posterior burrow lining	Arag.
Discovery Bay, Jamaica	Middle burrow lining	Arag.
	Anterior burrow lining	Arag.
Eufistulana mumia (Spengler)		
YPM 9589	Posterior tube	Arag.
Singapore	Middle tube	Arag.
Superfamily Pholadacea		
Family Pholadidae		
Barnea truncata (Say)		
YPM 10217a	Protoplax	Arag.
Great Sippowisset Marsh,		
Cape Cod, Massachusetts		
Diplothyra sp. cf. *D. smithii* Tryon		
YPM 9903, 9904, 9890, 9899	Mesoplax	Arag.
North of Fort Meyers, Florida,	Callum	Arag.
Pliocene (?)		
Martesia striata (Linnaeus)		
UNC 5199	Mesoplax	Arag.
Sanibel Island, Florida	Callum	Arag.
Family Teredinidae		
Bankia fimbriatula Moll and Roch		
UNC 5194	Posterior burrow lining	Calc.
Beaufort, North Carolina	Pallets	Calc.
Teredo navalis Linnaeus		
UNC 5195	Posterior burrow lining	Calc.
Beaufort, North Carolina	Pallets	Calc.
Psiloteredo megotara (Hanley)		
UNC 5196	Posterior burrow lining	Calc.
Shetland Islands, Scotland		
Uperotus clavus (Gmelin)		
UNC 5197	Posterior burrow lining	Calc.[21]
Australia	Anterior burrow lining	Arag.[21]
Kuphus arenaria Linnaeus		
YPM 9540	Posterior tube	Calc.
Australia		

Table III. (*Continued*)

Taxon[19]	Accessory structure	Mineralogy
Kuphus polythalamia (Linnaeus)		
UNC 5198	Posterior tube	Calc.
Solomon Islands		
Superfamily Clavagellacea		
Penicillus penis (Linnaeus)		
YPM 9588	Anterior tube	Arag.
Singapore	Posterior tube	Arag.

Notes and References

Notes

1. Shell mineralogy is inferred on the basis of analogy with modern shell microstructures.
2. Shell mineralogy was determined by X-ray diffraction.
3. Shell mineralogy is based on inference from differential preservation.
4. X-ray diffraction of whole-shell preparations revealed less than 5% calcite, suggesting that the shell may have been entirely aragonitic originally.
5. Calcite was also present, but was considered to be entirely secondary or a result of contamination.
6. Further analysis is required to verify whether the calcite is original.
7. Carter (personal observation).
8. Species authorship was not provided in the reference.
9. Kennedy and Taylor (1968) verified that aragonite occurs in the inner layers of another Upper Cretaceous species of this genus, i.e., *Plagioptychus aguilloni* (d'Orbigny) from Austria.
10. Feigl staining indicates a possible sparse intermittent aragonitic crust between the calcitic prismatic layer and the periostracum.
11. The outer calcitic prismatic layer covers all the juvenile postlarval shell (up to a shell length of about 2.5 mm), but may become restricted to the rib interspaces in the posterior of the later-formed shell. The rib crests in these later-formed parts of the shell may be aragonitic on the basis of Feigl staining. The anteroventral shell exterior remains largely calcitic in the postlarval shell, and this corresponds with a general lack of radial ribs in this portion of the shell.
12. Feigl staining indicates a possible aragonitic crust between the calcitic prismatic layer and the periostracum, localized to the posterodorsal shell surface and along a radial band from the umbones to the medial–ventral shell margin (i.e., along a radial sulcus).
13. Feigl staining indicates a possible aragonitic crust anteriorly between the calcitic prismatic layer and the periostracum.
14. Feigl staining indicates a possible aragonitic crust between the calcitic prismatic layer and the periostracum.
15. The inner aragonitic irregularly prismatic layer (myostracal?) resists Feigl staining longer than the nacreous layers but not as long as the outer calcitic prismatic layer, perhaps because of its coarse texture.
16. Feigl staining of the shell exterior with the periostracum removed by bleaching in Clorox (sodium hypochlorite) indicates that the larval shell and early postlarval shell exterior

(to a length of about 1.5–2.0 mm) are probably aragonitic. Thereafter, the shell exterior is predominantly calcitic, but the outer calcitic prismatic shell layer is occasionally absent, especially on shell margins secreted by a withdrawn mantle margin. In other places on the postlarval shell, aragonite (or fine-grained calcite?) may form the outermost shell layer in alternating and interfingering concentric bands.

17. Species range from Abbott, R. T., 1974, *American Seashells*, 2nd ed., Van Nostrand, New York.
18. Species range from Olsson, A.. A., 1961, *Panamic–Pacific Pelecypoda*, Paleontological Research Institution, Ithaca, New York.
19. YPM and UNC stand for Yale University Peabody Museum of Natural History and the University of North Carolina, respectively.
20. This high-magnesium calcite may represent contamination from encrusting calcareous algae (see Chapter 2).
21. Feigl staining reveals a distinct but patchy transition from aragonite to calcite toward the posterior of the burrow lining.

References

22. Bailey, J. B., 1975, Systematics, functional morphology, and ecology of Middle Devonian bivalves from the Solsville Member (Marcellus Formation), Chenango Valley, New York, Ph.D. dissertation, University of Illinois, Urbana, 290 pp.
23. Bathurst, R. G. C., 1964, The replacement of aragonite by calcite in the molluscan shell wall, in: *Approaches to Paleoecology* (J. Imbrie and N. D. Newell, eds.), pp. 357–376, John Wiley & Sons, New York.
24. Bathurst, R. G. C., 1975, *Carbonate Sediments and their Diagenesis*, 2nd Engl. ed., Elsevier Scientific, Amsterdam and New York, 658 pp.
25. Bøggild, O. B., 1930, The shell structure of the mollusks, *K. Dan. Vidensk. Selsk. Skr. Naturvidensk. Math. Afd. 9 Raekke* **2**:231–236.
26. Carter, J. G., and Eichenberger, N. L., 1977, Ontogenetic calcite in the Veneracea (Mollusca: Bivalvia), *Geol. Soc. Am. Abstr.*, 1977 Annual Meeting, Seattle, Washington, p. 922.
27. Carter, J. G., and Tevesz, M. J. S., 1978a, Shell microstructure of a Middle Devonian (Hamilton Group) bivalve fauna from central New York, *J. Paleontol.* **52**:859–880.
28. Carter, J. G., and Tevesz, M. J. S., 1978b, The shell structure of *Ptychodesma* (Cyrtodontidae; Bivalvia) and its bearing on the evolution of the Pteriomorphia, *Philos. Trans. R. Soc. London Ser. B* **284**:367–374.
29. Chave, K. E., and Schmalz, R. F., 1966, Carbonate–seawater interactions, *Geochim. Cosmochim. Acta* **30**:1037–1048.
30. Cheetham, A. H., Rucker, J. B., and Carver, R. E., 1969, Wall structure and mineralogy of the cheilostome bryozoan *Metrarabdotos*, *J. Paleontol.* **43**:129–135.
31. Feigl, F., 1937, *Qualitative Analysis by Spot Test*, Nordemann, New York, 400 pp.
32. Hall, A., and Kennedy, W. J., 1967, Aragonite in fossils, *Proc. R. Soc. London Ser. B* **168**:377–412.
33. Huegi, T., 1945, Gesteinsbildend wichtige Karbonate und deren Nachweis mittles Farbmethoden, *Schweiz. Mineral. Petrogr. Mitt.* **25**:114–140.
34. James, N. P., 1974, Diagenesis of scleractinian corals in the subaerial vadose environment, *J. Paleontol.* **48**:785–799.
35. Kennedy, W. J., and Taylor, J. D., 1968, Aragonite in rudists, *Proc. Geol. Soc. London*, No. 1645, pp. 325–331.

36. Kennedy, W. J., Morris, N. J., and Taylor, J. D., 1970, The shell structure, mineralogy and relationships of the Chamacea (Bivalvia), *Palaeontology* **13**:379–413.
37. MacClintock, C., 1967, Shell structure of patelloid and bellerophontoid gastropods (Mollusca), *Peabody Mus. Nat. Hist. Yale Univ. Bull.* **22**:1–140.
38. Newell, N. D., 1938, Late Paleozoic pelecypods: Pectinacea, *Kansas State Geol. Surv. Bull.* **10**:1–23 ("1937").
39. Newell, N. D., 1969, Various families in the Pectinacea, in: *Treatise on Invertebrate Paleontology, Part N, Mollusca 6* (R. C. Moore, ed.), pp. N332–N383, Geological Society of America and University of Kansas Press, Lawrence.
40. Philippon, J., 1975, Mise en évidence par coloration de la répartition calcite/aragonite sur les coquilles de gastropodes actuels et fossiles: Problèmes posés par ces colorations, *C. R. Acad. Sci. Ser. D* **281**:889–892.
41. Pojeta, J., Jr., 1962, The pelecypod genus Byssonychia as it occurs in the Cincinnatian at Cincinnati, Ohio, *Paleontogr. Am.* **4**:169–216.
42. Pojeta, J., Jr., 1966, North American Ambonychiidae (Pelecypoda), *Paleontogr. Am.* **5**(36):129–241.
43. Sandberg, P. A., 1975, Bryozoan diagenesis: Bearing on the nature of the original skeleton of rugose corals, *J. Paleontol.* **49**:587–606.
44. Schneidermann, N., and Sandberg, P. A., 1971, Calcite–aragonite differentiation by selective staining and scanning electron microscopy, *Trans. Gulf Coast Assoc. Geol. Soc.* **21**:349–352.
45. Skelton, P. W., 1974, Aragonitic shell structures in the rudist *Biradiolites*, and some palaeobiological inferences, *Geol. Mediterraneenne* **1**(2):63–74.
46. Skelton, P. W., 1976, Functional morphology of the Hippuritidae, *Lethaia* **9**:83–100.
47. Stasek, C., 1962, The form, growth, and evolution of the Tridacnidae (giant clams), *Arch. Zool. Exp. Gen.* **101**:1–40.
48. Stehli, F. J., 1956, Shell mineralogy in Paleozoic invertebrates, *Science* **123**:1031–1032.
49. Taylor, J. D., and Kennedy, W. J., 1969, The shell structure and mineralogy of *Chama pellucida* Broderip, *Veliger* **11**:391–398.
50. Taylor, J. D., Kennedy, W. J., and Hall, A., 1969, The shell structure and mineralogy of the Bivalvia: Introduction: Nuculacea–Trigonacea, *Bull. Br. Mus. (Nat. Hist.) Zool. Suppl.* **3**:1–125.
51. Taylor, J. D., Kennedy, W. J., and Hall, A., 1973, The shell structure and mineralogy of the Bivalvia. II. Lucinacea–Clavagellacea: Conclusions, *Bull. Br. Mus. (Nat. Hist.) Zool.* **22**(9):255–294.
52. Yochelson, E. L., White, J. S., Jr., and Gordon, M., Jr., 1967, Aragonite and calcite in mollusks from the Pennsylvanian Kendrick Shale (of Jillson) in Kentucky, *U.S. Geol. Surv. Prof. Pap.* **575D**:D76–D78.

Guide to Bivalve Shell Microstructures

JOSEPH GAYLORD CARTER

Introduction

Definitions of the following microstructure groups are provided in Chapter 2 (Section 2.2.1). The microstructure varieties in the Guide are arranged in outline form to simplify their definitions. References, indicated by numerical superscripts, are provided for amplification of these descriptions, or to indicate other interpretations of the terminology. Some of the descriptions used herein differ from those used in previous works in their independence from optical crystallographic criteria (e.g., in the case of composite prismatic and homogeneous structures). Using the terminology herein, shell microstructures are diagnosed on the basis of their major (i.e., first-order) structural arrangement. However, this same terminology can also be applied to substructures, i.e., to second-order configurations, if this finer level of observation is clearly indicated. For example, first-order simple prisms may show a homogeneous substructure, spherulitic prisms commonly show a fibrous prismatic substructure, and first-order lamellae of crossed foliated structure commonly show a regularly foliated substructure.

JOSEPH GAYLORD CARTER • Department of Geology, University of North Carolina, Chapel Hill, North Carolina 27514.

Microstructure Guide

Microstructure group	Microstructure varieties	Mineralogy	Figures
I. Prismatic	A. **Simple prismatic.**[1,3,10,16] Prisms lack a substructure of elongate subunits diverging toward the depositional surface. Prism boundaries are well defined and generally noninterdigitating. Prisms show a moderate length/width ratio.		
	1. **Regular simple prismatic.**[12] Prism cross sections are polygonal and later uniform.	Arag. + calc.	2
	2. **Irregular simple prismatic.** Prism cross sections are nonpolygonal to polygonal and highly variable along their lengths.	Arag. + calc.	3,4
	B. **Fibrous prismatic.**[3,8,9,12,13] As in simple prismatic, but prisms show a large length/width ratio.	Arag. + calc.	5–9
	C. **Spherulitic prismatic.**[3,5,7,11,20] Prisms show a substructure of elongate subunits radiating in three dimensions from a single nucleation site or spherulite toward the depositional surface.	Arag. + calc.	1
	D. **Composite prismatic.**[1,3,8,9,14,16] Prisms show a substructure of elongate subunits radiating in three dimensions from the central, longitudinal prism axis toward the depositional surface. The angle of divergence of these subunits is generally greater toward the edges of the prism than near the center.		
	1. **Denticular composite prismatic.** Three-dimensional divergence of the structural subunits is an artifact of deposition on a strongly curved and denticulated shell margin. See Bøggild[1] (his Plate II, Figs. 5 and 6) for examples in *Nucula*.	Arag.	
	Nondenticular composite prismatic. Three-dimensional divergence of the subunits	Arag. + calc.	11–15,19

	Description	Mineralogy	Ref.
	occurs independently of the shape of the depositional surface.		
3. **Compound composite prismatic.**	Nondenticular composite prisms (see 2 above) diverge in three dimensions from a central, longitudinal axis toward the depositional surface as an artifact of deposition on a strongly curved and denticulated shell margin.	Arag.	10
II. **Spherulitic**[11]	See Chapter 2.		
III. **Laminar**			
A. **Nacreous.**[1,3,15,16,21]	Laminae consist of polygonal to rounded tablets lying essentially parallel to the general depositional surface. Spiral growth of the tablets may locally disrupt the laminar arrangement.	Arag.	16–18
1. **Sheet nacreous.**[16]	Tablets show irregular, stairstep, and/or brick wall stacking modes in all vertical sections.	Arag.	23
2. **Row stack nacreous.**[3,21]	Parallel, elongate tablets show a vertical stacking mode in vertical sections perpendicular to their length, but nonvertical stacking in vertical sections parallel to their length.	Arag.	24
3. **Columnar nacreous.**[16,21]	Tablets show vertical stacking in all vertical sections.	Arag.	25
B. **Regularly foliated.**[1,8,9,10]	Laminae consist of parallel, elongate blades dipping uniformly over large portions of the depositional surface with a single predominant dip direction.	Calc.	26,27
IV. **Crossed**			
A. **Crossed lamellar.**[1,3,12,17,18]	Adjacent aggregations of numerous parallel, elongate subunits show only two predominant dip directions relative to the shell margin.	Arag. + calc.	28–31

(Continued)

Microstructure group	Microstructure varieties	Mineralogy	Figures
	Crossed foliated structure is a particular variety of crossed lamellar structure in which the subunits are calcitic blades or laths (Figs. 30 and 31).		
	B. **Crossed acicular.**[5,6] As in crossed lamellar, except that each aggregation consists of only a few parallel elongate subunits.	Arag.	32
	C. **Complex crossed lamellar.**[1,4,10,14] Adjacent aggregations of elongate subunits show three or more predominant dip directions, or they are arranged (ideally) on the surfaces of stacked cones. **Complex crossed foliated** structure[10] is a particular variety of complex crossed lamellar structure in which the elongate subunits are calcitic blades or laths (for example, see MacClintock[10]).		
	1. **Irregular complex crossed lamellar.**[1,3] Adjacent aggregations of numerous parallel, elongate subunits show three or more predominant dip directions. This may be called **irregular complex crossed foliated** if the subunits are calcitic blades or laths.	Arag. + calc.	32–35
	2. **Fine complex crossed lamellar.**[3] As in irregular complex crossed lamellar, except that the adjacent aggregations consist of only a few elongate subunits.	Arag.	41
	3. **Cone complex crossed lamellar.**[3] Adjacent aggregations of elongate subunits are arranged (ideally) on the surfaces of stacked cones. The angle of divergence of these subunits is generally constant between the center and the edge of each columnar aggregation [compare with composite prismatic structure (I.D above)].	Arag. + calc.	35–40

Unlike composite prismatic structure, the adjacent columns in many complex crossed lamellar structures are highly interdigitated, and they frequently intergrade with irregular complex crossed lamellar structure in the same general shell layer. Calcitic varieties of this structure with lath- or bladelike structural subunits may be called **cone complex crossed foliated**.

D. **Crossed-matted/lineated**.[22] Aggregations of elongate subunits of two particular types intimately mixed in the same shell layer, i.e.: (1) irregular aggregations of radiating subunits and (2) radially elongate aggregations of mutually parallel subunits. This second type of aggregation intergrades with crossed lamellar structure in the inner layer of many Arcacea, where the "lineated" aggregations are structurally similar to alternating first-order crossed lamellae.

Arag.   42–44

V. **Homogeneous**[1,3,8,9,16,18] A. **Homogeneous sensu stricto**.[16] A homogeneous structure with major structural units generally less than 5 microns in diameter.

Arag. + calc.   45–47

B. **Granular**. A homogeneous structure with major structural units generally greater than 5 microns in diameter.

Arag.   47,48

VI. **Isolated Spicules or Spikes**[2,3]

Arag.   49

VII. **Isolated Crystal Morphotypes**[19]

Arag. + calc.   50

References

1. Bøggild, O. B., 1930, The shell structure of the mollusks, *K. Dan. Vidensk. Selsk. Skr. Naturvidensk. Math. Afd. 9 Raekke* **2**:231–326.
2. Carter, J. G., and Aller, R. C., 1975, Calcification in the bivalve periostracum, *Lethaia* **8**:315–320.
3. Carter, J. G., and Tevesz, M. J. S., 1978, Shell microstructure of a Middle Devonian (Hamilton Group) bivalve fauna from central New York, *J. Paleontol.* **52**:859–880.
4. Denis, A., 1972, Essai sur la microstructure du test de Lamellibranches, *Trav. Lab. Paleontol. Fac. Sci. Orsay Univ. Paris,* pp. 1–89.
5. Erben, H. K., and Krampitz, G., 1970, Ultrastruktur und Aminosäuren-Verhältnisse in den Schalen der rezenten Pleurotomariidae (Gastropoda), *Biomineralisation* **6**:12–31.
6. Flajs, G., 1972, Die Ultrastruktur des Schlosses der Bivalvia I, *Biomineralisation* **6**:49–65.
7. Haas, W., 1972, Untersuchungen über die Mikro- und Ultrastruktur der Polyplacophorenschale, *Biomineralisation* **5**:3–52.
8. Kobayashi, I., 1969, Internal microstructure of the shell of bivalve molluscs, *A. Zool.* **9**:663–672.
9. Kobayashi, I., 1971, Internal shell microstructure of Recent bivalvian molluscs, *Sci. Rept. Niigata Univ. Ser. E* **2**:27–50.
10. MacClintock, C., 1967, Shell structure of patelloid and bellerophontoid gastropods (Mollusca), *Peabody Mus. Nat. Hist. Yale Univ. Bull.* **22**:1–140.
11. Mutvei, H., 1964, On the shells of *Nautilus* and *Spirula* with notes on the secretions in non-cephalopod molluscs, *Ark. Zool.* **16**:221–278.
12. Newell, N. D., 1938, Late Paleozoic pelecypods: Pectinacea, *Kansas State Geol. Surv. Bull.* **10**:1–23 ("1937").
13. Newell, N. D., 1942, Late Paleozoic pelecypods: Mytilacea, *Univ. Kansas Publ. State Geol. Surv.* **10**:1–115.
14. Oberling, J. J., 1964, Observations on some structural features of the pelecypod shell, *Mitt. Naturforsch. Ges. Bern* **20**:1–63.
15. Schmidt, W. J., 1923, Bau und Bildung der Perlmuttermasse, *Zool. Jahrb. Anat. Abt. Jena* **45**:1–148.
16. Taylor, J. D., Kennedy, W. J., and Hall, A., 1969, The shell structure and mineralogy of the Bivalvia: Introduction: Nuculacea–Trigonacea, *Bull. Br. Mus. (Nat. Hist.) Zool. Suppl.* **3**:1–125.
17. Uozumi, S., Iwata, K., and Togo, Y., 1972, The ultrastructure of the mineral in and the construction of the crossed-lamellar layer in molluscan shell, *J. Fac. Sci. Hokkaido Univ., Ser. 4* **15**:447–478.
18. Waller, T. R., 1978, Morphology, morphoclines, and a new classification of the Pteriomorphia (Mollusca: Bivalvia), *Philos. Trans. R. Soc. London Ser. B* **284**:345–365.
19. Wind, F. H., and Wise, S. W., Jr., 1976, Organic vs. inorganic processes in archaeogastropod shell mineralization, in: *The Mechanisms of Mineralization in the Invertebrates and Plants* (N. Watabe and K. M. Wilbur, eds.), pp. 369–387, Belle W. Baruch Library in Marine Science, No. 5, University of South Carolina Press, Columbia.
20. Wise, S. W., Jr., 1969, Study of molluscan shell ultrastructures, in: *Scanning Electron Microscopy* (O. Johari, ed.), pp. 205–216, IIT Research Institute, Chicago.
21. Wise, S. W., Jr., 1970, Microarchitecture and mode of formation of nacre (mother of pearl) in pelecypods, gastropods and cephalopods, *Eclogae Geol. Helv.* **63**:775–797.
22. Wise, S. W., Jr., 1971, Shell ultrastructure of the taxodont pelecypod *Anadara notabilis* (Röding), *Eclogae Geol. Helv.* **64**:1–12.

Microstructure Figures 1–50

Figure Captions

The abbreviations UNC and YPM stand for the Department of Geology, University of North Carolina, and the Yale University Peabody Museum of Natural History, respectively. The abbreviation SEM stands for scanning electron microscopy. In the descriptions, "first-order" refers to the major structural units, e.g., "first-order lamellae" of crossed lamellar structure.

Sample Preparations. The microstructure figures are based on four types of sample preparations indicated by number in the captions:

Preparation 1. SEM of gold/palladium-coated fracture surfaces.

Preparation 2. SEM of gold/palladium-coated acetate-peeled surfaces of epoxy-embedded, sectioned, polished, and HCl-etched surfaces. These electron micrographs show the peeled surface, not the acetate peel.

Preparation 3. Light microscopy of acetate peels of epoxy-embedded, sectioned, polished, and HCl-etched surfaces. These light photomicrographs show the acetate peel, not the peeled surface.

Preparation 4. Light microscopy of the natural surface of deposition viewed with unidirectional reflected light.

Figure Labels and Symbols

Abbreviations. The abbreviations used in the figures are as follows: (D) untreated depositional surface; (H) horizontal section or fracture; (O) oblique section or fracture; (R) radial section or fracture; (T) transverse section or fracture.

Arrows. Thin arrow: points toward the nearest shell margin. Thick arrow: points toward the shell interior or the depositional surface, or both.

Scale Bars. Scale bars are labeled in microns.

Figure 1. Aragonitic spherulitic prismatic structure immediately below the periostracum (P) in *Spisula solidissima* Dillwyn, UNC 7034, New Jersey. Preparation 2.

Figure 2. Calcitic regular simple prismatic outer layer of *Pinna bicolor* Gmelin, YPM 6964, Philippines. Preparation 1.

Figure 3. Aragonitic irregular simple prismatic inner layer of *Divalucina cumingi* (Adams and Angas), YPM 9065, New Zealand. Preparation 1.

Figure 4. Aragonitic irregular simple prismatic myostracum of *Cyrtodaria siliqua* (Spengler), YPM 4995, Martha's Vineyard, Massachusetts. Preparation 1.

Figure 5. Aragonitic fibrous prismatic outer layer of *Fragum subretusum* (Sowerby), YPM 9676, Philippines. Preparation 1. This fanlike arrangement occurs in only two dimensions; therefore, the structure is not composite prismatic. The periostracal and crossed lamellar layers appear at the top and bottom of the photograph, respectively.

Figure 6. Close-up of Fig. 5, showing the elongate structural units.

Figure 7. Calcitic fibrous prismatic middle part of the outer layer of *Mytilus californianus* (Conrad), YPM 9526, California. Preparation 1.

Figure 8. As in Fig. 7, but showing the inner part of the outer layer.

Figure 9. As in Fig. 7, but showing the outer part of the outer layer.

Figure 10. Three different sectional views showing the aragonitic compound composite prismatic outer layer of *Donax variabilis* Say, YPM 10074, Sanibel Island, Florida. Preparation 3. The underlying crossed lamellar layer is visible in the radial section.

Figure 11. Three different sectional views showing the aragonitic nondenticular composite prismatic outer layer of *Tapes literata* (Linnaeus), YPM 9679, Philippines. Preparation 3. The lower right photograph shows outer (upper left triangle) and inner (lower right triangle) portions of this shell layer.

Figures 12–15. Nondenticular composite prismatic and other structures in *Venerupis crenata* Lamarck, YPM 9748, New South Wales, Australia. Preparation 3. Figures 12–15 are of the same radial section and are progressively nearer the posterior shell margin. (i) Calcitic nondenticular composite prismatic; (ii) aragonitic nondenticular composite prismatic; (iii) aragonitic crossed lamellar; (iv) aragonitic fine complex crossed lamellar; (v) aragonitic irregular complex crossed lamellar. These four figures are of the same magnification; the bar scale in Fig. 15 represents 100 microns.

Figure 16. Three different sectional views showing the aragonitic spherulitic (not spherulitic prismatic) outer layer of *Anodontia bialata* Pilsbry, YPM 8963, Tsingtao, China. Preparation 3. The periostracum (P) is also visible in the radial section.

Figure 17. Another view of the spherulitic layer in Fig. 16, here showing the contact between a small patch of calcite (*left*) and part of an aragonitic spherulite (*right*). Preparation 2.

Figure 18. SEM of the aragonitic spherulitic structure in Fig. 16. Preparation 2.

Figures 19–22. Calcitic mineralogical aberrations in the outer and middle largely aragonitic shell layers of *Anodontia bialata* (same specimen as in Figs. 16–18). All Preparation 3. (i) Calcitic aberrations, showing nondenticular composite prismatic structure in the larger patch in Fig. 19; (ii) outer layer aragonitic spherulitic structure; (iii) middle layer aragonitic crossed lamellar and cone complex crossed lamellar structures; (iv) inner layer aragonitic irregular complex crossed lamellar structure.

Figure 23. Aragonitic sheet nacreous middle layer of *Pinctada radiata* Leach, YPM 6889, Bahamas. Preparation 1.

Figure 24. Aragonitic row stack nacreous inner layer of *Pinna bicolor* Gmelin, YPM 6964, Philippines. Preparation 1. The fracture is parallel to the direction of nacre tablet elongation.

Figure 25. Aragonitic columnar nacreous middle layer of *Neotrigonia gemma* Iredale, UNC 5427, Sydney, Australia. Preparation 1.

Figure 26. Calcitic regularly foliated inner layer of *Pecten diegensis* Dall, YPM 7856, California. Preparation 1.

Figure 27. Calcitic regularly foliated outer layer of *Anomia simplex* d'Orbigny, YPM 9715, New Haven, Connecticut. Preparation 1. Left valve.

Figure 28. Aragonitic crossed lamellar middle layer of *Barbatia obtusoides* (Nyst), YPM 6420, Japan. Preparation 1.

Figure 29. Close-up of Fig. 28.

Figure 30. Calcitic crossed foliated outer layer in the left valve of *Propeamussium dalli* (Smith), YPM 8387, west of Martinique, Windward Islands, with extremely large first-order lamellae. The branching, interdigitating pattern of the two first-order lamellae is not apparent at this high magnification. The periostracum appears in the lower part of the photograph. Preparation 1.

Figure 31. Close-up of Fig. 30, showing a regularly foliated substructure within a single first-order crossed lamella.

Figure 32. Aragonitic crossed acicular middle layer of *Arctica islandica* Linnaeus, UNC 7033, New Jersey. Preparation 2.

Figure 33. Aragonitic irregular complex crossed lamellar inner layer of *Geloina turgida* (Lea), YPM 9701, Philippines. Preparation 2.

Figure 34. Aragonitic irregular complex crossed lamellar inner layer of *Myrtea spinifera* (Montagu), YPM 9086, England. Only the shapes of the first-order lamellae are apparent at this low magnification. Preparation 4.

Figure 35. Intergradation of irregular and cone varieties of aragonitic complex crossed lamellar structure in the inner layer of *Glauconome virens* (Linnaeus), YPM 10080, Sumatra. Preparation 4. The depositional surface is planar; the apparent conical elevations are optical illusions.

Figure 36. Aragonitic cone complex crossed lamellar inner layer of *Semele purpurascens* (Gmelin), YPM 9646, Lake Worth, Florida. Preparation 4. As in Fig. 35, the depositional surface is planar; the apparent conical elevations are optical illusions.

Figures 37–39. Aragonitic cone complex crossed lamellar inner layer of *Crenella glandula* (Totten), YPM 9194, Cape Cod, Mass. All Preparation 2. The three thin horizontal bands in Fig. 37 are organic-rich layers of irregular simple prismatic structure. The columns that comprise the complex crossed lamellar structure are strongly interdigitated. Figures 38 and 39 show the boundary between two adjacent columns.

Figure 40. Aragonitic cone complex crossed lamellar structure between the outermost spherulitic prismatic layer and the underlying crossed lamellar layer (all exterior to the pallial line) in *Spisula solidissima* Dillwyn, UNC 7034, New Jersey. Preparation 1.

Figure 41. Aragonitic fine complex crossed lamellar layer immediately exterior to the pallial myostracum in *Arctica islandica* Linnaeus, UNC 7033, New Jersey. Preparation 1.

Figure 42. Three different sectional views showing the aragonitic crossed–matted/lineated inner layer of *Anadara notabilis* (Röding), YPM 6530, Naples, Florida. Preparation 3.

Figures 43 and 44. Lineated portion of the aragonitic crossed–matted/lineated inner layer of *Arca ventricosa* Lamarck, YPM 6251, Palau. Preparation 1. Note the tubules in Fig. 44. Figure 44 also shows two dip directions, indicating that the structure is nearly radial crossed lamellar in this part of the shell.

Figure 45. Aragonitic homogeneous *sensu stricto* inner layer of *Cooperella subdiaphana* (Carpenter), YPM 9645, Orange County, California. Preparation 1.

Figure 46. Close-up of Fig. 45.

Figure 47. Aragonitic homogeneous inner part of the outer layer of *Mya arenaria* Linneaus, YPM 10082, New Haven, Conn. Preparation 1. The structure near the top is homogeneous *sensu stricto*. The structure near the bottom is transitional between this and granular structure.

Figure 48. Close-up of Fig. 47.

Figure 49. Aragonitic periostracal spikes isolated from the posterior periostracum of *Spengleria rostrata* (Spengler), YPM 9473, Florida, by dissolution of the organic component of the periostracum in 5% NaOCl (Clorox). The spikes have been rinsed in tapwater prior to air-drying and have been gold/palladium-coated for SEM.

Figure 50. Aragonitic isolated crystal morphotypes in a void between the periostracum and the normal outer prismatic shell layer in the posterior of *Spengleria rostrata* (same specimen as Fig. 49). Preparation 1.

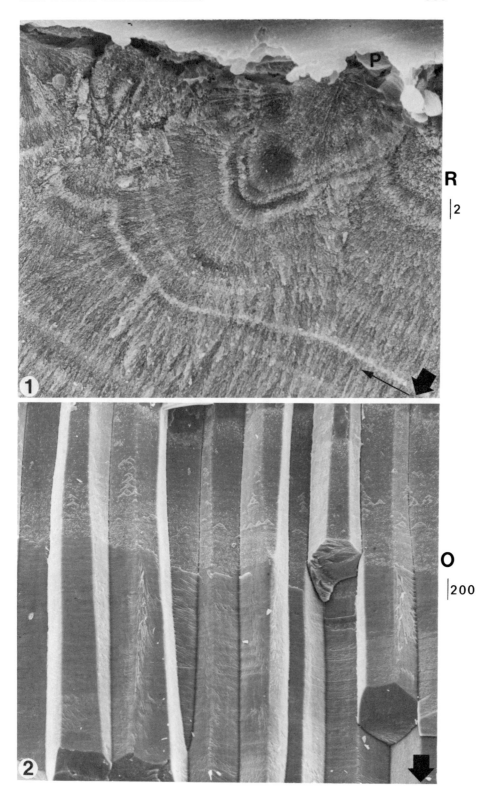

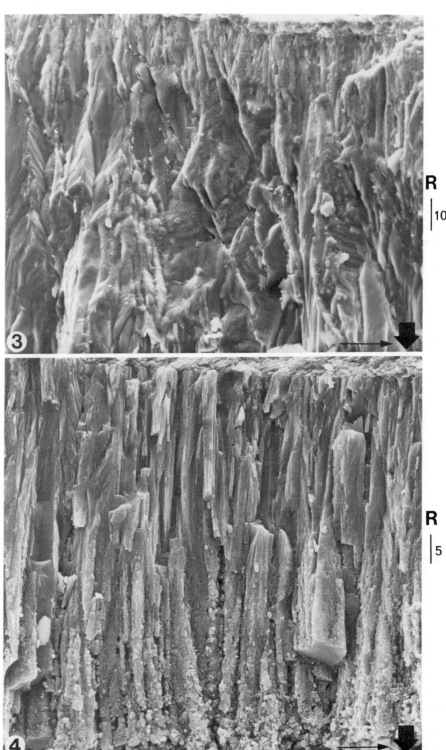

R
|10

R
|5

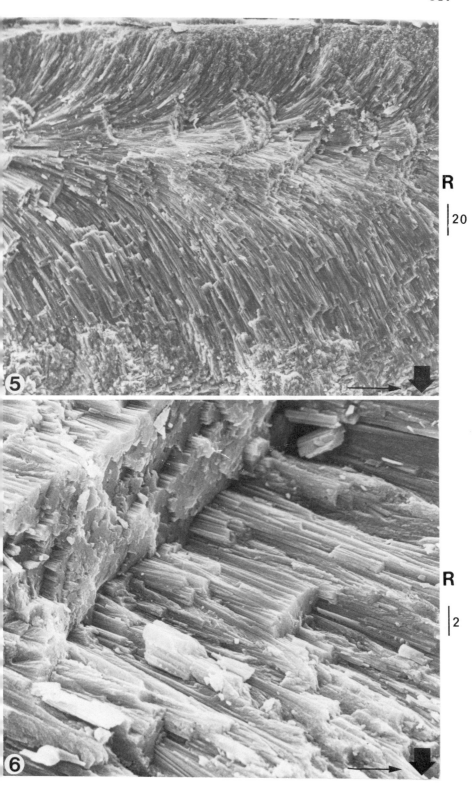

R

20

5

R

2

6

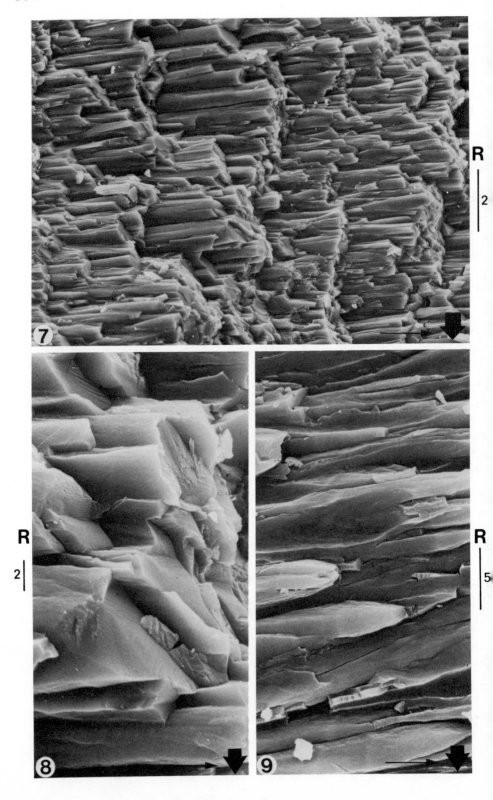

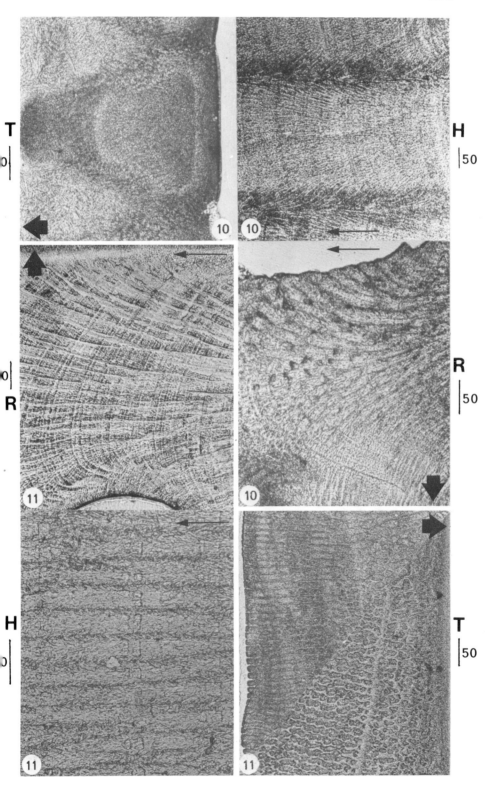

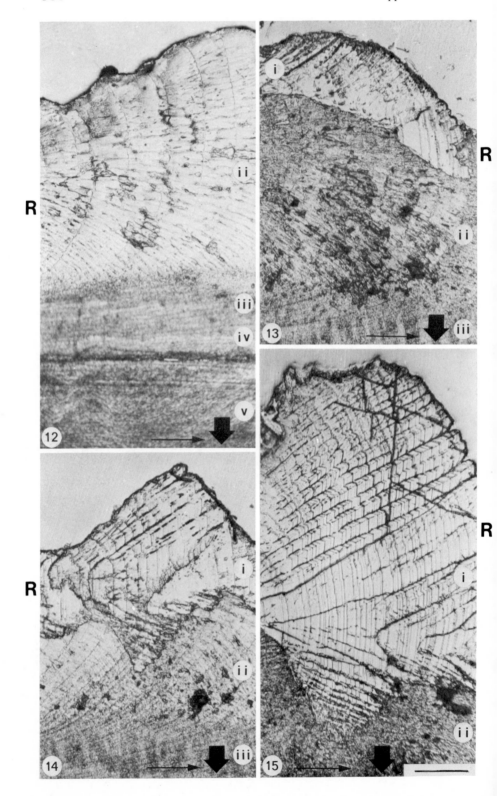

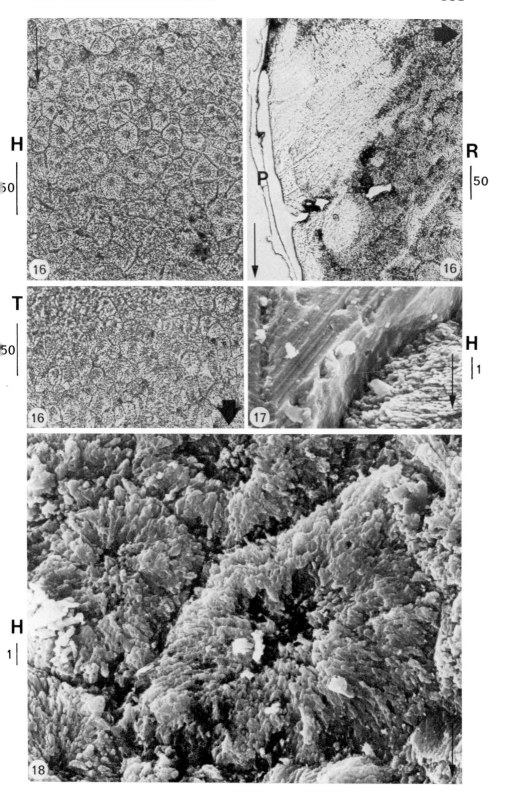

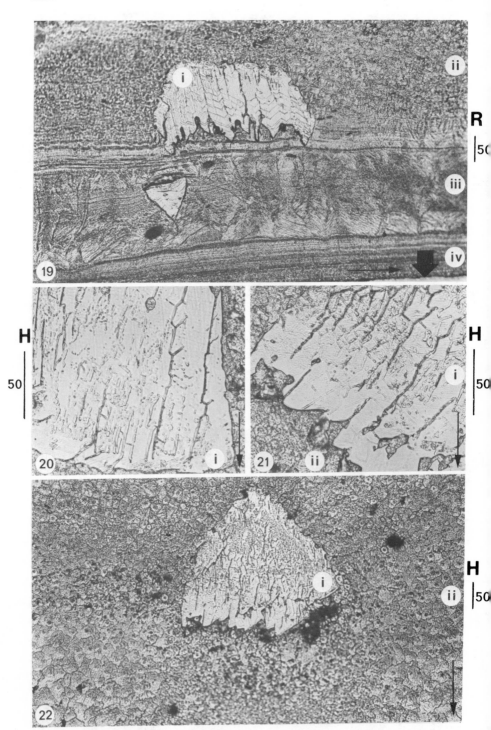

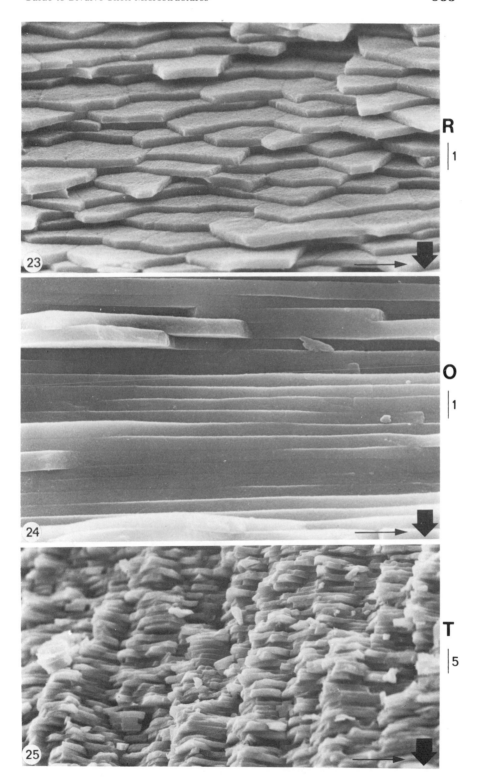

R

| 1

O

| 1

T

| 5

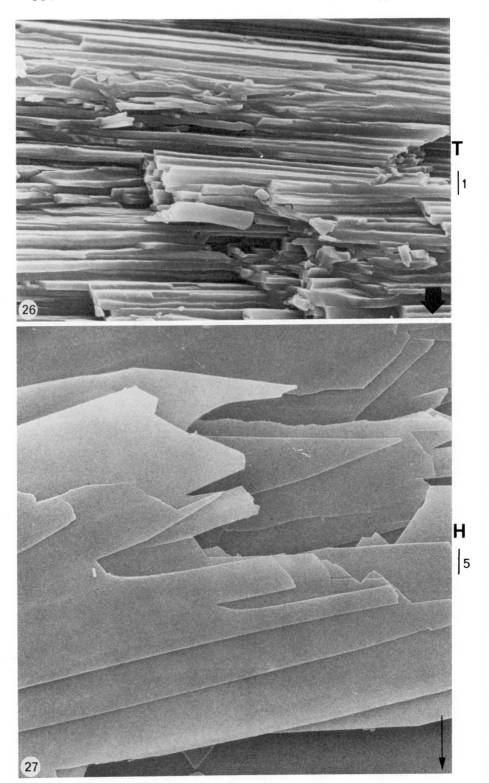

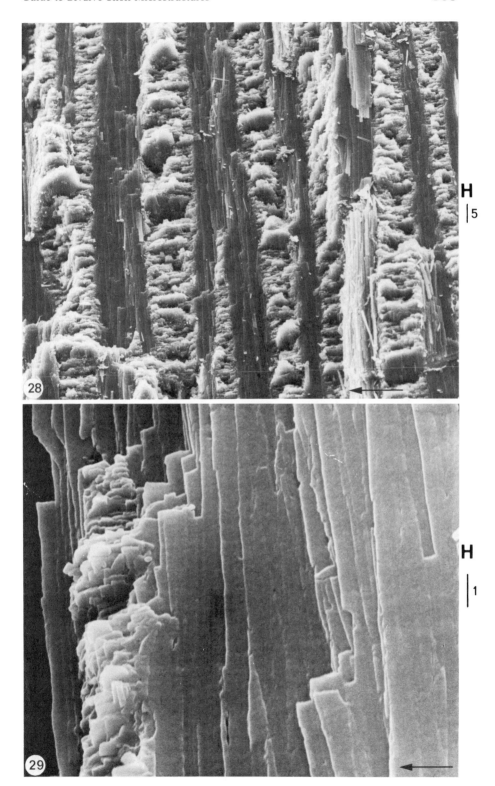

H

|5

H

|1

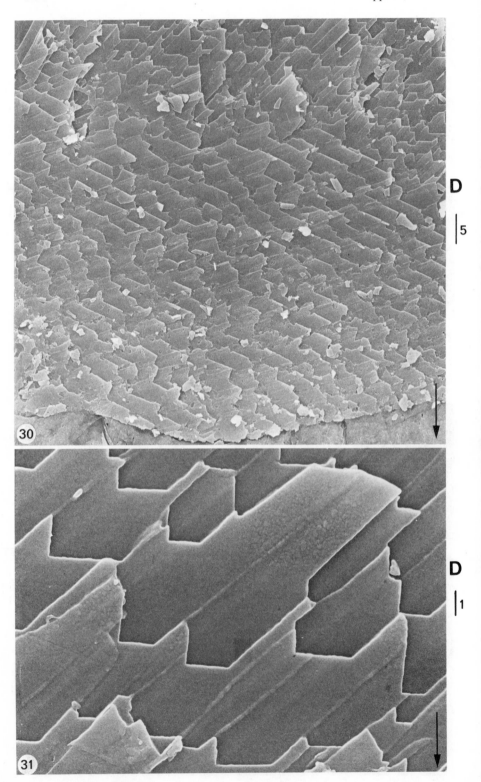

D

5

30

D

1

31

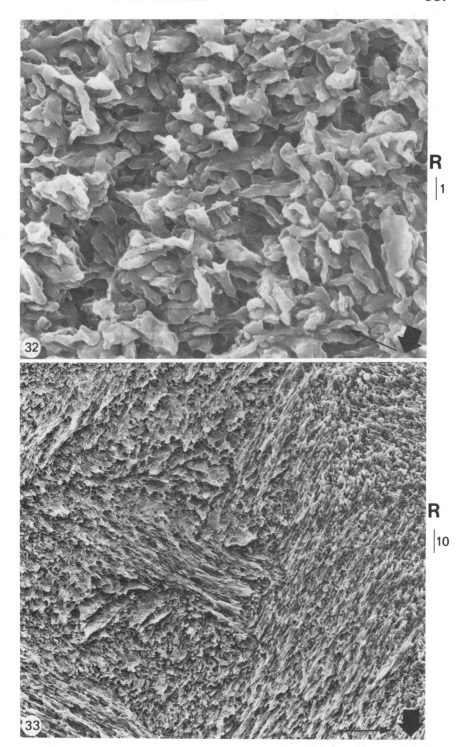

R

|1

R

|10

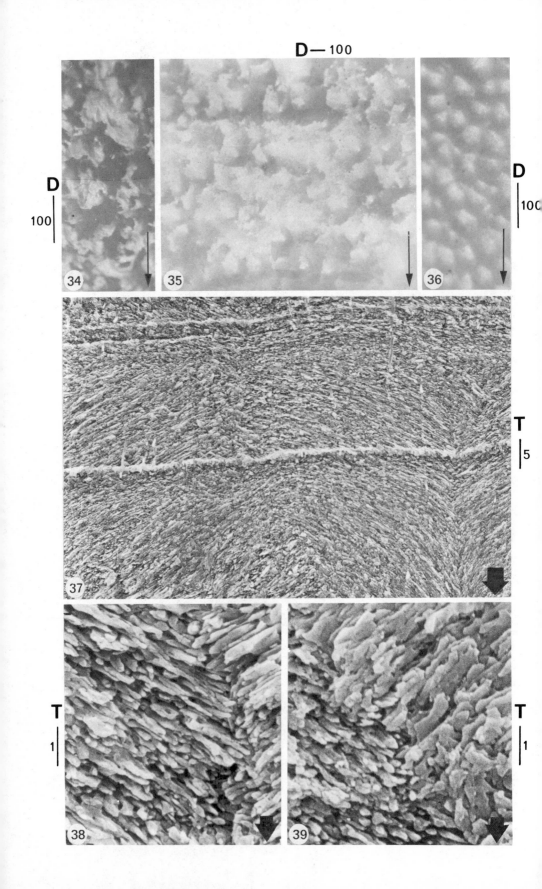

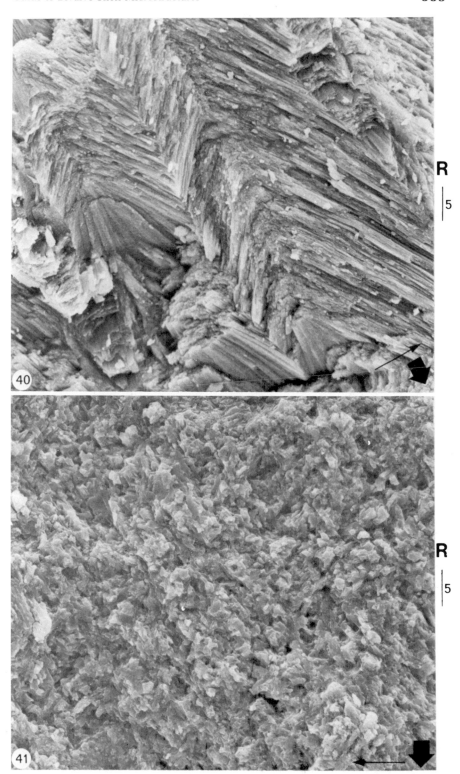

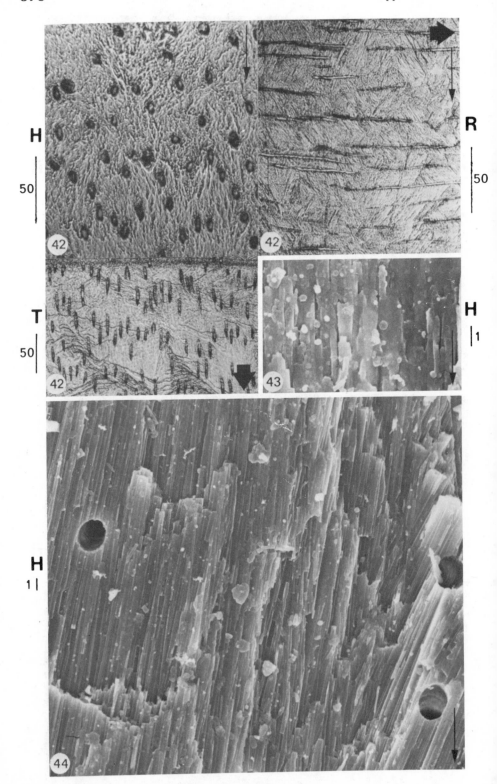

H
50
42

R
50
42

T
50
42

H
1
43

H
1

44

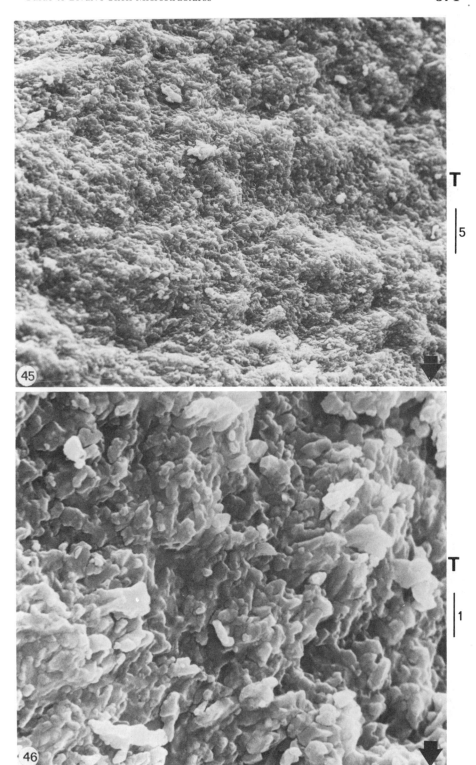

R

5

R

1

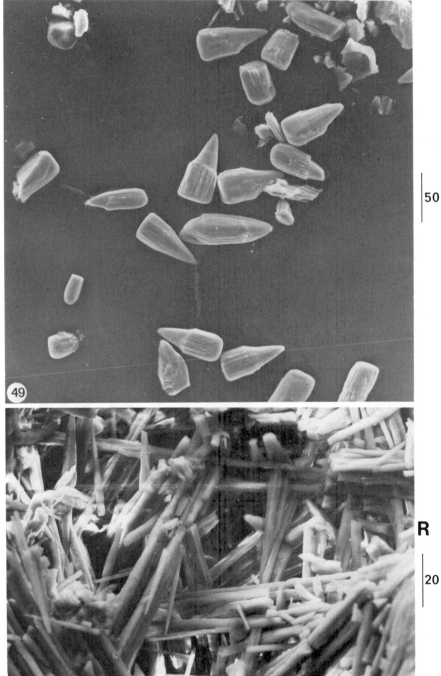

50

R

20

Quantitative Analysis
of Skeletal Growth Records

Application of Normalized Power Spectra to the Analysis of Chemical and Structural Growth Patterns

GARY D. ROSENBERG, M. ASHTON, R. HEWITT,
and D. J. SIMMONS

1. Introduction

This appendix is a modification of Dolman's (1975) method of evaluating chemical and structural growth periodicities with Fourier techniques. Mohr (1975) also applied Fourier analyses to the study of Precambrian stromatolite growth. The advantage of the present modification is that it determines amplitude (or power) spectra normalized to the sample variance, and thus facilitates comparison of spectra for allometric studies. Variations in growth rate in the time domain appear as variations in wavelength in the frequency domain.*

As an example of the difficulty of using nonnormalized spectra, note that the amplitude of a harmonic in the spectrum of shell growth patterns may vary with the size of the specimen. Specimens that grow to a relatively large size in one habitat may have apparently different spectra than spec-

---

* The term "wavelength," rather than "frequency," is used in growth studies because one seldom directly determines growth frequencies (cycles per unit time). As will become apparent below, one usually determines cycles per unit length of growth. The term "frequency" will be used in this sense in this appendix.

GARY D. ROSENBERG ● Geology Department, Michigan State University, East Lansing, Michigan 48824. *Present address:* Geology Department, Indiana University/Purdue University, Indianapolis, Indiana 46202.   M. ASHTON ● Geology Department, University of Newcastle Upon Tyne, Newcastle, England NE1 7RU.   R. HEWITT ● Geology Department, University of Birmingham, Birmingham, England B15 2TT.   D. J. SIMMONS ● Division of Orthopedic Surgery, Washington University Medical School, St. Louis, Missouri 63110.

675

imens that live as long, but do not grow as large, in another habitat. Thus, the parameter of interest is the relative power contained within each harmonic of the different spectra.

2. Normalized Power Spectra

The method of calculating normalized power spectra described here is based on Shimshoni's (1971) extension of Fisher's (1929) test. The method provides a test for the significance of each harmonic. However, we caution the reader that there are likely to be trends in skeletal growth patterns, in the mean, for example, that could cause distortion of signal and affect application of the test. These are matters that will affect all Fourier analyses, and despite such complicating growth trends, the normalization procedure still can be a valuable tool for facilitating comparison of different spectra, providing we keep these limitations in mind. Dolman (1975) provides a useful discussion of these kinds of problems.

A Fourier transform represents data in the frequency domain that were originally in the time domain. In the case of structural growth patterns, a growth increment is first defined, and then the widths of a contiguous series of these increments are measured throughout as much of the organism's growth span as is possible. No decision is made before measurement regarding the period of time represented by an increment, but the assumption that a given type of increment represents the same period wherever it is measured is tacit, if not stated outright.

Chemical distributions within the shell can be measured with the electron microprobe. A repeating chemical periodicity, analogous to a structural increment, cannot be defined prior to analysis. Rather, point source analyses are made at equally spaced intervals along a growth axis. However, one must remember that equally spaced sampling intervals may not represent equal time intervals because growth rate in bivalves decreases with time. This problem is especially critical if the growth series is long.

The Fourier representation of the data is a series of harmonics (sine or cosine waves) related to both the length of the series and the number of data points. Each harmonic has a frequency, amplitude, and phase. The graph of these values is termed a "periodogram." The periodicity that each harmonic represents is determined by dividing the total length of the data series by the harmonic number.

In the Fisher–Shimshoni test, the harmonics are ranked according to decreasing amplitude. Normalized periodogram estimates are compared with correspondingly ranked values of a periodogram that would be expected of a normally distributed, random data series of the same number of points (n, where n is an even number). If the $m[=(n-2)/2]$ harmonics down to rank r' exceed the random sample values, the probability that these harmonics arise by chance as a result of the randomness of the data is less than p. Normalization is related to the sample variance. That is, the $2m + 2(= n)$ terms of a time series $x(i) = 1, 2, \ldots, n$ can be written as a sum of their constituents:

$$x(i) = a_o/2 + \sum_{k=1}^{m} [a_k\cos(2\pi ki/n) + b_k\sin(2\pi ki/n)]$$

The values of a_o, a_k, and b_k are given by

$$a_o = (2/n) \sum_{i=1}^{n} x(i)$$

$$a_k = (2/n) \sum_{i=1}^{n} x(i)\cos(2\pi ki/n)$$

$$b_k = (2/n) \sum_{i=1}^{n} x(i)\sin(2\pi ki/n)$$

for $k = 1, 2, \ldots, m$.

The amplitude of each harmonic is defined by

$$c_k = (a_k^2 + b_k^2)^{1/2}$$

The normalized quantity g_k is defined by

$$g_k = c_k^2 \bigg/ \sum_{j=1}^{m} c_j^2$$

This is an estimate of the fractional variance contributed by the kth harmonic and is the quantity tabulated by Shimshoni for normal random samples. The relationship

$$\sum_{j=1}^{m} c_j^2 = (2/n) \sum_{i=1}^{n} [x(i) - a_o/2]^2$$

enables one to obtain the sum of squares without evaluating all the c_j values.

We apply the evaluation here to three examples of general interest to the readers of this volume. In the first case, allometric variations in chemical growth patterns are simply described in the laboratory rabbit incisor. This example has obvious applicability to analysis of growth periodicities in invertebrate exoskeletons. We will also consider structural and chemical rhythms in a Jurassic bivalve and belemnite.

3. Chemical Periodicities in the Rabbit Incisor

In this example, Ca distribution was measured (with the electron microprobe) parallel with the direction of growth on both the fast-growing labial and slow-growing lingual sides of the same lower incisor of a laboratory rabbit (Fig. 1). A recombination of high-ranking harmonics for both sides of the tooth shows that lower frequencies and longer wavelengths are more pronounced on the faster-growing side than on the slower-growing side. It is not necessary to evaluate the statistical significance of each harmonic here. It is sufficient to note that more than 80% of the power resides in harmonics representing wavelengths of 50–560 μm on the labial side. Less than half this amount of power resides in harmonics representing wavelengths of 56–700 μm on the lingual side.* This difference is expected, since the apposition rate on the lingual side is less than on the labial and the wavelengths of the patterns are consequently reduced.

4. Analysis of a Jurassic Mytilid

The method is next applied to a Jurassic mytilid, *Modiolus* sp., obtained from the Inferior Oolite (Bajocian) at Burton Cliff near Burton Bradstock, Dorset, England. Stratigraphic and paleoecological data suggest that the deposit may have accumulated in a turbulent environment tens of meters deep. The bivalve was studied using composite photos of acetate peels of acid-etched, polished shell sections cut along the axis of maximum growth (see Chapter 1 for illustration of this axis). Two of us (G.D.R. and M.A.) independently measured the widths of a series of 216 growth increments. Each increment was composed of a dark phase and a light phase as seen in the acetate peels (Fig. 2). The series of increment widths are graphed in Fig. 3A and B. Figure 3D shows the periodogram of the first 25 harmonics of the Fourier transform of the data in Fig. 3A. Four harmonics were determined to be significant ($p < 0.01$): harmonics 2, 3, 4, and 8. These represent periods of 108, 72, 54, and 27 increments, respectively, and are the only harmonics that are significant in both sets of measurements. In Fig. 3C, the significant harmonics have been recombined to approximate the original data series. The recombination is a markedly smoothed version of the original data.

---

* The highest frequency in both spectra is 1 cycle/2 μm.

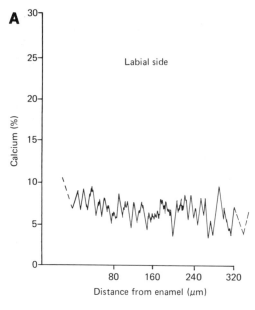

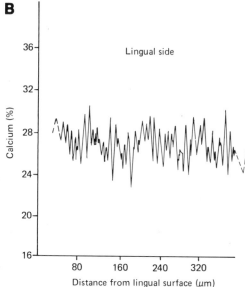

Figure 1. Distribution of Ca in a portion of the dentin of the laboratory rabbit incisor. (A) Labial dentin; (B) lingual dentin. Recombined harmonics from electron-microprobe series (1-μm spot size with measurements every 1 μm). Note that periodicities on the slower-growing lingual side are of a higher frequency than those on the labial side.

There are some uncertainties in the periods of the harmonics, since the spectrum we have calculated is not continuous. For example, the two terms on either side of harmonic 8 represent periods of 24–30.9 increments (= days?). We do not know how the power is distributed between 27 increments and the two extremes on either side, so our best estimate is

Figure 2. Acetate peel of growth increments in the outer shell layer of a Jurassic mytilid (*Modiolus* sp.) from the Inferior Oolite (Bajocian) at Burton Cliff near Burton Bradstock, Dorset, England. The shell was sectioned along the axis of maximum growth, polished, and acid-etched. Three increments are indicated by arrows and are some of the widest increments from the center of the growth series in Fig. 3. Entire growth series from posteriormost increments in a specimen that was approximately 2 years old at death. Growth to right. Each half of scale bar: 50 μm.

that the 27-increment term represents the summed power over the stated range.

Thus, harmonic 8 may represent a tidal periodicity. But it would be premature to conclude from these data that the length of the Jurassic month was 27 synodic days. We would need to evaluate additional specimens from the same facies, and from other habitats representing different tidal exposure.

5. Analysis of a Jurassic Belemnite

Finally, we studied a belemnite guard (*Cylindroteuthis puziosiana*) collected from a glauconitic sandstone associated with phosphatic nodules (Callovian) exposed on the shore at Port a' Brora, Sutherland, Scotland. This motile boreal species had a life span of several years (Urey *et al.*, 1951); its shell morphology varies with sedimentary facies. The 23-mm diameter of this specimen is typical of those from turbulent inshore subtidal environments. Figure 4A–C shows three electron-microprobe traverses of calcium concentration along a single line of section of the specimen (longitudinal section, adult portion of guard, dorsal to the phrag-mocone).

Ninety-six measurements were made at 50-μm intervals with a 25-μm spot size. Thus, the series is not continuous and probably does not record the highest-frequency rhythms. The lower three graphs in Fig. 4 (D–F) show recombined harmonics from each of the respective normalized power spectra (Fig. 5A–C). Each recombination uses at least 66% of the normalized power. The repeatability of the original data series, as well as the recombinations, can be judged from Fig. 4 by inspection. Although only one harmonic (transect 3) tested to be significant ($P < 0.01$), the similarity of the three spectra and the three recombinations suggests that other harmonics may be significant. That is, the significance test may be

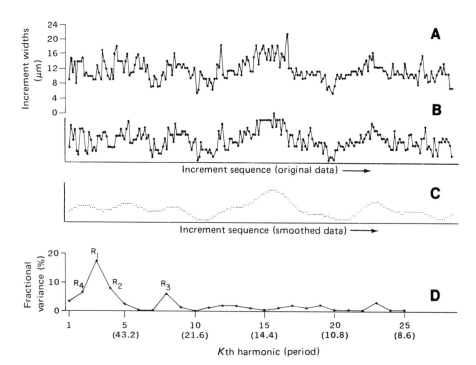

Figure 3. (A, B) Widths of 216 growth increments measured on acetate peels similar to those shown in Fig. 2. The two traverses (A and B) were made by different authors (M.A. and G.D.R.). Each data point corresponds to one increment. The direction of growth is from left to right. The average width of the increments is 11 μm. (C) A smoothed recombination of the four significant harmonics as determined in (D). (D) Normalized power spectrum of the first 25 harmonics of the transform of (A). The fractional variance (*ordinate*) is plotted against the kth harmonic (*abscissa*). The number in parentheses below each harmonic number indicates the period of the respective harmonic, which is obtained by dividing the length of the series (216 measurements) by the harmonic number. (R_n) Highest ranking harmonics.

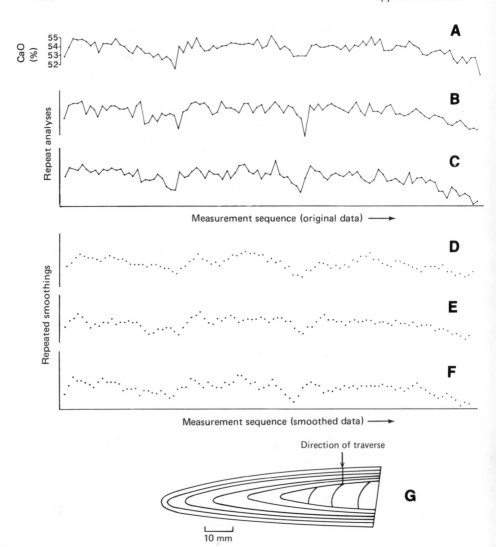

Figure 4. (A, B, C) Calcium distribution across the growth increments in the guard of a Jurassic belemnite, *Cylindroteuthis puziosiana* (Callovian age, from the shore at Port a' Brora, Sutherland, Scotland). The direction of growth is from left to right in each case. Three repeated traverses were made along the same line of section (longitudinal section, adult portion of guard, dorsal to the phragmocone). Measurements were made at 50-μm intervals using a 25-μm spot diameter. See (G) for the approximate location of these transects. (D, E, F) Three recombinations (summations) of high-ranking harmonics from the respective normalized power spectra (see Fig. 5). Each recombination uses 12 harmonics representing more than 66% of the total power. (G) Longitudinal section of C. *puziosiana*. The arrow shows the direction of the electron-microprobe traverse.

too severe; Shimshoni developed his test because of the severity of previous tests.

6. Comparison of the Jurassic Bivalve and Belemnite Data

We can average the three belemnite spectra (Fig. 5D) and arbitrarily select the four highest peaks (3, 10, 1, 7), which together contain 46% of the power. These harmonics and the highest-ranking harmonics from the *Modiolus* transform are listed in Table I. The power in each harmonic and the period of each harmonic are also listed. For *Modiolus*, the period is given first in terms of the number of increments per cycle and is then converted to the approximate number of microns per cycle by multiplying the number of increments by the average width of an increment (11 μm). For *Cylindroteuthis*, the period is given first in terms of the number of measurements per cycle and then converted to the number of microns per cycle by multiplying the number of measurements by 50 μm (the measurement interval) and subtracting 50 μm, because the number of intervals is one less than the number of measurements.

It is tempting to conclude that at least two of the high-ranking harmonics in both specimens are similar. Perhaps the 297-μm periodicity in *Modiolus* and the 430-μm periodicity in *Cylindroteuthis* represent similar (temporal) periodicities, as may the 594-μm period in *Modiolus* and the 635-μm period in *Cylindroteuthis*. Each of the four harmonics contains 6–8% of the power. The fact that the belemnite wavelengths are slightly longer than their bivalve counterparts may be explained by the 25-μm spot size and the 50-μm sampling interval for the belemnite. Because the chemical measurements were not contiguous, the periodicities could have been smoothed and shifted to longer wavelengths (for an explanation of aliasing, see Dolman, 1975).

In other words, one is prompted to ask whether the wavelengths of the rhythms in two different taxa are similar because both organisms responded to similar geophysical (day and tidal) periodicities in the Jurassic, or whether the similarities are coincidental. The answer is tempered with many uncertainties. The sample size is very small, the exact period of the harmonics is not known, the importance of other high-ranking harmonics in the spectra cannot be assessed without analyses of additional specimens, the sampling interval of the microprobe analyses may have introduced smoothing, and the use of the mean increment width to determine wavelength of structural cycles may have introduced errors.

Nevertheless, it should be clear from the three examples presented here that spectral normalization facilitates comparison of growth peri-

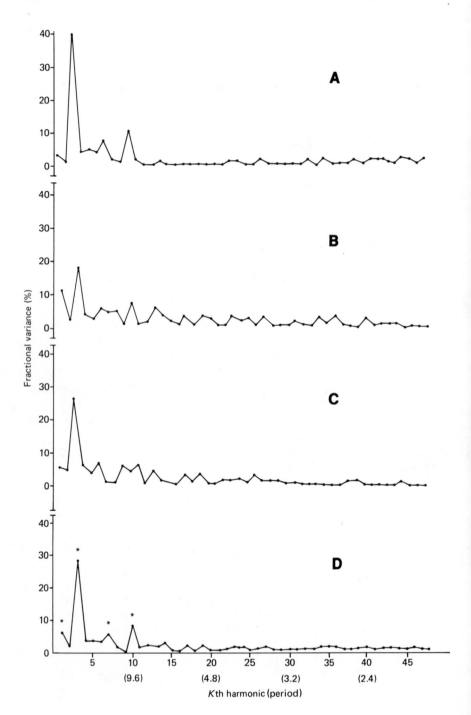

Table I. Highest-Ranking Harmonics from the Spectrum of Structural Rhythms in *Modiolus* sp. and Harmonics That Form Peaks in the Average Power Spectrum from Calcium Periodicities in *Cylindroteuthis puziosiana*[a]

	Modiolus				Cylindroteuthis			
Harmonic (No.)	3	4	8	2	3	10	1	7
Power (%)	18	8	6	6	27	7	6	6
t	72	54	27	108	32	9.6	96	13.7
τ	792	594	297	1188	1550	430	4750	635

[a] The table lists harmonic number, percentage power, and period in terms of number of measurements (t) of structural increments (*Modiolus*) or chemical periodicities (*Cylindroteuthis*), and microns per cycle (τ) for each major harmonic.

odicity power spectra. The problem that remains is the one of causal interpretation. In this case, it is a matter of obtaining additional independent data to determine the ontogenetic, physiological, and paleoecological significance of growth periodicities.

ACKNOWLEDGMENTS. We thank Dr. M. L. Richards for invaluable help with the principles of the Fisher–Shimshoni test. Mr. Bruce Spenner assisted with computer techniques. Mr. J. F. Rose kindly donated the *Modiolus* sp. specimen, and Dr. C. F. Parsons discussed the horizon of origin of the specimen. Dr. C. T. Scrutton and Dr. J. Dolman offered helpful comments on the manuscript. This research was supported by NSF Grant DES 75-00345 (Rosenberg), USPHS Grants AM16391 and DE04629 (Rosenberg and Simmons), and a N.E.R.C. (U.K.) studentship (Ashton).

References

Dolman, J. W., 1975, A method for the extraction of environmental and geophysical information from growth records in invertebrates and stromatolites, in: *Growth Rhythms and the History of the Earth's Rotation* (G. D. Rosenberg and S. K. Runcorn, eds.), pp. 191–222, John Wiley, London.

Fisher, R. A., 1929, Test of significance in harmonic analysis, *Proc. R. Soc. London Ser. A* **125:**54–59.

Figure 5. (A, B, C) Normalized power spectra for the respective data series in Fig. 4A, B, and C. (D) Average power spectrum obtained from the three spectra (A, B, and C). Starred harmonics represent peaks in the spectrum that are compared with ranking harmonics from the *Modiolus* transform in Table I. The number in parentheses below each harmonic number in the averaged spectrum is the period of the respective harmonic, which is obtained by dividing the length of the series (96 measurements) by the harmonic number.

Mohr, R. E., 1975, Measured periodicities of the Biwabik (Precambrian) stromatolites and their geophysical significance, in: *Growth Rhythms and the History of the Earth's Rotation* (G. D. Rosenberg and S. K. Runcorn, eds.), pp. 43–56, John Wiley, London.

Shimshoni, M., 1971, On Fisher's test of significance in harmonic analysis, *Geophys. J. R. Astron. Soc.* **23**:373–377.

Urey, H. D., Lowenstam, H. A., Epstein, S., and McKinney, C. R., 1951, Measurement of paleotemperatures of the Upper Cretaceous of England, Denmark, and the southeastern U.S.A., *Geol. Soc. Am. Bull.* **62**:399–416.

Study of Barnacle Shell Growth-Band Patterns Using Time-Series Analysis

CLIFFORD L. TRUMP and EDWIN BOURGET

1. Introduction

A basic conclusion drawn from the work described in Chapter 13 is that the width of a barnacle growth band is proportional to the duration of immersion. On the other hand, it is evident that there is a marked seasonal change in this proportionality, or growth rate, since growth bands are much thicker in early summer than in winter (Bourget and Crisp, 1975a). The molting cycle, which may be as short as 3–5 days, also affects the growth rate by 5–10%, and there may exist other regular changes related to tidal (e.g., fortnightly), solar, or weather cycles. Finally, aperiodic occurrences such as storms or heat-stress events can strongly affect the growth rate for a few tidal cycles. It is the purpose of this appendix to briefly describe some time-series techniques that can be used to examine how the growth rate varies with time. The basic working hypothesis we use is that the growth rate cannot be studied by analyzing the band-width record alone, since variations in the band widths are caused by the interaction of growth-rate variations and tidal-immersion variations, but the growth rate can be studied only by first removing from the band-width record any effects due to variations in the duration of the tidal immersions.

The model used here to describe the relationship between band

CLIFFORD L. TRUMP and EDWIN BOURGET ● Groupe Interuniversitaire de Recherches Océanographiques du Québec (GIROQ), Département de Biologie, Faculté des Sciences et de Génie, Université Laval, Québec, Canada G1K 7P4. Dr. Trump's present address is: Applied Hydrodynamics Branch, Environmental Sciences Division, Naval Research Laboratory, Washington, D.C. 20375.

widths and tidal immersions is

$$F(t) = G(t) \cdot T(t) \tag{1}$$

where $F(t)$ is the band thickness in μm, $T(t)$ is the tidal immersion period in hours, and $G(t)$ is the growth rate in μm/hr, all as functions of time. $T(t)$, and therefore $F(t)$, is a discrete function of time in that there is one value per tidal cycle. On the other hand, $G(t)$ is a continuous function of time that is felt to vary little over a tidal immersion, so that its value at high water can be considered the constant of proportionality for the immersion period. By using the high-water time and having one value of $F(t)$, $T(t)$, and $G(t)$ per tidal cycle as the basis for time-series analysis, we are implicitly assumming that high water occurs in the middle of an immersion and that the time intervals between successive high waters are all equal. Neither of these assumptions is completely correct, but both are close enough to reality for our purposes. The formulation of $G(t)$ is with t in hours for the sake of simplicity, but time-series analyses of band-width and growth-rate estimates must be with t in tidal cycles (1 tidal cycle = 12.42 hr) because we are limited to one estimate per tidal cycle. We then hypothesize that $G(t)$ takes the form

$$G(t) = C \cdot A(t) \cdot [1 + a_1 \sin(f_1 t)] \cdot [1 + a_2 \sin(f_2 t)] \tag{2}$$
$$\cdots \cdot [1 + a_n \sin(f_n t)] \cdot [1 + r(t)]$$

where C is some constant reflecting the relationship between a given animal and its environment and would be affected by such parameters as species, crowding, substratum, water, and climate characteristics. $A(t)$ represents the annual trend in the growth rate and varies by at least one order of magnitude over the year. The terms $[1 + a_i \sin(f_i t)]$ represent periodic oscillations in the growth rate of frequency f_i and proportionate effect a_i. The oscillatory terms take the form above because we feel that a periodic cycle (e.g., the molting cycle) would affect the basic growth rate by some percentage (e.g., 10%; $a = 0.1$), but that its absolute effect (in μm/hr) would depend on the absolute size of $C \cdot A(t)$. Finally, the term $[1 + r(t)]$ represents random changes in the growth rate possibly related to aperiodic environmental effects such as those mentioned earlier, or to random errors in estimating band widths and tidal-immersion periods.

2. A Specific Growth-Rate Model

We do not, as yet, have a time series of barnacle band widths and simultaneous immersion periods long enough to analyze. Therefore, we have generated an artifical series that will be used to demonstrate the proposed techniques. Using the 1976 tide tables for the St. Lawrence estuary, immersion periods were calculated for a point in the intertidal zone for the entire year. A time series of 707 immersion periods [ranging from 2.28 to 6.76 hr (see Fig. 1B)] was calculated over a total time scale of 8766 hr (1 year). The constant C was set at 1.0 μm/hr, and the annual cycle, $A(t)$, was defined by

$$A(t) = 8\{[(t/730.5) - 6]^2 + 4\}^{-1} \tag{3}$$

where t is in hours (730.5 hr $\approx$ 1 month). This formula yields a curve that resembles a normal curve with a maximum value in June when $t = 4383$ hr and $A(t) = 2.0$, and a minimum value in December when $t = 0$ or 8766 hr and $A(t) = 0.2$. Three oscillatory variations were included in the growth rate. One, analogous to a molting cycle, had a period of 80 hr ($f = 0.012$ cycle/hr = 0.155 cycle/tidal cycle) and an amplitude of 0.1 [10% of $C \cdot A(t)$]. Another, analogous to a fortnightly tidal cycle, had a period of 328 hr ($f = 0.003$ cycle/hr = 0.038 cycle/tidal cycle) and an amplitude of 0.5 (5% of $C \cdot A(t)$]. The third cycle, analogous to a diurnal cycle, had a period of 24 hr ($f = 0.042$ cycle/hr = 0.517 cycle/tidal cycle) with a maximum value at 13:30 and an amplitude of 0.10 [10% of $C \cdot A(t)$]. Finally, a uniformly random component simulating measurement error was introduced that affected the growth rate by up to 15% of $C \cdot A$ (t).

Using the tidal-immersion times as input and the growth-rate function [equation (2)] with the parameters above as the model, a band-width series was generated (see Fig. 1A). This series with one value per tidal cycle evaluated at high water will be used as the example for most of the techniques described hereafter.

3. Analysis of the Specific Band-Width Series

The tidal data set should consist of at least one full year of simultaneous band widths and immersion periods from a given location. It should contain no growth discontinuities or stress periods. Using the artificially created series of band widths and immersion periods described in Section 2, we will give an example of how various techniques can be used to (1) isolate and describe long-term variations in the growth rate, mainly the annual growth cycle, $A(t)$, perhaps including the reproductive cycle, and, if the record is long enough, the effect of barnacle age on $C \cdot A(t)$; (2) remove the random component from the record; and (3) isolate and describe residual periodic variations and aperiodic events.

Figure 1A and B shows the records of band widths and tidal im-

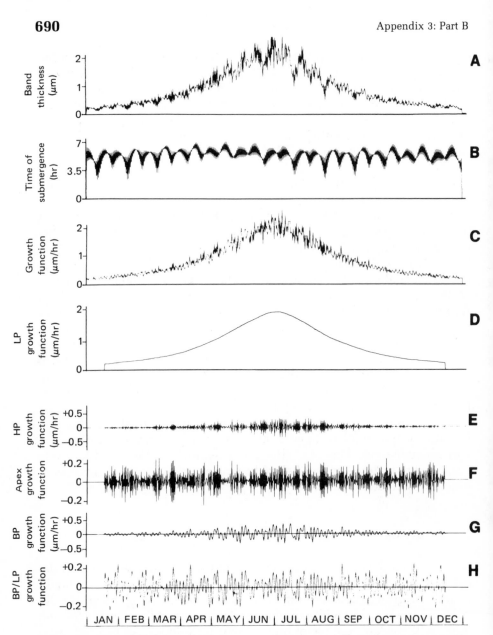

Figure 1. Components of the time-series analysis of a hypothetical barnacle band-width series. (A) Model band-width series $[F(t)]$; (B) simultaneous immersion periods $[T(t)]$; (C) growth function $[G(t) = F(t) \div T(t)]$; (D) low-pass (LP) filtered growth function; (E) high-pass (HP) filtered growth functions; (F) ratio of high-pass to low-pass filtered growth function (HP/LP); (G) band-pass (BP) filtered growth function; (H) ratio of band-pass to low-pass filtered growth function (BP/LP). There is one value for each of 707 tidal immersions covering one full year.

mersions. The basic assumption of the model is that the width of a band is proportional to the duration of the immersion. This assumption implies that any variation in immersion period will affect band width. Our main interest is to study the growth rate (the proportionality between the band width and the period), not solely the variations of the band widths. Therefore, the first step is to divide the band widths by the simultaneous immersion periods to produce a discrete estimate of the growth rate (in μm/hr) for each tidal cycle. The result of this first step is simpler than the band-width record (Fig. 1C). The annual trend $[C \cdot A(t)]$ is apparent, and at least one fairly regular oscillation becomes visible.

The next step is to isolate and describe the annual trend. To this end, a low-pass numerical filter* of length $n = 67$ and half-amplitude frequency of 0.01 cycle/tidal cycle (or $f = 0.0008$ cycle/hr, $T = 1240$ hr) was used on the growth function. The annual trend is quite clear in the resulting series (Fig. 1D). Note that this series is shorter than the original series. Numerical filtration results in a sacrifice of data points at each end of the series. The number of points sacrificed at each end equals $(n - 1)/2$, or, in this case, 33.

To isolate random noise and high-frequency oscillations, a high-pass filter of length $n = 67$ (the same as the first filter) and half-amplitude frequency of 0.25 cycle/tidal cycle (or $f = 0.02$ cycle/hr, $T \approx 50$ hr) was used. The resultant series is shown in Fig. 1E. This series contains information about oscillations with frequencies greater than 0.25 cycle/tidal cycle (periods < 50 hr), including all random measurement errors, and any information concerning possible 24-hr oscillations.

Examination of Fig. 1E indicates that there is more variance present in summer than in winter. The model of the growth-rate function predicts that it is a product of the annual trend $[C \cdot A(t)]$ and relative periodic $[l + a_i \sin(\omega_i t)]$ and random $[1 + r(t)]$ terms. Therefore, if the periodic and random terms are to be studied, the growth rate should be divided by the annual trend. In this case, the high-pass series (Fig. 1E) is divided by the low-pass series (Fig. 1D). The result is the series shown in Fig. 1F, which is nondimensional because it reflects the proportion of the annual trend present at frequencies greater than 0.25 cycle/tidal cycle (or $f > 0.02$ cycle/

---

* A numerical filter is analogous to a color filter in that it separates a time series into "color" groups. A low-pass filter will preserve all information about oscillations present with frequencies less than a predetermined frequency (half-amplitude frequency), a high-pass filter will preserve all information above a predetermined frequency, and a band-pass filter will preserve all information between two predetermined frequencies. Mathematically, a numerical filter is similar to a moving-average process, but where the "weights" in a moving-average process are all the same (e.g., 1/5, 1/5, 1/5, 1/5, 1/5, for $n = 5$), the "weights" in a numerical filter are shaped (e.g., 0.1, 0.2, 0.4, 0.2, 0.1, for $n = 5$). The length of a filter and its shape can be adjusted to yield the type of filter and half-amplitude frequency (or frequencies) desired (see Jenkins and Watts, 1968).

hr, $T < 50$ hr). Because the magnitudes in Fig. 1F range from -0.2 to $+0.2$, we conclude that high-frequency variations affect the growth rate by approximately 20%.

Twenty-four-hour oscillations cannot be examined using time-series analysis because at least two samples per oscillation are necessary for resolution. In this case, there is one sample (i.e., band width) per tidal period (12.4 hr) or 1.93 samples per 24-hr solar period. However, by plotting the data in Fig. 1F against the time of day that high water occurs for each immersion period, it is possible to observe directly any 24-hr cycle present. The results of this operation are shown in Fig. 2A. It is evident that a 24-hr cycle is present that peaks near 13:30. One could fit a sinusoidal curve to this plot to obtain a quantitative estimate of the 24-hr cycle. The variance about this fitted curve could then be used to estimate the magnitude of random error.

So far, the annual trend has been separated and examined using a low-pass numerical filter (Fig. 1D), and the high-frequency oscillations and random noise have been separated and examined using a high-pass filter (Figs. 1E and F and 2A). This leaves oscillations in the middle of the frequency band to be examined (periods of 4–100 tidal cycles). These oscillations could be extracted by the use of a band-pass filter with $\frac{1}{2}$-amplitude frequencies of 0.01 cycle/tidal cycle (or $f = 0.008$ cycle/hr, $T = 1240$ hr) and 0.25 cycle/tidal cycle (or $f = 0.02$ cycle/hr, $T = 50$ hr). This was done, and the result is displayed in Fig. 1G. This band-pass series was divided by the low-pass series to give a nondimensional series showing the proportionate effect of middle-frequency oscillations on the growth rate (Fig. 1H). A fairly regular oscillation is apparent in Fig. 1H. Its period is estimated by counting the number of cycles (100) and dividing this into the record length (641 tidal cycles, or 7961 hr) to give 79.6 hr, which is equivalent to the hypothetical molting cycle (80 hr). It also appears that this primary oscillation is modulated, possibly by another lower-frequency cycle. The best way to isolate and examine the various frequencies present in the band-pass series would be to calculate the frequency spectrum.[*] Two dominant frequencies can be identified: (1) the molting cycle, with frequency 0.155 cycle/tidal cycle ($T = 80$ hr); and (2) the fortnightly cycle, with frequency 0.038 cycle/tidal cycle ($T = 328$ hr) (Fig. 2B). The variance density present in the molting cycle is approximately 4 times greater than in the fortnightly cycle. Since the variance of an oscillation is proportional to the square of its amplitude, this relationship implies that the molting-cycle amplitude is twice as great as the fortnightly-cycle amplitude as was modeled. A careful treatment of

---

[*] The spectrum of a time series displays the distribution of variance as a function of frequency. In other words, it describes what oscillations are present in a time series and their magnitudes (see Jenkins and Watts, 1968).

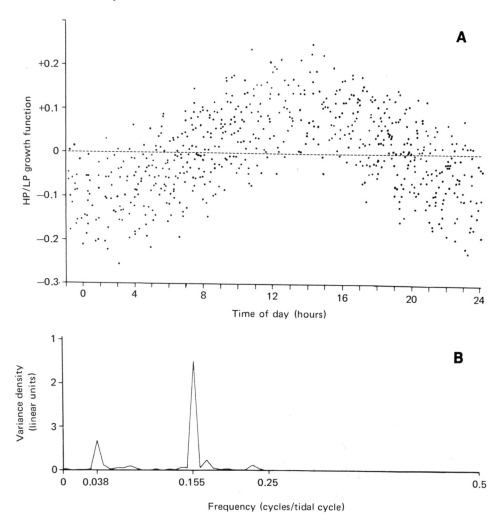

Figure 2. Analysis of subannual periodicities in the growth function. (A) Scatter diagram of the ratio of the high-pass (HP) to low-pass (LP) filtered growth function against the time of day of high water for each immersion period. (B) Spectrum of the ratio of the band-pass (BP) to low-pass (LP) filtered growth function. The molting cycle (1.55 cycles/tidal cycle) and the fortnightly tidal component (0.038 cycle/tidal cycle) are identified in this analysis.

the variance units will yield quantitative estimates of the magnitudes of the oscillations in the growth function. Notice that there is no variance energy present in Fig. 2B at frequencies less than 0.1 cycle/tidal cycle (or $f < 0.0008$ cycle/hr, $T > 1240$ hr) or greater than 0.25 cycle/tidal cycle (or $f > 0.02$ cycle/hr, $T < 50$ hr) because this energy was filtered out while producing the band-pass series shown in Figs. 1G and H. The smaller

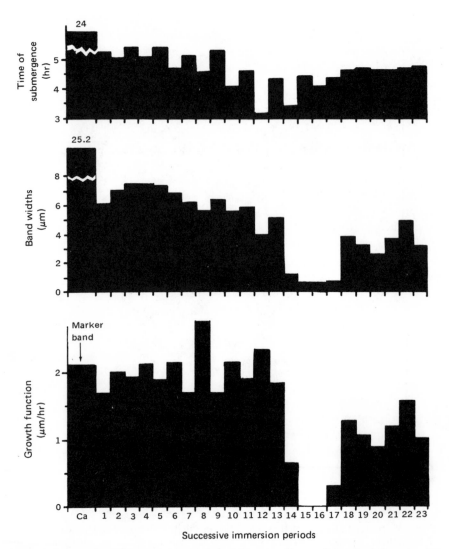

Figure 3. Components of a time-series analysis of an actual barnacle band-width series. (A) Immersion periods [$T(t)$]; (B) measured band widths [$F(t)$]; (C) calculated growth function [$G(t) = F(t) \div T(t)$]. Band Ca is a thick marker band induced in the barnacle by addition of $CaCl_2$ to the water.

peaks present in this spectrum are due to nonlinearities inherent in the formulation of the growth-rate model.

In reality, even with long-term, high-quality band-width series, realistic analyses will not be as precise as those presented above. Any cycles present will not be perfect. For instance, the molting cycle varies in period and amplitude from cycle to cycle and over the year (Crisp and Patel, 1958, 1960). Aperiodic events such as storms, stress events, or phytoplankton blooms will affect the time series, and some oscillations may be present only during parts of the year. However, we feel that the methods presented herein provide a firm basis for examining the relationship between band widths and immersion periods.

4. Analysis of Expected Band-Width Series

In practice, data series covering a full year will be uncommon. Good records of band widths and immersion periods covering about 50 tidal cycles are more likely, and in this case the methods described in the preceding section would have to be modified. The first step, and the most important one, is to divide the band widths by their immersion periods to obtain an estimate of the growth rate (in μm/hr). The next step is to identify trends in the growth rate. Numerical filtration is not a good technique to use on a short record because filtration demands a sacrifice of points on both ends of the time series (e.g., Fig. 1D). The combination of a short data set and a fairly long filter will result in a very small number of data points. The best process is to fit, by least-squares approximation, a polynomial to the growth rate, $G(t)$. A linear or parabolic fit would probably be sufficient.

After the trend is approximated, it can be subtracted from or divided into the growth function to obtain a time series that would best reveal information about shorter-period oscillations. Spectral analyses may not be usable because of the shortness of the records. However, simple visual examination and plotting of the values of the growth function against the time of day of high water may provide useful information.

5. Analysis of an Actual Band-Width Series

In Fig. 8 of Chapter 13, data representing 24 consecutive band widths and immersion periods are given. Figure 3 reproduces these data with the addition of the calculated growth rates. The absence of bands 15 and 16 and the thinness of bands 14 and 17 are attributed to heat stress caused by a long, hot emersion period between immersion periods 13 and 14 (Fig. 3B). It is not clear from the band-width record alone (bands 18–23) when the effects of the heat stress ceased and when the "normal" growth resumed. On the other band, when the growth rates (Fig. 3C) are examined, they appear quite consistent at near 2.0 μm/hr from the marker band to band 13, very small or zero for bands 14–17, and significantly less (mean

= 0.98 µm/hr) for bands 18–23. This pattern indicates that the barnacle had not fully recovered from the heat stress by period 23. By use of the growth rates, rather than just the band-thickness record, a better idea of the character of the stress event and its duration was obtained.

Next, we notice that the growth rates are alternately high and low for bands 1–13, and this pattern is not evident in the band-width record. This is because there were variations in growth rate that coincided with inverse variations in the immersion period and therefore were not apparent in the band-width record. If the growth rates estimated for these first 13 bands are plotted against the time of day in a manner similar to that done in Fig. 2A, there is an indication of a diurnal cycle (Fig. 4). The 7 nighttime immersions had a mean growth rate of 1.79 µm/hr, which was significantly less than the mean growth rate of 2.24 µm/hr during the 6 daytime immersions. Furthermore, all 7 nighttime growth-rate estimates were less than all 6 daytime growth-rate estimates. This series is too short to prove the existence of a 24-hr cycle, but this conclusion is indicated. Since it has been clearly shown, under controlled laboratory conditions, that the diurnal variation in light intensity has no significant effect on barnacle shell growth rate (Bourget and Crisp, 1975b), the day–night differences observed are probably related to other diurnally fluctuating environmental parameters such as surface water temperature, which is known to influence metabolic activity.

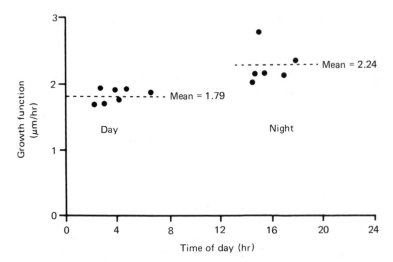

Figure 4. Time-series analysis of an actual barnacle band-width series (see Chapter 13, Fig. 8). The growth function is plotted against time of day of high water for immersion periods 1–13 of Fig. 3.

In this appendix, the formulation of the growth-rate model [equation (3)] is hypothetical and could be improved on through the analysis of real band-width series. The basic hypothesis, that the band widths are linearly related to the immersion periods, was demonstrated in Chapter 13, and its use in calculating the growth rates for a real band-width record (Section 5) proved to be of value in describing the effect of a heat-stress event and demonstrating the presence of a diurnal cycle in a barnacle's growth rate.

References

Bourget, E., and Crisp, D. J., 1975a, An analysis of the growth bands and ridges of barnacle shell plates, *J. Mar. Biol. Assoc. U. K.* **55:**439–461.

Bourget, E., and Crisp, D. J., 1975b, Factors affecting deposition of the shell in *Balanus balanoides* (L.), *J. Mar. Biol. Assoc. U. K.* **55:**231–249.

Crisp, D. J., and Patel, B. S., 1958, Relation between breeding and ecdysis in cirripedes, *Nature (London)* **181:**1078–1079.

Crisp, D. J., and Patel, B. S., 1960, The moulting cycle in *Balanus balanoides* (L.), *Biol. Bull.* **118:**31–47.

Jenkins, G. M., and Watts, D. B., 1968, *Spectral Analysis and Its Application,* Holden-Day, San Francisco, 525 pp.

Probabilistic Population Descriptions

J. RIMAS VAIŠNYS and RICHARD E. DODGE

1. Introduction

In this appendix we propose some probabilistic and statistical techniques for describing biological populations. The techniques are illustrated with actual data, obtained from studies of coral reefs in Bermuda. In our example, we use skeletal band counts, made for a small sample of corals, to construct an age-frequency description of the coral reef population. The techniques are quite general and are applicable whenever a quantifiable skeletal growth record is exhibited by an individual organism and when one is interested in estimating the frequency distribution of the same growth pattern in the species population.

Before we proceed with mathematical manipulations, it may be well to answer some possible questions. Why bother with a mathematical analysis when in many cases mathematical models are unreasonably artificial, and do not reflect the essential natural processes? We agree with the claim implied by the question, that natural science is often forced into the procrustean bed of available mathematical models. We believe that models should be adjusted to accommodate the science, rather than vice versa. This premise has prompted us to attempt to formulate techniques that can deal with real biological and ecological data, with actual field situations, and with hypotheses about real cause-and-effect relationships. Our goal has not been completely realized, but we believe that the results presented herein justify the effort, and we encourage further application and development of the approach.

J. RIMAS VAIŠNYS • Department of Geology and Geophysics, Yale University, New Haven, Connecticut 06520.   RICHARD E. DODGE • Nova University Ocean Sciences Center, Dania, Florida 33004.

One might further ask, why bother with such formal analyses when an experienced ecologist or paleontologist can draw the same conclusions intuitively? We cite two strong arguments in favor of formal analyses. First, even when an intuitive conclusion is thought to be correct, one rarely knows how right it is. Formal methods can have built-in techniques for evaluating the certainty of the conclusions. This feature is particularly important when one is dealing with probabilistic systems, statistical estimates, and small samples. These situations arise when the relationships between organisms and the environment are incompletely understood, when field observations are difficult and expensive, or when one simply wishes to preserve a local ecosystem by doing less than exhaustive sampling. Second, formal analyses, when appropriately applied, force us to state explicitly both our assumptions and our arguments. Such explicit statements may have the drawback of being lengthy, but they also can help avoid fruitless arguments and they encourage progress by serving as starting points for futher developments.

This appendix is organized as follows: In Section 2, coral collection and measurement procedures are summarized. This is done early so that the later mathematical development can be referred to the biological situation. Section 3 introduces the notions of probability distributions and hypotheses, examines possible classes of hypotheses to be considered for describing biological populations, reviews methods of relating observational data and hypotheses, and introduces methods for the comparison of hypotheses. Section 4 illustrates these methods by using them in the analysis of coral growth data obtained in the Bermuda field study. Section 5 presents conclusions and suggestions for further research.

2. Sampling Location, Sampling Methods, and Data Collection

Hermatypic corals are a major benthic faunal component in semi-tropical to tropical shallow-water marine habitats. In Bermuda, the predominant hermatypic corals are the brain corals *Diploria labyrinthiformis* and *Diploria strigosa*. This genus is not only important in terms of reef ecology, but also has been useful in studies of coral responses to environmental change (Dodge and Vaišnys, 1975, 1977).

Living specimens of both *Diploria* species were collected from reef sampling stations established in areas along the north and south coasts of Bermuda and in Castle Harbor, Bermuda. Dead coral heads, abundant in Castle Harbor, were also collected. At all stations, corals were obtained from no deeper than 6 m and were usually recovered from a depth range of 2–5 m. The age of each specimen was determined from annual density bands, revealed by X-radiography of medial sections of the skeleton (see Chapter 14 and Appendix 1.B).

Figure 1 shows the age distribution of live corals from Castle Harbor, the north and south coasts, and dead corals from Castle Harbor sites (hereafter designated respectively as CHL, NS, and CHD). In each case, the number of collected corals, c_k, in the kth age class is shown, where the kth age class includes all specimens having ages between k and $k + 1$ years. Included for each case is an inset figure giving the age-specific collection efficiency, ν_k. The estimates of the relative collection efficiency, given by the shape of the curves, are more certain than the absolute collection-efficiency values. In each population, the collection efficiency for

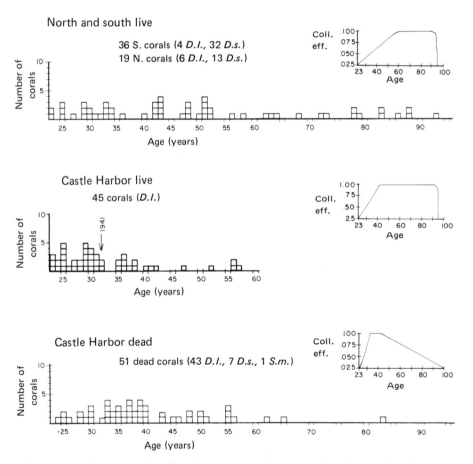

Figure 1. Number of corals collected vs. age at each site. (D.l.) *Diploria labyrinthiformis*; (D.s.) *Diploria strigosa*; (S.m.) *Stephanocoenia michilini*. The inset gives the collection efficiency vs. age.

corals older than approximately 90 years was zero; however, visual observations at each site yielded information about the presence and abundance of large corals, overall coral abundance, density, species composition, and other habitat conditions. A description of collection and measurement procedures, as well as discussion of the relevant coral biology, are given in Dodge and Vaišnys (1977), and this reference should be consulted for details.

3. Probability Distributions and Hypotheses

Sampling of a population is probabilistic, and because the inputs, responses, and states of reef ecosystems are incompletely known, probabilistic descriptions are appropriate. For our particular case, in which the age-frequency structure of the coral population is of interest, a collection site may be described by giving the probability of finding some stated number of corals in certain specified age classes. This is done by constructing probability distributions that give the probability that the site contains n_0 corals of the 0th age class, n_1 corals of the 1st age class, . . . , and finally $n_{\lambda-1}$ corals of the $(\lambda - 1)$st age class. This description is denoted by $p(n_0, n_1, \ldots, n_{\lambda-1})$, and will be abbreviated as $p(n)$. Lambda (λ) is the maximum longevity of the corals in the population. Such a probability distribution, describing the age structure of the reef, is a hypothesis about the reef population, and in the following discussion we will use the terms "probability distribution" and "hypothesis" interchangeably.

To illustrate some properties of such descriptions, Tables Ia–e present examples of probability distributions of hypothetical systems. The cases are constructed to be simple and yet to illustrate the basic nature and properties of a probability distribution. In all cases, we suppose that there are two age classes, labeled 1 and 2. We assume that the largest possible number of individuals in each age class is only 4. The cells at row and column intersections provide an exhaustive listing of all the possible states that might be observed, ranging from the completely empty system, zero individuals of class 1 and zero individuals of class 2, shown in the upper left-hand corner, to the most populated possible system, 4 individuals of class 1 and 4 individuals of class 2, corresponding to the lower right-hand corner. A numerical entry in each cell gives the probability assigned to the corresponding state.

Observe that for all tables, the sum of probabilities over all the possibilities is 1. The tables also have border columns and rows giving the marginal probabilities for the two age classes—in other words, the probability of observing the stated number of organisms in the given age class, without regard for the number of individuals in the other age class. Each marginal probability also sums to 1. Inspection of the entries of Tables Ia–d shows that the probability entry in any cell is equal to the product of the corresponding marginal probabilities. In these situations, the age class 1 and age class 2 populations are statistically independent. Not all probability distributions can be expressed as products of marginal probability distributions, as an inspection of Table Ie will show. At the bottom of each table are several entries that give expected values for different quantities and also give the entropy of the distribution. Observe that different distributions can have identical values for these quantities.

Tables Ia and Ib give examples of probability distributions that apply when the state of the system is certainly known; such distributions are referred to as "pure-state" distributions. In Tables Ia and Ib, all cells, except one, have the entry 0. Thus, Table Ia illustrates a probability dis-

Tables Ia–e. Probability Distributions Describing Five Hypothetical Systems of Organisms

Cell entries at row and column intersections give the probability assigned to the corresponding state. The expected number of organisms in each age class, as well as the entropy of the probability distribution, are given at the foot of each table.

Tables Ia and Ib illustrate pure-state distributions. These are the narrowest distributions possible for this system. Note that both these distributions have zero entropy. Table Ic illustrates the broadest possible distribution for this system, and it has the highest entropy. Tables Id and Ie give distributions of intermediate entropy. These last three distributions are mixed-state distributions.

Table Ia

Marginal probability for number in age class 2		0	0	1.0	0	0
Marginal probability for number in age class 1	Number in age class 2 / Number in age class 1	0	1	2	3	4
0	0	0	0	0	0	0
0	1	0	0	0	0	0
0	2	0	0	0	0	0
1.0	3	0	0	1.0	0	0
0	4	0	0	0	0	0

Expected number of organisms in age class 1: 3.00.
Expected number of organisms in age class 2: 2.00.
Entropy of illustrated probability distribution: 0.00.

Table Ib

Marginal probability for number in age class 2		0	1.0	0	0	0
Marginal probability for number in age class 1	Number in age class 2 / Number in age class 1	0	1	2	3	4
0	0	0	0	0	0	0
0	1	0	0	0	0	0
1.0	2	0	1.0	0	0	0
0	3	0	0	0	0	0
0	4	0	0	0	0	0

Expected number of organisms in age class 1: 2.00.
Expected number of organisms in age class 2: 1.00.
Entropy of illustrated probability distribution: 0.00.

Table Ic

Marginal probability for number in age class 2		0.20	0.20	0.20	0.20	0.20
Marginal probability for number in age class 1	Number in age class 2 / Number in age class 1	0	1	2	3	4
0.20	0	0.04	0.04	0.04	0.04	0l04
0.20	1	0.04	·0.04	0.04	0.04	0.04
0.20	2	0.04	0.04	0.04	0.04	0.04
0.20	3	0.04	0.04	0.04	0.04	0.04
0.20	4	0.04	0.04	0.04	0.04	0.04

Expected number of organisms in age class 1: 2.00.
Expected number of organisms in age class 2: 2.00.
Entropy of illustrated probability distribution: 3.22.

tribution that applies when it is certainly known that there are 3 individuals of age class 1 and 2 individuals of age class 2 in the system. Tables Ic–e illustrate what may be called mixed-state distributions, those in which two or more cells have non-zero entries. In biological applications, it is usual to consider pure-state descriptions when describing a single unique

Table Id

Marginal probability for number in age class 1	Marginal probability for number in age class 2	0.20	0.20	0.30	0.20	0.10
	Number in age class 2 Number in age class 1	0	1	2	3	4
0.10	0	0.02	0.02	0.03	0.02	0.01
0.20	1	0.04	0.04	0.06	0.04	0.02
0.20	2	0.04	0.04	0.06	0.04	0.02
0.40	3	0.08	0.08	0.12	0.08	0.04
0.10	4	0.02	0.02	0.03	0.02	0.01

Expected number of organisms in age class1: 2.20.
Expected number of organisms in age class 2: 1.80.
Entropy of illustrated probability distribution: 3.03.

Table Ie

Marginal probability for number in age class 1	Marginal probability for number in age class 2	0.11	0.20	0.33	0.26	0.10
	Number in age class 2 Number in age class 1	0	1	2	3	4
0.08	0	0.00	0.02	0.03	0.02	0.01
0.16	1	0.02	0.04	0.04	0.04	0.02
0.27	2	0.04	0.04	0.08	0.08	0.03
0.40	3	0.04	0.08	0.15	0.10	0.03
0.09	4	0.01	0.02	0.03	0.02	0.01

Expected number of organisms in age class 1: 2.26.
Expected number of organisms in age class 2: 2.04.
Entropy of illustrated probability distribution: 2.92.

system (e.g., a given tide pool) and mixed-state descriptions when considering an ensemble of systems (e.g., a collection of similar tide pools). When the primary goal is to obtain a system characterization at a given time in the absence of complete information (the usual case in biological and geological systems), a probability distribution will reflect more our state of knowledge than the nature of the real system. Then single systems are appropriately described by mixed-state distributions. In this case, mixed-state distributions can be viewed as superpositions of pure-state distributions.

To use the descriptions above, it is necessary to have a way of choosing the most applicable distribution (or distributions), out of all those that are possible, based on observational (or experimental) data and any other pertinent information. These eclectic choices are discussed below.

Descriptions by probability distributions in natural systems are almost never unique. To define a unique probability distribution for a given system, sampling must be exhaustive. To illustrate this point, we return to the probability distribution tabulated in Tables Ia–e. If our hypothetical tide pool is examined by taking a small, nonexhaustive, random sample in a single pool and one finds, for example, 1 individual of class 1 and 1 organism of class 2, this result is compatible with all the illustrated probability distributions. Only an exhaustive sample—i.e., finding all the organisms (and possibly examining all the pools)—could identify the uniquely applicable distribution. In biological sampling, it is expected that exhaustive sampling will be the exception rather than the rule.

In light of this discussion, one should search for ways to compare or rank hypotheses, in light of available, but usually incomplete, observational data, rather than demand that the probability distribution be unique. This task may be subdivided, for didactic purposes, into three subtasks: (1) specification of the class of hypotheses to be considered; (2) specification of procedures to be used in relating observational data to these hypotheses; and (3) specification of the hypothesis-comparison procedure itself. For reasons of brevity, it seems best to sequentially discuss these topics in the order 2, 3, and 1.

3.1. Propositional Calculus and Probabilities

An explicit procedure for relating observational data and hypotheses is provided by propositional calculus. In our coral population problem, this procedure is used because our observations are complicated by a collection procedure that has a varying and age-dependent collection efficiency. Because large numbers of variables and possibilities are involved, it is desirable to proceed in an explicit, stepwise fashion. In general, use of propositional calculus encourages a clear statement of assumptions, observational data, sampling procedure, and the hypotheses themselves, and specification of pertinent background information. In the following discussion, we will explain the construction of a few kinds of basic propositions and define ways of relating them to each other.

A proposition, symbolized by a letter (possibly subscripted), may be viewed as a statement about the systems under consideration. It is convenient to recognize two types of propositions: those pertaining to data and those pertaining to hypotheses. In the following examples, we use propositions illustrative of our Bermuda field problem. An example of a data proposition would be: $D_1 =$ "There are 3 corals in age class 23 in the Castle Harbor live coral collection." Another might be: $D_2 =$ "There are 2 corals in age class 24 in the Castle harbor live coral collection." A slightly different data proposition would be: $C_1 =$ "25% of the corals present on the inspected Castle Harbor reefs, having ages between and including 23 and 24 years, were collected." An example of a hypothesis proposition might be: $H_3 =$ "The probability that there exist 10 corals in age class 3 on the Castle Harbor reefs is 0.30." This latter proposition just restates in words part of the information specified mathematically by a probability distribution.

For such descriptions by propositions to be useful, it must be possible to combine, transform, and interrelate the propositions. The means for accomplishing this is provided by propositional algebra. For example, if proposition E corresponds to the statement, "There are 3 corals in age class 23 in the Castle Harbor live coral collection, and 25% of the corals present on the inspected Castle Harbor reefs, having ages between and including 23 and 24 years, were collected," then E may be derived from the individual propositions D_1 and C_1 stated above by: $E = D_1 \cdot C_1$, where the centered dot symbolizes the relationship *and*. The other basic combining relationship is *or*, denoted by $+$. For example if $G =$ "There are 3 corals in age class 23 in the Castle Harbor live coral collection or there are 2 corals in age class 24 in the Castle Harbor live coral collection or both," then $G = D_1 + D_2$. For the sake of completeness, we introduce the negation of a proposition, symbolized by $\sim$, so that, e.g., $\sim D_1 = d_1$. Finally, it is convenient to recall the following relationships when more complicated expressions involving *and* and *or* are encountered:

$$D_1 \cdot (D_2 + C_1) = D_1 \cdot D_2 + D_1 \cdot C_1$$

and

$$D_1 + (D_2 \cdot C_1) = (D_1 + D_2) \cdot (D_1 + C_1)$$

The foregoing operations make it possible to describe hypotheses conveniently. For example, the hypothesis Q: "The probability that there exist 15 corals in age class 1 on the Castle Harbor reefs is 0.20, and the probability that there exist 7 corals in age class 2 is 0.18, and the probability that there exist 10 corals in age class 3 is 0.30, etc.," is conveniently expressed as $Q = H_1 \cdot H_2 \cdot H_3 \cdot \ldots$. The reader will remember that as we have stated earlier, a hypothesis is equivalent to a specific probability distribution.

Rational numbers, lying between zero and one and called "probabilities," may be assigned to certain data and hypothesis propositions. Thus, the probability associated with proposition D_1 is denoted by $p(D_1)$, and this expression may be interpreted as claiming that the probability that there are 3 corals in age class 23 in the Castle Harbor collection is the number p. In sampling situations, this number may be given a relative frequency of occurrence interpretation. Once probabilities have been assigned to two propositions, say D_1 and D_2, the probability to be associated with the compound event $(D_1 \cdot D_2)$ is given by

$$p(D_1 \cdot D_2) = p(D_1/D_2)p(D_2) = p(D_2/D_1)p(D_1)$$

and with the compound event $(D_1 + D_2)$, it is given by

$$p(D_1 + D_2) = p(D_1) + p(D_2) - p(D_1 \cdot D_2)$$

The term $p(D_1/D_2)$ is a conditional probability, and indicates the probability of D_1 when D_2 is assumed true. If $p(D_1/D_2) = p(D_1)$, then D_2 is said to be irrelevant to D_1. Hypothesis propositions may also be assigned nonnegative real numbers, called "likelihoods," by a relationship of the form $L(H_i/Dx) = kp(D/H_i x)$, where k is a constant of proportionality and H_i and D are hypothesis and data propositions, respectively.

We have briefly summarized the propositional and probability concepts necessary for a discussion of the Bermuda coral observations and hypotheses. The interested reader should consult Cox (1961) for a more detailed description. The numerical propositional and probability assignments applicable to our field situation are given in Sections 4.1–4.3.

3.2. Comparison of Hypotheses

We now wish to discuss two methods for comparing and ranking hypotheses: Bayesian probability (Savage, 1972; Tribus, 1969) and likelihood (Edwards, 1972) procedures. In the Bayesian approach, probabilities are assigned to propositions specifying hypotheses, and the hypotheses are ranked according to these probabilities. Two sets of probabilities for any hypothesis, H_i, are involved: $p(H_i/x)$, often referred to as the "prior probability," describes the status of the hypothesis H_i in light of the background information specified by the proposition x, but before the outcome of the current observations on the system are known. In contrast, $p(H_i/Dx)$, often termed the "posterior probability," describes the status of the hypothesis, H_i, in light of the data arising from the observations symbolized by the propositions D. Bayes's rule relates the two probabilities as follows:

$$p(H_i/Dx) = p(H_i/x)[p(D/H_ix)/p(D/x)]$$

In practice, it is often interesting to compare some relatively small number of competing hypotheses two at a time. The comparison between hypotheses H_1 and H_2 is done by taking the ratio of the two posterior probabilities:

$$[p(H_1/Dx)/p(H_2/Dx)] = [p(H_1/x)/p(H_2/x)][p(D/H_1x)/p(D/H_2x)]$$

In the likelihood approach, hypotheses of interest are compared in terms of the likelihood ratio for pairs of hypotheses:

$$L(H_1,H_2/Dx) = L(H_1,H_2/x)[p(D/H_1x)/p(D/H_2x)]$$

where $L(H_1,H_2/x)$ is the prior likelihood ratio. It may be noted that if the ratio of prior hypothesis probabilities is numerically equal to the prior likelihood ratio, the Bayesian and likelihood comparisons will be numerically the same (though of course the interpretations may differ). Even if the above is not the case, the numerical results in both approaches are closely and simply related. For this reason, the explicit calculations given below are done only for the Bayesian approach.

3.3. Specification of Hypotheses

It will have been noted that to proceed with the analysis in any approach, it is necessary to define a class of hypotheses to be considered and to evaluate the prior hypothesis probabilities, $p(H_i/x)$, or the prior likelihood ratio, $L(H_1,H_2/x)$, for hypotheses of the given class. These choices must be made in light of prior information, and it is important to realize that the term "prior information" may include basic factual information, biological or ecological generalizations (having various degrees of generality, validity, or applicability), accepted assumptions, and even the goals of the given study. It is often the latter that defines the broad class of hypotheses to be considered. For example, it is the ecological orientation of our coral study that specifies that we consider probability distributions as a function of different numbers of corals in various age classes.

Before we present a specific hypothesis class or assign specific prior probabilities to members of the class, let us make it clear that the importance of such an assignment comes less from any specific selections and assignments *per se* but more from the explicit specification of what hypotheses, with what prior probabilities, are being considered—i.e., what and how prior information about the situation is being utilized. The nature of this information may range from a specific, more or less widely known biological fact to a certain accepted, but not necessarily uniformly valid, biological generalization, or to a convenient, and possibly debatable, assumption, and it is important that these facts and assumptions be explicitly spelled out. The detailed reason for stressing this explicitness will be seen later, but a summary of the conclusions may be in order at this point. If the method of analysis is independent of the specific initial assumptions, and if the initial assumptions are clearly stated so that the reader is aware of them, then any reader who disagrees with the assumptions or conclusions of the study may readily modify the overall analysis to incorporate any desired revised assumptions.

To illustrate the interplay of prior information and hypothesis choice, we return to our hypothetical tide pool system, discussed earlier. If our interest is simply in characterizing a specific, single pool at a given time, the natural class of hypotheses is that of pure-state distributions. Under the conditions of our example, there will be 25 such distributions or hypotheses in the class. If there is no additional biological information beyond that used in choosing the class of hypotheses, each of these hypotheses might be assigned the same prior probability, in this case 0.04. If we have additional biological information, the prior probability assignment would be different. For example, suppose we know that this pool has been under a constant environment for a long time and that its population reflects an appropriate steady state, and that there is age-dependent mortality. With this background information, it would be appropriate to assign a prior probability of zero for all the pure-state hypotheses (distributions) that describe a situation in which there are more class 2 organisms than class 1 organisms, thereby increasing the prior probability of the remaining hypotheses. If, on the other hand, our interest is in obtaining ecological generalizations—for example, we wish to specify those properties of an expected age-frequency distribution that would hold under a finite range of conditions and possibly for a variety of organisms— then the class of hypotheses would be that of mixed-state distributions, appropriate for describing ensembles of tidal-pool systems as mentioned above. With these goals, the assignment of prior probabilities would be more sensitive to the theoretical component of the prior information.

The class of hypotheses we choose to explore is the one made up of probability distributions that satisfy certain constraints, specified below for our coral system, on their associated expectation values. These hypotheses also have the highest possible entropy, with all members of this class being assigned the same prior probabilites (or likelihoods). The above choice is made for the following reasons: for a probabilistic description, biological and ecological generalizations refer not to the properties of individual systems, but to the properties of ensembles of systems. The requirement that a probability distribution satisfy expectation value constraints then ensures that the pertinent biology and ecology are reflected in the prior-probability assignments of the analysis. The requirement that the probability density have the largest possible entropy arises from the use of negative entropy as a measure of the amount of information required to specify the distribution (Cox, 1961; Tribus, 1969). Thus, among the distributions that are consistent with the available information and stated assumptions, the one that has the highest entropy will be the one that requires no additional information for its specification. A maximum-entropy probability distribution may be said to be an unprejudiced distribution in the sense that it makes no claims requiring additional information, beyond that already used as prior information, for its justification (Tribus, 1969). Tables Ia–e, in addition to displaying examples of probability distributions, discussed earlier, also give the entropy and the expected values for organism number in each age class for these distributions.

4. Description of Coral Populations

In this section, we analyze our Bermuda coral population data, which were summarized in Section 2, by the probability procedures we have just described. Section 4.1 develops the class of hypotheses to be evaluated, and Section 4.2 describes the relationship between the observational data and the hypotheses being considered. Section 4.3 discusses and compares various specific hypotheses, and Section 4.4 reviews some pertinent biological and ecological factors.

4.1. Specification of Population Hypotheses

In describing our coral reef populations, we will then work with probability distributions or hypotheses having the following properties: the distribution, $p(n)$, is chosen so that the entropy, $S = -\Sigma_n p(n)\ln p(n)$, is a maximum, $p(n)$ is normalized so that $\Sigma_n p(n) = 1$; and $p(n)$ also satisfies any constraints provided by background information and assumptions. The quantity Σ_n is a shorthand notation for the λ-fold sum over the permitted n_k's, each sum running from 0 to infinity. It is possible that we may have no relevant background information or prior knowledge. In such a case, maximum-entropy considerations, without additional constraints, lead to a broad distribution that makes all events equally probable. The assumption that absolutely nothing more is known about corals is probably too pessimistic. While it may be too much to expect very detailed statements about the biology of coral life history or environmental changes of a specific coral reef before observations are made, we feel that certain generalizations about population age distributions or corals can be made (Connell, 1974). This knowledge is reflected both in simple models that describe average population behavior (e.g., Chapter 10) and in accepted generalizations about life-history traits of other organisms (Cole, 1954). Much of this background information that is useful in describing coral reef populations can be stated, at least to a first approximation, by saying that the expected or average number of individuals varies with age in a specified manner. If the variation in the average number of corals with age class k is denoted by the function $f(k)$, then the probability distribution will have to satisfy the restriction $f(k) = \Sigma_n n_k\, p(n)$ for each age class.

To proceed further, we must make a specific choice for the function $f(k)$. In light of present knowledge, and our study of the Bermuda system, a reasonable form for determining a specific number of corals at a given age is

$$f(k) = \rho A F_k = \rho A\left(e^{-rk}\bigg/ \sum_{j=0}^{\lambda-1} e^{-rj}\right) \tag{1}$$

where ρ is the coral density (number per square meter), A is the reef area sampled in square meters, F_k is the fraction of the population aged between k and k + 1 years, r is the intrinsic rate of natural increase (or difference between instantaneous birth and death rates), and λ is the maximum longevity of the organisms in the population, i.e., the age of the oldest individual. The equation describes populations under steady-state conditions, as derived, for example, by Cole (1954). In addition, in an average sense, the expression will describe populations in many temporally variable environments, for example, when the differences between birth rates and death rates are approximately constant or when the birth–death rates vary on either a much shorter or a much longer time scale than mean longevity. We note that this specific choice for $f(k)$ is being made because of the general lack of life-history information about corals in general, and about the Bermuda sites in particular. In these circumstances, it seems best to begin with the simplest assumptions as long as they are not contradicted by available site-specific information.

The probability distribution that has the highest entropy and that also satisfies the additional restraint of the background information may be found by the method of Lagrange multipliers (Hildebrand, 1965). The desired probability function is the one that maximizes the expression

$$I = S - (\alpha - 1)\left[\sum_n p(n)\right] - \sum_{j=0}^{\lambda-1} \beta_j\left[\sum_n n_j p(n)\right] \tag{2}$$

where $\alpha, \beta_1, \ldots, \beta_{\lambda-1}$ are Lagrange multipliers. Differentiating I with respect to a typical probability term, $p(n')$, and setting the resulting expression equal to zero, we obtain the equation

$$-\ln p(n') - 1 - (\alpha - 1) - \sum_{j=0}^{\lambda-1} \beta_j n'_j = 0 \tag{3}$$

The desired probability may be expressed as $p(n) = e^{-\alpha - \Sigma_j \beta_j n_j}$. Thus, $p(n)$ may also be written as $\Pi_{j=0}^{\lambda-1} p(n_j)$, where $p(n_j) = e^{-\alpha_j} e^{-\beta_j n_j}$ and $e^{-\alpha_j} = (\sum_{n_j=0}^{\infty} e^{-\beta_j n_j})^{-1}$, which has the value of $1 - e^{-\beta_i}$ as a consequence of the normalization restriction. The probability of finding a certain number of corals in specific age classes, 0 through $\lambda - 1$, is now defined in terms of the Lagrange multipliers α and β_k. These multipliers may be determined, in principle, by substituting the above expression for the probability into the constraint equations. Alternatively, it is often more convenient, particularly in this case, to obtain an expression in which the multipliers are eliminated in favor of the value of $f(k)$. By using the above expression for $p(n)$, $f(k)$ may be written as

$$f(k) = \sum_{n_k} n_k e^{-\beta_k n_k}/e^{\alpha_k}$$

It may be observed that the numerator of this expression is simply minus the derivative of the denominator with respect to the multiplier β_k. Performing the summation defining the denominator and the indicated differentiation leads to

$$f(k) = 1/(e^{\beta_k} - 1)$$

With the help of this relationship, it is possible to express the probabilities $p(n_k)$ in terms of the $f(k)$:

$$p(n_k) = [f(k)]^{n_k}/[1 + f(k)]^{n_k+1} \tag{4}$$

Finally, if we introduce the value of $f(k)$ as specified by equation (1), the probabilities $p(n_k)$ are given by

$$p(n_k) = \left[\rho A\left(e^{-rk} \Big/ \sum_{j=0}^{\lambda-1} e^{-rj}\right)\right]^{n_k} \Bigg/ \left[1 + \rho A\left(e^{-rk} \Big/ \sum_{j=0}^{\lambda-1} e^{-rj}\right)\right]^{n_k+1} \qquad (5)$$

As developed above, a hypothesis (or a probability distribution) is specified by choosing three parameters: r, the intrinsic rate of natural increase (the difference between birth and death rates); λ, the maximum longevity of individuals in the population; and ρ, the coral density (number of corals/m^2). In the following discussion, such a hypothesis will be denoted by $H_{\lambda,r,\rho}$. Site-specific background information, obtained in the course of field observations, allows us to define ranges for these parameters before the results of the specimen collection are considered, and this information reduces the amount of numerical calculations that are required.

In Castle Harbor, an extensive examination revealed no living corals with an age greater than 60 years, and thus λ_{CHL} is taken as 60 years. On the north and south reefs, the oldest living individuals were at least 250 years old. This is also true for the Castle Harbor dead population, in which the oldest dead specimens were at least 250 years of age. These observations of λ maximum are supported by other reports of coral longevity (Scoffin and Garrett, 1974; Frazier, 1970; Johannes, 1972). We thus assign $\lambda_{NS} = 250$ yr and $\lambda_{CHD} = 250$ yr.

General biological considerations suggest that the overall expected r of a site will be greater than zero, though of course individual occurrences of sites with a negative r are possible. Visual inspection of the reefs gives a qualitative impression about the number of corals at various age classes and indicates that r is 0.2/yr or less at each site, but r is not negative. Visual inspection of the reefs also indicates that there are variations in coral density, ρ, between sites. For CHL, with an estimated observational area of 5000 m^2, it does not appear that ρ is less than 0.1/m^2 or greater than 6.0/m^2. For the NS corals, as well as for the CHD population, we estimate a minimum ρ of 1.0/m^2 and a maximum of no more than 60.0/m^2 over the estimated observational areas of 4000 m^2.

In concluding this section, we remind the reader that if he disagrees with the above assumptions, he may repeat the analysis, without change of form, substituting his own assumptions. If the revision is only in prior-probability assignments, the recalculation will be trivial; if a different class of hypotheses is used, more work will have to be done. Specifically, we wish to point out that different assumptions about the population age structure may be easily accommodated, simply by revising equation (1).

4.2. Data–Hypothesis Relationships

The discussion in the previous section showed that the quantitative relationship between the observational data and any hypothesis may be expressed by the conditional probability $p(D/H_ix)$. In our specific case, D is a compound proposition describing all the results of our age determinations for collected corals, as described in Section 2, and H_i is a short-hand notation for a hypothesis $H_{\lambda,r,\rho}$, as developed in the preceding section.

The evaluation of the terms of the form $p(D/H_ix)$ requires a further, more detailed, description of the collection procedure and its relationship to both the final data and the hypotheses. The evaluation is made easier by defining certain auxiliary propositions. Because later we will be dealing with several populations, these propositions are defined at this time more generally than is required in the analysis for a given site. Let K: "The kth age class consists of corals having ages between k and $k + 1$ years," C_{k,c_k}: "c_k corals of the kth age class are contained in the collection," A_{k,n_k}: "n_k corals of the kth age class exist at a given reef site," w: "Corals are collected independently of one another, with a certain collection efficiency that may be age- and site-dependent," x: "Other information and descriptive statements about corals, reefs, populations, environmental variables, etc."

It is also convenient to define explicitly several additional compound propositions. D_s: "c_0 corals of age class 0, and c_1 corals of age class 1, and . . . $c_{\lambda-1}$ corals of age class $\lambda - 1$ are contained in the collection." This may be expressed more briefly as $D_s = C_{0,c_0} \cdot C_{1,c_1} \cdot \; \cdots \; \cdot C_{\lambda-1,c_{\lambda-1}} = \Pi_{k=0}^{\lambda-1} C_{k,c_k}$, T_m: "n_0 corals of age class 0, and n_1 corals of age class 1, and . . . $n_{\lambda-1}$ corals of age class $\lambda - 1$ exist on the reef." This may be expressed as $T_m = \Pi_{k=0}^{\lambda-1} A_{k,n_k}$. A given T_m is but one possible statement from a set of mutually exclusive and exhaustive statements, so that $\Sigma_q p(T_q) = 1$, where Σ_q is shorthand for $\Sigma_{n_0=0}^{\infty} \Sigma_{n_1=0}^{\infty} \; . . . \; \Sigma_{n_{\lambda-1}}^{\infty}$. The term $p(D_s/H_iwx)$, where H_i is some specific hypothesis about coral reef populations, may now be evaluated with the help of the above auxiliary statements.

Because $\Sigma_q p(T_q) = 1$, we may, regardless of the meaning of D_s, H_i, w, and x (as long as none is the negation of T_q), write $\Sigma_q p(T_q/D_sH_iwx) = 1$. Multiplying both sides of this equation by $p(D_s/H_iwx)$, we obtain

$$p(D_s/H_iwx) = \sum_q p(D_s/H_iwx)p(T_q/D_sH_iwx)$$
$$= \sum_q p(D_sT_q/H_iwx)$$

We have already observed that $D_s = \Pi_{k=0}^{\lambda-1} C_{k,c_k}$ and $T_q = \Pi_{k=0}^{\lambda-1} A_{k,n_k}$, and hence the last expression may be written as

$$p(D_s/H_i wx) = \sum_q p\left(\prod_k C_{k,c_k} \cdot \prod_k A_{k,n_k}/H_i wx \right)$$

The terms C_{k,c_k} are independent of one another by statement w and the terms A_{k,n_k} are independent of one another by the nature of the maximum-entropy hypotheses [equations (3) and (4)]. Therefore, the above expression becomes $\sum_q \Pi_{k=0}^{\lambda-1} p(C_{k,c_k} \cdot A_{k,n_k}/H_i wx)$. Recalling that $\sum_q$ is shorthand for a λ-fold summation and expanding the enclosed term with the help of the equation describing compound events, we obtain

$$p(D_s/H_i wx) = \sum_{n_0=0}^{\infty} \sum_{n_1=0}^{\infty} \dots \sum_{n_{\lambda-1}=0}^{\infty} \prod_{k=0}^{\lambda-1} p(C_{k,c_k}/A_{k,n_k} wx) p(A_{k,n_k}/H_i wx)$$

An alternate form more convenient for numerical work may be obtained by expanding the terms and rearranging them, which gives the result

$$p(D_s/H_i wx) = \prod_{k=0}^{\lambda-1} \sum_{n_k=0}^{\infty} p(C_{k,c_k}/A_{k,n_k} wx) p(A_{k,n_k}/H_i wx)$$

The equivalence of the two forms may be demonstrated by expanding each expression directly and comparing the results. A saving in numerical work may be made by noting that the probability of having more corals in the collection than exist on the site is zero, so that the expression used in the computation is

$$p(D_s/H_i wx) = \prod_{k=0}^{\lambda-1} \sum_{n_k=c_k}^{\infty} p(C_{k,c_k}/A_{k,n_k} wx) p(A_{k,n_k}/H_i wx)$$

The evaluation of this expression for our coral observation in Bermuda and for the hypotheses introduced above proceeds as follows: The probability of collecting c_k corals, given there are n_k in existence, is, in accordance with the proposition w, given by

$$p(C_{k,c_k}/A_{k,n_k} wx) = [n_k!/c_k!(n_k - c_k)!][\nu_k^{c_k}(1 - \nu_k)^{(n_k - c_k)}]$$

where ν_k is the age-specific coral-collection efficiency. The term $p(A_{k,n_k}/H_i wx)$ is identical to the term $p(n_k)$ defined by equation (5). Combining these results and explicitly indicating that the hypothesis H_i refers to a specific choice of λ, r, ρ, we find that

$$p(D_s/H_{\lambda,r,\rho}wx)$$

$$= \prod_{k=0}^{\lambda-1} \sum_{n_k=c_k}^{\infty} [n_k!/c_k!(n_k - c_k)!][v_k^{c_k}(1 - v_k)^{(n_k-c_k)}]$$

$$\times \left[\rho{\cdot}A{\cdot}\left(e^{-rk} \middle/ \sum_{k=0}^{\lambda-1} e^{-rk}\right)\right]^{n_k} \middle/ \left[1 + \rho{\cdot}A{\cdot}\left(e^{-rk} \middle/ \sum_{k=0}^{\lambda-1} e^{-rk}\right)\right]^{n_k+1}$$

4.3. Comparison of Hypotheses

In this section, we examine hypotheses about coral populations from the Bayesian viewpoint (leaving a likelihood approach as an exercise for the reader). It will be recalled that in this approach, the quantity of interest is $p(H_{\lambda,r,\rho}/D_s x)$; we wish to know how this probability depends on λ, r, and ρ. For the Bermuda example, $p(H_{\lambda,r,\rho}/D_s x)$ is simply proportional to $p(D_s/H_{\lambda,r,\rho}wx)$ as specified above, and the following discussion is in terms of this latter quantity.

The amount of observational data is sufficiently large, and the number of required calculations sufficiently numerous, that the evaluation of hypotheses about the different sites is best done numerically. This was done by coding the appropriate expressions in FORTRAN and using an IBM-370/158 computer. Graphic printout was obtained by using SPEAKEASY. The calculations were performed with truncated summations and single-precision arithmetic. The adquacy of the approximation was checked by performing selected evaluations with twice the number of terms in the summation and double-precision arithmetic.

The simplest way to report the results of the calculations is to present contour diagrams of $p(D_s/H_{\lambda,r,\rho}wx)$ with respect to ρ and r, using the appropriate λ and D_s for each site. Figures 2A, B, and C present these contour diagrams for the CHL, NS, and CHD sites, respectively. The points at which the probability of the data, given the specific hypothesis, are at a maximum are: for CHL, $\lambda = 60$ yr, $r = 0.118$/yr, and $\rho = 0.25$/m^2; for NS, $\lambda = 250$ yr, $r = 0.068$/yr, and $\rho = 1.0$/m^2; and for CHD, $\lambda = 250$ yr, $r = 0.092$/yr, and $\rho = 1.0$/m^2. These maximum probability points are indicated by a 0 in each diagram. Hypotheses corresponding to these parameters are the ones favored most by the observations. We shall denote the hypotheses corresponding to the above values of the parameters by H_0^{CHL}, H_0^{NS}, and H_0^{CHD}, respectively. In each figure, the contour defined by the outer edge of the area denoted by 2's includes at least 95% of the total probability.

Inspection of Figs. 2A–C reveals that the CHL population is significantly different from either the NS or the CHD population; e.g., the 10^{-2} probability contour (denoted by the outer edge of the field of 2's in the figures) of CHL does not overlap with the 10^{-2} contour of either NS or CHD. It is only when the 10^{-5} probability contour is approached (which includes over 99.99% of the total probability) that the contours of CHL and CHD touch. It seems reasonable to conclude that the CHL population is very different biologically from the NS reef populations living outside the harbor or from the dead Castle Harbor (CHD) population. In contrast, it is interesting to note that the NS and CHD populations are quite similar, showing overlap even at the 10^{-2} probability contour.

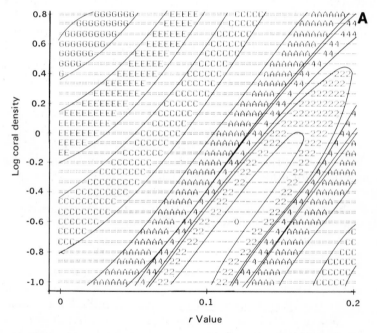

Figure 2. (A) Contours of $p(D_{CHL}/H_{\lambda=60,r,\rho}wx)$ vs. log ρ and r (intrinsic rate of increase) for the Castle Harbor live coral collection. (0) Point of maximum probability. Relative to this maximum, the probability contours of 1, 10^{-1}, 10^{-2}, 10^{-3}, 10^{-4}, 10^{-5} correspond to the outer edges of the fields indicated respectively by 0, $-$, 2, $-$, 4, $-$. The probability contours of 10^{10}, 10^{-20}, . . . , 10^{-90}, 10^{-100}, 10^{-150} correspond to the outer edges of the fields indicated by A, $=$, C, $=$, E, $=$, I, $=$, K, respectively. (B) Contours of $p(D_{NS}/H_{\lambda=250,r,\rho}wx)$ vs. log ρ and r for the North–South reef collection. The contours and symbols are as given for Fig. 2A. (C) Contours of $p(D_{CHD}/H_{\lambda=250,r,\rho}wx)$ vs. log ρ and r for the Castle Harbor dead coral collection. The contours and symbols are as given for Fig. 2A.

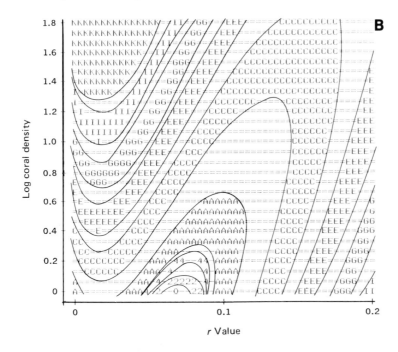

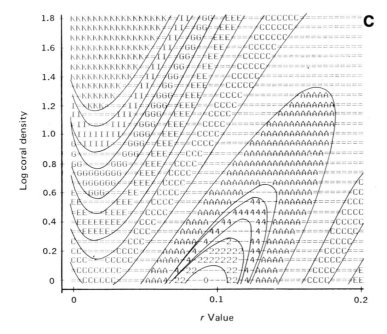

Figure 2. (Cont'd)

Another way of comparing hypotheses is by considering probability ratios for various hypotheses, an approach that is also closer to the likelihood method. One overall hypothesis deserving attention in such a comparison is $H_I = H_0^{CHL} \cdot H_0^{NS} \cdot H_0^{CHD}$, which claims that each of three populations is best described by its own distribution. A reasonable competing hypothesis is $H_{II} = (H_{\lambda=60, r^*, \rho^*}^{CHL}) \cdot (H_{\lambda=250, r^*, \rho^*}^{NS}) \cdot (H_{\lambda=250, r^*, \rho^*}^{CHD})$, which asserts that all three populations have identical age-distribution functions. If r^* and ρ^* are chosen to be 0.092/yr and 1.0/m², respectively, H_{II} is the most probable hypothesis in the class of such hypotheses. Even in this case, it compares unfavorably to H_1, as the following evaluation demonstrates [with $p(H_i/wx) = p(H_{II}/wx)$ and $D_t = D_{CHL} \cdot D_{NS} \cdot D_{CHD}$, the total data set]:

$$[p(H_I/D_t wx)/p(H_{II}/D_t wx)] = [p(D_{CHL}/H_0^{CHL} wx)/p(D_{CHL}/H_{60, r^*, \rho^*}^{CHL} wx)]$$

$$\cdot [p(D_{NS}/H_0^{NS} wx)/p(D_{NS}/H_{250, r^*, \rho^*}^{NS} wx)]$$

$$\cdot [p(D_{CHD}/H_0^{CHD} wx)/p(D_{CHD}/H_{250, r^*, \rho^*}^{CHD} wx)]$$

$$= (10^{15.47})(10^{+4.09})(10^{+0}) = (3 \times 10^{+19})$$

The results again indicate that it is more reasonable to consider each population as distinct rather than described by a common frequency-age distribution. The individual ratio terms above show that the population most significantly different from the others is the CHL population. In contrast, the difference in the parameters characterizing the CHD and NS populations is not large, and the evidence for a difference is not compelling.

Finally, having identified the most probable description of the coral population at each site, we may examine various features of the corresponding probability distributions. The distributions themselves may be explicitly computed from equation (5) by introducing the most probable values of λ, r, and ρ, as found above, for each site. The functions are Poisson at each age class. A summary feature of some interest, easily displayed graphically, is the dependence of the expected coral density on age. Figure 3 illustrates this for each of the hypotheses H_0^{CHL}, H_0^{NS}, and H_0^{CHD}, over an age range that corresponds approximately to that of the collection.

4.4. Biological and Ecological Factors

The preceding analysis of data describing the coral populations in Bermuda waters revealed that the most favored hypothesis describing Castle Harbor live (CHL) coral populations involves a much lower λ, a higher r, and a lower ρ than those appropriate for a dead population of

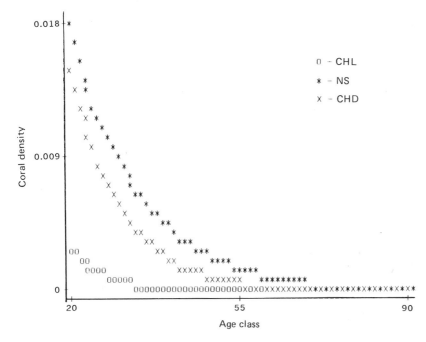

Figure 3. Average coral density vs. age for the three study sites. Each curve corresponds to the most probable hypothesis for the indicated site.

Castle Harbor (CHD) corals or for the reef coral populations now living external to the harbor (NS). In previous work (Dodge and Vaišnys, 1977), we indicated that large-scale dredging in Castle Harbor during 1941–1943 caused large-scale mass mortality of harbor corals. This conclusion was supported by the differences in species distributions and abundance between CHL collections and those from NS and CHD, by information about adverse effects of sedimentation on corals, by an analysis of coral growth patterns, and by the fact that most living corals in the harbor today postdate the dredging event. The quantitative analysis reported here also supports this conclusion. It indicates, first, that the NS and CHD populations are not too dissimilar in form. Such a situation is to be expected, since no major population disturbance is documented for NS or for Castle Harbor corals prior to 1941. The CHD population, being essentially a mass mortality or census, preserves the age structure it had in life. Second, the analysis shows not only that the CHL population differs from the NS and CHD populations, but also that it is relatively younger in average age and has a higher rate of natural increase and a lower coral density than the other populations. All these attributes are consistent with a new population recolonizing a disturbed habitat.

5. Conclusions and Suggestions for Further Research

We have presented methods for describing populations of organisms by means of quantitative hypotheses. The approach has a number of desirable properties. It can incorporate self-consistently both relevant background information about the system and the outcome of specific field observations on the system. A class of hypotheses, rather than a single hypothesis, can be considered, and hypotheses reflecting biological or ecological processes can be formulated and used. There is potential in this approach for dealing with realistic observational and sampling procedures. New information, arising either from further field observations or from more general theoretical arguments, can be incorporated into the analysis to update it without beginning the process all over again. The technique provides quantitative indicators that can be used to evaluate the performance of the analysis itself. The concrete utility of the approach has been demonstrated by analyzing data obtained in a field study of corals from Bermuda waters, and good agreement was found between the conclusions reached by qualitative arguments and by this quantitative approach.

In the early stage of development of this methodology, perhaps one of the most useful steps might be to begin designing field or laboratory observations and experiments within the framework presented above, before the work is actually started. This would probably increase the efficiency of the planned work as well as lead to specific improvements and further developments in each of the methods discussed in previous sections. With this methodology, real progress can be expected in formulating hypothesis classes based on biological and ecological generalizations, and also in improving field sampling procedures.

ACKNOWLEDGMENTS. Support of this research came from NSF grant No. OCE75-17618, the Geological Society of America, and The Woman's Seaman's Friend Society of Connecticut, Inc. We wish to thank Professor K. K. Turekian for discussion, the staff of the Bermuda Biological Station where field work was based, and Mr. Steven Dryer for able and energetic aid in collection.

References

Cole, L. C., 1954, The population consequences of life history phenomena, Q. Rev. Biol. **29**:103–137.

Connell, J. H., 1974, Population ecology of reef-building corals, in: Biology and Geology of Coral Reefs, Vol. 2 (O. A. Jones and R. Endean, eds.), pp. 205–245, Academic Press, New York.

Cox, R. T., 1961, *The Algebra of Probable Inference*, John's Hopkins Press, Baltimore, 114 pp.

Dodge, R. E., and Vaišnys, J. R., 1975, Hermatypic coral growth banding as environmental recorder, *Nature (London)* **258:**706–708.

Dodge, R. E., and Vaišnys, J. R., 1977, Coral populations and growth patterns: Responses to sedimentation and turbidity associated with dredging, *J. Mar. Res.* **35:**715–730.

Edwards, A. W. F., 1972, *Likelihood*, Cambridge University Press, Cambridge, p. 9.

Frazier, W. J., 1970, Description of reefs in Castle Harbor, Bermuda, *Bermuda Biol. Sta. Res. Spec. Publ.*, No. 6, pp. 63–72.

Hildebrand, F. B., 1965, *Advanced Calculus for Applications*, Prentice-Hall, Englewood Cliffs, New Jersey, p. 352.

Johannes, R. E., 1972, Coral reefs and pollution, *Proc. FAO Tech. Conf. Mar. Poll., Marine Pollution and Sea Life*, pp. 364–375.

Savage, L. J., 1972, *The Foundations of Statistics*, Dover Publications, New York, p. 45.

Scoffin, T. P., and Garrett, P., 1974, Processes in the formation and preservation of internal structures in Bermuda patch reefs, *Proc. 2nd Int. Coral Reef Symp.* **2:**429–448.

Tribus, M., 1969, *Rational Descriptions, Decisions, and Designs*, Pergamon Press, New York, 478 pp.

Index

Abrasion
  break, 260, 271
  of shell, 87–92
  of shell margin, 317
  of shell ornamentation, 317
Absolute growth, 33
Absolute mortality, 284
Acanthurus chirurgus, 545, 552
Accessory calcification
  in bivalves, 638–641
  mineralogy, 78–79
  types, 77–78
Accretive skeleton, 396
Acetate peel, 4, 391
  of bivalve shell, 597–601
  of fish sagitta, 621–622
  and power spectra analysis, 678
Acetazolamide, 120
Acidic sedimentary environments, 318
Acidic water, 302
Acmea, 385
Acmea digitalis, 382, 385
Acmea orbignyi, 382
Acmea persona, 382
Acmea scabra, 385, 389
Acropora palmata, 495
Acteocina canaliculata, 337
Acteocina senestra, 337
Adaptation, 384–387, 453–455
Adaptive strategy, 13–14, 347–354, 358
Adductor muscle, 27
Adhesive area, 384
Adventitious conchiolin patches, 315–316
Adventitious sublayers, 315–316
Aeolinites, 390–391

Aequipectin, 148
Aequipectin irradians, 144–145
Aerial exposure, 226
Aerobic metabolism, 211
After-the-fact pollution study, 13
Agariciella planulata, 496
Agarum, 389
Age
  and annuli on ligament, 308–309
  of barnacles, 485–488
  of bivalves, 273
  at death, 225, 273–274
  from fish sagittae, 520
  and growth bands, 307
  and growth rate, 261, 482
  and magnesium in shells, 144
  and periostracum color, 317
  of polychaetes, 584
  relative, 445–447
  and shell microstructure, 239, 241
  and size-frequency histograms, 581, 584
  and strontium in shells, 151
Age class, 426–440, 581, 584
Age dating, 155–156
Age-frequency histogram, 420–421, 581–586, 701
Age at maturity, 226
Aglaophamus verillii, 564, 584
Alae, 470
Albula vulpes, 548
Alectryonella, 347
Algae, 303, 494, 511–512
Allometry, 37–45
  and adaptation, 384–387
  and age, 39, 40, 43

725

Allometry (*cont.*)
  and color pattern, 382–383
  definition, 38
  doming, 382
  and ecological factors, 42–44
  flaring, 382
  and gastropod shell sculpture, 382–383
  and growth rate, 384–387
  and marginal growth, 381
Amblema plicata, 299
Ametamorphic development, 328
Amino acids in organic matrix, 122–124
Amplitude of harmonic, 677
Anaerobic glycolysis, 127
Anaerobic stress, 552
Anaerobiosis
  induction, 213
  and irregular nacreous laminae, 241
  and shell microgrowth increment
      formation, 208–209, 211–218
  and shell microstructural changes,
      231–235, 241
Angle of incidence, 49
Anguilla rostrata, 528
Anisomyarian condition of adductor
      muscle, 27
Annual event, 222, 241
Annual growth band, 306, 613–614
Annual growth increment, 397–398,
      495–496
Annual growth line, 564–566, 569, 585, 587
Annual growth ring, 401, 585
Annual microgrowth increment patterns,
      259, 269
Annulus, 546–548
Anodonta
  conchiolin patches, 315
  length/height proportions, 299
  periostracum color, 318
  shell regeneration, 118
  strontium content in shell, 151
Anodonta bialata, 76
Anodonta grandis, 298–300
Anodontoides ferussacianus, 299
Anterior–posterior axis, 27–28
Aperture rounding adaptations, 384
Apex theory, 330
Apical bosses, 313–315
Aragonite
  in accessory calcification, 78
  and bivalve shell microstructure
      categories, 81–82

Aragonite (*cont.*)
  and bivalve shell microstructure groups,
      79–81
  in bivalve shells, 72–73
  and Feigl solution, 632, 637–638
  and fish otoliths, 520
  inversion to calcite, 180
  and manganese substitutions, 141–142,
      145
  and oxygen isotopes, 175, 194
  and strontium substitutions, 154
  taxonomic distribution in bivalves, 73–77
  and temperature, 238–239
  See also Bivalve shell mineralogy
Aragonite–calcite oxygen isotope
      fractionation curve, 189, 194
Aragonite–calcite oxygen isotope
      thermometry, 190
Aragonite crystalline laminae, 241
Aragonitic complex crossed lamellar
      microstructure, 81
Aragonitic crossed acicular microstructure,
      81
Aragonitic crossed lamellar microstructure,
      81
Argonitic crossed-matted lineated
      microstructure, 81
Argonitic homogeneous microstructure, 81
Aragonitic nacreous microstructure, 79
Aragonitic prismatic microstructure, 81–82
Aragonitic prisms, 245–247
Aragonitic shell layers
  and environmental controls, 103
  evolution, 85–90
  inner, 87–90
  mechanical data, 87–90
  middle, 87–90
  outer, 86–87
Archaeoprion, 588
Area/volume ratio, 38
Argopecten irradians, 121, 136
Artica islandica, 245
Ascophyllum nodosum, 389
Astarte, 317, 341
Astarte castanea, 226
Astericus, 520
Autoradiography, 495–496
Axis of maximum growth, 28

Babinka, 28
Baculites scotti, 189, 191–199

Balanus balanoides, 461
    age and growth rate, 482
    age determination, 486
    environmental effects, 476–477
    external ridges, 484
    growth bands, 473, 479–480
    seasonal growth cycle, 481–482
    temperature effects, 479–480
Balanus crenatus, 482
Balcis, 380
Bankia gouldi, 124
Barnacle shell
    formation, 471–473
    growth, 471–473
        and coelomic pressure, 473
        formation of cuticle, 472
        and molting cycle, 472–473
        role of epithelium, 472
        seasonal cycle, 481–482
        secretion of calcium, 472
    growth patterns
        applications, 485–489
        endogenous control, 482
        and environment, 476–482
        external band, 473, 482–485
        and food supply, 477–479
        growth line, 473
        growth ridge, 473
        internal bands, 473–482
        model, 688–689, 691
        and photoperiod, 477
        and stress, 488
        striae, 473
        summary, 485
        and temperature, 479–480
        terminology, 473
        and tidal immersion, 477–480, 488,
            687–688
        time series analysis, 687–697
    morphology, 470–471
        alae, 470
        basal disk, 470
        cuticle, 470
        opercular membrane, 470
        opercular valves, 470
        paries, 470
        pseudosheath, 470
        radii, 470
        sheath, 470
        wall plate canals, 471
        wall plates, 470
Barnegat Bay, 274–291

Basal disk, 470
Bathylasma corraliforme, 482
Bayesian probability, 708–709, 717–720
Beak, 25
Belemnite guard, 680–684
von Bertalanffy growth equation, 34
Bicarbonate, 118–119
Bidaily microgrowth increment patterns,
    259, 262
Bioerosion, 506–507
Biometry, 589
Biospace, 3
Biostratigraphic index fossils, 359
Biradiolites angulosissimus, 74
Bivalve
    accessory calcification, 638–641
    development types, 345–347
    evolution, 30–31
    mineralogical evolution, 73
    origins, 28–29
    radiation, 30
Bivalve shell
    aberration, 303
    advantages, 29
    breakage, 303
    chemistry
        allometric variations, 136, 159
        boron, 143
        compositional gradations, 137
        diagenetic change, 155–156, 159
        and environmental cycles, 159
        and extrapallial fluid, 156–157
        lead, 143
        magnesium, 143–149
        manganese, 140–142
        ontogenetic variations, 136–139,
            144–146, 155–159
        phylogenetic trends, 136
        rhythms and body size, 159–161
        rhythms and lifespan, 159–161
        and shell geometry, 159
        strontium, 149–155
        sulfate, 142
        trace element variations, 139–155, 308
    deposition, 207
    dissolution
        adaptations, 310
        and conchiolin sublayers, 315
        and periostracum, 310
        and shell ornamentation, 317
    growth
        allometric techniques of analysis, 37–45

Bivalve shell (*cont.*)
  growth (*cont.*)
    multivariate techniques of analysis,
      45–52
    and temperature, 301–302
  growth curve, 35, 281
  growth equation, 34, 277
  growth pattern analysis, 678–680, 683–684
  growth rate
    and environment, 35
    genetic effects, 36
    latitudinal variations, 36
    and magnesium content, 144
    and strontium content, 150
  layer separation method, 308
  logrithmic spiral, 45
  margin, 316–317
  microgrowth increment patterns
    abrasion break, 260, 271
    and age, 273
    and age at death, 273–274
    annual, 259, 269
    bidaily, 259, 262
    daily, 259, 261–262
    and demographic analysis, 273–274
    fortnightly, 259, 262
    freeze-shock break, 260, 270
    growth breaks, 269–272
    and growth rate, 273
    heat-shock break, 260, 270
    lunar-month, 259, 262, 269
    and mass mortality, 274
    in *Mercenaria mercenaria*, 256–258
    and natural mortality, 274
    neap-tide break, 260, 272
    ontogenetic changes, 256–257
    and population dynamics, 273–274
    and recruitment, 273
    and season-of-death, 274
    seasonal changes, 262
    spawning break, 260, 271–272
    storm break, 261, 272
    subdaily, 259, 262
    thermal-shock break, 260, 270–271
    and water temperature, 261–262
    See also Molluscan shell microgrowth
      increment patterns
  microstructural sublayers
    and environmental controls, 100–103
    and respiration, 101–102
    and temperature, 101

Bivalve shell (*cont.*)
  microstructure
    adaptive significance, 105–106
    classification based on natural
      associations, 81–82
    classification based on structural
      arrangement, 79–81
    in diagenetically altered shells, 631–632
    and environment, 100–106
    evolution, 84–85, 104
    evolution of aragonitic shell layers,
      85–90
    evolution of calcitic shell layer, 90–95
    identification, 645–673
    and mechanical data, 87–90, 91–92
    and mechanical design, 85
    in unionaceans, 309–310
    See also Molluscan shell
      microstructure
  microstructure categories, 81–83
  microstructure groups, 79–81
  mineralogical–structural variations
    99–100
  mineralogy
    accessory calcification, 78–79
    analytical approaches, 632
    aragonite distribution, 72
    boron substitutions, 143
    calcite distribution, 72
    dahllite occurence, 72
    and environment, 96–99, 104
    evolution, 73, 84–85, 104
    evolution of aragonitic shell layers,
      85–90
    evolution of calcitic shell layers, 90–95
    general, 72–73
    in Hippuritacea, 74–75
    magnesium substitutions, 145
    manganese substitutions, 140–142
    microstructural intergradations, 72, 82
    in Modiomorphacea, 76
    in Mytilacea, 76–77, 632–638
    and paleontological data, 73
    in Pterioida, 73–74
    and salinity, 98–99
    strontium substitutions, 154
    and sulfates, 142
    superfamily distribution, 72
    taxonomic summary, 627–632
    and temperature, 96–99
    in Veneroida, 75–76

Bivalve shell (*cont.*)
  morphology
    aberrations, 303
    and acidic water, 302
    and competition, 304–305
    and environmental reconstruction,
      300–301
    and habitat, 297–301
    and hard water, 302–303
    height–habitat relationships, 298–301
    macroscopic features, 25–28
    obesity–habitat relationships, 297–301
    ontogenetic variations, 304
    parameters and models, 50–51
    and predation, 304–305
    and selection pressure, 305
    and sexual dimorphism, 303–304
    size–habitat relationships, 297–301
    shell outline–habitat relationships,
      298–301
    shell sculpture–habitat relationships,
      299
    and specialization, 304–305
    as a taxonomic character, 45
    and temperature, 301–302
    and trophic group, 303
  observation techniques, 613–614
  ornamentation, 317
  outline, 298–301
  preparation, 597–601, 609–611
  regeneration, 179
  sculpture, 299
  size, 36–37
Blood sinus, 116
Body size
  of adult
    and energy for reproduction, 351–353
    and larval mortality, 351–353
    and larval type, 351–353
  and mode of reproduction, 351–353
  and progenesis, 354
Boron substitution in shell, 143
Botula fusca, 76, 638
Brachiodontes recurvus, 149, 151
Breeding, 482
Brephic axials, 333
Brooded larvae, 328–329, 341
Buccinum undatum, 328
Burrow linings, 78
Burrowing adaptations, 57–61

Burrowing rate index, 59
Byssus gland, 30

Cabestana spengleri, 388
Calcite
  in accessory calcification, 78
  and bivalve shell microstructure
    categories, 81–82
  and bivalve shell microstructure groups,
    79–81
  in bivalve shells, 72–73
  and Feigl solution, 632, 637–638
  and magnesium substitutions, 145
  and manganese substitutions, 141–142
  oxygen isotope data for, 194
  and oxygen isotope fractionation, 175
  secretion, 92, 95
  and strontium substitutions, 154
  taxonomic distribution in bivalves, 73–77
  variation with temperature, 238–239
  See also Bivalve shell mineralogy
Calcite/aragonite ratio, 98
Calcitic complex crossed lamellar
    (=complex crossed foliated)
    microstructure, 81
Calcitic crossed lamellar (= crossed
    foliated) microstructure, 81
Calcitic prismatic microstructure, 81–82
Calcitic regularly foliated microstructure,
    79–81
Calcitic shell layers
  and active swimming, 94
  and bivalve shell density, 92
  and crack-stopping, 92
  evolution, 90–95
  and fracture localizations, 91–92
  mechanical data, 91–92
  and temperature, 96–99
Calcium
  concentration and shell dissolution, 125
  deposition in otoliths, 524–529
  ion and anaerobiosis, 211
  metabolism in fishes, 524–525, 527
  movement through mantle, 118–119
  secretion in barnacles, 472
  variations in bivalve shell, 135–139
Calcium carbonate
  as buffer, 208, 213
  deposition, 211, 532
  dissolution, 211

Calcium carbonate (*cont.*)
and molluscan shell microgrowth
increments, 206
and otoliths, 520
precipitation, 188
variations in shell, 206–207
Calcium-45, 208
Calliostoma, 383
Callista chione, 299
Calyptraea dilatata, 336
Cancellate ornamentation, 333
Cantharus ringens, 388
Canthyria spinosa, 317
Capitella capitata, 336
Carbon dioxide and oxygen isotope
paleothermometry, 171
Carbon-14 uptake, 509
Carbon isotope variations in corals, 509–510
Carbonic anhydrase, 120
Cardinal area, 25
Cardium, see Cerastoderma
Cephalopod, 221
Cerastoderma, salinity and shell shape, 44
Cerastoderma edule
annual microgrowth patterns, 222
coefficients of allometry, 39
daily microgrowth increments, 224
disturbance lines, 400
and environmental change, 226
fortnightly patterns, 221
growth strategies, 37
monthly microgrowth patterns 221
periodic microgrowth increments, 219
population attributes, 461
population census, 451–453
respiration, 127
strontium in shell, 150
substratum effects on shape, 43
sulfur variations, 142
thermally induced microstructural
changes, 240
Cerastostoma foliatum, 385, 388
Cerion uva, 390
Cerithidea, 383
Cerithium, 382
Cerithium columna, 388
Cerithium nodulosum, 387, 388
Chalky deposits within foliated layers, 102
Chama, 75, 76, 96
Chama macerophylla, 187–188
Chance influences on morphology, 53

Charonia, 386
Cheilosporum maximum, 387
Chemical growth patterns, 675–685
Chione undatella, 135–137
Chlamys, 142, 399
Chondrophones, 26
Chonetes laguessiana, 427
Chthamalus depressus, 486
Circadian rhythms, 160–161, 258
Circulation patterns, 308
Class frequency, 418
Class-frequency histogram, 418–420
Clausilia, 382
Climate and coral growth records, 509
Climatic isolation and speciation, 357
Clinocardium nuttalli, 221, 224
Cluster analysis, 411–413
Clyperomorus morus, 388
Coelomic pressure, 473
Coelorhynchus cilensis, 549
Cohort, 423
Collagen in fish sagittae, 525
Collection efficiency, 701
Colliculum, 522
Collisella, 357
Collum, 521
Colonization on hard surfaces, 30
Color pattern and allometry, 382–383
Columbella mercatoria, 388
Columnar nacreous sublayers, 101–102
Commissure, 28, 338
Competition
and morphologic diversity, 304–305
and nonplanktotrophy, 353
Compartment hypothesis, 121, 609
Complex cross foliated microstructure, 80
Complex crossed lamellar microstructure,
80, 245
Composite prismatic microstructure, 79
Concentric ridges, 317
Conchiolin
adventitious sublayers, 315–316
and ligament, 26
and molluscan shell microgrowth
increments, 206
and nacre formation, 217
patches, 315–316
sublayers, 315
Conchiolin/CaCO$_3$ ratio, 316
Concrescent shell, 471
Conditional probability, 708, 715

Conocardium, 220
Continental shelf and larval dispersal, 356
Convexity of larval shell, 340
Coral
  age-frequency histogram, 701
  growth
    and algal symbiosis, 511–512
    and bioerosion, 506–507
    and depth, 489–499
    environmental variations, 498–508
    general features, 494–498
    and geochronometry, 511
    and isotope variations, 509–511
    and light, 494, 498–499
    and nutrient supply, 506
    and radionuclide uptake, 495–496, 509
    and Ra-228, 495–496
    and sedimentation rate, 499–500
    and temperature, 494
    and turbidity, 494
    and wave activity, 508–509
    and wind energy, 508–509
  population
    age distribution model, 712
    probability distribution, 711–722
  population hypotheses, 711–722
  skeleton growth increments
    and bioerosion rates, 506–507
    high-density (HD) band, 497–498
    and latitudinal temperature, 505–506
    low-density (LD) band, 497–498
    methods of observation, 615–617
    and nutrient supply, 506
    and paleolatitude determination, 495
    preparation for study, 615–617
    and reef CaCO$_2$ budgets, 507
    sampling technique for chemical
      analysis, 617
    seasonal density variations, 498
    staining techniques, 617
    and temperature, 495
    and wave activity, 508–509
    width–density–mass characteristics,
      497
    and wind energy, 508–509
Corallina pilulifera, 387
Corbicula manillensis, 143
Correlation coefficient, 405–406, 410–413
Crassatella mississippiensis, 220
Crassostrea, 144, 148, 347
Crassostrea gigas, 157

Crassostrea virginica
  bicarbonate flux, 120
  carbonic anhydrase activity, 120
  egg production, 324
  larval size and temperature, 361
  lead concentration, 143
  magnesium concentration, 146
  substratum effects on shape, 43
Crenomytilus grayanus, 148
Crenulated shell margin, 317
Cross foliated microstructure, 80
Crossed acicular microstructure, 80
Crossed lamellar microstructure, 80. 240
Crossed-matted/lineated microstructures,
  80
Crossed microstructures, 80
Crowding and growth rate, 388
Cruciform muscle, 27
Crystal imprints, 124
Crystal morphotypes, 81
Crystal nucleation, 123–124, 213, 216
Cuticle, 470, 472
Cyclical growth patterns, 258–269
Cylindroteuthis puziosiana, 680, 684
Cymbium, 336
Cypraea lynx, 387
Cypraea moneta, 388
Cypraea tigris, 388

D-shaped veliger, 338
Daily growth increments
  in bivalves, 259, 261–262
  corals, 495
  in fish sagittae, 529, 531–532, 542–544, 554
  subdivision, 397
Daily growth lines in polychaete jaws, 568
Death assemblage
  census from, 6
  dynamic, 6
  mass mortality event, 6
  of Mercenaria mercenaria, 277–279
  and mortality estimation, 282, 444–451
  and population census, 451
  and shell microstructure study, 239–240
  size-frequency histogram, 276–277
Deep sea and larval dispersal types, 351
Demarcation line, 49
Demersal larvae, 327, 351
Demography, 3, 273–274
Density of shells
  and active swimming, 94

Density of shells (*cont.*)
  and calcite secretion, 92
  and chalky deposits, 102
  and percentage of organic matrix, 92
  and porosity, 92
  and sedentary life habits, 94
Denticular composite prisms, 87
Dentition of larvae, 341–435
Depth
  and adaptive strategy, 358
  and larval development type, 358
  and larval shell, 340
  and larval type, 351
  and molluscan shell, 228
  and reproductive type, 351
  and fish sagitta growth, 554
Detergents (oil spill removers), 240-241
Desiccation stress adaptations, 384
Determinate growth, 34, 381, 387–388
Development
  ametamorphic, 328
  bivalve, 345–347
  of brooded larvae, 328–329
  direct, 328
  and egg size, 330–332, 340–341
  and egg strings, 341
  gastropod, 335–337
  larval, 324–329
  lecithotrophic types, 327–328
  in oviparous species, 328
  strategies, 347–354
  type
    and adult body size, 351
    and depth, 351
    and dispersal, 348
    inferred from prodissoconch
      morphology, 330–335, 340–345
    latitudinal distribution, 348–350, 353
    and r–K model, 347–354
    taxonomic distribution of bivalves,
      345–347
    taxonomic distribution of gastropods,
      335–337
Diagenesis in shell, 155–156, 159, 631
Dicathais aegrota, 386
Dicathais orbita, 385
Dicathais textilosa, 386
Diet and gastropod shell pigment, 386–387
Dimyarian condition of adductor muscles,
  27
Diploria, 498, 507

Diploria labyrinthiformis
  age-frequency histogram, 700–701
  carbon isotopic variations, 510–511
  increment width–density–mass
    chareristics, 497
  radiometric growth determination, 496
Diploria stringosa
  age-frequency histogram, 700–701
  effect of light on growth, 498
Direct development, 328
Directive plane, 49
Directive spiral, 49
Dispersal
  in deep sea, 351
  and developmental type, 348
  and environmental tolerance, 354–355
  larval, 327
  and larval type, 356
  and natural selection, 356
  paleontological significance, 355–36
  and planktotrophy, 356
  and recolonization, 354–355
  and species turnover rates, 358
Dissoconch, 338
Diurnal events in molluscan shell, 219–220
Divaricate ornamentation, 317
Doming, 382, 384–385
Donax incarnatus, 424
Donax striatus, 43
Dredging and coral mortality, 500–505
Drupa, 382
Drupa ricinus, 388

Echinospiral larvae, 336
Ecological applications of skeletal growth
  records, 13–14
Ecophenotypic influences on morphology,
  53
Egg
  production, 324, 328
  size
    and development, 330–332, 340–341
    and development time, 331–332
    and hatching size, 331, 340–341
  strings and development, 341
Eisenia bicyclis, 386
Electron-dense lamellae, 213
Elliptio complanata, 298
Elminius modestus, 476–477, 482
Embryonic shell, 329

Endogenous rhythms, 399
Enteromorpha, 389
Entropy, 703, 711
Environment
  and allometry, 385–387
  and bivalve shell morphology, 297–301
  and bivalve shell outline, 298–301
  and bivalve shell sculpture, 299
  and fish sagitta growth discontinuity,
    552, 555
  and fish sagitta growth patterns, 553–556
  and height, 298–301
  and larval type, 348–351
  and obesity, 297–301
  and polychaete growth, 581
  and size, 297–301
Environmental change
  and bivalve shell morphology, 299–300
  and molluscan shell growth patterns,
    206
  and population attributes, 461
  and skeletal growth, 4
Environmental factors
  and bivalve shell microstructure, 105–106
  and bivalve shell mineralogy, 105–106
  and bivalve shell morphology, 55–61
Environmental monitoring
  and growth band information, 307
  and larval shell morphology, 355
Environmental quality and larval type, 355
Environmental reconstruction
  and adult size, 390
  and bivalve shell morphology, 299–300
  bivalve shell morphology–habitat
    relationships, 305
  and gastropod shell morphology, 390–391
  and growth lines, 391
Environmental stimuli and growth lines,
  398
Environmental stress
  and barnacle growth bands, 488
  and extinction, 358–359
  and fish sagittae, 549–552
  and growth band information, 308
  source recognition, 461
  and speciation, 358–359
Environmental tolerance
  and dispersal, 354–355
  and larval type, 356–357
  and recolonization, 354–355
  and speciation, 359

Environmental variability
  and bivalve shell microgrowth increment
    patterns, 228
  and fish otolith growth, 527–528
  and larval types, 348–351
Epibyssal attachment, 55–57
Epicontinental sea, 356
Epinephelus morio, 547
Epithelium and barnacle shell growth, 472
Equilibrium species, 348, 453
Escutcheon, 25
Eunice Kobiensis, 572
Eunice torquata, 580
Eunoe oerstedii, 566, 569
Euryoxic capacity, 127
Evolution of bivalves, 28–32
Excisura, 521
Expansion rate, 380
External growth increments in corals,
  494–495
External growth lines, 396
External growth patterns on barnacles, 485
External growth ridges on barnacles,
  483–485
External shell growth patterns, 204
Extinctions
  and environmental stress, 358–359
  and larval type, 356
Extrapallial fluid
  and bivalve shell composition variations,
    156–157
  chemical changes, 208
  composition, 119–120
  definition, 119
  levels of succinic acid, 211
  and molluscan shell dissolution, 125
  and internal molluscan shell growth
    patterns, 206
  and molluscan shell formation, 119–120

False annuli, 613–614
Favia speciosa, 496
Favites virens, 496
Fecundity and polychaete jaws, 586
Feeding
  and fish otolith growth, 528–529
  and fish sagitta growth, 528, 544
  and growth of fish scales, 528
  larval, 324, 327
  and polychaete growth, 581

Feeding habits and fish sagitta growth, 554
Feeding time and growth rate, 389
Feigl solution, 632, 637–638
Fibrous prismatic microstructure, 79
Filter-feeding adaptations, 384
Fine surface ridges on barnacles, 483–484
First-order lamellae, 80
Fish otolith
  fossil, 555–556
  growth
    and environmental variation, 527–528
    and feeding activity, 528–529
    and temperature, 528
    See also Fish sagitta growth
  methods of preparation, 619–623
  mineralogy, 520, 524–529
  morphology, 520
  and osteogenesis, 525
  and osteolysis, 529
  terminology, 520
Fish sagitta
  age estimates, 520
  comparison to bone, 525–526
  comparison to mollusc shells, 520
  growth
    discontinuities, 522, 549–552
    and feeding, 528, 544
    regulation of calcification rates, 528
    stages, 523
    taxonomic differences, 523
    See also Fish otolith growth
  growth discontinuity
    and anaerobic stress, 552
    and habitat, 552, 555
    with ontogeny, 552
    and physiological events, 551
    spawning break, 550
    and thermal stress, 552
    tidal breaks, 550
    and transplantation stress, 552
    winter disturbance, 550
  growth patterns, 522–523
    classification, 530
    daily, 529, 531–532, 542–544, 554
    and environmental stress, 549–552
    and feeding habits, 528, 554
    fortnightly, 545
    and habitat, 553–556
    inducing stress marks, 622–623
    larval increments, 543
    and latitude, 554
    lunar, 544–546

Fish sagitta (cont.)
  growth patterns (cont.)
    monthly, 545
    periodic spawning marks, 545–546
    seasonal, 546–549, 555
    subdaily, 530–531
    tidal, 544–546, 554–555
  growth rate and racial grouping, 522
  lamellae, 529–530
  margin and time of stress event, 550–551
  methods of observation, 619–620
  methods of preparation, 619–623
  mineralogy, 520, 524–529
  morphology, 520–524
    colliculum, 522
    collum, 521
    crenulations, 522
    excisura, 521
    growth patterns, 523
    growth phases evident in, 522
    kernel, 522
    nuclear area, 522
    opaque ring, 522
    postcaudal trough, 521
    of proximal side, 521–522
    and racial grouping, 522
    stages of growth, 523
    sulcus acusticus, 521
    taxonomic differences, 523
    in young fish, 522
  and sound perception, 524
  subseasonal rings, 529–530
  thin sectioning and acetate peel
    preparation, 621–622
Fish scales and feeding, 528
Fisheries management, 14
Flaring, 382, 384–385
Foliated microstructure, 81, 83
Food supply
  and allometry, 385
  and gastropod shell pigment, 386–387
  and gastropod shell size, 390
  and growth rate, 388–389
  and internal growth bands in barnacles,
    477–479
  and seasonal zones in sagitta, 547
Ford–Walford plot, 34, 426–427
Fordilla troyensis, 29
Foreign contaminants and secretion of
    conchiolin patches, 315–316
Form–habitat relationships in unionacean
    bivalves, 297–301

Fortnightly patterns
  in fish sagittae, 545
  in molluscan shell, 220–221, 259, 262
Fourier analysis, 675–676
Fractional variance, 677
Fractured section, 233, 601
Freeze-shock break, 260, 270
Frequency
  domain, 676
  histogram, 418
  spectrum, 692
  table, 418
Functional morphology, 52–61
Fusconaia flava, 299
Fucus serratus, 581
Fusinus, 383

Gastropod
  development types, 335–337
  growth increments, 391
Gastropod shell
  axis, 380
  growth, 379–383
    determinate, 381
    description, 380
    direction of axis, 380
    geometry, 379–383
    idealized, 380
    marginal 380
  growth rate
    and adult size, 387–390
    and allometry, 384–387
    and crowding, 388
    and feeding time, 389
    and food supply, 388–389
    and indeterminate growth, 388–389
    and population density, 389
    and tidal level, 389
    and translation rate, 384
  height adaptations, 384
  larval, 329–337
  microgrowth increment patterns, *see*
    Molluscan shell microgrowth
    increment patterns
  microstructure, *see* Molluscan shell
    microstructure
  morphology and environmental
    reconstruction, 390–391
  sculpture, 382–383
Gelidium, 387
Gemma gemma, 224–226
Generating curve, 46, 381–382

Genetic continuity, 351
Genetic divergence, 356
Geochronometry, 511
Geodesic coils, 380
Geodesics, 384
Geographic range
  and larval type, 356
  and planktotrophy, 357
Geukensia demissa
  fortnightly cycles in shells, 220
  shell dissolution 126
  shell microgrowth increment patterns,
    243–245
  shell microgrowth increments in
    nacreous layer, 216–217
  shell microstructure, 101–102
  shell microstructure change with
    latitude, 231–237
  shell microstructure change with season,
    231–235
Glycera convoluta, 571, 580
Glycera dibranchiata, 571
Glycera glaucopsammensis, 571
Glycymeris, 39, 46–47
Gnathampharete paradoxa, 561
Gradualism and nonplanktotrophic larvae,
  357–358
Granular microstructure, 80
Granulations, 329, 338
Gregariella coralliophaga, 76
Growth
  axis, 27, 379–380
  band
    and age determination, 307
    barnacle, 473
    definition, 397
    and environmental stress, 308
    and life-history parameters, 307
    origin, 305–306
    and seasonal changes, 309
    and temperature changes, 305–309
  biocheck, 4
  break, 4, 269–272
    description, 260–261, 269
    origin, 269–270
    thermally induced, 229
  check, 4
  center, 25
  coefficient in allometry, 38
  curve
    and barnacle winter rings, 488
    bivalve, 421

Growth (*cont.*)
  curve (*cont.*)
    method of comparison, 405
    and size-frequency histogram, 424
  increment
    annual, 398
    carbonate rich, 397–398
    color, 397
    definition, 397
    in gastropods, 391
    organic rich, 397–398
    problems in interpretation, 398–399
    subdivisions, 397
    layer, 71
  line
    barnacle, 473
    definition, 396
    and endogenous rhythms, 399
    and environmental reconstruction, 391
    and environmental stimuli, 398
    and environmental variability, 400
    expression, 396
    external, 396
    internal, 396
    patterns, 399–400
    periodic events, 206–207
    in polychaete jaws, 569–570
    position, 397
    problems in interpretation, 398–399
    on prodissoconch II, 338
  line analysis, 401–413
    cluster analysis, 411–413
    of contemporaneous growth records,
        407, 411–413
    filtering data, 406
    missing lines, 407–408
    of noncontemporaneous growth
        records, 410–411
    results of correlation study, 407–410
    seasonal effect on correlation, 408–410
    shell preparation, 404–405
    test for contemporaneity, 413–414
  rate
    and age in barnacles, 482
    and barnacle winter rings, 485–486
    from bivalve shell microgrowth
        increment patterns, 273
    and breeding in barnacles, 482
    and competition, 37
    in corals, 495–496
    of *Mercenaria mercenaria*, 277

Growth (*cont.*)
  rate (*cont.*)
    and molting in barnacles, 482
    and size-frequency histograms, 277,
        421
  rate curve, 400, 405–406
  ridge
    barnacle, 473, 482, 485
    coral, 495
    sublayer, 71
    vectors of bivalve shell, 46–48
Gryphaea, 387

Habitat, *see* Environment
Haliotis, 330
Haliotis cracherodii, 179
Haliotis kamchatkana, 386
Haliotis midae, 324
Haliotis rufescens, 179, 387
Halla parthenopeia, 572
Haloa japonica, 337
Halosydna brevisetosa, 566, 568
Hard water and bivalve shell morphology,
    302–303
Harmonic, 676
Harmothoe imbricata, 566, 568–569
Harmothoe sarsi, 569
Hatching size
  and egg size, 331, 340–341
  and predation, 353
  and physiological stress, 353
Heat-shock break, 260, 270
Height
  and habitat, 298–301
  larval shell, 338–340
Helcion pectunculus, 382
Heteromyarian condition of adductor
    muscles, 27
Heterodont condition of bivalve hinge, 26
Hinge, 26
Hinge teeth, 26
Hirsute ridge, 484
Historical influences on morphology, 52
Homeomorphology, 54
Homogeneous microstructure
  description, 80
  and oxygen levels, 245
  sensu stricto, 80
  in sublayers, 233–235
  variation with latitude, 239
Homogeneous sublayers, 101

Hyaline zone
  and migration, 549
  microstructure, 546–547
  spawning breaks, 549
  temporal interpretation, 548–549
  See also Seasonal zone
Hydrobia ulvae, 327
Hypotheses of coral populations, 711–722
Hypotheses choice and prior information,
    709–711
Hypotheses of population age structure,
    702–706
Hyridella, 299

Immersion period and barnacle shell,
    477–479
"Independent entities" concept, 51
Indeterminate growth, 388–389
Inner calcitic prismatic layer, 239
Inner homogeneous layer, 241
Inner layer conchiolin patches, 315
Inner nacreous layer, 241
Inoceramus sublaevis, 189–199
Intercellular junctions, 119
Internal growth increments, 3–5
  time series analysis, 500–508
Internal growth lines, 396
Internal shell growth patterns, 204–205
  See also Bivalve shell microgrowth
      increment patterns; Molluscan shell
      microgrowth increment patterns
Intracrystalline organic matrix, 210–211,
    216
Intraprismatic organic strands, 213
Ionotrophy, 124
Irregular nacreous laminae, 241
Irregular prismatic sublayers, 235
Irregular simple prismatic sublayers, 101
Isometric relationship in allometry, 38
Isomyarian condition of adductor muscles,
    27
Iteroparity in polychaetes, 587

K-selection, 348–354, 453–454
Kelletia kellettii, 179
Kernel, 522

Lachnolaimus maximus, 545, 552
Lacuna, 353–354
Lagena, 520
Lagrange multiplier, 713
Laminar microstructure, 79

Lampsilis excavata, 299
Lampsilis ovata, 299
Lampsilis radiata, 298–300, 303
Lampsilis ventricosa, 299–300
Lanthanum tracer, 119
Lapillus, 520
Larvae
  bivalve
    identification methods, 341–345
    shell morphology, 337–345
    taxonomic distribution, 345–347
  brooded, 328–329
  classification, 324
  demersal, 327
  dentition, 341–345
  development
    of brooded types, 328–329
    of lecithotrophic types, 327–328
    of pelagic types, 325–326
  development type
    and biostratigraphy, 359–360
    and body size, 351–353
    and depth, 358
    and dispersal, 356
    and environment, 348–351
    and environmental quality, 355
    and environmental tolerance, 356–357
    and extinctions, 356
    and genetic divergence, 356
    and geographic range, 356–357
    and mortality, 350
    and natural selection, 356
    and paleoclimatology, 361–363
    and palecommunity study, 360–361
    and paleogeography, 360
    and pollution, 355
    and population dynamics, 360–361
    and recolonization, 354–355
    and r–K model, 347–354
    and speciation, 356–359
    and tidal level, 351
  dispersal, 10
    of demersal types, 327
    of pelagic types, 327
  duration of pelagic stage, 326–328
  egg production
    of lecithotrophic types, 328
    of pelagic types, 324
  feeding
    of demersal types, 327
    of pelagic types, 324

Larvae (*cont.*)
  gastropod
    identification methods, 335
    shell mineralogy, 329–335
    taxonomic distribution, 335–337
  lecithotrophic, 327–328
  ligament, 345
  mortality
    in lecithotrophic types, 328
    in pelagic types, 325
  nonplanktotrophic, 327–329
  nourishment of lecithotrophic types,
    327–328
  of oviparous species, 328
  of ovoviviparous species, 328–329
  pelagic, 324–327
  planktotrophic, 324–327
  settlement of pelagic types, 326
  shell shape and taxonomy, 345
  of viviparous species, 329
Larval increments in fish sagittae, 543
Larval shell
  definition, 329, 338
  dimension terminology, 338–340
  size, 361–363
Larval shell morphology
  bivalves, 337–345
    brooded types, 341
    dissoconch, 338
    lateral hinge system, 341, 343–345
    ligament, 345
    ligament pits, 345
    metamorphic line, 338
    of nonpelagic types, 341
    of nonplanktotrophic types, 340–345
    of pelagic types, 340–341
    of planktotrophic types, 340–345
    prodissoconch, 338
    prodissoconch I, 338
    prodissoconch II, 338
    provinculum, 341–342
    terminology, 337–340
  dentition, 341–345
  and environmental monitoring, 355
  gastropod, 329–335
    of lecithotrophic types, 333–335
    of nonplanktotrophic types, 330–335
    ornamentation, 333–334
    of planktotrophic types, 330–335
    protoconch, 329
    protoconch I, 329

Larval shell morphology (*cont.*)
  gastropod (*cont.*)
    protoconch II, 330
    protoconch ultrastructure, 330
    terminology, 329–330
  and mode of life, 332–333
  and species turnover rates, 358
Lateral hinge system, 341, 343–345
Latitude
  and fish sagitta growth patterns, 554
  and homogeneous microstructure, 239
  and larvae development type, 348–350,
    353
  and molluscan shell microstructure,
    231–237
  and molluscan shell sublayers, 235–237
  and obesity in unionacean bivalves, 299
Lead in bivalve shells, 143
Lecithotrophic development, 351
Lecithotrophic larvae, 327–328
  and prodissoconch, 340
  and protoconch, 333–335
Length
  of larval shell, 338
  of shell, 27
  of geodesic, 380
Length/height ratio and habitat, 299
Lepidonotus clava, 566, 568
Lepomis cyanellus, 528
Lepomis gibbosus, 531
Leptasterias hexactis, 353
Life assemblage, 276, 279–281
Life history strategy, 350
Life table, 284, 440–443
Ligament, 26, 345
  annuli, 308–309
  pits, 345
Light and coral growth, 498–499, 510
Ligumia recta, 298
Likelihood, 708–709
Limopsis striatopunctatus, 220–221
Lithophaga, 78, 638
Littorina littorea, 382, 389
Littorina saxatilis, 336
Logarithmic conispiral, 380
Lopha, 347
Lunar patterns
  in fish sagitta, 544–546
  in molluscan shell, 206, 259, 262, 269
Lunula, 25
Lutjanus buccanella, 549

Lutjanus synagris, 546
Lutjanus vivanus, 531
Lutraria, 317
Lutz–Rhoads hypothesis of growth line
    formation, 211–218, 230–231
Lymnaea stagnalis, 389

Macoma, 179
Macrocystis pyrifera, 387
Macroscopic growth bands, 4
Maculae, 526
Magnesium
    in algal calcite, 78
    in bivalve shells, 78, 143–149, 157
        and salinity, 148–149
        and temperature, 145–148
Manganese, in bivalve shells, 140–142
Manicina areolata, 495
Mantle
    calcium and bicarbonate movement,
        118–119
    isthmus, 26
    and molluscan shell dissolution, 125
    retraction, 306
Margaritifera falcata, 315–316
Margaritifera mararitifera, 299
Margin growth in gastropods, 380, 381, 388
Marphysa belli, 572
Marphysa sanquinea, 580
Mass (census) mortality, 225–226, 274, 277
Matrix hypothesis of mollscan shell
        formation, 121
Maturity and polychaete jaw growth lines,
        585
Maximum attainable size, 34
Mechanical resiliency of shell, 310
Mediaster aequalis, 328
Megalops atlanticus, 527
Melongena, 382
Mentha aquatica, 389
Mercenaria
    and magnesium in shell, 148
    seasonal growth variations, 603
    subdaily growth lines, 398
Mercenaria campechiensis ochlockoneenis,
        220
Mercenaria mercenaria, 525
    abrasion break, 260, 271
    absolute mortality, 284
    age, 279
    age determination, 241

Mercenaria mercenaria (cont.)
    age and rate of shell growth, 261
    and anaerobic respiration, 208–209
    annual microgrowth patterns, 222, 259,
        269
    bidaily microgrowth increment patterns,
        259, 262
    biological rhythms, 207
    circadian rhythms, 258
    composition of organic matrix, 122–123
    corrosion of growth surfaces, 208
    cyclical growth patterns, 258–269,
    daily microgrowth increment patterns,
        259, 261–262
    death assemblage, 277–279
    diurnal periodicities, 219
    environmental effects on growth, 226
    fortnightly microgrowth increment
        patterns, 259, 262
    fortnightly cycles, 220
    freeze shock break, 260, 270
    growth, 281
    growth break, 260–261, 269–272
    growth line, 397
    growth rate, 277
    heat-shock break, 260, 270
    larval mortality, 284
    life assemblage, 279–281
    life table, 284
    lunar-month microgrowth increment
        patterns, 259, 262, 269
    monthly growth patterns, 221
    mortality patterns, 279, 282–291
    neap-tide break, 260, 272
    population dynamics, 273–274, 291
    recruitment patterns, 281
    seasonal growth rates, 262
    seasonal mortality, 283–284
    shell dissolution in, 125–126, 212
    shell growth equation, 277
    shell layer, 256
    shell microgrowth increment patterns,
        256–258
    shell microstructure change, 231
    size-frequency histogram, 276–282
    spawning break, 260, 271–272
    storm break, 261, 272
    subdaily microgrowth increment
        patterns, 259, 262
    temperature and rate of shell growth,
        261–262

Mercenaria mercenaria (*cont.*)
  thermal-shock break, 260, 270–271
  thermally induced effects in shell, 229,
    240
Merluccius bilinearis, 525, 544, 547, 552
Metamorphic line, 338
Metazoan evolution and circadian
    rhythms, 160–161
Meteoric waters and oxygen isotopes,
    181–183
Microscopic growth increments, 4–5
Microscopic growth interruption, 4
Mineralogical conversion of aragonite to
    calcite, 78–79
Mizihopecten yessoenisis, 148, 150
Modiolus
  acetate peel, 678
  and Feigl solution, 637
  normalized power spectra analysis,
    678–680, 683
  shell mineralogy, 97
Modiolus demissus, 43
Modiolus modiolus
  coefficients of allometry, 39
  crowding and shell shape, 43
  growth strategies, 37
Molluscan physiology and internal shell
    growth patterns, 206
Molluscan shell
  chemistry and shell dissolution, 237
  corrosion and shell microstructure,
    231–233
  damage and shell microgrowth patterns,
    223
  deposition, 207
  dissolution
    and anaerobic glycolysis, 127
    and anaerobiosis, 125–128, 212
    electron-dense lamellae, 213
    nacre formation, 213
    and organic-rich layers, 207–208
    and oxygen isotope thermometry, 237
    and shell chemistry, 237
    zonation, 126
  formation
    hydrogen ion removal, 120
    primary compartments, 116
    requirements of general theory, 121
    role of carbonic anhydrase, 120
    role of extrapallial fluid, 119–120
    role of organic matrix, 121–125

Molluscan shell (*cont.*)
  formation (*cont.*)
    role of outer mantle epithelium,
      116–119
  microgrowth increment formation
    and anaerobiosis, 208–209, 211,
      212–218
    electron dense lamellae, 213
    intracrystalline organic matrix, 216
    and intraprismatic organic strands, 213
    Lutz–Rhoads hypothesis, 211
    in nacre, 213. 216–218
    origin, 206–208
    prismatic layer, 213
    and shell valve movements, 206–207
  microgrowth increment patterns, 205
    and aerial exposure, 226
    and age at death, 225
    and age at sexual maturity, 226
    annual events, 222
    and crowding, 226
    ecological use, 223–230
    and environmental variations, 226, 228
    fortnightly events, 220–221
    lunar and solar cycles, 206
    and mass mortality, 225–226
    and monthly events, 221–222
    and paleobathymetry, 228
    paleocological use, 223–230
    and periodic environmental events,
      206–207
    and random events, 222–223
    season of death, 225–226
    season of reproduction, 226
    seasonal and climatic variations,
      228–229
    and seasonal environmental factors,
      224
    and sediment type, 226
    semidiurnal and diurnal events,
      219–220
    semiperiodic events, 222–223
    solar daily increments, 219
    and solar time, 206
    and temperature, 229
    temporal categories, 218–219
    and tides, 207
    transplantation effects, 226
    and water depth, 228
    See also Bivlave shell microgrowth
      increment patterns

Molluscan shell (*cont.*)
  microstructure, 205
    and age, 239, 241
    annual patterns, 241
    in death assemblages, 239–240
    and detergents, 240–241
    and latitude, 235–237
    and Lutz–Rhoads hypothesis,
      230–231
    mechanisms, 230–231
    and metabolic pathways, 243–244
    nomenclature, 230
    and organic material, 209–211
    phylogenetic implications, 243–247
    and shell mineralogy, 238–239
    temperature, 235–239
    as thermal stress indicator, 239
    See also Bivalve shell microstructure
  mineralogy
    and shell microstructure, 238–239
    See also Bivalve shell mineralogy
  organic–inorganic relationships, 209–211
  valve movements, 206–207
Molting in barnacles, 472–473
  and growth rate, 482
  and growth ridges, 485
  and hirsute ridges, 484
Monomyarian condition of adductor
    muscles, 27
Monotely, 587
Monastrea annularis
  increment width–density–mass
    characteristics, 497
  isotope variations, 509–510
  and light, 498
  and paleoclimatology, 509
  radiometric growth determination,
    495–496
  and sediment, 500
  spacing of septa and dissepiments, 498
  and temperature, 505
Monthly patterns
  in fish sagittae, 545
  in molluscan shell, 221–222
Mortality
  and accurate age estimates, 445–451
  and barnacle winter rings, 488
  and bivalve shell microgrowth increment
    patterns, 273–274
  of corals, 500–505
  data sources, 442

Mortality (*cont.*)
  and death assemblage interpretation,
    444–451
  of larvae, 325, 328, 351–353
  and larval type, 350
  and life tables, 440–443
  in *Mercenaria mercenaria*, 282–291
  problems with data interpretation,
    442–443
  in scolecodont assemblages, 588
  size dependent, 450–451
  and size-frequency histogram, 277, 421,
    447
  and transplantation, 284–288
Morula granulata, 388
Morula uva, 388
Mother of pearl, 79
Mulinia, 42
Mulinia lateralis
  as indicator species, 454–455
  larval development, 348
  survivorship, 447–449
  and trophic group amensalism, 462
Murex erinaceus, 461
Muscle scars on bivalve shell, 26–27
Musculus discors, 341
Musculus niger, 341
Mya arenaria
  larvae size and temperature, 361
  larvae survival, 325
  manganese substitutions in shell, 141–142
  substratum effects on shell form, 43
Myostracal prisms, 238
Mytilacea shell mineralogy, 632–638
Mytilus
  magnesium concentrations in shell, 146
  plasticity of shell shape, 45
  shell mineralogical–microstructural
    variations, 100
  shell mineralogy, 97–98
  strontium in shell, 150
  sulfates in shell, 142
Mytilus californianus
  and competition,37
  growth strategies, 37
  organic matrix 610
  and oxygen isotopes in shell, 179
  shell changes with temperature, 239
  shell inorganic concentrations, 135
Mytilus edulis, 56, 461
  age, 241

Mytilus edulis (*cont.*)
  allometry, 39–40
  and competition, 37
  crowding and shell form, 43
  growth, 33–35, 421
  growth curve, 425
  magnesium in shell, 148
  manganese in shell, 141–142
  periostracum, 313
  shell dissolution, 126
  survivorship, 444–445
Mytilus galloprovincialis, 56

Nacre
  formation, 213
  and growth-line formation, 216–218
  and molluscan shell microgrowth
    increments, 213–214
  and phylogeny, 243–244, 247
  sublayers, 235
Nacreous laminae, 241
Nacreous microstructure
  description, 81
  and mechanical resiliency, 310
  taxonomic distribution, 83
  and temperature, 239
Nacreous tablets, 231–235, 241
Nacroprismatic shell, 310
Nassarius vibex, 354–355
National Bureau of Standards reference
    water, 173
Natural mortality, 274
Nautilus pompilius, 124, 222
Neap-tide break, 260, 272
Negative allometry, 39
Nepthys, 566
Nepthys bucera, 565–566
Nepthys caeca
  age, 584
  annual growth rings, 585
  growth lines, 564–566
Nephtys caecoides, 565
Nephtys ciliata, 565–566
  age, 585
  growth, 581
Nephtys hombergii, 561
  age, 584
  annual growth rings, 585
  growth lines, 563–566
  jaw chemistry, 580
  size-frequency histogram, 581
Nephtys longosetosa, 565

Nephtys punctata, 565
Nereis diversicolor, 570, 585
Nereis virens, 584
Nereocystis leutkeana, 389
Nereocystus, 389
Nodilittorina, 382
  Nodilittorina picta, 391
Nonconcrescent shell, 471
Nonpelagic development in deep sea, 351
Nonpelagic larvae
  and genetic differentation, 356
  on oceanic islands, 350
  shell morphology, 341
Nonplanktotrophic larvae, 327–329
  and dispersal, 356
  and extinctions, 356
  and gradualism, 357–358
  and protoconch morphology, 333–335
  protoconch I, 331
  shell morphology, 330–335
  and speciation, 356–358
Nonreflected growth lines, 213–214
Normal distribution, 429
Normalized power spectra analysis
  advantages, 675–676
  comparison of examples, 638–685
  of Jurassic belemnite, 680–683
  of Jurassic mytilid, 678–680
  method of calculation, 676–677
  of rabbit incisor, 678
Northia, 383
Notovola meridionalis, 400
Nucella lapillus, 240–241
Nuclear area of fish sagitta, 522
Nucula
  allometry, 42
  larval development, 348
Nucula annulata
  egg production, 328
  as indicator species, 454–455
  and trophic group amensalism, 462
Nucula delphinodonta, 328
Nucula proxima
  egg production, 328
  monthly microgrowth patterns in shell,
    221–222
Numerical filter, 691–692
Nurse eggs, 328
Nutrient supply
  and coral growth, 506
  and planktotrophy 353
  and size, 10

Obesity
  and habitat, 297–301
  and latitude, 299
Ocean barriers, 356–357
Oceanic islands and larval types, 350
Ocyurus chrysurus, 548, 551
Olivella verreauxi, 327
Oncorhyncus nerka, 528
Ontogenetic growth, 3, 5
Ontogeny
  and bivalve shell morphology, 304
  and calcium variations, 136–139
  and fish sagitta growth discontinuities,
    552
  and fish sagitta growth increments, 543
  and gastroped shell axis, 380
  and magnesium in shell, 144–146
  and shell sculpture, 383
  and strontium in shell, 149–151
  and trace elements in shell, 140
Opaque ring, 522
Opaque zone in fish sagittae
  microstructure, 546–547
  temporal interpretation, 548–549
  See also Seasonal zone
Opercular membrane, 470
Opercular valves, 470
Ophisthogyrate umbo, 25
Ophryotrocha puerilis, 572
Opisthobranchia, 337
Opportunistic species, 348, 453
Organic material
  in otoliths, 520
  in shell, 135–139
Organic matrix
  composition, 122–124
  concentration
    and salinity, 103
    environmental control, 103–104
  as crystal nucleator, 123–124
  deposition in fish sagittae, 532
  increase in thickness, 217
  insoluble part, 121–122
  and molluscan shell microstructure, 210
  observation techniques, 609–611
  physical relation to other shell
    components, 607
  preparation for study, 609–611
  role in molluscan shell formation,
    121–125
  soluble part, 122–124
  structural relationships, 124

Organic-rich growth increments, 397–398
Ornamentation, 383
  of planktotrophic larvae, 333–334
  of protoconch, 333–334
Orthogyrate umbo, 25
Osteogenesis in fish, 525–526
Osteolysis, 529
Ostrea
  brooding, 347
  and paleotemperature determination,
    189, 192, 195
Otolin, 525
Outer mantle epithelium
  and calcium movement, 119
  functional zones, 117
  metabolic activity, 117, 119
  and rates of shell formation, 117
  ultrastructure and function, 117–118
Oviparity, 328
Ovoviviparity, 328–329
Oxygen isotopes
  atmospheric concentration, 169–170
  in corals, 509–510
  equilibrium
    constant, 176
    and precipitation of calcium
      carbonate, 188
  in foraminifera, 186
  fractionation, 170
    altitude effects, 181–182
    of aragonite–calcite, 189
    during condensation, 181–183
    curve, 179–180
    during evaporation, 182–185
    factors, 173, 175–176, 182
    kinetic effects in evaporation, 182–183
    latitudinal trends, 181–182
    during metabolism, 188
    during precipitation, 182
    and salinity, 183–185
    during snowfall, 182
    and temperature, 177–178
  and molluscan growth, 188
  nomenclature, 172
  in ocean water, 180, 181
    and melting ice sheets, 185–186
    and glaciations, 185–186
    and salinity, 184–185
    throughout geologic time, 185–186
  paleothermometry, 178–179
  reference standards, 172–175
  in shells, 171–172, 186–188

Oxygen isotopes (*cont.*)
  and temperature, 176
  thermometry, 177, 237
Oxygen-18 concentration and salinity,
    184–185
Oxygen-18/Oxygen-16 ratio, 171–172, 175
Oxygen levels, and aragonitic prisms,
    245–246

pH and periostracum color, 318
Pachymelania, 383
Paleobathymetry, 8–9, 228
Paleobiogeography, 10–11, 359–360
Paleoclimatology, 7–8, 228–229, 361–363
Paleocommunity studies, 360–361
Paleoecology
  and skeletal growth records, 7–12, 400
  test for contemporaneity, 413–414
Paleoenvironmental reconstruction, 11–12
Paleogeography, 360
Paleolatitude reconstruction, 7–8, 228–229,
    495
Paleosalinity, 9, 316
  in North American seaway during Late
    Campanian, 190, 196–199
Paleotemperature-gradient reconstruction,
    36, 363
Paleothermometry, 9
  and bivalve growth bands, 308
  equations, 178–179
  and internal growth increments of corals,
    495
  and late Cretaceous Pierre Shale, 190–195
  and molluscan shell microstructure
    changes, 239
  in North American seaway during Late
    Campanian, 190–196
  and prodissoconch II, 361
Pallets mineralogy, 78
Pallial line, 26
Pallial sinus, 26–27
Panorbarius corneus, 389
Papuina, 382
Paragnaths, 562
Paranereites, 587
Parasite infestations and bivalve shell
    aberrations, 303
Paries, 470
Patella, 385
Patella granularis
  doming 382
  growth rate, 388

Patella vulgata, 388
  growth rate, 389
  population attributes, 461
Patina pellucida, 386
Pavona, 505
Pavona gigantea, 496–498
Pecten, 145
Pecten diegensis, 397
Pecten maximus, 401–413
Pectinaria australis, 581
Pectinaria hyperborea, 581
Pediveliger, 338
Peedee Belemnite reference standard,
    172–175
Pelagic larvae
  and biostratigraphic index fossils, 359
  development, 324–327
  in deep sea, 351
  genetic differentation, 356
  on oceanic islands, 350
  shell morphology, 340–341
Penitella penita, 43, 224
Perinereis cultrifera, 570, 580
Periodogram, 676
Periostracum
  bonding of prismatic layer to, 313–317
  color, 317–318
  exterior surface texture, 318
  fluid-filled vacuoles, 313
  and prevention of bivalve shell
    dissolution, 310
  as hydrophobic barrier, 119
  resistance to peeling and cracking,
    311–313, 316–317
  as shell margin seal, 316
  structural bonding, 313
  thickness, 310–311
Peripheral isolates, 358
Perna, 97
Phenotypic plasticity, 387
Phloe minuta, 570
Phosphate–calcite thermometry, 189
Photoperiod and barnacle growth bands,
    477
Phylogeny and shell microstructure,
    243–247
Physiological response to trauma, 398
Physiological stress and hatching size, 353
Pinctada martensii, 43, 72, 145
Pinctada radiata, 213
Pisaster ochraceus, 353
Placopecten magellanicus, 400

Planaxis nucleus, 383
Planktotrophic larvae
   development, 324–327, 351, 353
   gastropod shell morphology, 330–335
   and ornamentation, 333–334
   prodissoconch I, 340
   prodissoconch II, 340
   and protoconch morphology, 333–335
   protoconch I, 330
   and punctuated equilibria, 357
   shell morphology, 340–345
Planktotrophy
   and biostratigraphy, 359–360
   and dispersal, 356
   and energy for reproduction, 351–353
   and genetic continuity, 351
   and geographic range, 357
   and nutrients, 353
   and progenesis, 354
   and speciation, 356
   and species longevity, 356–357
   and temperature, 353
Platygyra, 505
Poecilozonites, 387, 390
Pollution
   and larval type, 355
   and molluscan shell microgrowth
      patterns, 223
Polychaete growth
   and age class recognition, 581, 584
   and feeding, 581
   and segment number, 581
Polychaete jaw
   arrangement, 563
   biometry, 589
   chemical composition, 580
   and definition of standard animal, 586
   ecological and paleoecological
      application, 581–589
   and estimates of biochemical
      components, 586
   and fecundity, 586
   growth lines
      annual, 585
      daily, 568
      in Eunicidae, 571–572
      in Glyceridae, 570–571
      and maturation, 585
      in Nephtyidae, 564–566
      in Nereidae, 570
      in Polynoidae, 566–567
      and reproductive cycle, 571, 587

Polychaete jaw (*cont.*)
   growth lines (*cont.*)
      in Sigalionidae, 566–570
   length and body size relationship,
      585–586
   morphology, 562–563
      carrier piece, 563
      in Eunicidae, 571–572
      forceps, 563
      in Glyceridae, 570–571
      in Nephtyidae, 564–565
      in Nereidae, 570
      in Polynoidae, 566–568
      and reproductive strategy, 587
      in Sigalionidae, 566–569
      mandibles, 563
      maxillae, 563
   and population age structure, 581–586
   preparation for study, 625–626
   preservation, 572, 580
   and reproductive effort, 586
   secretion, 565
   as size reference, 586
   and taxonomy, 561–562
Polychaete reproductive strategy, 586–587
Polychaete size and jaw length, 585–586
Polytely, 587
Population, probabilistic description of,
   699–722
Population attributes
   and adaptation, 453–455
   categories, 417
   and growth band information, 307
   and environment, 455–4457, 461
Population density and growth rate, 389
Population dynamics
   and growth lines, 400
   and larval development types, 360–361
   of *Mercenaria mercenaria*, 291
   and bivalve shell microgrowth increment
      patterns, 273–274
Population hypotheses, 702–706, 711–722
Procelaneous microstructure
   description, 81
   taxonomic distribution, 83
Porites lobata, 497
Porosity of bivalve shells, 92, 94–95
Positive allometry, 39
Postcaudal trough, 521
Posterior encrustations on shells, 78
Posterior probability, 708–709
Posterior tubes, 78

Potamogeton, 389
Power function in allometry, 38
Praemytilus strathairdensis, 243
Predation
  and hatching size, 353
  and morphologic diverstiy, 304:305
  and nonplanktotrophy, 353
Prior information and hypothesis choice,
    709–711
Prior probability, 708
Prismatic layer
  bonding of periostracum to, 313–317
  formation and growth, 213
Prismatic microstructure, 79, 240
Probabilistic population description
    699–722
Probability distribution
  of coral population, 711–722
  maximum entrophy, 711, 713
Probability of proposition, 708
Probability paper, 427–440, 584
Proboscis, 562–563
Prodissoconch
  definition, 338
  of lecithotrophic larvae, 340
Prodissoconch I
  definition, 338
  morphology, 338
  of planktotrophic larvae, 340
Prodissoconch II
  growth lines, 338
  as paleotemperature indicator, 361
  of planktotrophic larvae, 340
Progenesis, 354
Proposition probability, 708
Propositional calculus, 706–708
Proptera alata, 298
Prosobranchia, 335–336
Prosogyrate umbo, 25
Protobranchia, 34
Protoconch
  definition, 329
  of lecithotrophic larvae, 333–335
  and mode of larval life, 332–333
  of nonplanktotrophic larvae, 333–335
  ornamentation, 333–334
  of planktotrophic larvae, 333–335
  ultrastructure, 330
  whorl number, 330
Protoconch I
  definition, 329
  morphology, 329–330

Protoconch I (*cont.*)
  of nonplanktotrophic larvae, 331
  of planktotrophic larvae, 330
Protoconch II, 330, 338
Protostracum, 338
Provinculum, 338, 341–342
Prunum apicinum, 355
Pseudoannual growth band, 306
Pseudopleuronectes americanus, 545, 552
Pseudosheath, 470
Pulmonata, 337
Punctae, 330, 338
Punctuated equilibria, 357
Pyrgoma, 471

Quadrula quadrula, 299

r–K selection model, 453–454
  and development type, 347–354
  and larval type, 348–354
  and polychaete reproductive strategies,
    586
Rabbit incisor chemical periodicities, 678
Racial grouping using fish sagittae, 522
Radial costae, 317
Radii of barnacles, 470
Radium-228 and coral growth, 495–496
Random events recorded in shell, 222–223
Rangia cuneata, 127
Rayleigh process, 181
Recolonization
  and dispersal capability, 354–355
  and environmental tolerance, 354–355
Recruitment
  and age-frequency histograms, 421
  and bivalve shell microgrowth increment
    patterns, 273
  of *Mercenaria mercenaria*, 281
  and size-frequency histograms, 277, 281,
    421, 425
Relative growth, 33
Reproductive energy
  and growth, 37, 571, 586
  and larval type, 351–353
  and planktotrophy, 351–353
Reproductive season recorded in
    molluscan shell, 226
Reproductive strategy
  and body size, 351–353
  and depth, 351
  and polychaete jaw morphology, 587

Reproductive strategy (*cont.*)
  of polychaetes
    inferred from scolecodonts, 587
    and r–K selection, 586
    types, 586–587
Resilium, 26
Respiration
  and bivalve shell microstructure, 101–102
  and cyclical growth patterns in shell,
    258
  and molluscan shell dissolution, 125–128
Reticulate ornamentation, 333
Retusa obtusa, 337
Rhinoclavis, 383
Rhomboplites aurorubens, 549
Rissoa membranacea, 336
Runcina, 337

Saccostrea, 347
Sacculus, 520
Salinity
  and allometry, 386
  and bivalve shell mineralogy, 98–99
  and boron in bivalve shells, 143
  and magnesium in bivalve shells, 148–149
  and manganese in bivalve shells, 142
  in North American seaway during Late
    Campanian, 190, 196–199
  and organic matrix concentration, 103
  and oxygen isotopes, 184–185
  and periostracum color, 317–318
Scanning densitometer, 616
Scaphoconch, 336
Scleroprotein/$CaCO_3$ ratio, 316
Scolecodonts, 561, 571, 580
  and mortality analysis, 588
  preparation for study, 626
  and reproductive strategy, 587
  size-frequency analysis, 588
Scomberomorus cavalla, 548
Season of death from microgrowth
    increment patterns, 225–226, 274, 488
Seasonal bands, 403
Seasonal environmental changes and
    growth records, 7–8, 309
Seasonal growth rates in *Mercenaria
    mercenaria*, 261–262
Seasonal growth record, 403
Seasonal mortality in *Mercenaria
    mercenaria*, 283–284
Seasonal patterns in fish sagittae, 546–549,
    555

Seasonal zones in fish sagittae, 546–547; *see
    also* Hyaline zone; Opaque zone
Sediment trapping in bivalves, 303
Sedimentation rate and coral growth,
    499–500
Segment number in polychaetes, 581
Selection pressure and shell morphology,
    305
Semelparity in polychaetes, 586–587
Semicontinuous reproduction, 587
Semidiurnal events recorded in shells,
    219–220, 222–223
Settlement of larvae, 326
Sexual dimorphism and bivalve shell
    morphology, 303–304
Sexual maturity recorded in shells, 226
Sheath, 470
Shell
  architecture, 71
  closure and periostracum, 316
  dissolution, 125
  figures, 470
  gape, 28
  layer, 71
  layer separation, 614
  margin repair, 315
  microstructural changes
    seasonal sequence, 231–235
    with temperature, 231–235
  microstructure, 71
  pigment and diet, 386–387
  pitting, 315–316
  shape and taxonomy, 345
  structure, 70
  sublayer, 71
  symmetry, 28
Shock breaks, 270–271
Simple prismatic microstructure, 79
Sinusigera lip, 333
Siphonaria gigas, 382
Siphonaria lessoni, 383
Size
  and food supply, 390
  and habitat, 297–301
  and progenesis, 354
  and temperature, 301–302
Size class interval, 419
Size-frequency distribution, *see* size-
    frequency histogram
Size-frequency histogram, 273
  age-class estimation, 426–440
  age-class shape, 423, 427

Size-frequency histogram (*cont.*)
  of cohort, 423
  factors controlling form, 277, 281, 421
  general form, 423–424
  growth curve estimation, 424
  of *Mercenaria mercenaria*, 276, 279
  and mortality, 447–449
  of polychaetes, 581, 584
  polymodal, 423
  and probability paper, 427–440
  problems in interpretation, 425–426
  recruitment estimation, 281, 425
  and survivorship, 444–445
Size of adult
  and determinate growth, 387–388
  and energy for reproduction, 351–353
  and environmental reconstruction, 390
  and growth rate, 387–390
  and larval types, 351–353
Size of larval shell and temperature,
      361–363
Skeletal growth records
  ecological applications, 13–14
  paleoecological applications, 7–12
Skeleton, 1
Solar cycles and molluscan microgrowth
      increments, 206
Solar daily increments, 219
Solenastra hyades, 496
Solution pits, 310
Sound perception and fish sagitta, 524
Spawning
  and magnesium in shells, 144
  and molluscan shell microgrowth
      patterns, 223
  break, 260, 271–272, 550
  mark, 545–546
  ring, 548
Speciation
  and climatic isolation, 357
  and environmental stress, 358–359
  and environmental tolerance, 359
  and larval type, 356–359
Specialization and bivalve shell
      morphology, 304–305
Species longevity and larval type, 356–357
Species turnover rates, 358
Spherulites, 313
Spherulitic microstructure, 79
Sphyraena barracuda, 547
Spicules or spikes in bivalves, 80
Spire height, 384

Spirorbis spirorbis, 581
Spisula solida, 421, 427, 458
Spisula solidissima, 241
Spring-tide events recorded in shell, 262
Staining method for corals, 617
Standard animal, 586
Standard mean ocean water reference
      standard, 172–175
Statoconia, 519
Statolith, 519
Stephanocoenia, 496
Stephanocoenia michilini, 701
Sthenelais boa, 566, 569, 580
Stochastic model of life history strategies,
      348
Storm break, 223, 261, 272
Straight-hinged veliger, 338
Strength of shells, *see* Bivalve shell
      microstructure, and mechanical data
Striae, 473
Striostrea, 347
Strombus, 382, 383
Strombus gigas, 179, 187, 387
Strongylocentrotus droebachiensis, 389
Strontium in bivalve shells, 149–155
Strontium-90 uptake in corals, 509.
Structural growth pattern analysis, 675–685
Structural influences on morphology, 53
Subdaily growth increments, 222–223, 397
Subdaily growth patterns in fish sagittae,
      530–531
Subdaily growth striations, 212
Subdaily microgrowth increment patterns,
      259, 262
Subdaily rings in fish sagittae, 531
Sublayer junctions, 243
Substratum type and morphologic
      adaptations, 57–61
Succinate levels during anoerobiosis 211
Succinic acid level in extrapallial fluid,
      211, 213
Sulcus acusticus, 521
Sulfates in bivalve shells, 142
Summer band, 397
Surface of accretion, 396
Surface of deposition, 396
Survivorship, 443–451
  and death assemblage, 444–451
  and life assemblage, 444
Survivorship curves
  and accurate age estimates, 445–451
  for *Mercenaria mercenaria*, 284

Survivorship curves (*cont.*)
  general shape, 444
  theoretical, 443
Swiftopecten swifti, 148
Synodic-month patterns, 221

Taphonomy, 5
Taxodont condition of bivalve hinge, 26
Taxonomy
  and developmental types, 335–337
  and phenotypic plasticity, 386–387
  and shell shape, 345
Tectarius, 382
Tectus pyramis, 383
Teleoconch, 330
Teleplanic larvae, 327, 334
Telescopium telescopium, 382
Temperature
  and allometry, 385–386
  and annual band formation, 306
  and bivalve shell growth and
    morphology, 301–302
  and bivalve shell microstructure
    sublayers, 101
  and bivalve shell mineralogy, 96–99
  and coral growth, 494–495, 505–506
  and daily microgrowth increments,
    261–262
  and fish otolith growth, 528
  and growth band information, 307–308
  and internal growth bands in barnacles,
    479–480
  and isotope variations in corals, 509–510
  and larval shell size, 361–363
  and magnesium in bivalve shells, 145–148
  and molluscan shell mineralogy and
    microstructure, 231–235, 237–239
  in North American seaway during Late
    Campanian, 190–196
  and planktotrophy, 335
  recorded in molluscan shell, 224
  and seasonal zones in fish sagittae, 547
  and shell sublayer type, 235–237
  and size, 301–302
  and skeletal growth records, 7
  and strontium in bivalve shell, 150
Template hypothesis, 607
Tensilium, 26
Terebra, 382, 383
Terebra crenulata, 383
Teredo, 42
Teredo navalis, 78

Tetraclita, 471
Thais haemastoma, 382
Thais lamellosa, 385, 388
Thais lapillus, 382, 461
Thais melones, 383
Theodoxus fluviatilis, 386
Thermal-shock break, 260, 270–271
Thermal stress
  and fish sagitta growth, 552
  and molluscan shell microstructural
    changes, 239–240
Thickness of larval shell, 340
Thin section preparation, 603–605
Thracia, 347
Thyasira gouldi, 341
Tibia, 383
Tidal breaks, 550
Tidal fluctuations and skeletal growth
    records, 10
Tidal immersion and internal growth
    bands, 488, 687–688
Tidal level
  and allometry, 385
  and growth rate, 389
  and internal growth bands in barnacles,
    479–480
  and manganese in bivalve shells, 141–142
Tidal patterns in fish sagittae, 544–545,
    554–555
Tides and molluscan shell microgrowth
    increments, 207
Tilapia, 528, 531, 546, 552
Tilapia melanotheron, 548
Tilapia mossambica, 543
Time domain, 676
Time series analysis
  of barnacle shell growth bands, 488–489,
    687–697
  of belemnite guard, 680–684
  of bivalve shell, 678–680, 683–684
  of coral growth increments, 500–508
  of growth patterns, 675–685, 687–697
  of rabbit incisor, 678
Tivela stultorum, 229, 241, 244
Trace elements in shell, 135–155, 308
Trace metals in bivalve soft tissue, 157–158
Transformation theory, 51
Transgressive–regressive cycles, 359
Translation rate, 380, 384
Transplantation
  and bivalve shell microgrowth increment
    patterns, 261

Transplantation (*cont.*)
  and fish sagitta growth, 552, 623
  of *Mercenaria mercenaria*, 284–288
  and molluscan shell microgrowth
        patterns, 233
Tressus nuttalli, 40
Trichomya hirsuta, 76
Tricolia, 383
Tridacna squamosa, 219–220
Triphora, 337
Trochus niloticus, 382–383
Trophic degree and shell morphology, 303
Trophic group amensalism, 462
Turbidity and coral growth, 494
Turbo cornutus, 386–387
Turritella, 384
Tympanotonus, 382
Tympanotonus radula, 383

Ulva, 389
Umbo
  description, 25
  dissolution, 316
  orientation, 25
Uperotus clavus, 78
Urophycis chuss, 552
Utriculus, 520

Valve closure
  and anaerobic respiration, 208–209
  periodic, 126
  and shell loss, 212
Valve-movement rhythmicity, 207
Valve thickness, 302

Vasum, 382
Veliconcha, 338
Veliger, 324–325, 328, 338
Veliger larvae size
  and latitudinal temperature gradient, 36
  and paleotemperature, 36
Venerupis pullastra, 43, 421
Venerupis rhomboides, 43, 45
Venerupis semidecussata, 43
Vital effect, 188
Viviparity, 329

Walford plot, 34, 426–427
Wall plate canals, 471
Water acidity and periostracum color, 317
Water depth and growth band information,
        308, 498–499
Water silt content and periostracum color,
        317
Wave activity and coral growth, 508–509
Wavelength, 675
Width of larval shell, 340
Width of shell, 28
Wind energy and coral growth, 508–509
Winter band, 397
Winter disturbance in fish sagittae, 550
Winter rings in barnacles, 485–488

X-radiograph of corals, 495–496, 616

Yoldia limatula, 447–448

Zooxanthellae, 494, 511–512